Biomaterials, Bioengineering and Sustainability

Volume 6

There is an urgent need to address the current paradigm shifts in bioengineering for Human Health aiming the creation of breakthrough tissue engineered products, manufacturing technologies and effective regenerative treatments for tackling different diseases/disorders in a personalized manner. Yet, the excessive costs and wastes related to the development and production of biomaterials and advanced therapy medicinal products, and the increasing use of plastics in cell culture methods and animal derived reagents has recognized the importance of decreasing the direct carbon footprint and thus, implement sustainable principles and solutions in the innovation ecosystem. The main goal of the volumes in Biomaterials, Bioengineering and Sustainability series is to catapult and consolidate new concepts and solutions towards the development of the next-generation of sustainable and eco-friendly biomaterials and tissue engineering and regenerative medicine approaches. Each volume will focus on the latest developments dealing with the identification of new sources of sustainable or recycled biomaterials, providing ideas for green technologies and methods that can be applied for biomaterials advanced processing and scaffolding strategies, and applications in biofabrication, tissue engineering, regenerative medicine, and drug delivery systems. It also aims to include the exploitation of renewable and sustainable source of human cells applied for cell therapies or in combination with sustainable biomaterials. The develop complex in vitro 3D/4D models and dynamic cell culture systems will be other subjects to be further explored from a sustainable perspective. This series aims to attract the contributions of leading experts in bioengineering, cell biology, materials engineering, and environmental sciences.

Zahid Hussain • Renjun Pei
Editors

Cell Membrane Engineering for Advancing Cell-Based and Nanotechnology-Based Therapies

Springer

Editors
Zahid Hussain
CAS Key Laboratory for Nano-Bio Interface
Suzhou Institute of Nano-Tech
and Nano-Bionics (SINANO)
Chinese Academy of Sciences (CAS)
Suzhou, China

Renjun Pei
CAS Key Laboratory for Nano-Bio Interface
Suzhou Institute of Nano-Tech
and Nano-Bionics (SINANO)
Chinese Academy of Sciences (CAS)
Suzhou, China

ISSN 2731-751X ISSN 2731-7528 (electronic)
Biomaterials, Bioengineering and Sustainability
ISBN 978-3-031-93424-7 ISBN 978-3-031-93425-4 (eBook)
https://doi.org/10.1007/978-3-031-93425-4

This Springer imprint is published by the registered company Springer Nature Switzerland AG
The registered company address is: Gewerbestrasse 11, 6330 Cham, Switzerland

With sincere respect and gratitude, we dedicate this book to the esteemed healthcare professionals and biomedical engineers whose unwavering commitment to patient care, public health, and medical innovation forms the foundation of this work. Your dedication inspires future generations and contributes to a healthier, more equitable world.

Foreword

Cell Membrane Engineering for Advancing Cell-Based and Nanotechnology-Based Therapies presents a timely and comprehensive review of how biomolecular engineering, nanotechnology, and biomaterials converge within the field of cell membrane engineering to drive innovative therapeutic advancements. This book provides insights into how cell membrane engineering can revolutionize therapeutic applications. With an increasing global demand for precision medicine, regenerative therapies, and targeted drug delivery, this book serves as a crucial resource for researchers including faculty and graduate students, clinicians, and industry professionals alike. It delves into cutting-edge methodologies that bridge biology and material sciences to address pressing challenges in healthcare. The carefully curated chapters, authored by distinguished experts, offer a holistic perspective on biomaterials, engineering strategies, and their translational potential in clinical applications. By bringing together key concepts in biomaterial science, nanotechnology, cellular engineering, and drug delivery, this book equips readers with the knowledge to navigate and contribute to the next frontier of biomedical advancements.

This book comprises 16 chapters, organized into five comprehensive sections, each focusing on a critical aspect of cell membrane engineering and its applications. It explores innovative strategies that are shaping the future of biomedical advancements, particularly in drug delivery, immunotherapy, tissue engineering, and regenerative medicine.

The first section introduces cell membrane engineering for nanoparticle development. It explores diverse membrane-based platforms, including extracellular vesicles, membrane-coated nanoparticles, and hybrid nanomaterials. Through detailed discussions, the authors highlight how surface engineering enhances biocompatibility, targeting efficiency, and therapeutic efficacy, addressing challenges in nanoparticle delivery and immune modulation.

The second section explores strategies in living cell surface engineering, focusing on functional biomolecules and biomaterials—such as aptamers, antibodies, peptides, and nanomaterials—that enable precise cellular modifications. It examines genetic and nongenetic engineering approaches, individually and in combination, highlighting their significant potential in enhancing drug loading efficiency, tissue penetration, and immunomodulation.

The third section focuses on transforming cellular therapies through nanoscale camouflage. It explores the integration of biomolecules and biomaterials into functional cell coatings, with applications in blood transfusion, insulin cell transplantation, and stem cell therapies. Advanced single-cell nanoencapsulation techniques are introduced, highlighting their role in enhancing therapeutic cell survival, targeting, and immune evasion.

The fourth and fifth sections of the book explores engineered membrane platforms for a variety of potential applications ranging, including cancer diagnosis, cancer immunotherapy, vaccine development, regenerative medicine, infectious disease treatment, neurodegenerative disorders, and inflammatory conditions. It also shows the potential of biomimetic cell membrane platforms in overcoming biological barriers, such as the blood-brain barrier, to improve drug delivery outcomes.

Each chapter in this book is meticulously crafted, providing not only biological insights but also practical applications of membrane engineering and biomaterials in the medical field. This book acts as a bridge between biology, engineering, and human healthcare. The diversity of topics ensures that this work remains relevant to a wide audience, from biomaterial scientists to clinicians and pharmaceutical innovators.

Cell Membrane Engineering for Advancing Cell-Based and Nanotechnology-Based Therapies is a vital contribution to biomedical engineering, offering innovative insights into biomaterials, biomolecular engineering, and nanotechnology in cell medicine. It paves the way for personalized healthcare and sustainable bioengineering solutions. For researchers and professionals in biotechnology, nanomedicine, and regenerative medicine, this book serves as a definitive guide to the latest advancements in cell therapy. Its in-depth exploration of membrane engineering makes it an invaluable resource for developing novel therapeutic strategies. As cell therapy evolves, this book highlights the power of interdisciplinary research in driving innovation. I highly recommend it to anyone interested in cell medicine.

Professor of Biomedical Engineering, Yong Wang
Dorothy Foehr Huck and J. Lloyd Huck Chair
in Cell Medicine, Department of Biomedical Engineering
The Pennsylvania State University,
University Park, PA, USA

Preface

Cells are the cornerstones of life and perform critical functions in complex microenvironments. The heterogeneous cell surface composition regulates numerous biological functions and represents a promising platform for diagnostic and therapeutic applications. Recent advancements in biology, chemistry, materials sciences, and imaging technology have enabled detailed exploration of cell membrane activities. On the other hand, advances in biomimetic nanotechnology and synthetic biology have led to the development of cell membrane-derived nanovesicles and cell membrane-coated nanomaterials. Genetic engineering, covalent conjugation, metabolic glycoengineering, bioorthogonal engineering, hydrophobic insertion, physical coating, and membrane fusion strategies have been used to integrate numerous bioactive materials onto cell membranes to effectively modulate cell membrane dynamics. Cell membrane engineering possesses multitasking capabilities and greatly enhances the biomedical potential of living cells and cell membrane-based nanoplatforms, as outlined in the following points:

- Therapeutic cells have potential for treating diseases and tissue defects; however, challenges like poor homing, low engraftment, decreased survival, undesired differentiation, and immune-related issues hinder their use in tissue modeling and cellular therapies. Recent studies suggest that cell surface engineering can enhance surface interactions, strengthen cell features, and add therapeutic functions to these therapies.
- Extracellular vesicles and bacterial membrane vesicles are membrane-bound structures containing bioactive molecules from parent cells, offering potential for biomedical applications. However, their clinical use is limited by low yield, heterogeneous cargo, poor targeting, and immune responses. Engineering strategies can enhance their bioactivity, tissue targeting, penetration, and applicability for disease treatment.
- Nanoparticles are used for drug delivery, chemotherapy, and diagnostics. However, their clinical application is limited by protein corona, rapid uptake by the reticuloendothelial system, and inefficient targeting. Recent studies have suggested that engineered cell membrane coatings can improve colloidal stability, immune evasion, barrier penetration, and targeting potential of modified nanoparticles.

- Single-antigen vaccines are less effective, and integrating multiple antigens is challenging. Purified mammalian and bacterial membranes show promise as nanovaccines, but low antigen expression limits their efficacy. Recent studies suggest that engineered hybrid membrane-based nanomaterials can express multiple antigens and ligands to induce immune responses and targeted delivery for cancer and infectious diseases.

This book explores advancements in surface-engineered cells, cell-derived extracellular vesicles, bacterial membrane vesicles, cell membrane-coated nanoparticles, and hybrid membrane-based nanomaterials. This book covers cell surface engineering strategies and explores biomaterials integrated onto cell membrane platforms. The integration of click groups, proteins, peptides, nucleic acids, drugs, polymers, and nanoparticles onto cell membrane can regulate cell-cell interactions, targeted homing, tracking, barrier penetration, targeted drug delivery, strengthening immune activation, conferring protection, immune shielding, and efficacy of cell membrane-based platform. This book explores the applications of engineered cells and cell membrane-based nanoplatforms in immune engineering, cancer treatment, gastrointestinal disease management, tissue engineering and regenerative medicine, universal blood transfusion, diabetes treatment, tissue modeling, circulating tumor cell capture and neutralization, biomedical imaging, and the treatment of infectious, neurodegenerative, and inflammatory diseases. These topics have been thoroughly explored, offering academics, clinicians, and researchers a detailed understanding of the latest advancements in cell membrane engineering and its biomedical potential. This book reviews recent discoveries in cell membrane therapeutics, offering instructional content for graduate students in chemical biology, biomaterials, biomedical engineering, immune engineering, nanobiotechnology, tissue engineering, and regenerative medicine. In summary, this interdisciplinary book brings together pioneering research to inspire a broad scientific audience and drive innovation in disease monitoring, precision treatment, and biomedical applications

Suzhou, China

Zahid Hussain
Renjun Pei

Acknowledgments

The editors would like to express their deep appreciation and indebtedness to all contributing authors for their hard work, dedication, and significant contributions to this book. We express our deep gratitude to Yong Wang for their insightful foreword. Our special thanks go to the Series Editors, J. Miguel Oliveira and Rui L. Reis, for initiating the Springer Reference Series on Biomaterials, Bioengineering and Sustainability. We sincerely appreciate the thoughtful feedback and constructive suggestions provided by J. Miguel Oliveira on our chapters, which have significantly enhanced the quality and impact of our work. Finally, we extend our heartfelt gratitude to the outstanding staff at Springer Nature for their generous assistance and exceptional support in publishing this book, especially Éva Lőrinczi (Publishing Editor), Sudha Kannan (Production Editor), Immaculate Sofiya Fernandus (Production Supervisor), and Gonzalo Cordova (Former Publishing Editor).

Contents

Editors and Contributors

About the Editors

Zahid Hussain CAS Key Laboratory for Nano-Bio Interface, Suzhou Institute of Nano-Tech and Nano-Bionics (SINANO), Chinese Academy of Sciences, Suzhou 215123, China

Zahid Hussain received his MSc degree (2013) in Biochemistry from Punjab University, Lahore, and his MS degree (2017) in Biochemistry and Molecular Biology from COMSATS University, Islamabad. He earned his PhD (2019–2023) in the same field at the University of Science and Technology of China (USTC). He then joined Suzhou Institute of Nano-Tech and Nano-Bionics (SINANO) as a Research Assistant Professor. His research interests focus on therapeutic nanomaterials, cell surface engineering, and fabrication of cell-instructive scaffolds. He has published up to 35 peer-reviewed academic papers in prestigious journals. https://orcid.org/0000-0002-2824-2672

Renjun Pei CAS Key Laboratory for Nano-Bio Interface, Suzhou Institute of Nano-Tech and Nano-Bionics (SINANO), Chinese Academy of Sciences, Suzhou 215123, China

Renjun Pei received his bachelor's and PhD degrees from the Department of Chemistry of Wuhan University in 1993 and 1998, respectively. He engaged in research at the Changchun Institute of Applied Chemistry of Chinese Academy of Sciences, Southern Illinois University, Kyoto University, and Columbia University from 1998 to 2011. Since 2011, he has been working as a professor at the Suzhou Institute of Nano-Tech and Nano-Bionics (SINANO), Chinese Academy of Sciences. Together with his research team, his current research interests include aptamer SELEX and applications, nano-based diagnosis and therapy, organoids, cell surface engineering, and stem cells for tissue regeneration. He has published up to 150 peer-reviewed academic papers in prestigious journals. https://orcid.org/0000-0002-9353-3935

Contributors

Nisar Ahmad College of Pharmacy, Dalian Medical University, Dalian, China

Wajiha Ahmed CAS Key Laboratory for Nano-Bio Interface, Suzhou Institute of Nano-Tech and Nano-Bionics (SINANO), Chinese Academy of Sciences (CAS), Suzhou, China

Derajram Benival National Institute of Pharmaceutical Education and Research (NIPER), Ahmedabad, Gujarat, India

Yun Chang Department of Biomedical Engineering, The Hong Kong Polytechnic University, Hong Kong, China

The Hong Kong Polytechnic University Shenzhen Research Institute, Shenzhen, China

Changchong Chen Department of Biomedical Engineering, National University of Singapore, Singapore

Hewan Dawit CAS Key Laboratory for Nano-Bio Interface, Suzhou Institute of Nano-Tech and Nano-Bionics (SINANO), Chinese Academy of Sciences (CAS), Suzhou, China

Nimeet Desai National Institute of Pharmaceutical Education and Research (NIPER), Ahmedabad, Gujarat, India

Cheng Dong Department of Biomedical Engineering, The Hong Kong Polytechnic University, Hunghom, Hong Kong, China

The Hong Kong Polytechnic University Shenzhen Research Institute, Shenzhen, China

Yavuz Nuri Ertas Department of Biomedical Engineering, Erciyes University, Kayseri, Türkiye

ERNAM–Nanotechnology Research and Application Center, Erciyes University, Kayseri, Türkiye

Tian Gao CAS Key Laboratory for Nano-Bio Interface, Suzhou Institute of Nano-Tech and Nano-Bionics (SINANO), Chinese Academy of Sciences (CAS), Suzhou, China

Ali Hasan Department of Zoology, Wildlife and Fisheries, University of Agriculture Faisalabad, Faisalabad, Pakistan

Maryam Heidari Department of Tissue Engineering and Regenerative Medicine, Faculty of Advanced Technologies in Medicine, Iran University of Medical Sciences, Tehran, Iran

Cellular and Molecular Research Center, Iran University of Medical Sciences, Tehran, Iran

Ali Hussain Institute of Zoology, University of the Punjab, Lahore, Pakistan

Eman Hussain Department of Pharmaceutics, University of Gezira, Wad Madani, Sudan

Zahid Hussain CAS Key Laboratory for Nano-Bio Interface, Suzhou Institute of Nano-Tech and Nano-Bionics (SINANO), Chinese Academy of Sciences (CAS), Suzhou, China

Lai Jiang School of Pharmaceutical Sciences, Zhejiang Chinese Medical University, Hangzhou, China

Mohammad Reza Khalili Blood Transfusion Research Center, High Institute for Research and Education in Transfusion Medicine, Tehran, Iran

Muhammad Kamran Khan Australian Regenerative Medicine Institute, Medical Nursing and Health Sciences (MNHS), Monash University, Clayton, VIC, Australia

Suzhou Industrial Park Monash Research Institute of Science and Technology, Suzhou, China

Department of Biotechnology, Southeast University-Monash University Joint Graduate School, Suzhou, China

Department of Microbiology, Biomedicine Discovery Institute, Medical Nursing and Health Sciences (MNHS), Monash University, Clayton, VIC, Australia

Mohammad Hassan Kheirandish Department of Medical Biotechnology, School of Medicine, Shahid Sadoughi University of Medical Sciences, Yazd, Iran

Selvaraj Kunjiappan Department of Biotechnology, Kalasalingam Academy of Research and Education, Virudhunagar, Tamil Nadu, India

Xiaoliu Liang College of Pharmacy, Guangxi Medical University, Nanning, China

Tong Li Department of Biomedical Engineering, The Hong Kong Polytechnic University, Hong Kong, China

Bin Liu Department of Biomedical Engineering, The Hong Kong Polytechnic University, Hong Kong, China

The Hong Kong Polytechnic University Shenzhen Research Institute, Shenzhen, China

Gang Liu State Key Laboratory of Vaccines for Infectious Diseases, Center for Molecular Imaging and Translational Medicine, Xiang An Biomedicine Laboratory, National Innovation Platform for Industry-Education Integration in Vaccine Research, School of Public Health, Xiamen University, Xiamen, China

College of Pharmacy, Guangxi Medical University, Nanning, China

Jiaxin Ma State Key Laboratory of Vaccines for Infectious Diseases, Center for Molecular Imaging and Translational Medicine, Xiang An Biomedicine Laboratory, National Innovation Platform for Industry-Education Integration in Vaccine Research, School of Public Health, Xiamen University, Xiamen, China

Shah Mehmood CAS Key Laboratory for Nano-Bio Interface, Suzhou Institute of Nano-Tech and Nano-Bionics (SINANO), Chinese Academy of Sciences (CAS), Suzhou, China

Arezou Mehrabi School of Medicine, Tonekabon Branch, Islamic Azad University, Tonekabon, Iran

Peiman Brouki Milan Department of Tissue Engineering and Regenerative Medicine, Faculty of Advanced Technologies in Medicine, Iran University of Medical Sciences, Tehran, Iran

Cellular and Molecular Research Center, Iran University of Medical Sciences, Tehran, Iran

Ethar Mohamed Faculty of Pharmacy, University of Khartoum, Khartoum, Sudan

Mahir Mohammed Faculty of Pharmacy, University of Khartoum, Khartoum, Sudan

ERNAM–Nanotechnology Research and Application Center, Erciyes University, Kayseri, Türkiye

Jeganpandi Senthamarai Pandi Department of Biotechnology, Kalasalingam Academy of Research and Education, Virudhunagar, Tamil Nadu, India

Griva Parmar A. R. College of Pharmacy and G. H. Patel Institute of Pharmacy, Anand, Gujarat, India

Parasuraman Pavadai Department of Pharmaceutical Chemistry, Faculty of Pharmacy, M.S. Ramaiah University of Applied Sciences, Bengaluru, Karnataka, India

Renjun Pei CAS Key Laboratory for Nano-Bio Interface, Suzhou Institute of Nano-Tech and Nano-Bionics (SINANO), Chinese Academy of Sciences (CAS), Suzhou, China

Zahra Poursalehi Department of Applied Cell Sciences, Faculty of Medicine, Urmia University of Medical Sciences, Urmia, Iran

Fazal Rahim Biomedical Sciences and Health Laboratory of Anhui Province, Institute of Health and Medicine, Hefei Comprehensive National Science Center, University of Science and Technology of China, Hefei, China

Ragini Rai National Institute of Pharmaceutical Education and Research (NIPER), Raebareli, Uttar Pradesh, India

Raja Rajeswari Rajeshkumar Department of Biotechnology, Kalasalingam Academy of Research and Education, Virudhunagar, Tamil Nadu, India

Dhwani Rana National Institute of Pharmaceutical Education and Research (NIPER), Ahmedabad, Gujarat, India

Saima Rani Molecular Virology Laboratory, Department of Biosciences, COMSATS University Islamabad (CUI), Islamabad, Pakistan

Murugesan Sankaranarayanan Medicinal Chemistry Research Laboratory, Department of Pharmacy, Birla Institute of Technology and Science Pilani, Pilani, Rajasthan, India

Nannan Shao CAS Key Laboratory for Nano-Bio Interface, Suzhou Institute of Nano-Tech and Nano-Bionics (SINANO), Chinese Academy of Sciences (CAS), Suzhou, China

Mohammad Ali Shokrgozar National Cell Bank of Iran, Pasteur Institute of Iran, Tehran, Iran

Meng Sun Key Laboratory of Molecular Medicine and Biotherapy, School of Life Sciences, Beijing Institute of Technology, Beijing, China

Panneerselvam Theivendren Department of Pharmaceutical Chemistry & Analysis, School of Pharmaceutical Sciences, Vels Institute of Science, Technology & Advanced Studies, Chennai, Tamil Nadu, India

Mohammad Tollabi Department of Tissue Engineering and Regenerative Medicine, Faculty of Advanced Technologies in Medicine, Iran University of Medical Sciences, Tehran, Iran

Cellular and Molecular Research Center, Iran University of Medical Sciences, Tehran, Iran

Ismat Ullah CAS Key Laboratory for Nano-Bio Interface, Suzhou Institute of Nano-Tech and Nano-Bionics (SINANO), Chinese Academy of Sciences (CAS), Suzhou, China

Salim Ullah The Fifth Affiliated Hospital of Wenzhou Medical University, Zhejiang, China

Lalitkumar K. Vora School of Pharmacy, Queen's University, Belfast, UK

Li Wang CAS Key Laboratory for Nano-Bio Interface, Suzhou Institute of Nano-Tech and Nano-Bionics (SINANO), Chinese Academy of Sciences (CAS), Suzhou, China

Yijia Xie State Key Laboratory of Vaccines for Infectious Diseases, Center for Molecular Imaging and Translational Medicine, Xiang An Biomedicine Laboratory, National Innovation Platform for Industry-Education Integration in Vaccine Research, School of Public Health, Xiamen University, Xiamen, China

Nourhan H. Zahran Department of Chemistry, Faculty of Science, Suez Canal University, Ismailia, Egypt

Qurat ul Ain Zahra Personalized Medicine Centre, Health Futures Institute, Murdoch University, Murdoch, WA, Australia

Jinfeng Zhang Key Laboratory of Molecular Medicine and Biotherapy, School of Life Sciences, Beijing Institute of Technology, Beijing, China

Ye Zhang CAS Key Laboratory for Nano-Bio Interface, Suzhou Institute of Nano-Tech and Nano-Bionics (SINANO), Chinese Academy of Sciences (CAS), Suzhou, China

Samira Zolfaghari Department of Tissue Engineering and Regenerative Medicine, Faculty of Advanced Technologies in Medicine, Iran University of Medical Sciences, Tehran, Iran

Cellular and Molecular Research Center, Iran University of Medical Sciences, Tehran, Iran

Part I

Advances in Cell Membrane Engineering for Biomedical Applications

Biological Membrane Engineering for Unlocking Therapeutic Potential of Membrane-Based Platforms

1

Wajiha Ahmed, Zahid Hussain, and Renjun Pei

Abstract

Advances in cell membrane research have enabled the development of innovative biological membrane-based platforms, including therapeutic cells, extracellular vesicles, bacterial membrane vesicles, membrane-coated nanoparticles, and hybrid membrane-based nanomaterials. These platforms leverage diverse membrane receptors for broad biomedical applications. However, several limitations continue to impede their full potential. Biological membrane engineering offers a promising approach to modifying these platforms, thereby enhancing treatment effectiveness. This chapter explores dynamic functionalities of cell membranes and practical characteristics of biological membranes sourced from various cell types, including red blood cells, platelets, white blood cells, stem cells, cancer cells, and bacteria. This chapter further explores the scope and limitations of membrane-based platforms and highlights membrane engineering strategies that enhance cell-cell and cell-biomaterial interactions, provide protection and immune shielding, improve bioactivity through cargo loading, modulate disease microenvironments, and incorporate capturing and tracking functions. It also addresses targeting the blood-brain barrier, improving homing to injured tissues, and strengthening the immune microenvironment. These advancements significantly improve the performance of membrane-based platforms, driving innovations in the treatment of neurodegenerative, infectious, cancerous, gastrointestinal, and other diseases. Finally, the chapter outlines future research directions for engineered membrane-based platforms, focusing on addressing unmet clinical needs and further expanding their biomedical impact.

W. Ahmed · Z. Hussain (✉) · R. Pei (✉)
CAS Key Laboratory for Nano-Bio Interface, Suzhou Institute of Nano-Tech and Nano-Bionics (SINANO), Chinese Academy of Sciences (CAS), Suzhou, China
e-mail: wanjia2024@sinano.ac.cn; zahid@mail.ustc.edu.cn; rjpei2011@sinano.ac.cn

Z. Hussain, R. Pei (eds.), *Cell Membrane Engineering for Advancing Cell-Based and Nanotechnology-Based Therapies*, Biomaterials, Bioengineering and Sustainability 6, https://doi.org/10.1007/978-3-031-93425-4_1

Keywords

Cell membrane · Surface markers · Membrane-based platforms · Engineered functionalities · Bioactivity enhancement · Biomedical applications

1.1 Introduction

Biological cells are the fundamental units of life, enclosed by a semi-permeable plasma membrane. This membrane functions as both a physical barrier and a dynamic organelle, maintaining stability and enabling essential activities. The cell membrane consists of a lipid bilayer embedded with proteins, carbohydrates, and glycoconjugates, such as glycoproteins, glycolipids, and proteoglycans. Phospholipids and cholesterol form a selective barrier, protecting the cell, regulating fluidity, and supporting lateral diffusion of membrane components (Cheng and Smith 2019; Shivatare et al. 2022). Proteins and glycoconjugates, linked to the lipid bilayer through hydrophobic or hydrophilic interactions, are organized across multiple structural scales. They function as receptors or ligands for signal exchange and play a crucial role in supporting various cellular functions (Chandler and Costello 2016). The cell membrane exhibits a heterogeneous composition, with distinct proteins and glycoconjugates collectively termed surfaceomes. These components vary across cell states and organisms (Bausch-Fluck et al. 2019), regulating numerous processes, including cell adhesion, migration, cell-niche communication, homing (Liu et al. 2025; Choi et al. 2023), pathogen recognition, and immune response (Borst et al. 2021; Gao et al. 2015). The expression and composition of cell surface receptors, ligands, and surface glycans in healthy and diseased states represent an emerging focus for diagnostic and therapeutic development. Notably, up to 50% of known drug targets are cell surfaceome components (Di Meo et al. 2024).

The cell membrane is essential in biological processes across all life forms, enabling specialized functions in dynamic environments and serving as a key target for drug development. Advancements in cell membrane research have driven the development of innovative strategies to address critical challenges in applied biomedical engineering (Sevencan et al. 2020). Over the past decade, numerous membrane-based platforms, including living cells (Park et al. 2018), mammalian extracellular vesicles (EVs) (Wiklander et al. 2019), bacterial membrane vesicles (BMVs) (Toyofuku et al. 2019), cell membrane-coated nanoparticles (NPs) (Kroll et al. 2017), and hybrid membrane-based nanomaterials (Peng et al. 2024), have been explored for disease diagnosis and therapy. The potential and limitations of these platforms are comprehensively examined in Sect. 1.3. Genetic and biomaterial-based engineering strategies have been employed to precisely modify the molecular landscape of the plasma membrane. These approaches enable the functionalization of cell surfaces with receptors, ligands, nanomaterials, antibodies, aptamers, nucleic acids, peptides, bioactive polymers, phenolic compounds, small molecules, and fluorescent probes. The overarching objective of cell membrane engineering is to enhance surface interactions, reinforce intrinsic cellular properties, optimize cell-microenvironment dynamics, enable targeted drug delivery, and integrate exogenous functionalities into membrane-based platforms to augment their biomedical potential (Csizmar et al. 2018; Cheng et al. 2023).

The cell membrane is essential for various biological processes, making it a key target for engineering to manipulate its functions for therapeutic purposes. This chapter explores the practical characteristics of cell membranes derived from various cell types, including red blood cells (RBCs), platelets (PLTs), white blood cells, stem cells, cancer cells, and bacteria. It discusses the scope and limitations of membrane-based platforms, laying the foundation for membrane engineering strategies aimed at enhancing cell-cell and cell-biomaterial interactions, providing protection and immune shielding, improving homing and targeting, strengthening the immune microenvironment, enhancing bioactivity through cargo loading and barrier penetration, and incorporating tracking, sensing, and in situ therapeutic functions. These functionalities significantly enhance the performance of cell-membrane-based platforms, driving innovations in numerous biomedical applications. The chapter also outlines future research directions for engineered membrane-based platforms to address unmet clinical needs. Overall, this chapter offers a comprehensive analysis of the specific properties modulated by cell membrane engineering strategies and their therapeutic potential in disease treatment, presenting a unique perspective not systematically covered in previous reviews (Fig. 1.1).

1.2 Cell Functions Based on Membrane Characteristics

Various cell types, including RBCs, PLTs, leukocytes, stem cells, cancer cells, and probiotic bacteria, have been recognized for their therapeutic benefits. Each cell type has unique functionalities and biological characteristics that contribute to its potential in medical applications (Table 1.1). This section discusses the fundamental properties of these cell types, with a particular focus on their surface characteristics.

1.2.1 Red Blood Cells and Platelets

RBCs are predominant blood cells, playing a vital role in oxygen and carbon dioxide transport to support metabolism. They exhibit remarkable flexibility, allowing them to withstand osmotic pressure and efficiently navigate narrow capillaries. Their biconcave shape provides a high surface-area-to-volume ratio and optimizes gas exchange. Their non-nucleated structure and lack of organelles give RBCs a high drug-loading capacity and simplified membrane purification. RBCs are easily isolated through centrifugation and are suitable for long-term storage. They circulate for approximately 120 days and undergo natural biodegradation without producing byproducts. Sialoglycoproteins on the RBC membrane create a negative charge, preventing aggregation and ensuring smooth blood flow (Zhang et al. 2024).

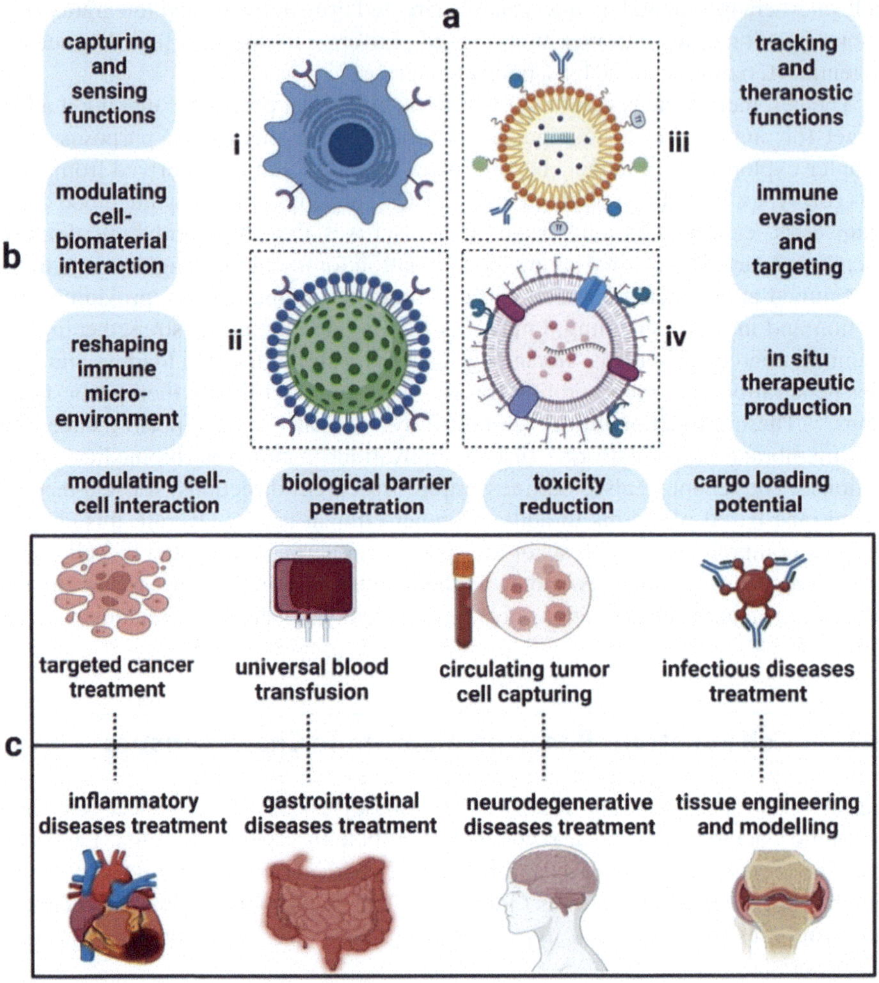

Fig. 1.1 (**a**) Overview of engineered cell membrane-based therapeutic platforms: (i) engineered living cells, (ii) engineered membrane coated nanoparticles, (iii) engineered bacterial membrane vesicles, and (iv) engineered mammalian extracellular vesicles. (**b**) Functional modulation and bioactivity enhancements properties of engineered cell membrane-based platforms. (**c**) Potential biomedical applications of engineered cell membrane-based platforms. (Source: Created using BioRender.com)

Membrane-embedded macromolecules carry blood group antigens and immunoregulatory proteins, which influence RBCs interactions within the immune system. CD47 functions as a "do not eat me" signal by engaging with SIRPα on phagocytes, inhibiting immune clearance. Other key membrane proteins, such as complement receptor 1, membrane cofactor protein, homologous restriction protein, CD55 (decay-accelerating factor), C8-binding protein, and CD59, contribute to immune

Table 1.1 Cell types, membrane receptors, and therapeutic applications based on membrane characteristics

Cell type	Key membrane receptors	Therapeutic applications
Red blood cells	Sialoglycoproteins, CD47, ABO blood group antigens, GPI-anchored proteins	Intravascular drug delivery, extended circulation, blood transfusion, compatibility
Platelets	Glycoprotein Ib-IX, αIIbβ3 integrins, CD47, CD55, CD59, GPIbα receptor	CTCs capture, tumor targeting, hemostasis enhancement, thrombosis inhibition, wound healing, immune modulation
Leukocytes	CCR2, VCAM-1, ICAM-1, MHC, CCR7, LFA-1, CXCR2, CD40, TLRs, β2, β7 integrins	Tumor/inflammation targeting, immune modulation, macrophage-mediated phagocytosis
Stem cells	CCR1, CCR2, CXCR4, VEGF, EGF receptors, α4β1, α5β1 integrins, Notch signaling proteins, CD34, CD73	Tissue repair, drug delivery, regenerative medicine, homing to injury sites, stem cell-based therapies for neurodegeneration
Cancer cells	E-cadherin, E-selectin, EpCAM, HER2, EGFR, αvβ3 integrins	Homotypic tumor targeting, cancer vaccines, immune checkpoint blockade, personalized cancer therapy
Bacteria	PAMPs (LPS, LTA), OmpA, pili, flagella	Drug delivery, immunotherapy, vaccine development, bacterial-mediated tumor targeting

regulation and protect RBCs from complement-mediated lysis. Overall, RBCs membrane-based platforms offer extended circulation, biocompatibility, immune privilege, and minimal toxicity, making them ideal for intravascular drug delivery (Xia et al. 2019; Rao et al. 2015).

PLTs are small anucleate cells and the second most abundant constituents of blood. Upon release from bone marrow megakaryocytes into the bloodstream, PLTs circulate for 7–10 days. Resting platelets have a smooth, disc-like shape, while in the activated state, they extend irregular pseudopods. The Glycoprotein Ib-IX receptor complex and integrins are essential for PLT adhesion to damaged endothelium, triggering activation, thrombus formation, and hemostasis. These functions enable PLTs to assess blood vessel integrity and serve as first responders to damage, bleeding, and inflammation (Periayah et al. 2017; Sekhon and Sen 2014). PLTs membrane possesses multiple receptors, enabling diverse functions. PLTs can bind to circulating tumor cells (CTCs) and aggregate at tumor sites through interactions with integrin αvβ3, P-selectin, and C-type lectin-like receptor 2. PLTs contain immunomodulatory receptors (CD47, CD55, and CD59) that signal "do not eat me," preventing macrophage clearance, suppressing the complement system, and prolonging their circulation. PLTs-membrane-based platforms can accumulate at the target site, reduce elimination of membrane-coated systems, and improve the pharmacokinetics of injected therapeutics. PLTs and PLTs membrane have been engineered for capturing CTCs, targeting metastatic niches, and targeting vascular inflammation and injured tissues (Hu et al. 2015; Li et al. 2016; Han et al. 2022a).

1.2.2 Leukocytes

Leukocytes are classified into two main categories: granulocytes and agranulocytes. Granulocytes differentiate into three subtypes: neutrophils, eosinophils, and basophils. Agranulocytes differentiate into monocytes and lymphocytes. Monocytes can further develop into dendritic cells (DCs), and macrophages, while lymphocytes are categorized into B-lymphocytes (B-cells), T-lymphocytes (T-cells), and natural killer (NK) cells. The next section details the membrane characteristics of key biomedical cells, including macrophages, DCs, and neutrophils.

Macrophages are key phagocytic cells of the innate immune system, serving as both effector cells and antigen-presenting cells. They can be categorized into pro-inflammatory (M1) and anti-inflammatory (M2) subtypes, both regulating immune responses and homeostasis. As phagocytes, macrophages engulf pathogens, damaged cells, and foreign particles while presenting antigens to T-cells to modulate adaptive immunity. Macrophages exhibit a natural tropism toward inflammatory sites and tumors, directed by cytokine and chemokine gradients (Chen et al. 2023). Leveraging this property, macrophage membrane-coated NPs target inflamed tissues via receptors-ligand interaction and adhesion molecules such as C-C chemokine receptor 2 (CCR2), vascular cell adhesion molecule-1 (VCAM-1), and intercellular adhesion molecule (ICAM). For instance, integrins present on macrophage membranes can bind to VCAM-1 on cancer cells, enabling efficient glioma-targeting delivery. In comparison to other phagocytes, macrophages have an extended lifespan, making them promising therapeutical candidates. Overall, macrophage offer prolonged circulation, pathogen sequestration, enhanced antigen presentation, T-cell activation, and targeted drug delivery for tumors and inflammation (Lopes et al. 2022; Wu et al. 2022).

Neutrophils are vital players in the innate immune system and play a key role in the inflammatory responses and can reach inflammatory sites faster than other immune cells. Upon detection of chemokines, they swiftly migrate to sites of infection or inflammation, adhering to endothelial cells through integrin-mediated interactions. Numerous neutrophil membrane coated nanoplatforms have been introduced to target endothelial cells. Subsequently, neutrophils traverse the endothelial barrier, infiltrating into tissues via intercellular gaps to execute their immune functions. Similarly, due to inherent cell membrane component, such as LFA-1, CXCR2 and CD11b, neutrophil membrane-coated nanosystems can inherit the biological properties and exert the ability of intercellular communication. Neutrophils are the most prevalent circulating leukocytes among immune cells; however, their limited circulation time may constrain their use in biomedical applications (Kang et al. 2017; Zhang et al. 2018).

DCs are antigen-presenting cells that capture tumor and pathogen antigens, present them on major histocompatibility complexes (MHC), and upregulate costimulatory markers on their surface. Adhesion and immune co-regulatory molecules, including ICAM-3, CD40, CD44, and integrins on DCs membranes, facilitate cell adhesion and enhance DC-T cell interactions, promoting T-cell activation and differentiation. Mature DC membrane-coated NPs retain antigen-presenting capabilities and can specifically activate T cells due to their broad spectrum of peptide/MHC complexes. Additionally, DCs membrane-coated NPs achieve lymph node targeting via the CCR7 receptor. DCs play a crucial role in mediating immunity, stimulating cytotoxic and helper T lymphocytes to combat infections and cancer (Harvey et al. 2022; Ma et al. 2023a). Overall, based on their diverse repertoire of receptors, leukocyte membranes serve as sophisticated biointerfaces for targeted drug delivery against tumors, inflammation, and infectious diseases. Additionally, leukocyte membrane-based platforms facilitate pathogen sequestration, immune modulation, and CTCs capture for cancer diagnostics. Their potential applications establish them as transformative tools in advanced therapeutics and precision medicine.

1.2.3 Stem Cells

Stem cells, including mesenchymal stem cells (MSCs), are multipotent differentiation, and therapeutic properties such as anti-inflammatory, anti-apoptotic, and immune-modulatory potential. These cells originate from sources like bone marrow, adipose tissue, liver, dental pulp, placenta, umbilical cord, and amniotic fluid and can differentiate into osteocytes, chondrocytes, hepatocytes, myocytes, cardiomyocytes, adipocytes, and neurons (Gopalarethinam et al. 2023). Stem cells possess innate targeting capabilities, enhancing their potential for treating degenerative diseases. Beyond regenerative medicine, stem cells are also employed as drug carriers, targeting injured tissues, and crossing the blood-brain barrier (BBB). Their diverse membrane composition, with receptors for cytokines, chemokines, growth factors, and cell-matrix components, likely underpins their biological functions. For example, CCR1 (CCL3, CCL5) and CCR2 (CCL2, CCL7) drive migration to inflamed sites, CCR3 (CCL5, CCL11) and CCR4 (CCL17, CCL22) aid homing and immune interactions, CCR5 (CCL3, CCL4) and CCR7 (CCL19, CCL21) support recruitment to inflammation, and CXCR4 (CXCL12) is key for bone marrow homing. Stem cells express diverse growth factor receptors that control proliferation, differentiation, and tissue regeneration. For example, fibroblast growth factors promote self-renewal capacity of stem cells, and platelet-derived growth factors drive mesenchymal stem cell differentiation into bone and fat cells (Wu and Zhao 2012; Han et al. 2022b).

1.2.4 Cancer Cells and Bacteria

Cancer cells exhibit angiogenesis, immortality, resistance to cell death, evasion of growth suppressors, uncontrolled proliferation, and metastasis, causing severe damage to tissues and organs. Tumor cells express surface proteins like cadherins, selectins, integrins, immunoglobulins, and homing receptors, which regulate invasion, metastasis, recognition, and immune evasion, driving cancer progression and therapy resistance (Zhang et al. 2020). Based on these characteristics, cancer cell membrane platforms have been utilized to evade immune recognition, selectively target homotypic tumors, and develop cancer vaccines. For example, membrane-associated markers such as Galectin-3, N-cadherin, the epithelial cell adhesion molecule (EpCAM), and E-cadherin facilitate tumor targeting, homotypic recognition, and adhesion. Cancer cell membrane-based platforms enhance nanomedicine endocytosis and tumor accumulation via homologous targeting. Enriched with tumor-specific antigens, they also activate antigen-presenting cells, stimulating cytotoxic T lymphocyte immunity and supporting cancer vaccine development. This targeted approach enhances the therapeutic effect on tumor cells while minimizing toxicity to healthy tissues. Unlike membranes derived from cultured cancer cells, those isolated from surgically removed primary tumors retain a comprehensive array of tumor antigens, which induce robust immune responses and hold significant potential for personalized immunotherapy (Ma et al. 2023b; Chen et al. 2021a).

Bacteria are distinguished as gram-positive or gram-negative types based on differences in their cell wall structure and lipid composition. Most bacteria exist as symbiotic organisms that support human health, while certain pathogenic species evade immune detection and cause disease. Both commensal and pathogenic bacteria, along with their membrane-derived components, have been extensively studied for therapeutic applications (Omer et al. 2022). Bacterial membrane vesicles (BMVs) closely resemble natural bacterial membranes, containing phospholipids, proteins, pathogen-associated molecular patterns (PAMPs), lipopolysaccharides (LPS) and lipoteichoic acids (LTA). BMVs have been utilized for drug delivery, leveraging bacterial membrane properties to enhance cellular uptake and immune system activation. Certain bacterial strains, including *Salmonella typhimurium*, *Bifidobacterium bifidum*, and *Clostridium beijerinckii,* produce BMVs with tumor-targeting capabilities, allowing them to selectively colonize hypoxic tumor regions for drug delivery. Probiotics and probiotic-derived BMVs function as adhesion mediators and are used in managing gastrointestinal diseases due to their immunomodulatory and targeting effects. Overall, bacterial membrane-based platforms offer significant potential for cancer and infectious disease drug delivery, vaccine development, and gastrointestinal disease management (Toyofuku et al. 2023; Sartorio et al. 2021).

1.3 Biological Membrane Platforms: Challenges and Solutions

1.3.1 Living Cells

Biological cells, as the fundamental building blocks of life, are intricate living machines capable of performing specialized functions within complex environments. Cell-based therapies provide integrated functions, including self-renewal, differentiation, paracrine stimulation, communication, tissue homing, barrier penetration, and therapeutic effects like anti-inflammatory, immune-modulatory, and pathological targeting. The unique functionalities of cell-based therapies are difficult to replicate with molecular medicine or scaffolds, which position living cells as a powerful option for treating intractable diseases. For instance, RBC transfusions are used to treat anemia (Du Pont-Thibodeau et al. 2014), chondrocytes help repair cartilage injuries (Akkiraju and Nohe 2015), Islets are employed to cure diabetes (Cayabyab et al. 2021), immune cells target and eradicate tumors (Irvine et al. 2022), stem cells contribute to tissue repair (Margiana et al. 2022), and probiotics are used in the treatment of gastrointestinal diseases (Milner et al. 2021). Natural cells have limitations that impede optimal outcomes, including susceptibility to harsh environments, reduced viability after transplantation, and loss of homing signals in diseased microenvironments. Surviving cells may experience undesired differentiation, immune complications, and off-target toxicity. Furthermore, these cells lack exogenous functions, exhibit low drug-loading efficiency, and have unclear biodistribution due to ineffective tracking. The success of cell-based therapies depends on the precise regulation of cell function within both physiological and pathological environments. These challenges emphasize the need for membrane engineering to improve targeting, differentiation, interactions, biodistribution tracking, immune modulation, and dynamic responsiveness for tissue regeneration and immunotherapy (Pires-Santos et al. 2024; Cubillos-Ruiz et al. 2021).

1.3.2 Mammalian Extracellular Vesicles

EVs are lipid-enclosed nanoparticles secreted during membrane turnover and exocytosis. Based on their biogenesis, they are categorized into exosomes and ectosomes. Exosomes originate through membrane invagination, endosomal maturation, intraluminal vesicle formation, and release via membrane fusion. Ectosomes, which include microvesicles and apoptotic bodies, differ in origin. Microvesicles are formed by the outward protrusion of the cell membrane, retaining characteristics of the parental cell. In contrast, apoptotic bodies form during programmed cell death, carrying organelles and nuclear fragments, though their complexity may limit their applications. The molecular composition of EVs is largely determined by their source cells. Their luminal cargo includes proteins, lipids, various nucleic acids species, carbohydrates, and metabolites, while their shell is composed of phospholipids and are enriched with lipid rafts, ceramides, receptors, cholesterol, and MHC molecules. These structural attributes enable EVs to evade immune detection, exhibit antigen specificity, cross the BBB, support targeted homing, and facilitate immunomodulation through bio-stimulatory effects (El Andaloussi et al. 2013). Natural EVs offer significant advantages; however, their clinical effectiveness is hindered by challenges such as inconsistent cargo composition, variability in surface markers, poor tissue targetability, limited efficiency in loading parent cell expressed cargo, and limited penetration, which result in variability in efficacy and inefficient targeting. For example, variability in surface markers across batches lead to uneven biodistribution and inefficient targeting. To confront these challenges, researchers are developing engineering strategies to enhance EV functionality, stability, targeting efficiency, real-time biodistribution monitoring, and therapeutic potential, expanding their applications in disease treatment (Zhang et al. 2021; Dooley et al. 2021).

1.3.3 Bacterial Membrane Vesicles

Gram-negative bacteria possess both an inner and outer membrane, with the latter enriched in LPS. A peptidoglycan layer lies within the periplasmic space between these membranes, and small projections from the outer membrane bud off to form nanovesicles. In contrast, gram-positive microorganisms possess a single lipid membrane surrounded by a thick peptidoglycan layer containing LTA, which limits vesicle production due to structural constraints. Notably, vesicles from gram-positive bacteria lack periplasmic components and LPS, distinguishing them from gram-negative bacterial vesicles. Bacterial membrane vesicles (BMVs) are classified based on their origin: outer membrane vesicles (OMVS) originate from gram-negative microorganisms, while membrane vesicles are derived from gram-positive microorganisms. BMVs contain key bacterial components, including LPS, phospholipids, peptidoglycan, membrane proteins, toxins, and nucleic acids. Enriched with PAMPs, BMVs elicit strong immune responses, making them valuable for immunotherapy. Their nanoscale size and structural stability prevent antigen degradation and minimize drug leakage, further supporting their potential as delivery vehicles (Toyofuku et al. 2023; Sartorio et al. 2021). Despite their significant potential, clinical use of membrane vesicles face challenges such as variability in surface markers and biochemical properties, leading to batch-to-batch inconsistencies, off-target effects, and drug delivery limitations. Additionally, BMVs from pathogenic bacteria may contain endotoxic LPS, compromising safety and efficacy. To address these challenges, engineering strategies enable functional modifications of BMVs during biogenesis or after isolation, enhancing targeting specificity and optimizing immunogenic potential. These advancements are crucial for vaccine development, adjuvant therapy, cancer immunotherapy, and antibacterial treatment (Qing et al. 2019; Long et al. 2022).

1.3.4 Membrane Coated Nanoparticles

Nanoparticles (NPs) have attracted significant interest in biomedical research due to their high surface area, drug-loading potential, and responsiveness to external triggers such as magnetic fields, and heat. Polymeric, metallic, metal oxide, inorganic, and supramolecular NPs have been developed for applications in chemotherapy, radiotherapy, immunotherapy, and photothermal therapy. However, systemic administration of bare NPs faces challenges such as protein corona formation, aggregation, rapid clearance through reticuloendothelial system, poor tissue target specificity, and off-target accumulation (Rampado et al. 2020; Dai et al. 2018). Significant advances optimize NPs-based delivery system through stealth coatings (e.g., polyethylene glycol) and ligand conjugation (e.g., sialic acid, aptamers, antibodies, peptides). However, recent studies report anti-PEG IgM antibody production, leading to unexpected clearance of PEGylated materials. Meta-analyses highlight insufficient NPs concentration at target sites, even with targeting ligands. The insufficient targeting potential and immunogenicity concerns highlight the need for biomimetic coatings to enhance the targeting efficiency of NPs (Menon et al. 2022; Chen et al. 2021b; Mima et al. 2015). Cell membranes inherit surface molecules from parent cells, replicating biological functions and inspiring their use in NPs coatings to address current challenges. The bio-interfacing properties of cell membranes depend on their source, influencing receptor expression and guiding their design and optimization. Cell membrane coatings improve NPs stability by evading immune clearance, prolonging circulation, crossing biological barriers, enabling homotypic targeting, and neutralizing biological responses. Additionally, tumor cell membranes and bacterial membrane vesicles carry immunogenic ligands that promote disease site accumulation, immune activation, and enhanced vaccine delivery and design (Fang et al. 2023). Despite their potential, single-cell membrane-coated NPs face limitations that hinder their translation into complex biological systems, often due to the absence of recognizable moieties. For example, RBCs coatings extend circulation time and reduce immune clearance but offer limited tissue targeting, while cancer cell membrane-coated NPs show strong tumor targeting but have reduced circulation time and blood compatibility. The inherent limitations of a single native membrane in targeting and efficacy have sparked interest in cell membrane engineering to enhance bio-interface interactions, adaptability, and biological performance (Ma et al. 2024; Ai et al. 2020).

1.3.5 Hybrid Membrane-Based Nanomaterials

Lipid membrane fusion plays a key role in cellular physiology and membrane engineering. Like natural vesicles, such as EVs and BMVs, liposomes have a spherical structure with an aqueous core and a synthetic phospholipid bilayer. However, liposomes face challenges like biofouling, short circulation times, immune recognition, and limited targeting. To address these limitations, there is increasing interest in integrating cell membrane fragments into liposomes to develop biomimetic nanomaterials. Combining membrane fragments from various cell sources or merging natural vesicles or liposome membranes can create vesicle-mimetic nanomaterials with superimposed functions and enhanced immune evasion, immunomodulation, tissue targeting, and drug-loading capacity (Wang et al. 2021; Sun et al. 2023). Unlike naturally derived vesicles, hybrid membrane-based nanomaterials can be produced in large quantities at lower costs. Recent studies have investigated the fusion of mammalian cell membranes and OMVs to create hybrid nanovesicles that combine functional characteristics from both sources. This approach inherits the immune-stimulatory properties of OMVs and tumor-specific antigens, making it essential for tumor targeting, drug delivery, and the development of anti-cancer vaccines (Zou et al. 2021). For instance, combining RBCs and PLTs membranes extends circulation time while improving affinity for pathogens and injured vasculatures (Liu et al. 2018). Hybrid membrane technology combines the functions of different cell types, enhancing biomedical potential over monotypic cell membrane platforms, but its clinical translation is limited and requires surface engineering with exogenous ligands, tracking agents, and therapeutic compounds. For example, fusing engineered liposomes, like photothermal and drug-loaded variants, with engineered vesicles enables phospholipid and cargo exchange, creating hybrid nanovesicles with enhanced stimulus-responsive properties. Modifying these hybrid membrane-based nanomaterials with targeting peptides, click moieties, therapeutic drugs, and tracing agents further enhances their functional versatility in biomedical applications (Jhan et al. 2020; Ducrot et al. 2023).

1.4 Modulating Performance of Membrane-Based Platforms Through Engineering

Biological membranes, composed of polysaccharides, proteins, and lipids, contain functional groups that enable chemical and physical modifications using synthetic materials. Rapid progress in biotechnology and materials science have led to the development of several strategies to enhance membrane functionality, including covalent conjugation, bioorthogonal glycoengineering, lipid insertion, membrane fusion, and genetic engineering. These approaches have enabled the functionalization of cell surfaces with receptors, nanomaterials, antibodies, aptamers, nucleic acids, peptides, polymers, phenolic compounds, small molecules, and fluorescent probes. Numerous reviews have been published on these strategies, and readers are referred to excellent reviews for more details (Csizmar et al. 2018; Cheng et al. 2023; Wang and Hu 2023; Adebowale et al. 2024). This section presents a detailed analysis of the specific properties modulated by cell membrane engineering strategies and their therapeutic potential in disease treatment.

1.4.1 Modulating Cell-Cell and Cell-Biomaterials Interaction

Scaffold-free strategies harness the self-organization potential of cells to form stable aggregates, such as cell sheets and spheroids, without exogenous materials. They offer high cell density, improved viability, and reduced cytotoxicity, supporting tissue engineering and ex vivo modeling. However, challenges remain in building multilayered architectures due to limited cell interactions and alignment issues in 3D constructs. The heterogeneity of human tissues makes replication difficult using traditional scaffold-free approaches, underscoring the need for cell surface engineering to enable precise patterning. Layer-by-layer (LbL) films are widely used to create ultrathin cell surface coatings by sequentially depositing oppositely charged polymers via electrostatic and intermolecular interactions. These films support single-cell survival, enhance cell clustering through surface adhesion, and enable scalable fabrication of complex 3D structures. For example, fibronectin (negatively charged) and gelatin (negatively charged) nano-coatings on single-cell surfaces enhance integrin-mediated adhesion, promoting rapid aggregation and and the formation of dense β-cell spheroids with improved insulin secretion. This simple and effective technique offers a promising approach for the rapid fabrication of multilayered tissues (Fukuda et al. 2018; Nishiguchi et al. 2011) (Fig. 1.2a). Click chemistry-based cell surface engineering enables precise cell placement and patterning for creating functional anatomical structures. For example, cationic liposomes have been used to deliver bioorthogonal click groups, such as ketone and oxyamine, to the surfaces of various cell types for targeted modification. This approach enables the formation of spheroid assemblies composed of interconnected cells through click ligation between ketone- and oxyamine-engineered cells. Unlike LbL approach, click chemistry-based engineering allows for the precise assembly of heterogeneous cell populations, enabling the fabrication of densely packed,

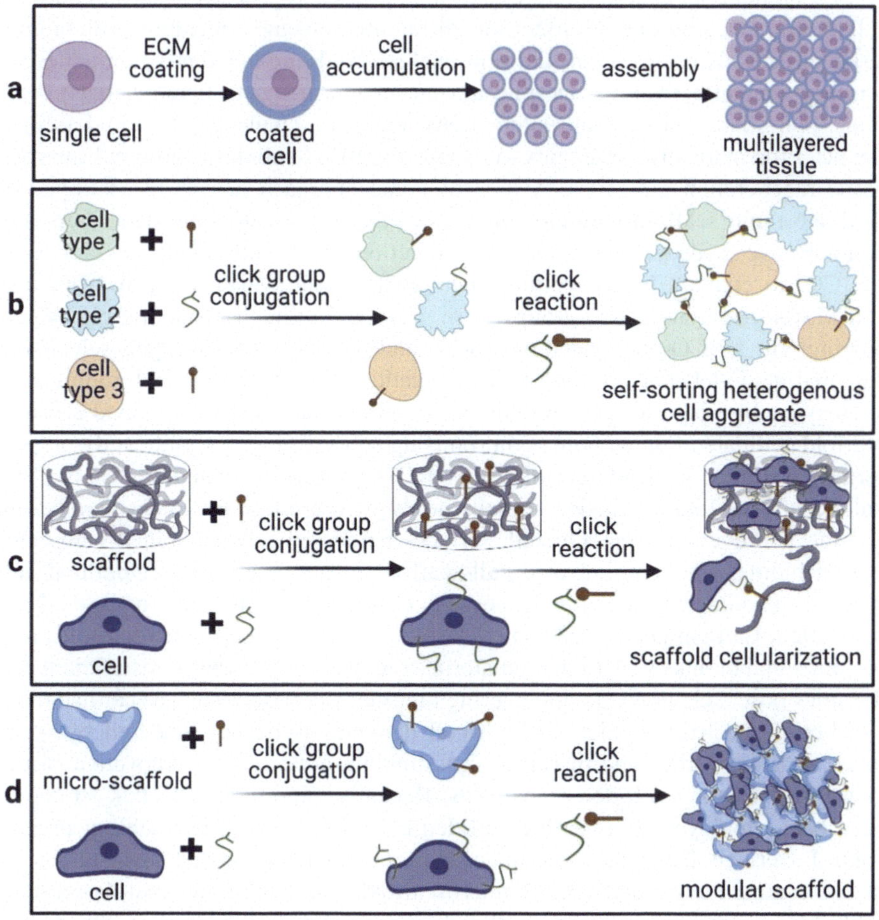

Fig. 1.2 Schematic illustration of: (**a**) Single-cell surface coating with biocompatible extracellular matrix materials for multilayered tissue assembly. (**b**) Single-cell surface engineering with different biorthogonal groups enables precise cellular interaction which could enable fabrication of self-sorting heterogenous cell aggregates. (**c**) Engineering 3D scaffolds and living cells with bioorthogonal groups enables precise cell-scaffold interactions via click reactions, enhancing cellularization. (**d**) Bioorthogonal functionalization of micro-scaffolds and cells enables precise interactions and high-density modular scaffold construction. (Source: Created with BioRender.com)

functional cardiac tissue. This method offers a promising strategy for constructing engineered tissues with enhanced structural and functional properties. Furthermore, the use of UV light-responsive photo-oxyamine lipids in click-crosslinked structures allows for the cleavage of photolabile oxime linkages upon UV exposure, enabling the controlled disassembly of microtissues. Overall, bio-orthogonal approach provides greater control over spheroid assembly and facilitates the creation of multilayer tissues with different cell types through specific interactions among distinct cell populations (Rogozhnikov et al. 2016; Luo et al. 2014) (Fig. 1.2b).

Cell-biomaterial interactions are essential for tissue regeneration, as they support cell attachment, adhesion, proliferation, differentiation, and integration with biomaterial surfaces. However, many conventional scaffolds are inherently non-adhesive, limiting their effectiveness. Despite advancements in ligand-functionalized scaffolds, optimizing cell-scaffold interactions remains a challenge. This underscores the need for innovative strategies to enhance scaffold cellularization and improve tissue regeneration outcomes. A bioorthogonal approach has been developed to facilitate rapid scaffold cellularization. For example, electrospun nanofibers and non-adhesive hydrogel surfaces were functionalized with dibenzocyclooctyne (DBCO) to create a biocompatible microenvironment. Meanwhile, macrophages and MSCs displaying azide groups were generated through metabolic glycoengineering. The DBCO-conjugated nanofibers and non-adhesive hydrogels selectively captured azide-labeled cells via a click reaction. This bioorthogonal conjugation strategy enables precise cell immobilization, enhanced cell spreading, and efficient scaffold cellularization without compromising cell viability. It significantly outperforms non-functionalized scaffolds and RGD peptide-decorated scaffolds. This unique cell-scaffold interaction strategy holds great promise for tissue engineering and regenerative medicine. Overall, surface-engineered cells combined with engineered biomaterials can enhance cell-scaffold interactions, support optimal cell densities, and improve the effectiveness of cell-seeded or cell-laden scaffolds (Mao et al. 2020; Battigelli et al. 2020) (Fig. 1.2c). Conventional cell-laden scaffolds rely on functional bioinks with high biomaterial-to-cell ratios. However, insufficient cell densities pose a challenge in engineering high-density tissues such as cardiac muscle, brain, and liver. Achieving high cell densities is further hindered by weak mechanical properties and disrupted crosslinking, necessitating innovative strategies for fabricating cell-dense scaffolds. Recently, bioorthogonal click strategies have been employed to assemble cell-dense modular scaffolds, enabling precise control over cell distribution and material properties for improved tissue engineering outcomes. For example, DNA microscaffolds and hyaluronic acid-based scaffolds have been functionalized with DBCO moieties, while azide groups were introduced into cell surface glycoproteins via metabolic glycoengineering. The click reaction between polymeric microscaffolds and azide-presenting cells enables rapid conjugation, forming cell-dense hydrogels. These modular hydrogels provide a larger surface area, customizable geometries, and controlled cell densities, allowing efficient entrapment of multiple cell types. Moreover, modular scaffolds reduce biomaterial use and effectively replicate the cell-rich nature of tissues compared to traditional cell-laden scaffolds. Consequently, they offer the potential for enhanced cellularization, integration, and transplantation outcomes, facilitating the reconstruction of cell-dense constructs (Nam et al. 2023; Lavrador et al. 2025) (Fig. 1.2d).

1.4.2 Conferring Protection and Immune Shielding

RBC transfusion is a life-saving treatment for patients with blood loss due to traumatic injuries, surgical hemorrhage, or anemia. However, the increasing demand is complicated by immune rejection caused by surface glycoprotein markers, leading to hemolysis and cross-type agglutination from mismatched blood. To address these challenges, researchers are developing strategies to mask RhD, A/B, or rare-type antigens, preventing antigen-antibody interactions and eliminating immune rejection from blood type mismatches. Gene editing to create antigen-free RBCs and enzyme-based antigen removal have been explored for universal blood. However, these strategies are costly, unsuitable for emergency transfusions, and may have side effects, limiting their practicality in critical situations. Enzymatic removal of RBC antigens can convert type A/B to type O but does not affect Rh antigens, as their removal disrupts the cell membrane. These limitations hinder the practical application of these strategies in universal blood development, highlighting the need for chemical-based cell surface engineering to shield RBC surface antigens. For example, ferric ions and tannic acid-based metal-phenolic networks, as well as polydopamine coatings, did not cause deterioration of the original shape and function of RBCs. Systemic assessment in a murine model demonstrated excellent survival profiles of coated RBCs, even with repeated transfusions. These coatings prevented coagulation reactions, reduced recognition of ABO and RhD-positive antigens by specific antibodies, attenuated agglutination, and could aid in developing antigenically shielded RBCs for universal emergency transfusions (Park et al. 2017; Wang et al. 2014) (Fig. 1.3a).

Probiotics are beneficial microorganisms that help maintain gut balance and support the treatment of gastrointestinal diseases by inhibiting pathogens, regulating immune responses, enhancing metabolism, neutralizing toxins, and strengthening the gut barrier. However, conventional probiotics often face limitations such as reduced functionality, poor colonization efficiency, and rapid clearance. Their viability and effectiveness are further compromised by environmental stressors, including temperature fluctuations, moisture, oxidation, and harsh gastrointestinal conditions such as acidic pH, digestive enzymes, bile salts, and rapid gut transit. To address these challenges, engineered probiotics have gained increasing attention. Compared to traditional microencapsulation methods, recent advancements have shifted the focus toward smart single-cell nanoencapsulation strategies. Supramolecular self-assembly and LbL coating around individual probiotic cells have been introduced to enhance shelf life, improve delivery efficiency, and enable targeted therapies for gastrointestinal diseases. For example, encapsulating probiotic within metal-phenolic network shells provides biocompatible protection against harsh gastrointestinal conditions and antibiotic-induced damage, enhancing their

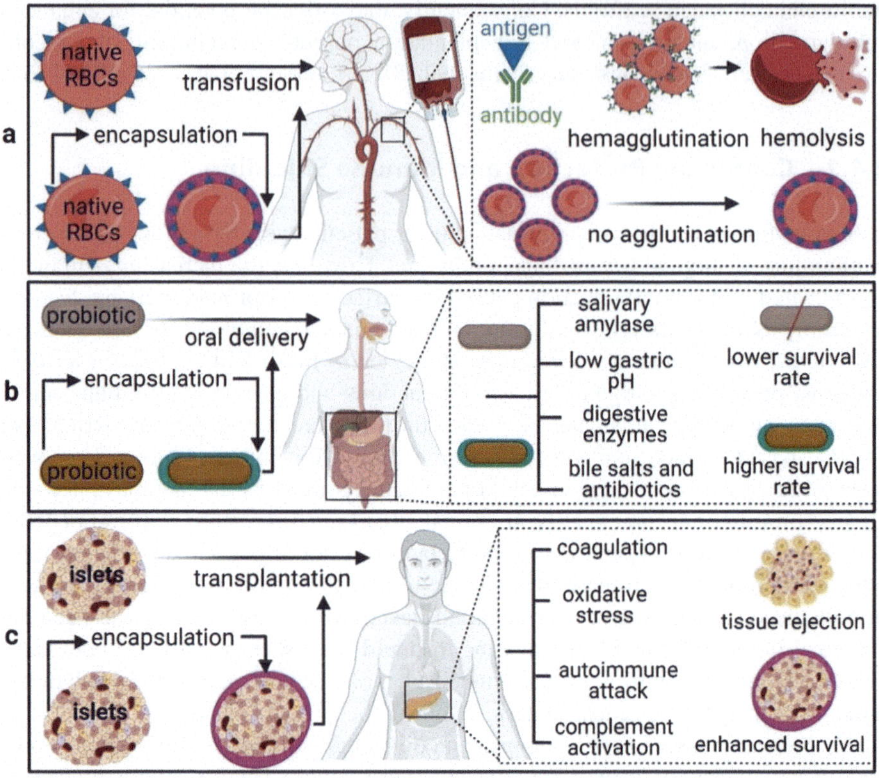

Fig. 1.3 Schematic illustration of: (**a**) Red blood cell surface engineering for universal blood transfusion. (**b**) Probiotic surface engineering to enhance survival in gastrointestinal conditions and ensure targeted delivery to inflamed gut sites. (**c**) Islet surface engineering to improve resistance against in vivo stressors and autoimmune attacks, promoting enhanced survival and functional longevity. (Source: Created with BioRender.com)

survival and colonization. Furthermore, mucin and tannic moieties into the coating enhance probiotic colonization in the mucus layer, reduces gut inflammation through ROS scavenging, and boosts intestinal microbiota, all essential for managing gastrointestinal diseases. Thus, the interaction between living probiotic surfaces and biomaterial coatings enhances protective and functional features, overcoming the limitations of native probiotics and traditional multicell encapsulation methods (Pan et al. 2022; Yang et al. 2022) (Fig. 1.3b).

Type 1 diabetes (T1D) is an immune-mediated disorder marked by the destruction of insulin producing pancreatic cells by self-reactive T cells. This leads to a deficiency in insulin production and impaired glucose regulation. Insulin therapy offers temporary glycemic control but may lead to numerous complications such as neuropathy, kidney issues, and cardiovascular disorders, while the daily injections and blood sugar monitoring can be stressful. Islet transplantation presents a promising alternative to insulin injections for T1D treatment; however, it faces challenges such as insufficient insulin secretion, high cell death, inflammation, thrombosis, immune activation, and the side effects of immunosuppression, often leading to graft failure. There is a need for strategies to protect transplanted islets from autoimmune rejection and harmful host responses, ensuring long-term graft acceptance without the use of immunosuppression. Single-cell nano-encapsulation and surface engineering strategies have been developed to protect transplanted cells, enhance graft acceptance, and enable glucose-responsive insulin release without needing immunosuppression. For example, dual-layer shielding strategy using multi-arm PEG-catechol and unfractionated heparin have been used to reduce inflammatory effects in transplanted islets. The double-layer coated islet surface preserved cell viability, provided anticoagulant effects, and showed significantly longer survival in an allograft model than unmodified islets. In combination with immunosuppressive drugs, shielded islets demonstrated significantly longer survival, highlighting their potential to prevent blood-mediated inflammatory reactions and immune attacks in islet transplantation. Furthermore, surface coating has been functionalized with the fibrinolytic enzyme urokinase and anticoagulant thrombomodulin that could suppress thrombogenic reactions on transplanted islets without insulin release, thereby enhancing graft survival. Overall, drug-coated insulin-producing cellular aggregates prevented thrombosis, reduced bleeding risk, showed no cytotoxicity, and mitigated innate immune responses, thereby preventing coagulation and tissue inflammation (Chen et al. 2011; Im et al. 2013)(Fig. 1.3c).

1.4.3 Enhanced Activity Through Cargo Loading

Traditional NPs-based drug delivery systems may cause side effects because they lack tissue-specific targeting. Given that a cell's targeting potential is largely governed by its membrane, there is growing interest in using cell membranes as vehicles for drug delivery. Cell membrane engineering for drug conjugation involves attaching active drugs or drug carriers to the cell surface, targeting the carriers to specific locations, and controlling drug release. Both artificial and intrinsic ligand-receptor interactions have been utilized for drug loading and targeted delivery to injury sites. For example, IL-12 nanostimulant-engineered chimeric antigen receptor T cell (CAR T) biohybrids (INS-CAR T) have been developed through chemical engineering to enhance the antitumor immunity of CAR T cells. In brief, IL-12-loaded human serum albumin NPs are attached to CAR T cells via click chemistry mechanism without affecting their anticancer potential. IL-12 is selectively released from INS-CAR T biohybrids when tumor antigens trigger an increase in thiol groups on the cell surface. The immune-boosting effects of the IL-12 nanochaperone substantially enhance the anticancer activity of CAR T cells, leading to the dramatic elimination of solid tumors (Fig. 1.4a) (Luo et al. 2022). Furthermore, EVs and BMVs have been engineered to load specific cargo both within their lumen and on their surface for enhanced biomedical applications. Active loading of recombinant and nucleic acid drug involves fusing recombinant proteins with integral proteins of EVs, such as lysosomal-associated membrane protein 2 (Lamp2) and

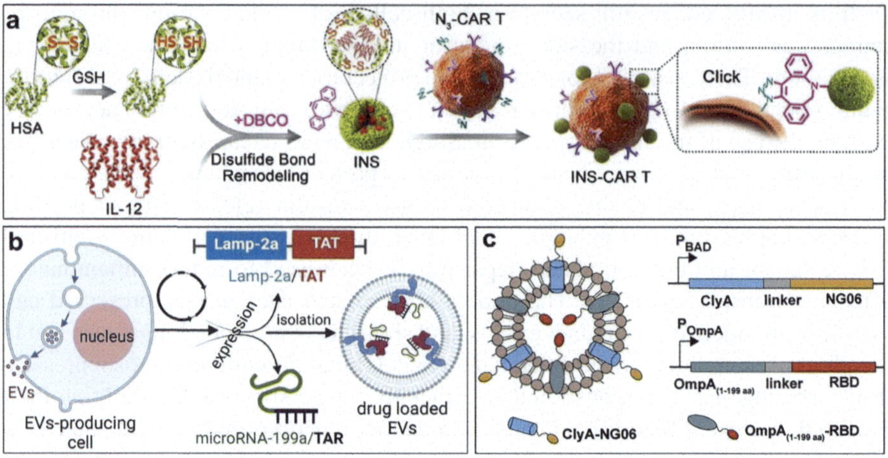

Fig. 1.4 Schematic illustration of: (**a**) Redox-sensitive nanoparticles loaded with IL-12 (INS) were fabricated through the remodeling of human serum albumin (HSA) disulfide bonds and dibenzocyclooctyne (DBCO) labeling. Azide-tagged CAR T cells were then conjugated with the INS nanoparticles using click chemistry, resulting in the creation of INS-CAR T biohybrids. This approach enhanced the effectiveness of immunotherapy in treating solid tumors. (Reproduced with permission from Luo et al. (2022), copyright 2022, Elsevier B.V.). (**b**) Parental cell-based active drug loading of recombinant drug in EVs through genetic engineering. Interaction of C-terminus of Lamp2a-TAT module with TAR-miR-199a facilitate microRNA loading in EVs lumen. (Source: Created with BioRender.com). (**c**) The RBD and NG06 of the spike protein, fused with the vesicle membrane-associated proteins OmpA and ClyA, are directed to the lumen and surface of the BMVs, respectively. (Reproduced with permission from (Wo et al. 2023), copyright 2023, Elsevier B.V.)

tetraspanin CD9. Such mechanisms could assist in specifically loading target proteins into specific site of EVs. For example, interaction of C-terminus of Lamp2a-TAT module with TAR-miR-199a facilitate microRNA loading in EVs lumen (Fig. 1.4b). Similarly, the C-terminus of CD9 has been fused with the RNA-binding human antigen R (HuR) protein using genetic engineering, enabling the CD9-HuR EVs to enrich microRNA-155 into its lumen (Sutaria et al. 2017; Li et al. 2018a). In BMVs, recombinant antigens can be fused with bacterial membrane proteins for targeted antigen presentation or molecular drug delivery. For instance, the receptor-binding domain (RBD) of the spike protein can be fused with the carrier outer membrane protein A (OmpA), while a core peptide of the RBD, referred to as NG06, can be fused with the carrier membrane-associated protein cytolysin A (ClyA). This specific fusion of recombinant proteins with particular carrier membrane proteins of BMVs enables the direct loading of the recombinant proteins into the lumen or their presentation on the surface of the bacterial vesicles (Fig. 1.4c) (Wo et al. 2023). Engineering mammalian and bacterial vesicles is valuable for presenting recombinant antigens or molecular drugs on their surfaces or within their lumens. This approach allows precise control over antigen or drug orientation, improving cargo loading and ensuring better protection for optimal function compared to random loading through physical methods. It also holds promises for targeted drug delivery, enhancing therapeutic efficacy for a broad range of diseases.

1.4.4 In Situ Therapeutics Production

Genetically modified probiotics act as micro-factories, engineered to secrete therapeutic molecules like antimicrobial peptides, anti-inflammatory peptides, and antioxidant enzymes in the gastrointestinal microenvironment to inhibit pathogens, reduce inflammation, reinforced of the intestinal barrier, and modulate the gut disease microenvironment. For example, *Escherichia coli* Nissle has been used as a beneficial probiotic strain and genetically engineered for sustained 3-hydroxybutyrate production in colitis-afflicted mice. The engineered probiotic significantly increased gut 3-hydroxybutyrate production and elevated short-chain fatty acid levels. As a key ketone body, the sustained presence of 3-hydroxybutyrate provides energy during starvation, shows therapeutic potential for inflammation, and promotes the growth of beneficial gut microorganisms, contributing to an improved colonic environment in the colitis mouse model (Yan et al. 2021). Furthermore, *E. coli* Nissle has been genetically modified to express interleukin-2 (IL-2), which can regulate the intestinal microenvironment and trigger tolerogenic immune responses. The successful delivery of IL-2-expressing *E. coli* Nissle through an enteric coating film enhances the accumulation of the engineered probiotic in the intestine and enables controlled IL-2 secretion. IL-2 regulates immune responses, modulates the gut microbiota, significantly reduces inflammation, and repairs the intestinal epithelial barrier in inflammatory bowel disease mouse model (Li et al. 2024). *E. coli* has been engineered as a probiotic strain, modified to express a specific tumor antigen fused with cytolysin A on its outer membrane surface, controlled by an arabinose-inducible promoter. When administered orally with arabinose in mice, the genetically modified probiotic triggered the generation of OMVs that successfully traversed the intestinal epithelium to reach the lamina propria. There, they promoted the maturation of dendritic cells. In mouse models of lung metastatic melanoma and subcutaneous colon tumors, the antigen-loaded OMVs effectively suppressed tumor growth and provided protection against tumor recurrence (Fig. 1.5) (Yue et al. 2022).

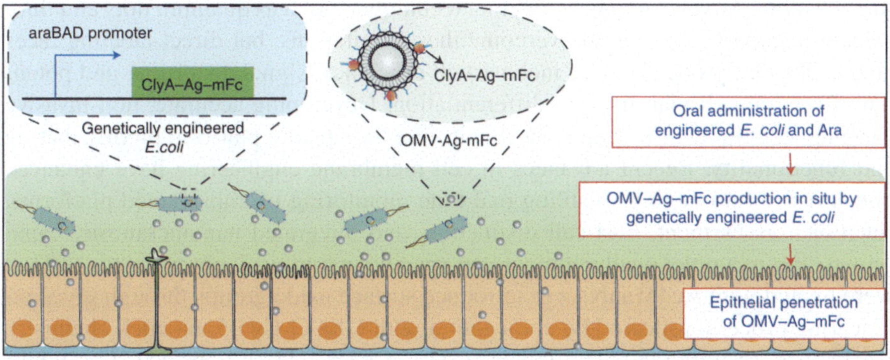

Fig. 1.5 Engineered *E. coli* was transformed with a plasmid expressing ClyA, fused with a tumor antigen and the Fc fragment of mouse IgG (ClyA–Ag–mFc), under the control of an arabinose-inducible promoter. Oral administration of arabinose and modified *E. coli* in mice triggered OMV production, which suppressed tumor growth, and provided protection against tumor re-challenge. (Reproduced with permission from Yue et al. (2022), copyright 2022, Springer Nature)

1.4.5 Capturing and Tracking Functions

Circulating tumor cells (CTCs) are cancerous cells released from primary tumors into the bloodstream. Metastasis is a complicated multi-stage process that contributes to approximately 90% of cancer-related mortality. As key liquid biopsy biomarkers, CTCs facilitate prognosis, metastatic risk evaluation, tumor burden estimation, and therapy monitoring. However, only a small fraction of CTCs survives in circulation, as the majority succumb to shear stress and immune surveillance. Furthermore, heterogeneous surface markers resulting from epithelial-to-mesenchymal transition and the presence of abundant background leukocytes further complicate their detection. Current CTCs separation techniques, including biophysical methods, biochemical approaches (e.g., antibody-, peptide-, and aptamer-based targeting), and microfluidic platforms, often suffer from low sensitivity, poor binding affinity, or inefficient release, underscoring the urgent need for sensitive and reproducible detection technologies (Ju et al. 2022). Recent studies demonstrate that engineered membrane platforms enhance capture efficiency, detection sensitivity, and minimize blood sample interference. For instance, cell-released magnetic vesicles, modified with DBCO, inherit the parent cell's membrane characteristics and are used to target and isolate metabolically labeled CTCs. This vesicle-based approach shows high reliability and broad potential for capturing various tumor cells, with excellent capture efficiency and low detection limits. Additionally, metabolic glycoengineering combined with click chemistry enables high-purity enrichment of CTCs, bypassing phenotypic fluctuations during cancer progression. This system enables efficient, sensitive, and specific CTCs capture while preserving cell viability for downstream analysis, making it a promising approach for rare cell detection and cancer monitoring (Fig. 1.6a) (Kang et al. 2021).

Living cells and membrane-based nanoplatforms offer promising disease treatments but face clinical challenges due to unclear in vivo fate and the need for effective therapeutic assessment. Traditional evaluations rely on postmortem histology and indirect metrics like tumor size, biomarkers, and survival rates, lacking real-time insights. Advancements in labeling techniques, such as quantum dots and fluorescent nanoparticles, aim to overcome these limitations, but direct labeling faces challenges like probe dilution, inconsistent efficiency, signal distortion, and potential effects on cell viability and differentiation. Developing accurate, non-invasive imaging techniques is essential for monitoring therapeutic platform biodistribution and functionality. Recent advances in cell membrane engineering have enhanced non-invasive bioimaging, enabling real-time monitoring of transplanted platforms, functional assessment, host-cell distinction, and integrated nanotheranostic capabilities. For example, ovalbumin antigen-specific cytotoxic T lymphocytes (CTLs) were treated with Ac4ManNAz to introduce surface azide groups through glycoengineering process, followed by conjugation with DBCO-Cy5.5 via click chemistry mechanism, resulting in Cy5.5-labeled CTLs (Cy5.5-CTLs). Labeled CTLs exhibited strong near-infrared fluorescence (NIRF) signals, maintained key functional properties, and antigen-specific cytolytic activity. In immunodeficient mice harboring ovalbumin-expressing tumors, the introduced Cy5.5-CTLs were easily

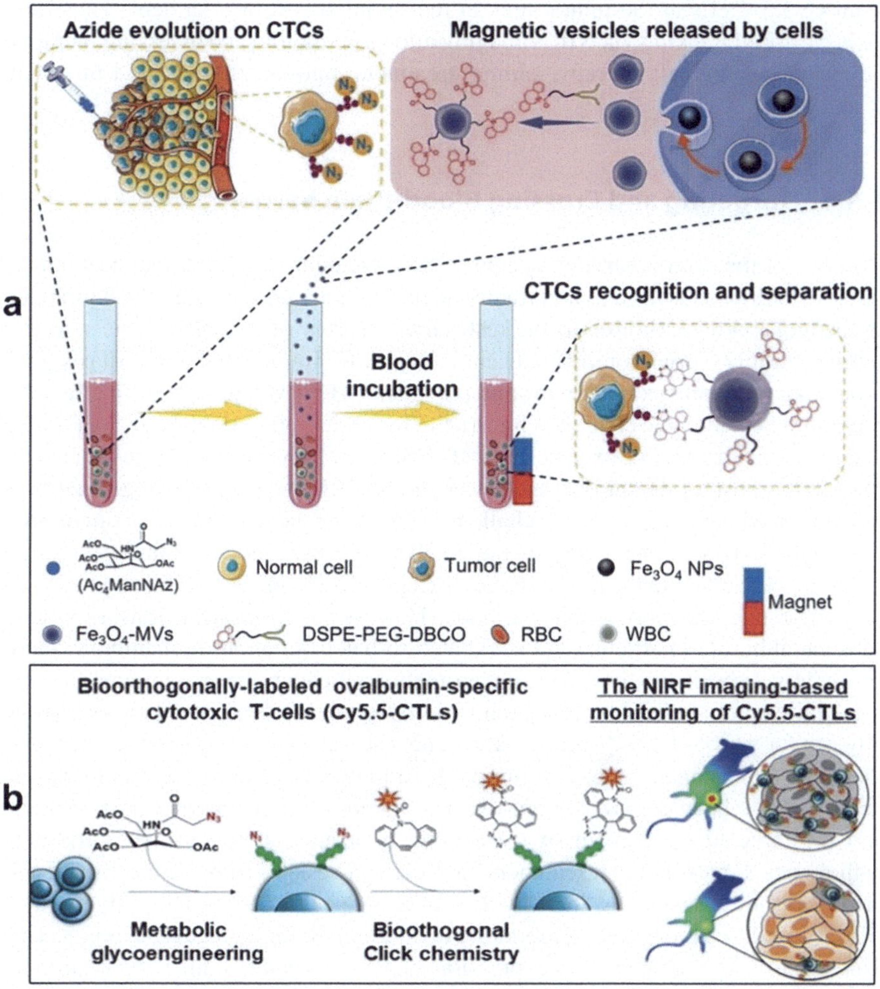

Fig. 1.6 (**a**) Magnetic vesicles, retaining their parent cells' membrane features, are modified with dibenzocyclooctyne (DBCO) to selectively target and isolate azide-labeled rare CTCs via click chemistry, ensuring high capture efficiency and ultra-sensitive cancer detection. (Reproduced with permission from Kang et al. (2021), copyright 2021, Wiley-VCH GmbH). (**b**) Schematic representation of click labeling of ovalbumin-specific cytotoxic T cells (CTLs). The labeled CTLs (Cy5.5-CTLs) enable near-infrared fluorescence (NIRF) imaging for tumor infiltration monitoring and therapy response prediction. (Reproduced with permission from (Kim et al. 2021), copyright 2021, Elsevier B.V.)

visualized at targeted tumors sites via minimally invasive NIRF imaging. Furthermore, the degree of tumor growth inhibition correlated closely with the intensity of NIRF signals from Cy5.5-CTLs at tumor sites. Cy5.5-CTLs demonstrated variable treatment outcome, based on tumor growth efficacy. These differing therapeutic outcomes were strongly associated with the intensity of NIRF signals from Cy5.5-CTLs at the tumor sites in the initial stages of treatment. Thus, noninvasive T-cell tracking via NIRF imaging allows predicting and assessing therapeutic responses in cell therapy, aiding treatment optimization (Fig. 1.6b) (Kim et al. 2021).

1.4.6 Targeting and Crossing Blood-Brain Barrier

Diseases of the central nervous system (CNS), including glioblastoma, Alzheimer's disease, Parkinson's disease, and traumatic brain injury, present significant treatment challenges, particularly due to the restrictive function of the BBB. This selective, semipermeable barrier controls nutrient transport for intracranial homeostasis while protecting the brain against toxins and pathogens. However, this protective function presents a considerable challenge for drug delivery, as approximately 98% of molecular medicines are unable to cross the BBB. Therefore, developing effective nanoscale carriers to transport theranostic agents across the BBB and enhance drug penetration into damaged tissues is a critical challenge. EVs have gained increasing attention as potential carriers to the CNS, owing to their intrinsic biocompatibility, reduced immunogenic effect in the bloodstream, high physicochemical stability, and capacity for long-range transport within the body. However, their clinical translation faces substantial hurdles, particularly due to batch-to-batch variations in membrane markers for target specificity, low cargo encapsulation efficiency, limited intracellular bioavailability, and minimal brain accumulation potential. To enhance their therapeutic efficacy for various CNS disorders, researchers are actively working on surface modifications and luminal cargo loading of these platforms. Numerous brain-targeting peptides have been engineered onto vesicle membranes, facilitating BBB crossing and the targeted enrichment of therapeutic molecules at specific sites for brain pathologies. For example, researchers have derived vesicles from stem cells and dendritic cells by transfecting them with a neuron-specific rabies virus glycoprotein (RVG) fused to the integral membrane protein Lamp2b through genetic engineering. This modification enabled the expression of RVG peptides onto vesicle surfaces. Furthermore, surface engineered has then been loaded with nucleic acid-based drugs. RVG-expressed nanovesicles cross the BBB, accumulate in ischemic regions of a mouse model, and effectively deliver their contents to neuronal cells. The engineered vesicles successfully modulated gene expression in various brain regions and promoted neurogenesis, highlighting their potential for treating brain diseases (Yang et al. 2017; Alvarez-Erviti et al. 2011).

1.4.7 Enhancing Homing Capabilities

Cells play a pivotal role in tissue healing and repair by responding to both chemical and mechanical signals from their microenvironment. These signals guide differentiation and regulate cellular functions within tissue-specific niches. One critical process in this regard is homing, where cells migrate to designated sites within the body. For example, T cells target inflamed tissues via integrins and selectins, and MSCs migrate to injury sites for tissue repair through CXC chemokine receptor 4 (CXCR4) and CCR2. Enhancing stem cell homing is essential for effective tissue repair, immune responses, and targeted therapies. However, cultured cells and cell-derived membranes often exhibit low expression of homing markers, and intravenous delivery presents challenges to therapeutic success. Since a cell's targeting abilities mainly stem from its cell membrane, efforts to leverage plasma membranes as a carrier for drug delivery are gaining importance for targeted therapies for treating cancer, infectious, inflammatory disease, neurodegenerative diseases, and tissue defects. To address these limitations, genetic engineering and biomaterials-based strategies have been developed to enhance ligand presentation, chemotactic signals, and adhesion molecules on cell surfaces and membrane platforms to allow precise control over cellular interactions. These strategies create a supportive "home" that improves cell homing, recruitment, fate determination, drug delivery, and overall therapeutic efficacy. For example, studies have shown that stromal-derived factor-1 (SDF-1) expression in the ischemic myocardium is upregulated after infarction and released into the bloodstream, where it reaches the bone marrow to recruit stem cells that express CXCR4. The CXCR4/SDF-1 axis is crucial for MSC homing to the infarcted myocardium, essential for cardiac recovery. However, MSC expansion during in vitro culture leads to the downregulation of CXCR4 expression, impairing their ability to migrate to the ischemic heart tissue. To address this, surface modification strategies have been developed to incorporate recombinant CXCR4 proteins into stem cells membranes using lipid-PEG insertion. Recombinant CXCR4 engineering on the cell membrane surface enhanced stem cell migration towards an SDF-1 concentration gradient, improving retention rates and homing, and, increasing their potential for cardiac repair (Won et al. 2014). Pretargeting and bioorthogonal engineering have been used to capture circulating stem cells and direct them to the damaged heart for repair. First, DBCO-PEG-CD41 antibodies were administered to the pretarget platelets at the myocardial infarction site. Pre-administration of granulocyte colony-stimulating factors stimulated the release of endothelial progenitor cells (EPCs) from the bone marrow into the bloodstream. After 48 h, azide-modified CD34 antibodies bound to EPCs, and their azide groups reacted with DBCO groups on platelets through a click chemistry mechanism. Platelets possess the inherent potential to accumulate in injured vessels and redirect engineered stem cells to the injured area. In a mice model of myocardial infarction, assessments confirmed effective heart repair facilitated by this pretargeting and bioorthogonal approach (Fig. 1.7) (Li et al. 2018b).

1.4.8 Reshaping Immune Microenvironment

1.4.8.1 Immune Checkpoint Blockade

Cancer immunotherapy enhances immune responses, reshapes the tumor microenvironment, and utilizes tumor-specific antibodies with immune effectors to combat malignancies. However, its efficacy is hindered by immunosuppressive checkpoint

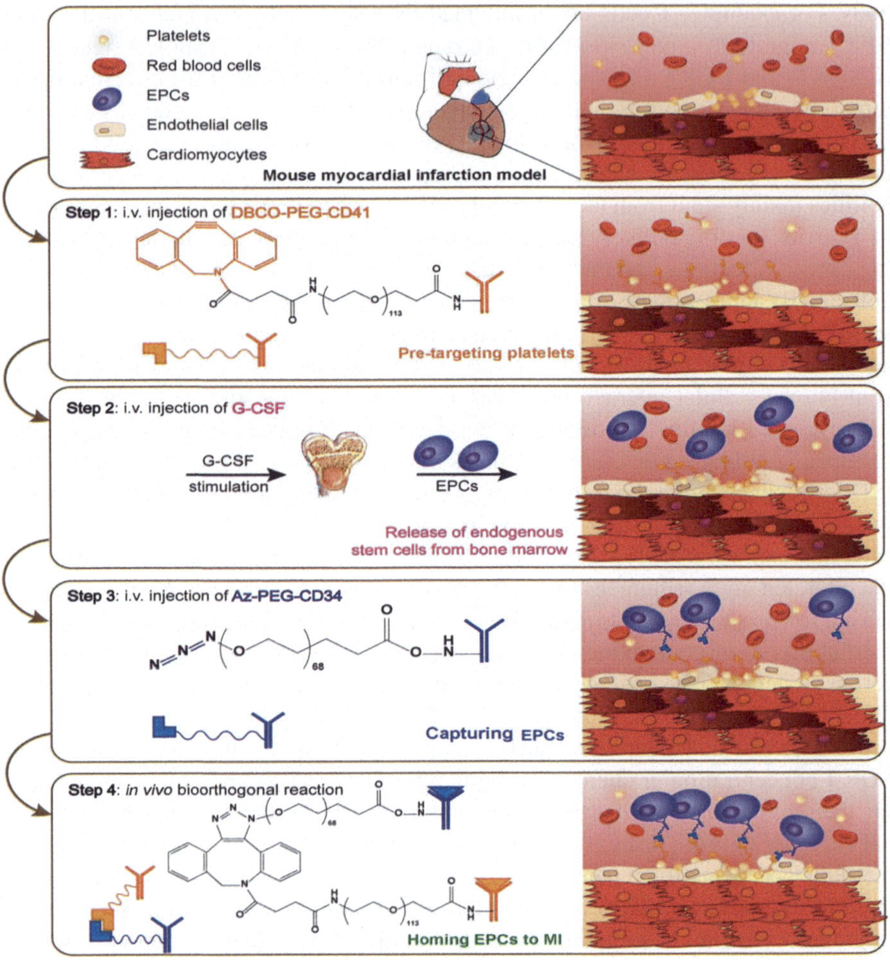

Fig. 1.7 Pretargeting and click engineering were used to direct stem cells to the wounded heart. DBCO-PEG-CD41 antibodies pretargeted platelets at the infarction site. Granulocyte colony-stimulating factors (G-CSF) stimulated the release of endothelial progenitor cells (EPCs) from the bone marrow. Azide-modified CD34 antibodies were administered to pretarget EPCs, and their azide groups reacted with DBCO groups on platelets. Platelets naturally accumulate in injured vessels, redirecting engineered EPCs for effective heart repair. (Reproduced with permission from Li et al. (2018b), copyright 2018, American Chemical Society)

ligands, including Cluster of Differentiation 47 (CD47), programmed cell death ligand-1 (PD-L1), and signal regulatory protein α (SIRPα), which enable tumors to evade immune surveillance and induce T-cell exhaustion. Fo example, PD-L1 inhibits T-cell activation by engaging with programmed cell death protein-1 (PD-1), while CD47 transmits a "don't eat me" signal via SIRPα, preventing macrophage-mediated phagocytosis. Recently, engineered membrane-based platforms offer a transformative approach to immune checkpoint blockade, restoring T-cell immunity and enhancing macrophage-mediated tumor clearance. Genetic engineering has enabled the modification of BMVs with the ectodomain of PD1, allowing their binding to PD-L1 on tumor cells, promote integration, and reduce PD-L1 levels. This protects T cells from immune suppression while simultaneously enhancing effector T cell infiltration through immune stimulation and checkpoint inhibition. As a result, bioengineered BMVs exhibit superior tumor growth inhibition compared to native BMVs and PD-L1 antibodies (Li et al. 2020). Nie et al. developed responsive exosome for cancer treatment using metabolic glycoengineering and click chemistry. They conjugated azide-modified BMVs from M1 macrophages with DBCO-modified CD47 and SIRPα antibodies via pH-sensitive linkers. In the acidic tumor microenvironment, the benzoic-imine bonds cleaved, releasing the antibodies, blocking SIRPα on macrophages and CD47 on cancer cells, enhancing macrophage activity and promoting immune-mediated cancer cell destruction (Nie et al. 2020). Genetically modified fusion cellular vesicles (Fus-CVs) expressing SIRPα variants and PD-1 trigger strong antitumor immune response by engaging both innate and adaptive immune responses. Dual blockades of CD47 and PD-L1 using Fus-CVs enhance macrophage-mediated phagocytosis, boost antigen presentation, and stimulates T-cell immunity. Their dual targeting potential ensures selective tumor cell binding, minimizing systemic side effects, improving therapeutic efficacy against malignant melanoma and carcinoma mouse models, and thereby act as promising platform for multi-targeted immune checkpoint inhibition therapy (Fig. 1.8a–c) (Meng et al. 2021).

1.4.8.2 Immune Tolerance Induction

Immune tolerance induction contrasts with immune blockade therapies by training the immune system to accept specific antigens without triggering a response. This approach is used in autoimmune diseases and transplantation to prevent immune rejection. Engineered membrane-based platforms have been developed to treat autoimmune conditions such as T1D and inflammation. For example, β-cell-targeted Ac4ManNAz-encapsulated NPs facilitate the delivery and establishment of β cells with high concentration of azide groups. The administration of a DBCO-functionalized PD-L1 immunoglobulin fusion protein (PD-L1-Ig) enables its direct conjugation to the surface of β cells. In a nonobese diabetic (NOD) mouse model, this strategy selectively conjugates PD-L1 onto β cells through click chemistry. Modified β cells present islet-specific antigens and PD-L1, effectively engaging and blocking autoreactive T-cell attacks on insulin-producing β cells. This process reverses T1D by decreasing IFN-γ-producing cytotoxic T cells and promoting antigen-specific immune tolerance (Fig. 1.8d) (Au et al. 2021). MSCs-derived EVs can be engineered to express PD-L1, enhancing their ability to target inflamed tissues. The PD-L1 receptors on engineered EVs interact with activated immune cells expressing PD-1, effectively inhibiting autoreactive T-cell activity and reshaping the local immune microenvironment. This cell-free PD-1/PD-L1 pathway-based treatment has shown significant efficacy in ulcerative colitis and psoriasis mouse models, restoring self-tolerance and suppressing autoimmunity, with potential for treating a range of autoimmune disorders (Xu et al. 2022).

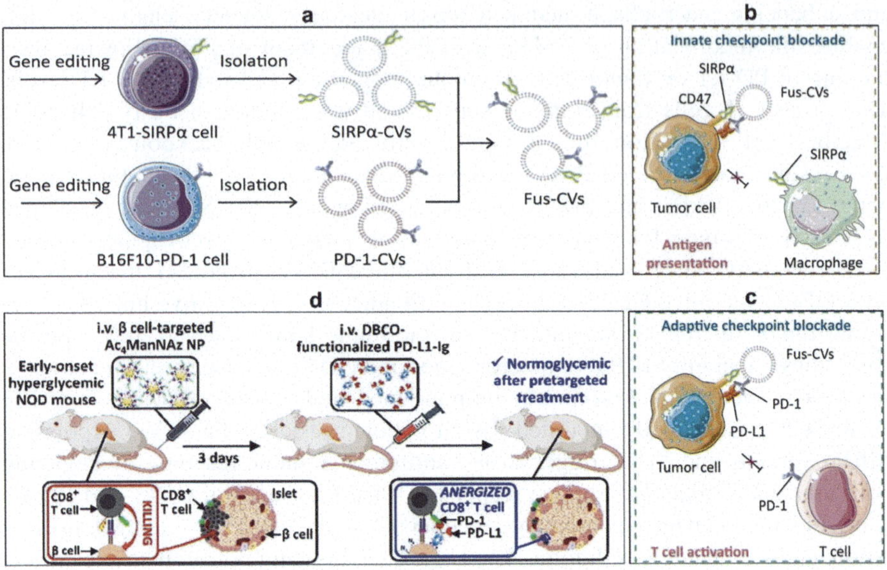

Fig. 1.8 (**a–c**) Preparation of genetically modified fusion cellular vesicles (Fus-CVs) displaying SIRPα and PD-1 for enhanced multi-target immune checkpoint blockade therapy. Fus-CVs simultaneously inhibit the CD47/SIRPα innate immune response and the PD-1/PD-L1 adaptive immune response, enhancing antigen presentation by dendritic cells and macrophages while amplifying antitumor T-cell responses. (Reproduced with permission from Meng et al. (2021), copyright 2021, Wiley-VCH GmbH). (**d**) Bioengineering of β cells to reverse type 1 diabetes. Ac4ManNAz nanoparticles deliver Ac4ManNAz to β cells, where metabolic glycoengineering generates azide-modified surface proteins. These enable click reactions with intravenously injected DBCO-PD-L1-Ig, promoting antigen presentation and PD-1/PD-L1 activation for self-tolerance. (Reproduced with permission from Au et al. (2021), copyright 2021, American Chemical Society)

1.5 Conclusion and Future Direction

Cell membrane engineering is rapidly advancing in translational medicine, leveraging the unique properties of cell membranes to develop therapeutic platforms, including engineered living cells, extracellular vesicles, bacterial membrane vesicles, membrane-coated nanoparticles, and hybrid nanomaterials. These platforms enhance cell-cell and cell-biomaterial interactions, supporting scaffold-free constructs, tissue modeling, and cell-dense structures for tissue engineering. Chemical modifications provide protection against harsh microenvironments, immune shielding, and therapeutic benefits, enabling applications such as engineered RBCs for universal blood transfusion, immune-protected islets for T1D treatment, and probiotics for gastrointestinal disease management. Membrane engineering also facilitates drug conjugation and loading onto living cells and vesicles, ensuring enhanced protection and targeted delivery for disease microenvironment modulation. Genetically modified probiotics function as micro-factories, efficiently inhibiting pathogens, reducing inflammation, and modulating the gut microbiome more effectively. Engineered membrane platforms improve CTCs capture for cancer monitoring and enhance non-invasive bioimaging, enabling real-time monitoring for various theranostic applications. Additionally, engineered membrane vesicles can traverse the BBB to deliver therapeutics precisely to specific brain regions for neurodegenerative disease treatment. By optimizing ligand presentation, chemotactic signals, and adhesion molecules, membrane engineering enhances cell recruitment, fate determination, drug delivery, and therapeutic efficacy in tissue engineering and targeted therapies. Moreover, engineered membrane-based platforms reshape the immune microenvironment, strengthening T-cell immunity, promoting macrophage-driven tumor clearance, and improving immune tolerance for autoimmune disease treatments.

Cell membrane-engineered platforms are essential for advancing precision therapies in complex diseases. However, further research is needed to precisely position mammalian cells and develop modular microtissues that replicate native tissue complexities. The heterogeneity of cell membranes across individuals and diseases presents opportunities for personalized medicine, enabling tailored membrane-based platforms through single-cell omics and surfaceome analysis. Engineering patient-derived membranes with specific antigens or targeting ligands can facilitate personalized vaccines and nanotherapeutics. Integrating artificial intelligence, bioinformatics, machine learning, and synthetic biology can refine bioengineering by optimizing ligand-receptor databases, balancing immunogenicity, and predicting patient-specific strategies. Additionally, engineered membranes capable of sensing microenvironmental cues and adapting their behavior will enhance precision medicine. In the future, the development of a multifunctional platform that seamlessly integrates targeting, tracking, immune modulation, and in situ therapeutic production holds significant promise for advancing membrane-based technologies.

References

Adebowale K, Liao R, Suja VC, Kapate N, Lu A, Gao Y, Mitragotri S (2024) Materials for cell surface engineering. Adv Mater 36(43):2210059

Ai X, Wang S, Duan Y, Zhang Q, Chen MS, Gao W, Zhang L (2020) Emerging approaches to functionalizing cell membrane-coated nanoparticles. Biochemistry 60(3):941–955

Akkiraju H, Nohe A (2015) Role of chondrocytes in cartilage formation, progression of osteoarthritis and cartilage regeneration. J Dev Biol 3(4):177–192

Alvarez-Erviti L, Seow Y, Yin H, Betts C, Lakhal S, Wood MJ (2011) Delivery of siRNA to the mouse brain by systemic injection of targeted exosomes. Nat Biotechnol 29(4):341–345

Au KM, Tisch R, Wang AZ (2021) In vivo bioengineering of beta cells with immune checkpoint ligand as a treatment for early-onset type 1 diabetes mellitus. ACS Nano 15(12):19990–20002

Battigelli A, Almeida B, Shukla S, Rocha AD, Shukla A (2020) Inducing mesenchymal stem cell attachment on non-cell adhesive hydrogels through click chemistry. Chem Comm 56(55):7661–7664

Bausch-Fluck D, Milani ES, Wollscheid B (2019) Surfaceome nanoscale organization and extracellular interaction networks. Curr Opin Chem Biol 48:26–33

Borst J, Busselaar J, Bosma DM, Ossendorp F (2021) Mechanism of action of PD-1 receptor/ligand targeted cancer immunotherapy. Eur J Immunol 51(8):1911–1920

Cayabyab F, Nih LR, Yoshihara E (2021) Advances in pancreatic islet transplantation sites for the treatment of diabetes. Front Endocrinol 12:732431

Chandler KB, Costello CE (2016) Glycomics and glycoproteomics of membrane proteins and cell-surface receptors: present trends and future opportunities. Electrophoresis 37(11):1407–1419

Chen H, Teramura Y, Iwata H (2011) Co-immobilization of urokinase and thrombomodulin on islet surfaces by poly (ethylene glycol)-conjugated phospholipid. J Contr Release 150(2):229–234

Chen Q, Zhang L, Li L, Tan M, Liu W, Liu S, Xie Z, Zhang W, Wang Z, Cao Y, Shang T (2021a) Cancer cell membrane-coated nanoparticles for bimodal imaging-guided photothermal therapy and docetaxel-enhanced immunotherapy against cancer. J Nanobiotechnol 19(1):449

Chen L, Hong W, Ren W, Xu T, Qian Z, He Z (2021b) Recent progress in targeted delivery vectors based on biomimetic nanoparticles. Signal transduction and targeted therapy. Signal Transduct Target Ther 6(1):225

Chen S, Saeed AF, Liu Q, Jiang Q, Xu H, Xiao GG, Rao L, Duo Y (2023) Macrophages in immunoregulation and therapeutics. Signal Transduct Target Ther 8(1):207

Cheng X, Smith JC (2019) Biological membrane organization and cellular signaling. Chem Rev 119(9):5849–5880

Cheng Q, Kang Y, Yao B, Dong J, Zhu Y, He Y, Ji X (2023) Genetically engineered-cell-membrane nanovesicles for cancer immunotherapy. Adv Sci 10(26):2302131

Choi A, Javius-Jones K, Hong S, Park H (2023) Cell-based drug delivery systems with innate homing capability as a novel nanocarrier platform. Int J Nanomed 18:509–525

Csizmar CM, Petersburg JR, Wagner CR (2018) Programming cell-cell interactions through non-genetic membrane engineering. Cell Chem Biol 25(8):931–940

Cubillos-Ruiz A, Guo T, Sokolovska A, Miller PF, Collins JJ, Lu TK, Lora JM (2021) Engineering living therapeutics with synthetic biology. Nat Rev Drug Discov 20(12):941–960

Dai Q, Bertleff-Zieschang N, Braunger JA, Björnmalm M, Cortez-Jugo C, Caruso F (2018) Particle targeting in complex biological media. Adv Healthc Mater 7(1):1700575

Di Meo F, Kale B, Koomen JM, Perna F (2024) Mapping the cancer surface proteome in search of target antigens for immunotherapy. Mol Ther 32(9):2892–2904

Dooley K, McConnell RE, Xu K, Lewis ND, Haupt S, Youniss MR, Martin S, Sia CL, McCoy C, Moniz RJ, Burenkova O (2021) A versatile platform for generating engineered extracellular vesicles with defined therapeutic properties. Mol Ther 29(5):1729–1743

Du Pont-Thibodeau G, Harrington K, Lacroix J (2014) Anemia and red blood cell transfusion in critically ill cardiac patients. Ann Intensive Care 4:1–11

Ducrot C, Loiseau S, Wong C, Madec E, Volatron J, Piffoux M (2023) Hybrid extracellular vesicles for drug delivery. Cancer Lett 558:216107

El Andaloussi S, Mäger I, Breakefield XO, Wood MJ (2013) Extracellular vesicles: biology and emerging therapeutic opportunities. Nat Rev Drug Discov 12(5):347–357

Fang RH, Gao W, Zhang L (2023) Targeting drugs to tumours using cell membrane-coated nanoparticles. Nat Rev Clin Oncol 20(1):33–48

Fukuda Y, Akagi T, Asaoka T, Eguchi H, Sasaki K, Iwagami Y, Yamada D, Noda T, Kawamoto K, Gotoh K, Kobayashi S (2018) Layer-by-layer cell coating technique using extracellular matrix facilitates rapid fabrication and function of pancreatic β-cell spheroids. Biomaterials 160:82–91

Gao W, Fang RH, Thamphiwatana S, Luk BT, Li J, Angsantikul P, Zhang Q, Hu CM, Zhang L (2015) Modulating antibacterial immunity via bacterial membrane-coated nanoparticles. Nano Lett 15(2):1403–1409

Gopalarethinam J, Nair AP, Iyer M, Vellingiri B, Subramaniam MD (2023) Advantages of mesenchymal stem cell over the other stem cells. Acta Histochem. 125(4):152041

Han H, Bártolo R, Li J, Shahbazi MA, Santos HA (2022a) Biomimetic platelet membrane-coated nanoparticles for targeted therapy. Eur J Pharm Biopharm 172:1–15

Han Y, Yang J, Fang J, Zhou Y, Candi E, Wang J, Hua D, Shao C, Shi Y (2022b) The secretion profile of mesenchymal stem cells and potential applications in treating human diseases. Signal Transduct Target Ther 7(1):92

Harvey BT, Fu X, Li L, Neupane KR, Anand N, Kolesar JM, Richards CI (2022) Dendritic cell membrane-derived nanovesicles for targeted T cell activation. ACS Omega 7(50):46222–46233
Hu CM, Fang RH, Wang KC, Luk BT, Thamphiwatana S, Dehaini D, Nguyen P, Angsantikul P, Wen CH, Kroll AV, Carpenter C (2015) Nanoparticle biointerfacing by platelet membrane cloaking. Nature 526(7571):118–121
Im BH, Jeong JH, Haque MR, Lee DY, Ahn CH, Kim JE, Byun Y (2013) The effects of 8-arm-PEG-catechol/heparin shielding system and immunosuppressive drug, FK506 on the survival of intraportally allotransplanted islets. Biomaterials 34(8):2098–2106
Irvine DJ, Maus MV, Mooney DJ, Wong WW (2022) The future of engineered immune cell therapies. Science 378(6622):853–858
Jhan YY, Prasca-Chamorro D, Zuniga GP, Moore DM, Kumar SA, Gaharwar AK, Bishop CJ (2020) Engineered extracellular vesicles with synthetic lipids via membrane fusion to establish efficient gene delivery. Int J Pharm 573:118802
Ju S, Chen C, Zhang J, Xu L, Zhang X, Li Z, Chen Y, Zhou J, Ji F, Wang L (2022) Detection of circulating tumor cells: opportunities and challenges. Biomark Res 10(1):58
Kang T, Zhu Q, Wei D, Feng J, Yao J, Jiang T, Song Q, Wei X, Chen H, Gao X, Chen J (2017) Nanoparticles coated with neutrophil membranes can effectively treat cancer metastasis. ACS Nano 11(12):1397–1411
Kang K, Zhou X, Zhang Y, Zhu N, Li G, Yi Q, Wu Y (2021) Cell-released magnetic vesicles capturing metabolic labeled rare circulating tumor cells based on bioorthogonal chemistry. Small 17(18):2007796
Kim W, Yoon HY, Lim S, Stayton PS, Kim IS, Kim K, Kwon IC (2021) In vivo tracking of bioorthogonally labeled T-cells for predicting therapeutic efficacy of adoptive T-cell therapy. J Control Release 329:223–236
Kroll AV, Fang RH, Zhang L (2017) Biointerfacing and applications of cell membrane-coated nanoparticles. Bioconjug Chem 28(1):23–32
Lavrador P, Moura BS, Almeida-Pinto J, Gaspar VM, Mano JF (2025) Engineered nascent living human tissues with unit programmability. Nat Mater 24:143–154
Li J, Ai Y, Wang L, Bu P, Sharkey CC, Wu Q, Wun B, Roy S, Shen X, King MR (2016) Targeted drug delivery to circulating tumor cells via platelet membrane-functionalized particles. Biomaterials 76:52–65
Li Z, Zhou X, Wei M, Gao X, Zhao L, Shi R, Sun W, Duan Y, Yang G, Yuan L (2018a) In vitro and in vivo RNA inhibition by CD9-HuR functionalized exosomes encapsulated with miRNA or CRISPR/dCas9. Nano Lett 19(1):19–28
Li Z, Shen D, Hu S, Su T, Huang K, Liu F, Hou L, Cheng K (2018b) Pretargeting and bioorthogonal click chemistry-mediated endogenous stem cell homing for heart repair. ACS Nano 12(12):12193–12200
Li Y, Zhao R, Cheng K, Zhang K, Wang Y, Zhang Y, Li Y, Liu G, Xu J, Xu J, Anderson GJ (2020) Bacterial outer membrane vesicles presenting programmed death 1 for improved cancer immunotherapy via immune activation and checkpoint inhibition. ACS Nano 14(12):16698–16711
Li M, Liu N, Zhu J, Wu Y, Niu L, Liu Y, Chen L, Bai B, Miao Y, Yang Y, Chen Q (2024) Engineered probiotics with sustained release of interleukin-2 for the treatment of inflammatory bowel disease after oral delivery. Biomaterials 309:122584
Liu Y, Wang X, Ouyang B, Liu X, Du Y, Cai X, Guo H, Pang Z, Yang W, Shen S (2018) Erythrocyte–platelet hybrid membranes coating polypyrrol nanoparticles for enhanced delivery and photothermal therapy. J Mater Chem B 6(43):7033–7041
Liu C, Gao J, Cheng Y, Zhang S, Fu C (2025) Homologous-adhering/targeting cell membrane- and cell-mediated delivery systems: a cancer-catch-cancer strategy in cancer therapy. Regen Biomater 12:rbae135
Long Q, Zheng P, Zheng X, Li W, Hua L, Yang Z, Huang W, Ma Y (2022) Engineered bacterial membrane vesicles are promising carriers for vaccine design and tumor immunotherapy. Adv Drug Deliv Rev 186:114321

Lopes J, Lopes D, Pereira-Silva M, Peixoto D, Veiga F, Hamblin MR, Conde J, Corbo C, Zare EN, Ashrafizadeh M, Tay FR (2022) Macrophage cell membrane-cloaked nanoplatforms for biomedical applications. Small Methods 6(8):2200289

Luo W, Pulsipher A, Dutta D, Lamb BM, Yousaf MN (2014) Remote control of tissue interactions via engineered photo-switchable cell surfaces. Sci Rep 4(1):6313

Luo Y, Chen Z, Sun M, Li B, Pan F, Ma A, Liao J, Yin T, Tang X, Huang G, Zhang B (2022) IL-12 nanochaperone-engineered CAR T cell for robust tumor-immunotherapy. Biomaterials 281:121341

Ma X, Kuang L, Yin Y, Tang L, Zhang Y, Fan Q, Wang B, Dong Z, Wang W, Yin T, Wang Y (2023a) Tumor–antigen activated dendritic cell membrane-coated biomimetic nanoparticles with orchestrating immune responses promote therapeutic efficacy against glioma. ACS Nano 17(3):2341–2355

Ma J, Dai L, Yu J, Cao H, Bao Y, Hu J, Zhou L, Yang J, Sofia A, Chen H, Wu F (2023b) Tumor microenvironment targeting system for glioma treatment via fusion cell membrane coating nanotechnology. Biomaterials 295:122026

Ma Y, Yi J, Ruan J, Ma J, Yang Q, Zhang K, Zhang M, Zeng G, Jin L, Huang X, Li J (2024) Engineered cell membrane-coated nanoparticles: new strategies in glioma targeted therapy and immune modulation. Adv Healthc Mater 13(20):2400514

Mao D, Zhang C, Liu J, Wang X, Li B, Yan H, Hu F, Kong D, Wang Z, Liu B (2020) Bio-orthogonal click reaction-enabled highly specific in situ cellularization of tissue engineering scaffolds. Biomaterials 230:119615

Margiana R, Markov A, Zekiy AO, Hamza MU, Al-Dabbagh KA, Al-Zubaidi SH, Hameed NM, Ahmad I, Sivaraman R, Kzar HH, Al-Gazally ME (2022) Clinical application of mesenchymal stem cell in regenerative medicine: a narrative review. Stem Cell Res Ther 13(1):366

Meng QF, Zhao Y, Dong C, Liu L, Pan Y, Lai J, Liu Z, Yu GT, Chen X, Rao L (2021) Genetically programmable fusion cellular vesicles for cancer immunotherapy. Angew Chem Int Ed 60(50):26320–26326

Menon I, Zaroudi M, Zhang Y, Aisenbrey E, Hui L (2022) Fabrication of active targeting lipid nanoparticles: challenges and perspectives. Mater Today Adv 16:100299

Milner E, Stevens B, An M, Lam V, Ainsworth M, Dihle P, Stearns J, Dombrowski A, Rego D, Segars K (2021) Utilizing probiotics for the prevention and treatment of gastrointestinal diseases. Front Microbiol 12:689958

Mima Y, Hashimoto Y, Shimizu T, Kiwada H, Ishida T (2015) Anti-PEG IgM is a major contributor to the accelerated blood clearance of polyethylene glycol-conjugated protein. Mol Pharmaceutics 12(7):2429–2435

Nam H, Jeon H, Kim H, Yoon HY, Kim SH, Lee JB (2023) Module-assembly of injectable cellular DNA hydrogel via clickable cells and DNA scaffolds. Chem Eng J 452:139492

Nie W, Wu G, Zhang J, Huang LL, Ding J, Jiang A, Zhang Y, Liu Y, Li J, Pu K, Xie HY (2020) Responsive exosome nano-bioconjugates for synergistic cancer therapy. Angew Chem Int Ed 59(5):2018–2022

Nishiguchi A, Yoshida H, Matsusaki M, Akashi M (2011) Rapid construction of three-dimensional multilayered tissues with endothelial tube networks by the cell-accumulation technique. Adv Mater 23(31):3506–3510

Omer R, Mohsin MZ, Mohsin A, Mushtaq BS, Huang X, Guo M, Zhuang Y, Huang J (2022) Engineered bacteria-based living materials for biotherapeutic applications. Front Bioeng Biotechnol 10:870675

Pan J, Gong G, Wang Q, Shang J, He Y, Catania C, Birnbaum D, Li Y, Jia Z, Zhang Y, Joshi NS (2022) A single-cell nanocoating of probiotics for enhanced amelioration of antibiotic-associated diarrhea. Nat Commun 13(1):2117

Park T, Kim JY, Cho H, Moon HC, Kim BJ, Park JH, Hong D, Park J, Choi IS (2017) Artificial spores: immunoprotective nanocoating of red blood cells with supramolecular ferric ion-tannic acid complex. Polymers 9(4):140

Park J, Andrade B, Seo Y, Kim MJ, Zimmerman SC, Kong H (2018) Engineering the surface of therapeutic "living" cells. Chem Rev 118(4):1664–1690

Peng X, Yang L, Yuan P, Ding X (2024) Hybrid cell membrane-based nanoplatforms for enhanced immunotherapy against cancer and infectious diseases. Adv Healthc Mater 13(19):2304477
Periayah MH, Halim AS, Saad AZ (2017) Mechanism action of platelets and crucial blood coagulation pathways in hemostasis. Int J Hematol-Oncol Stem Cell Res 11:319
Pires-Santos M, Nadine S, Mano JF (2024) Unveiling the potential of single-cell encapsulation in biomedical applications: current advances and future perspectives. Small Sci 4(5):2300332
Qing G, Gong N, Chen X, Chen J, Zhang H, Wang Y, Wang R, Zhang S, Zhang Z, Zhao X, Luo Y (2019) Natural and engineered bacterial outer membrane vesicles. Biophys Rep 5:184–198
Rampado R, Crotti S, Caliceti P, Pucciarelli S, Agostini M (2020) Recent advances in understanding the protein corona of nanoparticles and in the formulation of "stealthy" nanomaterials. Front Bioeng Biotechnol 8:166
Rao L, Bu LL, Xu JH, Cai B, Yu GT, Yu X, He Z, Huang Q, Li A, Guo SS, Zhang WF (2015) Red blood cell membrane as a biomimetic nanocoating for prolonged circulation time and reduced accelerated blood clearance. Small 11(46):6225–6236
Rogozhnikov D, O'Brien PJ, Elahipanah S, Yousaf MN (2016) Scaffold free bio-orthogonal assembly of 3-dimensional cardiac tissue via cell surface engineering. Sci Rep 6(1):39806
Sartorio MG, Pardue EJ, Feldman MF, Haurat MF (2021) Bacterial outer membrane vesicles: from discovery to applications. Annu Rev Microbiol 75(1):609–630
Sekhon UD, Sen GA (2014) Platelet-like nanoparticles: mimicking shape, flexibility, and surface biology of platelets to target vascular injuries. ACS Nano 8(11):11243–11253
Sevencan C, McCoy RS, Ravisankar P, Liu M, Govindarajan S, Zhu J, Bay BH, Leong DT (2020) Cell membrane nanotherapeutics: from synthesis to applications emerging tools for personalized cancer therapy. Adv Therap 3(3):1900201
Shivatare SS, Shivatare VS, Wong CH (2022) Glycoconjugates: synthesis, functional studies, and therapeutic developments. Chem Rev 122(20):15603–15671
Sun M, Yang J, Fan Y, Zhang Y, Sun J, Hu M, Sun K, Zhang J (2023) Beyond extracellular vesicles: hybrid membrane nanovesicles as emerging advanced tools for biomedical applications. Adv Sci 10(32):230361
Sutaria DS, Jiang J, Elgamal OA, Pomeroy SM, Badawi M, Zhu X, Pavlovicz R, Azevedo-Pouly ACP, Chalmers J, Li C, Phelps MA (2017) Low active loading of cargo into engineered extracellular vesicles results in inefficient miRNA mimic delivery. J Extracell Vesicles 6(1):1333882
Toyofuku M, Nomura N, Eberl L (2019) Types and origins of bacterial membrane vesicles. Nat Rev Microbiol 17(1):13–24
Toyofuku M, Schild S, Kaparakis-Liaskos M, Eberl L (2023) Composition and functions of bacterial membrane vesicles. Nat Rev Microbiol 21(7):415–430
Wang Y, Hu Q (2023) Bio-orthogonal chemistry in cell engineering. Adv Nanobiomed Res 3(3):2200128
Wang B, Wang G, Zhao B, Chen J, Zhang X, Tang R (2014) Antigenically shielded universal red blood cells by polydopamine-based cell surface engineering. Chem Sci 5(9):3463–3468
Wang J, Zhu M, Nie G (2021) Biomembrane-based nanostructures for cancer targeting and therapy: from synthetic liposomes to natural biomembranes and membrane-vesicles. Adv Drug Deliv Rev 178:113974
Wiklander OP, Brennan MÁ, Lötvall J, Breakefield XO, El Andaloussi S (2019) Advances in therapeutic applications of extracellular vesicles. Sci Transl Med 11(492):eaav8521
Wo J, Lv ZY, Sun JN, Tang H, Qi N, Ye BC (2023) Engineering probiotic-derived outer membrane vesicles as functional vaccine carriers to enhance immunity against SARS-CoV-2. Iscience 26(105772):1
Won YW, Patel AN, Bull DA (2014) Cell surface engineering to enhance mesenchymal stem cell migration toward an SDF-1 gradient. Biomaterials 35(21):5627–5635
Wu Y, Zhao RC (2012) Current understanding of stem cell mobilization: the roles of chemokines, proteolytic enzymes, adhesion molecules, cytokines, and stromal cells. Stem Cell Rev Rep 8:243–250

Wu Y, Wan S, Yang S, Hu H, Zhang C, Lai J, Zhou J, Chen W, Tang X, Luo J, Zhou X (2022) Macrophage cell membrane-based nanoparticles: a new promising biomimetic platform for targeted delivery and treatment. J Nanobiotechnol 20(1):542

Xia Q, Zhang Y, Li Z, Hou X, Feng N (2019) Red blood cell membrane-camouflaged nanoparticles: a novel drug delivery system for antitumor application. Acta Pharm Sin B 9(4):675–689

Xu F, Fei Z, Dai H, Xu J, Fan Q, Shen S, Zhang Y, Ma Q, Chu J, Peng F, Zhou F (2022) Mesenchymal stem cell-derived extracellular vesicles with high PD-L1 expression for autoimmune diseases treatment. Adv Mater 34(1):2106265

Yan X, Liu XY, Zhang D, Zhang YD, Li ZH, Liu X, Wu F, Chen GQ (2021) Construction of a sustainable 3-hydroxybutyrate-producing probiotic Escherichia coli for treatment of colitis. Cell Mol Immunol 18(10):2344–2357

Yang J, Zhang X, Chen X, Wang L, Yang G (2017) Exosome mediated delivery of miR-124 promotes neurogenesis after ischemia. Mol Ther Nucleic Acids 7:278–287

Yang X, Yang J, Ye Z, Zhang G, Nie W, Cheng H, Peng M, Zhang K, Liu J, Zhang Z, Shi J (2022) Physiologically inspired mucin coated Escherichia coli Nissle 1917 enhances biotherapy by regulating the pathological microenvironment to improve intestinal colonization. ACS Nano 16(3):4041–4058

Yue Y, Xu J, Li Y, Cheng K, Feng Q, Ma X, Ma N, Zhang T, Wang X, Zhao X, Nie G (2022) Antigen-bearing outer membrane vesicles as tumour vaccines produced in situ by ingested genetically engineered bacteria. Nat Biomed Eng 6(7):898–909

Zhang Q, Dehaini D, Zhang Y, Zhou J, Chen X, Zhang L, Fang RH, Gao W, Zhang L (2018) Neutrophil membrane-coated nanoparticles inhibit synovial inflammation and alleviate joint damage in inflammatory arthritis. Nat Nanotechnol 3(12):1182–1190

Zhang Z, Zhou L, Xie N, Nice EC, Zhang T, Cui Y, Huang C (2020) Overcoming cancer therapeutic bottleneck by drug repurposing. Signal Transduct Target Ther 5(1):113

Zhang X, Zhang H, Gu J, Zhang J, Shi H, Qian H, Wang D, Xu W, Pan J, Santos HA (2021) Engineered extracellular vesicles for cancer therapy. Adv Mater 33(14):2005709

Zhang X, Lin Y, Xin J, Zhang Y, Yang K, Luo Y, Wang B (2024) Red blood cells in biology and translational medicine: natural vehicle inspires new biomedical applications. Theranostics 14(1):220

Zou MZ, Li ZH, Bai XF, Liu CJ, Zhang XZ (2021) Hybrid vesicles based on autologous tumor cell membrane and bacterial outer membrane to enhance innate immune response and personalized tumor immunotherapy. Nano Lett 21(20):8609–8618

Engineering Cell Membrane-Derived Nanovesicles for Advanced Biomedical Applications

2

Zahid Hussain, Wajiha Ahmed, Tian Gao, Ye Zhang, and Renjun Pei

Abstract

Cell membrane-derived nanovesicles (CMNVs) represent a broad category of membrane-bound nano-sized particles. They encompass exosomes, ectosomes, and CMNV-mimetic nanoparticles. The lumen and surface of nanovesicles are enriched with bioactive biomolecules, including proteins, lipids, enzymes, metabolites, nucleic acids, and cell-specific receptors, depending on the source cell. Natural and mimetic CMNVs hold substantial promise in biomedical applications owing to their unique biochemical composition, bio-interfacing properties, circulation stability, and strong protection of cargo. Despite their promise, the clinical use of natural and mimetic CMNVs is challenged by heterogeneous cargo, limited tissue targeting, and the inability to harness all advantages from a single parent cell due to variations in surface markers and biochemical composition. These limitations reduce therapeutic efficacy, emphasizing the need for advanced surface modification and cargo engineering strategies to enhance the biomedical features of CMNVs. This chapter outlines methods for the purification of natural CMNVs and explores various pre-modification (genetic engineering, metabolic glycoengineering, pre-cargo engineering) and post-modification strategies (covalent conjugation, lipid insertion, membrane fusion, post-cargo engineering) to optimize the surface properties and drug-loading capabilities of engineered nanovesicles. Furthermore, this chapter discusses therapeutic potential of engineered CMNVs in addressing cancer, neurological disorders, inflammatory diseases, and promoting tissue regeneration.

Z. Hussain · W. Ahmed · T. Gao · Y. Zhang · R. Pei (✉)
CAS Key Laboratory for Nano-Bio Interface, Suzhou Institute of Nano-Tech and Nano-Bionics (SINANO), Chinese Academy of Sciences (CAS), Suzhou, China
e-mail: zahid@mail.ustc.edu.cn; wanjia2024@sinano.ac.cn; tgao2017@sinano.ac.cn; yezhang2017@sinano.ac.cn; rjpei2011@sinano.ac.cn

Z. Hussain, R. Pei (eds.), *Cell Membrane Engineering for Advancing Cell-Based and Nanotechnology-Based Therapies*, Biomaterials, Bioengineering and Sustainability 6, https://doi.org/10.1007/978-3-031-93425-4_2

Keywords

Cell membrane nanovesicles · Hybrid nanovesicles · Purification methods · Engineering strategies · Engineered nanovesicles · Biomedical applications

2.1 Introduction

Mammalian extracellular nanovesicles are heterogeneous lipid-bound spherical nanoparticles. These nanovesicles are produced by living cells as a part of normal membrane renewal and exocytosis. Natural nanovesicles have shown substantial promise in biomedical applications owing to their unique biochemical and bio-interfacing properties (Gill et al. 2019). In addition, nanovesicle-mimetic nanoparticles or artificial extracellular vesicles can be prepared from cell membranes by fusing cell membrane components from different cell membrane sources or by combining synthetic liposomes with cell membrane proteins or fragments (Sun et al. 2023).

Extracellular Nanovesicles The composition of extracellular nanovesicles varies depending on their source cells. Their lumen contains proteins, lipids, nucleic acids, carbohydrates, and metabolites, while their membrane consists of lipid rafts, ceramides, membrane proteins, receptors, cholesterol, and a major histocompatibility complex. Mammalian extracellular nanovesicles can be divided into exosomes and ectosomes (Pocsfalvi et al. 2016; Hallal et al. 2022). Exosomes (30–150 nm) are cup-shaped membrane-bound nanovesicles. They are formed through a multistep process that includes inward budding of the cell membrane, formation of endocytic nanovesicles, inward budding of the early endosome, generation of intraluminal nanovesicles, production of multivesicular bodies (containing numerous intraluminal vesicles), and their fusion with the plasma membrane to release intraluminal vesicles as exosomes. The sorting, chemical composition, and secretion of exosomes are regulated by their parent cells (Hessvik and Llorente 2018). Ectosomes can be subdivided into microvesicles and apoptotic bodies. Microvesicles (50–300 nm) originate from outward blebbing of the cell membrane, closely resembling the surface properties of the parent plasma membrane (Cocucci et al. 2009). In contrast, apoptotic bodies (500–5000 nm) are generated during apoptosis and are enriched in various cellular components, including organelles. Their large size and heterogeneous biochemical features may limit their application compared to other extracellular nanovesicles (Hu et al. 2024) (Fig. 2.1).

Nanovesicle-Mimetic Nanoparticles Like natural vesicles, liposomes have a spherical structure composed of an aqueous core enclosed by a synthetic phospholipid bilayer. Unlike biological nanovesicles, liposomes suffer from biofouling, low circulation time, immune recognition, and limited targeting (Moulahoum et al. 2021) due to their inability to replicate the complex bio-interface of natural vesicles. Polyethylene glycol (PEG) modified liposomes have shown prolonged circulation,

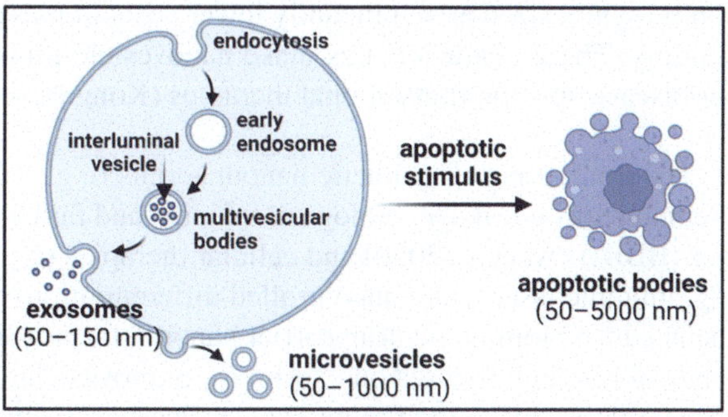

Fig. 2.1 Biogenesis and structure of exosomes, microvesicles, and apoptotic bodies. (Source: Created with BioRender.com)

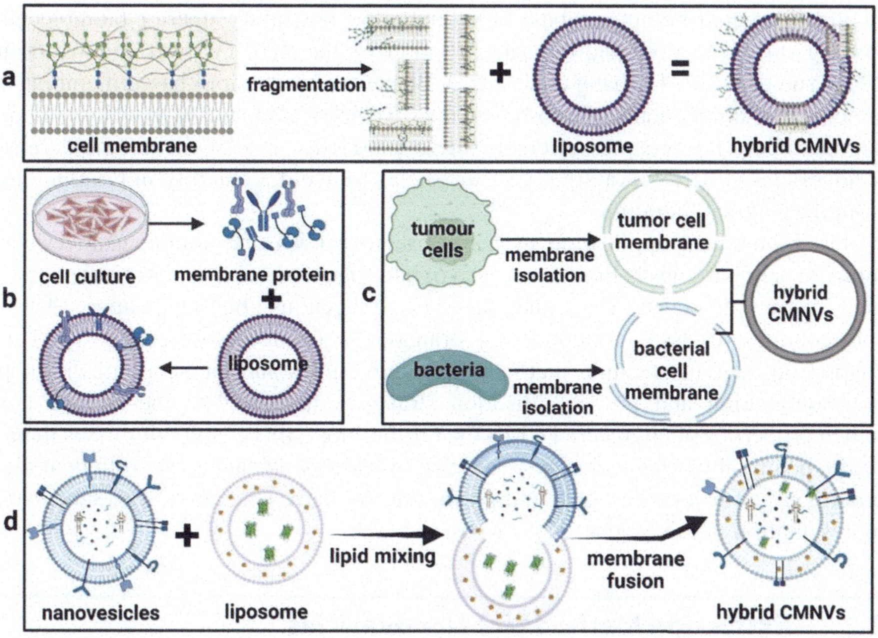

Fig. 2.2 (**a**) Fusion of cell membrane and liposomes. (**b**) Fusion of liposomes and cell membrane proteins. (**c**) Fusion of cell membrane from mammalian and bacterial cell sources. (**d**) Fusion of drug-loaded liposomes and natural nanovesicles. (Source: Created with BioRender.com)

but the generation of anti-PEG antibodies after repeated injections can cause hypersensitivity reactions (Zhang et al. 2016). There is significant interest in incorporating cell membrane fragments into liposomes for biomimetic applications. Furthermore, re-fusing cell membrane fragments from different cell sources can create nanovesicle-mimetic nanoparticles. Natural nanovesicles can be fused with liposomes to create tailored nanoparticles (Fig. 2.2). Nanovesicle-mimetic nanoparticles retain cell membrane properties, exhibit sufficient stability, and high drug-carrying potential. Unlike natural extracellular nanovesicles, these nanoparticles can be synthesized at significantly lower costs in mass quantities and possess good stability. These characteristics make nanovesicle-mimetic nanoparticles promising for disease-specific translational therapies (Kong et al. 2022; Le et al. 2021).

The limitations of synthetic nanomaterials (e.g., protein corona, poor targeting, premature drug leakage, biological toxicity, and failure to cross the blood-brain barrier (BBB)) (Ai et al. 2020) and cellular therapies (e.g., cell senescence, poor homing, immune issues, and uncontrolled differentiation) (Hoang et al. 2022) highlight the need for biomimetic drug carrier nanoplatforms (Shao et al. 2024). Current biomedical research focuses on exosomes, microvesicles, and natural-vesicle mimetic nanoparticles due to their remarkable properties, including bio-interfacing capabilities (such as immunological escape and specific antigen presentation), circulation stability, targeted homing, robust bio-stimulation potential, superior biomolecule loading and protection, and the capacity to cross the BBB (Witwer and Wolfram 2021; You et al. 2024; Zhang et al. 2023). Because phospholipid-based membranes are essential components of exosomes, microvesicles, and natural-vesicle mimetic nanoparticles, this chapter refers to these platforms as cell membrane-derived nanovesicles (CMNVs) or simply nanovesicles to avoid ambiguity in terminology from the original literature.

The biomedical application of natural nanovesicles and simple nanovesicle-mimetic nanoparticles is limited because of heterogeneous cargo, poor tissue targeting, restricted penetration, and lack of exogenous biotherapeutics. These shortcomings necessitate engineering strategies to create designer CMNVs. This chapter outlines purification methods for nanovesicles and the scope of different pre-modification and post-modification strategies to modulate the surface and lumen properties of engineered CMNVs. Furthermore, this chapter discusses therapeutic potential of engineered nanovesicles in addressing cancer, neurological disorders, inflammatory diseases, and promoting tissue regeneration. The schematic representation of this chapter is shown in Fig. 2.3.

2.2 Extraction Methods for Nanovesicles

Various cell sources can be cultured to produce large quantities of CMNVs for biomedical applications. Considering the existence of distinct subpopulations with similar morphologies and sizes, several approaches have been established to isolate CMNVs from biological samples. These methods include ultracentrifugation, size-based techniques, and microfluidic immunoaffinity-based isolation (Tauro et al. 2012).

2.2.1 Ultracentrifugation-Based Isolation

Differential and density gradient ultracentrifugation are widely used CMNVs purification strategies. Differential ultracentrifugation isolates nanovesicles based on

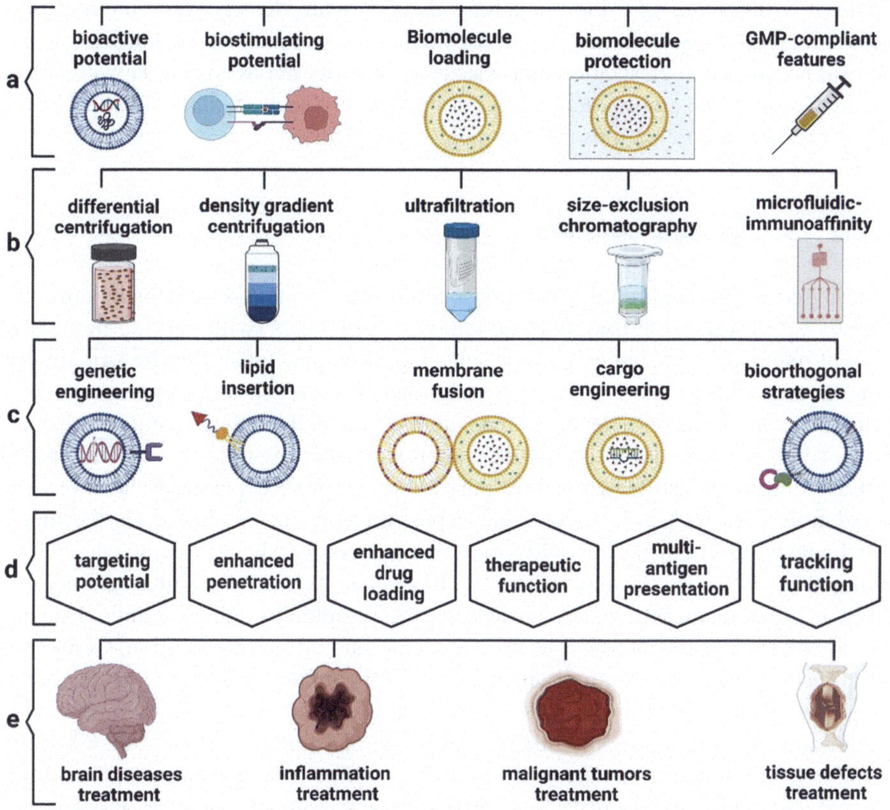

Fig. 2.3 (**a**) Characteristics features of CMNVs. (**b**) Extraction methods for CMNVs. (**c**) Strategies for CMNVs engineering. (**d**) Summary of enhanced functionalities in nanovesicles through engineered modifications for improved biomedical applications. (**e**) Biomedical potential of engineered CMNVs in addressing cancer, neurological disorders, inflammatory diseases, and promoting tissue regeneration. (Source: Created with BioRender.com)

their size and sedimentation rate through sequential centrifugation. Initially, samples undergo low-speed centrifugation to remove the cells (500 × g for 5 min) and large debris (3000× g for 10–20 min). The supernatant is then ultracentrifuged to collect ectosomes (10,000× g for 30–45 min) and exosomes (100,000× *g* for 50–100 min), though protocols may vary. This method is cost-effective, reagent-free, and considered the "gold standard". However, major disadvantages include long processing times, aggregation, structural damage to vesicle membranes or cargo from repeated centrifugation, and challenges in purifying nanovesicles from lipoproteins (Kverneland et al. 2023; Linares et al. 2015). Density gradient ultracentrifugation separates nanovesicles according to size, molecular mass, and sedimentation density by using continuous or discontinuous gradients of sucrose or iodixanol. The biological sample is added to the top of the medium and then centrifuged, allowing particles of different densities to migrate through the gradient and settle at equilibrium positions matching their buoyant density. This method processes larger sample volumes, separates CMNV subpopulations by buoyant densities for precise characterization, and yields high-purity nanovesicles. However, this method is time-consuming, requires prolonged centrifugation, and is unsuitable for high-throughput applications (Taylor and Shah 2015).

2.2.2 Size-Based Isolation

Size-based isolation methods encompass ultrafiltration and size-exclusion chromatography. Ultrafiltration uses pore-containing membranes (with varying molecular weight retention) to separate nanovesicles based on size, enabling the isolation of smaller and homogeneous particle populations. The ultrafiltration method is cost-effective, instrument-independent, and suitable for small and large sample volumes. However, larger particles can clog the membrane, resulting in lower recovery rates. Ultrafiltration may suffer from low purity due to limited specificity, as similarly sized molecules, such as lipoproteins, can pass through the membrane. Furthermore, mechanical stress during ultrafiltration can compromise the structural integrity of isolated nanovesicles (Vergauwen et al. 2017). Size exclusion chromatography utilizes an aqueous mobile phase to transport the sample through a column. The stationary phase consists of a porous polymer, enabling differential elution. Compared to ultracentrifugation, the main advantage of this approach is that it avoids harsh conditions (i.e., centrifugal forces and pressures), can handle various sample volumes, and better maintain the inherent structure and function of isolated CMNVs. However, this method has some drawbacks, such as time-consuming processing, lower yield, and reduced purity, as similar-sized contaminants, such as lipoproteins, can co-elute with nanovesicles (Monguió-Tortajada et al. 2019; Sidhom et al. 2020).

2.2.3 Microfluidic Immunoaffinity-Based Isolation

Microfluidics-immunoaffinity combines antigen-antibody interactions to selectively capture nanovesicles with microfluidics for fluid control in micrometer channels. Antibodies targeting nanovesicle surface markers are coated on microfluidic capillaries. Samples are allowed to pass through micrometer channels, and this design increases the surface area and incubation time, improving the capture of CMNVs. This method is suitable for diagnosis because it efficiently isolates specific CMNVs from complex biological samples and can handle small sample volumes. However, this method is expensive, applicable to small volumes of samples, and results in lower CMNVs yield. Isolating a broad population of CMNVs is challenging because not all nanovesicles carry adequate surface markers. Furthermore, incomplete antibody removal can potentially impair surface receptor functionality or targeting efficiency of nanovesicles (Yang et al. 2017a; Tian et al. 2019).

2.3 Strategies for Producing Engineered Nanovesicles

Natural CMNVs from different cell sources vary in surface markers, chemical composition, and physicochemical properties. These differences cause batch variability, poor biodistribution, limited targeting, and reduced effects. There is growing interest in modifying CMNV surfaces and lumen by adding therapeutic moieties and exogenous drugs to enhance tissue targeting, immune response, stimuli-responsiveness, and monitoring capabilities for targeted disease treatment (Ma et al. 2025). Various bioengineering approaches, including genetic engineering (GenEng), metabolic glycoengineering (MGEng) coupled click conjugation, covalent conjugation, lipid insertion, membrane fusion (MemFus), and cargo engineering (CarEng) have been introduced to engineer CMNVs. These methods can be categorized into pre-modification (modification of nanovesicle-producing cells) and post-modification strategies (direct modification of isolated nanovesicles). Pre-modification allows nanovesicle properties to be manipulated during biogenesis, ensuring consistent results, reducing batch variability, and enabling efficient therapeutic molecule incorporation with minimal disruption to intrinsic activities. Post-modification-based methods bypass parent cell modification and can be applied to isolated nanovesicles. They enable the incorporation of multiple exogenous molecules and allow for more extensive surface property modifications compared to pre-modification methods. The principles of various pre- and post-modification strategies are discussed in the following sub-section and are summarized in Table 2.1.

2.3.1 Pre-modification Strategies

2.3.1.1 Genetic Engineering

GenEng involves the transfection of donor cells with a target gene that integrates into the host genome and promotes recombinant protein expression through biogenesis. Changes in parent cells can be reflected in the secreted nanovesicles, enabling engineered CMNVs to promote the corresponding targeting, pharmaceutical, recognition, and immunological functions (Cheng et al. 2024) (Fig. 2.4a). Various receptors and ligands such as PD-1, PD-L1, CXCR4, TRAIL, iRGD peptide, GE11 peptide, GFE1 peptide, single-chain variable fragment (scFv), RVG (rabies virus glycoprotein), chondrocyte affinity peptide (CAP), cardiomyocyte-targeting peptide (CTP) have been expressed into CMNVs through GenEng. The recombinant gene encoding the targeting receptor has been fused with membrane-bound proteins of nanovesicles such as lysosome-associated membrane glycoprotein 2B (Lamp2b) (Bai et al. 2020), glycosylphosphatidylinositol (GPI)-anchor (Kooijmans et al. 2016), platelet-derived growth factor receptor (PDGFR) (Shi et al. 2020), CD63 (Rufino-Ramos et al. 2022), and CD9 (Yim et al. 2016). Specific presentation of targeted molecules fused with integral vesicle proteins can correct their spatial orientation and optimum function. For example, recombinant tags can be exposed on the nanovesicle surface or into their lumen when fused with the N-terminus or

Table 2.1 Comparative overview of different pre-modification and post-modification approaches for nanovesicle functionalization

Strategy	Subcategory	Key features	Advantages	Limitations
Pre-modification	Genetic engineering	Transfection of donor cells with recombinant genes	Precise control, scalable, maintains nanovesicle integrity	Limited to genetically modifiable cells, requires stable expression
	Metabolic glycoengineering and click conjugation	Incorporation of unnatural sugars with bioorthogonal handles for conjugation	Enables selective modifications, bioorthogonal tags facilitate click reactions	Limited by metabolic incorporation efficiency, complex synthesis
	Pre-cargo engineering	Passive or active drug loading in parent cells before vesicle secretion	Higher control over cargo loading, maintains vesicle integrity	Low efficiency for high molecular weight drugs, complex drug sorting mechanisms
Post-modification	Covalent conjugation	Surface functionalization via reactive biomolecules (e.g., NHS ester)	Stable modification, allows click chemistry-based targeting	May alter membrane proteins, potential loss of bioactivity
	Lipid insertion	Incorporation of ligand-conjugated lipid anchors	Simple, maintains membrane integrity, tunable ligand density	Potential impact on vesicle stability and function
	Membrane fusion	Fusion of natural vesicles with liposomes or other cell-derived membranes	Creates hybrid vesicles with enhanced functionalities	Complex optimization required, risk of vesicle instability
	Post-cargo engineering	Drug loading via electroporation, sonication, chemical transfection, or liposome fusion	Enables loading of hydrophilic therapeutics and nucleic acids	Risk of vesicle damage, aggregation, and loss of bioactivity

C-terminus of Lamp2b, respectively (Bai et al. 2020). Specific fusion of exogenous moieties with nanovesicles membrane proteins could result in enhanced loading, targeting specificity, immune recognition, or protection of inserted biomolecules. Overall, GenEng offers advantages like large-scale production of modified CMNVs once the engineered cell line is established, as well as the specific display of recombinant molecules on the surface or within the lumen of nanovesicles (Sterner and Sterner 2021; Cheng et al. 2023).

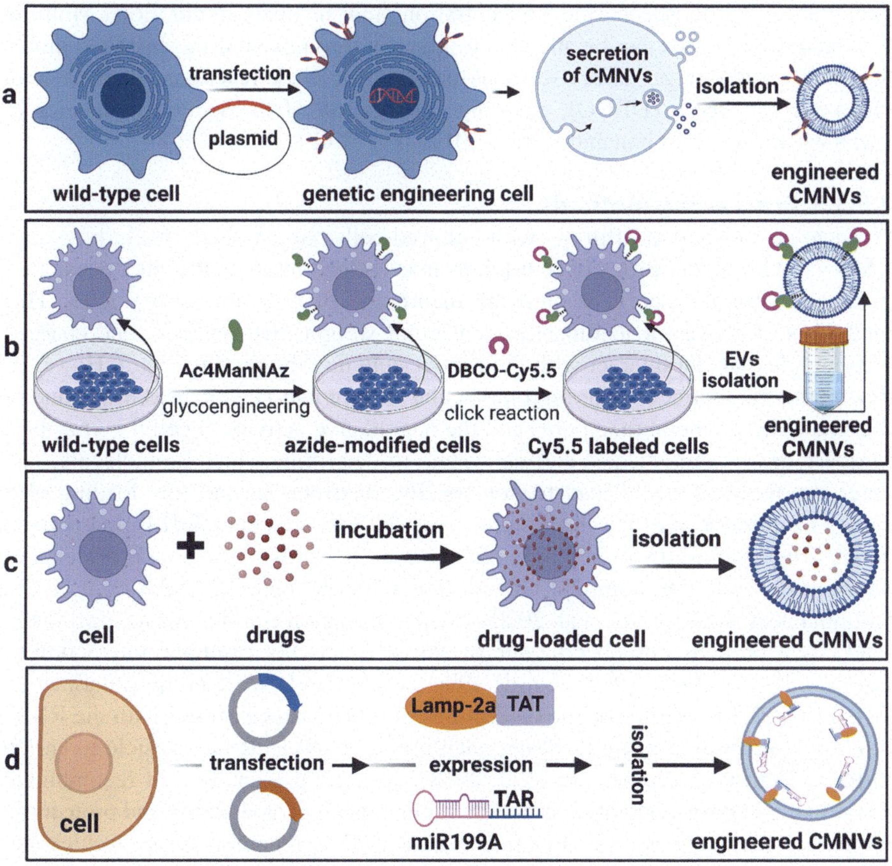

Fig. 2.4 (**a**) GenEng for pre-modification of nanovesicles. (**b**) MGEng coupled click conjugation-based pre-modification of nanovesicles. (**c**) Parental cell-based passive drug loading in nanovesicles. (**d**) Parental cell-based active drug loading in nanovesicles through GenEng. (Source: Created with BioRender.com)

2.3.1.2 MGEng and Click Conjugation

Glycosylation is a post-translational modification of more than 90% of cell-surface proteins. These modifications are essential for cellular adhesion, interactions, immune responses, and cell recognition. These significances have prompted research into MGEng for incorporating artificial agents into glycans for healthcare applications (Cheng et al. 2021). Unnatural monosaccharide analogs (mannosamine, galactosamine, glucosamine, sialic acid) bearing a bioorthogonal group (i.e., azide, terminal alkyne) can diffuse into cells, and are metabolized into nucleotide sugars for glycosylation. These modified glycoconjugates are then transported to endosomal and plasma membranes, increasing the presence of unnatural tags on sialoglycoproteins (Agatemor et al. 2019). Metabolic changes occurring in parent cells can be mapped on their secreted nanovesicles. Tag quantity can be controlled by manipulating monosaccharide concentration and incubation duration. Multiple bioorthogonal groups can be attached to glycans through MGEng. Installed bioorthogonal groups (e.g., azide) serve as handles for click reactions, such as with dibenzylcyclooctyne (DBCO), enabling conjugation of therapeutic molecules to CMNVs surface (Nischan and Kohler 2016) (Fig. 2.4b).

2.3.1.3 Pre-CarEng Methods

Physical co-incubation of drugs with cultured cells can passively pack drugs into CMNVs, with higher drug concentrations in cells leading to their higher packaging in the intraluminal space or onto the membrane of secreted nanovesicles. This method is effective for loading low molecular weight, hydrophobic, and charged compounds that can interact with the nanovesicle membrane (Fig. 2.4c). However, this method is ineffective for high molecular weight and nucleic acid drugs due to the small size of membrane pores and the degradative activity of cellular enzymes. Furthermore, a sophisticated sorting system to determine which biomolecules can enter nanovesicles could lead to non-specific distribution, and low loading efficiency of introduced drugs (Kalimuthu et al. 2018; Kalani et al. 2014). In contrast, active drug loading involves fusing recombinant proteins with integral proteins of nanovesicles, such as Lamp2b, GPI-anchor, PDGFR, and CD9 (Bai et al. 2020; Kooijmans et al. 2016; Shi et al. 2020). Such mechanism could assist to specifically load target proteins into specific site of nanovesicles. For example, interaction of Lamp2a-TAT module with TAR-miR-199a facilitate miRNA loading (Sutaria et al. 2017) (Fig. 2.4d). Similarly, the C-terminus of CD9 has been fused with the RNA-binding HuR protein using GenEng, enabling the CD9-HuR nanovesicle to enrich miRNA-155 into its lumen (Li et al. 2018). Specific presentation of recombinant drugs in the nanovesicle lumen is crucial for enhanced cargo loading and protection. However, this active pre-modification is restricted to recombinant molecules and may not be able to tag complex drugs.

2.3.2 Post-modification Strategies

2.3.2.1 Covalent Conjugation

The heterogeneous biomolecular landscape of the cell surface features reactive hotspots. Only a few moieties are reactive in complex environments, offering opportunities to modify biological activity. Functional groups such as primary amine, carboxylic acid, and thiol from membrane glycoproteins are potential targets for covalent bonding with conjugants. Carboxylic acid and thiol-based conjugation are rarely used for nanovesicle engineering compared to amine conjugation methods due to their low abundance on membrane surfaces (Yang et al. 2023). In contrast, amine groups ($-NH_2$) located on lysine residues of proteins and N-terminal end of polypeptide chains are widely used for surface engineering of nanovesicles through reaction with ester-labelled agents such as N-hydroxysuccinimide ester (NHS) labelled reagents (i.e., NHS-DBCO). NHS ester reacts with protein amino groups of membrane protein to form a stable amide bond, enabling its widespread use in

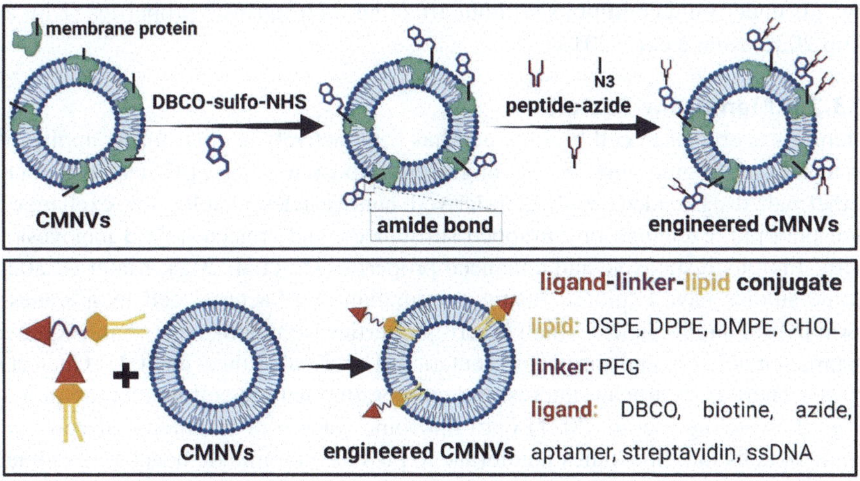

Fig. 2.5 (**a**) Covalent conjugation and click reaction-based attachment of targeted peptide onto nanovesicles. (**b**) Schematic lipid insertion methods for nanovesicle engineering. (Source: Created with BioRender.com). DSPE 1,2-distearoyl-sn-glycero-3-phosphoethanolamine, DPPE 1,2-dipal mitoyl-sn-glycero-3-phosphoethanolamine, DMPE 1,2-dimyristoyl-sn-glycero-3--phosphoethanolamine, CHOL cholesterol, PEG polyethylene glycol, DBCO dibenzocyclooctyne, NHS N-hydroxysuccinimide, CMNVs cell membrane-derived nanovesicles

bioconjugation for nanovesicle protein labeling. These reactions can be performed in neutral aqueous buffers. Primary bioorthogonal groups introduced via covalent conjugation can serve as handles for click conjugation, enabling the attachment of exogenous therapeutics to the CMNVs surface (Tian et al. 2018; Yang et al. 2024) (Fig. 2.5a).

2.3.2.2 Lipid Insertion

Lipid insertion facilitates the installation of non-natural moieties on nanovesicle through a lipid anchor. The arrangement of phospholipids, GPI, and cholesterol creates an attractive site for inserting ligand-conjugated lipophilic molecules. Studies have shown that incorporating phosphatidylcholine into membrane lipid layer of nanovesicles enhances their penetration and uptake to the recipient cells, underscoring the importance of lipid engineering for improved therapeutic delivery (Zhan et al. 2021). A range of synthetic materials can be conjugated to hydrophobic anchors (DSPE, DPPE, DMPE) and PEG-based spacer to form PEG-linked lipid conjugates (such as FITC-PEG-DPPE) for surface engineering (Cruje and Chithrani 2014; Antes et al. 2018) (Fig. 2.5b). For example, docetaxel (an acid-sensitive pro-drug) and CR8 peptide (to target vascular cellular adhesion molecule-1) have been conjugated to DSPE-PEG via an acid-cleavable ester bond. DSPE-PEG acted as a "flexible arm" for the surface modification of nanovesicles via lipid insertion (Zhong et al. 2021). Lipid insertion approach enables uniform presentation of introduced ligands under mild conditions without affecting their intrinsic properties (Kim and Kim 2022; Kolate et al. 2014).

2.3.2.3 Membrane Fusion

MemFus is crucial in cell biology and has been actively studied for its application in membrane engineering. Fusing engineered liposomes (e.g., pH-responsive, photothermal, drug-loaded) with cell-derived nanovesicles enables the exchange of phospholipid, exchange of entrapped substances, and created hybrid nanovesicles with stimulus-responsive and enhanced properties (Li et al. 2024; Ma et al. 2022). Some studies have explored fusing mammalian and bacterial cell membranes to form hybrid nanovesicles, which inherit properties from both sources and could aid in targeted delivery and antigen presentation (Fig. 2.6a) (Zou et al. 2021; Chen et al. 2021). There is significant interest in incorporating natural cell membrane proteins (Fig. 2.6b) (Zinger et al. 2021) onto liposome surface or fragments of cell membrane of cells into liposomes membrane to provide biomimetic benefits to chimeric nanovesicles (Fig. 2.6c) (Song et al. 2021; He et al. 2018). GenEng can enable nanovesicle-producing cells to express a single recombinant peptide onto nanovesicle surface, and fusing different engineered vesicles can create hybrids with dual peptides on their surface (Meng et al. 2021). Overall, MemFus facilitates rapid modification by integrating various parent membrane functionalities, enabling multi-targeting therapy and enhancing biomedical applications (Piffoux et al. 2018; Yu et al. 2022).

2.3.2.4 Post-CarEng Methods

Post-active drug loading can be performed using sonication, electroporation, chemical transfection, and liposome fusion. Electroporation uses an electric field, sonication uses an ultrasound probe, and transfection uses surfactants (saponin and Triton X-100) to create temporary pores in the nanovesicle membrane for drug loading (Fig. 2.7a). These methods enable sufficient loading of hydrophilic drugs, magnetic particles, and nucleic acid in nanovesicle. However, each active loading method has drawbacks: sonication can destabilize nanovesicle structure and may cause loss of endogenous components, ultrasound promotes nanovesicle aggregation and loss of surface proteins, and transfection may cause surfactant-induced hemolysis. These methods do not permit the loading of drugs to the nanovesicle interface. Furthermore, many biomolecules are susceptible to deactivation upon loading, necessitating careful optimization of drug loading protocols (Haney et al. 2015; Fu et al. 2020). Liposome-mediated drug loading involves incorporating drugs into liposomes during their formation. These drug-loaded liposomes can then fuse with natural nanovesicles to create hybrid nanovesicles with high drug content (Zheng et al. 2024) (Fig. 2.7b). However, compared to other drug-loading methods, this process alters the membrane fluidity of natural nanovesicles, which may limit its suitability for applications that require the preservation of their native characteristics.

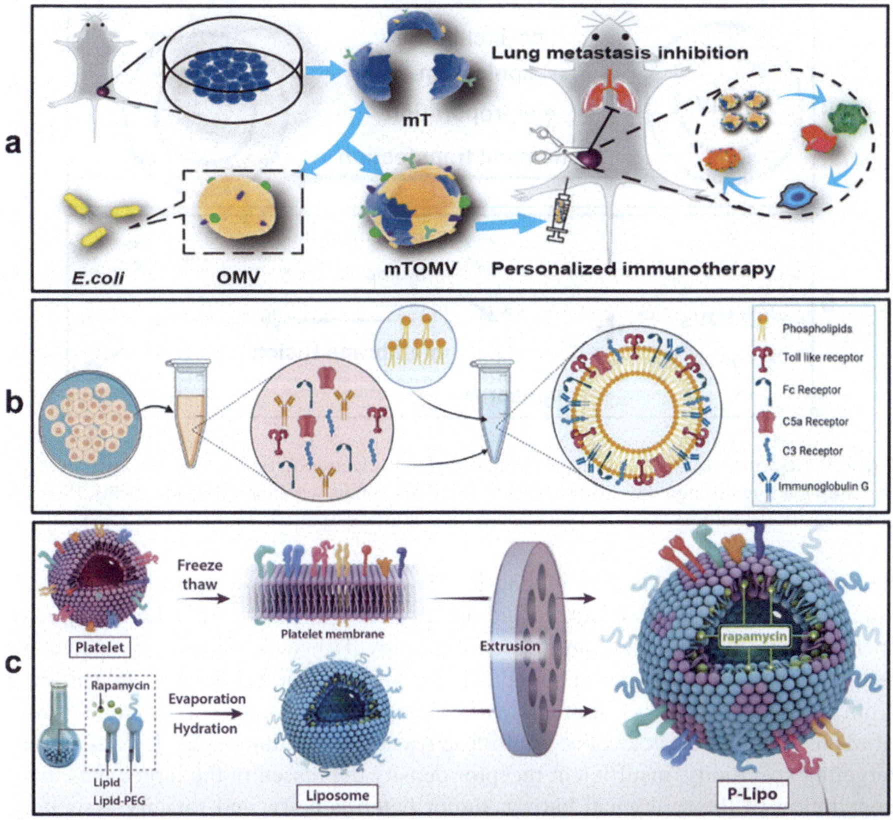

Fig. 2.6 (**a**) Tumor and bacterial membrane vesicles fusion enhance immune response for cancer immunotherapy. (Reprinted with permission from Zou et al. (2021), copyright 2021, American Chemical Society). (**b**) Cell membrane protein and liposome fusion to target brain lesions. (Reprinted with permission from Zinger et al. (2021), copyright 2021, Wiley-VCH GmbH). (**c**) Platelet membrane and liposome fusion to target atherosclerosis. (Reprinted with permission from Song et al. (2021), copyright 2021, Elsevier B.V.)

2.4 Potential Biomedical Applications of Engineered Nanovesicles

Engineered nanovesicles represent a frontier in biomedical applications, offering innovative solutions for cancer therapy, brain diseases, inflammation, and tissue engineering (Table 2.2).

2.4.1 Engineered Nanovesicles for Cancer Therapy

Cancer is characterized by abnormal cell growth that can spread to other body parts. This seriously threatens health and creates a substantial socio-economic burden

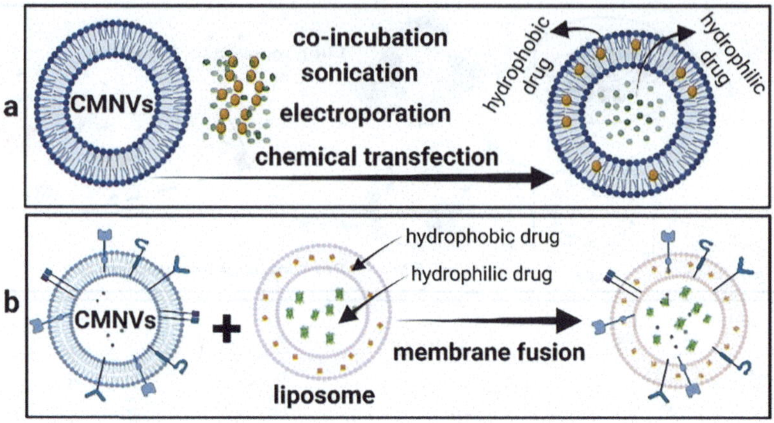

Fig. 2.7 (**a**) Drug loading via co-incubation, sonication, electroporation, or chemical transfection. (**b**) Drug loading through the fusion of drug-loaded liposomes with natural nanovesicles. (Source: Created with BioRender.com)

worldwide. Various treatment modalities have been developed from bench to bedside at an unprecedented rate, including surgical excision, chemotherapy, phototherapy, radiotherapy, and combinatorial therapy. However, despite advancements in these treatment strategies and medical technologies, mortality rates and suffering from many cancer types remain essentially unchanged. Issues such as the concern of metastasis, tumor recurrence, multidrug resistance of tumor cells, lack of tumour targeting specificity, insufficient receptor density expressed in the target sites, poor penetration of the biological barrier, tumor heterogeneity, and toxicity associated with traditional therapies pose significant hurdles in satisfactory cancer treatment. These challenges highlight the need for improved immune modulation and targeted nanosized drug delivery strategies. Engineered nanovesicles have been introduced for cancer treatment through targeted drug therapy and immunotherapy (Yong et al. 2022; Ruan et al. 2022).

2.4.1.1 Cancer Treatment Through Targeted Drug Therapy

Several studies have used transfected HEK293T cells to display specific targeting peptides, such as GE11, IL-3, and iRGD, on nanovesicle surfaces. This was achieved by fusing these peptides with integral nanovesicle proteins like PDGFR or Lamp2b. This approach enables targeted delivery to specific cell types, e.g., GE11 targets EGFR-expressing breast cancer cells, IL-3 targets IL-3 receptors in myeloid leukemia, and iRGD targets solid tumors with high integrin $\alpha v\beta 3$ levels. The engineered nanovesicles were loaded with various nucleic acid drugs, including miRNA, BCR-ABL siRNA, and KRAS siRNA, using a pre-loading method. This loading approach enhanced nucleic acid stability and prevented uncontrolled diffusion. Results showed that peptide-engineered nanovesicles effectively delivered nucleic acids to cancer cells through specific ligand-receptor interactions and allowed for adequate tumor tissue penetration. In mouse models, these engineered nanovesicles

Table 2.2 Overview of cell sources, engineering strategies, surface display agents, incorporated drugs, mechanisms, and biomedical applications of engineered nanovesicles

Cell source	Loaded drug (method)	Surface display agent	Engineering strategies	Working mechanism and key results	Ref.
Engineered CMNVs for cancer therapy					
HEK293 cells	miRNA (pre-loading)	PDGFR-GE11 peptide	GenEng, CarEng	GE11 peptide-decorated, and miRNA-encapsulated nanovesicles enable targeted delivery to EGFR-expressing breast cancer in mouse models, suggesting a potential platform for cancer treatment.	Ohno et al. (2013)
HEK293T cell	BCR-ABL siRNA (pre-loading)	Lamp2b-IL3	GenEng, CarEng	Nanovesicles were engineered with IL3 and siRNA. Intraperitoneal injection of engineered nanovesicles showed potential to home cancer tissue and inhibit cancer growth in mice models.	Bellavia et al. (2017)
HEK293T cells	KRAS siRNA (pre-loading)	Lamp2b-iRGD peptide	GenEng, CarEng	Cells were programmed to express tumor-penetrating iRGD on nanovesicle surfaces and then loaded with siRNA. Intravenous injection of these nanovesicles targeted tumor and suppressed tumor growth in a mouse model.	Zhou et al. (2019)
Immature dendritic cells	Doxorubicin (electroporation)	Lamp2b-iRGD peptide	GenEng, CarEng	iRGD-engineered nanovesicles were loaded with doxorubicin drug. These nanovesicles targeted integrin-positive cancer cells and delivered the drug to tumors in mice, resulting in tumor growth inhibition.	Tian et al. (2014)
BM-MSCs	PI3Kγ inhibitor (pre-loading)	Lamp2b-GFE1 protein	GenEng, CarEng	GFE1-engineered nanovesicles accumulate in the lungs of melanoma mice, releasing a PI3Kγ inhibitor that targets the premetastatic niche to prevent tumor metastasis and enhance immune stimulation against cancer.	Han et al. (2022)
M1 macrophage, platelets, B16F10 cells	cGAMP (sonication)	SIRPα variants	GenEng, MemFus, CarEng	SIRPα-engineered hybrid nanovesicles target tumor cells, repolarize M2-to-M1, enhance antigen presentation, and inhibit melanoma in mice. cGAMP loading further boosts tumor inhibition in breast cancer models.	Rao et al. (2020)
M1 macrophages	–	aCD47, aSIRPα	MGEng coupled click conjugation	Azide-modified nanovesicles were conjugated with DBCO-aCD47 and DBCO-aSIRPα using sensitive linkers to target tumors, reprogram macrophages, and release antibodies, thereby enhancing tumoricidal activity.	Nie et al. (2020)

(continued)

Table 2.2 (continued)

Cell source	Loaded drug (method)	Surface display agent	Engineering strategies	Working mechanism and key results	Ref.
Dendritic cells	–	aCD19, PD1	GenEng	Dendritic cell-derived nanovesicles contain co-stimulatory molecules for antigen presentation. Engineered vesicles can penetrate huCD19-expressing tumors and reverse immunosuppressive microenvironment.	Xu et al. (2023)
RAW264.7 cells	CPI-444 (incubation)	aPD1-scFv	GenEng, CarEng	Nanovesicles displaying aPD-1 enhance T cell anti-tumor activation and infiltration. CPI-444-loaded nanovesicles antagonize immune-suppressive metabolites and restrain tumor progression in mice.	Xue et al. (2022)
BM-MSCs	ICG, R837 (sonication)	aPD-L1, ovalbumin	GenEng, CarEng	Nanovesicles expressing aPD-L1, immune adjuvants and photosensitizers enabled targeted photothermal cancer ablation, enhanced dendritic cell maturation and anti-tumor responses in melanoma mice.	Chen et al. (2022)
Engineered CMNVs for brain diseases treatment					
BM-MSC	Curcumin (incubation)	c(RGDyK) peptide	Covalent coupled click conjugation	Intravenous administration of peptide-conjugated and curcumin-loaded nanovesicles resulted in targeted delivery to the lesion site, suppressing inflammation and inducing cellular apoptosis to treat cerebral ischemia.	Tian et al. (2018)
BM-MSCs	miRNA-124 (electroporation)	Lamp2b-RVG peptide	GenEng, CarEng	Systemic administration of RVG-exposed and miRNA-124-loaded nanovesicles crosses the BBB, accumulates in the ischemic region, and promotes cortical neurogenesis by modulating gene expression.	Yang et al. (2017b)
Immature dendritic cells	siRNA (electroporation)	Lamp2b-RVG peptide	GenEng, CarEng	Intravenous injection of RVG-engineered nanovesicles delivers siRNA to neuronal cells, leading to targeted and long-term gene silencing in brain regions, indicating potential for treating brain diseases.	Alvarez-Erviti et al. (2011)
Neural stem cells	Bryostatin-1 (sonication)	PDGFRα peptide	GenEng, CarEng	Systemic administration of engineered nanovesicles crosses the endothelial barrier, enhances the targeting to brain lesions, and modulates neuroinflammation in autoimmune encephalomyelitis mice.	Wu et al. (2022)

M2 microglia	–	DA7R peptide, SDF-1 peptide	Covalent coupled click conjugation	DA7R attracts endothelial cells, and SDF-1 recruits neural stem cells. Engineered nanovesicles targeted injured vasculature, recruited neural stem cells, and induced neurogenesis after ischemic stroke.	Ruan et al. (2023)
Bone-marrow macrophages	GDNF (pre-loading)	–	GenEng	Engineered nanovesicles protect GDNF, target inflamed brain tissue, improve neuronal survival, and reduce inflammation and neurodegeneration in a Parkinson's disease mouse model.	Zhao et al. (2022)
BM-MSCs	Tyrosine phosphatase-2 (pre-loading)	–	GenEng	MSCs-derived nanovesicles modified with tyrosine phosphatase-2 penetrate the BBB, inducing mitophagy and inhibiting neuroinflammation, thereby rescuing synaptic loss and cognitive decline in an Alzheimer's mode.	Xu et al. (2022a)
Engineered CMNVs for inflammatory diseases treatment					
Numerous cancer cell lines	–	DBCO-PHA	MGEng coupled click conjugation	Unnatural azide groups were expressed on donor cells, and later, DBCO-PHA labelled the nanovesicle's surface. PHA target CD44-overexpressing tissues, allowing vesicles to accumulate in collagen-induced arthritis mice.	Lim et al. (2021)
Adipose-derived stem cells	–	DBCO-dextran sulfate	MGEng coupled click conjugation	Stem cells expressing azide enabled DBCO-dextran sulfate attachment to the nanovesicle surface. Engineered vesicles targeted M1 macrophages and reduced inflammation in collagen-induced arthritis mice.	You et al. (2021)
RAW264.7 Cells, HUVEC	Curcumin (sonication)	–	MemFus, CarEng	Hybrid nanovesicles with dual cell membrane proteins showed high uptake by inflammatory cells. Curcumin loading reduced pro-inflammatory cytokine expression and alleviated lipopolysaccharides-induced organ damage in mice.	He et al. (2024)
Activated neutrophil	Keratinocyte growth factor (liposome fusion)	–	MemFus, CarEng	Keratinocyte growth factor-loaded neutrophil/liposome nanovesicles exhibit sufficient stability and target inflamed bowel tissue in sodium sulfate-induced mice, promoting the recovery of bowel function.	Zhao et al. (2019)

(continued)

Table 2.2 (continued)

Cell source	Loaded drug (method)	Surface display agent	Engineering strategies	Working mechanism and key results	Ref.
MC3T3 cells	Curcumin (co-incubation)	CXCR4	GenEng, CarEng	CXCR4-expressing nanovesicles target inflammatory cells. Curcumin release enhances macrophage polarization. Engineered vesicles show anti-inflammatory effects for treating ulcerative colitis in mice.	Wang et al. (2021)
RBCs	–	Docetaxel, CR8 peptide	Lipid insertion	Nanovesicles were conjugated with docetaxel and CR8 peptide to target inflammatory endothelial cells, allowing for responsive drug release to manage atherosclerosis in apolipoprotein E-deficient mice.	Zhong et al. (2021)
HEK293T cells	srIkB (pre-loading)	–	GenEng	Parent cells were engineered to express CIBN-EGFP-CD9 and srIκB-mCherry-CRY2 for loading srIκB into nanovesicles. Intraperitoneal injection of these vesicles reduced organ injury and inflammation in a septic mouse model.	Choi et al. (2020)
Engineered CMNVs for tissue engineering					
Cardiosphere-derived cells	–	Lamp2b-CTP	GenEng	Peptide-engineered nanovesicles enhanced cardiomyocyte uptake, reduced apoptosis, stimulated angiogenesis, and improved cardiac retention following injection in mice.	Mentkowski and Lang (2019)
HEK293 cells	–	Lamp2b-CTP	GenEng	CTP expression on nanovesicles improve their targeting of heart cells in a wild-type mouse. CTP-enriched nanovesicles could serve as a therapeutic tool for heart disease.	Kim et al. (2018)
BM-MSCs, platelets	–	–	MemFus	Membrane fusion imparted pro-angiogenic properties and targeted injured endothelium from source cells. These hybrid nanovesicles reduce nonspecific uptake and improve retention in ischemic heart tissue.	Li et al. (2021) and Hu et al. (2021a)
Chondrocytes cells	miRNA-140 (electroporation)	Lamp2b-CAP	GenEng, CarEng	CAP-expressing and miRNA-140-loaded nanovesicles targeted chondrocytes, penetrated the cartilage matrix, inhibited cartilage-degrading proteases, and reversed osteoarthritis in rats.	Liang et al. (2020)

Dendritic cells	Cas9 sgMMP-13 (transfection)	Lamp2b-CAP	GenEng, CarEng MemFus	CAP-expressing, plasmid-loaded hybrid nanovesicles showed potential in targeting chondrocytes, penetrating deep into cartilage, knocking down MMP-13 expression, and reducing cartilage degradation in mouse models.	Liang et al. (2022)
NIH-3 T3 cells	Antagomir-188 (liposome fusion)	CXCR4	GenEng, CarEng, MemFus	CXCR4-expressing nanovesicles fused with antagomir-188 liposomes selectively targeted bone marrow, promoting bone growth and reducing bone loss in mice, and can treat pathological bone loss.	Hu et al. (2021b)
BM-MSCs	–	PD-L1	GenEng	PD-L1-expressing nanovesicles target inflamed tissues in colitis and psoriasis. They activate immune cells and restore tissue lesions by inhibiting autoreactive T-cell activity and reshaping the immune microenvironment.	Xu et al. (2022b)
Human umbilical cord-derived MSCs	5-aminolevulinate hydrochloride	–	CarEng	Cytokine-primed and drug-loaded nanovesicles enhance immunosuppression by promoting anti-inflammatory macrophages, increasing the Treg/effector T-cell ratio, targeting inflamed pancreas, and supporting β-cell survival.	Wang et al. (2023)

GenEng genetic engineering, *CarEng* cargo engineering, *MemFus* membrane fusion, *MGEng* metabolic glycoengineering, *BM-MSCs* bone marrow-derived mesenchymal stem cells, *HEK293* human embryonic kidney 293, *Lamp2b* lysosome-associated membrane glycoprotein 2B, *PDGFR* platelet-derived growth factor receptor, *iRGD* internalizing RGD peptide, *RVG* rabies virus glycoprotein, *CAP* chondrocyte affinity peptide, *CTP* cardiomyocyte-targeting peptide, *CXCR4* C-X-C chemokine receptor type 4, *PD-1* programmed death-1, *PD-L1* programmed death-ligand 1, *SIRPα* signal-regulatory protein alpha, *miRNA* microRNA, *siRNA* small interfering RNA, *BBB* blood-brain barrier, *ICG* indocyanine green significantly inhibited cancer cell growth, highlighting their potential for nucleic acid-based cancer therapies (Ohno et al. 2013; Bellavia et al. 2017; Zhou et al. 2019). Other research studies have adopted similar peptide engineering strategies but loaded synthetic drugs to inhibit cancer growth. For example, Tian et al. produced nanovesicles from immature mouse dendritic cells, genetically engineering them to express the αv integrin-specific iRGD peptide fused with Lamp2b. The resulting CMNVs were incorporated with doxorubicin via electroporation. These engineered nanovesicles effectively targeted αv integrin-positive breast cancer cells and delivered the drug specifically to tumor tissues in mice, inhibiting tumor growth (Tian et al. 2014). Han et al. engineered BM-MSCs-derived nanovesicles to express the GFE1 peptide by fusing it with the Lamp2b protein. These nanovesicles were loaded with a PI3Kγ inhibitor to inhibit melanoma lung metastasis. In the melanoma mouse model, the nanovesicles effectively accumulated in pulmonary vascular endothelial cells. The gradual release of the PI3Kγ inhibitor blocked the migration and recruitment of granulocytic myeloid-derived suppressor cells. This transformation changed the lung microenvironment from immunosuppression to immune stimulation, ultimately inhibiting melanoma lung metastasis (Han et al. 2022).

2.4.1.2 Cancer Treatment Through Immunotherapy

The immune system defends against external invasions and eliminates transformed cells. Cancer immunotherapy enhances immune responses, modulates the tumor microenvironment, and integrates tumor-specific antibodies with cellular immune effectors for therapy. However, cancer cells express immunosuppressive checkpoint ligands such as CD47, PD-L1, and SIRPα. These ligands lead to T-cell exhaustion and allow cancer cells to escape immune surveillance. Tumor microenvironment promotes the polarization of tumor-associated macrophages into a pro-tumorigenic M2 phenotype. PD-L1 inhibits T-cell activation by interacting with PD-1. CD47 sends a "don't eat me" signal to macrophages through SIRPα, preventing phagocytosis. Unfortunately, cancer immunotherapy faces limitations due to heterogeneous antigen expression, reduced tumor-specific antigens, immunosuppressive tumors microenvironment, and increased expression of immune checkpoint inhibitors. Immune checkpoint blockade and the disruption of the CD47-SIRPα axis can enhance T-cell immunity and macrophage response in cancer therapy (Lian et al. 2019; Jin et al. 2024). Recently, engineered nanovesicles have emerged as a promising approach for immunotherapy of solid tumors and hematological malignancies.

Rao et al. demonstrate that hybrid CMNVs retain unique properties from their source cells. Platelet-derived nanovesicles can bind to cancer cells. M1 macrophage-derived nanovesicles promote M2-to-M1 repolarization in the tumor-associated microenvironment. Cancer cell-derived nanovesicles expressing SIRPα can block CD47-SIRPα signaling. These features enhance cancer targeting potential, improve antigen presentation, stimulate T-cell responses, and boost antitumor M1 macrophage activity. Furthermore, loading the stimulator of interferon genes agonist cGAMP into hybrid nanovesicles enhances tumor inhibition in melanoma mouse models and triple-negative breast cancer models (Rao et al. 2020). Nie et al.

developed responsive CMNVs for cancer therapy using MGEng-coupled click chemistry. They conjugated azide-modified CMNVs from M1 macrophages with DBCO-modified CD47 and SIRPα antibodies via pH-sensitive linkers. These nanovesicles specifically target tumors, and in the acidic tumor microenvironment, the benzoic-imine bonds cleaved, releasing the antibodies. This blocked SIRPα on macrophages and CD47 on cancer cells, enhancing macrophage activity and promoting immune-mediated cancer cell destruction (Nie et al. 2020). Xu et al. engineered dendritic cells to express anti-CD19 single-chain variable fragments (scFv) for targeting tumor antigens and the immune checkpoint protein PD-1 to reshape the immunosuppressive tumor microenvironment. The resulting nanovesicles retained co-stimulatory molecules from the parent source for effective antigen presentation and T-cell activation. When administered in tumor-bearing mice, these nanovesicles effectively penetrated huCD19-expressing solid tumors and reversed the immunosuppressive microenvironment by blocking PD-L1 expression (Xu et al. 2023).

Xue et al. engineered RAW264.7 macrophages to express anti-PD-1 single-chain variable fragments (aPD-1-scFv) on the surface of nanovesicles. They loaded CPI-444 onto these nanovesicles to target the adenosine A2A receptor. Intravenous administration of these engineered nanovesicles in tumor-bearing mice enhanced T cell anti-tumor activation and infiltration through PD-1 blockade. CPI-444 antagonized immune-suppressive metabolites in the tumor microenvironment, significantly restraining tumor progression in the mouse model (Xue et al. 2022). Chen et al. developed a photoimmunotherapy strategy by engineering BM-MSCs to express anti-PD-L1 antibodies (aPD-L1) on nanovesicles and loading them with indocyanine green photosensitizers. Intravenous injection of these nanovesicles selectively targeted PD-L1-overexpressing tumor cells, enabling effective photothermal ablation in melanoma mouse models. In another approach, aPD-L1 and ovalbumin were expressed on the nanovesicles, which were loaded with the immune adjuvant Imiquimod R837. This immunization targeted dendritic cells, promoting their maturation and enhancing the anti-tumor immune response. This strategy integrates ligand-targeted delivery with photoimmunotherapy using versatile nanoplatforms (Chen et al. 2022). Overall, these studies demonstrate that engineered nanovesicles have significant potential as an effective immunotherapeutic strategy.

2.4.2 Engineered Nanovesicles for Brain Disorders Treatment

Brain-related disorders are challenging to treat, leading to long-term disabilities. The selective permeability of BBB and complex brain tumor microenvironment hinder the effective treatment of brain diseases (Arvanitis et al. 2020). BBB consists of tightly connected brain endothelial cells, pericytes, an extracellular basal membrane, and astrocytic perivascular endfeet. It serves as a selective semipermeable barrier between the bloodstream and the brain, controlling nutrient transport while protecting the brain against toxins and pathogens. BBB maintains intracranial homeostasis but also hinders the delivery of synthetic drugs, with ~98% for molecular medicines cannot cross BBB (Terstappen et al. 2021). Complex brain tumor microenvironments further restrict nanomaterial penetration, raising uncertainty regarding whether drugs that cross the BBB can achieve therapeutic levels in damaged tissues. Thus, developing an effective nanoscale carrier for transporting theranostic agents across the BBB and enhancing drug penetration into damaged tissues remains a critical challenge (Arvanitis et al. 2020; Terstappen et al. 2021).

Natural nanovesicles from specific source cells possess the potential to penetrate hard-to-reach tissues, sparking interest in their use for brain therapies. However, nanovesicles derived from cultured cells often show limited therapeutic efficacy and poor brain targeting capabilities. Therefore, modifying CMNVs is an exciting research area to enhance their effectiveness in treating brain diseases. Researchers have derived nanovesicles from BM-MSCs and dendritic cells by transfecting them with a neuron-specific rabies virus glycoprotein (RVG) fused to the integral membrane protein Lamp2b through GenEng. This modification enabled the expression of RVG peptides onto nanovesicle surfaces. Electroporation has loaded engineered nanovesicles with nucleic acid drugs, such as miRNA-124 or siRNA. RVG-expressed nanovesicles cross the BBB, accumulate in ischemic regions of a mouse model, and effectively deliver their contents to neuronal cells. The engineered nanovesicles successfully modulated gene expression in various brain regions and promoted neurogenesis, highlighting their potential for treating brain diseases (Yang et al. 2017b; Alvarez-Erviti et al. 2011). Using a similar strategy, neural stem cells were engineered to express PDGFRα on nanovesicles for targeted delivery to brain lesions. The engineered nanovesicles were loaded with Bryostatin-1 (an anti-inflammatory drug), and systemic administration allowed them to target brain lesions in a mouse model of autoimmune encephalomyelitis. This approach modulated inflammation at lower drug dosages, reduced myelin loss, and showed promise for treating neuroinflammatory diseases (Wu et al. 2022).

Tian et al. grafted c(RGDyK) peptide to the surface of nanovesicles derived from BM-MSCs using covalent conjugation followed by a click reaction. This peptide has a high affinity for αvβ3 integrin, enabling precise delivery to cerebral vascular endothelial cells in the ischemia-affected brain. Curcumin was loaded into the lumen of the nanovesicles to suppress inflammatory responses in the affected area. In vivo studies using a cerebral ischemia mouse model demonstrated that the engineered nanovesicles served as a targeted delivery vehicle for curcumin, offering a potential treatment strategy for cerebral ischemia (Tian et al. 2018). Ruan et al. extracted CMNVs from M2 microglia to enhance the directed differentiation of neural stem cells into neurons. The extracted nanovesicles were modified with a vascular endothelial cell-targeting peptide (DA7R) and a local neural stem cell-recruiting factor (SDF-1) using covalent interactions and click chemistry. These engineered nanovesicles demonstrated homing potential toward injured vasculature, facilitated the recruitment of local neural stem cells, and improved their directed differentiation. Following intravenous injection, the engineered nanovesicles accumulated in the ischemic area of a transient middle cerebral stroke mouse model, resulting in notable improvements in neurogenesis at the injured site (Ruan et al. 2023).

Some researchers have leveraged the intrinsic brain-targeting potential of native nanovesicles to deliver drugs to damaged brain tissues. Studies indicate that nanovesicles derived from bone marrow macrophages and MSCs can cross the BBB and accumulate in the brains of mouse models. Molecular drugs such as glial cell line-derived neurotrophic factor (GDNF) and tyrosine phosphatase-2 have been incorporated into these sourced cell-derived nanovesicles through pre-loading methods. Engineered nanovesicles protect the drugs and demonstrate an intrinsic ability to target inflamed brain tissues in transgenic mouse models of Parkinson's or Alzheimer's disorders. Targeted drug accumulation enhances neuronal survival, improves locomotor function, mitigates synaptic loss, and reduces neuroinflammation, offering potential benefits for neurodegenerative disorders (Zhao et al. 2022; Xu et al. 2022a). Overall, numerous brain-targeting peptides have been engineered onto CMNVs surfaces, enabling them to cross the BBB and serving as promising nanocarriers for brain pathologies. However, the comparative analysis of the innate brain tropism properties of CMNVs from different cell sources has not been specifically studied. Choosing suitable cell sources and engineering strategies could further enhance brain tropism.

2.4.3 Engineered Nanovesicles for Inflammation Treatment

Inflammation is a protective response to trauma, infection, and autoimmune injury, aimed at restoring tissue homeostasis. However, uncontrolled inflammation can harm the host and contribute to diseases like rheumatoid arthritis, bowel disease, myocardial ischemia, pancreatitis, and atherosclerosis. Macrophages are crucial in this process, displaying two primary phenotypes: M1 (pro-inflammatory) and M2 (anti-inflammatory). Although strategies exist to modulate macrophage activity and employ anti-inflammatory drugs, targeted delivery to specific inflammatory sites remains a challenge. This has led to increased interest in using engineered nanovesicles for treating inflammatory diseases (Lei et al. 2023; Tang et al. 2020).

MGEng-coupled click conjugation strategies were used to attach PEGylated hyaluronic acid (PHA) and dextran sulfate to nanovesicles for targeted inflammatory tissue binding. Cultured cells were treated with azide-containing unnatural sialic acids to express azide groups on the nanovesicles. DBCO-modified PHA or dextran sulfate labelled these azides, allowing incorporation into the nanovesicles. PHA nanovesicles target CD44-overexpressing tissues, while dextran sulfate nanovesicles bind to inflammatory macrophages. Systemic administration of engineered nanovesicles in collagen-induced arthritis mouse models led to targeted accumulation in inflamed joint tissues, promoting macrophage polarization and reducing joint inflammation, indicating a promising anti-inflammatory strategy for rheumatoid arthritis (Lim et al. 2021; You et al. 2021). Drug-loaded hybrid nanovesicle strategies have been introduced to address inflammation challenges. He et al. developed hybrid nanovesicles by fusing RAW264.7 macrophage membranes with HUVEC membranes and loading them with curcumin. These nanovesicles demonstrated high uptake in lipopolysaccharides-induced inflammatory cells and effectively targeted inflammatory sites in the heart and thoracic aorta. The curcumin-loaded nanovesicles reduced the expression of pro-inflammatory cytokine and alleviated lipopolysaccharides-induced organ damage, indicating their potential for treating cardiovascular inflammation (He et al. 2024). On the other hand, Zhao et al. developed hybrid nanovesicles by fusing activated neutrophil membranes with keratinocyte growth factor-loaded liposomes. The drug-loaded hybrid nanovesicles demonstrated sufficient stability and effectively targeted inflamed bowel tissue in a sodium sulfate-induced ulcerative colitis mouse model (Zhao et al. 2019).

GenEng and CarEng strategies have been used to engineer nanovesicles for targeted delivery to inflammatory tissues. Wang et al. transfected MC3T3 cells to express CXCR4 on the nanovesicle surface, which were then loaded with curcumin. These engineered nanovesicles showed improved targeting to inflammatory tissues, enhanced M2 macrophage polarization, and reduced inflammation in a mouse model of ulcerative colitis, indicating their potential for treating ulcerative colitis and apical periodontitis (Wang et al. 2021). Zhong et al. developed stimulus-responsive nanovesicles for the management of atherosclerosis. They conjugated docetaxel (an acid-sensitive prodrug) and CR8 peptide (to target vascular cellular adhesion molecule-1) to DSPE-PEG via an acid-cleavable ester bond. DSPE-PEG acted as a "flexible arm" for the surface modification of nanovesicles. This CR8-mediated active targeting and pH-sensitive release of the prodrug in acidic environments show promise for treating atherosclerosis in an apolipoprotein E-deficient mouse model (Zhong et al. 2021). Choi et al. pioneered an optogenetically engineered nanovesicle system to load super-repressor IκBα (srIκB) into CMNVs. HEK293T cells were engineered to express CIBN-EGFP-CD9 and srIκB-mCherry-CRY2, enabling the production and loading of srIκB into the nanovesicle lumen. SrIκB inhibits nuclear factor κB translocation, crucial for inflammatory responses. Intraperitoneal injection of these nanovesicles reduced proinflammatory cytokines, systemic inflammation, and organ injury in septic mice (Choi et al. 2020). Overall, the ongoing development of nanovesicle-based approaches holds great potential for improving treatment outcomes in inflammatory diseases by enhancing targeted delivery and minimizing systemic side effects.

2.4.4 Engineered Nanovesicles for Promoting Tissue Regeneration

Engineered nanovesicles have been used for the management of heart diseases, cartilage diseases, bone diseases, and diabetes. Mentkowski et al. and Kim et al. introduced cardiomyocyte-targeting peptide (CTP) onto the nanovesicle surface through fusion with Lamp2b protein. These studies found that CTP was effectively displayed on nanovesicle surfaces with native properties. CTP-enriched nanovesicles better target heart cells and heart tissue in mice models (Mentkowski and Lang 2019; Kim et al. 2018). Some studies have leveraged intrinsic regenerative and targeting potential of nanovesicles through membrane fusion strategies for myocardial infarction treatment. Hybrid nanovesicles were created by fusing BM-MSC and platelet membrane-sourced nanovesicles. This fusion combines the pro-angiogenic properties of stem cells with the natural tropism of platelet membranes. Hybrid nanovesicles demonstrated improved macrophage clearance evasion and increased uptake by endothelial cells and cardiomyocytes. These nanovesicles showed increased retention in ischemic hearts and significantly improved cardiac function compared to unmodified nanovesicles in mouse models of myocardial infarction (Li et al. 2021; Hu et al. 2021a).

Osteoarthritis is a chronic joint disease that predominantly affects the elderly and creates a significant socioeconomic burden. Knocking down matrix metalloproteinase-13 (MMP-13) has emerged as a promising strategy to reverse cartilage degeneration, but the dense and avascular nature of cartilage complicates drug delivery. Recent studies have concentrated on conjugating chondrocyte affinity peptide (CAP) onto nanovesicles through fusion with Lamp2b using GenEng. These CAP-enriched nanovesicles were then loaded with miRNA-140, capable of inhibiting MMP-13 (Liang et al. 2020). CAP-expressed nanovesicles were fused with liposomes and Cas9 sgMMP-13 plasmid to create drug-loaded hybrid nanovesicles (Liang et al. 2022). The results indicated that engineered nanovesicles penetrated the matrix of cartilage, specifically targeting chondrocytes. This targeted delivery inhibited cartilage-degrading proteases, reduced extracellular matrix degradation, and could reverse osteoarthritis symptoms in a rat model (Liang et al. 2020, 2022). Hu et al. demonstrated the fusion of CXCR4-expressing nanovesicles with liposomes containing antagomir-188 to form drug-loaded hybrid nanovesicles. These hybrid nanovesicles are effectively localized in the bone marrow. The released antagomir-188 significantly boosted osteogenic differentiation and suppressed adipogenesis in BM-MSCs. This approach could effectively reverse trabecular bone density and reduce cortical bone porosity in mice model (Hu et al. 2021b).

Recent research has shown that MSCs-derived engineered nanovesicles can serve as therapeutic tools for diabetes. Xu et al. transfected MSCs to express PD-L1, resulting in nanovesicles with enhanced targeting ability in inflamed tissues. The PD-L1 receptors on these nanovesicles interacted with activated immune cells expressing PD-1, effectively inhibiting autoreactive T-cell activity and reshaping the local immune microenvironment. This approach shows promise for autoimmune diseases, including diabetes (Xu et al. 2022b). Cytokine-primed MSCs-derived nanovesicles can enhance immunosuppression in diabetes by promoting macrophage conversion to an anti-inflammatory phenotype and increasing the Treg/effector T-cell ratio. These nanovesicles upregulate PD-L1 expression, inducing apoptosis in activated T cells while boosting Treg proportions. Additionally, nanovesicles were loaded with hexyl 5-aminolevulinate hydrochloride for fluorescence imaging and anti-inflammatory benefits. These engineered nanovesicles target the inflamed pancreas, induce macrophage transition, suppress T cells, and promote islet β-cell survival, highlighting their potential for diabetes treatment (Wang et al. 2023).

Overall, engineering strategies can enhance targeting capabilities and tracking functions, provide exogenous therapeutic functions, increase drug loading efficiency, boost penetration potential, and enable the presentation of multiple antigens on nanovesicles for disease-specific treatments. These unique biointerfacing and functional potentials are rarely present in naturally derived nanovesicles from specific sources. The engineered functionalities in nanovesicles are crucial for maximizing their potential in addressing cancer, neurological disorders, inflammatory diseases, promoting tissue regeneration, and treating numerous other diseases.

2.5 Conclusions and Future Directions

Natural nanovesicles inherit attributes from parent cells but their clinical applications are limited owing to variations in surface markers and internal cargo. Recent research has engineered nanovesicles using numerous modification strategies to enhance bio-interfacing, tissue targeting, tissue penetration, immunological potential, antigen presentation, and monitoring potential. Unique endogenous features, modifiable functions, exogenous properties, and GMP compliance increase the pharmaceutical spectrum of engineered nanovesicles in cancer, brain disease, inflammation, and tissue regeneration. Designer nanovesicles eliminate the limitations of natural nanovesicles, advancing from in vitro promise to preclinical studies, and may replace current cell therapies. However, unclear biochemical components of nanovesicles may lead to unforeseen side effects. Numerous studies lack data on clearance or biodistribution rates of engineered nanovesicles, highlighting the need for scalable methods to produce consistent engineered nanovesicles, such as high-yield designer cell lines. Integrating natural nanovesicles' features with engineered functionalities is crucial for biomedical use, but standardized data on source cell-specific nanovesicles is lacking. Collaboration among cell biology, chemical engineering, GenEng, immunology, microbiology, and nanomedicine could bridge gaps and unlock the biomedical potential of designer nanovesicles for treating refractory diseases and advancing precision medicine.

References

Agatemor C, Buettner MJ, Ariss R, Muthiah K, Saeui CT, Yarema KJ (2019) Exploiting metabolic glycoengineering to advance healthcare. Nat Rev Chem 3(10):605–620

Ai X, Wang S, Duan Y, Zhang Q, Chen MS, Gao W, Zhang L (2020) Emerging approaches to functionalizing cell membrane-coated nanoparticle. Biochemistry 60(13):941–955

Alvarez-Erviti L, Seow Y, Yin H, Betts C, Lakhal S, Wood MJ (2011) Delivery of siRNA to the mouse brain by systemic injection of targeted exosomes. Nat Biotechnol 29(4):341–345

Antes TJ, Middleton RC, Luther KM, Ijichi T, Peck KA, Liu WJ, Valle J, Echavez AK, Marbán E (2018) Targeting extracellular vesicles to injured tissue using membrane cloaking and surface display. J Nanobiotechnol 16:1–15

Arvanitis CD, Ferraro GB, Jain RK (2020) The blood–brain barrier and blood–tumour barrier in brain tumours and metastases. Nat Rev Cancer 20(1):26–41

Bai J, Duan J, Liu R, Du Y, Luo Q, Cui Y, Su Z, Xu J, Xie Y, Lu W (2020) Engineered targeting tLyp-1 exosomes as gene therapy vectors for efficient delivery of siRNA into lung cancer cells. Asian J Pharm Sci 15(4):461–471

Bellavia D, Raimondo S, Calabrese G, Forte S, Cristaldi M, Patinella A, Memeo L, Manno M, Raccosta S, Diana P, Cirrincione G (2017) Interleukin 3-receptor targeted exosomes inhibit in vitro and in vivo Chronic Myelogenous Leukemia cell growth. Theranostics 7(5):1333

Chen L, Qin H, Zhao R, Zhao X, Lin L, Chen Y, Lin Y, Li Y, Qin Y, Li Y, Liu S (2021) Bacterial cytoplasmic membranes synergistically enhance the antitumor activity of autologous cancer vaccines. Sci Transl Med 13(601):eabc2816

Chen H, Zhang P, Shi Y, Liu C, Zhou Q, Zeng Y, Cheng H, Dai Q, Gao X, Wang X, Liu G (2022) Functional nanovesicles displaying anti-PD-L1 antibodies for programmed photoimmunotherapy. J Nanobiotechnol 20(1):61

Cheng B, Tang Q, Zhang C, Chen X (2021) Glycan labeling and analysis in cells and in vivo. Annu Rev Anal Chem 14:363–387

Cheng Q, Kang Y, Yao B, Dong J, Zhu Y, He Y, Ji X (2023) Genetically engineered-cell-membrane nanovesicles for cancer immunotherapy. Adv Sci 10(26):2302131

Cheng Q, Li R, He Y, Zhu Y, Kang Y, Ji X (2024) Genetically engineered cellular nanovesicles: theories, design and perspective. Adv Funct Mater 34(46):2407842

Choi H, Kim Y, Mirzaaghasi A, Heo J, Kim YN, Shin JH, Kim S, Kim NH, Cho ES, In Yook J, Yoo TH (2020) Exosome-based delivery of super-repressor IκBα relieves sepsis-associated organ damage and mortality. Sci Adv 6(15):eaaz6980

Cocucci E, Racchetti G, Meldolesi J (2009) Shedding microvesicles: artefacts no more. Trends Cell Biol 2:43–51

Cruje C, Chithrani DB (2014) Polyethylene glycol density and length affects nanoparticle uptake by cancer cells. J Nanomed Res 1(00006):8

Fu S, Wang Y, Xia X, Zheng JC (2020) Exosome engineering: current progress in cargo loading and targeted delivery. NanoImpact 20:100261

Gill S, Catchpole R, Forterre P (2019) Extracellular membrane vesicles in the three domains of life and beyond. FEMS Microbiol Rev 43(3):273–303

Hallal S, Tűzesi Á, Grau GE, Buckland ME, Alexander KL (2022) Understanding the extracellular vesicle surface for clinical molecular biology. J Extracell Vesicles 11(10):e12260

Han X, Bi L, Wu Y, Yan J, Wu X, Zheng R, Sun Y, Zhang H, Wang Z, Wang Y, Zhang H (2022) Genetically engineered exosomes for targetedly preventing premetastatic niche formation and suppressing postoperative melanoma lung metastasis. Nano Today 46:101597

Haney MJ, Klyachko NL, Zhao Y, Gupta R, Plotnikova EG, He Z, Patel T, Piroyan A, Sokolsky M, Kabanov AV, Batrakova EV (2015) Exosomes as drug delivery vehicles for Parkinson's disease therapy. J Control Release 207:18–30

He H, Guo C, Wang J, Korzun WJ, Wang XY, Ghosh S, Yang H (2018) Leutusome: a biomimetic nanoplatform integrating plasma membrane components of leukocytes and tumor cells for remarkably enhanced solid tumor homing. Nano Lett 18(10):6164–6174

He J, Wu X, Li L, Chen J, Liao J, Wu A, Zhang M, Chen Y, Mao X, Shen X, Tao L (2024) Construction of curcumin-loaded macrophage and HUVECs membrane-derived vesicles for drug delivery in cardiovascular inflammatory. J Drug Deliv Sci Technol 95:105611

Hessvik NP, Llorente A (2018) Current knowledge on exosome biogenesis and release. Cell Mol Life Sci 75:193–208

Hoang DM, Pham PT, Bach TQ, Ngo AT, Nguyen QT, Phan TT, Nguyen GH, Le PT, Hoang VT, Forsyth NR, Heke M (2022) Stem cell-based therapy for human diseases. Signal Transduct Target Ther 7(1):1–41

Hu S, Wang X, Li Z, Zhu D, Cores J, Wang Z, Li J, Mei X, Cheng X, Su T, Cheng K (2021a) Platelet membrane and stem cell exosome hybrids enhance cellular uptake and targeting to heart injury. Nano Today 39:101210

Hu Y, Li X, Zhang Q, Gu Z, Luo Y, Guo J, Wang X, Jing Y, Chen X, Su J (2021b) Exosome-guided bone targeted delivery of Antagomir-188 as an anabolic therapy for bone loss. Bioact Mater 6(9):2905–2913

Hu Z, Qian S, Zhao Q, Lu B, Lu Q, Wang Y, Zhang L, Mao X, Wang D, Cui W, Sun X (2024) Engineering strategies for apoptotic bodies. Smart Med 3:e20240005

Jin S, Wang H, Li Y, Yang J, Li B, Shi P, Zhang X, Zhou X, Zhou X, Niu X, Wu M (2024) Discovery of a novel small molecule as CD47/SIRPα and PD-1/PD-L1 dual inhibitor for cancer immunotherapy. Cell Commun Signal 22(1):173

Kalani A, Kamat PK, Chaturvedi P, Tyagi SC, Tyagi N (2014) Curcumin-primed exosomes mitigate endothelial cell dysfunction during hyperhomocysteinemia. Life Sci 107(1–2):1–7

Kalimuthu S, Gangadaran P, Rajendran RL, Zhu L, Oh JM, Lee HW, Gopal A, Baek SH, Jeong SY, Lee SW, Lee J (2018) A new approach for loading anticancer drugs into mesenchymal stem cell-derived exosome mimetics for cancer therapy. Front Pharmacol 9:1116

Kim S, Kim K (2022) Lipid-mediated ex vivo cell surface engineering for augmented cellular functionalities. Biomater Adv 140:213059

Kim H, Yun N, Mun D, Kang JY, Lee SH, Park H, Park H, Joung B (2018) Cardiac-specific delivery by cardiac tissue-targeting peptide-expressing exosomes. Biochem Biophys Res Commun 499(4):803–808

Kolate A, Baradia D, Patil S, Vhora I, Kore G, Misra A (2014) PEG—a versatile conjugating ligand for drugs and drug delivery systems. J Control Release 192:67–81

Kong H, Yi K, Zheng C, Lao YH, Zhou H, Chan HF, Wang H, Tao Y, Li M (2022) Membrane-fusogenic biomimetic particles: a new bioengineering tool learned from nature. J Mater Chem B 10(36):6841–6858

Kooijmans SA, Aleza CG, Roffler SR, van Solinge WW, Vader P, Schiffelers RM (2016) Display of GPI-anchored anti-EGFR nanobodies on extracellular vesicles promotes tumour cell targeting. J Extracell Vesicles 5(1):31053

Kverneland AH, Østergaard O, Emdal KB, Svane IM, Olsen JV (2023) Differential ultracentrifugation enables deep plasma proteomics through enrichment of extracellular vesicles. Proteomics 23(7–8):2200039

Le QV, Lee J, Lee H, Shim G, Oh YK (2021) Cell membrane-derived vesicles for delivery of therapeutic agents. Acta Pharm Sin B 11(8):2096–2113

Lei P, Yu H, Ma J, Du J, Fang Y, Yang Q, Zhang K, Luo L, Jin L, Wu W, Sun D (2023) Cell membrane nanomaterials composed of phospholipids and glycoproteins for drug delivery in inflammatory bowel disease: a review. Int J Biol Macromol 249:126000

Li Z, Zhou X, Wei M, Gao X, Zhao L, Shi R, Sun W, Duan Y, Yang G, Yuan L (2018) In vitro and in vivo RNA inhibition by CD9-HuR functionalized exosomes encapsulated with miRNA or CRISPR/dCas9. Nano Lett 19(1):19–28

Li Q, Song Y, Wang Q, Chen J, Gao J, Tan H, Li S, Wu Y, Yang H, Huang H, Yu Y (2021) Engineering extracellular vesicles with platelet membranes fusion enhanced targeted therapeutic angiogenesis in a mouse model of myocardial ischemia reperfusion. Theranostics 11(8):3916

Li Y, Cai Z, Wang Z, Zhu S, Liu W, Wang C (2024) Construction of biomimetic hybrid nanovesicles based on M1 macrophage-derived exosomes for therapy of cancer. Chin Chem Lett 36:109942

Lian S, Xie R, Ye Y, Xie X, Li S, Lu Y, Li B, Cheng Y, Katanaev VL, Jia L (2019) Simultaneous blocking of CD47 and PD-L1 increases innate and adaptive cancer immune responses and cytokine release. EBioMedicine 42:281–295

Liang Y, Xu X, Li X, Xiong J, Li B, Duan L, Wang D, Xia J (2020) Chondrocyte-targeted microRNA delivery by engineered exosomes toward a cell-free osteoarthritis therapy. ACS Appl Mater Interfaces 12(33):36938–36947

Liang Y, Xu X, Xu L, Iqbal Z, Ouyang K, Zhang H, Wen C, Duan L, Xia J (2022) Chondrocyte-specific genomic editing enabled by hybrid exosomes for osteoarthritis treatment. Theranostics 12(11):4866

Lim GT, You DG, Han HS, Lee H, Shin S, Oh BH, Kumar EP, Um W, Kim CH, Han S, Lee S (2021) Bioorthogonally surface-edited extracellular vesicles based on metabolic glycoengineering for CD44-mediated targeting of inflammatory diseases. J Extracell Vesicles 10(5):e12077

Linares R, Tan S, Gounou C, Arraud N, Brisson AR (2015) High-speed centrifugation induces aggregation of extracellular vesicles. J Extracell Vesicles 4(1):29509

Ma Y, Zhang Y, Han R, Li Y, Zhai Y, Qian Z, Gu Y, Li S (2022) A cascade synergetic strategy induced by photothermal effect based on platelet exosome nanoparticles for tumor therapy. Biomaterials 282:121384

Ma Y, Dong S, Grippin AJ, Teng L, Lee AS, Kim BY, Jiang W (2025) Engineering therapeutical extracellular vesicles for clinical translation. Trends Biotechnol 43:61–82

Meng QF, Zhao Y, Dong C, Liu L, Pan Y, Lai J, Liu Z, Yu GT, Chen X, Rao L (2021) Genetically programmable fusion cellular vesicles for cancer immunotherapy. Angew Chem Int Ed 60(50):26320–26326

Mentkowski KI, Lang JK (2019) Exosomes engineered to express a cardiomyocyte binding peptide demonstrate improved cardiac retention in vivo. Sci Rep 9(1):10041

Monguió-Tortajada M, Gálvez-Montón C, Bayes-Genis A, Roura S, Borràs FE (2019) Extracellular vesicle isolation methods: rising impact of size-exclusion chromatography. Cell Mol Life Sci 76:2369–2382

Moulahoum H, Ghorbanizamani F, Zihnioglu F, Timur S (2021) Surface biomodification of liposomes and polymersomes for efficient targeted drug delivery. Bioconjug Chem 32(8):1491–1502

Nie W, Wu G, Zhang J, Huang LL, Ding J, Jiang A, Zhang Y, Liu Y, Li J, Pu K, Xie HY (2020) Responsive exosome nano-bioconjugates for synergistic cancer therapy. Angew Chem Int Ed 59(5):2018–2022

Nischan N, Kohler JJ (2016) Advances in cell surface glycoengineering reveal biological function. Glycobiology 26(8):789–796

Ohno SI, Takanashi M, Sudo K, Ueda S, Ishikawa A, Matsuyama N, Fujita K, Mizutani T, Ohgi T, Ochiya T, Gotoh N (2013) Systemically injected exosomes targeted to EGFR deliver antitumor microRNA to breast cancer cells. Mol Ther 21(1):185–191

Piffoux M, Silva AK, Wilhelm C, Gazeau F, Tareste D (2018) Modification of extracellular vesicles by fusion with liposomes for the design of personalized biogenic drug delivery systems. ACS Nano 12(7):6830–6842

Pocsfalvi G, Stanly C, Vilasi A, Fiume I, Capasso G, Turiák L, Buzas EI, Vékey K (2016) Mass spectrometry of extracellular vesicles. Mass Spectrom Rev 35(1):3–21

Rao L, Wu L, Liu Z, Tian R, Yu G, Zhou Z, Yang K, Xiong HG, Zhang A, Yu GT, Sun W (2020) Hybrid cellular membrane nanovesicles amplify macrophage immune responses against cancer recurrence and metastasis. Nat Commun 11(1):4909

Ruan S, Greenberg Z, Pan X, Zhuang P, Erwin N, He M (2022) Extracellular vesicles as an advanced delivery biomaterial for precision cancer immunotherapy. Adv Healthc Mater 11(5):2100650

Ruan H, Li Y, Wang C, Jiang Y, Han Y, Li Y, Zheng D, Ye J, Chen G, Yang GY, Deng L (2023) Click chemistry extracellular vesicle/peptide/chemokine nanocarriers for treating central nervous system injuries. Acta Pharm Sin B 13(5):2202–2218

Rufino-Ramos D, Lule S, Mahjoum S, Ughetto S, Bragg DC, de Almeida LP, Breakefield XO, Breyne K (2022) Using genetically modified extracellular vesicles as a non-invasive strategy to evaluate brain-specific cargo. Biomaterials 281:121366

Shao M, Rodrigues J, Sousa-Oliveira I, Moradialvand M, Asadollahi P, Veiga F, Hameed H, Jha NK, Sillanpää M, Sethi G, Paiva-Santos AC (2024) Revolutionizing cancer treatment via bioengineered extracellular vesicles: exploring nanovesicles to fully synthetic solutions. Appl Mater Today 40:102395

Shi X, Cheng Q, Hou T, Han M, Smbatyan G, Lang JE, Epstein AL, Lenz HJ, Zhang Y (2020) Genetically engineered cell-derived nanoparticles for targeted breast cancer immunotherapy. Mol Ther 28(2):536–547

Sidhom K, Obi PO, Saleem A (2020) A review of exosomal isolation methods: is size exclusion chromatography the best option? Int J Mol Sci 21(18):6466

Song Y, Zhang N, Li Q, Chen J, Wang Q, Yang H, Tan H, Gao J, Dong Z, Pang Z, Huang Z (2021) Biomimetic liposomes hybrid with platelet membranes for targeted therapy of atherosclerosis. Chem Eng J 408:127296

Sterner RC, Sterner RM (2021) CAR-T cell therapy: current limitations and potential strategies. Blood Cancer J 11(4):69

Sun M, Yang J, Fan Y, Zhang Y, Sun J, Hu M, Sun K, Zhang J (2023) Beyond extracellular vesicles: hybrid membrane nanovesicles as emerging advanced tools for biomedical applications. Adv Sci 10(32):2303617

Sutaria DS, Jiang J, Elgamal OA, Pomeroy SM, Badawi M, Zhu X, Pavlovicz R, Azevedo-Pouly ACP, Chalmers J, Li C, Phelps MA (2017) Low active loading of cargo into engineered extracellular vesicles results in inefficient miRNA mimic delivery. J Extracell Vesicles 6(1):1333882

Tang TT, Wang B, Lv LL, Liu BC (2020) Extracellular vesicle-based nanotherapeutics: emerging frontiers in anti-inflammatory therapy. Theranostics 10(18):8111

Tauro BJ, Greening DW, Mathias RA, Ji H, Mathivanan S, Scott AM, Simpson RJ (2012) Comparison of ultracentrifugation, density gradient separation, and immunoaffinity capture methods for isolating human colon cancer cell line LIM1863-derived exosomes. Methods 56(2):293–304

Taylor DD, Shah S (2015) Methods of isolating extracellular vesicles impact down-stream analyses of their cargoes. Methods 87:3–10

Terstappen GC, Meyer AH, Bell RD, Zhang W (2021) Strategies for delivering therapeutics across the blood–brain barrier. Nat Rev Drug Discov 20(5):362–383

Tian Y, Li S, Song J, Ji T, Zhu M, Anderson GJ, Wei J, Nie G (2014) A doxorubicin delivery platform using engineered natural membrane vesicle exosomes for targeted tumor therapy. Biomaterials 35(7):2383–2390

Tian T, Zhang HX, He CP, Fan S, Zhu YL, Qi C, Huang NP, Xiao ZD, Lu ZH, Tannous BA, Gao J (2018) Surface functionalized exosomes as targeted drug delivery vehicles for cerebral ischemia therapy. Biomaterials 150:137–149

Tian F, Li C, Lin L, Chen Q, Sun J (2019) Microfluidic analysis of circulating tumor cells and tumor-derived extracellular vesicles. Trends Analyt Chem 117:128–145

Vergauwen G, Dhondt B, Van Deun J, De Smedt E, Berx G, Timmerman E, Gevaert K, Miinalainen I, Cocquyt V, Braems G, Van den Broecke R (2017) Confounding factors of ultrafiltration and protein analysis in extracellular vesicle research. Sci Rep 7(1):1–12

Wang D, Jiang S, Zhang F, Ma S, Heng BC, Wang Y, Zhu J, Xu M, He Y, Wei Y, Zhang X (2021) Cell membrane vesicles with enriched CXCR4 display enhances their targeted delivery as drug carriers to inflammatory sites. Adv Sci 8(23):2101562

Wang L, Qi C, Cao H, Zhang Y, Liu X, Qiu L, Wang H, Xu L, Wu Z, Liu J, Wang S (2023) Engineered cytokine-primed extracellular vesicles with high PD-L1 expression ameliorate type 1 diabetes. Small 19(38):2301019

Witwer KW, Wolfram J (2021) Extracellular vesicles versus synthetic nanoparticles for drug delivery. Nat Rev Mater 6(2):103–106

Wu WC, Tian J, Xiao D, Guo YX, Xiao Y, Wu XY, Casella G, Rasouli J, Yan YP, Rostami A, Wang LB (2022) Engineered extracellular vesicles encapsulated Bryostatin-1 as therapy for neuroinflammation. Nanoscale 14(6):2393–2410

Xu F, Wu Y, Yang Q, Cheng Y, Xu J, Zhang Y, Dai H, Wang B, Ma Q, Chen Y, Lin F (2022a) Engineered extracellular vesicles with SHP2 high expression promote mitophagy for Alzheimer's disease treatment. Adv Mater 34(49):2207107

Xu F, Fei Z, Dai H, Xu J, Fan Q, Shen S, Zhang Y, Ma Q, Chu J, Peng F, Zhou F (2022b) Mesenchymal stem cell-derived extracellular vesicles with high PD-L1 expression for autoimmune diseases treatment. Adv Mater 34(1):2106265

Xu F, Jiang D, Xu J, Dai H, Fan Q, Fei Z, Wang B, Zhang Y, Ma Q, Yang Q, Chen Y (2023) Engineering of dendritic cell bispecific extracellular vesicles for tumor-targeting immunotherapy. Cell Rep 42(10):113138

Xue T, Zhang Z, Fang T, Li B, Li Y, Li L, Jiang Y, Duan F, Meng F, Liang X, Zhang X (2022) Cellular vesicles expressing PD-1-blocking scFv reinvigorate T cell immunity against cancer. Nano Res 15(6):5295–5304

Yang F, Liao X, Tian Y, Li G (2017a) Exosome separation using microfluidic systems: size-based, immunoaffinity-based and dynamic methodologies. Biotechnol J 12(4):1600699

Yang J, Zhang X, Chen X, Wang L, Yang G (2017b) Exosome mediated delivery of miR-124 promotes neurogenesis after ischemia. Mol Ther Nucleic Acids 7:278–287

Yang H, Yao L, Wang Y, Chen G, Chen H (2023) Advancing cell surface modification in mammalian cells with synthetic molecules. Chem Sci 14(46):13325–13345

Yang X, Xu Z, Shu F, Xiao J, Zeng Y, Lu X, Yu F, Xi L, Cheng F, Gao B, Chen H (2024) Bioorthogonal targeted cell membrane vesicles/cell-sheet composites reduce postoperative tumor recurrence and scar formation of melanoma. J Control Release 372:372–385

Yim N, Ryu SW, Choi K, Lee KR, Lee S, Choi H, Kim J, Shaker MR, Sun W, Park JH, Kim D (2016) Exosome engineering for efficient intracellular delivery of soluble proteins using optically reversible protein–protein interaction module. Nat Commun 7(1):1–9

Yong T, Wei Z, Gan L, Yang X (2022) Extracellular-vesicle-based drug delivery systems for enhanced antitumor therapies through modulating the cancer-immunity cycle. Adv Mater 34(52):2201054

You DG, Lim GT, Kwon S, Um W, Oh BH, Song SH, Lee J, Jo DG, Cho YW, Park JH (2021) Metabolically engineered stem cell–derived exosomes to regulate macrophage heterogeneity in rheumatoid arthritis. Sci Adv 7(23):eabe0083

You Q, Liang F, Wu G, Cao F, Liu J, He Z, Wang C, Zhu L, Chen X, Yang Y (2024) The landscape of biomimetic nanovesicles in brain diseases. Adv Mater 36(7):2306583

Yu Y, Cheng Q, Ji X, Chen H, Zeng W, Zeng X, Zhao Y, Mei L (2022) Engineered drug-loaded cellular membrane nanovesicles for efficient treatment of postsurgical cancer recurrence and metastasis. Sci Adv 8(49):eadd3599

Zhan Q, Yi K, Li X, Cui X, Yang E, Chen N, Yuan X, Zhao J, Hou X, Kang C (2021) Phosphatidylcholine-engineered exosomes for enhanced tumor cell uptake and intracellular antitumor drug delivery. Macromol Biosci 21(8):2100042

Zhang P, Sun F, Liu S, Jiang S (2016) Anti-PEG antibodies in the clinic: current issues and beyond PEGylation. J Control Release 244:184–193

Zhang E, Phan P, Zhao Z (2023) Cellular nanovesicles for therapeutic immunomodulation: a perspective on engineering strategies and new advances. Acta Pharm Sin B 13(5):1789–1827

Zhao YZ, ZhuGe DL, Tong MQ, Lin MT, Zheng YW, Jiang X, Yang WG, Yao Q, Xiang Q, Li XK, Xu HL (2019) Ulcerative colitis-specific delivery of keratinocyte growth factor by neutrophils-simulated liposomes facilitates the morphologic and functional recovery of the damaged colon through alleviating the inflammation. J Control Release 299:90–106

Zhao Y, Haney MJ, Fallon JK, Rodriguez M, Swain CJ, Arzt CJ, Smith PC, Loop MS, Harrison EB, El-Hage N, Batrakova EV (2022) Using extracellular vesicles released by GDNF-transfected macrophages for therapy of Parkinson disease. Cells 11(22):1933

Zheng L, Hu B, Zhao D, Liu W, Liu Q, Huang Y, Ruan S (2024) Recent progresses of exosome–liposome fusions in drug delivery. Chin Chem Lett 35(2):108647

Zhong Y, Qin X, Wang Y, Qu K, Luo L, Zhang K, Liu B, Obaid EAMS, Wu W, Wang G (2021) "Plug and play" functionalized erythrocyte nanoplatform for target atherosclerosis management. ACS Appl Mater Interfaces 13(29):33862–33873

Zhou Y, Yuan Y, Liu M, Hu X, Quan Y, Chen X (2019) Tumor-specific delivery of KRAS siRNA with iRGD-exosomes efficiently inhibits tumor growth. ExRNA 1:28

Zinger A, Soriano S, Baudo G, De Rosa E, Taraballi F, Villapol S (2021) Biomimetic nanoparticles as a theranostic tool for traumatic brain injury. Adv Funct Mater 31(30):2100722

Zou MZ, Li ZH, Bai XF, Liu CJ, Zhang XZ (2021) Hybrid vesicles based on autologous tumor cell membrane and bacterial outer membrane to enhance innate immune response and personalized tumor immunotherapy. Nano Lett 21(20):8609–8618

Engineered Bacterial Membrane Vesicles for Advanced Biomedical Applications

3

Saima Rani and Muhammad Kamran Khan

Abstract

Bacterial membrane vesicles are nanoscale lipid bilayer biomaterials (30–250 nm), obtained from both gram-positive and gram-negative bacteria, formed through distinct biogenesis mechanisms. These non-replicating vesicles comprise lipopolysaccharides, phospholipids, peptidoglycan, membrane proteins, and cargoes such as enzymes, toxins, and nucleic acids. Bacterial membrane vesicles inherit key attributes from their parent bacteria, reflecting their membrane composition, and play critical roles in pathogenesis, quorum sensing, nutrient acquisition, intercellular communication, and immune activation. Their unique biochemical and physicochemical properties position them as promising therapeutic nanomaterials. However, variability in surface markers and composition can cause batch-to-batch inconsistencies, toxicity, off-target effects, and weak immunogenicity, limiting their efficacy and safety. To address these challenges, careful selection of cell sources and bacterial membrane vesicles engineering is essential for enhancing immune-reactive determinants,

S. Rani
Molecular Virology Laboratory, Department of Biosciences, COMSATS University Islamabad (CUI), Islamabad, Pakistan

M. K. Khan (✉)
Australian Regenerative Medicine Institute, Medical Nursing and Health Sciences (MNHS), Monash University, Clayton, VIC, Australia

Suzhou Industrial Park Monash Research Institute of Science and Technology, Suzhou, China

Department of Biotechnology, Southeast University-Monash University Joint Graduate School, Suzhou, China

Department of Microbiology, Biomedicine Discovery Institute, Medical Nursing and Health Sciences (MNHS), Monash University, Clayton, VIC, Australia
e-mail: muhammad.khan2@monash.edu

Z. Hussain, R. Pei (eds.), *Cell Membrane Engineering for Advancing Cell-Based and Nanotechnology-Based Therapies*, Biomaterials, Bioengineering and Sustainability 6, https://doi.org/10.1007/978-3-031-93425-4_3

minimizing toxins, optimizing payload loading, and improving targeted effects. This chapter emphasizes the need for bacterial membrane vesicles modifications, provides an overview of characterization techniques, and outlines genetic engineering, click conjugation, lipid insertion, and membrane fusion strategies for producing engineered vesicles. Additionally, this chapter explores recent advances in bioengineered bacterial membrane vesicles for antimicrobial applications, including bacterial/viral vaccine development, anti-adhesion strategies, and potential in cancer therapy, gastrointestinal disease treatment, and osteoporotic bone defect repair.

Keywords

Outer membrane vesicles · Vesicles modification · Vaccine development · Cancer immunotherapy · Gastrointestinal treatment · Bone defects treatment

3.1 Introduction

Bacteria are categorized as gram-positive and gram-negative, depending on the differences in their structure and the composition of their cell wall lipid. Gram-negative bacteria exhibit a distinctive structure composed of inner membrane (IM) and outer membrane (OM). The outer membrane is characterized by lipopolysaccharides (LPS), while the region between these membranes, referred to as the periplasmic space, includes a layer of peptidoglycan (PG). Small projections from the outer membrane can pinch off to form spherical nanovesicles (30–250 nm) (Fig. 3.1a)

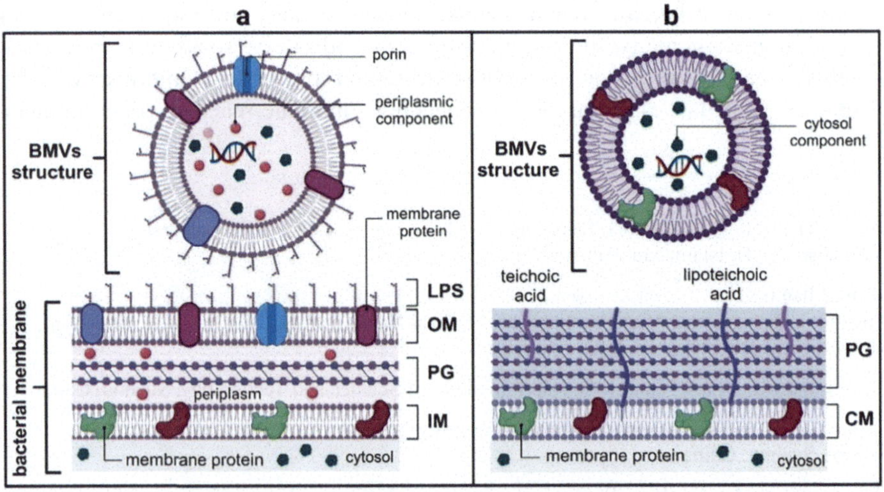

Fig. 3.1 Diagrammatic depiction of biogenies, structures, & compositions of: (**a**) BMVs from gram-negative microorganisms, and (**b**) BMVs from gram-positive microorganisms. IM inner membrane, OM outer membrane, PG peptidoglycan layer, LPS lipopolysaccharides, CM cell membrane, BMVs bacterial membrane vesicles. (Source: Created with BioRender.com)

(Avila-Calderón et al. 2021; Furuyama and Sircili 2021). In contrast, gram-positive bacteria possess a single lipid membrane encased by a robust cell wall made up of a thick peptidoglycan layer interwoven with lipoteichoic acid, which is embedded in the cytoplasmic membrane via diglyceride linkage. They are less known for producing vesicles, as only a few species generate them in limited quantities due to the dense peptidoglycan layer surrounding the cytoplasmic membrane. Notably, nanovesicles produced from gram-positive bacteria show substantial differences from gram-negative nanovesicles, primarily due to their lack of periplasmic components and LPS (Fig. 3.1b) (Cao and Lin 2021; Briaud and Carroll 2020).

Gram-positive and gram-negative bacteria produce nanovesicles through various propagation mechanisms. Their production is driven by defects influenced by environmental and genetic factors, including disruptions in cell envelope components (e.g., peptidoglycan and outer membrane) and 'bilayer coupling' effects caused by curvature-inducing molecules, as well as the aggregation of misfolded polypeptides in the periplasmic space (Toyofuku et al. 2023). Nanovesicles are named according to their origin or preparation, e.g., nanovesicles naturally released by gram-negative bacteria are outer membrane vesicles (OMVs), while nanovesicles produced by gram-positive microorganisms are known as membrane vesicles (Sangiorgio et al. 2024). Nanovesicles formed through the breakdown of bacterial cells and the merging of bacterial vesicles with other vesicles, like liposomes or extracellular vesicles, via a self-assembly mechanism. These are known as bacterial biomimetic vesicles (Fazal and Lee 2021; Zheng et al. 2019). Additionally, bacterial cells can undergo lysis to produce a variety of membrane vesicles, such as inner membrane vesicles, outer-inner membrane vesicles, explosive OMVs, and bacterial protoplast-derived nanovesicles (Toyofuku et al. 2019; Wang et al. 2018; Kim et al. 2015). Unless specified otherwise, this study refers to these structures as bacterial membrane vesicles (BMVs) to maintain consistency and avoid ambiguity in terminology as used in the original literature.

BMVs are non-replicating, circular and nanoscale lipid bilayer entities that are composed of LPS, phospholipids, peptidoglycan, porin, lipoprotein, periplasmic components, membrane-bound proteins, pathogen-associated molecular patterns (PAMPs), and cargoes that are encapsulated (like enzymes, toxins, nucleic acids, and signaling molecules) (Sartorio et al. 2021; Nagakubo et al. 2020). They are considered a bacterial 'sample pack' because they retain the membrane composition and properties of their parent bacterial cells and are crucial in several biological mechanisms e.g., pathogenesis, quorum sensing, nutrient utilization, and communication (e.g., horizontal gene transfer among different types of bacteria). Furthermore, BMVs engage in numerous biological processes, can function as a biological shuttle system to transport virulence factors into host cells, and possess inherent adjuvanticity to provoke host immune responses (Zhu et al. 2021; Jalalifar et al. 2023). From a biomedical perspective, bacterial membrane vesicles offer several significant advantages (Table 3.1).

Natural BMVs inherit various attributes from their parent cells. Their unique compositional, cargo loading, biological, and physicochemical characteristics make them promising therapeutic nanomaterials and drug delivery platforms (Fig. 3.2a).

Table 3.1 Key features and advantages of natural bacterial membrane vesicles for biomedical applications

BMVs feature	Description
Efficient uptake and tissue tropism	Their nanoscale size enables rapid uptake by antigen-presenting cells, cellular penetration, immune activation, and passive accumulation in target tissues via retention effect (Mehanny et al. 2020; Liang et al. 2022)
Adjuvant properties	BMVs are enriched with PAMPs, which elicit robust immune responses, making them effective as pathogen vaccines, adjuvants, and immunotherapy tools (Prior et al. 2021; Sanders and Feavers 2011)
Enhanced stability and protection	Stable and rigid membrane structure minimizes loaded antigen degradation and drug leakage during circulation (Alves et al. 2016)
Engineering versatility	Genetic, chemical, and physical modifications can enhance targetability and enable them to carry desired payloads and expand applications in biomedical fields (Luo et al. 2024; Gao et al. 2022)
Improved safety	Unlike live-attenuated bacteria, BMVs cannot replicate autonomously in vivo, offering a safer profile (Jiang et al. 2019)
GMP compliance	BMVs are nanoscale in size and can be easily sterilized using filtration strategies, aligning with Good Manufacturing Practice standards (Lui and Leung 2022)

However, these properties vary among BMVs, as those derived from different bacterial types possess distinct surface markers and biochemical compositions. These variations are influenced by endogenous factors (e.g., bacterial strain, biogenesis pathways) and exogenous factors (e.g., growth stage, environmental conditions, and extraction methods). For example, BMVs from different types of bacteria, such as pathogenic and symbiotic bacteria, can have entirely distinct functions, and the presence of endotoxic LPS and danger signals on the surface or in the interior of BMVs may lead to toxicity and can compromise therapeutic efficacy (Huang et al. 2016a; Van de Waterbeemd et al. 2010). BMVs may lack tissue-targeting surface markers for specific applications, resulting in off-target effects, particularly outside tumor tissues (Gao et al. 2021). These variations limit the clinical utility of natural BMVs, requiring careful bacterial strain selection and engineered modifications to improve safety, antigenic completeness, immune reactivity, functionality, and therapeutic potential of vesicles (Ellis and Kuehn 2010; Qing et al. 2019; Bai et al. 2024). Engineering versatility is an important feature of nanovesicles. Pre-engineering (modification of parent cells) through genetic engineering and post-engineering (modification after BMVs isolation) through click conjugation, lipid insertion, and membrane fusion strategies have been used to overcome undesirable characteristics of BMVs and to improve targeting potential, promotion of safety, antigen delivery, and delivery of therapeutics. BMVs can be used as efficient vaccines for bacterial infections, adjuvants, agents for tumor immunotherapy, drug transport systems, and anti-bacterial adherence compounds (Fig. 3.2b–c) (Krishnan et al. 2022; Gilmore et al. 2021).

This chapter offers an in-depth understanding of the need for BMVs modifications, summarizes the techniques used for morphological, biophysical, biochemical, and functional characterization of BMV, and outlines strategies for BMVs engineering. The chapter further explores the advances in bioengineered bacterial membrane

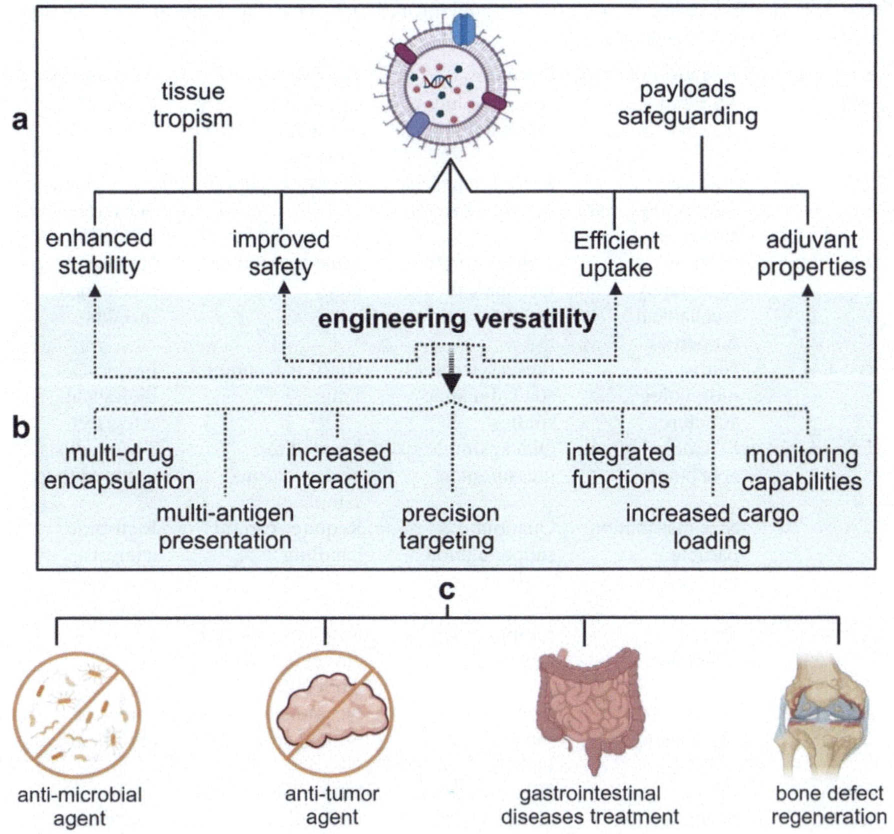

Fig. 3.2 Schematic diagram illustrates (**a**) Advantages of natural bacterial membrane vesicles, including their inherent properties and functionalities, (**b**) Engineering functionalities to enhance intrinsic properties and introduce diverse exogenous functions into bacterial membrane vesicles, and (**c**) Potential biomedical applications of engineered bacterial membrane vesicles. (Source: Created with BioRender.com)

vesicles in various applications, including antibacterial agents, viral vaccine design, targeted tumor therapy, gastrointestinal disease treatment, and osteoporotic bone defects treatment.

3.2 Characterization Techniques for Bacterial Membrane Vesicles

BMVs possess distinct properties and have been characterized through various morphological, biophysical, biochemical composition, and functional analyses (Table 3.2).

Table 3.2 General strategies used for morphological, biophysical, biochemical, and functional characterization of nanovesicles

Technique	Key insights	Strengths	Limitations	Advantages
TEM	Morphology, size, internal structure	High-resolution internal imaging	Complex preparation, 2D imaging	Detailed structural insights
SEM	Surface morphology, 3D structures	Surface analysis, 3D visualization	Requires coating	High-quality 3D surface images
AFM	Surface topography, mechanical properties	Non-destructive, mechanical insights	Time-consuming prep	Nanoscale mechanical insights
Cryo-EM	Native morphology, 3D structures	Preserves natural state, dynamic studies	High cost, complex setup	Preserves biological integrity
DLS	Size distribution, zeta potential	Quick, simple size measurement	Limited for heterogeneous samples	Fast analysis, minimal prep
NTA	Size distribution, particle concentration	Quantitative, subpopulation analysis	Requires careful handling	Real-time analysis, complex samples
BCA assay	Protein quantification	Simple, fast, sensitive	Requires standard curve; interference from buffers	Fast, buffer-compatible
SDS-PAGE	Protein size, expression profiling	Size-based separation, visualizes proteins	Denaturation alters protein structure	Effective protein analysis
Western Blotting	Protein detection, quantification	High specificity, low-abundance detection	Requires quality antibodies, time-consuming	High specificity, sensitive detection
ELISA	Quantifies specific proteins/ antigens	High sensitivity, adaptable for high throughput	Requires antibodies, limited for low abundance	High-throughput, specific
Mass Spectrometry	Protein identification, quantification	High sensitivity, analyzes complex mixtures	Expensive, requires skilled personnel	Sensitive, detailed protein analysis
Confocal Microscopy	Cellular imaging, vesicle tracking	High resolution, live-cell imaging	Limited by photobleaching	Real-time, non-invasive imaging

3.2.1 Morphological and Biophysical Characterization

The morphology of BMVs can be examined using different types of optical microscopes, including transmission electron microscopy (TEM), scanning electron microscopy (SEM), atomic force microscopy (AFM), and cryo-electron microscopy (Cryo-EM). The biophysical characteristics like size distribution, particle density and surface charge of extracted or engineered BMVs have been characterized utilizing dynamic light scattering (DLS), nanoparticle tracking analysis (NTA), and zeta potential measurement.

TEM is a widely used imaging technique for characterizing BMVs. TEM captures high-resolution images that enable the detailed visualization of vesicle morphology, size, and structural integrity at the nanoscale. It identifies vesicle shape, uniformity, and surface characteristics, such as spherical or elongated forms. TEM is beneficial for examining membrane structures and internal contents, as well as detecting vesicle aggregation and deformities. By employing specific staining methods, TEM enhances contrast and facilitates the distinction of membrane components. Additionally, TEM enables the evaluation of vesicle interactions and the monitoring of engineering modifications on vesicle shape and composition. Although sample preparation may be complex, TEM remains a vital tool in BMV characterization (Pascucci and Scattini 2021). SEM is a powerful tool for characterizing BMVs at the micrometer to nanometer scale. SEM provides high-resolution surface images that reveal vesicle shape, size, and texture. Unlike TEM, SEM offers 3D imaging, helping to analyze vesicle morphology and surface features. It can identify external structures such as pores, protrusions, or irregularities. SEM is valuable for evaluating vesicle aggregation and clumping, giving insights into vesicle stability. The technique also enables the visualization of vesicle interactions with materials or surfaces. Collaborating with energy-dispersive X-ray spectroscopy (EDX), SEM can recognize surface elemental compositions. It also helps assess the effects of engineering modifications on vesicle properties. Although sample preparation involves coating the vesicles, SEM remains essential for characterizing BMVs' surface characteristics and interactions (Pascucci and Scattini 2021; Malenica et al. 2021).

AFM is a high-resolution tool for analyzing BMVs, enabling examining their surface and mechanical characteristics. AFM provides 3D topographical images to analyze vesicle morphology, roughness, and elasticity. It is useful for evaluating the stiffness and flexibility of BMVs, critical for their stability in biological environments. AFM also measures interactions between BMVs and surfaces or biomolecules, offering insights into binding affinities and drug delivery potential. It can assess structural changes from engineering modifications, such as protein incorporation. AFM is sensitive to small variations in surface topography and is non-destructive, preserving vesicle integrity. Combining AFM with other techniques like fluorescence microscopy allows real-time monitoring of BMVs. Despite the need for careful sample preparation, AFM is essential for understanding nanoscale features and modifications in BMVs (Parisse et al. 2017; Sharma et al. 2018). Cryo-EM is an advanced technique for studying nanovesicles at high resolution in their original condition. By quickly freezing samples in vitreous ice, Cryo-EM preserves vesicle structure without chemical fixation or staining. This method enables detailed visualization of their size, shape, membrane characteristics, and internal contents. Cryo-EM provides 3D reconstructions, revealing how engineering modifications, like protein incorporation or lipid changes, affect vesicle form. It also allows the observation of dynamic behavior and interactions with biomolecules in real-time. It is a non-destructive technique that preserves the native structure of nanovesicles, making it essential for biomedical applications. Despite its cost and complex preparation, Cryo-EM is crucial for advancing nanovesicle engineering in drug delivery and vaccine development (Morandi et al. 2022).

DLS is a widely used procedure for characterizing the dimensional characteristics and distribution of BMVs in the nanometer range by detecting light intensity fluctuations caused by their Brownian motion in suspension. It provides information on the average size of vesicles and their size distribution, offering insights into the uniformity and stability of vesicle populations. DLS is non-invasive and does not require labeling of samples, making it ideal for analyzing biological vesicles like extracellular vesicles. The results are typically represented as a size distribution profile, which can be used to assess polydispersity or aggregation within the vesicle sample. DLS is highly sensitive to changes in vesicle size, allowing it to monitor processes such as vesicle fusion or aggregation. However, it is limited in its ability to characterize vesicle morphology or internal structure. Despite this, DLS remains an essential tool for the preliminary analysis of vesicles in pharmaceutical applications, where size uniformity is crucial for therapeutic performance (Lyu et al. 2021). NTA is a widely utilized technique for characterizing vesicles, including bacterial membrane vesicles. By tracking the movement of individual vesicles in a liquid sample through video microscopy, NTA analyzes Brownian motion to determine particle size and size distribution. It provides size profiles for heterogeneous populations, measures particle concentration, and offers insights into vesicle stability, aggregation, and engineering-induced size variations. Fluorescent labeling enhances the technique's specificity, enabling the analysis of distinct subpopulations. While requiring careful sample preparation, NTA remains a robust and reliable method for studying vesicle properties and concentrations, contributing significantly to biomedical research (Vestad et al. 2017).

3.2.2 Biochemical and Functional Characterization

The qualitative and quantitative analysis of proteins and engineered moieties in bacterial membrane vesicles have been primarily determined using methods such as the bicinchoninic acid (BCA) assay, western blotting, sodium dodecyl sulfate (SDS)-polyacrylamide gel electrophoresis (PAGE), enzyme-linked immunosorbent assay (ELISA), mass spectrometry (MS), and laser scanning confocal microscopy (LSCM).

BCA assay is a commonly utilized technique for measuring proteins in vesicle preparations. This assay involves the depletion of Cu^{2+} to Cu^{1+} in an alkaline medium, which subsequently interacts with the BCA reagent to produce a complex with a purple color. The color intensity is directly proportional to protein concentration and can be accurately quantified using spectrophotometer. The BCA assay offers high sensitivity, enabling the detection of proteins in small vesicle samples, making it ideal for the identification of protein content in vesicles, including exosomes, microvesicles, and bacterial membrane vesicles. SDS-PAGE is a commonly utilized method for the analysis and separation of proteins, including those associated with vesicles. In SDS-PAGE, proteins are initially treated with SDS, a detergent that denatures them and gives them a negative charge proportional to their size. The proteins are subsequently separated by molecular weight through an electric field applied across a polyacrylamide gel matrix. This method allows us to identify and characterize protein profiles in vesicle preparations, according to their size and migration pattern. SDS-PAGE is often coupled with western blotting to further identify specific proteins (Sódar et al. 2017).

ELISA is a highly sensitive immunoassay used to determine specific proteins and measure the quantity of proteins or other biomolecules within vesicles. The process involves immobilizing an antibody related to the desired protein target on a solid surface (such as a microplate), followed by adding vesicle samples. After the binding of the protein of interest, an enzyme-conjugated secondary antibody is then added. The enzyme catalyzes the reaction of the substrate to generate a colorimetric, fluorescent, or chemiluminescent signal, which can be measured to determine the concentration of the target molecule. ELISA is particularly valuable in vesicle engineering for profiling protein modifications, surface marker analysis, and assessing the success of vesicle-based therapeutic strategies (Yang et al. 2022). Mass spectrometry (MS) is a potent analytical method used in vesicle engineering for the detailed characterization of proteins, lipids, and other biomolecules present in extracellular vesicles. MS allows for the identification, quantification, and structural characterization of molecules with high sensitivity and specificity. In vesicle engineering, MS is often used to analyze the protein and lipid content of engineered vesicles, identify post-translational modifications, and uncover potential biomarkers for disease diagnosis or therapeutic applications. The process includes ionizing the sample and recognizing them to produce a spectrum. Techniques like tandem mass spectrometry (MS/MS) can further accommodate sequence information, enabling the analysis of complex molecular mixtures. MS is invaluable for profiling engineered vesicles in therapeutic research, ensuring that vesicles are successfully loaded with cargo and functionalized for specific biomedical applications (Burton et al. 2023).

LSCM is a high-resolution imaging technique used to visualize and identify the morphology, distribution, and interactions of engineered vesicles. In bacterial membrane vesicle engineering, LSCM offers advantages over traditional microscopy methods by providing optical sectioning, allowing for detailed 3D imaging of vesicle structures and their interactions with cells or other biomolecules. Using fluorescence-based markers, LSCM can investigate the surface properties, ingestion mechanisms, and intracellular trafficking of engineered vesicles. The technique is beneficial for assessing the effectiveness of vesicle surface modifications, protein or nucleic acid loading, and targeting efficiency in cellular environments. LSCM also allows for real-time monitoring of vesicle behavior, making it an essential tool in the progression of vesicle-based drug delivery process and therapeutic applications (Reclusa et al. 2020).

Each characterization technique offers unique advantages, and combining imaging with analytical methods ensures a comprehensive understanding of BMVs properties. Bacterial membrane vesicle characterization methods are selected based on parameters such as protein content, surface modifications, and morphology. Protein quantification, identification, and surface modification analysis are essential for understanding the biological properties and functional attributes of BMVs. The integration of these techniques supports their advancement in different biomedical applications (Bitto et al. 2021). Quality control of BMVs remains a challenge, requiring standardized characterization methods. The International Society for Extracellular Vesicles recommends detecting at least three vesicle protein markers via western blotting, flow cytometry, or proteomic analysis. Vesicle heterogeneity should be assessed using at least two analytical methods, with high-resolution imaging providing both close-up and wide-field views. Size distribution can be evaluated using NTA, DLS, or resistive pulse sensing. To maintain consistency, BMV yield should be normalized to bacterial count using optical density or colony-forming units. Additionally, batch-to-batch variations must be carefully managed in BMV-based product development (Lötvall et al. 2014; Théry et al. 2018).

3.3 Strategies for Bacterial Membrane Vesicles Engineering

3.3.1 Genetic Engineering

Specific targeting ligands can be introduced onto the surfaces of bacterial membrane vesicles through genetic engineering. This approach involves modifying parent bacterial cells, which subsequently produce BMVs with enhanced functionalities. In bacterial membrane nanovesicles, recombinant antigens have been fused with hemoglobin protease (Hbp) (van den Berg van Saparoea et al. 2020), cytolysin A (ClyA) (Wo et al. 2023) and outer membrane protein A (OmpA) (Alves et al. 2015). For example, mCherry can be fused with carrier OmpA, and directed recombinant protein loading into the lumen of bacterial membrane vesicle (Fig. 3.3a). Two different antigens can be co-expressed in the same vesicle, with one antigen displayed on the surface and the other in the vesicle lumen (Wo et al. 2023). This approach is useful for targeted drug delivery and presentation of foreign antigens onto vesicle surfaces. Since bacteria have advantages such as rapid proliferation, and genetically engineered probiotic strains can be regarded as a micro-factory designed to efficiently create engineered BMVs with durable modifications and it has been introduced for intestinal disease management that exert their function through the release of engineered BMVs (Gujrati et al. 2019; Yue et al. 2022).

The LPS in BMVs makes them powerful vaccines and adjuvants, but it can also trigger a significant inflammatory response, septic shock, and cause tissue damage. The structure of LPS is composed of core oligosaccharides, acyl chains, lipid A moieties, and O-specific polysaccharide chain. Lipid A components are the main factor responsible for the toxicity of LPS. Reducing LPS-mediated toxicity in bacterial membrane vesicles using chemical detergents has partially succeeded. This approach may enhance BMVs aggregation tendency, potentially compromise product quality. Furthermore, Chemical detergents can deplete lipoproteins and other surface-exposed proteins, such as outer membrane proteins and pattern recognition receptors, thereby reducing the adjuvanticity of BMVs. Genetic engineering has

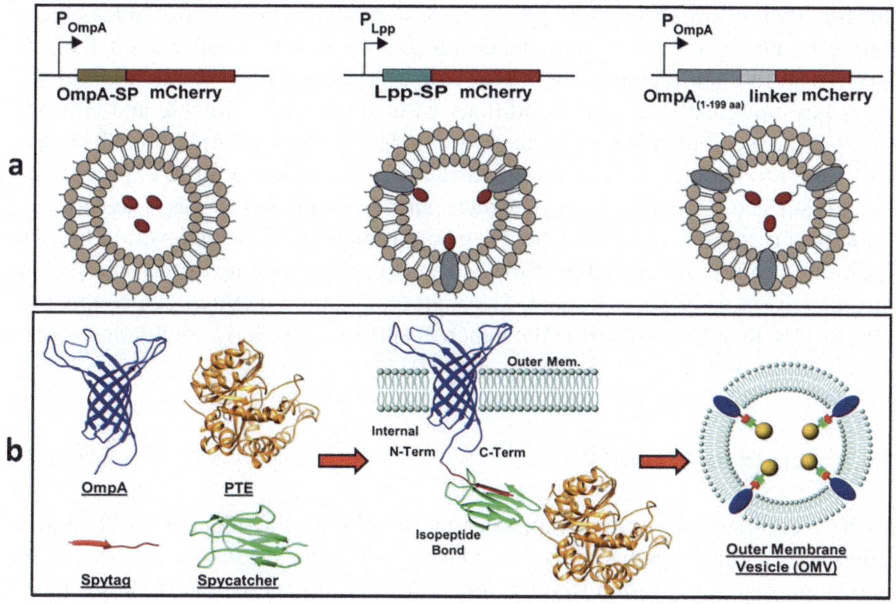

Fig. 3.3 (**a**) mCherry fused with OmpA is directed in the lumen of bacterial membrane vesicles. (Reproduced with permission from Wo et al. (2023), copyright 2023, Elsevier B.V.). (**b**) Schematic pairing among SpyTag and SpyCatcher onto bacteria membrane vesicle. (Reproduced with permission from Alves et al. (2015), copyright 2015, American Chemical Society)

been employed to minimize LPS-mediated pathogenicity by knocking out or modifying genes responsible for LPS synthesis or genes associated with the synthesis of core oligosaccharide and O-antigen are organized in an operon, such as msbA, msbB, lpxL1, and lpxM. This general method has the potential to reduce the endotoxicity caused by LPS and result in the production of safer membrane vesicles compared to their natural counterparts. The detoxification facilitated by genetic engineering may be more effective approach compared to other methods and extracted BMVs from these engineered bacteria did not provoke a noticeable inflammatory response and systemic toxicity, indicating the biocompatibility of these BMVs (Chen et al. 2016; Kay et al. 2019; Zariri et al. 2016; Zhuang et al. 2021; Needham et al. 2013). Overall, the advantages of genetic engineering methods in functionalizing BMVs are evident and enable their use to deliver vaccines, drugs, and immunotherapy agents.

3.3.2 Click Conjugation

SpyTag is a peptide that plays a pivotal role in specific protein-protein interactions by forming a stable isopeptide linkage with its complementary protein partner, SpyCatcher. This interaction enables the formation of a stable SpyCatcher/SpyTag pair for efficient coupling. The SpyTag-SpyCatcher ligation system enables precise conjugation between these components and is biologically analogue to click chemistry. This bio-click reaction offers stable, irreversible, and rapid reaction kinetics under physiologically relevant conditions without cofactors (Keeble and Howarth 2020). From a cell membrane engineering standpoint, SpyTag peptide can be genetically fused to integral cell membrane carrier protein, allowing for its expression on the cell surface. In contrast, SpyCatcher can be conjugated to target agents, and libraries of recombinant SpyCatcher have been rapidly introduced. SpyTag (fused to a cell surface protein) encounters SpyCatcher (fused to targeting agents), they spontaneously form an isopeptide bond. These protein ligation methods can couple proteins to present target antigens in the lumen or surface of bacterial membrane vesicle (Fig. 3.3b) (Alves et al. 2015; Cheng et al. 2021).

3.3.3 Lipid Insertion

BMVs membrane primarily comprises a phospholipid bilayer and embedded molecules. The natural organization of the vesicle membrane creates attractive sites for membrane functionalization through the insertion of ligand-conjugated molecules. Lipid insertion facilitates the spontaneous incorporation of functional moieties onto the BMVs via lipid anchors. Various synthetic materials with hydrophobic anchors, including phospholipids, alkanes, alkyl chains, oleyl chains, and similar moieties, are commonly employed for BMVs surface functionalization. A variety of synthetic targeting materials, including peptides, aptamers, saccharides, nanoparticles, and fluorescent molecules, can be conjugated to hydrophobic anchors, such as DPPE (1,2-dipalmitoyl-sn-glycero-3-phosphoethanolamine) via PEG-based spacers. This process forms PEG-linked lipid conjugates, which are commonly utilized for surface functionalization. For example, by conjugating a pH-sensitive drug delivery system to DSPE-PEG, it can be conveniently integrated into the plasma membrane phospholipid bilayer, resulting in a low rate of intercellular migration (Fig. 3.4a) (Guo et al. 2021). The lipid insertion technique facilitates the consistent incorporation of functional biomaterials into BMVs, ensuring their uniform presentation under mild conditions. The hydrophobic insertion method has a minimal impact on the intrinsic bioactivity of modified BMVs. However, several challenges persist, including the variable retention period of engineered molecules (ranging from 4 to 72 h) and the difficulty in maintaining a consistent density of artificial molecules on the EV membrane. This issue arises due to the potential translocation of inserted molecules to the inner leaflet of the nanovesicle. Optimizing the PEG length and the distance between the EV surface and the functional moiety is essential for enhancing the functionality of inserted molecules. This optimization ensures optimal performance by preventing inward penetration into the lumen while maintaining flexibility for terminal moieties (Kim and Kim 2022; Kolate et al. 2014).

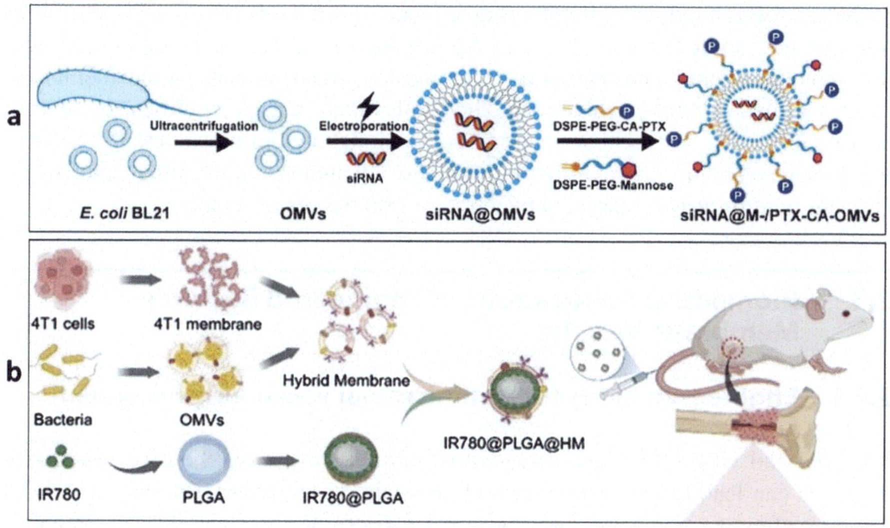

Fig. 3.4 (**a**) A pH-sensitive DSPE-PEG-conjugated drug delivery system can be integrated into the plasma membrane and minimize intercellular migration. (Reproduced with permission from Guo et al. (2021), copyright 2021, American Chemical Society). (**b**) Hybrid membranes were created by fusing breast tumor cell membranes and outer membrane vesicles (OMVs), which were then used to encapsulate IR780-loaded poly(lactic-co-glycolic acid) (PLGA) to create hybrid nanoparticles to effectively target and inhibit tumors. (Reproduced with permission from Wang et al. (2024a), copyright 2024, Springer Nature)

3.3.4 Membrane Fusion

Lipid membrane fusion plays a crucial role in cell biology and has attracted significant attention for its potential applications in membrane engineering. Hybrid membrane formation occurs as BMVs fuse with eukaryotic cell membranes such as those originating from tumor cells are being highly studied for their superior functionalities (Zhao et al. 2024). For example, hybrid membranes were created by fusing breast tumor cell membranes with bacterial OMVs, which were then utilized to encapsulate IR780-loaded poly(lactic-co-glycolic acid) (PLGA) to produce nanoparticles. The experiments demonstrated that the hybrid cell membrane-coated nanoparticles exhibited strong biocompatibility, successfully targeted tumors, enhanced polarization of macrophages, activated dendritic cells, and displayed the capability to eliminate tumors effectively in both vitro and in vivo settings (Fig. 3.4b) (Wang et al. 2024a). Additionally, there is growing interest in incorporating bacterial membrane proteins, membrane fragments, or fused BMVs into liposomes to confer biomimetic properties to hybrid nanovesicles. For example, siRNA@PLOV, a combination of photothermal-sensitive liposomes and weakened Salmonella OMVs, served to target cancer tissues and T lymphocytes in cancer treatment. The combined PLOVs kept the biological properties of OMVs while improving drug-loading capacity. The outcomes showed that the OMVs have immunogenic properties, combined with photothermal effects, significantly boosted T cell infiltration into the tumor (Zhai et al. 2021). Membrane fusion facilitates rapid modification, precise biomolecule encapsulation, and the integration of parent membrane functionalities, thereby enhancing the multifunctional capabilities of hybrid nanovesicles. These hybrid structures have demonstrated effectiveness in prolonging blood circulation, providing biomimetic advantages, presenting heterotypic antigens, and improving targeted penetration into tissues of interest.

3.4 Biomedical Applications of Engineered Bacterial Membrane Vesicles

3.4.1 Engineered BMVs for Antimicrobial Vaccine Development

Bacterial and viral infections activate host immune responses; however, excessive reactions can lead to tissue damage and, in severe cases, death. Engineered BMVs serve as effective adjuvants and microbial carriers due to their diverse antigens, PAMPs, and virulence factors, which enhance and modulate immune responses when delivered in a controlled manner. Engineered BMV-based vaccines, featuring specific antigens, trigger strong humoral and cellular immune responses, offering a promising approach for immune system training against pathogens. This section explores the potential of engineered BMVs in bacterial and viral vaccine development.

3.4.1.1 Bacterial Vaccine Design

Antibiotics are commonly used in clinics to cure bacterial infections; however, widespread resistance has emerged across all classes. The efficacy of antibiotics has declined significantly due to antimicrobial resistance developed in bacteria, and the rising incidence of infectious diseases highlights the urgent need for vaccines (Brüssow 2024; Shyr et al. 2021). Vaccines have been developed from attenuated microbes, subunit antigens, inactivated toxins, and nucleic acids, however, traditional vaccines do not protect against all infectious diseases. This gap is often attributed to antigenic variation among serogroups (Micoli et al. 2021). Traditional strategies face notable shortcomings: nucleic acid vaccines risk genome integration, live attenuated vaccines can potentially return to a virulent form, and subunit vaccines need adjuvants to boost their immune response. Furthermore, selecting and integrating multiple antigens remains challenging, highlighting the urgent need for developing multicomponent vaccines (Khalid and Poh 2023; Hussain et al. 2023).

BMVs are replication-defective, and they are crucial in the transportation of toxic substances and pathogenicity factors to target cells. They are rich in multiple PAMPs and bacteria-derived antigens, making them practical for inducing immune responses as opposed to their parent bacteria. The exploration of BMVs has been undertaken to develop vaccines against *Neisseria meningitidis* and numerous other pathogens. However, wild-type BMVs can exhibit inherent cytotoxicity; for instance, the presence of endotoxic LPS may trigger septic shock (Kim et al. 2009).

Several studies have focused on attenuating endotoxin and surface glycan engineering of BMVs-producing strains to create safe vaccines (Gerritzen et al. 2017). Furthermore, advancement has been made to decorate heterologous antigens and targeting ligands onto nanovesicles' surface via fusing them with integral carrier membrane proteins of nanovesicles such as hemoglobin protease (Hbp) (van den Berg van Saparoea et al. 2020), cytolysin A (ClyA), and outer membrane protein A (OmpA) (Alves et al. 2015) for specific biomedical applications.

Heterologous bacterial antigens have been expressed on bacterial membrane vesicles, allowing these membrane vesicles to be leveraged as vaccines against different infections. For example, Sun et al. developed a multitargeting heterologous click vaccine. The engineered BMVs consisting of trOmpA fused with SpyCatcher, can bind to SpyTag-fused *Staphylococcus aureus* recombinant antigens (SpA, Sbi, or EsxA). Subcutaneous injection of this click-based vaccine elicited both humoral and cellular immune responses that are particular to antigens, providing immunity against lethal *S. aureus* complications in a mouse model study (Sun et al. 2023). Huang et al. utilized GenEng to display Omp22 of *Acinetobacter baumannii* on *Escherichia coli* DH5α-derived BMVs by fusing Omp22 with the ClyA carrier protein. The resulting ClyA-Omp22 engineered membrane vesicles elicited high titers of Omp22-specific humoral antibodies and significantly protected mice against lethal *A. baumannii* infections. This approach improved survival rates in a murine sepsis model, reduced bacterial burdens in tissues, and suppressed serum inflammatory cytokine levels (Fig. 3.5a) (Huang et al. 2016b). Li et al. engineered BMVs via the lipoprotein transport machinery (Lpp) to display the SaoA antigen, focusing on developing a vaccine against *Streptococcus suis*. Immunization with these Lpp-SaoA-engineered bacterial membrane vesicles induced robust antibody responses, enhanced the opsonophagocytic uptake of *Streptococcus suis* by macrophages, and improved bacterial reduction in a mouse infection model (Li et al. 2023). Overall, these results indicated that heterologous antigen-expressed BMVs can be used as adjuvant-free vaccines for significant bacterial pathogens.

3.4.1.2 Viral Vaccine Design

The treatment of viral diseases primarily relies on antiviral drugs, vaccines, and supportive care, with specific strategies varying by virus. Significant progress has been made in combating viral infections, challenges remain, including viral resistance, rapid mutation rates, limited treatments for emerging threats, and the need for global vaccine accessibility (Hussain et al. 2023). Developing engineered BMVs, with their ability to mimic viral structures, enhance immune responses, and rapidly adapt to emerging threats, represents a cost-effective and promising advancement in vaccine technology for controlling both existing and future viral diseases. For example, a bivalent antigen display platform targeting the receptor binding domain (RBD) of the spike protein of *SARS-CoV-2*, which is associated with angiotensin-converting enzyme 2 in target cells, has been introduced. They fused the foreign proteins NG06 and RBD with carrier membrane proteins ClyA and OmpA, respectively, allowing their optimal expression onto the nanovesicle surface and lumen. Engineered nanovesicles elicited synergistic immunogenicity in mice and showed

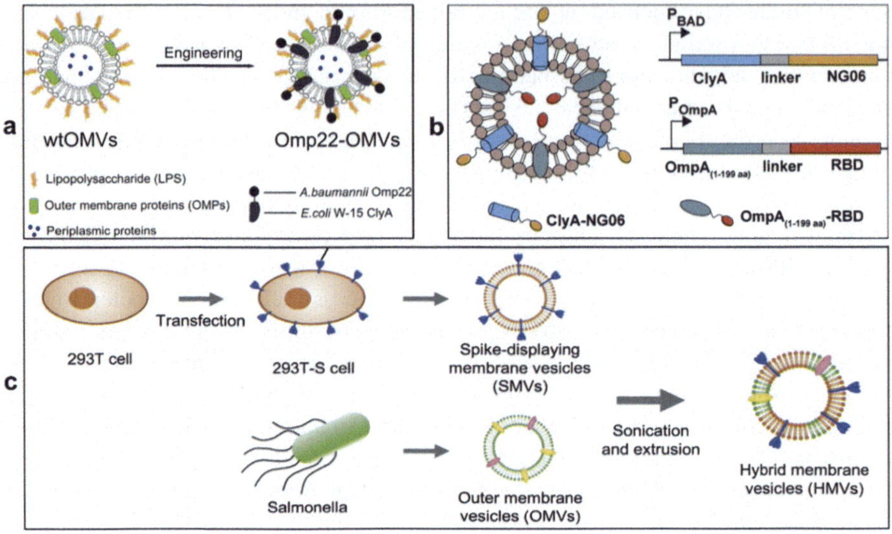

Fig. 3.5 (**a**) Fusion of *A. baumannii* Omp22 protein with the ClyA carrier protein, allowing ClyA-Omp22 to be displayed on nanovesicles. (Reproduced with permission from (Huang et al. 2016b), copyright, 2016 Springer Nature). (**b**) Expression of ClyA-NG06 onto nanovesicle surface and ClyA-RBD in the nanovesicle lumen in their natural conformation. (Reproduced with permission from Wo et al. (2023), copyright, 2023 Elsevier B.V.). (**c**) Envelope virus-mimetic vaccines were prepared by hybridizing viral antigens expressing mammalian cell membranes with bacterial outer membrane nanovesicles. (Reproduced with permission from Zhang et al. (2022), copyright 2022, Elsevier B.V.)

promise as a vaccine against *SARS-CoV-2* variants (Fig. 3.5b) (Wo et al. 2023). Furthermore, recombinant plasmids containing ClyA-RBD have been transformed into *E. coli* to express the ClyA-RBD protein. High-pressure homogenization has been used to create and enhance the presentation of recombinant antigens onto nanovesicle surfaces. Subcutaneous injections of these engineered nanovesicles resulted in accumulation in lymph nodes, promoting antigen presentation and eliciting *SARS-CoV-2* distinctive immune responses were studied in mice (Yang et al. 2021). Zhang et al. developed nano-vaccines mimicking envelope viruses by hybridizing mammalian cell membranes expressing viral spike proteins with bacterial nanovesicles from *Salmonella typhimurium*. The resulting hybrid nanovesicles retained viral proteins on their lipid bilayer and incorporated bacterial PAMPs in native conformation. This engineering enhanced antigen presentation by dendritic cells, resulting in strong immune responses in mice, demonstrating their potential for developing vaccines against *SARS-CoV-2* (Fig. 3.5c) (Zhang et al. 2022). Overall, engineered BMVs exhibit a better safety profile than natural bacterial vesicles. Recent advancements in expressing viral antigens on BMVs improve their ability to modulate immunity against viral infections. Bioengineering further improves their interaction, penetration, and targeted delivery, making them promising candidates for bacterial and viral vaccines.

3.4.2 Engineered BMVs in Targeted Cancer Therapy

Traditional cancer treatments like surgery, chemotherapy, and radiotherapy are helpful but still often cause serious impacts and have limited efficacy against metastatic tumors, highlighting the need for safer and more effective therapies. Cancer immunotherapy offers a promising alternative by stimulating the immune system to cancer and is effective in reducing tumor metastasis and preventing recurrence. Approaches like tumor-specific vaccines, chimeric antigen receptor T lymphocyte therapy, and immune checkpoint barrier harness immune cells to create potent adaptive responses, showing significant potential in treating malignant tumors. However, the complexity and unpredictability of cell-based immunotherapies can lead to severe adverse reactions from immune overactivity, highlighting the need for new strategies to mitigate these risks. Cancer cell membranes, abundant in cancer-related antigens and neoantigens, are promising tumor vaccine candidates. However, their low immunogenicity limits robust systemic immune responses. Engineered BMVs offer promise in delivering antigens, enhancing low-immunogenic antigens, and boosting antitumor innate immunity, and this cell-free immunotherapy strategy has gained considerable attention as a potential approach.

3.4.2.1 Membrane Fusion for Multi-antigen Presentation

Tumor heterogeneity, often driving metastasis, poses significant challenges for effective therapy. A vesicle system (mTOMV) was developed by hybridizing OMVs derived from bacteria and cell membranes obtained from tumors. The resulting mTOMV nanoplatforms enhanced innate immune cell activation and improved T cell-mediated lysis in homogeneous tumors. It effectively targeted inguinal lymph nodes, inhibiting lung metastasis. Furthermore, it elicited adaptive immune responses in homologous tumors, underscoring its potential for personalized immunotherapy to suppress tumor growth and metastasis (Zou et al. 2021). Cell membrane-camouflaged nanoparticles are increasingly utilized for biomedical applications because of their capability to mimic natural cellular functions. A hybrid membrane known as OMV-CC is created by combining bacterial OMVs with cancer cell membrane, successfully coating hollow polydopamine nanoparticles. This system integrates bacterial vesicle-based immunotherapy to enhance antitumor efficacy against melanoma through photothermal therapy. Intravenous administration of engineered cell membrane-coated nanoparticles targets melanoma and rapidly stimulates dendritic cells in lymph nodes and elicits strong antitumor immune responses. The synergy between immune activation and photothermal therapy effectively eliminates melanoma without significant side effects and could offer adaptability for diverse therapeutic and imaging applications (Fig. 3.6a) (Wang et al. 2020).

3.4.2.2 Checkpoint Inhibitor Therapy

Therapy that targets programmed cell death protein-1 (PD-1) and its ligand programmed cell death ligand-1 (PD-L1) through checkpoint inhibitors has demonstrated promoting clinical outcomes in particular tumors. Genetic engineering has

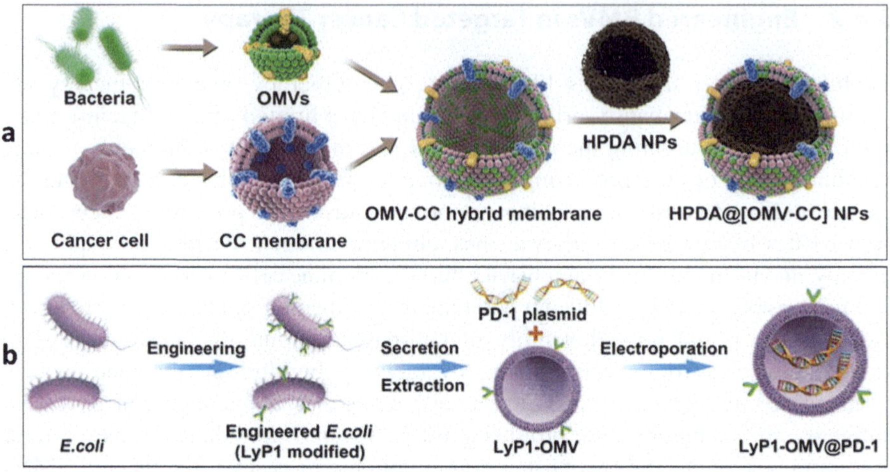

Fig. 3.6 (**a**) A schematic illustrating the fusion of outer membrane vesicles (OMVs) with the cancer membrane, leading to the creation of fused membrane wrapped nanoparticles for synergistic photothermal therapy and immunotherapy against melanoma. (Reproduced with permission from Wang et al. (2020), copyright 2020, American Chemical Society). (**b**) LOMV@PD-1 nanoparticles were prepared by genetically engineering attenuated *E. coli* to express the tumor-targeting peptide LyP-1. Extracted OMVs were then loaded with the PD-1 plasmid via electroporation for tumor immunotherapy. (Reproduced with permission from Pan et al. (2022), copyright 2022, Wiley-VCH GmbH)

been used to modify BMVs with the ectodomain of PD1. The potential of BMVs to initiate immune stimulation remains unaffected by this modification. Significantly, vesicles modified with PD1 bind to PD-L1 on tumor cells, facilitating their integration and reducing PD-L1 levels. This helps protect T cells from the immune-suppressive effects of the PD1/PD-L1 axis. Effector T cell collection in tumors is enhanced by the combined impact of immune stimulation and checkpoint inhibition, leading to enhanced tumor growth inhibition compared to native BMVs and PD-L1 antibodies. Bioengineered BMVs act as an effective immunotherapeutic agent for regulating the immune microenvironment of tumors and enhancing antitumor efficacy (Li et al. 2020). LyP1 polypeptide-modified OMVs (LOMVs) loaded with a PD-1 plasmid come into being for tumor-specific self-blockade of PD-L1. LOMVs accumulate in tumor tissues via bacterial membrane vesicles targeting potential and LyP1-mediated internalization, delivering the PD-1 plasmid into tumor cells. This induces PD-1 expression within tumor cells and selectively binds to PD-L1 on both neighboring and autologous cells, resulting in localized self-blockade. Additionally, LOMV membrane proteins attract cytotoxic lymphocytes and natural killer cells to the tumor site, promoting IFN-γ secretion and enhancing antitumor activity (Fig. 3.6b) (Pan et al. 2022).

3.4.2.3 Neoantigens Presentation and Personalized Therapies

Therapeutic mRNA vaccination offers a promising strategy to activate antitumor immunity, yet effective mRNA delivery technologies for customized cancer vaccines remain limited. OMVs engineered with lysosomal escape protein listeriolysin O and RNA-binding protein L7Ae efficiently adsorb mRNA antigens via L7Ae binding. Through listeriolysin O-mediated endosomal escape, the resultant nanoplatforms enable dendritic cells to take up and cross-present the delivered molecules. Engineered vesicles effectively impeding melanoma progression and resulting in complete tumor regression were observed in a colon cancer model and induced long-term immune memory (Li et al. 2022). Neoantigens from mutant tumor cell proteins can generate strong, customized anti-cancer immunity. However, the vaccine vehicle composition and formation of neo-antigens are critical for strong responses. By harnessing synthetic biology, genetically engineered bacteria have been created for personalized therapeutic vaccines. Fusion of neo-antigens and generation of vesicles from these bacteria effectively present these neo-antigens (BDVs-Neo). BDVs-Neo, combined with granulocyte-macrophage colony-stimulating factor (GM-CSF) were delivered subcutaneously within a hydrogel, enabling the sustained release of BDV-LPS and GM-CSF to recruit dendritic cells, and this approach successfully halted relapse of tumors and their metastasis (Meng et al. 2022). The intestinal epithelial barrier and gastrointestinal environment pose challenges for oral tumor vaccines. Cytolysin A protein of OMVs was fused with tumor antigens, and genetically engineered bacteria produced in the intestine have demonstrated strong efficacy in mouse models. Oral ingestion of engineered bacteria led to engineered OMV production, which crossed the intestinal epithelium, promoting dendritic cell development. In mouse models of metastatic melanoma and subdermal colon cancer, tumor growth was inhibited and protection against tumor rechallenge was provided by these antigens bearing OMVs. This approach highlights the potential of in situ OMV creation by using modified bacteria for oral vaccines and therapeutics (Yue et al. 2022). Overall, combined with cancer cell membrane and cancer antigen, engineered BMVs can safely activate immune cells, and produce strong antitumor immune responses through the existence of bacterial proteins effectively eradicating tumor tissues with significant results in laboratory and preclinical studies.

3.4.3 Engineered BMVs in Gastrointestinal Disease Treatment

Persistent inflammation of the gastrointestinal tract is a key indicator of inflammatory bowel disease (IBD). IBD has no cure and is driven by complex interactions among environmental, genetic, and microbial factors. Additionally, *Helicobacter pylori* (*H. pylori*) infection is a major contributor to gut diseases, including gastritis, peptic ulcers, and gastric cancer. Advancements from the Human Microbiome Project have significantly enhanced our insight into gut microbiota and its functional impact on health. The gut microbiota is crucial for sustaining intestinal homeostasis through its interactions with mucosal cells. However, both IBD and

H. pylori infection disturbs this microbial balance, resulting in dysbiosis. This condition is marked by a decline in bacterial diversity and a higher presence of proinflammatory bacteria. This disruption compromises the integrity of the intestinal barrier and impairs immune function, further exacerbating gastrointestinal inflammation and disease progression (Shan et al. 2022; Wang et al. 2024b; Nash et al. 2017). Substantial evidence has demonstrated the BMVs originating from probiotics and commensal bacteria play a crucial role in mediating communication between bacterial communities and target cells (epithelial and immune cells), influencing microbial homeostasis and host intestinal health by delivering bioactive molecules (Wang et al. 2023). This section emphasizes the possible therapeutic applications of engineered BMVs in restoring intestinal health through two key mechanisms: (a) serving as innovative therapeutic tools and drug delivery vehicles and (b) acting as anti-bacterial adhesion agents.

3.4.3.1 Encapsulation of Therapeutic Agents

BMVs-mediated pathways can be harnessed to deliver active proteins to treat intestinal diseases. Liang et al., develop a comprehensive engineering strategy for the large-scale production of probiotic membrane vesicles that encapsulate fucoxanthin for the purpose of intervening in colitis. Engineered membrane vesicles present a significant boost in yield, enhanced polypeptide content, and improved fucoxanthin stability. They effectively scavenged free radicals, mitigating H_2O_2-induced oxidative damage. Results demonstrated that engineered vesicles promoted macrophage polarization to the M2 type, reduced colon tissue injury, alleviated inflammation, and significantly suppressed proinflammatory cytokines. Engineered membrane vesicles in the colon reshaped gut microbiome communities and increased short-chain fatty acid production, highlighting their potential as a therapeutic approach for colitis (Liang et al. 2023). *Bacteroides thetaiotaomicron*, a major intestinal microbiota member, has been engineered to incorporate bacterial, viral, and human-derived proteins into its BMVs. Engineered BMVs delivered these therapeutics to the gut and respiratory tracts, offering protection against infection, tissue injury and inflammation. For example, BMVs have been engineered to incorporate vaccine antigens from *Influenza A virus* and *Salmonella enterica* serovar Typhimurium. These BMVs elicit antigen-specific immune and antibody responses and can provide heterotypic immunity in mice against a virulent dose of *Influenza A virus*. Furthermore, BMVs modified to produce keratinocyte growth factor-2 demonstrated significant stability. When delivered orally, they mitigated colitis severity and supported intestinal epithelial healing in a dextran sodium sulfate-induced colitis model (Gujrati et al. 2019).

3.4.3.2 Anti-bacterial Adhesion Strategies

Bacterial infections continue to be a leading cause of morbidity and mortality. The increasing prevalence of antibiotic resistance underscores the critical need for novel therapeutic strategies to effectively manage these infections (Chinemerem Nwobodo et al. 2022). Host-pathogen interaction is a ubiquitous biological event, and infections typically start with bacterial interaction to target cells and tissues, which enables adhered pathogens to avoid the body's clearance mechanisms, enabling them to attain a sufficient mass to initiate infection. Anti-adhesion therapies offer a hopeful alternative to antibiotics by preventing infection without inducing antibiotic resistance. Furthermore, inhibiting bacterial adhesion enhances their uptake, recognition, and clearance by immune cells. These advantages have driven the production of various anti-adhesion agents that target bacterial adhesins to combat bacterial infection without inducing bacterial resistance. *H. pylori* is highly adapted to colonize the human stomach, causing gastric disorders such as chronic gastritis, ulcers, and gastric adenocarcinoma. The recent investigation demonstrated that *H. pylori* BMV-coated nanoparticles inherit natural bacterial adhesins and competitively inhibit the adhesion of pathogenic *H. pylori* to mouse gastric tissues in a concentration-dependent manner. These results highlight the potential of BMVs to prevent bacterial infections by blocking the attachment of pathogenic bacteria to host tissues. Thus, nanotherapeutic strategies that integrate bacterial and host cell membranes offer a promising approach to reducing infection risk and treating infectious diseases (Fig. 3.7) (Zhang et al. 2019). Due to the complexity of pathogen attachment, multi-target interventions are crucial for competing with bacterial

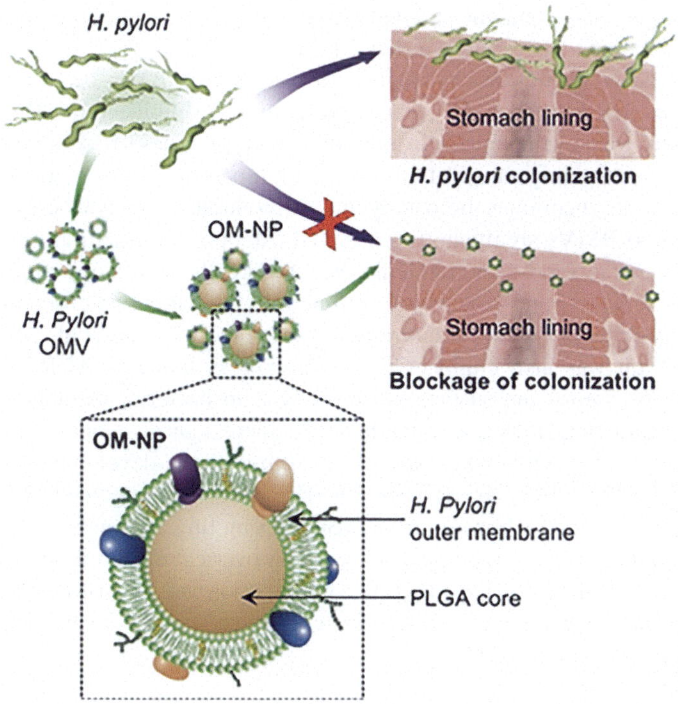

Fig. 3.7 The schematic illustrates the application of *H. pylori* outer membrane vesicle-entrapped nanoparticles, enabling specific attachment to gastric epithelial cells and significantly decreasing the adhesion of pathogenic *H. pylori* to the mouse stomach lining. (Reproduced with permission from Zhang et al. (2019), copyright 2019, Wiley-VCH GmbH)

adhesion sites. Future engineered BMVs could mimic bacteria and be loaded with antibiotics, providing a dual-action strategy to inhibit bacterial adhesion and improve therapeutic delivery against infections.

3.4.4 Engineered BMVs in Osteoporotic Bone Defect Treatment

Osteoporosis is a disorder that affects the skeletal system. It is marked by a reduction in bone mineral density, weakened bone strength, and microstructural deterioration due to an imbalance between osteoblasts and osteoclasts. This abnormal remodeling increases fracture risks, particularly in the elderly and postmenopausal women (Rachner et al. 2011; Khosla and Hofbauer 2017). Osteoporotic fractures lower life quality, and significantly burden public health, and are projected to increase due to the aging population trend. In China, around 3.17 million cases of osteoporotic bone fracture were recorded in 2020, with an estimated surge of 5.33 million cases by 2040 (Cui et al. 2021). The demand for repair materials is high, but osteoporotic fractures respond poorly to traditional therapies. Effective bone reconstruction requires inhibiting osteoclastogenesis, promoting osteogenesis, and incorporating biomimetic complexities. Tissue engineering has gained momentum but developing microenvironment-targeted strategies against osteoporotic bone defects are still challenging (Patel and Wairkar 2023).

Engineered probiotics were designed to overexpress BMP-2 and CXCR4 and both proteins fused to the BMV surface protein ClyA. Engineered BMVs-BMP-2-CXCR4 were then evaluated for osteoporosis therapy. These BMVs demonstrated excellent bone-targeting capability in vivo, high biocompatibility, and an extraordinary capacity to encourage the osteogenic differentiation of MSCs. Furthermore, the engineered BMVs significantly mitigated osteoporotic fracture progression in an ovariectomized mouse model. This one-step synthetic biology approach integrates bone-targeting and bone-forming functions, providing an efficient and scalable strategy for osteoporotic fracture treatment (Fig. 3.8) (Liu et al. 2024). *E. coli* probiotic strain was also engineered to overexpress human CXCR4 (hCXCR4) fused with the membrane surface protein ClyA on bacterial membrane vesicles. These bone-targeting BMVs, which inherently possess anti-osteoporotic properties, were further loaded with SOST siRNA, resulting in the development of BMVs-hCXCR4-SOST siRNA. The engineered BMVs effectively targeted bone tissue and were readily absorbed by bone marrow MSCs, promoting osteogenic differentiation while maintaining high cytocompatibility. In an ovariectomized mouse model, the bioengineered vesicles successfully reversed osteoporosis. This BMVs-based therapeutic approach presents a novel, safe, and effective strategy for treating osteoporotic fractures (Liu et al. 2023).

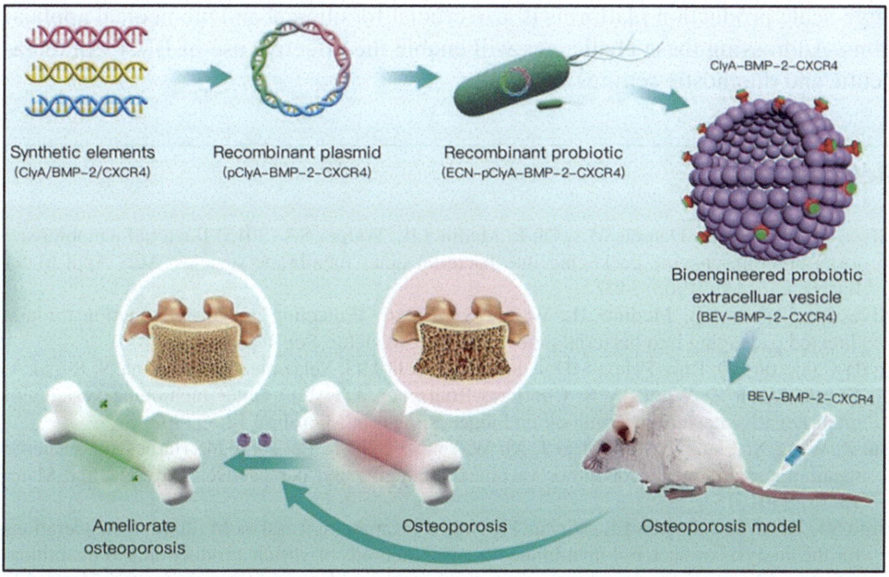

Fig. 3.8 Engineered *E. coli* Nissle 1917 (ECN) produces membrane vesicles displaying bone morphogenetic protein 2 (BMP-2) and C-X-C chemokine receptor type 4 (CXCR4) on their surface, demonstrating strong bone-targeting and osteogenesis-inducing features. (Reproduced with permission from Liu et al. (2024), copyright 2024, Wiley-VCH GmbH)

3.5 Conclusion and Future Directions

Bacterial membrane vesicles are non-replicating entities that hold significant promise for biomedical applications. Genetic engineering, click conjugation, lipid insertion, and membrane fusion have been employed to modify the properties of BMVs, enhancing their potential for various biomedical applications. These advancements have facilitated their use in antimicrobial vaccine development, cancer therapy, treatment of gastrointestinal diseases, and osteoporotic bone defect repair. However, several challenges must be addressed to fully unlock their potential. BMVs derived from different sources exhibit varied functions, necessitating a deeper understanding of their mechanisms and structures to optimize their use in biomedical contexts. A major concern is the presence of LPS and virulence factors in BMVs, which can induce toxicity or undesirable immunogenicity. Balancing the elicitation of an effective immune response while minimizing adverse effects is critical for their successful application in biomedicine. Standardizing BMVs research is essential to overcome challenges such as selecting strain, membrane vesicle diversity, accurate antigen loading, immunogenicity, safety, and scalable manufacturing for targeted therapies. Currently, methodologies vary across laboratories, leading to inconsistencies in the characterization and interpretation of BMVs. To advance the field, improving reproducibility and refining techniques for isolating, purifying, and understanding bacterial membrane vesicle safety is essential. Establishing robust large-scale production platforms is also crucial for clinical and biomedical applications. Addressing these challenges will enable the effective use of BMVs in therapeutic and diagnostic settings.

References

Alves NJ, Turner KB, Daniele MA, Oh E, Medintz IL, Walper SA (2015) Bacterial nanobioreactors–directing enzyme packaging into bacterial outer membrane vesicles. ACS Appl Mater Interfaces 7(44):24963–24972

Alves NJ, Turner KB, Medintz IL, Walper SA (2016) Protecting enzymatic function through directed packaging into bacterial outer membrane vesicles. Sci Rep 6(1):24866

Avila-Calderón ED, Ruiz-Palma MD, Aguilera-Arreola MG, Velázquez-Guadarrama N, Ruiz EA, Gomez-Lunar Z, Witonsky S, Contreras-Rodríguez A (2021) Outer membrane vesicles of gram-negative bacteria: an outlook on biogenesis. Front Microbiol 12:557902

Bai Z, Wang X, Liang T, Xu G, Cai J, Xu W, Yang K, Hu L, Pei P (2024) Harnessing bacterial membrane components for tumor vaccines: strategies and perspectives. Adv Healthc Mater 13(26):2401615

Bitto NJ, Zavan L, Johnston EL, Stinear TP, Hill AF, Kaparakis-Liaskos M (2021) Considerations for the analysis of bacterial membrane vesicles: methods of vesicle production and quantification can influence biological and experimental outcomes. Microbiol Spectr 9(3):e01273–e01221

Briaud P, Carroll RK (2020) Extracellular vesicle biogenesis and functions in gram-positive bacteria. Infect Immun 88(12):10–1128

Brüssow H (2024) The antibiotic resistance crisis and the development of new antibiotics. Microb Biotechnol 17(7):e14510

Burton JB, Carruthers NJ, Stemmer PM (2023) Enriching extracellular vesicles for mass spectrometry. Mass Spectrom Rev 42(2):779–795

Cao Y, Lin H (2021) Characterization and function of membrane vesicles in Gram-positive bacteria. Appl Microbiol Biotechnol 105:1795–1801

Chen L, Valentine JL, Huang CJ, Endicott CE, Moeller TD, Rasmussen JA, Fletcher JR, Boll JM, Rosenthal JA, Dobruchowska J, Wang Z (2016) Outer membrane vesicles displaying engineered glycotopes elicit protective antibodies. Proc Natl Acad Sci 113(26):E3609–E3618

Cheng K, Zhao R, Li Y, Qi Y, Wang Y, Zhang Y, Qin H, Qin Y, Chen L, Li C, Liang J (2021) Bioengineered bacteria-derived outer membrane vesicles as a versatile antigen display platform for tumor vaccination via Plug-and-Display technology. Nat Commun 12(1):2041

Chinemerem Nwobodo D, Ugwu MC, Oliseloke Anie C, Al-Ouqaili MT, Chinedu Ikem J, Victor Chigozie U, Saki M (2022) Antibiotic resistance: the challenges and some emerging strategies for tackling a global menace. J Clin Lab Anal 36(9):e24655

Cui L, Jackson M, Wessler Z, Gitlin M, Xia W (2021) Estimating the future clinical and economic benefits of improving osteoporosis diagnosis and treatment among women in China: a simulation projection model from 2020 to 2040. Arch Osteoporos 1-10:16

Ellis TN, Kuehn MJ (2010) Virulence and immunomodulatory roles of bacterial outer membrane vesicles. Microbiol Mol Biol Rev 74(1):81–94

Fazal S, Lee R (2021) Biomimetic bacterial membrane vesicles for drug delivery applications. Pharmaceutics 13(9):1430

Furuyama N, Sircili MP (2021) Outer membrane vesicles (OMVs) produced by gram-negative bacteria: structure, functions, biogenesis, and vaccine application. Biomed Res Int 2021(1):1490732

Gao J, Wang S, Dong X, Wang Z (2021) RGD-expressed bacterial membrane-derived nanovesicles enhance cancer therapy via multiple tumorous targeting. Theranostics 11(7):3301

Gao J, Su Y, Wang Z (2022) Engineering bacterial membrane nanovesicles for improved therapies in infectious diseases and cancer. Adv Drug Deliv Rev 186:114340

Gerritzen MJ, Martens DE, Wijffels RH, van der Pol L, Stork M (2017) Bioengineering bacterial outer membrane vesicles as vaccine platform. Biotechnol Adv 35(5):565–574
Gilmore WJ, Johnston EL, Zavan L, Bitto NJ, Kaparakis-Liaskos M (2021) Immunomodulatory roles and novel applications of bacterial membrane vesicles. Mol Immunol 134:72–85
Gujrati V, Prakash J, Malekzadeh-Najafabadi J, Stiel A, Klemm U, Mettenleiter G, Aichler M, Walch A, Ntziachristos V (2019) Bioengineering commensal bacteria-derived outer membrane vesicles for delivery of biologics to the gastrointestinal and respiratory tract. Nat Commun 10(1):1114
Guo Q, Li X, Zhou W, Chu Y, Chen Q, Zhang Y, Li C, Chen H, Liu P, Zhao Z, Wang Y (2021) Sequentially triggered bacterial outer membrane vesicles for macrophage metabolism modulation and tumor metastasis suppression. ACS Nano 15(8):13826–13838
Huang W, Wang S, Yao Y, Xia Y, Yang X, Li K, Sun P, Liu C, Sun W, Bai H, Chu X (2016a) Employing Escherichia coli-derived outer membrane vesicles as an antigen delivery platform elicits protective immunity against Acinetobacter baumannii infection. Sci Rep 37242(1):6
Huang W, Wang S, Yao Y, Xia Y, Yang X, Li K, Sun P, Liu C, Sun W, Bai H, Chu X (2016b) Escherichia coli-derived outer membrane vesicles as an antigen delivery platform elicits protective immunity against Acinetobacter baumannii infection. Sci Rep 6(1):37242
Hussain Z, Rani S, Ma F, Li W, Shen W, Gao T, Wang J, Pei R (2023) Dengue determinants: necessities and challenges for universal dengue vaccine development. Rev Med Virol 33(2):e2425
Jalalifar S, Morovati Khamsi H, Hosseini-Fard SR, Karampoor S, Bajelan B, Irajian G, Mirzaei R (2023) Emerging role of microbiota derived outer membrane vesicles to preventive, therapeutic and diagnostic proposes. Infect Agents Cancer 18(1):3
Jiang L, Schinkel M, van Essen M, Schiffelers RM (2019) Bacterial membrane vesicles as promising vaccine candidates. Eur J Pharm Biopharm 145:1–6
Kay E, Cuccui J, Wren BW (2019) Recent advances in the production of recombinant glycoconjugate vaccines. NPJ Vaccines 4(1):16
Keeble AH, Howarth M (2020) Power to the protein: enhancing and combining activities using the Spy toolbox. Chem Sci 11(28):7281–7291
Khalid K, Poh CL (2023) The promising potential of reverse vaccinology-based next-generation vaccine development over conventional vaccines against antibiotic-resistant bacteria. Vaccines 11(7):1264
Khosla S, Hofbauer LC (2017) Osteoporosis treatment: recent developments and ongoing challenges. Lancet Diabetes Endocrinol 5(11):898–907
Kim S, Kim K (2022) Lipid-mediated ex vivo cell surface engineering for augmented cellular functionalities. Biomater Adv 140:213059
Kim SH, Kim KS, Lee SR, Kim E, Kim MS, Lee EY, Gho YS, Kim JW, Bishop RE, Chang KT (2009) Structural modifications of outer membrane vesicles to refine them as vaccine. Biochim Biophys Acta Biomembr 1788(10):2150–2159
Kim OY, Choi SJ, Jang SC, Park KS, Kim SR, Choi JP, Lim JH, Lee SW, Park J, Di Vizio D, Lötvall J (2015) Bacterial protoplast-derived nanovesicles as vaccine delivery system against bacterial infection. Nano Lett 15(1):266–274
Kolate A, Baradia D, Patil S, Vhora I, Kore G, Misra A (2014) PEG—A versatile conjugating ligand for drugs and drug delivery systems. J Control Release 192:67–81
Krishnan N, Kubiatowicz LJ, Holay M, Zhou J, Fang RH, Zhang L (2022) Bacterial membrane vesicles for vaccine applications. Adv Drug Deliv Rev 185:114294
Li Y, Zhao R, Cheng K, Zhang K, Wang Y, Zhang Y, Li Y, Liu G, Xu J, Xu J, Anderson GJ (2020) Bacterial outer membrane vesicles presenting programmed death 1 for improved cancer immunotherapy via immune activation and checkpoint inhibition. ACS Nano 14(12):16698–16711
Li Y, Ma X, Yue Y, Zhang K, Cheng K, Feng Q, Ma N, Liang J, Zhang T, Zhang L, Chen Z (2022) Rapid surface display of mRNA antigens by bacteria-derived outer membrane vesicles for a personalized tumor vaccine. Adv Mater 34(20):2109984
Li Q, Zhou G, Fei X, Tian Y, Wang S, Shi H (2023) Engineered bacterial outer membrane vesicles with lipidated heterologous antigen as an adjuvant-free vaccine platform for Streptococcus suis. Appl Environ Microbiol 89(3):e02047–e02022

Liang X, Dai N, Sheng K, Lu H, Wang J, Chen L, Wang Y (2022) Gut bacterial extracellular vesicles: important players in regulating intestinal microenvironment. Gut Microbes 14(1):2134689
Liang D, Liu C, Li Y, Wu C, Chen Y, Tan M, Su W (2023) Engineering fucoxanthin-loaded probiotics' membrane vesicles for the dietary intervention of colitis. Biomaterials 297:122107
Liu H, Zhang H, Wang S, Cui J, Weng W, Liu X, Tang H, Hu Y, Li X, Zhang K, Zhou F (2023) Bone-targeted bioengineered bacterial extracellular vesicles delivering siRNA to ameliorate osteoporosis. Compos Part B Eng 255:110610
Liu H, Song P, Zhang H, Zhou F, Ji N, Wang M, Zhou G, Han R, Liu X, Weng W, Tan H (2024) Synthetic biology-based bacterial extracellular vesicles displaying BMP-2 and CXCR4 to ameliorate osteoporosis. J Extracell Vesicles 13(4):e12429
Lötvall J, Hill AF, Hochberg F, Buzás EI, Di Vizio D, Gardiner C, Gho YS, Kurochkin IV, Mathivanan S, Quesenberry P, Sahoo S (2014) Minimal experimental requirements for definition of extracellular vesicles and their functions: a position statement from the International Society for Extracellular Vesicles. J Extracell Vesicles 3(1):26913
Lui PP, Leung YT (2022) Practical considerations for translating mesenchymal stromal cell-derived extracellular vesicles from bench to bed. Pharmaceutics 14(8):1684
Luo Z, Cheng X, Feng B, Fan D, Liu X, Xie R, Luo T, Wegner SV, Ma D, Chen F, Zeng W (2024) Engineering versatile bacteria-derived outer membrane vesicles: an adaptable platform for advancing cancer immunotherapy. Adv Sci 11(33):2400049
Lyu TS, Ahn Y, Im YJ, Kim SS, Lee KH, Kim J, Choi Y, Lee D, Kang E, Jin G, Hwang J (2021) The characterization of exosomes from fibrosarcoma cell and the useful usage of Dynamic Light Scattering (DLS) for their evaluation. PLoS One 16(1):e0231994
Malenica M, Vukomanović M, Kurtjak M, Masciotti V, Dal Zilio S, Greco S, Lazzarino M, Krušić V, Perčić M, Jelovica Badovinac I, Wechtersbach K (2021) Perspectives of microscopy methods for morphology characterisation of extracellular vesicles from human biofluids. Biomedicines 9(6):603
Mehanny M, Koch M, Lehr CM, Fuhrmann G (2020) Streptococcal extracellular membrane vesicles are rapidly internalized by immune cells and alter their cytokine release. Front Immunol 11:80
Meng F, Li L, Zhang Z, Lin Z, Zhang J, Song X, Xue T, Xing C, Liang X, Zhang X (2022) Biosynthetic neoantigen displayed on bacteria derived vesicles elicit systemic antitumour immunity. J Extracell Vesicles 11(12):12289
Micoli F, Bagnoli F, Rappuoli R, Serruto D (2021) The role of vaccines in combatting antimicrobial resistance. Nat Rev Microbiol 19(5):287–302
Morandi MI, Busko P, Ozer-Partuk E, Khan S, Zarfati G, Elbaz-Alon Y, Abou Karam P, Napso Shogan T, Ginini L, Gil Z, Regev-Rudzki N (2022) Extracellular vesicle fusion visualized by cryo-electron microscopy. PNAS Nexus 1(4):pgac156
Nagakubo T, Nomura N, Toyofuku M (2020) Cracking open bacterial membrane vesicles. Front Microbiol 10:3026
Nash AK, Auchtung TA, Wong MC, Smith DP, Gesell JR, Ross MC, Stewart CJ, Metcalf GA, Muzny DM, Gibbs RA, Ajami NJ (2017) The gut mycobiome of the Human Microbiome Project healthy cohort. Microbiome 5:1–3
Needham BD, Carroll SM, Giles DK, Georgiou G, Whiteley M, Trent MS (2013) Modulating the innate immune response by combinatorial engineering of endotoxin. Proc Natl Acad Sci 110(4):1464–1469
Pan J, Li X, Shao B, Xu F, Huang X, Guo X, Zhou S (2022) Self-blockade of PD-L1 with bacteria-derived outer-membrane vesicle for enhanced cancer immunotherapy. Adv Mater 34(7):2106307
Parisse P, Rago I, Ulloa Severino L, Perissinotto F, Ambrosetti E, Paoletti P, Ricci M, Beltrami AP, Cesselli D, Casalis L (2017) Atomic force microscopy analysis of extracellular vesicles. Eur Biophys J 46:813–820
Pascucci L, Scattini G (2021) Imaging extracelluar vesicles by transmission electron microscopy: coping with technical hurdles and morphological interpretation. Biochim Biophys Acta Gen Subj 1865(4):129648

Patel D, Wairkar S (2023) Bone regeneration in osteoporosis: opportunities and challenges. Drug Deliv Transl Res 13(2):419–432

Prior JT, Davitt C, Kurtz J, Gellings P, McLachlan JB, Morici LA (2021) Bacterial-derived outer membrane vesicles are potent adjuvants that drive humoral and cellular immune responses. Pharmaceutics 13(2):131

Qing G, Gong N, Chen X, Chen J, Zhang H, Wang Y, Wang R, Zhang S, Zhang Z, Zhao X, Luo Y (2019) Natural and engineered bacterial outer membrane vesicles. Biophys Rep 5:184–198

Rachner TD, Khosla S, Hofbauer LC (2011) Osteoporosis: now and the future. Lancet 377(9773):1276–1287

Reclusa P, Verstraelen P, Taverna S, Gunasekaran M, Pucci M, Pintelon I, Claes N, de Miguel-Pérez D, Alessandro R, Bals S, Kaushal S (2020) Improving extracellular vesicles visualization: from static to motion. Sci Rep 10(1):6494

Sanders H, Feavers IM (2011) Adjuvant properties of meningococcal outer membrane vesicles and the use of adjuvants in Neisseria meningitidis protein vaccines. Expert Rev Vaccines 10(3):323–334

Sangiorgio G, Nicitra E, Bivona D, Bonomo C, Bonacci P, Santagati M, Musso N, Bongiorno D, Stefani S (2024) Interactions of gram-positive bacterial membrane vesicles and hosts: updates and future directions. Int J Mol Sci 25(5):2904

Sartorio MG, Pardue EJ, Feldman MF, Haurat MF (2021) Bacterial outer membrane vesicles: from discovery to applications. Ann Rev Microbiol 75(1):609–630

Shan Y, Lee M, Chang EB (2022) The gut microbiome and inflammatory bowel diseases. Annu Rev Med 73(1):455–468

Sharma S, LeClaire M, Gimzewski JK (2018) Ascent of atomic force microscopy as a nanoanalytical tool for exosomes and other extracellular vesicles. Nanotechnology 29(13):132001

Shyr ZA, Cheng YS, Lo DC, Zheng W (2021) Drug combination therapy for emerging viral diseases. Drug Discov Today 26(10):2367–2376

Sódar BW, Kovács Á, Visnovitz T, Pállinger É, Vékey K, Pocsfalvi G, Turiák L, Buzás EI (2017) Best practice of identification and proteomic analysis of extracellular vesicles in human health and disease. Expert Rev Proteomics 14(12):1073–1090

Sun J, Lin X, He Y, Zhang B, Zhou N, Huang JD (2023) A bacterial outer membrane vesicle-based click vaccine elicits potent immune response against Staphylococcus aureus in mice. Front Immunol 14:1088501

Théry C, Witwer KW, Aikawa E, Alcaraz MJ, Anderson JD, Andriantsitohaina R, Antoniou A, Arab T, Archer F, Atkin-Smith GK, Ayre DC (2018) Minimal information for studies of extracellular vesicles 2018 (MISEV2018): a position statement of the International Society for Extracellular Vesicles and update of the MISEV2014 guidelines. J Extracell Vesicles 7(1):1535750

Toyofuku M, Nomura N, Eberl L (2019) Types and origins of bacterial membrane vesicles. Nat Rev Microbiol 17(1):13–24

Toyofuku M, Schild S, Kaparakis-Liaskos M, Eberl L (2023) Composition and functions of bacterial membrane vesicles. Nat Rev Microbiol 21(7):415–430

Van de Waterbeemd B, Streefland M, Van der Ley P, Zomer B, Van Dijken H, Martens D, Wijffels R, Van der Pol L (2010) Improved OMV vaccine against Neisseria meningitidis using genetically engineered strains and a detergent-free purification process. Vaccine 28(30):4810–4816

van den Berg van Saparoea HB, Houben D, Kuijl C, Luirink J, Jong WS (2020) Combining protein ligation systems to expand the functionality of semi-synthetic outer membrane vesicle nanoparticles. Front Microbiol 11:890

Vestad B, Llorente A, Neurauter A, Phuyal S, Kierulf B, Kierulf P, Skotland T, Sandvig K, Haug KB, Øvstebø R (2017) Size and concentration analyses of extracellular vesicles by nanoparticle tracking analysis: a variation study. J Extracell Vesicles 6(1):1344087

Wang S, Gao J, Li M, Wang L, Wang Z (2018) A facile approach for development of a vaccine made of bacterial double-layered membrane vesicles (DMVs). Biomaterials 28-38:187

Wang D, Liu C, You S, Zhang K, Li M, Cao Y, Wang C, Dong H, Zhang X (2020) Bacterial vesicle-cancer cell hybrid membrane-coated nanoparticles for tumor specific immune activation and photothermal therapy. ACS Appl Mater Interfaces 12(37):41138–41147

Wang X, Lin S, Wang L, Cao Z, Zhang M, Zhang Y, Liu R, Liu J (2023) Versatility of bacterial outer membrane vesicles in regulating intestinal homeostasis. Sci Adv 9(11):eade5079

Wang J, Liang S, Chen S, Ma T, Chen M, Niu C, Leng Y, Wang L (2024a) Bacterial outer membrane vesicle-cancer cell hybrid membrane-coated nanoparticles for sonodynamic therapy in the treatment of breast cancer bone metastasis. J Nanobiotechnol 22(1):328

Wang X, Peng J, Cai P, Xia Y, Yi C, Shang A, Akanyibah FA, Mao F (2024b) The emerging role of the gut microbiota and its application in inflammatory bowel disease. Biomed Pharmacother 179:117302

Wo J, Lv ZY, Sun JN, Tang H, Qi N, Ye BC (2023) Engineering probiotic-derived outer membrane vesicles as functional vaccine carriers to enhance immunity against SARS-CoV-2. iScience 26(1):105772

Yang Z, Hua L, Yang M, Liu SQ, Shen J, Li W, Long Q, Bai H, Yang X, Ren Z, Zheng X (2021) RBD-modified bacterial vesicles elicited potential protective immunity against SARS-CoV-2. Nano Lett 21(14):5920–5930

Yang Z, Atiyas Y, Shen H, Siedlik MJ, Wu J, Beard K, Fonar G, Dolle JP, Smith DH, Eberwine JH, Meaney DF (2022) Ultrasensitive single extracellular vesicle detection using high throughput droplet digital enzyme-linked immunosorbent assay. Nano Lett 22(11):4315–4324

Yue Y, Xu J, Li Y, Cheng K, Feng Q, Ma X, Ma N, Zhang T, Wang X, Zhao X, Nie G (2022) Antigen-bearing outer membrane vesicles as tumour vaccines produced in situ by ingested genetically engineered bacteria. Nat Biomed Eng 6(7):898–909

Zariri A, Beskers J, van de Waterbeemd B, Hamstra HJ, Bindels TH, van Riet E, van Putten JP, van der Ley P (2016) Meningococcal outer membrane vesicle composition-dependent activation of the innate immune response. Infect Immun 84(10):3024–3033

Zhai Y, Ma Y, Pang B, Zhang J, Li Y, Rui Y, Xu T, Zhao Y, Qian Z, Gu Y, Li S (2021) A cascade targeting strategy based on modified bacterial vesicles for enhancing cancer immunotherapy. J Nanobiotechnol 19:1–18

Zhang Y, Chen Y, Lo C, Zhuang J, Angsantikul P, Zhang Q, Wei X, Zhou Z, Obonyo M, Fang RH, Gao W (2019) Inhibition of pathogen adhesion by bacterial outer membrane-coated nanoparticles. Angew Chem Int Ed 58(33):11404–11408

Zhang M, Wang L, Liu J, Pang Y (2022) Envelope virus-mimetic nanovaccines by hybridizing bioengineered cell membranes with bacterial vesicles. iScience 25(6):104490

Zhao J, Yu B, Li L, Guo S, Sha X, Ru W, Du GQ, Xue JY (2024) Outer membrane vesicle–cancer hybrid membrane coating indocyanine green nanoparticles for enhancing photothermal therapy efficacy in tumors. ACS Biomater Sci Eng 10(12):7619–7631

Zheng B, Xu J, Chen G, Zhang S, Xiao Z, Lu W (2019) Bacterium-mimicking vector with enhanced adjuvanticity for cancer immunotherapy and minimized toxicity. Adv Funct Mater 29(33):1901437

Zhu Z, Antenucci F, Villumsen KR, Bojesen AM (2021) Bacterial outer membrane vesicles as a versatile tool in vaccine research and the fight against antimicrobial resistance. MBio 12(4):10–1128

Zhuang Q, Xu J, Deng D, Chao T, Li J, Zhang R, Peng R, Liu Z (2021) Bacteria-derived membrane vesicles to advance targeted photothermal tumor ablation. Biomaterials 268:120550

Zou MZ, Li ZH, Bai XF, Liu CJ, Zhang XZ (2021) Hybrid vesicles based on autologous tumor cell membrane and bacterial outer membrane to enhance innate immune response and personalized tumor immunotherapy. Nano Lett 21(20):8609–8618

Hybrid Cell Membrane Engineered Platform for Theranostic Application

4

Meng Sun and Jinfeng Zhang

Abstract

In recent years, hybrid cell membrane (HCM)-engineered platforms have recently emerged as powerful tools in advancing disease diagnosis and therapy, owing to their enhanced biocompatibility, intrinsic targeting capabilities, improved in vivo stability, and prolonged circulation time. These platforms are poised to revolutionize the fields of drug delivery and disease management, due to their ability to mimic the natural characteristics of biological membranes. Notably, they have significantly propelled the progress of nanomedicine by integrating cellular features, such as cell-specific targeting and immune modulation, into engineered nanosystems. This chapter provides an overview of various HCM-engineered platforms derived from different cell membrane sources and reviews the fabrication techniques for these biomimetic systems. The applications of HCM-based platforms in disease diagnosis and therapy are also discussed, focusing on conditions such as cancer, inflammation, bacterial infections, and tissue regeneration. Additionally, the chapter outlines the future prospects and current challenges in this rapidly evolving field, aiming to inspire further research across multiple disciplines and to facilitate the advancement of personalized and precision medicine.

Keywords

Hybrid cell membrane · Extracellular vesicles · Liposomes · Biomimetic nanoparticles · Diagnosis and treatment

M. Sun · J. Zhang (✉)
Key Laboratory of Molecular Medicine and Biotherapy, School of Life Sciences, Beijing Institute of Technology, Beijing, China
e-mail: jfzhang@bit.edu.cn

Z. Hussain, R. Pei (eds.), *Cell Membrane Engineering for Advancing Cell-Based and Nanotechnology-Based Therapies*, Biomaterials, Bioengineering and Sustainability 6, https://doi.org/10.1007/978-3-031-93425-4_4

4.1 Introduction

With the rapid development of nanotechnology, there has been a significant push toward using nanomaterials to diagnose and treat various diseases, such as cancer (Guan et al. 2024; Lu et al. 2025), cardiovascular disease (Gu et al. 2024a; Nankivell et al. 2024), brain disease (Han et al. 2025; Quan et al. 2025), inflammatory diseases (Feng et al. 2024; Huang et al. 2024), *etc.* Conventional medicines suffer from low permeability, poor bioavailability, short half-life, and severe side effects, which cannot sufficiently exert the desired therapeutic effect. Nanodrugs, with reduced toxicity, improved drug stability, and outstanding pharmacokinetics, have become one of the most promising strategy to improving the properties of traditional drugs (Fang et al. 2023; Zhang et al. 2024). Although promising, nanodrugs' clinical use is still limited by their poor biocompatibility, weak targeting ability, and limited immune evasion (Lopes et al. 2023; Liao et al. 2020; Pan et al. 2024), highlighting the urgent needs to develop novel and efficient strategies to optimize the performance of nanomedicines and accelerate their clinical translation and application.

Biomimetic nanodrugs with cell membranes isolated from stem cells (Tian et al. 2024), red blood cells (Gu et al. 2024b), platelets (Xu et al. 2024a), immune cells (Duan et al. 2023; Li et al. 2024a), and cancer cells (Jiang et al. 2023) have been of great concern to meet therapeutic needs. Various cell membranes endow nanodrugs superior biological functionalities such as enhanced biocompatibility, intrinsic targeting, increases *in vivo* stability, and extended circulation time. In addition, the platform with a hybrid cell membrane (HCM) derived from different cells, extracellular vesicles, or multifunctional liposomes, can endow nanodrugs with flexibility and controllable functions (Yu et al. 2024; Sun et al. 2023; Mondal et al. 2023). For example, Zhang et al. reported a leukocyte/platelet hybrid membrane-camouflaged nanoparticle, with dual-targeting ability for triple-negative breast cancer (TNBC) therapy, which could naturally enter cancer cells and release the loaded phototherapeutic agent (IR780) and chemotherapeutics (DOX) to obtain synergistic and pleiotropic effects against TNBC (Zhang et al. 2021a). Therefore, HCM provides a promising platform equipped with diagnostic and therapeutic functions for diseases (Fig. 4.1).

Building on recent advancements in this emerging field, this chapter aims to summarize the representative works and current advances of the HCM-engineered platform in nanodrugs for theranostic applications. Firstly, we describe the different types of HCM-engineered platforms based on a fusion of cell membrane and cell membrane, cell membrane and extracellular vesicles, cell membrane and liposome, and cell membrane and bacterial outer membrane. Then we review the fabrication methods of biomimetic platform. Next, a specific emphasis on the theranostic applications of the biomimetic nanostructures towards various diseases with representative examples. Finally, this chapter discusses the current challenges and prospects of the biomimetic platform in clinical applications. We hope to provide researchers and valuable insights from diversified disciplines with HCM-engineered platforms in biomedicine and spark broad interest among researchers from this promising nanoplatforms research field.

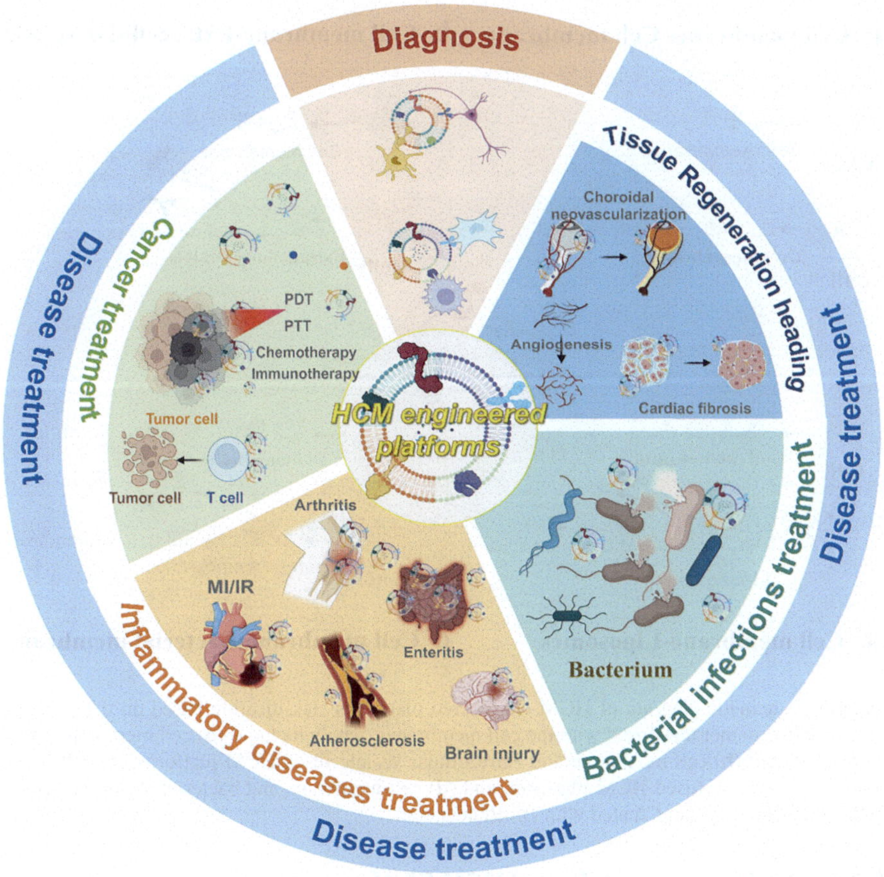

Fig. 4.1 Schematic representation of HCM engineered platforms for various theranostic applications. (Source: Created with BioRender.com)

4.2 Classification of HCM Engineered Platforms Based on Membrane Sources

Different types of cell membranes play a critical role in the diagnosis and therapeutic management of diseases in biomimetic nanostructures. Therefore, based on the different origins of the membranes fused with the cell membranes, HCM-engineered platforms can be classified into four main categories: (1) cell membrane-fused HCM platform; (2) cell membrane and extracellular vesicle-fused HCM platform; (3) cell membrane and liposome-fused HCM platform; and 4) cell membrane and bacterial membrane-fused HCM platform (Fig. 4.2).

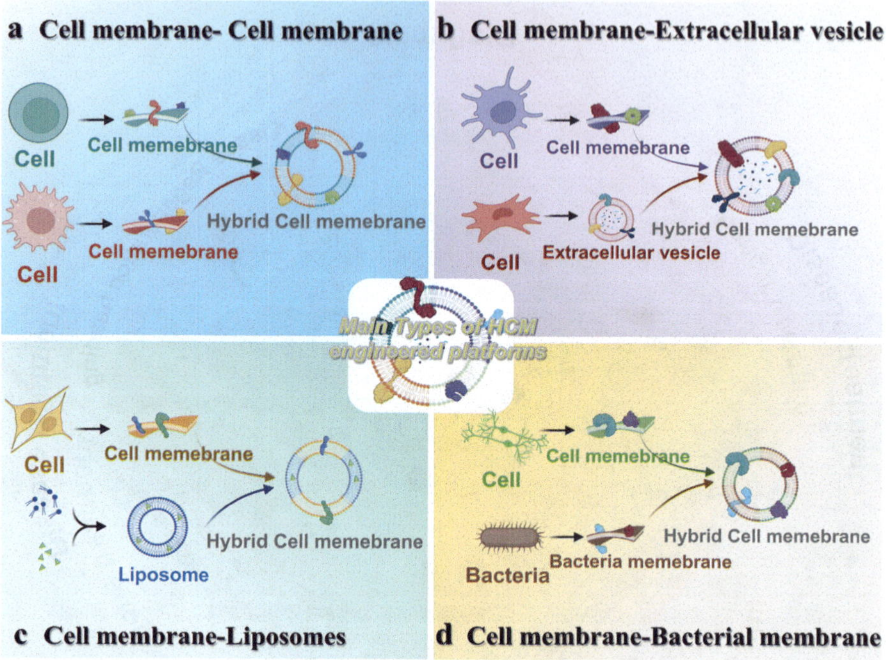

Fig. 4.2 Schematic diagram of HCM engineered platforms classification based on the different origins of the membranes fused with the cell membranes, which includes (**a**) cell membrane-fused HCM platform; (**b**) cell membrane and extracellular vesicle-fused HCM platform; (**c**) cell membrane and liposome-fused HCM platform; and (**d**) cell membrane and bacterial membrane-fused HCM platform. (Source: Created with BioRender.com)

4.2.1 Cell Membrane-Fused HCM Platform

The cell membrane inherits the functional characteristics of the parent cell, endowing biomimetic nanomedicines with diverse biological activities (Dehaini et al. 2017; Bu et al. 2019; Dong et al. 2023; Chen et al. 2024a). Numerous studies have demonstrated that red blood cells (RBCs) membranes can evade immune detection and extend the circulation time of drugs in the body; macrophage membranes can target sites of inflammatory damage and modulate the disease microenvironment; stem cell membranes exhibit both homing and inflammation-targeting capabilities; and platelet membranes tend to aggregate at injury sites, while also targeting tumor cells through P-selectin-mediated recognition of the CD44 protein, which is highly expressed on tumor cell surfaces. Therefore, the fusion of different cell membranes not only enhances the biocompatibility of nanomedicines but also boosts their targeting capabilities to further improve therapeutic efficacy. Jiang et al. fused RBCs membranes, which prolong circulation time, with MCF-7 cell membranes, which have homologous targeting effects, to prepare melanin-loaded nanomedicines (Melanin@RBC-M) for enhancing cancer therapeutic efficacy in photothermal therapy (PTT) (Jiang et al. 2019). Additionally, this study found that by adjusting the ratio of the two types of cell membranes, the nanomedicine's tendency toward enhancing homologous targeting and improving circulation time could be modulated. This approach endowed the nanomedicine with flexibility and controllability, synergistically improving its tumor-killing effect in PTT therapy. Recently, Chen et al. synthesized a red blood cell-platelet fusion membrane (RPM) camouflaged Ga-based ICD inducer, which exhibited high atomic utilization and excellent Ga^{3+} on-demand release capabilities, demonstrating significant antitumor therapeutic effects and preventing tumor recurrence and metastasis (Chen et al. 2024b).

4.2.2 Cell Membrane and Extracellular Vesicle-Fused HCM Platform

Extracellular vesicles (EVs) are nanoscale membrane-bound entities secreted by living cells, carrying various bioactive molecules (DNA, RNA, lipids, proteins, *etc.*) from the parent cells. Compared to cell membranes, EVs also inherit the beneficial biological properties of their parent cells, such as low immunogenicity, high biocompatibility, strong targeting ability, and ease of modification (Li et al. 2022a; Sun et al. 2024; Kalluri and LeBleu 2020; Chavda et al. 2024). Notably, EVs hold significant potential as therapeutic agents for drug delivery, gene therapy, immune modulation, and disease diagnostics. The cell membrane-EVs fusion platform enhances the biocompatibility and targeting of nanoparticles, further facilitating their therapeutic effects (Ducrot et al. 2023; Lai et al. 2022). Xie et al. fused platelet membranes with M2 macrophage derived EVs (M2-EVs) to create hybrid nanovesicles (P-M2EV) for targeting and treating atherosclerosis (AS). The platelet membrane, through glycoprotein Ib (GPIb), targets macrophages in atherosclerotic plaques, while M2-derived EVs (M2-EVs) regulate macrophage polarization by delivering miRNAs and other bioactive molecules, thereby alleviating disease progression (Xie et al. 2024a). Additionally, studies have shown that platelet membranes retain adhesive proteins and the natural ability to target damaged vascular systems. Li et al. fused platelet membranes with EVs from stem cells promoting angiogenesis, creating a biomimetic nanomedicine (P-EV) that targeted myocardial ischemia-reperfusion injury sites and promoted vascular regeneration (Li et al. 2022b).

4.2.3 Cell Membrane and Liposome-Fused HCM Platform

Liposomes are artificially engineered lipid bilayers that mimic cell membranes. They offer advantages such as high stability, excellent drug-carrying capacity, outstanding biocompatibility, and scalable production, which make them widely used in the preparation of membrane fusion platforms (Wang et al. 2015; Lv et al. 2020; Liu et al. 2023; Guo et al. 2024). Li et al. developed a nanosystem, H-MnO2- DOX-Col NPs, to enhance drug penetration, alleviate hypoxia, and improve chemotherapy in breast cancer. The researchers fused the cell membrane of RAW264.7 cells with pH-responsive liposomes to enable inflammation-targeting and drug-responsive release. This nanosystem reduced cardiac toxicity and promoted the penetration and therapeutic efficacy of the loaded chemotherapeutic agent, DOX, at the tumor site (Li et al. 2023). Gao and Li's team developed a pH-responsive liposomal membrane that undergoes charge inversion and fused it with cancer cell membranes with homologous targeting ability. This system was utilized for the co-delivery of paclitaxel (DTX) and phosphoglycerate mutase 1 (PGAM1) siRNA, enabling synergistic RNA interference and achieving efficient management of non-small cell lung cancer (Zhang et al. 2021b).

4.2.4 Cell Membrane and Bacterial Membrane-Fused HCM Platform

Prokaryotic bacterial membranes are also significant components for the preparation of cell membrane fusion platforms, playing a key role in enhancing and activating the immune function especially (Ding et al. 2023; Peng et al. 2024b). Chen et al. prepared a eukaryotic-prokaryotic vesicle (EPV) nanoplatform loaded with poly (lactic-co-glycolic acid)-indocyanine green (ICG) moiety (PI) to combine local PTT with antitumor effects. This nanoplatform leverages the natural adjuvant properties of melanoma cell membrane antigens and attenuated *Salmonella* membrane vesicles, facilitating potent immunotherapy and offering a novel approach for tumor vaccine design (Chen et al. 2020). Additionally, bacterial outer membrane vesicles (OMVs) can also fuse with plant cell membranes to produce cancer vaccines. As reported by Zhuang et al., hybrid vesicles (BPN) formed by the fusion of plant thylakoid membranes with OMVs could activate immune cells and improve the immunosuppressive microenvironment (Zhuang et al. 2022). Besides, BPN could synergize with photodynamic therapy to effectively inhibit tumor progression and metastasis. This study provides a new approach for developing multifunctional membrane-based hybrid systems for efficient cancer treatment.

Beyond the types mentioned above, there are additional biomimetic membrane fusion mechanisms that do not directly involve the cell membrane, such as the fusion of EVs and EVs as well as the fusion between EVs and liposomes (Zhang et al. 2023a; Kang et al. 2023; Xie et al. 2024b; Song et al. 2021; Hu et al. 2021a; Liang et al. 2022). These processes further expand the biomedical use of biomimetic nanoparticles. The HCM-engineered platforms exhibit promising potential for a wide array of biomedical applications.

4.3 Different Strategies for Preparation of HCM Engineered Platforms

It is essential to establish robust nanoplatform preparation strategies that ensure standardization, reproducibility, scalability, and productivity to advance the translational potential of biomimetic nanomedicine. Therefore, we analyzed the research methods in the current study and categorized the preparation strategies into three primary types, including physical strategies, chemical strategies, and biological strategies.

4.3.1 Physical Preparation Strategy

Co-extrusion represents one of the most widely employed physical preparation methods. This technique involves physically mixing different types of membrane materials in a specific ratio, which are then sequentially extruded through filter membranes with varying pore sizes, facilitating membrane fusion (Patty and Frisken 2003). The pore size of the filter membrane is directly related to the size of the resulting biomimetic nanomedicine. For instance, Yu et al. used co-extrusion to fuse TIGIT-expressing cell membranes with platelet membranes, while simultaneously loading the chemotherapy drug oxaliplatin (OXA), resulting in the creation of drug-loaded fusion nanovesicles (O-TPNV) (Yu et al. 2022). These nanodrugs specifically target postoperative wounds, interact with circulating tumor cells, block the CD155/TIGIT pathway, activate immune cells, and effectively suppress recurrence and metastasis of TNBC following surgery.

Sonication is often used as an auxiliary method for membrane fusion preparation. For example, in the preparation of biomimetic nanodrugs to inhibit B16F10 and 4T1 tumor growth, Han et al. first used sonication on RBCs membranes and tumor cell membranes and then employed the co-extrusion method to prepare a fusion membrane system for antigen delivery (Han et al. 2019). This system triggered the activation of immune cells around the tumor, achieving a highly effective anti-tumor response.

In addition to the commonly used membrane fusion methods mentioned above, freeze-thaw and co-incubation are also often employed as supplementary techniques to efficiently promote membrane fusion and facilitate the preparation of biomimetic nanodrugs. In the study by Li et al., they first prepared platelet membranes using the freeze-thaw method and then co-extruded them with mesenchymal stem cells (MSCs) exosomes to create a fusion system (P-EV) for vascular regeneration therapy (Li et al. 2021a).

4.3.2 Chemical Preparation Strategy

Chemical methods typically involve the use of chemical reagents to induce interactions between membrane materials and promote membrane fusion. Currently, polyethylene glycol (PEG) is widely used not only as an inducer in cell fusion but also for membrane fusion. Sato et al. investigated the fusion potential of exosomes and liposomes induced by PEG with different molecular weights and demonstrated the effective membrane fusion capability of PEG (Sato et al. 2016). They found that PEG8000 was the most efficient under these conditions, possibly because the larger PEG molecules can form more stable bridging structures between membranes, thereby enhancing the fusion efficiency.

Additionally, PEG is commonly used to facilitate the preparation of fusion membranes. Hu et al. first treated a mixture of human umbilical cord MSCs exosomes and platelet membranes with 5% PEG, followed by co-extrusion to produce mixed vesicles with a particle size of around 200 nm (Hu et al. 2021b). This treatment enhanced the uptake of the fusion membrane by endothelial and cardiomyocyte cells, promoting anti-inflammatory repair effects in myocardial injury.

4.3.3 Biological Preparation Strategy

Biological methods often utilize biomolecules to achieve membrane fusion. For example, viral fusion proteins or recombinant proteins possess inherent membrane fusion capabilities. Gao et al. employed specific viral membrane fusion proteins co-expressed in cells, along with lectin-neuraminidase (HN) proteins, to obtain engineered membranes (Gao et al. 2019). Subsequently, membrane fusion mediated by viral membrane proteins facilitated the fusion of biological membranes with exosomes, thereby enabling the rapid detection of specific miRNAs in tumor-derived exosomes.

Moreover, biological methods are commonly employed to impart diverse functionalities to fusion membrane components, enhancing their targeting ability, biocompatibility, and ultimately improving their therapeutic effects. For instance, Rao and Chen's group utilized genetic engineering strategy to fabricate tumor cell membrane vesicles expressing high levels of PD-1 and SIRPα, which were fused with PD-L1 and CD47 (Meng et al. 2021). This resulted in gene-programmable fusion cell vesicles (Fus-CV) that simultaneously block the innate immune checkpoint CD47 and the adaptive immune checkpoint PD-L1, thus activating tumor immunotherapy and demonstrating significant antitumor efficacy. In a similar vein, Liang et al. used gene editing to produce exosomes expressing a cartilage cell-affinity peptide (CAP) on their surface (Liang et al. 2022). These exosomes were subsequently fused with liposomes to target and deliver CRISPR/Cas9 plasmids, specifically knocking down matrix metalloproteinase 13 (MMP-13) in chondrocytes, effectively alleviating osteoarthritis. While biosynthetic methods are widely used to fuse liposomes with exosomes, they also offer a novel fabrication strategy for HCM-engineered platforms.

Overall, more novel and efficient preparation strategies are continuously emerging, further expanding the applicability of the HCM-engineered platforms and advancing their research progress in biomedical and clinical translation applications.

4.4 HCM Engineered Platform for Theranostic Application

4.4.1 HCM Engineered Platform for Diagnosis

Biomimetic membrane-coated nanoparticles play a crucial role in enhancing the specificity and sensitivity of disease diagnosis. Reports demonstrated that circulating tumor cells (CTCs), which are cancerous cells shed from solid tumors into the circulation, are critical for early cancer detection, tracking, and prognosis. In the capture and detection of CTCs, both specificity and sensitivity are paramount challenges. To address that, Rao et al. developed a hybrid membrane composed of platelets and leukocytes to coat immunomagnetic beads (IMBs), with surface modification of CTC-specific antibodies (Rao et al. 2018). This approach aimed to enhance the binding affinity of platelets to cancer cells, minimize interactions with leukocytes, and improve targeting, purity, and sensitivity in CTC capture (Fig. 4.3a). As shown

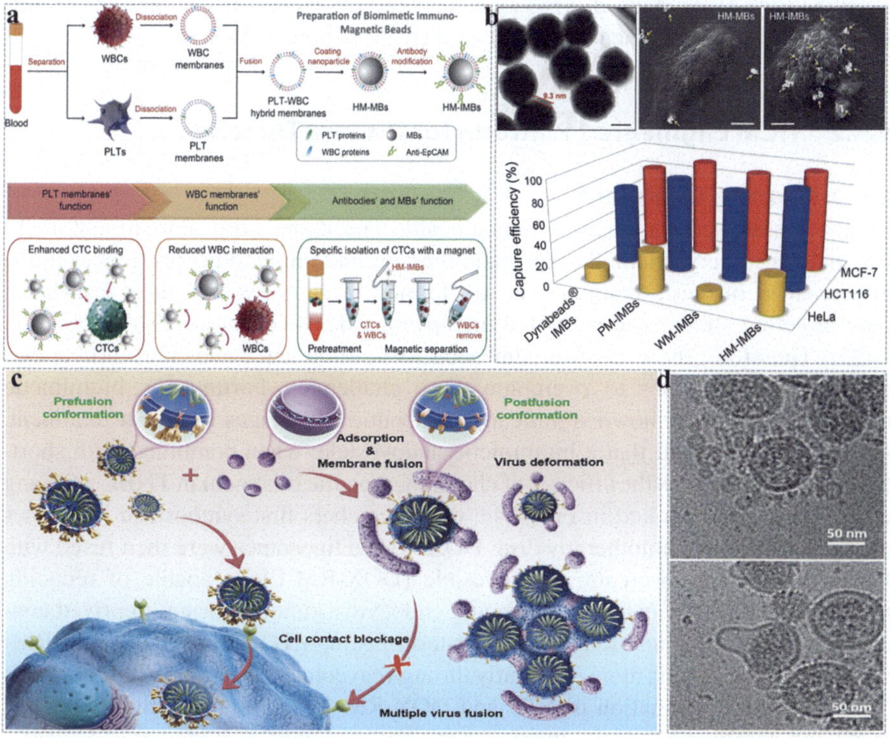

Fig. 4.3 (**a**) Schematic representation of the preparation process and capture mechanism of cell membrane-fusion nanoparticle HM-IMB for fast detection of circulating tumor cells (CTCs). (**b**) TEM morphology of HM-IMB and a bar chart showing the capture efficiency of different types of tumor cells. (Reproduced with permission from Rao et al. (2018), copyright 2018, Wiley-VCH GmbH). (**c**) Schematic illustration of the antiviral mechanism of nanodecoys. (**d**) TEM morphology of nanodecoys after fusion with the virus. (Reproduced with permission from Li et al. (2024b), copyright 2024, Elsevier B.V.)

in Fig. 4.3b, following antibody modification on the hybrid membrane surface, HM-IMBs demonstrated a significantly increased ability to bind CTCs, resulting in a marked improvement in capture efficiency. Moreover, compared to single-component membrane-coated particles, HM-IMBs exhibited enhanced CTC capture rates across various cancer cell types (HeLa, HCT116, and MCF-7), advancing the clinical translation of bionic cell membrane-coated systems.

In addition, Li et al. reported an artificial nanodecoy that combines membrane fusion strategies to bind and inactivate viruses, offering a new approach to the development of antiviral therapeutic and diagnostic nanosystems (Fig. 4.3c) (Li et al. 2024b). The researchers fabricated the artificial cell membrane glycocalyx nanodecoy by using sulfate liposomes loaded with hydrogenated soy phosphatidylcholine (HSPC) and cholesterol sulfate (CS). This nanodecoy could interact with the spike protein of the coronavirus, blocking viral-cell interactions. Furthermore, the nanodecoys, mimicking the structure of the cell membrane, fuse with surrounding viruses or other nanodecoys, leading to conformational changes and viral inactivation (Fig. 4.3d). This in vivo system, which captures, fuses, and inactivates viruses, expands the clinical potential of antiviral nanomaterials.

4.4.2 HCM Engineered Platform for Disease Therapy

4.4.2.1 Cancer Treatment

Cancer poses a serious threat to human health. Traditional clinical treatment methods include surgery, chemotherapy, radiotherapy, and immunotherapy, however, the effectiveness of these treatments is still limited by high toxicity, drug resistance, poor targeting ability, and restricted therapeutic effects (He et al. 2024; Zhu et al. 2023). Therefore, there is an urgent need to develop safer and more efficacious therapeutic approaches to overcome these challenges. Fortunately, biomimetic nanomedicines have shown significant therapeutic advantages in cancer treatment. Chiang et al. reported that a biomimetic nanovesicle, when combined with short-term fasting, enhances the efficacy of chemotherapeutic treatment in TNBC. (Chiang et al. 2024) As illustrated in Fig. 4.4a, the researchers first synthesized liposomes encapsulating the chemotherapy drug DOX. These liposomes were then fused with RBCs membranes to create a nanovesicle (DOX-RM-LPs) capable of reducing macrophage phagocytosis and enhancing endocytosis under nutrient-deprived conditions. The study found that after short-term fasting (STF) treatment, DOX-RM-LPs entered tumor cells more efficiently through caveolae/raft-mediated endocytosis. Moreover, the combination of STF and DOX-RM-LPs suppressed the tumor cell cycle and aerobic glycolysis, increased the production of reactive oxygen species (ROS) and autophagosomes, and significantly improved the anti-tumor effect (Fig. 4.4b). Besides, Liu et al. fused the membranes of breast cancer cells with platelet membranes to create hybrid vesicles (CPMV), which were further loaded with therapeutic miRNAs to enhance the tumor's sensitivity to chemotherapy (Fig. 4.4c) (Liu et al. 2021). The study showed that CPMV maintained prolonged circulation and enhanced specific accumulation in tumor models. This approach

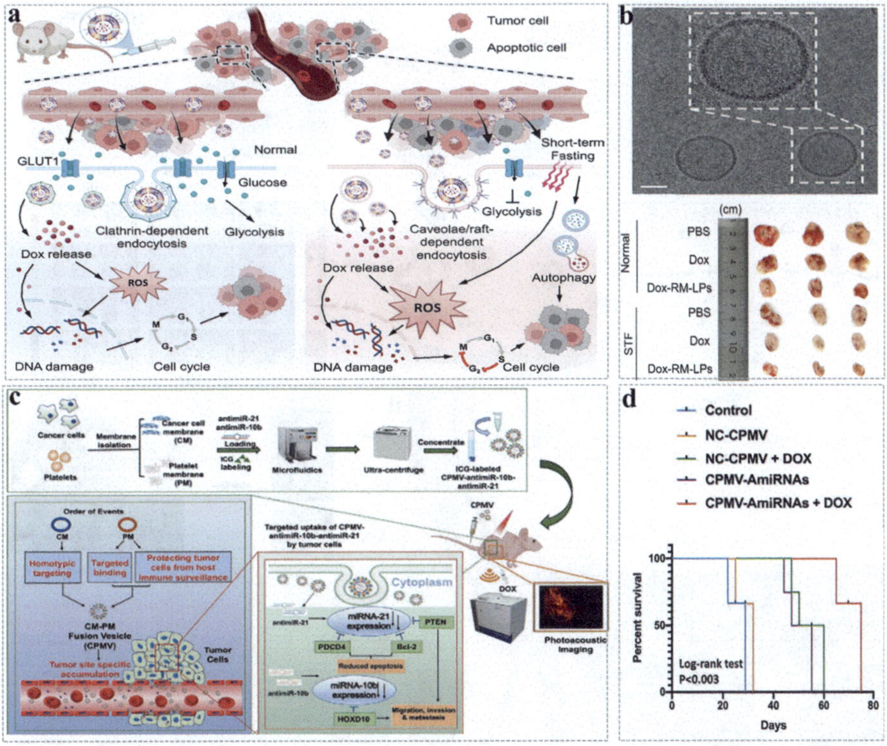

Fig. 4.4 (**a**) Schematic representation of the synthesis of biomimetic nanovesicles DOX-RM-LPs and the mechanism for enhanced therapeutic efficacy against triple-negative breast cancer. (**b**) TEM morphology and anticancer effects of DOX-RM-LPs. (Reproduced with permission from Chiang et al. (2024), copyright 2024, American Chemical Society). (**c**) Schematic representation of the preparation of drug-loaded hybrid vesicles (CPMV) and their mechanism of anticancer activity. (**d**) Survival curve after CPMV treatment. (Reproduced with permission from Liu et al. (2021), copyright 2021, Wiley-VCH GmbH)

increased the sensitivity of TNBC to DOX (Fig. 4.4d), promoting anti-tumor effects and improving the survival rates in the model group. These biomimetic nanovesicles have significant clinical therapeutic potential and hold promise for improving personalized treatments for malignant tumors.

The blood-brain barrier (BBB) presents a critical challenge in brain tumor treatment. To address this challenge, a research team developed a hybrid membrane-coated biomimetic nanocomposite for the efficient treatment of early-stage glioma. As shown in Fig. 4.5a, the team prepared a HCM composed of membranes from brain metastatic breast cancer and glioma cells to load nanoparticles containing gallic acid (GA) and indocyanine green (ICG), resulting in a biomimetic nanocomposite (HMGINP) with homologous tumor-targeting capabilities and great BBB penetration (Chi et al. 2023). Combined with PTT, the HMGINP efficiently crossed the BBB, reduced toxicity, and enhanced the tumor's sensitivity to heat, enabling

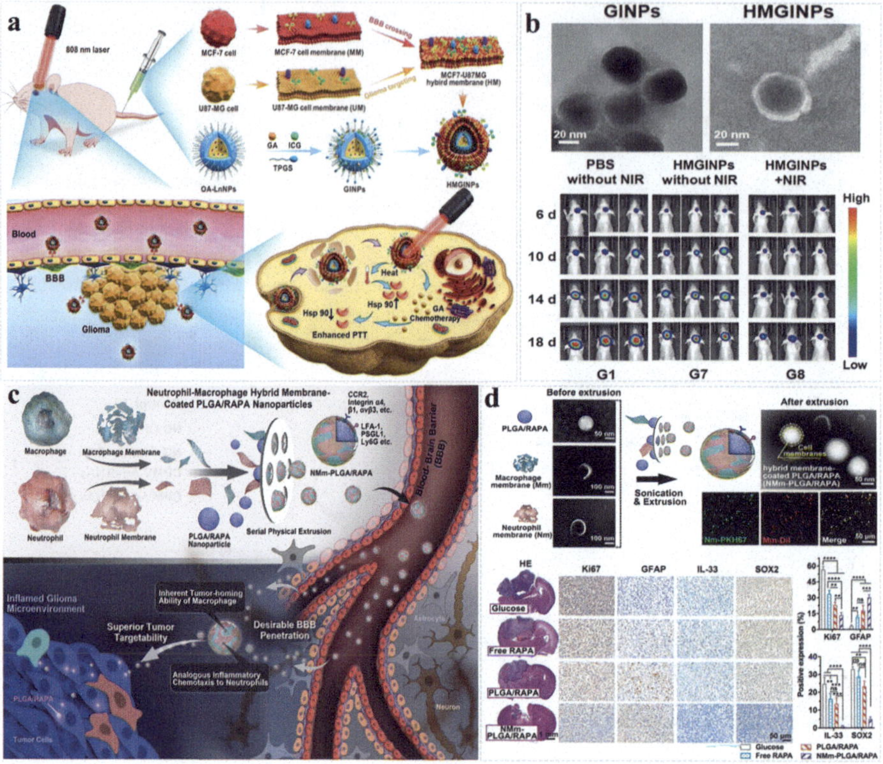

Fig. 4.5 (**a**) Schematic illustration of the preparation of biomimetic nanovesicles HMGINP and their mechanism for penetrating the BBB and therapeutic efficacy against cancer. (**b**) TEM morphology and fluorescence imaging of brain tumor treatment of HMGINP. (Reproduced with permission from Chi et al. (2023), copyright 2023, Wiley-VCH GmbH). (**c**) Schematic illustration of the preparation and the targeting and anticancer mechanisms of fusion membrane PLGA/RAPA nanoparticle platform. (**d**) TEM morphology and the anticancer tissue sectioning effects of PLGA/RAPA. (Reproduced with permission from Yin et al. (2022), copyright 2022, Elsevier B.V.)

the safe and efficient treatment of early-stage glioma (Fig. 4.5b). In the tumor-inflammation microenvironment, glioma cells often recruit peripheral macrophages, which promote tumor growth and metastasis. Based on this characteristic, Yin et al. selectively fused macrophage membranes with neutrophil membranes to combine homing targeting and inflammatory chemotaxis (Yin et al. 2022). Additionally, they loaded the nanocomposite with rapamycin (RAPA) loaded in poly (lactic-co-glycolic acid) (PLGA), a polymer that passively crosses the BBB. The nanocomposite responds to the tumor's heterogeneous microenvironment to deliver drugs, significantly enhancing anti-tumor therapy (Fig. 4.5c, d).

Fusion membrane biomimetic nanoparticles have found widespread applications in PTT for tumors. A team of researchers prepared fusion vesicles composed of RBCs membranes and melanoma cell membranes, along with hollow copper nanoparticles loaded with DOX (Wang et al. 2018). These fusion membrane-coated nanoparticles (DCuS@[RBC-B16]) exhibited excellent biocompatibility, long circulation times, and specific targeting capabilities. Under 1064 nm laser exposure, the biomimetic nanoparticles demonstrated exceptional synergistic photothermal and chemotherapeutic effects, significantly inhibiting melanoma growth (Fig. 4.6a, b). This development further advanced the use of nanoparticle-based personalized cancer therapies. To further enhance the penetration of the light source, Yang et al. developed iron-gallic acid coordination polymer nanodots (FeCNDs) that exhibit excellent photothermal effects and Fenton reaction catalytic activity (Yang et al. 2022). The surface of FeCNDs was functionalized with a hybrid membrane composed of platelet membranes and pH-responsive membranes, resulting in biomimetic nanoparticles with significant anti-tumor efficacy. Compared to the control group, these nanoparticles improved penetration efficiency by 55%, demonstrating excellent targeting properties and effectiveness in breast cancer treatment (Fig. 4.6c, d).

Immunotherapy aims to boost the body's immune system, relying on its inherent immune functions to eliminate cancer cells. However, the immune-suppressing tumor microenvironment hinders the effectiveness of immune responses. Research

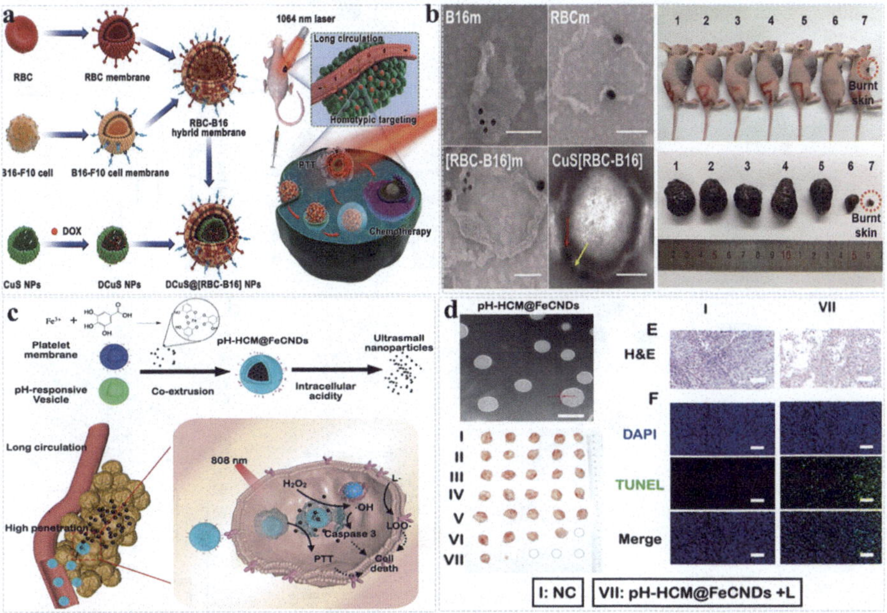

Fig. 4.6 (**a**) Schematic representation of the preparation and the anti-tumor effects with PTT of biomimetic nanoparticles (DCuS@[RBC-B16]). (**b**) TEM morphology of DCuS@[RBC-B16] and the changes in tumor size during treatment. (Reproduced with permission from Wang et al. (2018), copyright 2018, American Chemical Society). (**c**) Schematic illustration of the preparation of pH-responsive hybrid nanodrug (FeCNDs) and their anticancer mechanism in combination with PTT. (**d**) TEM morphology of FeCNDs and the changes in tumor morphology, tissue sectioning, as well as cell proliferation and apoptosis in anticancer activity. (Reproduced with permission from Yang et al. (2022), copyright 2022, Ivyspring International Publisher)

has shown that tumor-associated macrophages (M2 macrophages) play a critical role in immune suppression, and autophagy further promotes the polarization of macrophages to adopt M2 phenotype. As shown in Fig. 4.7a, Chen's team utilized the FDA-approved antidote, Prussian blue, to load the autophagy inhibitor Hydroxychloroquine (HCQ), and surface-modified the particles with mannose (Hou et al. 2022). The resulting nanoparticles (Man-HMPB/HCQ) can release iron ions to induce the re-polarization of M2 macrophages to the M1 macrophages, inhibit autophagy, and extend circulation time in the body. Based on this, the team fused macrophage membranes with thylakoid membranes and coated them on the nanoparticles. This further endowed the biomimetic nanoparticles with homing-targeting properties, alleviated hypoxia, promoted T-cell activation, and inhibited tumor growth (Fig. 4.7b). OMVs have become an important tool for immune activation in tumor treatment. As reported by Zhang, when Escherichia coli (*E. coli*) OMVs were fused with tumor cell membranes, the resulting mTOMVs efficiently

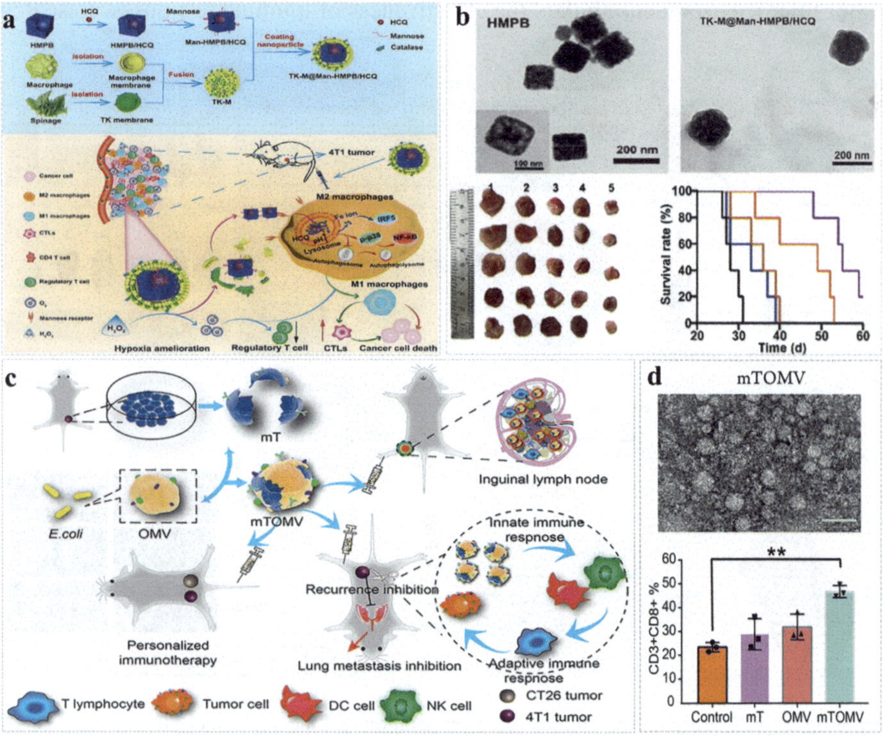

Fig. 4.7 (**a**) Schematic illustration of biomimetic nanoparticles Man-HMPN/HCQ for tumor immunotherapy. (**b**) TEM morphology of each component and their anticancer effects. (Reproduced with permission from Hou et al. (2022), copyright 2022, Wiley-VCH GmbH). (**c**) Schematic diagram of the preparation of mTOMV and the various anticancer mechanisms. (**d**) TEM morphology of mTOMV and their therapeutic effects after immunotherapy. (Reproduced with permission from Zou et al. (2021), copyright 2021, American Chemical Society)

accumulated in homologous tumors (Zou et al. 2021). These mTOMVs inhibited lung metastasis and activated adaptive immune responses within the homologous tumor, providing a personalized immunotherapy strategy for cancer treatment (Fig. 4.7c, d).

In addition, Xiong et al. used fusion vesicles composed of ID8 ovarian cancer cell membranes and RBCs membranes, and loaded them with Fe_3O_4 nanoparticles conjugated with ICG (Xiong et al. 2021). The resulting biomimetic nanoparticles, Fe_3O_4-ICG@IRM, not only induced photothermal tumor cell death but also activated CD8+ T cell activity. This combination of photothermal and immune therapy enhanced the treatment of both primary and metastatic tumors. Zhang' s team reported a hybrid membrane composed of OMVs and B16-F10 melanoma cells, which encapsulated hollow poly (dopamine) (HPDA) nanoparticles (Wang et al. 2020). These nanoparticles targeted the tumor while synergizing with PTT and antitumor immune responses, leading to the efficient and safe elimination of melanoma. Fusion cell membrane engineering platforms have made significant contributions to tumor treatment, advancing the clinical translation of nanoparticle-based therapies for cancer.

4.4.2.2 Inflammatory Diseases Treatment

Inflammation is a critical defense mechanism of the body to injury. Moderate inflammation helps eliminate damaging factors and promote tissue repair. However, excessive inflammation can exacerbate injury, leading to a cytokine storm, an overproduction of ROS, and lipid peroxidation, all of which can adversely affect health. Biomimetic nanomedicines are gaining prominence in the regulation of inflammation. Lin et al. developed a biomimetic nanoparticle based on Prussian blue particles and the clinical drug methotrexate (MTX), which is commonly used for managing rheumatoid arthritis (RA) (Lin et al. 2022). The researchers loaded Prussian blue nanoparticles with MTX to create the therapeutic nanoparticles PB@SIN NPs. Then, they prepared fusion membranes made from RBCs membranes and macrophage cell membranes. These fusion membranes not only encapsulated PB@SIN NPs but also incorporated hyaluronic acid (DSPE-PEG2000-HA) on the outer surface to target fibroblast-like synoviocytes (FLSs), enhancing the macrophage-targeting effect. This biomimetic nanoparticle significantly accumulated at the site of inflammatory damage in the joints, inhibited the production of inflammatory cytokines by synovial macrophages, and significantly suppressed the progression of arthritis (Fig. 4.8a, b). Gouty arthritis (GA) is typically marked by elevated urate accumulation in the joints and the development of an inflamed immune microenvironment around the joints. To address the clinical challenge where separately reducing uric acid or using anti-inflammatory treatments alone does not effectively treat GA, Xu et al. developed a unique biomimetic nanosystem (D-N[EM2]), which was made from a membrane fused from M2 macrophage membranes and M2 macrophage-derived exosomes (Xu et al. 2024b). This membrane encapsulated a combination of uricase, hyaluronic acid platinum/poly(dopamine) nanozyme, and resveratrol liposomes. Specifically, D-N[EM2] was designed to target the inflammatory sites, where it boosted the anti-inflammatory effects of resveratrol and

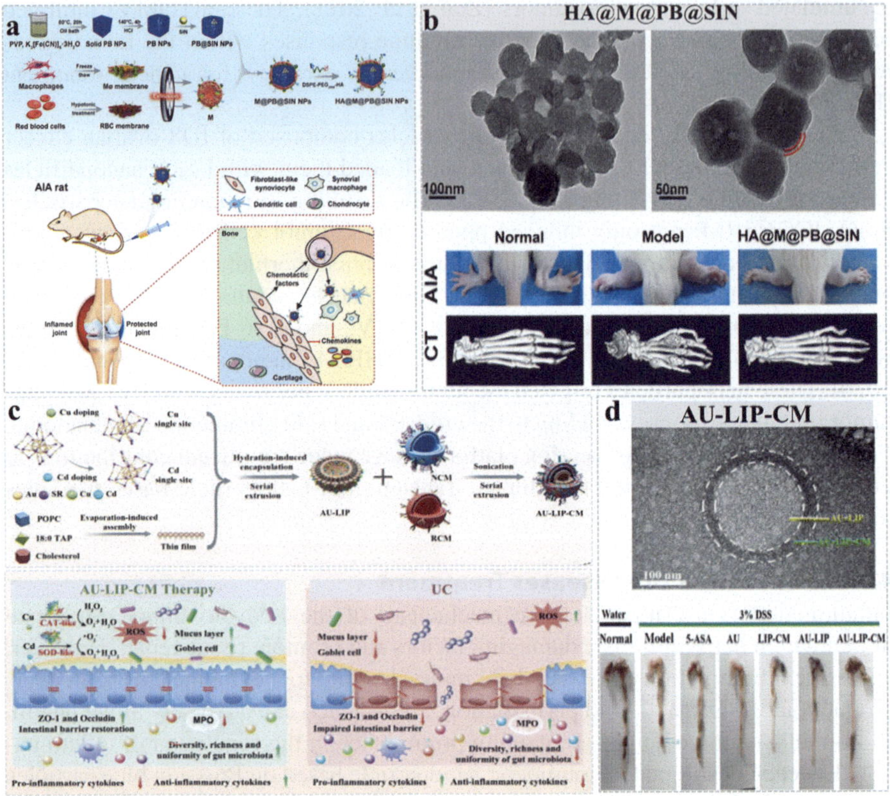

Fig. 4.8 (**a**) Schematic representation of the preparation of biomimetic nanoparticles (HA@M@PB@SIN) for the treatment of rheumatoid arthritis and their anti-inflammatory effects. (**b**) TEM morphology of HA@M@PB@SIN and photographs and CT images of mouse joints after drug treatment. (Reproduced with permission from Lin et al. (2022), copyright 2022, Elsevier B.V.). (**c**) Schematic illustration of the preparation of the biomimetic nanoparticle platform (AU-LIP-CM) and its therapeutic effects in UC. (**d**) TEM morphology of AU-LIP-CM and the colon repair morphology. (Reproduced with permission from Wang et al. (2023), copyright 2023, Wiley-VCH GmbH)

promoted transition of M1 macrophages from a pro-inflammatory state to an anti-inflammatory M2 phenotype. Additionally, D-N[EM2] released uricase at the inflammatory sites to break down uric acid. The platinum nanoparticles in the system also supported PTT to induce tissue repair at the injury sites, alleviate pain, and further treat GA in a synergistic manner.

Ulcerative colitis (UC) is characterized by high incidence and recurrence rates, often accompanied by excessive ROS production, macrophage activation, and continuous inflammation (Zhang et al. 2023b; Lu et al. 2024). Wang et al. developed a bioinspired nanoparticle platform (AU-LIP-CM) for the efficient treatment of UC by fusing liposomes containing gold clusters (AU) with neutrophil (NC) and RBCs membranes (Wang et al. 2023). AU-LIP-CM inherits inflammation-targeting and immune evasion properties, and the released gold (AU) demonstrates superoxide dismutase and catalase-like activities, clearing excess ROS, modulating the immune system, and repairing the intestinal barrier. Additionally, AU can emit fluorescence at >1200 nm, which, in combination with near-infrared (NIR-II) imaging, allows for real-time tracking of the material's aggregation, achieving excellent therapeutic effects in UC treatment (Fig. 4.8c, d).

Myocardial infarction (MI)/ischemia-reperfusion (IR) damage is often accompanied by myocardial cell apoptosis, fibrosis, and cardiac impairment, making the repair of myocardial cells crucial. Zhou et al. designed a platelet-macrophage hybrid membrane (HM) platform that incorporates membrane-penetrating peptide (P-Ben) and siSav1-loaded nanoparticles, aiming to inhibit the Hippo pathway and induce myocardial cell regeneration (Zhou et al. 2023). Under the inflammatory homing targeting effect, BSPC@HM efficiently accumulated at the site of IR-induced cardiac damage, significantly downregulated Sav1, inhibited myocardial cell apoptosis, and promoted cardiac function recovery. Additionally, the acidic microenvironment induced a charge reversal of the aconitine (PC) in siSav1, enhancing nanoparticle penetration and promoting heart repair. Additionally, Zhang et al. fused stem cell EVs with monocyte membranes to prepare a hybrid membrane nanoparticle platform that targets damaged myocardial cells while promoting angiogenesis, endothelial cell maturation, and macrophage subset regulation to repair heart damage (Zhang et al. 2020). Similarly, Hu et al. developed fusion vesicles composed of platelet membranes and stem cell exosomes (P-XOs), which reduced macrophage uptake, enhanced targeting ability to the damaged heart, and showed promising therapeutic effects for myocardial injury (Hu et al. 2021b).

AS is a major cause of cardiovascular diseases, characterized by endothelial damage, inflammation, and lipid deposition. Studies have shown that macrophages interact with lipoproteins, exacerbating lipid accumulation, triggering cellular dysfunction, and amplifying inflammation. Based on this, researchers fused anti-inflammatory M2 macrophage membranes with lipid-derived, anti-inflammatory peptides (HBSP), which could target plaques to promote the restoration of endothelial cell function (Guo et al. 2022). Additionally, the hybrid membrane vesicles were loaded with the classic cholesterol-lowering drug simvastatin (ST), which mediates cholesterol efflux and promotes plaque regression, offering an effective biomimetic nanoplatform strategy for treating AS (Fig. 4.9a, b).

Inflammatory brain diseases often manifest as damage to the BBB, an impaired brain microenvironment, and severe brain injury. Therefore, repairing the BBB is crucial in treating inflammatory brain conditions. Brain microvessel endothelial cells (BMECs) form the structural basis of the BBB, and vascular cell adhesion molecule-1 (VCAM1) is a key target for treating brain inflammation. As shown in Fig. 4.9c, researchers fused neural stem cell (NSC) membranes with simple liposomes to create hybrid bioresponsive vesicles (NSC-Lipo) that were loaded with the anti-inflammatory drug Metformin (Wu et al. 2023). These vesicles facilitated efficient treatment in ischemic stroke mouse models. The NSC membrane expressed VLA-4, enabling precise targeting of VCAM1, which is highly expressed on BMECs. The liposome enhanced the drug loading efficiency of Metformin, which

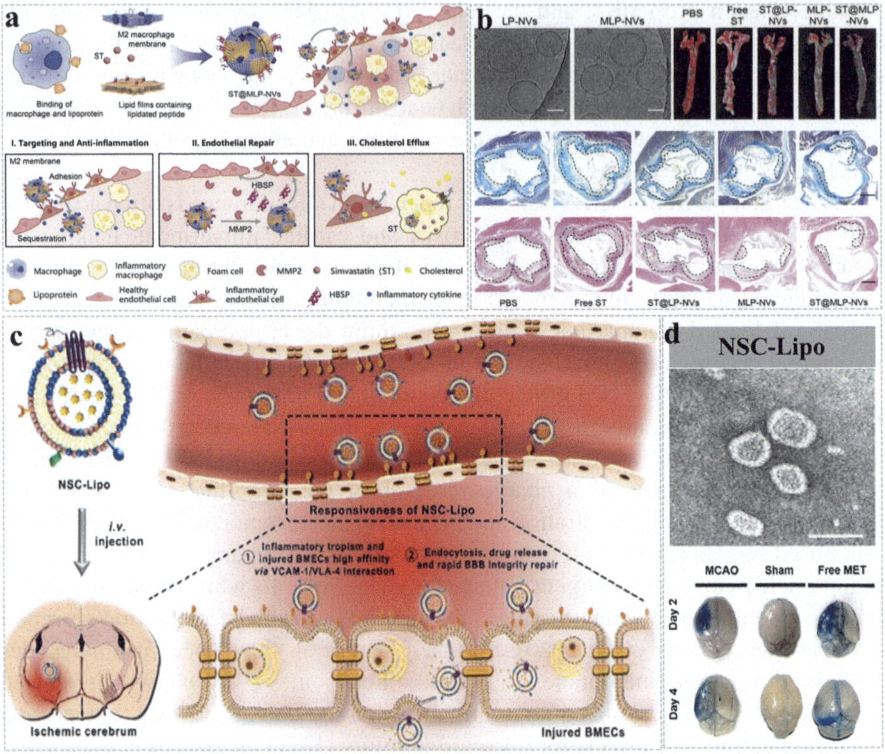

Fig. 4.9 (**a**) Schematic representation of the fabrication and therapeutic actions of fusion membrane nanovesicles (MLP-NV) for multi-modal treatment of AS. (**b**) TEM morphology of MLP-NV and the changes in vascular morphology and histological sections after treatment. (Reproduced with permission from Guo et al. (2022), copyright 2022, Wiley-VCH GmbH). (**c**) Schematic illustration of the mechanism of hybrid bioresponsive vesicles (NSC-Lipo) for crossing the BBB and treating brain injury. (**d**) TEM morphology and BBB repair effects of NSC-Lipo. (Reproduced with permission from Wu et al. (2023), copyright 2023, Elsevier B.V.)

in turn reduced inflammation in BMECs, promoting the restoration of BBB integrity (Fig. 4.9d). To reduce ischemia-reperfusion brain injury and promote neuronal repair, researchers developed a modular satellite-like structured nanoparticle platform (MEps), using SiO_2 nanoparticles as the core. These nanoparticles were electrostatically adsorbed with stem cell-derived exosomes and macrophage membranes, allowing targeted and rapid adhesion to damaged brain endothelial surfaces (Zhou et al. 2024). This nanoparticle inhibited neutrophil infiltration into the damaged brain tissue, further releasing growth factors to promote neuronal repair and alleviate ischemia-reperfusion brain damage.

4.4.2.3 Bacterial Infections Treatment

Bacterial infections pose a serious risk to public health. Traditional antibiotic treatments are prone to inducing drug resistance and adverse side effects, highlighting the critical need to develop targeted antibacterial strategies that mitigate resistance. Biomimetic nanoparticles prepared using hybrid membrane platforms have demonstrated significant potential in multimodal and targeted therapies. Black phosphorus quantum dots (BPQDs), with their excellent photothermal and photodynamic characteristics, can act as effective antibacterial agents under near-infrared (NIR) laser irradiation, significantly enhancing bacterial sensitivity to antibiotics. Liu et al. first utilized liposomal membranes to encapsulate BPQDs and the classical antibiotic amikacin (AM), forming AB@Lip nanoparticles with excellent stability (Liu et al. 2024). Subsequently, the researchers fused macrophage and RBCs membranes, coating them onto the surface of AB@Lip to fabricate AB@LRM, which exhibited inflammation-targeting capabilities and prolonged retention (Fig. 4.10a). As shown in Fig. 4.10b, under laser irradiation, AB@LRM demonstrated outstanding anti-inflammatory and antimicrobial effects in models of *Pseudomonas aeruginosa*-induced pneumonia and wound infections.

To combat infections caused by *Pseudomonas aeruginosa*, Ding's research team developed a multi-antigen antibacterial nanovaccine, expanding the application of hybrid membranes in antibacterial therapy (Peng et al. 2024a). The researchers first fused macrophage membranes with OMVs to obtain hybrid vesicles (HMVs). The surface of HMV was decorated with secreted toxins (SPs) from *Pseudomonas aeruginosa* and encapsulated gold nanoparticles (AuNPs) as an adjuvant. This design reduced the toxicity of SPs while eliciting an anti-toxin immune response. Notably, the AuNP@HMV@SP nanovaccine exhibited excellent preventive effects, achieving a 100% survival rate in a sepsis model, providing an effective strategy for the development of antibacterial nanovaccines (Fig. 4.10c, d).

In addition, researchers developed a nanorobot by coating gold nanowires with hybrid vesicles composed of RBCs membranes and platelet membranes. (Esteban-Fernández de Ávila et al. 2018) The nanorobot, capable of moving under ultrasound stimulation, can adhere to pathogens and neutralize toxins in the body, enabling the simultaneous removal of pathogens and toxins. Pore-forming toxins (PFTs) are regarded as one of the most potent types of protein involved in bacterial infections and represent a promising target for antibacterial therapy. He et al. developed hybrid nanovesicles (RM-PL) capable of neutralizing PFTs by fusing biological RBCs membranes with synthetic lipid bilayer membranes (He et al. 2019). Both *in vitro* and *in vivo* experimental studies demonstrated that RM-PL exhibits safe and effective antibacterial and tissue repair properties.

4.4.2.4 Tissue Regeneration

The regulation of tissue regeneration, both inhibition and promotion, plays a crucial role in disease progression. Choroidal neovascularization (CNV) is a major contributor to vision loss in many blinding conditions. Li et al. developed a HCM by combining retinal endothelial cell membranes with red blood plasma membranes, loaded with PLGA nanoparticles, as a substitute for traditional anti-angiogenic agents for non-invasive targeted CNV therapy (Li et al. 2021b). The biomimetic anti-angiogenic nanodrug demonstrated prolonged circulation time and effective

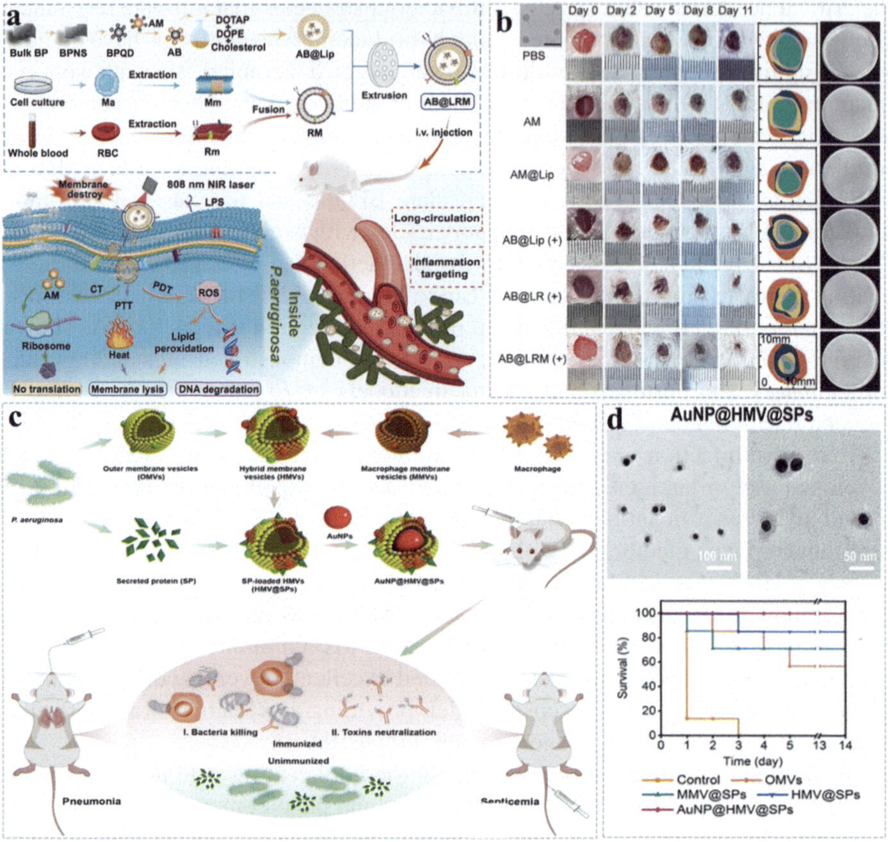

Fig. 4.10 (**a**) Schematic illustration of the construction of biomimetic nanoparticles AB@LRM and their antibacterial mechanism. (**b**) TEM morphology of AB@LRM and photographs of wound infection repair. (Reproduced with permission from Liu et al. (2024), copyright 2024, Wiley-VCH GmbH). (**c**) Schematic illustration of the preparation of biomimetic vaccines and their antibacterial therapeutic mechanism. (**d**) TEM morphology of AuNP@HMV@SP and the survival rate curve after treatment. (Reproduced with permission from Peng et al. (2024a), copyright 2024, Wiley-VCH GmbH)

anti-angiogenic activity, offering a promising non-invasive therapeutic approach for CNV.

Cardiac fibrosis significantly increases the risk of sudden heart failure. JQ1, a BRD4 inhibitor, exhibits potent anti-fibrotic properties. Li et al. developed a multi-functional biomimetic nanoparticle drug delivery nanoplatform by fusing platelet and RBCs membranes while co-loading PLGA and JQ1. These nanoparticles can specifically target cardiac fibroblasts and collagen, offering prolonged circulation time and efficient delivery of JQ1 (Li et al. 2024c). This system expands the potential of hybrid membrane-based nanodrugs in treating cardiac fibrosis and minimizing drug side effects. Overall, HCM-engineered platforms provide a safer and more efficient strategy for theranostic application.

Promoting angiogenesis is a promising therapeutic approach for treating ischemic heart disease (IHD). Li et al. ingeniously integrated MSC-EVs with angiogenic potential and platelet cell membranes with vascular injury-targeting properties to prepare fusion vesicles (P-EVs) that exhibit efficient targeting of MI sites and promote angiogenesis (Li et al. 2021a). Notably, P-EVs demonstrated significant regenerative effects, expanding the application and development of biomimetic fusion vesicles for managing vascular inflammatory disorders such as IHD, AS and MI/IR.

4.5 Conclusions and Perspectives

The engineered platforms based on fused cell membranes are playing an increasingly important role in various aspects such as the development of nanomedicines, disease diagnosis, and therapy. This progress has significantly advanced membrane-based nanomedicine in clinical research and translational applications, offering tremendous market potential and application value. However, this emerging field still faces several challenges that require further investigation and refinement: (1) Development of standardized production methods. The diversity of cell types and their distinct functions pose challenges, as complex preparation methods may disrupt the structure and composition of functional proteins. Additionally, varying membrane ratios can affect the final performance of hybrid membranes, thereby limiting the extended applications of the biomimetic platform. (2) Establishment of long-term stable storage methods. The application of biomimetic hybrid membrane platforms depends on effective transportation and storage solutions. Developing storage techniques that maintain the long-term functionality of biomimetic membranes is crucial. (3) Modification of multifunctional membranes. The biomedical applications of biomimetic nanomaterials are limited by the inherent functions of the cell membrane. By utilizing techniques such as genetic engineering and chemical modification, specific proteins can be expressed on the cell membrane surface, enhancing targeting efficiency and biocompatibility. This approach imparts multifunctionality to the HCM engineered platforms, further improving the diagnostic and therapeutic effects of nanomaterials. (4) Investigation of superior biosafety. Biosafety and prolonged toxicity should be systematically evaluated, particularly for hybrid membranes. One of the major challenges in this regard lies in the complexity of fusion membranes, as their interactions with biological systems may lead to unforeseen immunological responses or chronic toxicity. Therefore, understanding the underlying mechanisms and investigating biosafety are crucial for developing and applying HCM engineered platforms. With ongoing contributions from researchers across disciplines, advancements in nanotechnology, and a deeper understanding of fused cell membranes in vivo and in vitro, we believe that the engineered platform of fused cell membranes will exhibit even broader application prospects, driving the realization of personalized medicine and precision therapy.

References

Bu LL, Rao L, Yu GT et al (2019) Cancer stem cell-platelet hybrid membrane-coated magnetic nanoparticles for enhanced photothermal therapy of head and neck squamous cell carcinoma. Adv Funct Mater 29(10): 1807733

Chavda VP, Luo G, Bezbaruah R et al (2024) Unveiling the promise: exosomes as game-changers in anti-infective therapy. Exploration 4:20230139

Chen Q, Huang G, Wu W et al (2020) A hybrid eukaryotic-prokaryotic nanoplatform with photothermal modality for enhanced antitumor vaccination. Adv Mater 32(16):e1908185

Chen Z, Wei D, Wang Q et al (2024a) Multifunctional biomimetic nanosystem for retinoblastoma treatment. Chem Eng J 499: 156017

Chen Z, Guo W, Tan L et al (2024b) Biomimetic MOF-based nano-immunoactivator via disruption of ion homeostasis for strengthened tumor microwave-immunotherapy. Adv Funct Mater 34(36): 2401359

Chi S, Zhang L, Cheng H et al (2023) Biomimetic nanocomposites camouflaged with hybrid cell membranes for accurate therapy of early-stage glioma. Angew Chem Int Ed Engl 62(29):e202304419

Chiang ST, Chen Q, Han T et al (2024) Biomimetic nanovesicles synergize with short-term fasting for enhanced chemotherapy of triple-negative breast cancer. ACS Nano 18(50):33875–33889

Dehaini D, Wei X, Fang RH et al (2017) Erythrocyte-platelet hybrid membrane coating for enhanced nanoparticle functionalization. Adv Mater 29(16): 1606209

Ding C, He R, Cheng T et al (2023) Bacterial outer membrane-based biomimetic immune adaptors: mild immunomodulatory and bacterial targeted delivery strategy against implant-related infections. Adv Funct Mater 33(42): 2304168

Dong X, Yang Q, Wang H et al (2023) Targetedly attenuating cancer stemness and plasticity by homologous cancer stem cell-inherited fusion membrane nanoeffectors against cancer metastasis. Small Sci 4(2): 2300111

Duan Y, Zhou J, Zhou Z et al (2023) Extending the in vivo residence time of macrophage membrane-coated nanoparticles through genetic modification. Small 19(52): 2305551

Ducrot C, Loiseau S, Wong C et al (2023) Hybrid extracellular vesicles for drug delivery. Cancer Lett 558:216107

Esteban-Fernández de Ávila B, Angsantikul P, Ramírez-Herrera DE et al (2018) Hybrid biomembrane–functionalized nanorobots for concurrent removal of pathogenic bacteria and toxins. Science. Robotics 3(18):eaat0485

Fang F, Wang S, Song Y et al (2023) Continuous spatiotemporal therapy of a full-API nanodrug via multi-step tandem endogenous biosynthesis. Nat Commun 14(1):1660

Feng Y, Wei C, Gu Y et al (2024) pH-sensitive cationic nanoparticles for endosomal cell-free DNA scavenging against acute inflammation. J Control Release 369:88–100

Gao X, Li S, Ding F et al (2019) Rapid detection of exosomal microRNAs using virus-mimicking fusogenic vesicles. Angew Chem Int Ed Engl 58(26):8719–8723

Gu Z, Liu T, Rao W et al (2025) Innovation of lipid-based nanoparticles: the design and applications in the cardiovascular disease management. Adv Funct Mater 35(2): 2412335

Gu J, Yan C, Yin S et al (2024b) Erythrocyte membrane-coated nanocarriers modified by TGN for Alzheimer's disease. J Control Release 366:448–459

Guan M, Cheng K, Xie X-T et al (2024) Regulating copper homeostasis of tumor cells to promote cuproptosis for enhancing breast cancer immunotherapy. Nat Commun 15(1):1807733

Guo L, Miao Y, Wang Y et al (2022) Biomimetic macrophage membrane and lipidated peptide hybrid nanovesicles for atherosclerosis therapy. Adv Funct Mater 32(52): 2204822

Guo Q, Xu X, Lai X et al (2025) Antigen/adjuvant-free liposome induces adjuvant effects for enhancing cancer immunotherapy. Exploration 5(2): 20230115

Han X, Shen S, Fan Q et al (2019) Red blood cell–derived nanoerythrosome for antigen delivery with enhanced cancer immunotherapy. Sci Adv 5(10):eaaw6870

Han S, Yoo W, Carton O et al (2025) PEGylated multimeric RNA nanoparticles for siRNA delivery in traumatic brain injury. Small 21:e2405806

He Y, Li R, Li H et al (2019) Erythroliposomes: integrated hybrid nanovesicles composed of erythrocyte membranes and artificial lipid membranes for pore-forming toxin clearance. ACS Nano 13(4):4148–4159

He Y, Zhang S, She Y et al (2024) Innovative utilization of cell membrane-coated nanoparticles in precision cancer therapy. Exploration 4(6): 20230164

Hou L, Gong X, Yang J et al (2022) Hybrid-membrane-decorated prussian blue for effective cancer immunotherapy via tumor-associated macrophages polarization and hypoxia relief. Adv Mater 34(14):e2200389

Hu Y, Li X, Zhang Q et al (2021a) Exosome-guided bone targeted delivery of Antagomir-188 as an anabolic therapy for bone loss. Bioact Mater 6(9):2905–2913

Hu S, Wang X, Li Z et al (2021b) Platelet membrane and stem cell exosome hybrid enhances cellular uptake and targeting to heart injury. Nano Today 39: 101210

Huang H, Zheng S, Wu J et al (2024) Opsonization inveigles macrophages engulfing carrier-free bilirubin/JPH203 nanoparticles to suppress inflammation for osteoarthritis therapy. Adv Sci 11(22): 2400713

Jiang Q, Liu Y, Guo R et al (2019) Erythrocyte-cancer hybrid membrane-camouflaged melanin nanoparticles for enhancing photothermal therapy efficacy in tumors. Biomaterials 192:292–308

Jiang X, Zhang X, Guo C et al (2023) Genetically engineered cell membrane-coated magnetic nanoparticles for high-performance isolation of circulating tumor cells. Adv Funct Mater 34(7): 2304426

Kalluri R, LeBleu VS (2020) The biology, function, and biomedical applications of exosomes. Science 367(6478):aau6977

Kang K, Zhang Y, Zhou X et al (2023) Hybrid extracellular vesicles-liposomes camouflaged magnetic vesicles cooperating with bioorthogonal click chemistry for high-efficient melanoma circulating tumor cells enrichment. Adv Healthc Mater:e2202825

Lai J, Huang C, Guo Y et al (2022) Engineered extracellular vesicles and their mimics in cardiovascular diseases. J Control Release 347:27–43

Li Q, Song Y, Wang Q et al (2021a) Engineering extracellular vesicles with platelet membranes fusion enhanced targeted therapeutic angiogenesis in a mouse model of myocardial ischemia reperfusion. Theranostics 11(8):3916–3931

Li M, Xu Z, Zhang L et al (2021b) Targeted noninvasive treatment of choroidal neovascularization by hybrid cell-membrane-cloaked biomimetic nanoparticles. ACS Nano 15(6):9808–9819

Li M, Fang F, Sun M et al (2022a) Extracellular vesicles as bioactive nanotherapeutics: an emerging paradigm for regenerative medicine. Theranostics 12(11):4879–4903

Li Q, Huang Z, Wang Q et al (2022b) Targeted immunomodulation therapy for cardiac repair by platelet membrane engineering extracellular vesicles via hitching peripheral monocytes. Biomaterials 284:121529

Li J, Gong C, Chen X et al (2023) Biomimetic liposomal nanozymes improve breast cancer chemotherapy with enhanced penetration and alleviated hypoxia. J Nanobiotechnology 21(1):123

Li Q, Byun J, Kim D et al (2024a) Cell membrane-coated mRNA nanoparticles for enhanced delivery to dendritic cells and immunotherapy. Asian J Pharm Sci 19(6): 100968

Li X, Cheng N, Shi D et al (2024b) Sulfated liposome-based artificial cell membrane glycocalyx nanodecoys for coronavirus inactivation by membrane fusion. Bioact Mater 33:1–13

Li Y, Yu J, Cheng C et al (2024c) Platelet and erythrocyte membranes coassembled biomimetic nanoparticles for heart failure treatment. ACS Nano 18(39):26614–26630

Liang Y, Xu X, Xu L et al (2022) Chondrocyte-specific genomic editing enabled by hybrid exosomes for osteoarthritis treatment. Theranostics 12(11):4866–4878

Liao Y, Zhang Y, Blum NT et al (2020) Biomimetic hybrid membrane-based nanoplatforms: synthesis, properties and biomedical applications. Nanoscale Horiz 5(9):1293–1302

Lin Y, Yi O, Hu M et al (2022) Multifunctional nanoparticles of sinomenine hydrochloride for treat-to-target therapy of rheumatoid arthritis via modulation of proinflammatory cytokines. J Control Release 348:42–56
Liu Y, Sukumar UK, Kanada M et al (2021) Camouflaged hybrid cancer cell-platelet fusion membrane nanovesicles deliver therapeutic microRNAs to presensitize triple-negative breast cancer to doxorubicin. Adv Funct Mater 31(41): 2103600
Liu C, Xi L, Liu Y et al (2023) An inhalable hybrid biomimetic nanoplatform for sequential drug release and remodeling lung immune homeostasis in acute lung injury treatment. ACS Nano 17(12):11626–11644
Liu H, Tang L, Yin Y et al (2024) Photoresponsive multirole nanoweapon camouflaged by hybrid cell membrane vesicles for efficient antibacterial therapy of pseudomonas aeruginosa-infected pneumonia and wound. Adv Sci 11(35):e2403101
Lopes D, Lopes J, Pereira-Silva M et al (2023) Bioengineered exosomal-membrane-camouflaged abiotic nanocarriers: neurodegenerative diseases, tissue engineering and regenerative medicine. Mil Med Res 10(1):19
Lu J, Shen X, Li H et al (2024) Recent advances in bacteria-based platforms for inflammatory bowel diseases treatment. Exploration 4(5): 20230142
Lu H, Liang B, Hu A et al (2025) Engineered biomimetic cancer cell membrane nanosystems trigger gas-immunometabolic therapy for spinal-metastasized tumors. Adv Mater 37(1):e2412655
Lv Q, Cheng L, Lu Y et al (2020) Thermosensitive exosome-liposome hybrid nanoparticle-mediated chemoimmunotherapy for improved treatment of metastatic peritoneal cancer. Adv Sci (Weinh) 7(18):2000515
Meng QF, Zhao Y, Dong C et al (2021) Genetically programmable fusion cellular vesicles for cancer immunotherapy. Angew Chem Int Ed Engl 60(50):26320–26326
Mondal J, Pillarisetti S, Junnuthula V et al (2023) Hybrid exosomes, exosome-like nanovesicles and engineered exosomes for therapeutic applications. J Control Release 353:1127–1149
Nankivell V, Vidanapathirana AK, Hoogendoorn A et al (2024) Targeting macrophages with multifunctional nanoparticles to detect and prevent atherosclerotic cardiovascular disease. Cardiovasc Res 120(8):819–838
Pan S, Zhong W, Lan Y et al (2024) Pathology-guided cell membrane-coated polydopamine nanoparticles for efficient multisynergistic treatment of periodontitis. Adv Funct Mater 34(23)
Patty PJ, Frisken BJ (2003) The pressure-dependence of the size of extruded vesicles. Biophys J 85(2):996–1004
Peng X, Luo Y, Yang L et al (2024a) A multiantigenic antibacterial nanovaccine utilizing hybrid membrane vesicles for combating Pseudomonas aeruginosa infections. J Extracell Vesicles 13(10):e12524
Peng X, Chen J, Gan Y et al (2024b) Biofunctional lipid nanoparticles for precision treatment and prophylaxis of bacterial infections. Sci Adv 10(14):eadk9754
Quan Z, Wang S, Xie H et al (2025) ROS regulation in CNS disorder therapy: unveiling the dual roles of nanomedicine. Small 21:e2410031
Rao L, Meng Q-F, Huang Q et al (2018) Platelet-leukocyte hybrid membrane-coated immunomagnetic beads for highly efficient and highly specific isolation of circulating tumor cells. Adv Funct Mater 28(34): 1803531
Sato YT, Umezaki K, Sawada S et al (2016) Engineering hybrid exosomes by membrane fusion with liposomes. Sci Rep 6:21933
Song Y, Zhang N, Li Q et al (2021) Biomimetic liposomes hybrid with platelet membranes for targeted therapy of atherosclerosis. Chem Eng J 408: 127296
Sun M, Yang J, Fan Y et al (2023) Beyond extracellular vesicles: hybrid membrane nanovesicles as emerging advanced tools for biomedical applications. Adv Sci 10:2303617
Sun M, Li M, Hu M et al (2024) Fully bioactive nanodrugs: stem cell-derived exosomes engineered with biomacromolecules to treat CCl(4)- and extreme hepatectomy-induced acute liver failure. ACS Nano 18:33907–33921

Tian Y, Sun D, Liu N et al (2024) Biomimetic mesenchymal stem cell membrane-coated nanoparticle delivery of MKP5 inhibits hepatic fibrosis through the IRE/XBP1 pathway. J Nanobiotechnol 22(1): 741

Wang Q, Ren Y, Mu J et al (2015) Grapefruit-derived nanovectors use an activated leukocyte trafficking pathway to deliver therapeutic agents to inflammatory tumor sites. Cancer Res 75(12):2520–2529

Wang D, Dong H, Li M et al (2018) Erythrocyte-cancer hybrid membrane camouflaged hollow copper sulfide nanoparticles for prolonged circulation life and homotypic-targeting photothermal/chemotherapy of melanoma. ACS Nano 12(6):5241–5252

Wang D, Liu C, You S et al (2020) Bacterial vesicle-cancer cell hybrid membrane-coated nanoparticles for tumor specific immune activation and photothermal therapy. ACS Appl Mater Interfaces 12(37):41138–41147

Wang X, Jiao M, Tian F et al (2023) A biomimetic nanoplatform with improved inflammatory targeting behavior for ROS scavenging-based treatment of ulcerative colitis. Adv Healthc Mater 12(29):e2301450

Wu H, Jiang X, Li Y et al (2023) Hybrid stem cell-derived bioresponsive vesicles for effective inflamed blood-brain barrier targeting delivery. Nano Today 49: 101800

Xie L, Chen J, Hu H et al (2024a) Engineered M2 macrophage-derived extracellular vesicles with platelet membrane fusion for targeted therapy of atherosclerosis. Bioact Mater 35:447–460

Xie M, Wu Y, Zhang Y et al (2024b) Membrane fusion-mediated loading of therapeutic siRNA into exosome for tissue-specific application. Adv Mater 36(33):e2403935

Xiong J, Wu M, Chen J et al (2021) Cancer-erythrocyte hybrid membrane-camouflaged magnetic nanoparticles with enhanced photothermal-immunotherapy for ovarian cancer. ACS Nano 15(12):19756–19770

Xu X, Zhang X, Li R et al (2024a) Platelet membrane-coated curcumin-PLGA nanoparticles promote astrocyte-neuron transdifferentiation for intracerebral hemorrhage treatment. Small 20(42): 2311128

Xu J, Wu M, Yang J et al (2024b) Multimodal smart systems reprogramme macrophages and remove urate to treat gouty arthritis. Nat Nanotechnol 19(10):1544–1557

Yang H, Ding Y, Tong Z et al (2022) pH-responsive hybrid platelet membrane-coated nanobomb with deep tumor penetration ability and enhanced cancer thermal/chemodynamic therapy. Theranostics 12(9):4250–4268

Yin Y, Tang W, Ma X et al (2022) Biomimetic neutrophil and macrophage dual membrane-coated nanoplatform with orchestrated tumor-microenvironment responsive capability promotes therapeutic efficacy against glioma. Chem Eng J 433: 133848

Yu Y, Cheng Q, Ji X et al (2022) Engineered drug-loaded cellular membrane nanovesicles for efficient treatment of postsurgical cancer recurrence and metastasis. Sci Adv 8(49):eadd3599

Yu Y, Peng Y, Shen WT et al (2024) Hybrid cell membrane-coated nanoparticles for biomedical applications. Small Struct 5(5): 2300473

Zhang N, Song Y, Huang Z et al (2020) Monocyte mimics improve mesenchymal stem cell-derived extracellular vesicle homing in a mouse MI/RI model. Biomaterials 255:120168

Zhang T, Liu H, Li L et al (2021a) Leukocyte/platelet hybrid membrane-camouflaged dendritic large pore mesoporous silica nanoparticles co-loaded with photo/chemotherapeutic agents for triple negative breast cancer combination treatment. Bioact Mater 6(11):3865–3878

Zhang W, Gong C, Chen Z et al (2021b) Tumor microenvironment-activated cancer cell membrane-liposome hybrid nanoparticle-mediated synergistic metabolic therapy and chemotherapy for non-small cell lung cancer. J Nanobiotechnology 19(1):339

Zhang W, Ngo L, Tsao SC et al (2023a) Engineered cancer-derived small extracellular vesicle-liposome hybrid delivery system for targeted treatment of breast cancer. ACS Appl Mater Interfaces 15(13):16420–16433

Zhang Y, Wang T, Sun M et al (2023b) Advanced nanomedicine: redefining therapeutic paradigms for inflammatory bowel disease. Adv Healthc Mater:e2300069

Zhang C, Zhao D, Fang F et al (2024) Quintuple Free-radical therapy: an ultralong-retention FAND for NIR-involved multiple site-acting hypoxic tumor therapy. Adv Funct Mater 34(34):2401840

Zhou Y, Liang Q, Wu X et al (2023) siRNA delivery against myocardial ischemia reperfusion injury mediated by reversibly camouflaged biomimetic nanocomplexes. Adv Mater 35(23):e2210691

Zhou H, Wu Y, Li M et al (2024) Modular satellite nanoparticles for remedying primary and secondary injury in cerebral ischemia-reperfusion. Adv Funct Mater 34(30): 2315274

Zhu L, Luo M, Zhang Y et al (2023) Free radical as a double-edged sword in disease: deriving strategic opportunities for nanotherapeutics. Coord Chem Rev 475: 214875

Zhuang WR, Wang Y, Lei Y et al (2022) Phytochemical engineered bacterial outer membrane vesicles for photodynamic effects promoted immunotherapy. Nano Lett 22(11):4491–4500

Zou MZ, Li ZH, Bai XF et al (2021) Hybrid vesicles based on autologous tumor cell membrane and bacterial outer membrane to enhance innate immune response and personalized tumor immunotherapy. Nano Lett 21(20):8609–8618

Part II

Biomaterials and Strategies for Cell Membrane Engineering

Functional Biomaterials Used for Cell Membrane Engineering and Their Functional Aspects

5

Salim Ullah, Fazal Rahim, Nisar Ahmad, and Qurat ul Ain Zahra

Abstract

Natural cell membranes from various sources vary in surface markers, chemical composition, and physicochemical properties, which can contribute to batch variability, poor biodistribution, limited tissue targeting, restricted penetration, and reduced therapeutic effects. Interest is growing in engineering cell membrane of living cells and cell membrane-based platforms such as extracellular vesicles, bacterial membrane vesicles, vesicle mimetic nanoparticles, and cell membrane coated nanoparticles. Functional biomaterials such as recombinant receptors, peptides, aptamers, antibodies, liposomes, bioorthogonal groups, fluorescent probes, nanomaterials, polymeric materials, hydrophobic anchors, and synthetic drugs have emerged as a transformative tool in the field of cell surface engineering, offering novel capabilities to modulate cellular functions for various biomedical applications. These biomaterials are designed to interact with

S. Ullah
The Fifth Affiliated Hospital of Wenzhou Medical University, Zhejiang, China
e-mail: salimbio@mail.ustc.edu.cn

F. Rahim
Biomedical Sciences and Health Laboratory of Anhui Province, Institute of Health and Medicine, Hefei Comprehensive National Science Center, University of Science and Technology of China, Hefei, China
e-mail: fazal@mail.ustc.edu.cn

N. Ahmad
College of Pharmacy, Dalian Medical University, Dalian, China
e-mail: nisarahmadahmad@creighton.edu

Q. u. A. Zahra (✉)
Personalized Medicine Centre, Health Futures Institute, Murdoch University, Murdoch, WA, Australia
e-mail: quratulain.zahra@murdoch.edu.au

Z. Hussain, R. Pei (eds.), *Cell Membrane Engineering for Advancing Cell-Based and Nanotechnology-Based Therapies*, Biomaterials, Bioengineering and Sustainability 6, https://doi.org/10.1007/978-3-031-93425-4_5

cell membrane to enhance homing potential, cell-cell recognition, targeted accumulation, drug loading potential, monitoring capabilities, antigen presentation, immunological escape, or immune modulation for disease-specific treatment. These engineered functions paving the way to membrane-engineered platform to enhance tissue engineering, regenerative medicine, drug delivery, immunotherapy, vaccine development, diagnostics, and theranostics application. This chapter provides an overview of the various functional biomaterials employed in cell surface engineering, highlighting their functional aspects, applications in therapeutics, and the potential for future innovation in engineered cell membrane-based biotechnologies.

Keywords

Cell membrane engineering · Antibodies · Aptamers · Peptides · Nanomaterials · Polymeric materials

5.1 Introduction

As the foundational building blocks of life, biological cells govern the architectural integrity, functional coordination, and metabolic processes of mammalian systems. Each cell is enclosed by a dynamic, thick phospholipid bilayer called the plasma membrane, which serves as a selectively permeable interface crucial for preserving intracellular equilibrium and mediating bidirectional molecular traffic (Campos Muñiz and Fernández Perrino 2023). This lipid matrix, enriched with cholesterol to optimize fluidity and mechanical resilience, is interspersed with transmembrane proteins, receptors, and glycoconjugates (e.g., glycoproteins and glycolipids). These embedded macromolecules orchestrate specialized functions, such as ligand-specific signaling, selective ion transport, and intercellular communication, thereby underpinning the adaptability and survival of the cell. These components not only regulate the fluidity and flexibility of the membrane but also serve as receptors and ligands, forming the basis of cellular recognition and signaling (Vafaei 2017). The surface composition of the plasma membrane, often referred to as the surfaceome, displays a complex arrangement of proteins and glycoconjugates that control critical processes such as cell-cell communication, adhesion, migration, tissue-specific targeting, and immune response. The expression and behavior of surface receptors and ligands differ in healthy versus diseased states, making them crucial targets for diagnostics and therapeutic interventions. In fact, a significant portion of the drug candidates cataloged in the Drug Bank database are designed to interact with cell membrane proteins (Pietrobono and Stecca 2021).

The cell surface is a complex structure that governs numerous processes critical to biological functions. These potential characteristics of cell membrane have empowered researchers to consider cell membrane-based platforms for treating intractable diseases and tissue defects. Despite the tremendous therapeutic potential of natural cells, certain limitations such as low survival rates, undesired differentiation, limited targeting ability, immune rejection, and insufficient cell-to-cell interactions limit their application. Furthermore, properties of cell membrane based nanoplatform such as extracellular vesicles and bacterial membrane vesicles depend on various cell membrane sources which differ in surface markers, chemical composition, and physicochemical properties, leading to batch variability, poor biodistribution, limited tissue targeting, restricted penetration, and reduced therapeutic effects. These limitations have led to growing interest in bioengineering strategies to enhance cellular membrane function.

The molecular landscape of the cell membrane presents a unique platform for such engineering efforts (Stephan and Irvine 2011). Genetic engineering, chemical conjugation, lipid insertion, and physical methods have all been developed to modulate the cell surface. Genetic engineering involves transfection of donor cells with a target gene that integrates into the host genome and promotes recombinant protein expression through biogenesis. Changes in parent cells can be reflected to their cell membrane as well as in the secreted nanovesicles, enabling engineered living cells and cell membrane-based platforms to promote the corresponding targeting, pharmaceutical, recognition, and immunological functions. However, gene editing faces challenges, including reliance on genetic material uptake for protein expression, limited recombinant protein display on membrane surface, and inability to modify cell membrane with nonbiological agents. These shortcomings underscore need for chemical conjugation, lipid insertion, and physical methods. These non-genetic methods allowing for the attachment of various biomaterials, including receptors, peptides, antibodies, aptamers, nanomaterials, polymeric materials, and fluorescent probes in order to improve cellular targeting, immunogenicity, biodistribution, and therapeutic efficacy, thereby enabling advanced applications in tissue engineering, regenerative medicine, drug delivery, diagnostics, and theranostic applications (Jia et al. 2022). Functional biomaterials have emerged as transformative tools in cell surface engineering (CSE), offering new ways to modulate cellular functions and interactions. These materials, such as recombinant receptors, peptides, aptamers, antibodies, liposomes, nanomaterials, and polymeric materials can be used to enhance cell-cell recognition, improve drug delivery, and elicit specific immune responses (Othman et al. 2018).

Cell membrane engineering has broadened the therapeutic spectrum of living cells and cell membrane-based platforms. They offer numerous therapeutic benefits and play an important role in treating diverse diseases. These functional biomaterials or targeting ligands once attached to cell membrane can modulate cell-cell surface interaction, cell-microenvironment interactions (can enhance immune escape), and strengthening cell features (cell homing and confer cell survival) and improving their biological function (cell proliferation and differentiation). Furthermore, surface engineering strategies can render engineered living cells and cell-based platforms with additional unnatural functionalities such as directed therapy, targeted drug delivery, in vivo tracking, and biosensing potential. Surface modification techniques such as hydrophobic insertion, covalent conjugation, and layer-by-layer attachment have expanded the capabilities of these materials, allowing for precise control over cellular interactions and functions. For example, surface-engineered nanoparticles (NPs) have shown great promise in improving drug delivery efficiency by overcoming biological barriers and enhancing the accumulation of therapeutic agents at target sites (Guo et al. 2017). Despite these advances, challenges remain, such as the formation of protein coronas on NPs, immune system clearance, and non-specific accumulation in tissues. However, innovations in biomimetic and hybrid cell membrane-coated nanomaterials are addressing these issues by leveraging the functional benefits of natural cell membranes (Sushnitha et al. 2020). Such strategies are pushing the boundaries of what is possible in CSE, offering exciting opportunities for next-generation biomedical applications.

The simplest technique that uses the membrane proteins' surface-exposed functional groups as grafting sites is chemical conjugation. The wide range of applications is the main benefit of chemical conjugation. Cross-linker-functionalized biomaterials can be utilized to alter a range of cells. Enzymatic and metabolic conjugation techniques offer a more specific adhesion of biomaterials than random alteration by chemical labeling. Enzymatic mediated conjugation offers the covalent binding of functional biomaterials to a specific location on the cell membrane, just like metabolic conjugation does. For example, *Staphylococcus aureus* transpeptidase sortase A effectively conjugates peptides or proteins with the LPETG-motif to the N-terminal glycine that can be visible on the surface of several cell types (Fig. 5.1) (Lee et al. 2018). In contrast to other non-genetic surface membrane modification techniques, alterations made via covalent conjugation are permanent even when conjugated biomaterials eventually fade over time. Furthermore, covalent conjugation makes it challenging to regulate the degree of modification, and a greater degree of modification involving bioactive molecules—no matter how big or small—may result in important physiological changes, like decreased membrane fluidity and diffusion kinetics to the altered cells (Swee et al. 2015).

This chapter provides a comprehensive overview of the various functional biomaterials employed in CSE, emphasizing their role in enhancing cellular functions and their potential to revolutionize cell-membrane based therapeutic approaches across to address unmet clinical needs across wide range of medical fields.

5.2 Functional Material Used to Modulate Cell Membrane

5.2.1 Receptors and Peptides

Endocytosis, pinocytosis, and phagocytosis are all ways used to absorb lipids and proteins from cell membranes. Biomaterials, depending on their nature, properties, and size, may internalize primarily through endocytosis when chemically associated, electrostatically adsorbed, or hydrophobically embedded in the membrane. This process is driven by ligand-receptor interactions or the uptake of bioactive substances across the cell membrane (Tang et al. 2021). These processes result in the formation of small pockets containing receptors and damaged portions of the cell membrane. The pockets are then sealed, and the newly created nanoscale vesicles are transported to internal cellular compartments (Cooper and McNeil 2015).

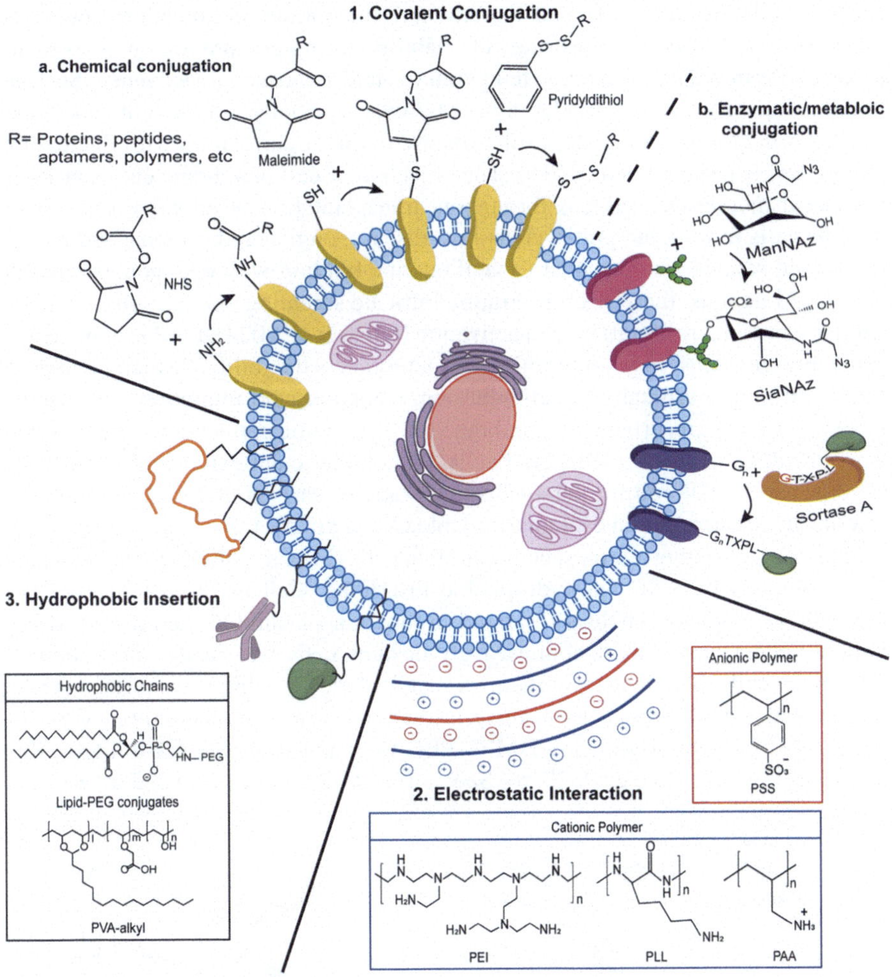

Fig. 5.1 Strategies for biomaterial-based cell surface engineering. (**1**) Through chemical covalent conjugation, cross-linkers enable cell surface remodeling using modified biomaterials. Enzymatic processes and the metabolism of artificial sugars within cells can be used to affix functional groups on the cell surfaces. (**2**) Cells may be modified layer by layer via ionic interactions between the cell membrane and charged polymers. (**3**) Through hydrophobic contact, long alkyl-chain polymers or lipid-conjugated bioactive compounds can be incorporated into the cell membrane. (Reprinted with permissions from Lee et al. (2018), copyright 2018, Springer Nature). NHS N-hydroxyl-succinimidyl ester, PSS poly(styrene) sulfate, PEG poly(ethylene glycol), PVA poly(vinyl alcohol), PEI poly(ethyleneimine), SiaNAz N-α-azidoacetyl sialic acid), PAA poly(acrylic acid), PLL poly-L-lysine, PSS poly(styrene) sulfate

Endocytosis allows cells to absorb materials from the invaginated membrane and surrounding media, potentially compromising surface engineering functionality. Thus, strategies must account for cell membrane dynamics to extend the retention of targeted functional materials (Lee et al. 2018).

Since cells are central to cell therapy, any membrane modifications must not compromise cell survival. Ensuring cell viability during preparation involves regulating pH, osmolality, pressure, temperature, and exposure to solvents. Surface modifications should not hinder nutrient flow, especially for insulin-producing β-cells, as they must modulate insulin release in direct correlation with fluctuating glucose levels (Ahmed et al. 2016). The objective of cell membrane engineering is to conceal membrane antigens throughout transplantation or adoptive transfer of immune cells to mitigate graft-versus-host disease, engineered proteins and receptors should remain unobstructed to facilitate the binding with endogenous growth factors and ligands, thereby orchestrating intricate signaling cascades that regulate cellular survival, proliferation, and activation (Ullah et al. 2024). Furthermore, engineered proteins must not compromise the membrane's flexibility elasticity and flexibility, attributing critically to cell adhesion, signaling, and migration. Ultimately, the expense of engineering cell membrane for therapeutic applications must remain economical. Genetically engineered cells, like chimeric antigen receptor T (CAR-T) cells, can be tailored for personalized therapy in various diseases, but the high cost of each treatment makes it unaffordable (Xu et al. 2021).

A great way to improve mesenchymal stem cells (MSCs) homing to the damaged myocardium is by CSE via hydrophobic insertions. Cell modification involving hydrophobic insertion enables the rapid production of tailored therapeutic MSCs without causing harm as it noninvasively designs cells and easily alters the cell membrane with therapeutic compounds that include lipophilic anchoring (Abdal Dayem et al. 2024). Stem cells were surface-engineered with recombinant CXCR4 (rCXCR4) hydrophobized using DMPE-PEG to demonstrate the feasibility of this approach (Fig. 5.2) (Lee et al. 2018). Specialized MSCs were produced in less than

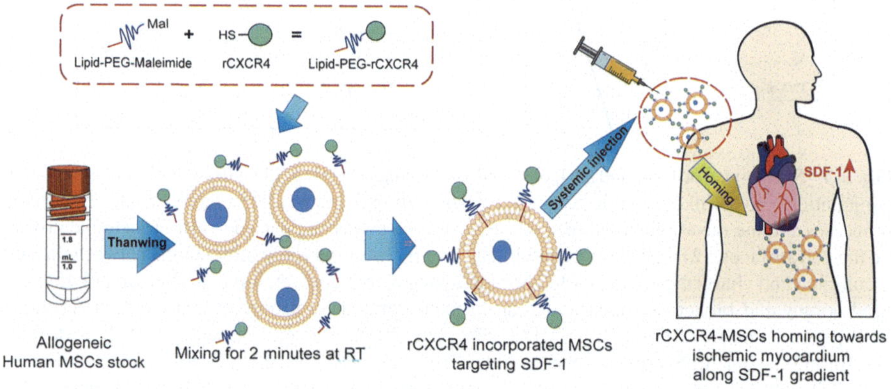

Fig. 5.2 Schematic depiction of surface-modified stem cells to support cardiac repair. *Ex vivo* grown and multiplied MSCs surface-functionalized via hydrophobic insertion strategies to integrate rCXCR4 into their membrane surface. Systemically administered rCXCR4-modified MSCs might move to the ischaemic myocardium through the utilization of the SDF-1 concentration gradient. (Reprinted with permissions from Lee et al. (2018), copyright 2018, Springer Nature). SDF-1 stromal-derived factor-1, PEG poly (ethylene glycol), MSCs mesenchymal stem cells, rCXCR4 recombinant CXC chemokine receptor 4

ten minutes after pre-expanded stem cells were incubated with rCXCR4-PEG-DMPE. These cells showed a twofold increase in their capacity to migrate along the stromal cell-derived factor 1 (SDF-1) concentration gradient, indicating a restored responsiveness to SDF-1.

To enhance certain cell adhesion or alter cell signaling pathways, researchers have been adding bioactive peptides or proteins to the cell surface. Cell-based biosensors have been developed using this method, in which cells are modified to have certain surface receptors that are able to recognize and react to target compounds. CSE is crucial for regenerative medicine, especially in producing programmable cell assemblies and cell-dense scaffolds materials. Researchers are exploring functionalized cells as building blocks for creating microtissues, leveraging existing toolkits. Scientists can control how cells self-assemble into intricate patterns by creating complimentary chemical interactions on various cell surfaces. Multicellular structures with regulated spatial organization, including cell sheets, clusters, or tissue-like structures, have been produced using this method. Additionally, the building of artificial cellular systems with sophisticated functions is made possible by the dynamic management of cell-cell interactions made possible by the integration of genetic circuits onto cell surfaces. In general, the kind of surface modification and assembly method may be used to adjust engineered cell assemblies. Various receptors, such as PD-1, PD-L1, CXCR4, TRAIL, iRGD peptide, GE11 peptide, GFE1 peptide, single-chain variable fragment (scFv), RVG (rabies virus glycoprotein), chondrocyte affinity peptide (CAP), cardiomyocyte-targeting peptide (CTP), DA7R peptide, and SDF-1 peptide has been discussed briefly in Table 5.1.

5.2.2 Aptamers

Aptamers and antibodies, as biomolecular tools, frequently modify cell surface characteristics due to their high binding affinities and specificity. Both are useful in fields like bioengineering, medicines, and diagnostics since they can target certain cell surface indicators. Despite the widespread use of antibodies for decades, aptamers, a more recent type of ligand, offer several distinct benefits. Aptamers are small, single-stranded DNA or RNA sequences that that fold into unique three-dimensional structures to attach to specific molecule, like proteins, small compounds, or even whole cells (Zahra et al. 2022). From the standpoint of how they work, antibodies and aptamers are comparable in that they both use spatial conformations that match to detect their ligands. Aptamers are frequently referred to as chemical antibodies. Aptamer-based biomedical applications, particularly in CSE, have garnered a lot of interest recently due to the materials' numerous appealing benefits. First, aptamers may be easily found using an iterative procedure known as Systematic Evolution of Ligands by Exponential Enrichment (SELEX), and they are compatible with large-scale in vitro synthetic technologies.

After many selection rounds, SELEX enriches aptamers through a randomized library with at least 1 trillion distinct sequences. Cell-SELEX is a variant approach that allows aptamer selection for an entire target cell. There is plenty of space for

Table 5.1 An overview of peptides and receptors associated in membrane engineering

No.	Peptide/ receptor	Role	Applications	Ref.
1	PD-L1	PD-1 Ligand; impedes T-cell activation to suppress immune response.	Immune checkpoint blockade therapies for cancer.	Liu et al. (2024)
2	PD-1	Immune checkpoint receptor that regulates immune response.	Cancer immunotherapy; immune evasion resistance.	Funauchi et al. (2024)
3	DA7R peptide	Tumor-homing peptide targeting glioblastoma cells.	Targeted drug and gene therapy for brain tumors.	Raucher (2019)
4	CXCR4	Regulating cell migration and homing, predominantly in cancer and immune cells.	Targeted cancer therapeutics; stem cell mobilization.	Choi et al. (2012)
5	GE11 peptide	EGFR-binding peptide for targeted therapeutics.	Cancer-targeted drug delivery.	Hailing et al. (2022)
6	CTP	Targeting cardiomyocytes for therapeutic applications.	Heart disease treatment; targeted drug delivery to the heart.	Das et al. (2024)
7	TRAIL	Induces apoptosis in cancer cells by attaching to death receptors.	Cancer therapy through selective tumor cell apoptosis.	Guerrache and Micheau (2024)
8	GFE1 peptide	EGFR-binding peptide for cancer therapeutics and imaging.	Tumor targeting in drug delivery and diagnostics.	Cheng et al. (2024)
9	iRGD peptide	Tumor-penetrating peptides which attaches to integrins and neuropilin-1 receptors.	Targeted drug delivery to tumors; higher permeability and retention.	Nikitovic et al. (2024)
10	CAP	coupling to chondrocytes for targeting cartilage.	Osteoarthritis treatment; regeneration of cartilage.	Zhang et al. (2024)
11	RVG	Targeting neural cells via coupling to nicotinic acetylcholine receptors.	Drugs and gene delivery to the CNS.	Nguyen et al. (2024)
12	SDF-1 peptide	Stem cell homing factor that regulates migration and adhesion.	Regenerative medicine; tissue repair and wound healing.	Wei et al. (2024)
13	scFv	short antibody fragments which retain antigen-binding specificity.	Targeted immunotherapy; cancer and disease-specific targeting.	Gezehagn Kussia and Tessema (2024)

PD-1 programmed cell death protein 1, *PD-L1* programmed death-ligand 1, *scFv* single-chain variable fragment, *RVG* rabies virus glycoprotein, *CAP* chondrocyte affinity peptide, *CTP* cardiomyocyte-targeting peptide, *EGFR* epidermal growth factor receptor, *CNS* central nervous system, *CXCR4* C-X-C Chemokine Receptor 4, *TRAIL* tumor necrosis factor-related apoptosis-inducing ligand

identifying possible biomarkers on the cell surface because the cell-SELEX approach doesn't demand prior understanding of the target. Furthermore, because aptamer molecules are smaller than antibodies, they could potentially be used to equip the cell surface without significantly altering the structural integrity of the native proteins. Additionally, aptamer chemical modification is stoichiometrically exact, site-specific, and versatile. Consequently, the improved programmability of aptamer-based CSE enables DNA nanotechnology to facilitate "Lego-like" modular building. Because of these benefits, aptamers will definitely be used a lot in research that focuses on the cell surface. For instance, they might be used to create sensing devices that can explore the microenvironment that is localized on the cell surface, to selectively image biomarkers that are linked to tumors, and to change how plasma membrane proteins dimerize and phosphorylate.

Antibodies and aptamers can be conjugated to create antibody-aptamer conjugates for cell surface engineering (CSE), enabling receptor binding. Aptamers form bivalent or polyvalent nanostructures that act as receptor agonists or antagonists, modulating cell signaling by blocking or promoting receptor clustering. They can be substituted with peptides or proteins to create DNA-protein or DNA-peptide conjugates. Due to their small size, stability, and chemical modification tolerance, aptamers are used to guide fluorescent probes to target points on the plasma membrane, aiding in the development of membrane probes. Aptamer-functionalized fluorescent probes are commonly used for in situ mapping of various membrane-bound proteins in living cells. These proteins commonly include mucin 1 (MUC1), protein tyrosine kinase 7, and epithelial cell adhesion molecule (EpCAM). An aptamer-conjugated polydiacetylene (PDA) liposomal sensor was designed by Wang et al. for detecting MUC1 on the surface of breast cancer cells via "turning-on" the fluorescence. In this approach, a cyanine 3 (Cy3) fluorophore was used to chemically label a MUC1-binding aptamer. On the basis of intermolecular energy transfer between the PDA conjugated backbone and Cy3, the aptamer's fluorescence was considerably reduced upon assembly with the PDA liposome (Fig. 5.3) (Wang et al. 2020). Fluorescence signals were recovered when the aptamer recognized and binds to its target MUC1, causing a change in aptamer configuration that simultaneously separates the associated Cy3 from PDA. Such a probe made it possible to selectively and sensitively profile MUC1 on the plasma membrane of breast cancer cells.

A novel class of site-selective protein conjugation technologies combines aptamers with click chemistry, characterized by "recognition-then-reaction." Robinson and colleagues illustrated this by employing a modified aptamer attached to a cyclooctyne and reporter group on the cell surface. This approach allowed for the metabolic tagging of sialic acids on glycoproteins with azido groups (Fig. 5.4). Site-selective chemical attachment of reporters was enabled by the precise positioning of cyclooctyne near azido groups, facilitated by the aptamer's high affinity for its target glycoprotein.

Aptamers offer several advantages for cell-surface engineering, including high stability against nuclease degradation and the ability to be chemically synthesized and modified. Being nucleic acids, aptamers can be easily designed with longer DNA tails. A recent bifunctional DNA molecule with a nucleic acid tail and aptamer can hybridize with other DNA oligonucleotides while also recognizing cells. Aptamers' complementary sequences allow them to be activated and detached from the cell surface, offering flexibility and reversibility for adjusting cell-surface engineering as needed. Monovalent surface engineering may easily functionalize immune cells using single aptamers (Shi and Wang 2021). For instance,

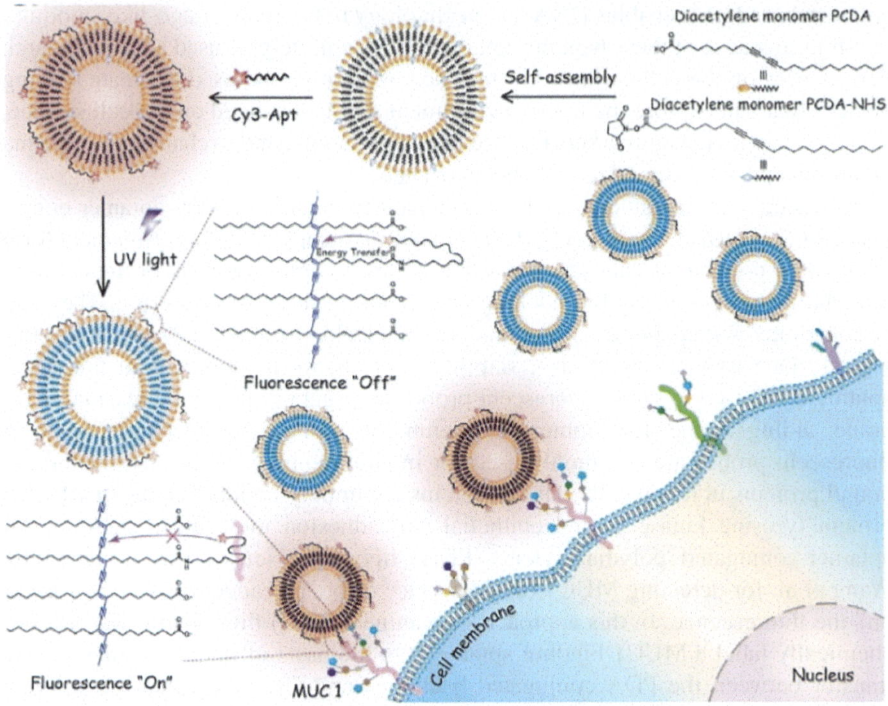

Fig. 5.3 Schematic depicting the construction of a fluorescence resonance energy transfer system including a PDA liposome and MUC1-binding aptamer (Cy3-labeled) for the targeted imaging of MUC1 expressed on plasma membranes. (Reprinted with permissions from Wang et al. (2020), copyright 2020, Elsevier)

hydrophobic insertions have been used to modify CD8+ cytotoxic T lymphocytes (CTLs) to exhibit TD05 aptamers (Fig. 5.5a) (Xiong et al. 2012). Within six hours, the altered CTLs destroyed 30% of the Ramos cells, but the unaltered CTLs only killed 5% of the Ramos cells. Two aptamer types with distinct functionalities can also be used to functionalize the cell surface. A basic requirement for many cell treatments is cell–cell recognition. For instance, when immune cells are used in cancer immunotherapy, potent and targeted interactions between the immune system and cancer cells must be realized (Coulie et al. 2014) in order to increase therapeutic efficacy while minimizing adverse effects and utilizing fewer therapeutic cells. Recently, Zhang et al. functionalized natural killer (NK) cells using the PDL1 aptamer and the aptamer TLS11a (Fig. 5.5b) (Zhang et al. 2020). The latter could control PD1/PDL1 signaling, whereas the former might target cancer cells. The modified NK cells demonstrated improved effectiveness in eliminating cancer cells by demonstrating the ability to target tumors and inhibit PD1/PDL1 checkpoints (Shi and Wang 2021). A densely branched DNA nanostructure with multivalent interactions and biomolecular recognition potential was designed and utilized for NK surface engineering, inspired by natural polyvalent interactions (Mammen et al.

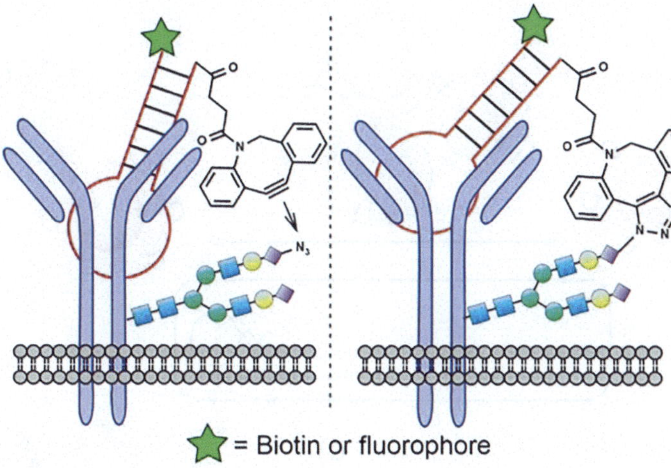

Fig. 5.4 The operational mechanism of an aptamer-based labeling approach for the targeted alteration of glycoproteins on the plasma membrane. (Reprinted with permissions from Robinson et al. (2015), copyright 2015, American Chemical Society)

1998). An aptamer for identifying the matching cancer antigen was present in each branch (Fig. 5.5c). The apparent K_d value of a polyvalent aptamer may be one order of magnitude lower than that of a monovalent aptamer due to polyvalent interactions. Therefore, compared to natural NK cells or those functionalized using monovalent aptamers, the NK cells designed with the polyvalent nanostructure could identify and eliminate target cancer cells with more efficacy (Shi et al. 2020). In this proof-of-concept work, the supramolecular DNA nanostructure hybridized with a single aptamer type; however, for enhanced cell recognition, this nanostructure can be functionalized with several aptamer types or any other ligands, including peptides, antibodies (Shi et al. 2020).

DNA aptamers have gained significant interest in cell-surface engineering, but only a few can target cell-membrane proteins despite many aptamers available for soluble proteins. Identifying aptamers with high specificity and affinity for cell-surface proteins remains a challenge, leading most research to focus on in vitro rather than in vivo studies. A detection unit exhibiting an exceptional target binding affinity is the essential function of aptamers utilized in CSE. For functional units attached to aptamers to induce site-specific changes, aptamers are typically required to selectively bind to their targets on the cell surface (Elskens et al. 2020). This design approach is appropriate for researchers who want to address concerns about the mobility, quantity, and distribution pattern of certain plasma membrane components in their natural habitat. Aptamer-based probes are often linked to the cell membrane via chemical conjugation or hydrophobic interaction to target extracellular components (Ni et al. 2020). Aptamer-based cell membrane modification encounters challenges, especially regarding the limited quantity and target specificity of aptamers compared to antibodies. However, aptamers show potential

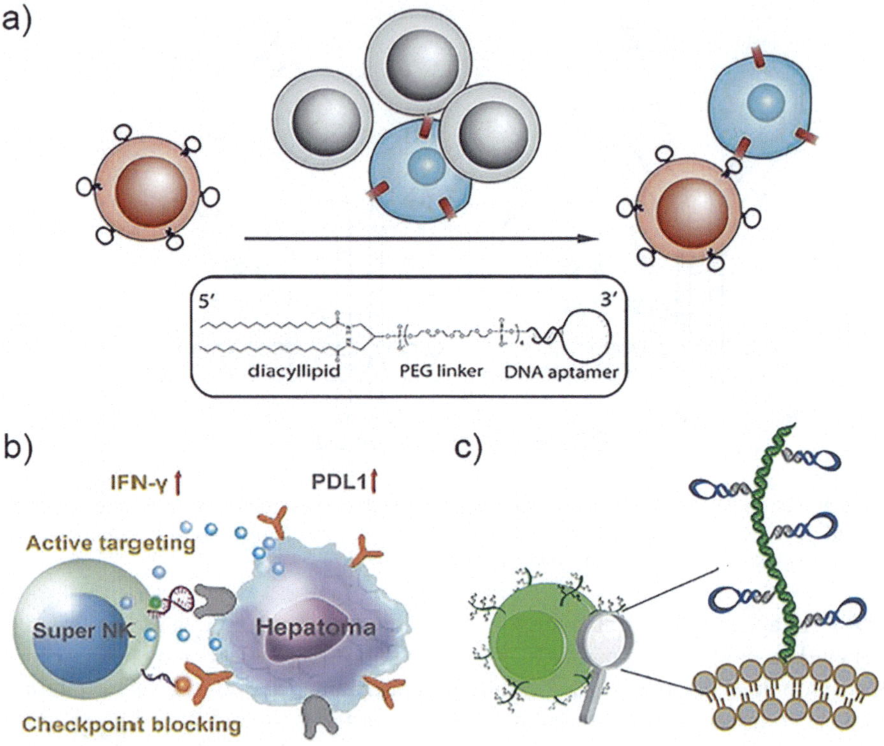

Fig. 5.5 CSE to enhance interactions between immune cells and cancer cells. (**a**) Monovalent surface engineering using a single aptamer type. (Reprinted with permissions from Xiong et al. (2012)), copyright 2012, Wiley-VCH GmbH). (**b**) Two kinds of aptamers combined with monovalent surface engineering. (Reprinted with permissions from Zhang et al. (2020), copyright 2020, Wiley-VCH GmbH). (**c**) Several aptamers combined with polyvalent surface engineering. The strategy makes use of aptamers, although any ligand, including antibodies and peptides, can be substituted. (Reprinted with permissions from Shi et al. (2020), copyright 2020, Wiley-VCH GmbH)

advantages in molecular design, chemical modification, and steric action. To overcome these limitations, efforts should focus on improving current methods to increase aptamer availability and discovering new aptamer targets using advanced technologies like cell-based SELEX (Zhou and Rossi 2017). This could enable biologists and chemists to develop aptamers into a toolkit similar to antibodies. While aptamer-based sensors monitor cell surface environments, they often require extra washing to remove unbound sensors and struggle to distinguish between analytes in the bulk phase and plasma membrane microenvironments. Future research may focus on creating membrane-binding-responsive modules to temporarily deactivate sensors before surface contact. Additionally, many aptamer-based DNA nanomachines work slowly through strand displacement or conformational transition, failing to capture rapid biological processes or short-term concentration changes of target analytes (Andrianova and Kuznetsov 2020).

The ability to use aptamer-mediated cell surface modifications in living things is a huge step forward in personalized medicine and drug development. For example, NK cells and cytotoxic T lymphocytes adorned with aptamers exhibit enhanced selectivity and target effectiveness in recognizing tumor cells. The chemical engineering of probiotic bacteria using aptamers is being rigorously studied to effectively counteract the immune-suppressive milieu within malignancies. A significant technological issue is sustaining the prolonged exposure of cell surface-localized aptamers (Zhang and Wang 2024). Cell surface-localized aptamers diminish over time due to spontaneous dissociation or rapid translocation, despite chemical or physical modifications. Solutions include (i) improving hydrophobic membrane anchors for stable retention and (ii) presenting aptamers in a multivalent manner. A major challenge remains enhancing the in vivo efficacy of aptamers, including their stability in complex matrices and cell-targeting capacity (Zhang et al. 2021). Only a small number of aptamers exhibit affinity for targets in animal models, likely due to their selection under ideal conditions (e.g., pure buffers, low temperatures) and target-enriched environments, which differ from the complex physiological contexts of cells. Enhancing aptamer selection, structural design, and functional optimization will enhance cell membrane modification, making aptamer-based biomaterials more applicable in biomedical fields (Urmi et al. 2024).

5.2.3 Antibodies

The immune system's B cells develop the Y-shaped, highly specialized proteins known as antibodies, which have a strong affinity for antigens. Their selectivity is used in CSE to precisely target and modify cells. Certain indicators on the cell membrane can be recognized and bound to by antibodies. For additional engineering or therapeutic interventions, this selectivity makes it possible to target certain cell types—such as immune cells, cancer cells, or stem cells—selectively. Functional molecules, nanoparticles, and other biomolecules can be attached to the cell surface using antibodies as anchors (Aribi 2020). For example: To develop bio-catalytic systems, enzymes can be immobilized on cell surfaces. It is possible to attach fluorescent markers for diagnostic and imaging applications. Cells are labeled using antibodies conjugated with radionuclides, fluorescent dyes, or nanoparticles. Tracking cell motions, interactions, or localization within biological systems is made easier by this labeling, which is crucial for research on immune response and metastasis (Feng et al. 2022).

It is possible to design antibodies to either activate or inhibit particular cell surface receptors. For instance, checkpoint inhibitors and other antibodies disrupt inhibitory receptors (such PD-1/PD-L1) in cancer immunotherapy to boost the immune system's ability to fight malignancies. In order to support desired signaling pathways, monoclonal antibodies can activate receptors. It is possible to build antibodies to transport medications, therapeutic substances, or gene-editing tools (like

CRISPR-Cas9) straight to the cytoplasm or cell surface. This improves treatment efficacy and reduces off-target effects (Dermani et al. 2019).

Monoclonal antibodies (mAbs) are widely utilized in targeted therapies. Their capacity to identify individual epitopes renders them optimal for accurate cell surface modification. Bispecific antibodies are designed to concurrently bind two distinct antigens, enabling the formation of cell-cell contacts or dual-functionalized surfaces. Single-chain variable fragments (scFvs) and Fab fragments are diminutive and exhibit enhanced tissue penetration, rendering them appropriate for densely packed cellular environments. Antibody conjugates are antibodies attached to therapeutic agents (e.g., antibody-drug conjugates) or imaging molecules for integrated therapy and diagnostics (Madsen et al. 2024).

Unintentional interactions might happen even with their great specificity. Computational modeling and sophisticated screening techniques enhance antibody design and lessen off-target effects. Under physiological circumstances, antibodies may deteriorate or cease to function. Stabilization techniques that increase their endurance include glycosylation and PEGylation (Oliveira et al. 2021). The high cost of large-scale antibody production drives advancements in cell-free expression systems and recombinant DNA technologies aimed at reducing production costs. More sophisticated applications are being made possible by the combination of nanotechnology and synthetic biology. For instance, smart cell engineering with responsive or programmable behaviors can be made possible by antibodies coupled with nanomaterials (Hunt et al. 2025). Antibody-functionalized cells have the ability to either directly target and destroy cancer cells or provide immune-stimulating signals to malignancies. Stem cells may be guided to certain organs by antibodies, which improves their integration and therapeutic potential. Diagnostic gadgets employ surfaces or cells coated with antibodies to find biomarkers or infections. Systemic toxicity is decreased by antibody-based targeting systems, which enhance medication delivery to certain tissues or cells. Cancer cells express immunosuppressive checkpoint ligands such as CD47, CD41, CD19, PD-L1, anti-VCAM1 antibodies, and SIRPα. The importance of employing the antibodies against these ligands for cancer immunotherapy has been given in Table 5.2.

5.2.4 Liposomes

Liposomes are sphere shaped synthetic vesicles with an aqueous core surrounded by one or more phospholipid bilayers. They are extensively utilized in medicine, biochemistry, and biophysics, especially for drug administration and as models for biological membrane research. When phospholipids with hydrophilic heads and hydrophobic tails form bilayers in an aqueous medium, liposomes are formed. Preparation techniques like sonication or extrusion can influence the size of these structures, which range in size from tiny unilamellar vesicles with a single bilayer to giant multilamellar vesicles with many layers (Norling 2020).

Zhou et al. established a novel technique to increase MSCs for the therapy of radiation-induced lung damage. Antioxidant liposomes (Lipo-OPC) were coupled

Table 5.2 An overview of antibodies used in cancer immunotherapy that target immunosuppressive checkpoint ligands

Antibody target	Role	Therapeutic applications	Ref.
SIRPα	Suppresses macrophage activity and inhibits phagocytosis by interacting with CD47.	Enhancing macrophage-mediated tumor clearance by disrupting the interaction between macrophages.	Sakano et al. (2024)
CD19	B-cell marker frequently overexpressed in B-cell malignancies.	Treatments for lymphoma and B-cell leukemia involve the use of anti-CD19 antibodies, such as CAR-T therapies and monoclonal antibodies.	Al-Amodi et al. (2025)
CD41	Participates in platelet aggregation and might be a factor in tumor metastasis.	Antibodies that target CD41 may lessen metastases and thrombosis linked to cancer.	Borsig (2008)
PD-L1	Inhibits immunological responses and promotes tumor immune evasion via binding to PD-1 on T cells.	Immune checkpoint inhibitors, such as anti-PD-L1 antibodies, increase T-cell activation against malignancies.	Blank et al. (2005)
CD47	The immunosuppressive ligand CD47 inhibits macrophage-mediated phagocytosis by interacting with SIRPα.	Anti-CD47 antibodies facilitate immune-mediated tumor removal by blocking this interaction.	Hao et al. (2023)
Anti-VCAM1	VCAM-1 interacts with integrins to promote tumor cell adherence and metastasis.	By blocking this adherence, anti-VCAM1 antibodies may slow the spread of metastases.	Schlesinger and Bendas (2015)

to MSC membrane via click reactions, resulting in MSCs-Lipo-OPC without affecting cell functions. This transformation allowed MSCs-Lipo-OPC to effectively localize to lung tissues. The transformed MSCs impacted innate immune cells including Tregs, neutrophils, and macrophages causing them to polarize towards anti-inflammatory actions. This technique successfully halted the advancement of radiation-induced pulmonary fibrosis. This technique enhanced the therapeutic value of both the drugs and the stem cells through the combination of drug-loaded NPs and MSCs, improving their use in radiation-induced damage management (Zhou et al. 2023).

Stroke is a serious disorder with a high death rate. M2 microglia have shown potential in decreasing inflammation and increasing neuronal healing. Nevertheless, their targeted distribution to stroke locations is difficult due to their confinement in the reticuloendothelial system. Platelet-hybrid microglia-based frameworks have already been created to address these issues and improve adherence to ischemic brain tissues. IL-4 has been embedded in ultrasound-responsive liposomes and coupled to platelet-hybrid microglia through click reactions involving azide-labelled microglia and DBCO-labeled liposomes. This approach allowed platelet-hybrid microglia to attach to damaged brain tissues, wherein ultrasound stimulated IL-4 and triggered M2 polarization across both designed and native microglia. This technique enhanced neuroprotection, vascular repair, and neurogenesis, demonstrating its promise for recovery from stroke (Li et al. 2021a). To facilitate nanoparticle attachment, ferrocene-coupled liposomes and red blood cells (RBCs) modified with β-cyclodextrin were incubated together. In the presence of high ROS levels in inflamed lungs, ferrocene-coupled liposomes detach from β-cyclodextrin-modified RBCs, releasing therapeutic agents to reduce inflammation significantly (Li et al. 2022).

Two lipid bilayers combine to form a single, continuous membrane during membrane fusion, a crucial biological process that is necessary for vesicle trafficking, virus entry, fertilization, and neurotransmitter release. Liposomes are used as models to investigate this intricate process. Membrane contact usually initiates fusion, which is followed by hemifusion, in which the bilayers' outside leaflets combine but their inner layers stay distinct. Following the formation of a fusion pore, which permits the contents to mingle, the pore eventually enlarges to finish the fusion (Marsden et al. 2011). Environmental factors, membrane curvature, fusion proteins, and lipid composition all have an impact on the fusion process. While fusion proteins like SNAREs or viral glycoproteins propel the process in biological systems, certain lipids encourage fusion by disrupting the bilayer. Because liposomes may be designed to merge with biological membranes and transfer medications or genetic material straight into cells, they have useful uses in drug delivery. They are also employed in the investigation of viral fusion processes and in the administration of vaccines (An 2023).

5.2.5 Bioorthogonal Groups

Click chemistry refers to a specific type of irreversible conjugation between two molecules characterized by strong orthogonality and rapid second-order reaction rate constants, unaffected by surrounding molecules. A variety of click reactions have been investigated for surface engineering, including Staudinger ligation, copper(I)-catalyzed azide-alkyne cycloaddition, strain-promoted alkyne-azide cycloaddition (SPAAC), and inverse-electron-demand Diels–Alder (IEDDA) reaction (Thirumurugan et al. 2013). A multitude of reviews have been written on various click-reaction mechanisms, and readers are referred to the subsequent research. The recent advancement of copper-free click methodologies, such as SPAAC and IEDDA, has facilitated their utilization in medicine (Rigolot et al. 2021). These bioorthogonal methods possess multiple roles in cell membrane engineering. Bioorthogonal groups can conjugate drug-loaded nanoparticles or liposomes to the membrane for targeted drug delivery. They are utilized in cell surface modification to investigate receptor-ligand interactions or immunological recognition. Moreover, in theranostics, bioorthogonal processes enable the conjugation of imaging agents to cell membranes, allowing for real-time monitoring of treatment responses (Wu et al. 2024).

5.2.5.1 Metabolic Glycoengineering-Based Bioorthogonal Group's Introduction

Over 90% of proteins on the cell membrane undergo glycosylation as a modification that occurs after translation. Cell adhesion, immunological responses, identification and cell interactions all are dependent on such modifications. These implications have led to studies on metabolic glycoengineering, which involves adding synthetic molecules to glycans for medical purposes. Bioorthogonal groups (such as terminal alkynes or azides) on artificial monosaccharide analogs (such as sialic acid, mannosamine, glucosamine, and galactosamine) can permeate into cells and undergo conversion into nucleotide sugars for glycosylation (Costa et al. 2025). Elevated unnatural tags on sialoglycoproteins are made possible by the subsequent transport of these altered glycoconjugates to plasma and endosomal membranes. On their secreted nanovesicles, parent cells' metabolic alterations may be charted. By adjusting the monosaccharide concentration and incubation time, tag amount may be managed. By using metabolic glycoengineering, glycans can have many bioorthogonal groups bonded to them. The conjugation of medicinal compounds to the surface of CMNVs is made possible by preinstalled bioorthogonal groups (for example, azide) acting as handles for click reactions, such as those with dibenzylcyclooctyne (DBCO).

5.2.5.2 Liposomes Fusion Based Bioorthogonal Group's Introduction

The liposomal fusion-based transport (LiFT) method enables the targeted attachment of functional molecules, including DNA, to the inner surface of the cell membrane (Lee et al. 2016). This is accomplished by integrating cholesterol-DNA conjugates into liposomes and enzymatically destroying external DNA, illustrating the possibility for precise localization control. In addition to DNA anchoring, LiFT functions as a multifaceted platform for the targeted delivery of functional moieties, especially metal catalysts, to designated membrane areas (Emmetiere et al. 2013). Figure 5.6 demonstrates the MAC-LiFT method's capability to facilitate the deposition of metal catalytic species on the surfaces of membranes (Elahipanah et al. 2016). The resultant membrane-anchored catalysts (MACs) demonstrate improved stability and activity, enabling two catalytic centers to function simultaneously with spatial-temporal accuracy. This innovation facilitates effective abiotic reactions on the cytoplasmic membrane and has prospective applications in intracellular catalysis and drug administration, involving the building of bioactive molecules that contravene traditional molecular design principles (Chen et al. 2023).

Despite its potential, obstacles such as response efficiency, off-target effects, and in vivo stability remain. Future study will focus on optimizing reaction kinetics and improving biocompatibility in order to increase their therapeutic and diagnostic uses.

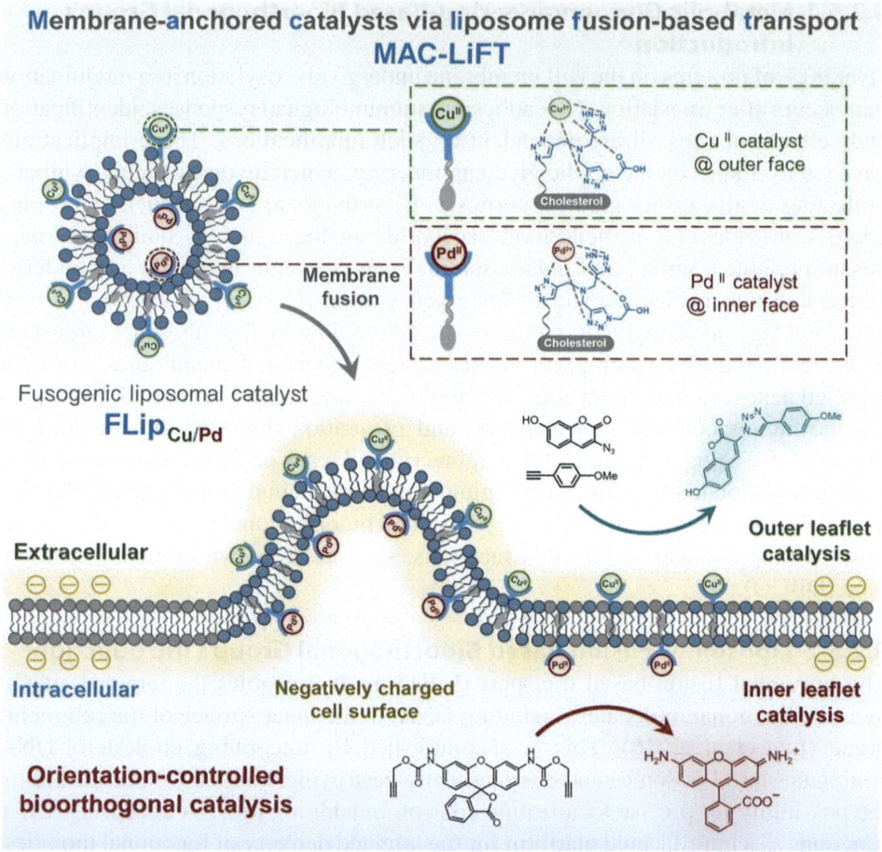

Fig. 5.6 Diagram illustrating the use of two different kinds of bioorthogonal metal catalysts in the MAC-LiFT technique to design the plasma membrane. (Reprinted with permissions from Chen et al. (2023), copyright 2023, American Association for the Advancement of Science)

5.2.6 Fluorescent Probes

Specialized compounds known as fluorescent probes glow when exposed to particular light wavelengths. They are essential resources for understanding, observing, and modifying the cell surface in CSE. These probes offer vital information on molecular interactions, cell activity, and the results of programmed changes.

The lipid architecture of biomembranes exhibits tremendous variety, as shown by sensing and imaging. As a result, the plasma membrane contains the bulk of sphingomyelin and cholesterol, which preferentially cluster at the outer leaflet and form microdomains that are in charge of several cell processes (Casares et al. 2019). Cell biologists and biophysicists have advanced chemical tools for studying biological membranes, providing deeper insights into their structures and functions. The exceptional sensitivity of fluorescence methods, enhanced by notable advancements in fluorescence microscopy, makes it especially relevant to fluorescent probes (Drummen 2012). Given the vast array of fluorescent membrane probes that have been generated, we were compelled to categorize and further arrange them in this account according to their optical response, sensing profile, and targeting principles. There are several probes for organelles and plasma membrane in terms of targeting. The most researched target is most likely the cell membrane. In terms of the sensing profile, we find probes for chemical and biophysical sensing as well as basic membrane markers with continuous emission (Fig. 5.7). Traditional membrane markers are made from fluorophores that are unaffected by the environment and used solely for biomembrane imaging. In contrast, environment-sensitive fluorophores, which change characteristics like fluorescence intensity, emission color, or lifetime based on viscosity, polarity, and other factors, form the basis of membrane probes for biophysical sensing. Solvatochromic membrane probes, designed to detect lipid order and local polarity, have been developed (Demchenko et al. 2009).

Other teams devised flippers to deal with the membrane tension, lipid order and molecular rotors to track local viscosity. Localized chemical species including ions, redox species, protons (pH), as well as biomolecules including proteins and lipids, are detected by the chemically sensitive probes at the biomembranes. Lastly, the response profile of the membrane probes may be used to classify them. The

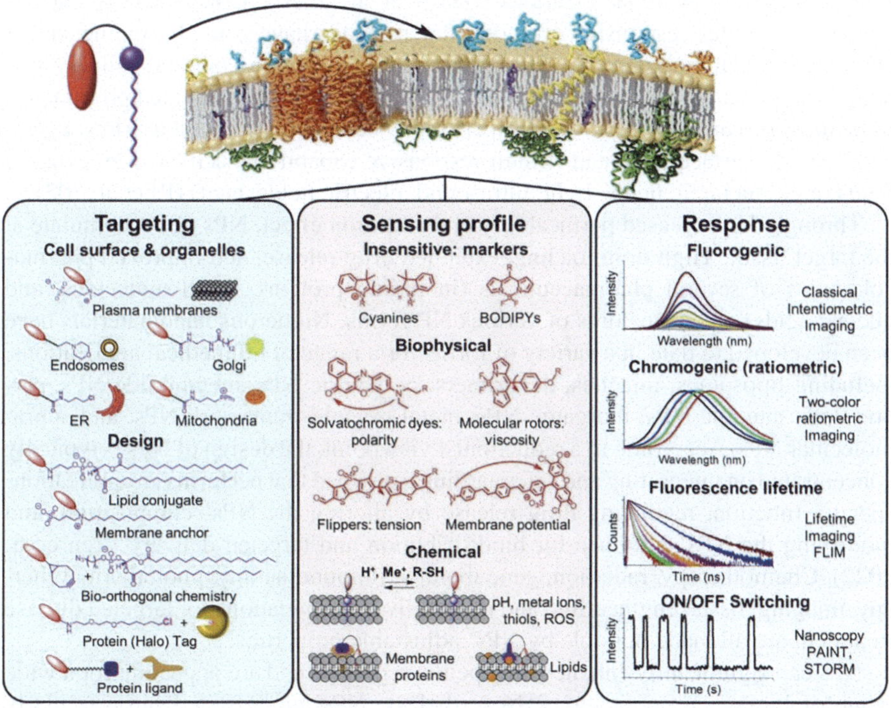

Fig. 5.7 Generic framework for membrane probe categorization. (Reprinted with permissions from Klymchenko (2022), copyright 2022, American Chemical Society)

traditional microscopy techniques are utilized with fluorogenic probes, which illuminate their fluorescence, enabling background-free bioimaging (Specht et al. 2017).

Ratiometric probes react by altering the color of their emission or excitation; we have studied solvatochromic probes in great detail. Their emission is separated into red and blue components under a microscope, and this information is then further examined as a ratio or as a "generalized polarization" (GP) characteristic (Klymchenko 2017). Fluorescence lifetime imaging (FLIM) can then be used with fluorescence lifetime-responsive probes, including flippers and fluorescence molecular rotors. Lastly, single-molecule localization microscopy (SMLM) may be performed with probes that blink. It is important to distinguish between SMLM probes that undergo photo-switching (STORM) and those that function by reversible binding to biomembranes (PAINT) (Klymchenko 2022).

5.2.7 Functional Nanomaterials

Drug delivery pertains to the provision of medicinal molecules to specific cells or tissue, including processes of absorption, distribution, administration, and elimination. Conventional methods depend on the systemic circulation of the "naked drug," which has significant limits, since over 90% of the medication in systemic circulation is removed by renal clearance (Dash et al. 2020). Consequently, carrier-mediated stimulus responsive drug delivery methods have gained prominence for enabling optimal drug deposition at disease sites and creating tailored delivery systems. Nanotechnology-derived solutions, particularly NPs, have gained prominence in medical research owing to their superior physicochemical characteristics, extensive specific surface area, and stimuli-responsive capabilities activated by external fields (e.g., magnetic fields, light, ultrasound, electric fields, heat) (PJ et al. 2021).

Through the increased permeability and retention effect, NPs can accumulate at the target tissue. High drug loading, extended drug release, and improved pharmacokinetics of several pharmaceuticals (including proteins, small molecules, and nucleic acids) are all features of various NP forms. Numerous nanomaterials have been developed to date in a variety of forms for a range of biomedical applications, including liposomes, micelles, dendrimers, polymeric NPs, metal oxide NPs, carbon dots, metallic NPs, inorganic NPs, metal-organic framework NPs, and supramolecular NPs. According to a reductionist viewpoint, the design of NPs is typically concentrated on integrating and safeguarding a payload that performs an application-specific function, regulating drug release by altering the NPs' composition, and modifying the NPs' interface for biodistribution and targeted delivery (Ren et al. 2022). Chemotherapy, radiation, gene therapy, immunotherapy, photothermal therapy, imaging, tissue engineering, and drug delivery applications for targeted disease treatment are all made possible by NPs' adjustable properties.

NPs can significantly enhance therapeutic regimens and are appealing for a wide range of biological applications. Nevertheless, NPs administered intravascularly face a complicated circulatory environment, and their practical use is limited by a number of NPs' limitations. For instance, a process known as "opsonization" occurs in the circulation when different macromolecular components create a protein corona on NPs surface (Yao et al. 2014). The physicochemical characteristics of NPs (larger particle size), circulation time, targeting efficiency, penetrating potential, non-specific tissue accumulation, and poor colloidal stability are all impacted by the development of protein corona. Additionally, the immune system can identify NPs because of their "foreign" components, and the reticuloendothelial system can quickly remove them before they reach their target. Because Pristine NPs often lack the potential to perceive the disease microenvironment and are unable to pass through biological barriers, there is less drug accumulation at the target site (Li et al. 2021b). Unpredictable treatment results and severe side effects are caused by NPs' non-specific dispersion and off-target effects. Furthermore, because of structural changes brought on the protein contact with the NPs surface, proteins may get denatured and lose their ability to function as intended. These drawbacks limit their clinical use and highlight the necessity of surface engineering techniques (Nienhaus and Nienhaus 2019).

The main focus of notable developments in nanotechnology is on improving the characteristics of NPs by means of chemical surface modifications (e.g., targeted ligand conjugation and stealth surface modification). These tactics were developed to overcome the difficulties posed by bare NPs and to allow for site-specific medication delivery. In order to achieve anti-biofouling, immune escape, and extended systemic circulation time, stealth-based modification involves coating various polymeric materials, such as polyethylene glycol (PEG), polyethylene oxide (PEO), zwitterionic phospholipids, and PEG-based copolymers (poloxamines, polysorbates, and poloxamers), onto the surface of nanoparticles. However, targeting ligand conjugation onto NP surfaces (e.g., sialic acid, aptamers, antibodies, peptides, phage) has been utilized to increase the EPR effect and improve targeting efficiency and preferential accumulation at a target region (Lowe et al. 2015).

Enhancing their effectiveness for biological applications may require combining targeted ligand conjugation with stealth surface modification. However, a number of studies have shown the unexpected removal of PEGylated nanomaterials in vivo following repeated and prolonged administration, as well as the formation of anti-PEG antibodies in some situations. Additionally, even when single targeting ligands are present, meta-analyses have demonstrated the ongoing limits of inadequate NPs concentration at the target locations. For improved clinical translation, biomimetic and functional coating techniques on the NP surface are required due to immunogenicity issues with PEGylated NPs and the limited targeted capability of ligand-conjugated material (Chehelgerdi et al. 2023).

A common technique for creating cell-mimicking nanoplatforms is to create a core-shell nanostructure by covering synthetic nanomaterials with a layer of natural cell membrane. Due to its ability to evade immune clearance, increase blood circulation time, penetrate biological barriers, improve targeting efficiency, and provide targeted homing potential, cell membrane coating has attracted significant attention. Because of the unique functional expression of cell surface proteins, these prospective bio interfacing and functional capabilities of cell membranes rely on cell sources. In the past 10 years, camouflaging synthetic NPs with a particular type of cell membrane has received a lot of attention. Single cell membrane-coated NPs, however, show certain limitations in their functionality. For example, RBCs membrane-coated NPs have the potential to increase circulation time, but their limited therapeutic benefits are due to their poor targeting capacity. Similarly, NPs coated with cancer cell membranes show great effectiveness in targeting tumors, but they have limited circulation capacity and don't appear to be sufficiently covert to avoid immune monitoring (Liu et al. 2019).

Recently, coating hybrid cell membranes onto NPs has gained significant attention. By fusing the cell membranes of two distinct cell sources, a hybrid cell membrane can be created. This success has led to the development of various "hybrid membranes," including those between osteosarcoma cells and macrophages, RBCs and cancer cells, RBCs and platelets, and platelets and leukocytes. In comparison to original single parent membrane coated nanoplatforms, hybrid cell membranes may conduct complicated tasks in concert, have protein markers from both cell sources, inherit several source cell-relevant functionalities, and are adaptable to complex biological systems. Furthermore, because tailored cell membrane coating onto NPs offers novel targeting properties, multitasking capabilities, and additional therapeutic potential, a great deal of study has been done on the subject. Ultimately, the distinct bio interfacing capabilities, biomimetic design, and synergistic benefits of both natural cell membranes and synthetic nanomaterials have drawn attention to cell membrane camouflaged NPs in the field of biomedicine. In a variety of biomedical applications, such as cancer immunotherapy, targeted drug delivery for cancer treatment, brain disease treatment, inflammation control, disease diagnosis/tracking, detoxification, and the treatment of infectious diseases, engineered cell membrane camouflaged NPs offer enormous potential (Liu et al. 2023).

5.2.8 Polymeric Materials and Hydrophobic Anchors

Functional polymeric materials are synthetic or natural polymers designed to interact with or modify cell surfaces. Their tunability, versatility, and ability to incorporate various functionalities make them powerful tools in CSE, enabling applications in biomedical research, tissue engineering, immunotherapy, and diagnostics. Functional polymeric materials surface engineering refers to the deliberate modification of polymer surfaces to enhance or impart specific functional characteristics, improving their interaction with the environment, other materials, or biological systems. This process is crucial in applications such as biomedicine, electronics, environmental technologies, and packaging, where surface properties like wettability, adhesion, biocompatibility, and durability are key. Standard methods of surface engineering include plasma treatment, which introduces functional groups (e.g., -OH, -COOH) to improve wettability and adhesion; chemical functionalization, where molecules are covalently bonded to the surface via techniques like graft polymerization or click chemistry; and layer-by-layer (LbL) assembly, which deposits alternating layers of polymers or nanomaterials to create thin films with desired properties like anti-fouling or bioactivity. Self-assembled monolayers form single molecular layers on polymer surfaces, enhancing properties like hydrophobicity or biocompatibility, while laser treatment induces surface morphology changes to improve adhesion or roughness (Xu et al. 2009). Coating and deposition techniques, such as spray coating or physical vapor deposition, add protective or functional layers to polymers, imparting qualities like UV resistance or antimicrobial effects. Nanomaterial incorporation further improves polymer surfaces by enhancing conductivity, mechanical strength, or antibacterial properties, and surface etching or roughening increases surface area or promotes cell attachment for tissue engineering applications. UV and ozone treatments can oxidize polymer surfaces, introducing reactive groups for enhanced reactivity and interaction. These surface modifications are vital for applications in biomedical fields, such as implants, drug delivery systems, and antimicrobial surfaces, as well as in electronics, where improved adhesion and conductivity are needed (Neděla et al. 2017). In environmental applications, functionalized polymer surfaces can prevent microbial fouling and improve water repellency, making them ideal for water filtration, self-cleaning surfaces, or energy-harvesting devices. Overall, functional polymer surface engineering allows for the precise tuning of surface properties of biological entities without altering the intrinsic functions, enabling the creation of highly specialized materials for various technological, medical, and environmental uses (Das and Das 2023).

To lessen the inflammatory consequences in transplanted islets, Im et al. suggested a double-layer shielding technique utilizing N-hydroxy succinimidyl-linked unfractionated heparin (UFH) and multi-arm PEG-catechol (PEG8). In an allograft model, the PEG8/UFH-coated islet surface demonstrated much longer survival than unmodified islets, maintained cell viability, and exhibited anticoagulant activity. Shielded islets exhibited prolonged survival and function when combined with immunosuppressive drugs, indicating their ability to stop immune assaults and blood-mediated inflammatory responses in islet transplantation (Im et al. 2013). Employing a polymerized network of dopamine molecules, Wang et al. hypothesized a method based on muscle sticky chemistry to protect antigenic epitopes on RBCs. pDA-coated RBCs (pDA-RBCs) can be used in blood transfusions and can stop coagulation responses. Even after many transfusions, pDA-RBCs show outstanding survival characteristics, according to systemic evaluation in a mouse model. pDA coating can create universal RBCs that are insulated from antigens and has no negative impact on RBCs morphology (Wang et al. 2014). A "cell-in-shell" strategy was proposed to conceal immune-stimulating epitopes on RBCs surfaces, using a supramolecular iron ion (Fe^{III}) and tannic acid (TA) based supramolecular metal-phenolic network coating. The galloyl groups of TA form a multivalent coordination with Fe^{III} ions, giving the coating its black color. The TA/Fe^{III} nanofilm exhibits strong surface adhesion to various substrates, including live cells, facilitating the creation of immune-camouflaged RBCs. The thick MPN shell does not alter the RBCs natural shape or oxygen-carrying capacity and effectively reduces immunoglobulin-induced agglutination. This immune-camouflaged RBCs strategy may aid in the advancement of universal blood transfusion options (Park et al. 2017).

Hydrophobic anchors such as DSPE (1,2-distearoyl-sn-glycero-3--phosphoethanolamine),DPPE (1,2-dipalmitoyl-sn-glycero-3-phosphoethanolamine), DMPE (1,2-dimyristoyl-sn-glycero-3-phosphoethanolamine), and CHOL (cholesterol) are pivotal in cell membrane engineering due to their potential to spontaneously integrate into the lipid bilayer of cell membranes. These molecules possess hydrophobic tails that insert into the membrane, while their functionalized heads can be chemically modified with reactive groups (e.g., alkyne, azide, or maleimide). This allows for the covalent conjugation of a broad spectrum of functional materials, including biomolecules, fluorescent probes, and nanoparticles, directly to the cell surface through bio-orthogonal reactions or other chemistries. This lipid insertion mechanism provides a powerful and versatile tool for non-invasively functionalizing cell surfaces for applications in targeted delivery, imaging, and cellular manipulation.

5.2.9 Synthetics Drugs and Microenvironment Modulating Agents

Although regular monitoring is difficult and may not be able to manage blood sugar levels, conventional insulin injections offer momentary control. Recent research has focused on glucose-responsive insulin administration utilizing RBCs, indicating that glucose-derivative-modified insulin (Glc-insulin) attaches to RBCs efficiently through interactions that involve glucose transporter (GLUT) molecules on their surface. Because free glucose competes with GLUT, this binding can be reversed in hyperglycemia, releasing insulin from RBCs. The infusion of Glc-insulin-conjugated RBCs extended the therapeutic impact of insulin, according to in vivo investigations conducted on a diabetic mouse model. The blood sugar levels were successfully kept within the standard range using this method. The glucose-sensitive insulin delivery device replicates the action of pancreatic β-cells and could revolutionize the management of diabetes (Wang et al. 2017).

More than 80% of patients undergoing joint revision surgery have heterotopic ossification, which may be extremely problematic for medical professionals and is common in burn and trauma situations. Although naïve osteoclasts display restricted bioactivity and adherence to ectopic locations, they are unable to eradicate calcified plaques. Osteoclasts (OCs) can remodel bone by removing organic bone matrix and mineralized bone component throughout resorption of bone. Tetracycline (TC) has been used to chemically modify osteoclasts in order to overcome this issue, producing engineered osteoclasts (TC-osteoclasts). TC attaches itself on hydroxyapatite to target calcium minerals. TC-osteoclasts shown more potential for cell therapy than naïve osteoclasts because of their wide calcium mineral affinity, which allowed them to efficiently target ectopic calcified regions and eliminate heterotopic ossification (Fig. 5.8) (Jin et al. 2021).

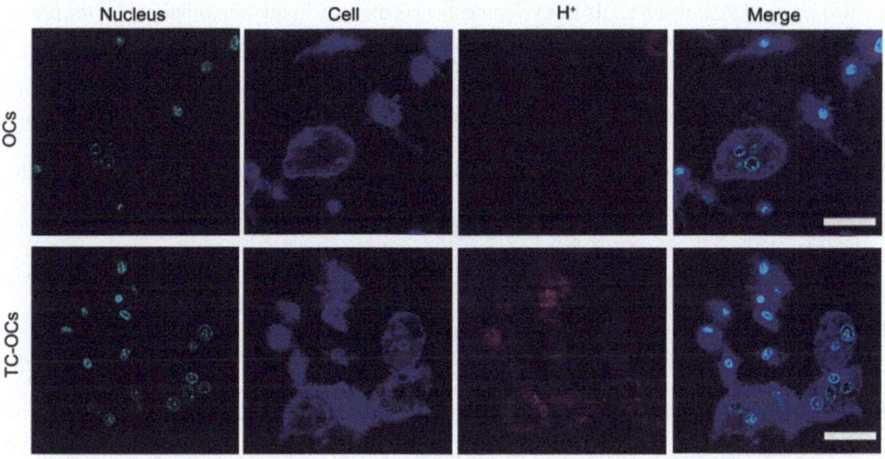

Fig. 5.8 Confocal laser scanning microscopy images demonstrate ectopic calcified tissue resorption potential of osteoclasts (OCs) and tetracycline-OCs (TC-OCs) seeded into cultured plates. Scale bar: 20 μm. (Reprinted with permissions from Jin et al. (2021), copyright 2021, Springer Nature)

5.3 Conclusions

This chapter explores the transformative potential of functional biomaterials in cell membrane engineering to address complex biomedical challenges. Researchers have leveraged materials such as recombinant receptors, peptides, aptamers, antibodies, and nanomaterials to develop innovative strategies aimed at improving cellular function, overcoming immune rejection, enhancing targeting mechanisms, and advancing therapeutic applications. Despite the promising advancements, tailoring these materials for clinical use presents challenges that highlight the need for continuous improvements in biocompatibility, precision, and cost-effectiveness. The integration of these technologies holds significant potential to revolutionize fields such as regenerative medicine, tissue engineering, drug delivery, and diagnostics, thereby facilitating the development of more efficient and personalized biomedical solutions. Combining natural cell membrane features with synthetic materials, along with advancing surface modification techniques to control the degree of modification and address batch-to-batch variability, is crucial for overcoming existing shortcomings and enhancing the effectiveness of engineered cell membrane-based platforms in biomedical application.

References

Abdal Dayem A, Yan E, Do M, Kim Y, Lee Y, Cho S-G et al (2024) Engineering extracellular vesicles for ROS scavenging and tissue regeneration. Nano Converg 11(1):24

Ahmed I, Qazi IM, Jamal S (2016) Developments in osmotic dehydration technique for the preservation of fruits and vegetables. Innovative Food Sci Emerg Technol 34:29–43

Al-Amodi HS, Bedair HM, Gohar S, Mohamed DA-W, Abd El Gayed EM, Nazih M et al (2025) FCRL1 and BAFF mRNA expression as novel diagnostic and prognostic biomarkers in diffuse large B-cell lymphoma: expression signatures predict R-CHOP therapy response and survival. Int J Mol Sci 26(3):1269

An D (2023) Membrane rupture, membrane fusion and the regulation of exocytosis. Columbia University

Andrianova M, Kuznetsov A (2020) Logic gates based on DNA aptamers. Pharmaceuticals 13(11):417

Aribi M (2020) Introductory chapter: B-cells. In: Normal and malignant B-cell. IntechOpen, London, p 3

Blank C, Gajewski TF, Mackensen A (2005) Interaction of PD-L1 on tumor cells with PD-1 on tumor-specific T cells as a mechanism of immune evasion: implications for tumor immunotherapy. Cancer Immunol Immunother 54:307–314

Borsig L (2008) The role of platelet activation in tumor metastasis. Expert Rev Anticancer Ther 8(8):1247–1255

Campos Muñiz C, Fernández Perrino FJ (2023) Evolution of the concepts of architecture and supramolecular dynamics of the plasma membrane. Membranes 13(6):547

Casares D, Escribá PV, Rosselló CA (2019) Membrane lipid composition: effect on membrane and organelle structure, function and compartmentalization and therapeutic avenues. Int J Mol Sci 20(9):2167

Chehelgerdi M, Chehelgerdi M, Allela OQB, Pecho RDC, Jayasankar N, Rao DP et al (2023) Progressing nanotechnology to improve targeted cancer treatment: overcoming hurdles in its clinical implementation. Mol Cancer 22(1):169

Chen Y, Wu T, Xie S, Bai Y, Xing H (2023) Orientation-controlled membrane anchoring of bioorthogonal catalysts on live cells via liposome fusion–based transport. Science. Advances 9(19):eadg2583

Cheng Q, Li R, He Y, Zhu Y, Kang Y, Ji X (2024) Genetically engineered cellular nanovesicles: theories, design and perspective. Adv Funct Mater 34(46):2407842

Choi W-T, Duggineni S, Xu Y, Huang Z, An J (2012) Drug discovery research targeting the CXC chemokine receptor 4 (CXCR4). J Med Chem 55(3):977–994

Cooper ST, McNeil PL (2015) Membrane repair: mechanisms and pathophysiology. Physiol Rev 95(4):1205–1240

Costa AF, Teixeira A, Reis CA, Gomes C (2025) Novel anticancer drug discovery efforts targeting glycosylation: the emergence of fluorinated monosaccharides analogs. Expert Opin Drug Discov 20(2):193–203

Coulie PG, Van den Eynde BJ, Van Der Bruggen P, Boon T (2014) Tumour antigens recognized by T lymphocytes: at the core of cancer immunotherapy. Nat Rev Cancer 14(2):135–146

Das P, Das S (2023) Novel polymeric materials for environmental applications. World Scientific, Singapore

Das D, Jothimani G, Banerjee A, Dey A, Duttaroy AK, Pathak S (2024) A brief review on recent advances in diagnostic and therapeutic applications of extracellular vesicles in cardiovascular disease. Int J Biochem Cell Biol 173:106616

Dash P, Piras AM, Dash M (2020) Cell membrane coated nanocarriers-an efficient biomimetic platform for targeted therapy. J Control Release 327:546–570

Demchenko AP, Mély Y, Duportail G, Klymchenko AS (2009) Monitoring biophysical properties of lipid membranes by environment-sensitive fluorescent probes. Biophys J 96(9):3461–3470

Dermani FK, Samadi P, Rahmani G, Kohlan AK, Najafi R (2019) PD-1/PD-L1 immune checkpoint: potential target for cancer therapy. J Cell Physiol 234(2):1313–1325

Drummen GP (2012) Fluorescent probes and fluorescence (microscopy) techniques—illuminating biological and biomedical research. Molecules 17(12):14067–14090

Elahipanah S, Radmanesh P, Luo W, O'Brien PJ, Rogozhnikov D, Yousaf MN (2016) Rewiring gram-negative bacteria cell surfaces with bio-orthogonal chemistry via liposome fusion. Bioconjug Chem 27(4):1082–1089

Elskens JP, Elskens JM, Madder A (2020) Chemical modification of aptamers for increased binding affinity in diagnostic applications: Current status and future prospects. Int J Mol Sci 21(12):4522

Emmetiere F, Irwin C, Viola-Villegas NT, Longo V, Cheal SM, Zanzonico P et al (2013) 18F-labeled-bioorthogonal liposomes for in vivo targeting. Bioconjug Chem 24(11):1784–1789

Feng Y, Xu Y, Liu S, Wu D, Su Z, Chen G et al (2022) Recent advances in enzyme immobilization based on novel porous framework materials and its applications in biosensing. Coord Chem Rev 459:214414

Funauchi M, Serada S, Hiramatsu K, Funajima E, Kanda M, Nagase Y et al (2024) Tumor cell-expressed lipolysis-stimulated lipoprotein receptor negatively regulates T-cell function. Int J Cancer 154(3):425–433

Gezehagn Kussia G, Tessema TS (2024) The Potential of single-chain variable fragment antibody: role in future therapeutic and diagnostic biologics. J Immunol Res 2024(1):1804038

Guerrache A, Micheau O (2024) TNF-related apoptosis-inducing ligand: non-apoptotic signalling. Cells 13(6):521

Guo S, Zhu X, Loh XJ (2017) Controlling cell adhesion using layer-by-layer approaches for biomedical applications. Mater Sci Eng C 70:1163–1175

Hailing T, Yonghong P, Yufeng Z, Haitao T (2022) Challenges for the application of EGFR-targeting peptide GE11 in tumor diagnosis and treatment. J Control Release 349:592–605

Hao Y, Zhou X, Li Y, Li B, Cheng L (2023) The CD47-SIRPα axis is a promising target for cancer immunotherapies. Int Immunopharmacol 120:110255

Hunt AC, Rasor BJ, Seki K, Ekas HM, Warfel KF, Karim AS et al (2025) Cell-free gene expression: methods and applications. Chem Rev 25(1):91–149

Im B-H, Jeong J-H, Haque MR, Lee DY, Ahn C-H, Kim JE et al (2013) The effects of 8-arm-PEG-catechol/heparin shielding system and immunosuppressive drug, FK506 on the survival of intraportally allotransplanted islets. Biomaterials 34(8):2098–2106

Jia H-R, Zhang Z, Fang X, Jiang M, Chen M, Chen S et al (2022) Recent advances of cell surface modification based on aptamers. Mater Today Nano 18:100188

Jin W, Lin X, Pan H, Zhao C, Qiu P, Zhao R et al (2021) Engineered osteoclasts as living treatment materials for heterotopic ossification therapy. Nat Commun 12(1):6327

Klymchenko AS (2017) Solvatochromic and fluorogenic dyes as environment-sensitive probes: design and biological applications. Acc Chem Res 50(2):366–375

Klymchenko AS (2022) Fluorescent probes for lipid membranes: from the cell surface to organelles. Acc Chem Res 56(1):1–12

Lee J, Lee H, Goh U, Kim J, Jeong M, Lee J et al (2016) Cellular engineering with membrane fusogenic liposomes to produce functionalized extracellular vesicles. ACS Appl Mater Interfaces 8(11):6790–6795

Lee DY, Cha B-H, Jung M, Kim AS, Bull DA, Won Y-W (2018) Cell surface engineering and application in cell delivery to heart diseases. J Biol Eng 12:1–11

Li Y, Teng X, Yang C, Wang Y, Wang L, Dai Y et al (2021a) Ultrasound controlled anti-inflammatory polarization of platelet decorated microglia for targeted ischemic stroke therapy. Angew Chem Int Ed 60(10):5083–5090

Li H, Wang Y, Tang Q, Yin D, Tang C, He E et al (2021b) The protein corona and its effects on nanoparticle-based drug delivery systems. Acta Biomater 129:57–72

Li J, Ding Y, Cheng Q, Gao C, Wei J, Wang Z et al (2022) Supramolecular erythrocytes-hitchhiking drug delivery system for specific therapy of acute pneumonia. J Control Release 350:777–786

Liu B, Wang W, Fan J, Long Y, Xiao F, Daniyal M et al (2019) RBC membrane camouflaged Prussian blue nanoparticles for gamabutolin loading and combined chemo/photothermal therapy of breast cancer. Biomaterials 217:119301

Liu H, Su Y-Y, Jiang X-C, Gao J-Q (2023) Cell membrane-coated nanoparticles: a novel multifunctional biomimetic drug delivery system. Drug Deliv Transl Res 13(3):716–737

Liu Q, Guan Y, Li S (2024) Programmed death receptor (PD-) 1/PD-ligand (L) 1 in urological cancers: the "all-around warrior" in immunotherapy. Mol Cancer 23(1):183

Lowe S, O'Brien-Simpson NM, Connal LA (2015) Antibiofouling polymer interfaces: poly (ethylene glycol) and other promising candidates. Polym Chem 6(2):198–212

Madsen AV, Pedersen LE, Kristensen P, Goletz S (2024) Design and engineering of bispecific antibodies: insights and practical considerations. Front Bioeng Biotechnol 12:1352014

Mammen M, Choi SK, Whitesides GM (1998) Polyvalente Wechselwirkungen in biologischen Systemen: Auswirkungen auf das Design und die Verwendung multivalenter Liganden und Inhibitoren. Angew Chem 110(20):2908–2953

Marsden HR, Tomatsu I, Kros A (2011) Model systems for membrane fusion. Chem Soc Rev 40(3):1572–1585

Neděla O, Slepička P, Švorčík V (2017) Surface modification of polymer substrates for biomedical applications. Materials 10(10):1115

Nguyen Q-A, Klein PM, Xie C, Benthall KN, Iafrati J, Homidan J et al (2024) Acetylcholine receptor based chemogenetics engineered for neuronal inhibition and seizure control assessed in mice. Nat Commun 15(1):601

Ni S, Zhuo Z, Pan Y, Yu Y, Li F, Liu J et al (2020) Recent progress in aptamer discoveries and modifications for therapeutic applications. ACS Appl Mater Interfaces 13(8):9500–9519

Nienhaus K, Nienhaus GU (2019) Towards a molecular-level understanding of the protein corona around nanoparticles–recent advances and persisting challenges. Curr Opin Biomed Eng 10:11–22

Nikitovic D, Kukovyakina E, Berdiaki A, Tzanakakis A, Luss A, Vlaskina E et al (2024) Enhancing tumor targeted therapy: the role of iRGD peptide in advanced drug delivery systems. Cancers 16(22):3768

Norling K (2020) Liposomes for mucosal vaccine delivery: physicochemical characterization and biological application. Chalmers Tekniska Hogskola, Gothenburg

Oliveira LM, Gasser T, Edwards R, Zweckstetter M, Melki R, Stefanis L et al (2021) Alpha-synuclein research: defining strategic moves in the battle against Parkinson's disease. npj Parkinson's Dis 7(1):65

Othman Z, Pastor BC, van Rijt S, Habibovic P (2018) Understanding interactions between biomaterials and biological systems using proteomics. Biomaterials 167:191–204

Park T, Kim JY, Cho H, Moon HC, Kim BJ, Park JH et al (2017) Artificial spores: immunoprotective nanocoating of red blood cells with supramolecular ferric ion-tannic acid complex. Polymers 9(4):140

Pietrobono S, Stecca B (2021) Aberrant sialylation in cancer: biomarker and potential target for therapeutic intervention? Cancers 13(9):2014

Raucher D (2019) Tumor targeting peptides: novel therapeutic strategies in glioblastoma. Curr Opin Pharmacol 47:14–19

Ren R, Lim C, Li S, Wang Y, Song J, Lin T-W et al (2022) Recent advances in the development of lipid-, metal-, carbon-, and polymer-based nanomaterials for antibacterial applications. Nanomaterials 12(21):3855

Rigolot V, Biot C, Lion C (2021) To view your biomolecule, click inside the cell. Angew Chem 133(43):23268–23289

Robinson PV, de Almeida-Escobedo G, de Groot AE, McKechnie JL, Bertozzi CR (2015) Live-cell labeling of specific protein glycoforms by proximity-enhanced bioorthogonal ligation. J Am Chem Soc 137(33):10452–10455

PJ JC, Saigeetha S, Samrot AV, Ponniah P, Chakravarthi S (2021) Overview on toxicity of nanoparticles, it's mechanism, models used in toxicity studies and disposal methods–a review. Biocatalysis and Agricultural. Biotechnology 36:102117

Sakano Y, Sakano K, Hurrell BP, Shafiei-Jahani P, Kazemi MH, Li X et al (2024) SIRPα engagement regulates ILC2 effector function and alleviates airway hyperreactivity via modulating energy metabolism. Cell Mol Immunol 21(10):1158–1174

Schlesinger M, Bendas G (2015) Vascular cell adhesion molecule-1 (VCAM-1)—an increasing insight into its role in tumorigenicity and metastasis. Int J Cancer 136(11):2504–2514

Shi P, Wang Y (2021) Synthetic DNA for cell-surface engineering. Angew Chem 133(21):11684–11695

Shi P, Wang X, Davis B, Coyne J, Dong C, Reynolds J et al (2020) In situ synthesis of an aptamer-based polyvalent antibody mimic on the cell surface for enhanced interactions between immune and cancer cells. Angew Chem Int Ed 59(29):11892–11897

Specht EA, Braselmann E, Palmer AE (2017) A critical and comparative review of fluorescent tools for live-cell imaging. Annu Rev Physiol 79(1):93–117

Stephan MT, Irvine DJ (2011) Enhancing cell therapies from the outside in: cell surface engineering using synthetic nanomaterials. Nano Today 6(3):309–325

Sushnitha M, Evangelopoulos M, Tasciotti E, Taraballi F (2020) Cell membrane-based biomimetic nanoparticles and the immune system: immunomodulatory interactions to therapeutic applications. Front Bioeng Biotechnol 8:627

Swee LK, Lourido S, Bell GW, Ingram JR, Ploegh HL (2015) One-step enzymatic modification of the cell surface redirects cellular cytotoxicity and parasite tropism. ACS Chem Biol 10(2):460–465

Tang S, Davoudi Z, Wang G, Xu Z, Rehman T, Prominski A et al (2021) Soft materials as biological and artificial membranes. Chem Soc Rev 50(22):12679–12701

Thirumurugan P, Matosiuk D, Jozwiak K (2013) Click chemistry for drug development and diverse chemical–biology applications. Chem Rev 113(7):4905–4979

Ullah S, Xu H-T, Shen J (2024) ROS-degradable polythioketal urethane scaffold for porcine wound repair. Acta Pharm Sin B 14(10):4628

Urmi R, Banerjee P, Singh M, Singh R, Chhillar S, Sharma N et al (2024) Revolutionizing biomedicine: aptamer-based nanomaterials and nanodevices for therapeutic applications. Biotechnol reports 42:e00843

Vafaei S (2017) Solid-supported lipid bilayer as a platform to study cell-extracellular matrix interactions

Wang B, Wang G, Zhao B, Chen J, Zhang X, Tang R (2014) Antigenically shielded universal red blood cells by polydopamine-based cell surface engineering. Chem Sci 5(9):3463–3468

Wang C, Ye Y, Sun W, Yu J, Wang J, Lawrence DS et al (2017) Red blood cells for glucose-responsive insulin delivery. Adv Mater 29(18):1606617

Wang D-E, Gao X, You S, Chen M, Ren L, Sun W et al (2020) Aptamer-functionalized polydiacetylene liposomes act as a fluorescent sensor for sensitive detection of MUC1 and targeted imaging of cancer cells. Sensors Actuators B Chem 309:127778

Wei F-L, Zhai Y, Wang T-F, Zhao J-W, Wang C-L, Tang Z et al (2024) Stem cell–homing biomimetic hydrogel promotes the repair of osteoporotic bone defects through osteogenic and angiogenic coupling. Sci Adv 10(44):eadq6700

Wu X, Hu JJ, Yoon J (2024) Cell membrane as a promising therapeutic target: from materials design to biomedical applications. Angew Chem 136(18):e202400249

Xiong X, Liu H, Zhao Z, Altman MB, Lopez-Colon D, Yang CJ et al (2012) DNA aptamer-mediated cell targeting. Angew Chem 52(5):1472

Xu Z-K, Huang X-J, Wan L-S (2009) Surface engineering of polymer membranes. Springer Science & Business Media, Hangzhou/Berlin

Xu J, Wang B, Li S (2021) Gene and cell therapy. Elsevier, Remington, pp 463–488

Yao J, Yang M, Duan Y (2014) Chemistry, biology, and medicine of fluorescent nanomaterials and related systems: new insights into biosensing, bioimaging, genomics, diagnostics, and therapy. Chem Rev 114(12):6130–6178

Zahra QA, Ullah S, Shahzad F, Qiu B, Fang X, Ammar A et al (2022) MXene-based aptasensors: advances, challenges, and prospects. Prog Mater Sci 129:100967

Zhang Y, Wang J (2024) Research progress of cell membrane biomimetic nanoparticles for circulating tumour cells. Front Oncol 14:1389775

Zhang D, Zheng Y, Lin Z, Liu X, Li J, Yang H et al (2020) Equipping natural killer cells with specific targeting and checkpoint blocking aptamers for enhanced adoptive immunotherapy in solid tumors. Angew Chem 132(29):12120–12126

Zhang B, Wang C, Du Y, Paxton R, He X (2021) A 'smart'aptamer-functionalized continuous label-free cell catch–transport–release system. J Mater Chem B 9(35):7196–7204

Zhang H, Yan W, Wang J, Xie S, Tao WA, Lee C-W et al (2024) Surface functionalization of exosomes for chondrocyte-targeted siRNA delivery and cartilage regeneration. J Control Release 369:493–505

Zhou J, Rossi J (2017) Aptamers as targeted therapeutics: current potential and challenges. Nat Rev Drug Discov 16(3):181–202

Zhou H, Zhang Y, Pei P, Shen W, Yi X, Yang K (2023) Liposome-anchored mesenchymal stem cells for radiation pneumonia/fibrosis treatment. Biomaterials 300:122202

Genetically Mediated Cell Membrane Engineering Strategies and Their Scope in Biomedical Applications

6

Mohammad Tollabi, Mohammad Hassan Kheirandish, Arezou Mehrabi, Zahra Poursalehi, Maryam Heidari, and Peiman Brouki Milan

Abstract

Genetically mediated cell surface engineering is widely explored for modifying both eukaryotic and prokaryotic membranes. These approaches show promise in utilizing engineered membrane-based platforms, such as living cells, membrane-coated nanoparticles, and membrane-derived vesicles, for various biomedical applications. Genetic engineering aims to modify cell membranes with recombinant molecules to enhance biological interactions, improve targeting efficiency, boost immunological responses, and reduce the toxicity of membrane-based therapeutic platforms. Additionally, it can enable the loading of recombinant nucleic acids into the vesicle lumen and facilitate the fusion of recombinant molecules with vesicle surface proteins. This approach enhances drug loading, membrane surface functionalization, and significantly improves the bioactivity,

M. Tollabi · M. Heidari · P. B. Milan (✉)
Department of Tissue Engineering and Regenerative Medicine, Faculty of Advanced Technologies in Medicine, Iran University of Medical Sciences, Tehran, Iran

Cellular and Molecular Research Center, Iran University of Medical Sciences, Tehran, Iran
e-mail: boroukimilan.t@iums.ac.ir

M. H. Kheirandish
Department of Medical Biotechnology, School of Medicine, Shahid Sadoughi University of Medical Sciences, Yazd, Iran
e-mail: mh.kheirandish@ssu.ac.ir

A. Mehrabi
School of Medicine, Tonekabon Branch, Islamic Azad University, Tonekabon, Iran
e-mail: arezoumehrabbi@iau.ac.ir

Z. Poursalehi
Department of Applied Cell Sciences, Faculty of Medicine, Urmia University of Medical Sciences, Urmia, Iran
e-mail: poursalehi.za@umsu.ac.ir

Z. Hussain, R. Pei (eds.), *Cell Membrane Engineering for Advancing Cell-Based and Nanotechnology-Based Therapies*, Biomaterials, Bioengineering and Sustainability 6, https://doi.org/10.1007/978-3-031-93425-4_6

targeting efficiency, and pharmaceutical function of engineered membrane-derived vesicles. This chapter explores advanced gene delivery systems, including viral vectors (e.g., retroviral, adenoviral, lentiviral) and non-viral vectors (e.g., liposomes, polymers, nanoparticles). It discusses genetic engineering strategies for improved drug loading, membrane surface modification, and reduced toxicity of membrane-based platforms. Furthermore, this chapter discusses the application of genetically modified cell membrane platforms for diverse biomedical applications, including cancer therapy, therapeutic production, gene delivery, and targeted disease treatment. This chapter also discusses challenges and future research directions for genetically engineered membrane-based platforms to address unmet clinical needs.

Keywords

Genetic engineering · Cell membrane modification · Pre-drug loading · Targeted drug delivery · Bioproduction · Cancer immunotherapy

6.1 Introduction

Understanding the crucial role of cell membrane in cell-environment interactions and their impact on cell fate and function led to the emergence of cell membrane engineering as a strategy to functionalize cells by modifying their surface (Custódio and Mano 2016). Cell surface engineering is a cutting-edge method that boosts the therapeutic capabilities of cell-based treatments by altering the surfaces of cells with different materials to enhance their functionality, longevity, and efficacy for regenerative medicine applications (Lee et al. 2018). The cell surface has been modified generally through two major strategies: non-genetic and genetic engineering. Non-genetic engineering approaches include covalent chemical conjugation and noncovalent physical bioconjugation (Liu et al. 2019). Genetic engineering is an ideal methodology for targeted modification of cell surface proteins by employing the biosynthetic machinery of the cells. Due to the importance of cell membrane receptors as primary conduits for collecting and transmitting environmental signals and the direct effect of these foreign signals on cellular functions, genetically mediated cell membrane engineering is a valuable approach to improve cell-based therapeutic approaches (Custódio and Mano 2016). Genetic engineering enables regulation of cell membrane proteins by introducing or knocking down strategies to increase the potential therapeutics of the cells and promote their *in vivo* function (Amaral and Pasparakis 2019). Various strategies are currently used to achieve high-efficiency cell transfection. These strategies are broadly classified into two main categories: viral delivery methods, including retroviral, lentiviral, and adenoviral systems, and non-viral delivery methods, such as liposomes, polymers, and nanoparticles (NPs).

This chapter explores advanced gene delivery systems, including viral vectors (e.g., retroviral, adenoviral, lentiviral) and non-viral vectors (e.g., liposomes, polymers, NPs). It discusses genetic engineering strategies for improved drug loading, membrane surface modification, and reduced toxicity of membrane-based platforms. Furthermore, this chapter discusses the application of genetically modified cell membrane platforms for diverse biomedical applications, including cancer immunotherapy, biomanufacturing, gene delivery, and targeted disease treatment. This chapter also discusses challenges and future research directions for genetically engineered membrane-based platforms to address unmet clinical needs.

6.2 Therapeutic Membrane Features of Different Cells and Need for Engineering

Cells, the fundamental units of life, rely on the cell membrane to define their structure and function. This membrane, composed primarily of a bilayer of phospholipids, functions as a protective barrier that shields the cell's interior from external disturbances. Beyond its structural role, the membrane enables selective exchange of substances and facilitates signal transmission from the extracellular environment. In addition to its structural function, the membrane controls the selective exchange of substances and facilitates signaling through its specialized transmembrane proteins (Huang et al. 2024). It has been found that cell membranes have the intrinsic properties of their original cells that must be considered in their biomedical application. For example, red blood cells (RBCs) membranes due to their low immunogenicity and long half-life in blood circulation can be used as an ideal carrier for drug delivery (Yu et al. 2021). Platelet cell membranes have been used for therapeutic purposes and the production of anti-tumor structures due to their strong cell-cell connections (Di Huang et al. 2024). Cancer cell membranes are commonly utilized in tumor therapy due to their similar structure to tumor cells and the presence of surface antigens. In this context, cancer cell membranes were coupled with poly(lactic-co-glycolic acid) NPs (Fang et al. 2014), or drug-loaded NPs (Sun et al. 2016; Chen et al. 2016a) for cancer treatment approaches. This strategy not only improved NPs accumulation at the tumor site (Sun et al. 2017), but also increased drug uptake, and optimal utilization of antigen-presenting cells (Kroll et al. 2017; Zhao et al. 2022a; Li et al. 2022). Macrophages are specialized cells tasked with identifying, engulfing, and eliminating "invaders." Due to the expression of the CCR2 (C—C chemokine receptor type 2) on the macrophage surface, the macrophage membrane shows unique properties such as tumor-targeting capabilities via CCR2-CCL2 axis, and drug-targeted delivery to the inflamed site (Lim et al. 2016; Zhao et al. 2018). Macrophage membrane-coated NPs are widely used in regenerative medicine for multiple applications (Gao et al. 2020; Li et al. 2024a; Gu et al. 2024).

Cell-based treatments have gained popularity as a promising treatment for various diseases, especially incurable diseases. The demand for cell therapy products is forecast to increase more rapidly in the next decades due to increasing investment from industry (De Pieri et al. 2021). Beyond recent clinical and commercial successes, there are numerous challenges that limit the widespread clinical use of cell-based therapeutics (Mount et al. 2015). Low cell survival, retention, and tissue integration at the target site are among the most important challenges after cell transplantation that decrease the therapeutic impact of cell-based therapy. Moreover, untargeted cell therapy-related side effects should be considered before clinical using of cell-based therapeutics (De Pieri et al. 2021). To overcome these challenges and decrease off-target effects, genetic engineering strategies are used to manipulate native cells, especially on their plasma membrane by expressing specific proteins or receptors (Ding et al. 2022; Oh et al. 2021; Hassanzadeh et al. 2022; Wu et al. 2022a).

6.3 Methods for Genetic Engineering

To protect genetic materials against biological fluid enzymes and prevent immune responses genetic materials are not used in naked form, and they are used in vector structure (Planul and Dalkara 2017). The choice of gene-carrying vector should be based on the type and amount of genetic material to be transfected, and the genetic engineering strategy. Effective vectors should possess key characteristics, including targeted genetic material transfer, low immune activation, cost-effectiveness, and minimal genotoxicity risk (Butt et al. 2022). Gene-carrying vectors used in genetic engineering approaches generally include viral and non-viral vectors.

6.3.1 Viral Vectors

Viral vectors are applicable tools for introducing specific DNA sequences into host cells through their inherited infectious ability. To date, different DNA and RNA-based viral vectors have been designed to carry and deliver genetic materials for basic researches, gene therapy, and vaccine development (Park and Lee 2024). Viral vectors are preferred over non-viral ones due to their high transduction efficiency, and their expression can be tailored for both short- and long-term applications (Lundstrom 2023). The main reason for utilizing viruses as gene-carrying vectors is their natural ability to infect cells and efficient gene delivery into host cells. Retroviruses, adenoviruses, and lentiviruses are widely used viral vectors in gene therapy and cell engineering (Fig. 6.1) (Park and Lee 2024; Zhao et al. 2022b).

6.3.1.1 Retroviral

Retroviruses are characterized by enzymes like reverse transcriptase and integrase, which play key roles in the retroviral cell cycle. Retroviral vectors, derived from natural retroviruses, are commonly used in gene therapy for cell modification (Yoder et al. 2021). For example, Moloney murine leukemia virus is commonly used in gene delivery due to its high gene transfer and integration efficiency, leading to permanent genetic modification. Successful gene therapy with these vectors requires high production of infectious retroviral particles, effective transfection, and sufficient transgene expression to achieve therapeutic effects without toxicity (Morgan et al. 2021). High gene delivery capacity and prolonged gene expression are among

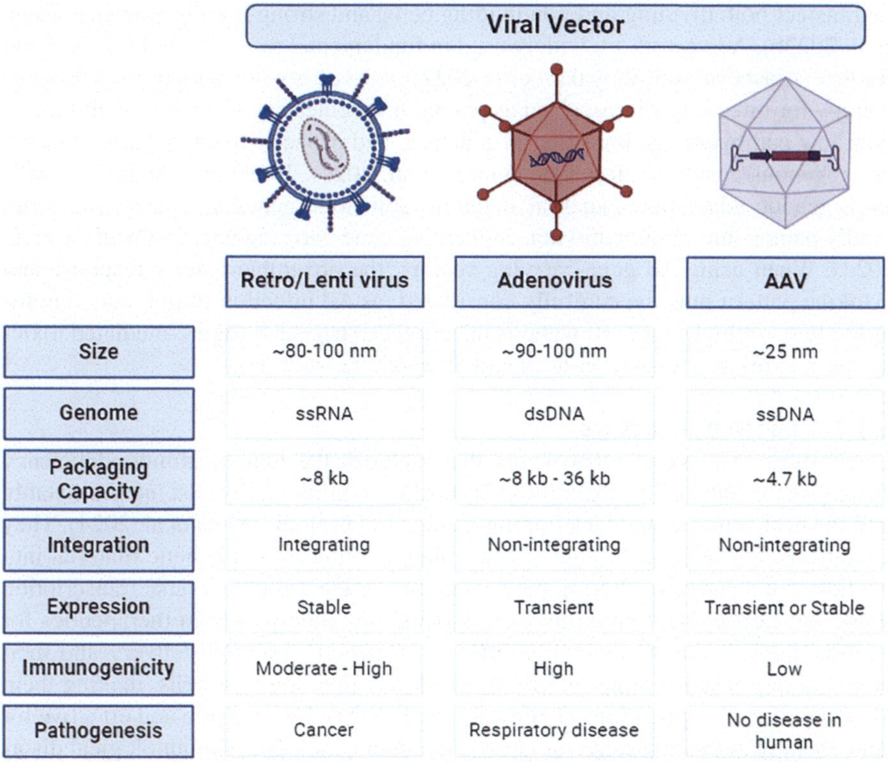

Fig. 6.1 The main features of typical viral gene-carrying vectors for gene delivery and therapeutics. (Reprinted with permission from Park and Lee (2024), copyright 2024, Springer Nature)

the most important advantages of these types of vectors (Yi et al. 2011). These vectors are ideal for gene delivery in biomedical applications, such as chimeric antigen receptor (CAR)-T cell production (Pinto et al. 2025). In addition to gene therapy and vaccine development, retroviral vectors are used as display systems for protein engineering (Urban and Merten 2011). Early retroviral vectors faced limitations due to insertional mutagenesis risks, but modern variants like lentiviruses are widely used clinically. However, retroviral vectors are limited to infecting dividing cells and carry a high risk of insertional mutagenesis due to off-target effects (Zhao et al. 2022b).

6.3.1.2 Adenoviral

Adenoviruses (Ad) are non-enveloped viruses having a double-stranded DNA genome that can be isolated from several vertebrate hosts, including humans, mice, and dogs (Sakurai et al. 2022). They are among the earliest gene-carrying viral vectors explored for potential in vivo gene therapy applications. Ad vectors are popular genetic engineering tools for various biomedical applications, such as cancer treatment and vaccine formulation, due to their short-term high gene expression, ability to transfect both dividing and nondividing cells, and strong immunogenicity (Zhao et al. 2022b). Ad vectors are widely used in fundamental research, and translational studies (Atasheva and Shayakhmetov 2022). Ad vectors are widely used in gene therapy for infectious diseases and as potential vaccines due to their easy modification, low pathogenicity, high-titer production, and capacity to carry large amounts of exogenous genetic material (Salauddin et al. 2024). To enhance safety and efficacy, genetic adaptations, such as deleting replication genes, transform Ads from mildly pathogenic viruses into non-replicating gene-carrying vectors (Wallace et al. 2024). When using Ad gene-carrying vectors, the pro-inflammatory response and cytokine pattern must be carefully considered, as Ad infection in immunocompromised hosts or high-dose intravenous injection can trigger cytokine-mediated toxicity and a cytokine storm (Atasheva and Shayakhmetov 2022).

6.3.1.3 Lentiviral Vectors

Lentiviruses, a class of retroviruses that includes the human immunodeficiency virus, possess enhanced capabilities to transduce non-dividing cells, integrate stably into the host genome, and lack immunogenic viral proteins (Maia et al. 2024). They are crucial in molecular biology as they integrate transported genetic material into the host cell's genome, altering the genetic landscape through reverse transcription (Park and Lee 2024). Lentiviral vectors (LVs) are widely used in therapeutics for various diseases due to their high efficacy in genetic material delivery and their potential to transduce nonproliferating or slowly dividing host cells, making them ideal for clinical applications (Jadlowsky et al. 2024). LVs are safe and effective for gene therapy in treating range of genetic disorders, including immunological disorders, and neurometabolic diseases. Their ability to transduce both proliferating and quiescent cells, coupled with stable genome integration, makes them ideal for diverse cell-based therapies (Poletti and Mavilio 2021).

6.3.2 Nonviral Vectors

Nonviral vectors are essential in cell-membrane engineering, offering a safer, more adaptable alternative to viral methods. They deliver genetic material without integration risks by forming DNA complexes taken up via endocytosis. Using cationic lipids or polymers, they protect DNA from degradation, enhance uptake, and improve transfection efficiency (Mintzer and Simanek 2009). Nonviral vectors employ a range of strategies to improve cellular uptake, including surface functionalization with biomimetic components like mannitol, which facilitates caveolae-mediated endocytosis (Ma et al. 2020). Recent advances in nonviral gene delivery systems, including liposomes, lipid NPs, and polymers, have enhanced mesenchymal stem cells (MSCs) transfection efficiency. Integrating these systems with effective uptake strategies enables precise phenotype engineering for therapeutic application (Hamann and Pannier 2022; Wang et al. 2023). Nonviral vectors are increasingly favored for gene delivery due to their safety over viral vectors, which pose risks like immunogenicity and toxicity. Innovations such as supramolecular

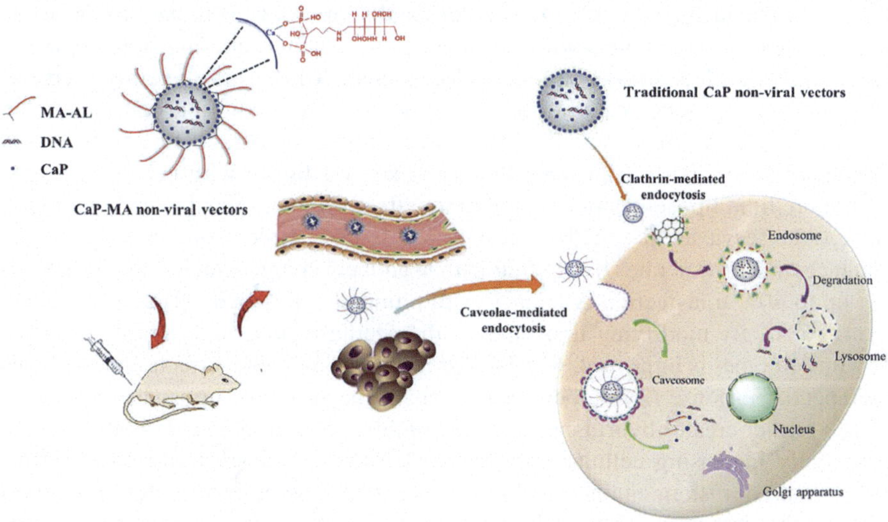

Fig. 6.2 Schematic illustration of the cellular internalization and intracellular transport of bio-inspired CaP-MA non-viral vectors. CaP calcium phosphate, MA-AL mannitol-alendronate. (Reprinted with permission from Ma et al. (2020), copyright 2020, BioMed Central Ltd)

self-assembled nanocomplexes enhance their effectiveness by efficiently crossing cellular and endosomal membranes, significantly improving gene delivery efficiency (Yin et al. 2013. Nonviral vectors are valued in gene therapy for their ease of preparation, high gene loading, and low immunogenicity. However, challenges such as poor intracellular bioavailability, rapid gene degradation, and lower transfection efficiency, particularly in vivo, limit their clinical use and hinder therapeutic success (Ma et al. 2020; Romano 2007; Marcinkowski et al. 2019). Unlike viral vectors, which naturally overcome cellular barriers, nonviral vectors must be strategically designed to achieve efficient delivery. Traditional nonviral vectors primarily rely on clathrin- and caveolae-mediated pathways for cellular uptake (Akita and Harashima 2008; Xiang et al. 2012), leading to different intracellular trafficking results: clathrin-mediated endocytosis typically directs cargo to lysosomes for degradation, whereas caveolae-mediated uptake circumvents this degradative environment, thereby preventing gene loss (Wattiaux et al. 2000; Luzio et al. 2009). Effectively managing cellular uptake and intracellular fate is a promising strategy to enhance the transgene efficiency of conventional non-viral delivery vectors (Fig. 6.2).

6.3.2.1 Liposomes

Liposomes are gaining recognition as versatile tools in genetic engineering, particularly for gene delivery. Their ability to encapsulate nucleic acids and enhance cellular uptake makes them a viable alternative to traditional viral vectors. These structures can encapsulate various genetic materials, including plasmid DNA, siRNA, and antisense oligonucleotides, improving stability and delivery efficiency

(Nordling-David and Golomb 2013). Additionally, liposome formulation can be tailored to modify interactions with cell membranes, such as altering lipid composition to influence NPs uptake pathways (Xiang et al. 2018). This alteration is crucial as it enhances genetic material delivery efficiency by modifying cell membrane properties, like stiffness and fluidity, which impact membrane fusion processes (Bompard et al. 2020). Liposome fluidity is key for fusion with cell membranes, with more fluid liposomes interacting more effectively with flexible and metastatic cells (Alekseeva et al. 2015). Cationic liposomes, made from cationic lipids, enhance transfection rates by binding with negatively charged nucleic acids, achieving up to 40% transfection efficiency with minimal cytotoxicity (Bravo-Estupiñan et al. 2024). By modifying liposomes with specific ligands, their ability to target certain cell types is improved, which helps minimize off-target effects and enhances therapeutic response (Zylberberg et al. 2017). The fusion of liposomes with exosomes further creates hybrid carriers that combine the advantages of both systems, potentially increasing cellular uptake and delivery efficiency (Sato et al. 2016). Besides direct fusion, engineered liposomes can also be utilized to develop hybrid systems that improve genetic delivery. Sato et al. outlined a technique for creating proteoliposomes capable of delivering hydrophilic peptides to cells through gap junctions, highlighting the potential of liposomes to promote intercellular communication and the transfer of genetic material (Sato et al. 2016). This approach highlights liposomes' adaptability in engineering cellular interactions and improving genetic payload delivery.

6.3.2.2 Polymers

Genetic engineering utilizes polymer techniques to enhance gene delivery and expression for biomedical applications. Key advancements include gene synthesis platforms, nonviral delivery systems, and biocompatible polymer carriers. Notably, seamless cloning of rolling-circle amplicons enables efficient synthesis of repetitive genes, facilitating diverse protein polymer libraries, improving transformation efficiency, and reducing costs (Homma et al. 2024) In the realm of nonviral delivery systems, polyplexes use polycationic compounds to condense nucleic acids, providing a safer alternative to viral methods. Nonviral vectors like cationic polymers have shown low toxicity and high stability, making them effective for the transient expression of genome-editing nucleases such as CRISPR/Cas9 (Germer and Wagner 2022). The advancement of nanoscale polymer carriers also enhances gene delivery by increasing bioavailability and reducing off-target effects (Rai et al. 2019; Schaffert and Wagner 2008). Despite advancements in polymer-based genetic engineering, challenges in cytotoxicity and delivery efficiency remain. Dendritic polymers are emerging as promising gene delivery systems, designed to enhance transfection efficiency, which is often low in synthetic non-viral vectors. Modifying their surface properties allows researchers to develop gene carriers with minimal toxicity and high target specificity, improving their ability to transport genetic material across cell membranes (Paleos et al. 2009). Another approach is cell encapsulation, which offers a secure and scalable method for delivering therapeutic proteins from genetically modified cells. This technique employs membrane polymers that protect the engineered cells from the immune system, enabling a controlled release of therapeutic proteins *in vivo*. The encapsulation process helps preserve cell viability while allowing the cells to react to external stimuli, such as specific inducers (Serguera et al. 1999). Additionally, Kalfopulos et al. developed a chitosan-based bone substitute with genetically modified periosteum cells to improve autologous grafts. Adenoviral infection with angiogenin and GFP genes enhanced bone regeneration, showing promise in a rabbit model for spine surgery (Zarate-Kalfopulos et al. 2006). Serguera et al. showed that engineered allogeneic cells implanted in membrane polymers enable sustained therapeutic protein delivery. These polymers provide immunoprotection, preventing rejection and enhancing long-term genetic engineering applications (Serguera et al. 1999).

6.3.2.3 Nanoparticles

Genetic engineering enhances cell membrane-camouflaged NPs by modifying their membranes to improve drug delivery, targeting, and therapeutic effectiveness. One approach involves altering the genetic material of the source cells from which the biological membrane is derived. By genetically engineering these cells, specific proteins or receptors can be overexpressed on the cell membrane. For instance, overexpression of Toll-like receptor 4 has been shown to enhance the NPs ability to recognize to bacterial components, thereby improving their antibacterial properties (Deng et al. 2023). Moreover, nonviral vectors, like polyethylenimine, aid in gene delivery by compacting DNA into NPs, which allows for receptor-mediated endocytosis. They improve intracellular transport and facilitate escape from endosomes through the 'proton sponge' effect, ultimately enabling the nuclear import of genetic material (Belguise-Valladier and Behr 2001). Cationic lipid-based liposomes also serve as nonviral vectors that self-assemble with nucleic acids to create lipoplexes. These vectors enhance gene delivery by regulating cellular gene expression, maximizing transfection efficiency while reducing cytotoxic effects (Pezzoli et al. 2013). Polyelectrolyte particles act as non-viral vectors for delivering anionic substances into cells, employing star-shaped or spherical micelle structures to support gene therapy and siRNA applications. This approach improves techniques for genetically mediated cell membrane engineering (He et al. 2010). Moreover, these nanomaterials can replicate the characteristics of natural cell membranes, serving as a platform for targeted drug delivery and immunotherapy (Li et al. 2018a). By modifying the membranes of coated NPs to display specific ligands or receptors, researchers can improve the targeting efficiency of these systems, leading to more effective treatments for diseases like cancer. Another technique is membrane fusion, which allows for the integration of various cellular membranes, combining different functionalities. By fusing membranes from engineered cells with synthetic NPs, a hybrid system can be created that leverages both the natural characteristics of cell membranes and the customizable features of synthetic materials (Zhang et al. 2021a).

6.4 Genetic Engineering Strategies for Cell-Membrane Engineering

Genetic engineering of cell membranes is gaining interest in therapeutic delivery, cellular interaction modulation, and synthetic biology. Both genetic and non-genetic methods enhance membrane functionality, including DNA-based scaffolds that attach molecules via hydrophobic interactions, avoiding complex genetic modifications and harmful chemicals (Guo et al. 2022). This approach facilitates the development of adjustable DNA scaffold networks that can effectively influence cell-cell interactions (Guo et al. 2022). Furthermore, integrating DNAzymes into cell membranes has demonstrated the ability to mediate specific interactions through metal ion regulation, thereby increasing the adaptability of DNA-based engineering methods (Qian et al. 2021). DNA nanotechnology enables programmable cell surfaces, while CRISPR/Cas9 revolutionizes genetic engineering by allowing precise membrane modifications to enhance targeting and interaction (Mali et al. 2013; Ma et al. 2014). For example, engineering MSCs to display homing ligands on their membranes has been shown to enhance their targeting abilities in therapeutic applications (Sarkar et al. 2010). This system enables precise genomic modifications, allowing for the expression of specific proteins on cell membranes that can improve cellular functions, such as targeting capabilities in CAR T cell therapy (Xiong et al. 2024). However, the complexity and expense associated with genetic engineering can be a barrier, prompting researchers to investigate alternative methods that may provide simpler and more efficient solutions (Roy et al. 2020). Additionally, CRISPR-based systems have been shown to aid in the creation of genetically engineered exosomes, which can be used for site-specific drug delivery (Olson et al. 2024). Moreover, CRISPR technology can be integrated with other genetic tools to optimize metabolic pathways and boost the functionality of engineered cells (Lian et al. 2017). Alongside genetic techniques, non-genetic methods like chemical cross-linking and lipid anchoring enable direct molecule attachment to cell membranes, complementing genetic engineering (Wang et al. 2024a). These methods allow for quick and reversible modifications of cell surfaces, which can be particularly beneficial in therapeutic settings where dynamic responses are necessary (Gabrielse et al. 2014). Exploring living biointerfaces, where engineered cells interact through modified membranes, expands possibilities in cell-membrane engineering by utilizing cells that express specific membrane proteins to enhance adhesion and direct differentiation. These methods aid tissue engineering by controlling cell fate and creating dynamic interfaces that respond to signals, enhancing the integration of engineered cells into host tissues (Custódio and Mano 2016). Additionally, the advancement of high-throughput screening methods, like fluorescence-activated cell sorting, has made it easier to identify the best genetic modifications for improving membrane properties (Menasche et al. 2019). These techniques enable rapid evaluation of engineered cells by surface traits, facilitating the selection of optimal groups for research and therapy, crucial for advancing cell-membrane engineering and clinical applications (Csizmar et al. 2018).

6.4.1 Genetic Engineering for Pre-drug Loading

Genetic engineering approaches in cell membrane engineering have become essential methods for optimizing pre-drug loading in therapeutic uses. These techniques utilize the distinctive characteristics of cell membrane-derived vesicles (CMVs) to improve the efficiency and targeting abilities of drug delivery systems. CMVs are membrane particles produced or secreted by nearly all living cells, featuring a phospholipid bilayer structure such as extracellular vesicles (EVs), and bacterial membrane vesical (BMVs) (El Andaloussi et al. 2013). EVs play a crucial role in sorting and delivering biomacromolecules such as proteins and nucleic acids to recipient cells, contributing to normal physiological functions, immune responses, and disease progression (El Andaloussi et al. 2013; Rädler et al. 2023; Jeppesen et al. 2023). Although the processes behind cargo sorting in EVs are not fully understood, they have been effectively utilized to encapsulate endogenous biomacromolecules (Zhang et al. 2021b). Artificially created EVs (aEVs) are generally produced using manual techniques like ultrasonic extrusion while maintaining the properties of EVs. In comparison to polymeric NPs, membrane vesicles provide benefits such as stable lipid bilayer structures, excellent biocompatibility, and reduced toxicity, positioning them as promising drug delivery vehicles for therapeutic applications (Zhang et al. 2021b; Munagala et al. 2016).

Natural CMVs face challenges like low yield and heterogeneity due to variations in source cells, biogenetic pathways, and purification methods, limiting their research and therapeutic applications (Jeppesen et al. 2023). This variability complicates the understanding of their components and functional properties, ultimately diminishing their therapeutic efficacy (Jeppesen et al. 2019; Buzas 2023). Furthermore, although CMVs contain biological elements from their originating cells, they often fall short in achieving specific therapeutic outcomes due to poor tissue targeting, inadequate signal modulation, and absence of external therapeutic enhancements (Raposo and Stahl 2023). Additional research has shown that engineered membrane vesicles are a promising form of nanomedicine with a wide array of biomedical applications (Dooley et al. 2021; Liang et al. 2021). By combining genetic engineering with bioorthogonal chemistry and techniques like membrane hybridization, both the internal and membrane surface functionalities of CMVs can be optimized, thereby enhancing or introducing new therapeutic functions. This approach has sparked increasing interest in engineered CMVs for biotherapeutic applications (Zhang et al. 2021b). Due to the presence of various EVs subpopulations (including exosomes, ectosomes, and apoptotic bodies) with similar morphologies and overlapping sizes, effective isolation and purification strategies are essential for elucidating their mechanisms and enhancing their biomedical applications. Unique purification techniques have been developed for EVs isolation, which can also be applied to genetically engineered donor cells (Fig. 6.3).

In addition to EV extraction methods, aEVs can be produced by mechanically isolating the cell membrane. This process begins with converting cells into a lysis solution through mechanical disruption (Cheng et al. 2023a). In Method I: extrusion, the original cells are lysed and subsequently forced multiple times through

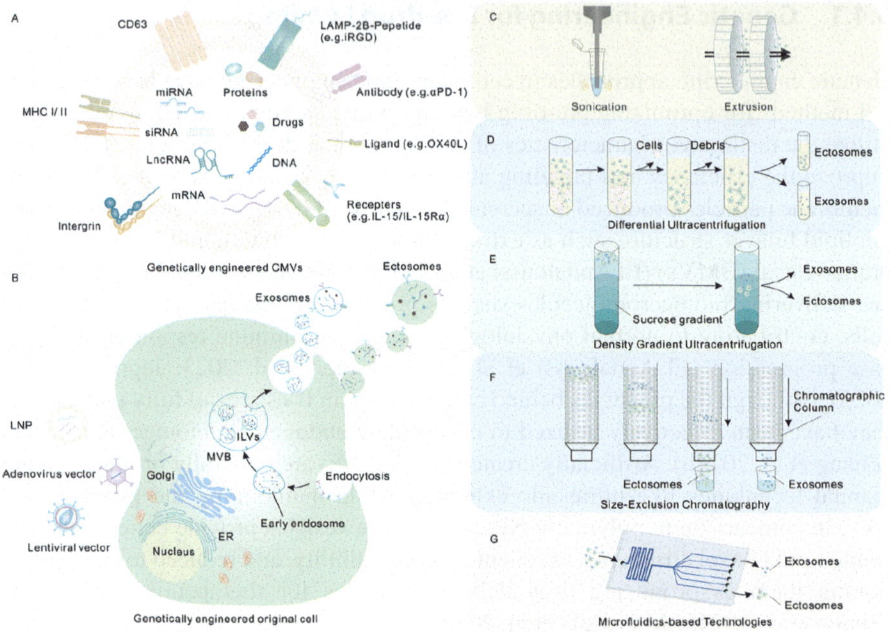

Fig. 6.3 Schematic illustrates the mechanism for genetically modifying CMVs and the strategies for their separation and purification. (**a**) Methods for genetic modification of CMVs; (**b**) Transfection and delivery of genetic material via LNPs are crucial for cell genetic modification, allowing donor cells to produce specific biomacromolecules. Modified CMVs can then transport exogenous biotherapeutics for disease treatment; (**c**) aEVs are obtained through sonication and extrusion; (**d–e**) Ultracentrifugation techniques, including differential and density gradient ultracentrifugation, separate EVs based on size or molecular weight; (**f–g**) Size-exclusion chromatography and microfluidics-based strategies. Abbreviations: LNP, lipid nanoparticles; ILVs, intraluminal vesicles; MVBs, multivesicular bodies; ER, endoplasmic reticulum; CMVs, cell membrane-derived vesicles. (Reprinted with permission from Li et al. (2024b), copyright 2024, OAE Publishing Inc)

polycarbonate membrane filters of varying pore sizes using a handheld extruder, resulting in core-shell structured aEVs (Jang et al. 2013). During this process, therapeutic molecules can be incorporated with the cells, enabling the collection of cargo-loaded aEVs. A previous study showed that a cell suspension was passed through an extruder three times to obtain either drug-free or drug-encapsulated aEVs for targeted delivery of chemotherapeutic agents directly to tumors (Jang et al. 2013). Likewise, other studies have documented the use of this method to create aEVs capable of drug delivery (Fuhrmann et al. 2015; Haney et al. 2015). Jhan et al. described the sequential extrusion of mixtures containing cell suspension and lipids through openings of varying diameters vesicles (Jhan et al. 2020). In each stage, the mixtures were manually squeezed and subsequently, the engineered EVs were collected to develop an effective gene delivery platform (Jhan et al. 2020). In method II: sonication, the homogenized cell mixture is sonicated at 20% power with

4-second pulses followed by 2-second pauses, repeated six times, and then cooled on ice for 2 min. Gradient centrifugation is then used to isolate high-purity cell membranes (Wen et al. 2022). The expression of recombinant proteins in cultured cells can facilitate passive drug loading into EVs, with higher concentrations of these proteins leading to increased packaging within the EVs. However, a complex sorting system determines which molecules can enter EVs, and random expression may result in insufficient packaging. To address this issue, active fusion of recombinant molecules with EVs through their constitutive proteins is necessary. Research has demonstrated that the specific interaction between the C-terminus of the Lamp2a-TAT module and TAR-miR-199a can enhance miRNA loading into the lumen of EVs (Sutaria et al. 2017). In a similar manner, the C-terminus of the EVs membrane protein CD9 has been combined with the RNA-binding human antigen R (HuR) protein, allowing the CD9-HuR nanovesicles to concentrate miRNA-155 within their lumen (Li et al. 2018b).

EVs offer therapeutic potential due to their genetic engineering-based loading capabilities. Researchers have leveraged native EVs' brain-targeting properties to deliver therapeutics to damaged brain tissues. Studies show that EVs from bone marrow macrophages and MSCs can cross the blood-brain barrier and accumulate in brain tissues in mouse models. Therapeutic molecules, such as glial cell line-derived neurotrophic factor and tyrosine phosphatase-2, have been integrated into these cell-derived EVs. Engineered EVs protect drugs from degradation and naturally target brain inflammation. In transgenic models of Parkinson's and Alzheimer's, they enhance neuronal survival, improve locomotion, reduce synaptic loss, and lower neuroinflammation. This makes them a promising strategy for overcoming drug delivery challenges in neurodegenerative diseases (Zhao et al. 2022c; Xu et al. 2022). A novel optogenetically engineered nanovesicle system was created to encapsulate super-repressor IκBα (srIκB) for therapeutic purposes. HEK293T cells were genetically modified to express two essential fusion proteins: CIBN-EGFP-CD9 and srIκB-mCherry-CRY2. This modification facilitated the generation of nanovesicles that contained srIκB within their interior. The srIκB protein acts as a powerful inhibitor of nuclear factor κB (NF-κB) translocation, which is a crucial pathway in inflammatory responses. In preclinical studies, these engineered nanovesicles were administered through intraperitoneal injection in mouse models of LPS-induced sepsis. The treatment significantly reduced the release of proinflammatory cytokines, alleviated systemic inflammation, and provided protection against organ damage caused by sepsis (Choi et al. 2020). Researchers have successfully used genetic engineering to load biotherapeutics molecules into original parent cells, transferring them to secreted EVs via natural biogenesis or aEVs through artificial processes (Fig. 6.4). Table 6.1 shows strategies for modifying CMVs to incorporate exogenous biotherapeutics for therapeutic purposes.

Biotherapeutic proteins are typically conjugated with intrinsic membrane vesicle surface proteins, such as lysosomal-associated membrane protein 2B (LAMP-2B), to facilitate their surface presentation (Raposo and Stahl 2023; Dooley et al. 2021). The presentation of bio-functional proteins on CMVs surface improves targeting and therapeutic efficacy for a range of diseases (Lewis et al. 2021). Wu et al.

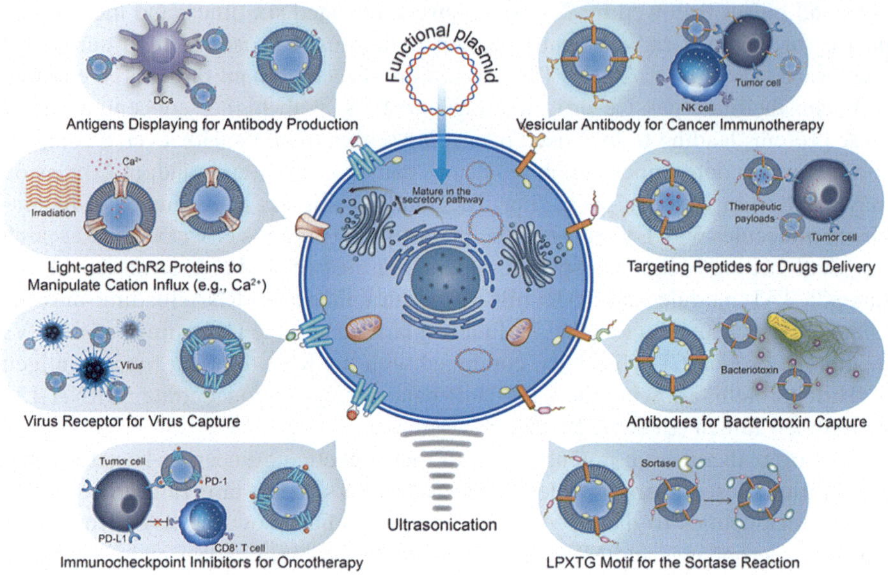

Fig. 6.4 This schematic illustrates the design of genetically engineered membrane vesicles using advanced genetic techniques. These methods enable precise modification of artificial proteins, such as antibodies, antigens, and peptides on the cell membrane. The resulting vesicles can also act as "tailorable shells" with internal compartments for incorporating bioactive components, offering great potential for therapeutic applications. (Reprinted with permission from Ren et al. (2021), copyright 2021, Wiley-VCH GmbH)

Table 6.1 Features and benefits of engineering approaches for CMVs

Therapeutic modality	Biomolecular categories	Modification strategies	Key benefits
Nucleic acid cargo	Small interfering RNA, microRNA, messenger RNA, and antisense oligonucleotide	Passive incubation, genetic engineering	Enables precise gene silencing or activation, effective delivery, extends systemic circulation
Chemical compounds	Chemotherapeutics, photosensitizers, immunomodulators, inhibitors	Passive incubation, surface functionalization	Enhances bioavailability, minimizes systemic toxicity, enables controlled and targeted drug release,
Protein	Receptor, ligand, antibodies, peptide	Genetic engineering, chemical conjugation	Natural mechanisms, safe delivery, precise targeting, enhances immune evasion

developed a stable cell line that overexpresses the brain lesion-targeting ligand PDGFRα and harvested the EVs. When used alongside Bryostatin-1 (Bryo-1), this method notably changes the pro-inflammatory phenotype of microglia in the central nervous system and leads to significant improvements in disease progression in mice (Wu et al. 2022b). CMVs derived from MSCs, which are rich in tyrosine phosphatase-2, were employed to cross the blood-brain barrier, promote mitophagy in neuronal cells, and reduce their apoptosis, thereby contributing to the therapeutic management of Alzheimer's disease (Xu et al. 2022). Oncolytic adenoviruses camouflaged with cell membrane vesicles containing targeting ligands effectively evade immune recognition and enhance their targeted delivery efficiency (Lv et al. 2019; Zhang et al. 2018).

6.4.2 Genetic Engineering for Membrane Surface Modification

Gene engineering for membrane surface modification encompasses various techniques aimed at enhancing the functionality of cell membranes, particularly for therapeutic applications. One innovative approach is glycosyl-phosphatidylinositol-anchored protein transfer, which enables the expression of new proteins on cell membranes without gene transfer. This method utilizes purified glycosyl-phosphatidylinositol-anchored proteins that can spontaneously integrate their lipid tail into cell membranes, showing promise in developing cancer vaccines and immunotherapies by directly modifying tumor cell membranes with immunological molecules (Jiyun et al. 2003). Another key approach is surface modification of retroviral or lentiviral vectors for gene delivery. This includes covalent chemical conjugation, membrane-topic reagents, and adaptor systems, enabling post-production alterations to enhance delivery efficiency and safety (Metzner et al. 2013). Additionally, a novel approach using amphiphilic poly-γ-glutamate (APG) has demonstrated the reconstitution of membrane proteins (MPs) into live cells. This method stabilizes certain receptors, such as G protein-coupled receptors (GPCRs), in mammalian cell plasma membranes without compromising cell viability. The interaction between APG and the membrane surface charge is crucial in this reconstitution process, facilitating functional modifications of cell membranes (Kim et al. 2024). Kim et al. successfully reconstituted GPCRs and 5-Hydroxytryptamine 3A receptor subunit (5HT3A) into mammalian cell membranes using APG while maintaining cell viability. Following reconstitution, 5HT3A exhibited ligand-dependent Ca^{2+} ion transport activity, demonstrating functional integration. The electrostatic interaction between APG and the membrane surface charge played a critical role in the reconstitution process (Kim et al. 2024). Recent advances in gene engineering have significantly improved membrane surface modification, leading to enhanced biomaterial performance. Key techniques such as ion implantation and covalent binding are essential for creating customized surface properties. The integration of these methods can accelerate the development of innovative biomaterials, making them more suitable for biomedical applications (Wu et al. 2015).

6.4.3 Genetic Engineering for Reducing Toxicity of Membrane-Based Platforms

Outer membrane vesicles (OMVs) are natural, non-replicative nanoparticles, secreted by Gram-negative bacteria and are associated with pathogenesis and transportation of virulence factors. In addition to the major components of the outer membrane, OMVs contain the periplasmic contents of the bacterium from which they originated (Carvalho et al. 2019). Due to their particulate nature and innate composition, OMVs have wide biomedical applications. To date, they have been used as anti-tumor agents, delivery platforms, adjuvants, and vaccines (Carvalho et al. 2019; Chen et al. 2016b; Zariri et al. 2016; Zhuang et al. 2021). For example, glycoconjugate vaccines can be produced using genetically engineered OMVs, offering new therapeutic approaches for treating microbial diseases (Chen et al. 2016b; Kay et al. 2019). Although OMVs have shown tremendous potential for biomedical applications, like any other therapeutic approach, OMVs-based approaches have their drawbacks and face challenges. Biosafety is the main concern of OMVs, as they contain lipopolysaccharide (LPS). To solve this problem, bacteria are modified by genetically engineered methods to produce OMVs with reduced LPS content and low toxicity (Van Der Pol et al. 2015; Chen et al. 2025). For example, Kim et al. modified *Escherichia coli* (*E. coli*) outer membrane by mutational inactivation of the MsbB (LpxM) lipid A acyltransferase and developed a platform technology to yield low-endotoxic OMVs (Kim et al. 2009). Low efficiency and unspecific functionality are other concerns of OMVs-based therapeutic approaches. To overcome these problems, OMV proteins are modified mainly by genetic engineering to develop functionalized bioengineered types. In this context, Cheng et al. developed a tumor vaccine platform with specific anti-tumor immune response by expressing tumor antigens on the surface of OMVs by utilizing genetic engineering and "Plug-and-Display" technology (Cheng et al. 2021).

6.5 Biomedical Applications of Genetically Engineered-Membrane-Based Platforms

6.5.1 Targeted Cancer Immunotherapy

CAR-T cell therapy is one of the best examples of clinically successful cell-based therapeutics based on genetic engineering. In this approach, immune cells especially lymphocytes are genetically modified to express CARs on their surface to recognize target antigens independent from MHC receptor and evoke vigorous antitumor response (Sterner and Sterner 2021). Due to their efficient therapeutic outcomes, the US Food and Drug Administration approved autologous T cells genetically engineered to target the CD19 B lymphocyte molecule as a reliable treatment for B-cell malignancies (June et al. 2018). CAR structure which is made up of an extracellular single-chain variable fragment (scFv), a transmembrane spacer, and intracellular activation domains expressed on the cell surface after transfecting the target immune cell with different gene delivery methods such as mRNA, viral vector, and nanoparticles (Newick et al. 2017). Researchers are exploring clinically applicable methods, including systemically injected viral vectors and polymer- or lipid-based nanocarriers loaded with CAR-encoding genes, to simplify CAR T-cell manufacturing and address challenges in ex vivo production (Pinto et al. 2025). Viral vectors have limited gene delivery capacity and require costly, time-consuming production, restricting their clinical use. To overcome these challenges, viral vector-free strategies have emerged for transfecting and engineering immune cells. For example, transposon systems like *Sleeping Beauty* and *piggyBac* have been explored as cost-effective alternatives for CAR T-cell engineering (Larson and Maus 2021). CAR-T cell therapy faces challenges like toxicity, poor solid tumor efficacy, antigen escape, immunosuppression, and limited infiltration. To overcome these, complementary anticancer therapies and innovative engineering strategies have been developed to enhance safety and clinical efficacy (Sterner and Sterner 2021).

6.5.2 Targeted Drug Delivery System

Targeted drug delivery platforms are designed to transport defined doses of active pharmaceutical ingredients directly to specific sites of the body, thereby reducing off-target drug accumulation and minimizing associated side effects (Manzari et al. 2021). To achieve this, various targeting moieties, including antibodies (Le Joncour et al. 2019), peptides (Kheirandish et al. 2019), and polymers (Chen et al. 2019), have been utilized in recent years to develop diverse drug conjugates, enhancing the localized accumulation and therapeutic efficacy of the drugs. Certain types of cells inherently exhibit a natural tendency to home to the specific site of the body. For example, platelets prefer homing to inflammatory or wound sites due to collagen receptors (Clemetson and Clemetson 2007). T lymphocytes and monocytes migrate to the tumor sites by specific cytokines (Imhof and Aurrand-Lions 2004) and neutrophils and macrophages move toward inflammatory sites (Butterfield et al. 2006). Dendritic cells also can attract to lymph nodes (Förster et al. 2012). Moreover, erythrocytes target splenic and hepatic tissue (Ukidve et al. 2020).

Genetic engineering can enable cell membranes to anchor drugs by expressing specific surface proteins. For example, dendritic cells transfected with a HaloTag-encoding plasmid used a chloroalkane linker to bind glycopolymers after protein expression (Yu et al. 2020). Proper localization of the protein to the membrane requires a localization sequence. In the aforementioned study, the immunoglobulin κ-chain signal peptide and a transmembrane domain derived from the platelet-derived growth factor receptor were incorporated into the HaloTag sequence to ensure correct membrane targeting. Additionally, protein payloads, such as therapeutic agents or targeting moieties, can be directly expressed on the cell membrane following gene delivery to target cells. However, challenges associated with this approach, including efficient transfection of hard-to-transfect cell types, must be carefully addressed (Li et al. 2016). Many diseases are associated with heightened local inflammation, making it essential to deliver drugs specifically to these inflamed areas for effective treatment (Chen et al. 2021). Inflamed endothelial cells often express elevated levels of vascular cell adhesion molecule-1 (VCAM-1), which can be targeted therapeutically (Pickett et al. 2023). Park et al. developed a biomimetic NPs formulation by utilizing the specific interaction between very late antigen-4

(VLA-4) and VCAM-1. They coated polymeric NPs cores with plasma membranes derived from genetically modified cells that continuously express VLA-4. This coating improved the nanoparticles' ability to bind to target cells that overexpress VCAM-1. The nanoparticles were also loaded with the anti-inflammatory drug dexamethasone, which enhanced drug delivery to inflamed lung tissues and demonstrated significant therapeutic effects in vivo (Park et al. 2021).

As mentioned earlier, RBCs are being explored as drug delivery carriers due to their ability to circulate in the body for extended periods (Zhang et al. 2022). However, traditional internal drug-loading methods may compromise RBCs membrane integrity and lead to uncontrolled drug release (Li et al. 2021). To address these issues, by considering the large surface area and functional groups on RBCs, researchers have investigated cell surface engineering as a complementary approach (Bashor et al. 2022). Additionally, researchers have loaded chemotherapeutic agents like doxorubicin into RBCs and attached iron oxide NPs for imaging-guided combined therapies, achieving synergistic antitumor effects (Wang et al. 2014). In another study, Park et al. developed genetically modified NPs coated with cell membranes for the efficient delivery of dexamethasone to inflamed lungs. In this research, they engineered these NPs to precisely target areas of inflammation (Fig. 6.5).

6.5.3 Optimized Targeted Gene Delivery

Gene delivery is a crucial step in genome editing, essential for both gene therapy and cell line development. It involves transferring nucleic acids across the cell membrane and ensuring their safe release into the cytoplasm or nucleus for genomic or transcriptomic modifications, such as knock-down, knock-out, or knock-in alterations. The advancement of genome editing tools like CRISPR/Cas9 has underscored in vivo gene delivery as a major challenge, driving extensive research efforts to enhance its efficiency and effectiveness (Ay and Reinisch 2025; Kheirandish et al. 2023; Sung and Kim 2019; Kohn et al. 2023).

Extracellular vesicles (EVs) are considered as efficient delivery vehicles for transporting diverse payloads, which can be loaded into them either endogenously or exogenously (Sun et al. 2024). Among different types of EVs, exosomes have emerged as a leading option for drug delivery, owing to their small size (30–150 nm), biocompatibility, stability, and well-understood biogenesis process (Wang et al. 2024b). Their membrane proteins and receptors provide an opportunity for functionalization with targeting moieties, making them particularly suitable for in vivo targeted gene therapy (Cheng et al. 2023b). EVs are often engineered to enhance their efficiency and targeting capabilities. These engineering strategies can be categorized into two main types: engineering the parent cells from which the EVs are derived (before EVs isolation) and direct modification of EVs (after EVs isolation) (Table 6.2) (Ge et al. 2023; An et al. 2019). The following sections provide a detailed overview of these EVs engineering strategies employed for targeted gene delivery. By introducing a vector construct into the parent cells of EVs, which contains an expression cassette for a desired peptide or protein fused to the coding sequence of

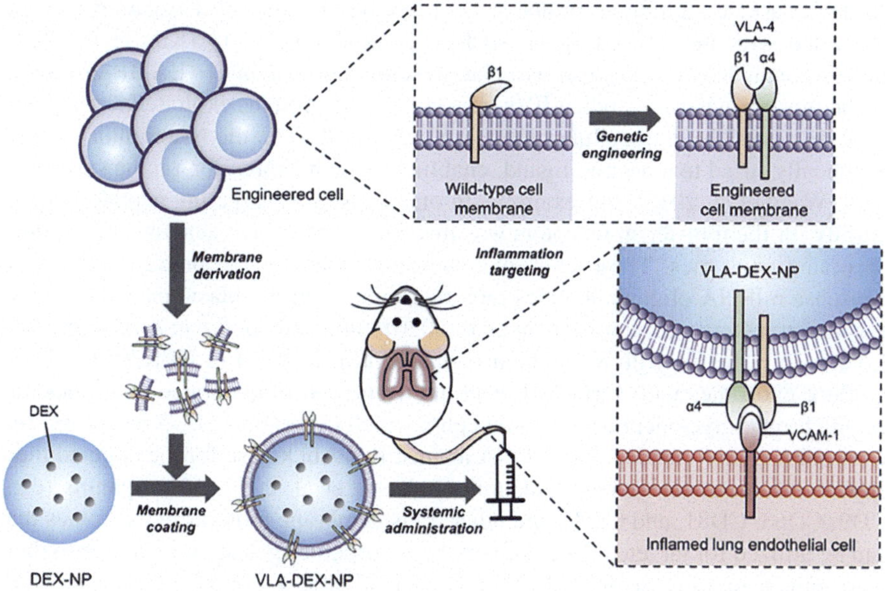

Fig. 6.5 A schematic depicts genetically modified plasma membrane-coated nanoparticles designed for precise drug delivery to inflamed lungs. Cells were engineered to express VLA-4, and their plasma membranes were used to coat dexamethasone-loaded nanoparticles (DEX-NP). The resulting VLA-DEX-NP specifically targets VCAM-1 on inflamed lung endothelial cells, ensuring precise drug delivery. (Reprinted with permission from Park et al. (2021), copyright 2021, American Association for the Advancement of Science)

Table 6.2 Different strategies for membrane engineering of EVs

Engineering strategy	Cargo type	Surface decoration	Targeted cell	Ref.
LAMP-2b fusion	miRNA	T7 targeting peptide	Glioblastoma	Kim et al. (2020)
LAMP-2b fusion	siRNA	Interleukin 3	chronic myelogenous leukemia	Bellavia et al. (2017)
CD63 fusion	miRNA	Apolipoprotein A1	Hepatocellular carcinoma	Liang et al. (2018)
LAMP-2b fusion	miRNA	HER2-binding affibody	Colorectal cancer	Liang et al. (2020)
LAMP-2b fusion	siRNA	iRGD peptide	Colorectal cancer	Lin et al. (2021)
LAMP-2b fusion	CRISPR-Cas9 Plasmid	HN3 antibody	Hepatocellular carcinoma	He et al. (2020)
LAMP-2b fusion	siRNA	tLyp-1 targeting peptide	Human non-small cell lung cancer	Bai et al. (2020)
LAMP-2b fusion	siRNA	iRGD peptide	Lung cancer	Zhou et al. (2019)

the EVs surface membrane proteins, EVs can be obtained post-isolation that are decorated with the desired agent, such as a targeting moiety (Kim et al. 2020; Bellavia et al. 2017). Lysosome-associated membrane protein-2b (LAMP-2b) is one of the most extensively studied EVs transmembrane proteins exploited for targeted gene delivery. The extracellular domain of LAMP-2b (i.e., the N-terminus) can be genetically fused to a desired ligand, enabling its high expression on the surface of EVs (Wang et al. 2018). For example, in one study, a T7 targeting peptide with an affinity for the transferrin receptors was fused to LAMP-2b to generate T7 peptide-decorated exosomes. These exosomes were designed for the precise delivery of antisense miRNA oligonucleotides targeting miR-21 to glioblastoma cells overexpressing transferrin receptors (Kim et al. 2020). Similarly, *Bellavia et al.* expressed LAMP-2b in fusion with a fragment of Interleukin 3 (IL-3) in HEK293T cells to produce exosomes with surface IL-3 protein. These IL-3-decorated exosomes targeted chronic myelogenous leukemia cells, which overexpress IL-3 receptors, and were subsequently loaded with siRNA against Bcr-Abl kinase for the targeted therapy of chronic myelogenous leukemia (Bellavia et al. 2017). Tetraspanins (e.g., CD9, CD63, CD81, and CD82) are other transmembrane proteins of exosomes that can be utilized for targeted gene delivery by incorporating desired ligands into their extracellular loops (Umeda et al. 2020; Khan et al. 2021). For example, *Liang et al.* engineered Apo-A1-CD63 exosomes by fusing CD63 tetraspanin with apolipoprotein A1 (Apo-A1). This fusion enabled the exosomes to specifically bind to the SR-B1 receptor, which is highly expressed on the surface of liver cancer cells. After loading the exosomes with miR-26a, a microRNA known to induce tumor-specific apoptosis, they effectively targeted HepG2 liver cancer cells for therapeutic delivery (Liang et al. 2018).

6.5.4 Therapeutic Bioproduction

Prokaryotic and eukaryotic cells serve as versatile cell factories in industrial biotechnology, enabling the cost-effective and sustainable biomanufacturing of pharmaceuticals, biofuels, bulk chemicals, and natural ingredients. However, optimizing these systems often requires extensive modifications to enhance production yield and adaptability to diverse growth conditions (Navarrete et al. 2020; Qi et al. 2019). Besides well-understood metabolic pathway engineering and synthetic biology to improve the performance of microbial cell factories (Lee et al. 2019), membrane engineering has recently gained attention as an innovative strategy to address the limitations of these traditional approaches. This strategy aims to enhance bioproduct production by modifying membrane composition and functionality. As a protective barrier, the cell membrane shields the cytoplasm from external stresses, enabling host cells to function efficiently even in harsh environments (Robertson 2018). Beyond its protective role, the membrane is deeply involved in physiological processes such as cell growth, secretion, and energy metabolism. Thus, maintaining membrane homeostasis is essential for the survival and productivity of host cell

Table 6.3 Different membrane engineering strategies for enhanced biomanufacturing

Engineering strategy	Bio-product	Species	Result	Ref.
Changing unsaturated fatty acids from cis to trans isomer using cis-trans isomerase from *P. aeruginosa*	Octanoic acid	*E. coli*	Alleviated toxicity, 41% enhanced octanoic acid production	Tan et al. (2016)
Increasing unsaturated fatty acid content by overexpressing stearoyl-CoA desaturase	Lycopene	*S. cerevisiae*	Alleviated toxicity, enhanced lycopene production	Hong et al. (2019)
Increasing the net negative charge of membrane by depleting membrane phosphatidylethanolamine and cardiolipin	α-amylase	*B. subtilis*	Increased product secretion, 47% enhanced production	Cao et al. (2017)
Regulating the expression of the endogenous aromatic transporter CouP	Catechol	*E. coli*	Alleviated toxicity, enhanced catechol production	Wu et al. (2018a)
Recombinant expression of an efflux pump from *Alcanivorax borkumensis*	Limonene	*E. coli*	Alleviated toxicity, enhanced limonene production	Dunlop et al. (2011)
Expanding membrane surface and improving the storage capacity by overexpressing *almgs, plsB* and *plsC* genes	Lycopene	*E. coli*	Enhanced lycopene storage capacity and production	Wu et al. (2018b)
Overexpressing ABC transporter and changing membrane lipid composition by overexpressing the stearoyl-CoA desaturase and	β-carotene	*S. cerevisiae*	Alleviated toxicity, enhanced membrane fluidity, and improved β carotene production	Bu et al. (2020)

factories (Ray et al. 2016). Different membrane engineering strategies for enhancing the synthesis of bio-products were summarized in Table 6.3.

High-level production of certain compounds can stress the host cell membrane, as their accumulation often leads to toxicity, disrupts growth and physiological functions, and ultimately reduces production efficiency and yield (Boyarskiy et al. 2016; Dunlop et al. 2011). To address these challenges, membrane remodeling strategies have been developed, including modifying lipid composition to enhance stability and stress tolerance. For example, studies on *Escherichia coli* show that adjusting unsaturated fatty acid content improves membrane fluidity and resistance to bulk chemicals (Royce et al. 2013). In this context, a study demonstrated that the overexpression of the cis-trans isomerase gene from *Pseudomonas aeruginosa* in *E. coli* which acts as a convertor of unsaturated fatty acids from cis to trans, enhanced membrane integrity and reduced membrane fluidity. This modification improved the strain's tolerance to octanoic acid, resulting in a 41% increase in production titer (Tan et al. 2016). Moreover, *Hong et al.* increased the unsaturated fatty acid content of the *Saccharomyces cerevisiae* membrane by overexpressing stearoyl-CoA desaturase, encoded by the *OLE1* gene. This modification alleviated the toxic effects of lycopene and, in combination with other strategies, significantly enhanced lycopene production (Hong et al. 2019). Similarly, research by Cao et al. on *Bacillus subtilis* revealed that the depletion of two major membrane phospholipids; phosphatidylethanolamine and cardiolipin, increased the net negative charge of the cell surface. This alteration significantly improved the secretion efficiency of recombinant α-amylase, resulting in a 47% increase in enzyme production (Cao et al. 2017).

Another effective membrane engineering strategy enhances the efflux of stress-induced compounds by overexpressing native transporters or introducing heterologous systems. This reduces toxicity from accumulated products while improving host cell productivity. For example, modulating the expression of the endogenous aromatic transporter CouP in *Escherichia coli* enhanced substrate uptake and conversion, mitigating toxicity and increasing catechol production by 30–40% (Wu et al. 2018a). Furthermore, the heterologous expression of an efflux pump from *Alcanivorax borkumensis* in *E. coli* enhanced the strain's tolerance to limonene, leading to a 68% improvement in production yield (Dunlop et al. 2011). Alternatively, expanding the surface area of the cell membrane directly increases the storage capacity of cells for synthesized products, thereby improving overall productivity. *Wu et al.* reported that overexpressing the *almgs* gene from *Acholeplasma laidlawii* in *E. coli* strain, which is involved in membrane bending, increased the storage capacity for lycopene, leading to a 12% improvement in its production. Also, the overexpression of *plsB* and *plsC* genes, which are associated with the synthesis of phosphatidylethanolamine, further enhanced membrane synthesis, resulting in an additional 13% increase in lycopene production (Wu et al. 2018b). Despite advances in membrane engineering, many strategies are limited to specific bioproducts or hosts due to membrane complexity and sensitivity. Even minor changes can impact viability and function, requiring careful optimization to enhance cell factory performance with minimal disruption.

6.6 Challenges in Membrane Engineering

Despite its advantages, genetic engineering faces clinical challenges, including safety concerns, low efficacy, poor cell survival, and limited control over cell phenotype (Amaral and Pasparakis 2019). Using genetically engineered cells for therapy requires costly and time-consuming clinical-grade vector preparation and testing, along with specialized production facilities (Jadlowsky et al. 2024). Genetic engineering-based cell surface modification faces challenges like inefficient gene delivery, oncogene activation risks, vector size limits, unintended cellular changes, and inability to modify membranes with non-biological materials or nucleic acids (Stephan and Irvine 2011; Yano and Matsuzaki 2009). Viral vectors offer high transfection efficiency but pose tumorigenesis risks, while non-viral vectors are safer but less effective. Cell surface modification relies on unpredictable epitope expression, requiring normalization for reliable data (Pai et al. 2015). Maintaining long-term stability of surface modifications is challenging, as membrane renewal can cause molecule loss. Systematic studies are needed to assess their impact on intravascular transport, cell interactions, and phenotype. Additionally, surface modifications may trigger immune responses, leading to rapid cell clearance (Liu et al. 2019; Tan et al.

2015). The high expression of targeted proteins depends on the amount of genetic material internalized by cells, while the limited capacity of EVs to display recombinant proteins restricts molecule density. Consequently, most studies engineer EVs for a single function, as integrating multiple genes is challenging. Additionally, displaying complex antigens is difficult due to uncertainties in protein folding.

6.7 Conclusion

Despite its significant popularity and therapeutic efficacy, cell-based therapy faces multiple challenges that limit its widespread clinical use and commercialization. Among cell-manipulation strategies, genetic-mediated cell membrane engineering strategies have been developed vigorously over the last decades to overcome native cell-based therapy challenges by improving the targeting, functionality, and therapeutic efficacy of the cells through modifying cells to express and/or secrete specific proteins. To achieve this, multiple gene-carrying vectors are developed to transfect host cells. These cells are engineered to have promoted functionalities for widespread applications such as cancer immune therapy, targeted drug delivery, bioproduction, and improving gene delivery efficacy. Despite the encouraging findings from research and preliminary preclinical studies, this strategy remains in its early stages and encounters several challenges that hinder its advancement into clinical applications.

References

Akita H, Harashima H (2008) Nonviral gene delivery. In: Gene Therapy for Renal Diseases and Transplantation, vol 159. Karger Publishers, Basel, pp 13–29

Alekseeva A, Kapkaeva M, Shcheglovitova O, Boldyrev I, Pazynina G, Bovin N et al (2015) Interactions of antitumour Sialyl Lewis X liposomes with vascular endothelial cells. Biochim Biophys Acta (BBA)-Biomembranes 1848(5):1099–1110

Amaral AJ, Pasparakis G (2019) Cell membrane engineering with synthetic materials: applications in cell spheroids, cellular glues and microtissue formation. Acta Biomater 90:21–36

An Y, Jin T, Zhu Y, Zhang F, He P (2019) An ultrasensitive electrochemical aptasensor for the determination of tumor exosomes based on click chemistry. Biosens Bioelectron 142:111503

Atasheva S, Shayakhmetov DM (2022) Cytokine responses to adenovirus and adenovirus vectors. Viruses 14(5):888

Ay C, Reinisch A (2025) Gene therapy: principles, challenges and use in clinical practice. Wien Klin Wochenschr 137:261–271

Bai J, Duan J, Liu R, Du Y, Luo Q, Cui Y et al (2020) Engineered targeting tLyp-1 exosomes as gene therapy vectors for efficient delivery of siRNA into lung cancer cells. Asian J Pharm Sci 15(4):461–471

Bashor CJ, Hilton IB, Bandukwala H, Smith DM, Veiseh O (2022) Engineering the next generation of cell-based therapeutics. Nat Rev Drug Discov 21(9):655–675

Belguise-Valladier P, Behr J-P (2001) Nonviral gene delivery: towards artificial viruses. Cytotechnology 35:197–201

Bellavia D, Raimondo S, Calabrese G, Forte S, Cristaldi M, Patinella A et al (2017) Interleukin 3- receptor targeted exosomes inhibit in vitro and in vivo Chronic Myelogenous Leukemia cell growth. Theranostics 7(5):1333–1345

Bompard J, Rosso A, Brizuela L, Mebarek S, Blum LJ, Trunfio-Sfarghiu A-M et al (2020) Membrane fluidity as a new means to selectively target cancer cells with fusogenic lipid carriers. Langmuir 36(19):5134–5144

Boyarskiy S, Davis López S, Kong N, Tullman-Ercek D (2016) Transcriptional feedback regulation of efflux protein expression for increased tolerance to and production of n-butanol. Metab Eng 33:130–137

Bravo-Estupiñan DM, Montaño-Samaniego M, Mora-Rodríguez RA, Ibáñez-Hernández M (2024) Cationic lipid derived from a basic amino acid: design and synthesis. Appl Sci 14(23):10892

Bu X, Lin JY, Cheng J, Yang D, Duan CQ, Koffas M et al (2020) Engineering endogenous ABC transporter with improving ATP supply and membrane flexibility enhances the secretion of β-carotene in Saccharomyces cerevisiae. Biotechnol Biofuels 13:168

Butt MH, Zaman M, Ahmad A, Khan R, Mallhi TH, Hasan MM et al (2022) Appraisal for the potential of viral and nonviral vectors in gene therapy: a review. Genes 13(8):1370

Butterfield TA, Best TM, Merrick MA (2006) The dual roles of neutrophils and macrophages in inflammation: a critical balance between tissue damage and repair. J Athl Train 41(4):457

Buzas EI (2023) The roles of extracellular vesicles in the immune system. Nat Rev Immunol 23(4):236–250

Cao H, van Heel AJ, Ahmed H, Mols M, Kuipers OP (2017) Cell surface engineering of Bacillus subtilis improves production yields of heterologously expressed α-amylases. Microb Cell Fact 16(1):56

Carvalho AL, Fonseca S, Miquel-Clopés A, Cross K, Kok K-S, Wegmann U et al (2019) Bioengineering commensal bacteria-derived outer membrane vesicles for delivery of biologics to the gastrointestinal and respiratory tract. J Extracell Vesicles 8(1):1632100

Chen Z, Zhao P, Luo Z et al (2016a) Cancer cell membranebiomimetic nanoparticles for homologous-targeting dualmodal imaging and photothermal therapy. ACS Nano 10:10049–10057. https://doi.org/10.1021/acsnano.6b04695

Chen L, Valentine JL, Huang C-J, Endicott CE, Moeller TD, Rasmussen JA et al (2016b) Outer membrane vesicles displaying engineered glycotopes elicit protective antibodies. Proc Natl Acad Sci 113(26):E3609–E3E18

Chen K, Cai H, Zhang H, Zhu H, Gu Z, Gong Q et al (2019) Stimuli-responsive polymer-doxorubicin conjugate: antitumor mechanism and potential as nano-prodrug. Acta Biomater 84:339–355

Chen Y-C, Gad SF, Chobisa D, Li Y, Yeo Y (2021) Local drug delivery systems for inflammatory diseases: status quo, challenges, and opportunities. J Control Release 330:438–460

Chen M-Y, Cheng T-W, Pan Y-C, Mou C-Y, Chiang Y-W, Lin W-C et al (2025) Endotoxin-free outer membrane vesicles for safe and modular anticancer immunotherapy. ACS Synth Biol 14(1):148–160

Cheng K, Zhao R, Li Y, Qi Y, Wang Y, Zhang Y et al (2021) Bioengineered bacteria-derived outer membrane vesicles as a versatile antigen display platform for tumor vaccination via plug-and-display technology. Nat Commun 12(1):2041

Cheng Q, Kang Y, Yao B, Dong J, Zhu Y, He Y et al (2023a) Genetically engineered-cell-membrane nanovesicles for cancer immunotherapy. Adv Sci 10(26):2302131

Cheng W, Xu C, Su Y, Shen Y, Yang Q, Zhao Y et al (2023b) Engineered Extracellular Vesicles: a potential treatment for regeneration. iScience 26(11):108282

Choi H, Kim Y, Mirzaaghasi A, Heo J, Kim YN, Shin JH et al (2020) Exosome-based delivery of super-repressor IκBα relieves sepsis-associated organ damage and mortality. Sci Adv 6(15):eaaz6980

Clemetson KJ, Clemetson JM (2007) Collagen receptors as potential targets for novel anti-platelet agents. Curr Pharm Des 13(26):2673–2683

Csizmar CM, Petersburg JR, Wagner CR (2018) Programming cell-cell interactions through non-genetic membrane engineering. Cell Chemi Biol 25(8):931–940

Custódio CA, Mano JF (2016) Cell surface engineering to control cellular interactions. ChemNanoMat 2(5):376–384
De Pieri A, Rochev Y, Zeugolis DI (2021) Scaffold-free cell-based tissue engineering therapies: advances, shortfalls and forecast. NPJ Regen Med 6(1):18
Deng Y, Ren M, He P, Liu F, Wang X, Zhou C et al (2023) Genetically engineered cell membrane-coated nanoparticles for antibacterial and immunoregulatory dual-function treatment of ligature-induced periodontitis. Front Bioeng Biotechnol 11:1113367
Di Huang XW, Wang W, Li J, Zhanga X, Xia B (2024) Cell-membrane engineering strategies for clinic guided design of nanomedicine. Biomater Sci 12:2865
Ding Y, Wang Y, Hu Q (eds) (2022) Recent advances in overcoming barriers to cell-based delivery systems for cancer immunotherapy. Exploration. Wiley Online Library
Dooley K, McConnell RE, Xu K, Lewis ND, Haupt S, Youniss MR et al (2021) A versatile platform for generating engineered extracellular vesicles with defined therapeutic properties. Mol Ther 29(5):1729–1743
Dunlop MJ, Dossani ZY, Szmidt HL, Chu HC, Lee TS, Keasling JD et al (2011) Engineering microbial biofuel tolerance and export using efflux pumps. Mol Syst Biol 7:487
El Andaloussi S, Mäger I, Breakefield XO, Wood MJ (2013) Extracellular vesicles: biology and emerging therapeutic opportunities. Nat Rev Drug Discov 12(5):347–357
Fang RH, Hu CM, Luk BT et al (2014) Cancer cell membrane-coated nanoparticles for anticancer vaccination and drug delivery. Nano Lett 14:2181–2188. https://doi.org/10.1021/nl500618u
Förster R, Braun A, Worbs T (2012) Lymph node homing of T cells and dendritic cells via afferent lymphatics. Trends Immunol 33(6):271–280
Fuhrmann G, Serio A, Mazo M, Nair R, Stevens MM (2015) Active loading into extracellular vesicles significantly improves the cellular uptake and photodynamic effect of porphyrins. J Control Release 205:35–44
Gabrielse K, Gangar A, Kumar N, Lee JC, Fegan A, Shen JJ et al (2014) Reversible re-programing of cell–cell interactions. Angew Chem Int Ed 53(20):5112–5116
Gao C, Huang Q, Liu C, Kwong CHT, Yue L, Wan J-B et al (2020) Treatment of atherosclerosis by macrophage-biomimetic nanoparticles via targeted pharmacotherapy and sequestration of proinflammatory cytokines. Nat Commun 11(1):2622
Ge L, Wang K, Lin H, Tao E, Xia W, Wang F et al (2023) Engineered exosomes derived from miR-132-overexpresssing adipose stem cells promoted diabetic wound healing and skin reconstruction. Front Bioeng Biotechnol 11:1129538
Germer J, Wagner E (2022) Polymer-based nonviral nucleic acid delivery and genome editing. In: Molecular analyses. CRC Press, Boca Raton, pp 165–182
Gu C, Geng X, Wu Y, Dai Y, Zeng J, Wang Z et al (2024) Engineered macrophage membrane-coated nanoparticles with enhanced CCR2 expression promote spinal cord injury repair by suppressing neuroinflammation and neuronal death. Small 20(10):2305659
Guo Z, Zhang L, Yang Q, Peng R, Yuan X, Xu L et al (2022) Manipulation of multiple cell–cell interactions by tunable DNA scaffold networks. Angew Chem Int Ed 61(7):e202111151
Hamann A, Pannier AK (2022) Innovative nonviral gene delivery strategies for engineering human mesenchymal stem cell phenotypes toward clinical applications. Curr Opin Biotechnol 78:102819
Haney MJ, Klyachko NL, Zhao Y, Gupta R, Plotnikova EG, He Z et al (2015) Exosomes as drug delivery vehicles for Parkinson's disease therapy. J Control Release 207:18–30
Hassanzadeh A, Shamlou S, Yousefi N, Nikoo M, Verdi J (2022) Genetically-modified stem cell in regenerative medicine and cancer therapy; a new era. Curr Gene Ther 22(1):23–39
He C-X, Tabata Y, Gao J-Q (2010) Non-viral gene delivery carrier and its three-dimensional transfection system. Int J Pharm 386(1-2):232–242
He C, Ali DJ, Li Y, Zhu Y, Sun B, Xiao Z (2020) Engineering of HN3 increases the tumor targeting specificity of exosomes and upgrade the anti-tumor effect of sorafenib on HuH-7 cells. PeerJ 8:e9524
Homma T, Yamamoto R, Ang LZP, Fehaid A, Ebara M (2024) A novel gene synthesis platform for designing functional protein polymers. bioRxiv 2024.09. 01.610679

Hong J, Park SH, Kim S, Kim SW, Hahn JS (2019) Efficient production of lycopene in Saccharomyces cerevisiae by enzyme engineering and increasing membrane flexibility and NAPDH production. Appl Microbiol Biotechnol 103(1):211–223

Huang D, Wang X, Wang W, Li J, Zhang X, Xia B (2024) Cell-membrane engineering strategies for clinic-guided design of nanomedicine. Biomater Sci 12(11):2865–2884

Imhof BA, Aurrand-Lions M (2004) Adhesion mechanisms regulating the migration of monocytes. Nat Rev Immunol 4(6):432–444

Jadlowsky JK, Leskowitz R, McKenna S, Karar J, Ma Y, Dai A et al (2024) Long-term stability of clinical-grade lentiviral vectors for cell therapy. Mol Ther Methods Clin Dev 32(1)

Jang SC, Kim OY, Yoon CM, Choi D-S, Roh T-Y, Park J et al (2013) Bioinspired exosome-mimetic nanovesicles for targeted delivery of chemotherapeutics to malignant tumors. ACS Nano 7(9):7698–7710

Jeppesen DK, Fenix AM, Franklin JL, Higginbotham JN, Zhang Q, Zimmerman LJ et al (2019) Reassessment of exosome composition. Cell 177(2):428–445. e18

Jeppesen DK, Zhang Q, Franklin JL, Coffey RJ (2023) Extracellular vesicles and nanoparticles: emerging complexities. Trends Cell Biol 33(8):667–681

Jhan Y-Y, Prasca-Chamorro D, Zuniga GP, Moore DM, Kumar SA, Gaharwar AK et al (2020) Engineered extracellular vesicles with synthetic lipids via membrane fusion to establish efficient gene delivery. Int J Pharm 573:118802

Jiyun Y (2003) Cell-surface engineering technique and cancer vaccine. Mian yi xue za zhi = J Immunol 19(6):478–479, 481

June CH, O'Connor RS, Kawalekar OU, Ghassemi S, Milone MC (2018) CAR T cell immunotherapy for human cancer. Science 359(6382):1361–1365

Kay E, Cuccui J, Wren BW (2019) Recent advances in the production of recombinant glycoconjugate vaccines. npj Vaccines 4(1):16

Khan H, Pan JJ, Li Y, Zhang Z, Yang GY (2021) Native and bioengineered exosomes for ischemic stroke therapy. Front Cell Dev Biol 9:619565

Kheirandish MH, Jaliani HZ, Rahmani B, Nikukar H (2019) Specific targeting of a pore-forming toxin (listeriolysin O) to LHRH-positive cancer cells using LHRH targeting peptide. Toxicon 164:82–86

Kheirandish MH, Rahmani B, Zarei Jaliani H, Barkhordari F, Mazlomi MA, Davami F (2023) Efficient site-specific integration in CHO-K1 cells using CRISPR/Cas9-modified donors. Mol Biol Rep 50(7):5889–5899

Kim S-H, Kim K-S, Lee S-R, Kim E, Kim M-S, Lee E-Y et al (2009) Structural modifications of outer membrane vesicles to refine them as vaccine delivery vehicles. Biochim Biophys Acta (BBA)-Biomembranes 1788(10):2150–2159

Kim G, Kim M, Lee Y, Byun JW, Hwang DW, Lee M (2020) Systemic delivery of microRNA-21 antisense oligonucleotides to the brain using T7-peptide decorated exosomes. J Control Release 317:273–281

Kim NH, Shim G, Park GH, Yu YG (2024) A nondestructive membrane engineering method using an amphiphilic polymer. Protein Sci 33(9):e5143

Kohn DB, Chen YY, Spencer MJ (2023) Successes and challenges in clinical gene therapy. Gene Ther 30(10):738–746

Kroll AV, Fang RH, Jiang Y et al (2017) Nanoparticulate delivery of cancer cell membrane elicits multiantigenic antitumor immunity. Adv Mater 29:1703969–1703977. https://doi.org/10.1002/adma.201703969

Larson RC, Maus MV (2021) Recent advances and discoveries in the mechanisms and functions of CAR T cells. Nat Rev Cancer 21(3):145–161

Le Joncour V, Martins A, Puhka M, Isola J, Salmikangas M, Laakkonen P et al (2019) A novel anti-HER2 antibody–drug conjugate XMT-1522 for HER2-positive breast and gastric cancers resistant to trastuzumab emtansine. Mol Cancer Ther 18(10):1721–1730

Lee DY, Cha B-H, Jung M, Kim AS, Bull DA, Won Y-W (2018) Cell surface engineering and application in cell delivery to heart diseases. J Biol Eng 12:1–11

Lee SY, Kim HU, Chae TU, Cho JS, Kim JW, Shin JH et al (2019) A comprehensive metabolic map for production of bio-based chemicals. Nat Catal 2(1):18–33

Lewis ND, Sia CL, Kirwin K, Haupt S, Mahimkar G, Zi T et al (2021) Exosome surface display of IL12 results in tumor-retained pharmacology with superior potency and limited systemic exposure compared with recombinant IL12. Mol Cancer Ther 20(3):523–534

Li J, Sharkey CC, Wun B, Liesveld JL, King MR (2016) Genetic engineering of platelets to neutralize circulating tumor cells. J Control Release 228:38–47

Li PY, Fan Z, Cheng H (2018a) Cell membrane bioconjugation and membrane-derived nanomaterials for immunotherapy. Bioconjug Chem 29(3):624–634

Li Z, Zhou X, Wei M, Gao X, Zhao L, Shi R et al (2018b) In vitro and in vivo RNA inhibition by CD9-HuR functionalized exosomes encapsulated with miRNA or CRISPR/dCas9. Nano Lett 19(1):19–28

Li Y, Raza F, Liu Y, Wei Y, Rong R, Zheng M et al (2021) Clinical progress and advanced research of red blood cells based drug delivery system. Biomaterials 279:121202

Li J, Huang D, Cheng R et al (2022) Multifunctional biomimetic nanovaccines based on photothermal and weak-immunostimulatory nanoparticulate cores for the immunotherapy of solid tumors. Adv Mater 34:2108012–2108023. https://doi.org/10.1002/adma.202108012

Li B, Wang W, Zhao L, Wu Y, Li X, Yan D et al (2024a) Photothermal therapy of tuberculosis using targeting pre-activated macrophage membrane-coated nanoparticles. Nat Nanotechnol 1–12

Li X, Wei Y, Zhang Z, Zhang X (2024b) Harnessing genetically engineered cell membrane-derived vesicles as biotherapeutics. Extracell Vesicles Circ Nucl Acids 5(1):44

Lian J, HamediRad M, Hu S, Zhao H (2017) Combinatorial metabolic engineering using an orthogonal tri-functional CRISPR system. Nat Commun 8(1):1688

Liang G, Kan S, Zhu Y, Feng S, Feng W, Gao S (2018) Engineered exosome-mediated delivery of functionally active miR-26a and its enhanced suppression effect in HepG2 cells. Int J Nanomedicine 13:585–599

Liang G, Zhu Y, Ali DJ, Tian T, Xu H, Si K et al (2020) Engineered exosomes for targeted co-delivery of miR-21 inhibitor and chemotherapeutics to reverse drug resistance in colon cancer. J Nanobiotechnol 18:1–15

Liang Y, Duan L, Lu J, Xia J (2021) Engineering exosomes for targeted drug delivery. Theranostics 11(7):3183

Lim SY, Yuzhalin AE, Gordon-Weeks AN, Muschel RJ (2016) Targeting the CCL2-CCR2 signaling axis in cancer metastasis. Oncotarget 7(19):28697

Lin D, Zhang H, Liu R, Deng T, Ning T, Bai M et al (2021) iRGD-modified exosomes effectively deliver CPT1A siRNA to colon cancer cells, reversing oxaliplatin resistance by regulating fatty acid oxidation. Mol Oncol 15(12):3430–3446

Liu L, He H, Liu J (2019) Advances on non-genetic cell membrane engineering for biomedical applications. Polymers 11(12):2017

Lundstrom K (2023) Viral vectors in gene therapy: where do we stand in 2023? Viruses 15(3):698

Luzio JP, Parkinson MD, Gray SR, Bright NA (2009) The delivery of endocytosed cargo to lysosomes. Biochem Soc Trans 37(5):1019–1021

Lv P, Liu X, Chen X, Liu C, Zhang Y, Chu C et al (2019) Genetically engineered cell membrane nanovesicles for oncolytic adenovirus delivery: a versatile platform for cancer virotherapy. Nano Lett 19(5):2993–3001

Ma Y, Zhang L, Huang X (2014) Genome modification by CRISPR/Cas9. FEBS J 281(23):5186–5193

Ma X-X, Xu J-L, Jia Y-Y, Zhang Y-X, Wang W, Li C et al (2020) Enhance transgene responses through improving cellular uptake and intracellular trafficking by bio-inspired non-viral vectors. J Nanobiotechnol 18:1–14

Maia A, Tarannum M, Romee R (2024) Genetic manipulation approaches to enhance the clinical application of NK cell-based immunotherapy. Stem Cells Transl Med 13(3):230–242

Mali P, Esvelt KM, Church GM (2013) Cas9 as a versatile tool for engineering biology. Nat Methods 10(10):957–963

Manzari MT, Shamay Y, Kiguchi H, Rosen N, Scaltriti M, Heller DA (2021) Targeted drug delivery strategies for precision medicines. Nat Rev Mater 6(4):351–370
Marcinkowski B, Stevanović S, Helman SR, Norberg SM, Serna C, Jin B et al (2019) Cancer targeting by TCR gene-engineered T cells directed against Kita-Kyushu Lung Cancer Antigen-1. J Immunother Cancer 7:1–9
Menasche BL, Crisman L, Gulbranson DR, Davis EM, Yu H, Shen J (2019) Fluorescence activated cell sorting (FACS) in genome-wide genetic screening of membrane trafficking. Curr Protoc Cell Biol 82(1):e68
Metzner C, Kochan F, Dangerfield JA (2013) Postexit surface engineering of retroviral/lentiviral vectors. Biomed Res Int 2013(1):253521
Mintzer MA, Simanek EE (2009) Nonviral vectors for gene delivery. Chem Rev 109(2):259–302
Morgan MA, Galla M, Grez M, Fehse B, Schambach A (2021) Retroviral gene therapy in Germany with a view on previous experience and future perspectives. Gene Ther 28(9):494–512
Mount NM, Ward SJ, Kefalas P, Hyllner J (2015) Cell-based therapy technology classifications and translational challenges. Philos Trans R Soc B Biol Sci 370(1680):20150017
Munagala R, Aqil F, Jeyabalan J, Gupta RC (2016) Bovine milk-derived exosomes for drug delivery. Cancer Lett 371(1):48–61
Navarrete C, Jacobsen IH, Martínez JL, Procentese A (2020) Cell factories for industrial production processes: current issues and emerging solutions. Processes 8(7):768
Newick K, O'Brien S, Moon E, Albelda SM (2017) CAR T cell therapy for solid tumors. Annu Rev Med 68(1):139–152
Nordling-David MM, Golomb G (2013) Gene delivery by liposomes. Isr J Chem 53(9–10):737–747
Oh J, Son YS, Kim WH, Kwon O-K, Kang B-J (2021) Mesenchymal stem cells genetically engineered to express platelet-derived growth factor and heme oxygenase-1 ameliorate osteoarthritis in a canine model. J Orthop Surg Res 16:1–16
Olson C, Ivanov K, Boyes D, Bengford D, Ku J, Flojo R et al (2024) Dual-omics approach unveils novel perspective on the quality control of genetically engineered exosomes. Pharmaceutics 16(6):824
Pai AA, Pritchard JK, Gilad Y (2015) The genetic and mechanistic basis for variation in gene regulation. PLoS Genet 11(1):e1004857
Paleos CM, Tziveleka L-A, Sideratou Z, Tsiourvas D (2009) Gene delivery using functional dendritic polymers. Expert Opin Drug Deliv 6(1):27–38
Park A, Lee JY (2024) Adenoviral vector system: a comprehensive overview of constructions, therapeutic applications and host responses. J Microbiol 62(7):491–509
Park JH, Jiang Y, Zhou J, Gong H, Mohapatra A, Heo J et al (2021) Genetically engineered cell membrane–coated nanoparticles for targeted delivery of dexamethasone to inflamed lungs. Sci Adv 7(25):eabf7820
Pezzoli D, Kajaste-Rudnitski A, Chiesa R, Candiani G (2013) Lipid-based nanoparticles as nonviral gene delivery vectors. Nanomater Interfaces Biol Methods Protocols:269–279
Pickett JR, Wu Y, Zacchi LF, Ta HT (2023) Targeting endothelial vascular cell adhesion molecule-1 in atherosclerosis: drug discovery and development of vascular cell adhesion molecule-1–directed novel therapeutics. Cardiovasc Res 119(13):2278–2293
Pinto E, Lione L, Compagnone M, Paccagnella M, Salvatori E, Greco M et al (2025) From ex vivo to in vivo chimeric antigen T cells manufacturing: new horizons for CAR T-cell based therapy. J Transl Med 23(1):10
Planul A, Dalkara D (2017) Vectors and gene delivery to the retina. Annu Rev Vis Sci 3(1):121–140
Poletti V, Mavilio F (2021) Designing lentiviral vectors for gene therapy of genetic diseases. Viruses 13(8):1526
Qi Y, Liu H, Chen X, Liu L (2019) Engineering microbial membranes to increase stress tolerance of industrial strains. Metab Eng 53:24–34
Qian R-C, Zhou Z-R, Guo W, Wu Y, Yang Z, Lu Y (2021) Cell surface engineering using DNAzymes: metal ion mediated control of cell–cell interactions. J Am Chem Soc 143(15):5737–5744
Rädler J, Gupta D, Zickler A, Andaloussi SE (2023) Exploiting the biogenesis of extracellular vesicles for bioengineering and therapeutic cargo loading. Mol Ther 31(5):1231–1250

Rai R, Alwani S, Badea I (2019) Polymeric nanoparticles in gene therapy: new avenues of design and optimization for delivery applications. Polymers (Basel) 11:745–749
Raposo G, Stahl PD (2023) Extracellular vesicles-on the cusp of a new language in the biological sciences. Extracell Vesicles Circ Nucl Acids 4(2):240
Ray S, Kassan A, Busija AR, Rangamani P, Patel HH (2016) The plasma membrane as a capacitor for energy and metabolism. Am J Physiol Cell Physiol 310(3):C181–C192
Ren E, Liu C, Lv P, Wang J, Liu G (2021) Genetically engineered cellular membrane vesicles as tailorable shells for therapeutics. Adv Sci 8(21):2100460
Robertson JL (2018) The lipid bilayer membrane and its protein constituents. J Gen Physiol 150(11):1472–1483
Romano G (2007) Current development of nonviral-mediated gene transfer. Drug News Perspect 20(4):227–231
Roy S, Cha JN, Goodwin AP (2020) Nongenetic bioconjugation strategies for modifying cell membranes and membrane proteins: a review. Bioconjug Chem 31(11):2465–2475
Royce LA, Liu P, Stebbins MJ, Hanson BC, Jarboe LR (2013) The damaging effects of short chain fatty acids on Escherichia coli membranes. Appl Microbiol Biotechnol 97:8317–8327
Sakurai F, Tachibana M, Mizuguchi H (2022) Adenovirus vector-based vaccine for infectious diseases. Drug Metab Pharmacokinet 42:100432
Salauddin M, Saha S, Hossain MG, Okuda K, Shimada M (2024) Clinical application of adenovirus (AdV): a comprehensive review. Viruses 16(7):1094
Sarkar D, Vemula PK, Zhao W, Gupta A, Karnik R, Karp JM (2010) Engineered mesenchymal stem cells with self-assembled vesicles for systemic cell targeting. Biomaterials 31(19):5266–5274
Sato YT, Umezaki K, Sawada S, Mukai S-a, Sasaki Y, Harada N et al (2016) Engineering hybrid exosomes by membrane fusion with liposomes. Sci Rep 6(1):21933
Schaffert D, Wagner E (2008) Gene therapy progress and prospects: synthetic polymer-based systems. Gene Ther 15(16):1131–1138
Serguera C, Bohl D, Rolland E, Prevost P, Heard JM (1999) Control of erythropoietin secretion by doxycycline or mifepristone in mice bearing polymer-encapsulated engineered cells. Hum Gene Ther 10(3):375–383
Stephan MT, Irvine DJ (2011) Enhancing cell therapies from the outside in: cell surface engineering using synthetic nanomaterials. Nano Today 6(3):309–325
Sterner RC, Sterner RM (2021) CAR-T cell therapy: current limitations and potential strategies. Blood Cancer J 11(4):69
Sun H, Su J, Meng Q et al (2016) Cancer-cell-biomimetic nanoparticles for targeted therapy of homotypic tumors. Adv Mater 28:9581–9588. https://doi.org/10.1002/adma.201602173
Sun HP, Su JH, Meng QS et al (2017) Cancer cell membrane-coated gold nanocages with hyperthermiatriggered drug release and homotypic target inhibit growth and metastasis of breast cancer. Adv Funct Mater 27:1604300–1604308. https://doi.org/10.1002/adfm.201604300
Sun M, Zhang H, Liu J, Chen J, Cui Y, Wang S et al (2024) Extracellular vesicles: a new star for gene drug delivery. Int J Nanomedicine 19:2241–2264
Sung YK, Kim S (2019) Recent advances in the development of gene delivery systems. Biomater Res 23(1):8
Sutaria DS, Jiang J, Elgamal OA, Pomeroy SM, Badawi M, Zhu X et al (2017) Low active loading of cargo into engineered extracellular vesicles results in inefficient miRNA mimic delivery. J Extracell Vesicles 6(1):1333882
Tan S, Wu T, Zhang D, Zhang Z (2015) Cell or cell membrane-based drug delivery systems. Theranostics 5(8):863
Tan Z, Yoon JM, Nielsen DR, Shanks JV, Jarboe LR (2016) Membrane engineering via trans unsaturated fatty acids production improves Escherichia coli robustness and production of biorenewables. Metab Eng 35:105–113
Ukidve A, Zhao Z, Fehnel A, Krishnan V, Pan DC, Gao Y et al (2020) Erythrocyte-driven immunization via biomimicry of their natural antigen-presenting function. Proc Natl Acad Sci USA 117(30):17727–17736

Umeda R, Satouh Y, Takemoto M, Nakada-Nakura Y, Liu K, Yokoyama T et al (2020) Structural insights into tetraspanin CD9 function. Nat Commun 11(1):1606

Urban JH, Merten CA (2011) Retroviral display in gene therapy, protein engineering, and vaccine development. ACS Chem Biol 6(1):61–74

Van Der Pol L, Stork M, van der Ley P (2015) Outer membrane vesicles as platform vaccine technology. Biotechnol J 10(11):1689–1706

Wallace R, Bliss CM, Parker AL (2024) The immune system—a double-edged sword for adenovirus-based therapies. Viruses 16(6):973

Wang C, Sun X, Cheng L, Yin S, Yang G, Li Y et al (2014) Multifunctional theranostic red blood cells for magnetic-field-enhanced in vivo combination therapy of cancer. Adv Mater (Deerfield Beach, Fla) 26(28):4794–4802

Wang X, Chen Y, Zhao Z, Meng Q, Yu Y, Sun J et al (2018) Engineered exosomes with ischemic myocardium-targeting peptide for targeted therapy in myocardial infarction. J Am Heart Assoc 7(15):e008737

Wang C, Pan C, Yong H, Wang F, Bo T, Zhao Y et al (2023) Emerging non-viral vectors for gene delivery. J Nanobiotechnol 21(1):272

Wang D, Zhou X, Huang M, Duan J, Qiu Y, Yi H et al (2024a) Cascade enzymes confined in DNA nanoanchors for antitumor therapy. ACS Appl Mater Interfaces 16(38):50295–50304

Wang J, Yin B, Lian J, Wang X (2024b) Extracellular vesicles as drug delivery system for cancer therapy. Pharmaceutics 16(8):1029

Wattiaux R, Laurent N, Wattiaux-De Coninck S, Jadot M (2000) Endosomes, lysosomes: their implication in gene transfer. Adv Drug Deliv Rev 41(2):201–208

Wen Y, Fu Q, Soliwoda A, Zhang S, Zheng M, Mao W et al (2022) Cell-derived nanovesicles prepared by membrane extrusion are good substitutes for natural extracellular vesicles. Extracell Vesicle 1:100004

Wu G, Li P, Feng H, Zhang X, Chu PK (2015) Engineering and functionalization of biomaterials via surface modification. J Mater Chem B 3(10):2024–2042

Wu W, Liu F, Singh S (2018a) Toward engineering E. coli with an autoregulatory system for lignin valorization. Proc Natl Acad Sci 115(12):2970–2975

Wu T, Ye L, Zhao D, Li S, Li Q, Zhang B et al (2018b) Engineering membrane morphology and manipulating synthesis for increased lycopene accumulation in Escherichia coli cell factories. 3 Biotech 8(6):269

Wu H, Peng Z, Xu Y, Sheng Z, Liu Y, Liao Y et al (2022a) Engineered adipose-derived stem cells with IGF-1-modified mRNA ameliorates osteoarthritis development. Stem Cell Res Ther 13(1):19

Wu W-C, Tian J, Xiao D, Guo Y-X, Xiao Y, Wu X-Y et al (2022b) Engineered extracellular vesicles encapsulated Bryostatin-1 as therapy for neuroinflammation. Nanoscale 14(6):2393–2410

Xiang S, Tong H, Shi Q, Fernandes JC, Jin T, Dai K et al (2012) Uptake mechanisms of non-viral gene delivery. J Control Release 158(3):371–378

Xiang S, Sarem M, Shah S, Shastri VP (2018) Liposomal treatment of cancer cells modulates uptake pathway of polymeric nanoparticles by altering membrane stiffness. Small 14(14):1704245

Xiong T, Chen Y, Peng Q, Li M, Lu S, Chen X et al (2024) Pyrazolone-protein interaction enables long-term retention staining and facile artificial biorecognition on cell membranes. J Am Chem Soc 146(34):24158–24166

Xu F, Wu Y, Yang Q, Cheng Y, Xu J, Zhang Y et al (2022) Engineered extracellular vesicles with SHP2 high expression promote mitophagy for Alzheimer's disease treatment. Adv Mater 34(49):2207107

Yano Y, Matsuzaki K (2009) Tag–probe labeling methods for live-cell imaging of membrane proteins. Biochim Biophys Acta (BBA)-Biomembranes 1788(10):2124–2131

Yi Y, Jong Noh M, Hee LK (2011) Current advances in retroviral gene therapy. Curr Gene Ther 11(3):218–228

Yin L, Song Z, Kim KH, Zheng N, Gabrielson NP, Cheng J (2013) Non-viral gene delivery via membrane-penetrating, mannose-targeting supramolecular self-assembled nanocomplexes. Adv Mater (Deerfield Beach, Fla). 25(22):3063

Yoder KE, Rabe AJ, Fishel R, Larue RC (2021) Strategies for targeting retroviral integration for safer gene therapy: advances and challenges. Front Mol Biosci 8:662331

Yu L, Feng R, Zhu L, Hao Q, Chu J, Gu Y et al (2020) Promoting the activation of T cells with glycopolymer-modified dendritic cells by enhancing cell interactions. Sci Adv 6(47):eabb6595

Yu X, Sha L, Liu Q, Zhao Y, Fang H, Cao Y, Zhao J (2021) Recent advances in cell membrane camouflage-based biosensing application. Biosens Bioelectron 194(113623)

Zarate-Kalfopulos B, Sanchez AR, Gonzalez D, Lecona H (2006) P2. Formation and characterization of genetically modified periosteum cells with overexpressed angiogenin incorporated into chitosan polymer for potential bone tissue engineering. Spine J 6(5):83S–84S

Zariri A, Beskers J, van de Waterbeemd B, Hamstra HJ, Bindels TH, van Riet E et al (2016) Meningococcal outer membrane vesicle composition-dependent activation of the innate immune response. Infect Immun 84(10):3024–3033

Zhang P, Zhang L, Qin Z, Hua S, Guo Z, Chu C et al (2018) Genetically engineered liposome-like nanovesicles as active targeted transport platform. Adv Mater 30(7):1705350

Zhang M, Cheng S, Jin Y, Zhang N, Wang Y (2021a) Membrane engineering of cell membrane biomimetic nanoparticles for nanoscale therapeutics. Clin Transl Med 11(2):e292

Zhang X, Zhang H, Gu J, Zhang J, Shi H, Qian H et al (2021b) Engineered extracellular vesicles for cancer therapy. Adv Mater 33(14):2005709

Zhang E, Phan P, Algarni HA, Zhao Z (2022) Red blood cell inspired strategies for drug delivery: emerging concepts and new advances. Pharm Res 39(11):2673–2698

Zhao H, Li L, Zhang J, Zheng C, Ding K, Xiao H, et al. C–C chemokine ligand 2 (CCl2) recruits macrophage-membrane-camouflaged hollow bismuth selenide nanoparticles to facilitate photothermal sensitivity and inhibit lung metastasis of breast cancer. ACS Appl Mater Interfaces 2018;10(37):31124-31135

Zhao Z, Dong S, Liu Y et al (2022a) Tumor microenvironment activable manganese-boosted catalytic immunotherapy combined with PD-1 checkpoint blockade. ACS Nano 16:20400–20418. https://doi.org/10.1021/acsnano.2c06646

Zhao Z, Anselmo AC, Mitragotri S (2022b) Viral vector-based gene therapies in the clinic. Bioeng Transl Med 7(1):e10258

Zhao Y, Haney MJ, Fallon JK, Rodriguez M, Swain CJ, Arzt CJ et al (2022c) Using extracellular vesicles released by GDNF-transfected macrophages for therapy of Parkinson disease. Cells 11(12):1933

Zhou Y, Yuan Y, Liu M, Hu X, Quan Y, Chen X (2019) Tumor-specific delivery of KRAS siRNA with iRGD-exosomes efficiently inhibits tumor growth. ExRNA 1(1):1–7

Zhuang Q, Xu J, Deng D, Chao T, Li J, Zhang R et al (2021) Bacteria-derived membrane vesicles to advance targeted photothermal tumor ablation. Biomaterials 268:120550

Zylberberg C, Gaskill K, Pasley S, Matosevic S (2017) Engineering liposomal nanoparticles for targeted gene therapy. Gene Ther 24(8):441–452

Biomaterial-Based Cell Membrane Engineering Strategies and Their Scope in Biomedical Applications

7

Mohammad Tollabi, Arezou Mehrabi, Samira Zolfaghari, Mohammad Reza Khalili, Mohammad Ali Shokrgozar, and Peiman Brouki Milan

Abstract

Cell surface engineering strategies aim to functionalize living cells and cell membrane-based platforms, such as membrane vesicles and membrane-coated nanoparticles, by modifying their membranes. These modifications enhance targeted homing, tissue penetration, selective accumulation, drug-loading efficiency, real-time monitoring, precise antigen presentation, and immune modulation for disease-specific therapies. Various ligands and biomolecules, such as peptides, biopolymers, nanoparticles, nucleic acids, and bioactive proteins, have been attached to membrane-based platforms using diverse methods. Biomaterial-based cell membrane engineering strategies, including covalent coupling, hydrophobic interaction, electrostatic interaction, membrane fusion, and metabolic glycoengineering, along with integrated approaches, have been introduced to enhance the functionality of cell membrane-based platforms. Biomaterial-based strategies are increasingly utilized because of their low cost, low toxicity, broad applications, and ease of engineering compared to genetic

M. Tollabi · S. Zolfaghari · P. B. Milan (✉)
Department of Tissue Engineering and Regenerative Medicine, Faculty of Advanced Technologies in Medicine, Iran University of Medical Sciences, Tehran, Iran

Cellular and Molecular Research Center, Iran University of Medical Sciences, Tehran, Iran
e-mail: zolfaghari.sa@iums.ac.ir; boroukimilan.t@iums.ac.ir

A. Mehrabi
School of Medicine, Tonekabon Branch, Islamic Azad University, Tonekabon, Iran
e-mail: arezoumehrabbi@iau.ac.ir

M. R. Khalili
Blood Transfusion Research Center, High Institute for Research and Education in Transfusion Medicine, Tehran, Iran

M. A. Shokrgozar
National Cell Bank of Iran, Pasteur Institute of Iran, Tehran, Iran

Z. Hussain, R. Pei (eds.), *Cell Membrane Engineering for Advancing Cell-Based and Nanotechnology-Based Therapies*, Biomaterials, Bioengineering and Sustainability 6, https://doi.org/10.1007/978-3-031-93425-4_7

engineering-based strategies. These methods have been extensively utilized in cancer therapy, precision drug delivery, tissue engineering and other biomedical applications. This chapter describes cell membrane features and the need for membrane engineering to enhance biomedical applications. It emphasizes chemically driven engineering strategies using biomaterials, highlighting their potential to improve outcomes for living cells and membrane-based platforms in biomedical applications, while also discussing the ongoing challenges of these strategies.

Keywords

Membrane engineering · Biomaterials · Covalent conjugation · Hydrophobic insertion · Membrane fusion · Metabolic glycoengineering · Biomedical application

7.1 Introduction

With the rise of translational science, cells and their derivatives, such as extracellular vesicles (EVs) and membrane-coated nanoparticles, have become a promising approach in regenerative medicine. The effectiveness of cell-based therapeutics is mainly due to the inherent properties of living cells and their capability to sense the surrounding environments (Fischbach et al. 2013; Hoang et al. 2022). Using native cells with naked plasma membrane in cell-based therapies and membrane-based platforms demonstrated many challenges that decreased their efficacy and restricted clinical applications. These challenges include high immunogenicity, poor homing, insufficient tissue engraftment, unwanted differentiation, and reduced viability during transplantation and *ex vivo* manipulation (Lanza et al. 2019; Naderi-Meshkin et al. 2015; Han et al. 2024; Koh et al. 2020). To address these challenges and enhance stability, specific action, and effectiveness of cell-based therapeutics, various cell-manipulating methods such as preconditioning, genetic engineering, and utilizing biomaterials as a synthetic extracellular matrix (ECM) to recreate a favorable microenvironment for cells are developed. With the advent of genetic engineering methods, long-lasting cell engineering became possible by modifying the expression of cell membrane proteins to enhance their therapeutic properties. Despite their advantages, genetic engineering-based cell membrane strategies have several limitations, including poor safety, challenges in controlling cell phenotype, and a lack of assured efficacy, as this process is limited to peptide conjugation due to the constraints of the central dogma. Genetic-based cell membrane engineering methods may influence normal cell function by hindering biological processes inside the cell (Qian et al. 2021; Ying et al. 2022). Moreover, the high expression of desired proteins relies on the amount of genetic material that cells internalize. Furthermore, only a limited number of recombinant proteins can be displayed on EVs, making it challenging to achieve high molecule density. As a result, most studies focus on engineering EVs for a single function, as incorporating multiple genes into EVs remains challenging (Richter et al. 2021; Murphy et al. 2019; Ivanova et al. 2023). Displaying complex antigens on EVs is challenging due to uncertainties in protein folding, EVs heterogeneity, and limitations in scalable production. These important challenges, alongside high risks of tumorigenesis and immunogenicity, restrict the broad-spectrum clinical use of this modification strategy and convince researchers to look for cheaper and safer alternative cell membrane engineering methods (Amaral and Pasparakis 2019; Lee et al. 2018).

To enhance the effectiveness of cell-based therapies, various chemical-based approaches are being developed to modify the cell surface with bioactive molecules, bioorthogonal chemical groups, bioactive peptides, nucleic acids, and nanomaterials, imparting new characteristics to the cells (Almeida-Pinto et al. 2023; Jin et al. 2025). Usually, cell membrane engineering technology is divided into two main categories, including non-biological or chemically-driven modifications (including covalent interaction, electrostatic functionalization, hydrophobic insertion, and biomolecular recognition) and biologically-driven modification (including genetic engineering and enzymatic modification) (Fig. 7.1) (Almeida-Pinto et al. 2023). Each modification method has its application, benefit, and limitation, but generally, non-biological membrane engineering methods are more reliable and reversible modifications than biological methods to control cellular functions (Hebb 2005; Liu et al. 2019a). Table 7.1 summarizes some important points of non-biological engineering methods and the genetic engineering approach (as a member of the biological membrane engineering approach). The aim of developed biological and non-biological membrane engineering approaches is to modify the cells and cell membrane-based platforms to have controllable functions and enhanced specific potential for targeted drug delivery (Almeida-Pinto et al. 2023), biosensing application (Tan et al. 2024), bioimaging (Guan et al. 2024), targeted cell-based therapies (Stephan and Irvine 2011), and tissue engineering (Gaspar et al. 2020).

Membrane fusion technology is a cell membrane engineering approach developed to assemble hybrid membrane structures and fuse them by different methods, such as electricity-induced and virus-induced methods (Sun et al. 2023). These methods address clinical challenges associated with cell-based platforms, particularly EVs, and enhance their therapeutic potential (Rankin-Turner et al. 2021). The surfaces of most mammalian cells are covered by glycocalyx, a layer of carbohydrates that have variable thicknesses in different types of cells. Cell membrane glycans play a crucial role in determining a cell's function and defining its interactions with the environment (Du and Yarema 2010). By manipulating the set of cell surface glycans, it is possible to gain insight into cellular behavior modulation for different biomedical applications (Ying et al. 2022). Due to the unique production and installation of glycans on the cell surface, their manipulation is not possible using genetic engineering methods (Kufleitner et al. 2023). To address this limitation, metabolic glycoengineering (MGE) is created to manipulate glycosylation by increasing glycans level and to attach non-natural monosaccharides into glycoconjugates (Agatemor et al. 2019).

Gathering on this potential, this chapter describes cell membrane features and necessities for membrane engineering for enhanced biomedical applications.

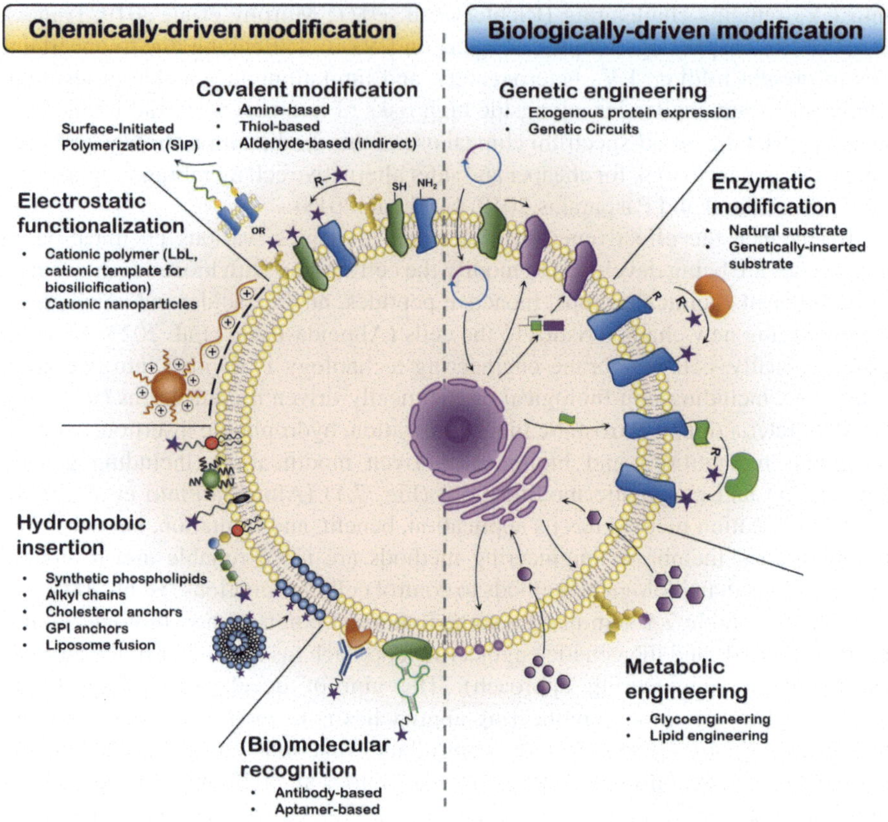

Fig. 7.1 Schematic of currently available cell surface modifying strategies, including chemically or non-biologically driven modifications and biologically-driven modification. (Reprinted with permission from Almeida-Pinto et al. (2023), copyright 2023, Wiley-VCH GmbH)

Subsequently, a distinct emphasis is placed on highlighting chemically driven engineering strategies by utilizing biomaterials and a brief overview of their biomedical applications. Finally, the ongoing challenges of these engineering strategies are briefly presented. Overall, this chapter emphasizes the potential of biomaterial-based cell membrane engineering in significantly improving the outcomes of living cells and cell membrane-based platforms for different biomedical applications.

7.2 Membrane Features and Necessities for Engineering

Cells are the fundamental units of life, with the cell membrane serving as a pivotal element in defining their structure and function. This membrane, composed primarily of a bilayer of phospholipids, acts as a protective barrier that shields the cell's interior from external disturbances. Beyond its structural role, the membrane

Table 7.1 Summarization of non-biological and genetic engineering approach for cell membrane modification

Modification approach	Explanation	Challenge	Benefits	Ref
Non-biological Engineering	Involves physical or covalently conjugating therapeutics to membrane-associated proteins	Requires precise control of reaction conditions to prevent unwanted side reactions Complexity can lead to variability in product quality	Enables specific and stable attachment of functional groups Enables tailored modifications for diverse applications.	Tan et al. (2024), Coronado et al. (2024), Wu et al. (2023), and Zhang et al. (2021)
Genetic Engineering	Enables multicomponent modifications via gene editing. Utilizes lipid fusion techniques to attach therapeutic agents into cell membranes	Gene editing challenges can complicate modifications Regulatory hurdles may arise in clinical use of genetically modified cells Achieving efficient lipid fusion can be challenging due to membrane instability.	Allows for sophisticated functionalities Can lead to long-lasting changes in cell behavior and therapeutic outcomes Direct incorporation of therapeutics enhances targeted delivery	Liu et al. (2019a), Tan et al. (2024), Kang and Tullman-Ercek (2018), and Wu et al. (2024a)

enables selective exchange of substances and facilitates signal transmission from the extracellular surroundings. The presence of specialized transmembrane proteins and receptors further enhance the membrane's functionality, allowing cells to dynamically adapt to and interact with their surroundings (Ying et al. 2022; Huang et al. 2024).

Cell surface engineering is a rapidly evolving field that provides valuable tools for regulating cellular functions and enhancing cell-based therapeutics' efficacy. As mentioned earlier, cell surface modification holds great promise for advancing cell-based therapies and tissue engineering. The membrane of mammalian cells is a complex and refined structure that precisely controls cellular function and physiological properties in natural biosystems (Amaral and Pasparakis 2019). It consists of different components and is known as a biological barrier that encompasses all organisms and delimits the extracellular environment (Wu et al. 2024b). The cell membrane can be used as a proper chemical substrate for engineering due to the negative charge and hydrophobicity of its lipid parts, as well as the chemical groups of proteins and carbohydrates. In cell surface engineering methods the cell membrane is decorated by different types of biological, chemical, and physical approaches to refine and improve cell membrane functionalities (Almeida-Pinto et al. 2023). Cell membrane building blocks including proteins, and glycans are ideal docking sites for the conjugation of biopolymers or therapeutic elements through direct chemical modification. In this strategy, functional molecules are conjugated to the hydroxyl (-OH) residues, amine ($-NH_2$) groups, carboxyl (-COOH) groups, and sulfhydryl (-SH) groups of cell membrane proteins or polysaccharides by covalent bonds (Almeida-Pinto et al. 2023; Abbina et al. 2017). As mentioned earlier, native cells have several clinical and commercial challenges that limit their wide-spectrum clinical application (Lanza et al. 2019; Naderi-Meshkin et al. 2015; Han et al. 2024), though several types of cells including red blood cells (RBCs), platelets, mesenchymal stem cells (MSCs), and some types of white blood cells are functionalized through cell membrane engineering methods to address these clinical limitations. Like cells, the clinical use of cell membrane-derived nanovesicles faces unresolved challenges, including low targeting efficiency, poor stability, limited tissue penetration, scalability issues, and difficulties in molecular characterization for design and regulation. Multiple bio-engineering approaches have been innovated to meet these challenges and enhance the therapeutic potency of cells and cell-based platforms (Claridge et al. 2021). RBCs are mainly used as drug carriers in cell-based platforms, but this approach faces some challenges, such as poor drug loading capacity, poor targeting, low stability after storage, and potential for immunogenicity (Rossi et al. 2019; Vincy et al. 2022). Multiple membrane engineering strategies are developed based on RBCs membrane functional groups such as -NH2, -COOH, and -SH (Fu et al. 2018). Zhang et al. fabricated cholesterol-modified RBCs nanovesicles by incubating free cholesterols with RBCs ghosts in ammonium sulfate buffer, exposed to physical disruption to induce vesiculation. These modified nanovesicles not only increased their structural stability but also improved their pH gradient maintenance ability (Zhang et al. 2017a). Despite all the progress, biomaterial-based cell membrane engineering strategies face formidable challenges related to the intrinsic properties of cell membranes, including the dynamic structure of the membrane, continuous flip-flop movement, and relatively high rate of degradation and replacement (McMahon and Gallop 2005).

7.3 Biomaterial-Based Cell Membrane Engineering Strategies

As previously mentioned, despite the magnificent advances in increasing the effectiveness of cell-based therapeutics, there is still quite a long way to adopt the cell-based therapy approach for clinical use as a completely safe and effective treatment method (Liu et al. 2019a). To circumvent the main challenges of cell-based therapeutics, including lack of control over the transport fate of the cells *in vivo*, decreasing the therapeutic potential of injected cells because of immune surveillance, and inability to completely control the differentiation as well as cell-cell and/or cell-matrix interactions, different types of cell membrane engineering methods are developed (Amaral and Pasparakis 2019; Teramura and Iwata 2010). For instance, non-covalent cell surface engineering strategies have been utilized to present synthetic glycopolymers that imitate cell surface mucins, which play roles in regulating cell–cell interactions (Rabuka et al. 2008).

In contrast to genetic engineering-based surface modification, non-biological modifications not only manipulate the cell's surface proteins but also surface lipids and glycans (Custódio and Mano 2016). By using non-biological modifications including covalent conjugation, electrostatic interaction, hydrophobic interaction, membrane fusion, and metabolic glycoengineering, various types of biopolymers and/or biomaterials have been tethered on the cell surface (Teramura and Iwata 2010). In this context, it is crucial to consider choosing materials compatible with the physiological conditions of the cells, designing the appropriate density and distribution of surface functional sites, and applying a suitable mechanism for macromolecule immobilization to ensure optimal performance and biocompatibility (Amaral and Pasparakis 2019; Stephan and Irvine 2011), and selecting stable material along with suitable engineering methods (Custódio and Mano 2016).

7.3.1 Covalent Conjugation Methods

The covalent chemical conjugation methods are developed to conjugate synthetic molecules to the cell surface directly and stably by utilizing this method (Guan et al. 2024). This strategy involves modifying the -NH_2 groups of lysine residues, as well as -SH groups of cell membrane peptides through acylation reactions, which are common methods for covalent conjugation to enhance cell therapies (Stephan and Irvine 2011; Chen et al. 2011), cell labeling, and imaging applications (Xi et al. 2022). To address the challenge of reduced cell viability in cell therapy, Koh et al. developed an ECM-coating method that reduces disulfide bonds around the cell membrane, enhancing the viability and engraftment potential of human iPSC-derived mesenchymal stem cells (Fig. 7.2) (Koh et al. 2020). In this context, masking the cell surface antigens to reduce the immunogenic responses and improve the efficacy of cell therapy is a relatively common and popular approach (Nacharaju et al. 2005; Wilson and Chaikof 2008). For example, Lee et al. modified the pancreatic islets by conjugating succinimidyl ester-conjugated polyethylene glycol (PEG) into the surface amines to mask immunogenic antigens, thus increasing the therapeutic efficacy and survival time of transplanted cells (Lee et al. 2006). Using a similar strategy, -SH groups on cysteine residues of mammalian cell membranes can be modified with various labeling groups. For example, maleimide reactive functional group, can be used to immobilize nanoparticles or polymers on the cell surface (Stephan et al. 2010). By this chemical method, RBCs were functionalized by conjugation of maleimide-functionalized PEG to hide the blood group antigens to reduce immunogenic reactions (Nacharaju et al. 2005). Cai et al. engineered umbilical vein endothelial cells and skin fibroblasts with succinimide ester-methoxy PEG to enhance their migration capability (Cai et al. 2023). PEGylation method is also being widely utilized to achieve effective cell-cell interaction by modifying proteins, enzymes, artificial surfaces, liposomes, and drugs (Chung et al. 2004). Recent advancements have demonstrated that functionalizing cells with short oligonucleotides can enhance their targeted adhesion properties (Custódio and Mano 2016). For instance, the integration of complementary DNA sequences conjugated

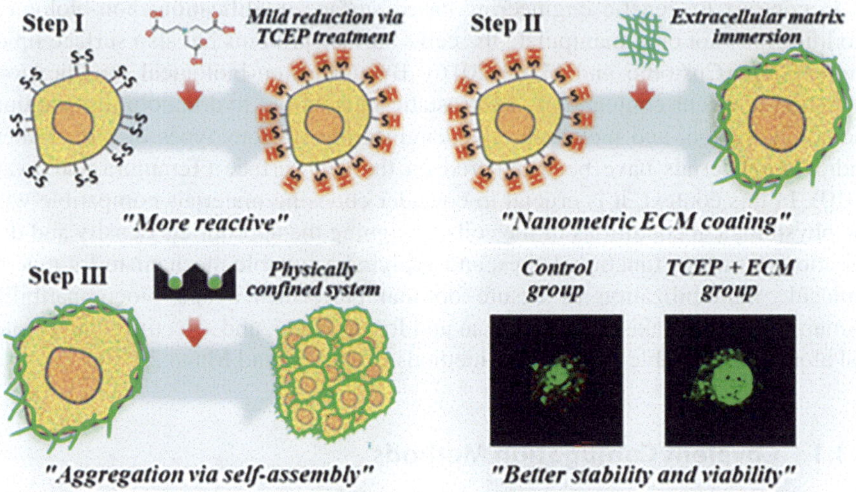

Fig. 7.2 A schematic illustrates the new nanometric ECM-coating method onto stem cells. (step I) To achieve the mild reduction of stem cells surface sulfide bonds, cells were treated with tris(2-carboxyethyl)phosphine (TCEP). (step II) Exposed free thiol groups on the cell membrane surface improve adhesion of ECM to the stem cells membrane. (step III) Formation of three-dimensional spheroids via self-assembly of the cells within a confined microenvironment. (Reprinted with permission from Koh et al. (2020), copyright 2020, American Chemical Society)

to amphiphilic PEG-lipids facilitates the orchestration of cell–cell and cell–biomaterial interactions through sequence-specific hybridization, thereby enabling precise biointerface modulation. Iwata and colleagues developed amphiphilic PEG-lipid polymers that were connected to specific oligonucleotide sequences (Li et al. 2019).

Biotinylation of the cell membrane occurs through a chemical reaction between cell surface amine groups and amine-reactive biotin. The biotin-labeled cells can then be functionalized by biotinylated molecules via a streptavidin linker (Stephan and Irvine 2011). Based on this, Krishnamachari and co-workers modify the human embryonic kidney 293 cell membrane by covalent conjugation of biotin to create a cell-microparticle hybrid (Krishnamachari et al. 2008). Moreover, there are many -COOH groups present at the C-terminus of membrane peptides, along with aspartic acid and glutamic acid residues, which can be functionalized through the amide coupling reaction mediated by 1-ethyl-3-(3-dimethylaminopropyl) (EDC) (Lutterotti et al. 2013). This method has proved to be completely efficient and safe to couple the antigen with cells in autoimmune and/or allergic diseases experimental models. Furthermore, maleimide derivatives can establish thioether bonds with the sulfhydryl groups on the peptides through the Michael addition reaction (Jones et al. 2017). In addition to -NH_2 or -SH groups, aldehydes and ketones can be used for covalent conjugation of different molecules. However, chemical reactive groups must first be introduced through enzymatic or chemical treatment of carbohydrates on the cell surface (Holden et al. 2010).

7.3.2 Hydrophobic Interactions

Hydrophobic interactions are developed as a noncovalent physical bioconjugation engineering method to change the surface of cell-based therapeutics by conjugating different types of materials (Liu et al. 2019a). Cell membrane as an amphipathic physical bilayer (Peng and Chen 2024), enables the conjugation of modifying substances to its surface through complementary interactions between the hydrophobic regions of the molecules and the hydrophobic region of the membrane (Liu et al. 2019a). In this strategy, lipophilic molecules are anchored in the hydrophobic region that is created by cell membrane phospholipids and modify the cell membrane with minimal effect on cell viability and bioactivity, compared to other direct chemical modification strategies (Davis et al. 2019). In this strategy, cell membrane modification by unnatural molecules is mainly mediated by lipid anchors such as alkyl chains (e.g., C18 chain), synthetic phospholipids, cholesterol, and glycosylphosphatidylinositol (GPI) anchors through hydrophobic interactions. To achieve optimal anchoring efficacy, it is crucial to consider the dynamic structure of the cell membrane when designing and selecting the hydrophobic anchor (Almeida-Pinto et al. 2023). Lipid insertion strategy makes possible the attachment of non-natural moieties on cell membrane surface and nanovesicles by a lipid anchor (Huang et al. 2024). The arrangement of phospholipid, glycosylphosphatidylinositol, and cholesterol creates an ideal engineering site for inserting ligand-conjugated lipophilic molecules.

This strategy has been widely used to modify RBC membranes, particularly for creating nanovesicles or masking nanomaterials. Zhang et al. applied a lipid-insertion technique to modify RBC membranes by tethering a tumor-targeting ligand-linker-lipid into RBC membrane ghosts. They then coated a polymeric nanoparticle core with the modified membrane to create functional cell membrane-coated nanoparticles for cancer treatment (Fang et al. 2013; Guo et al. 2015). Moreover, this strategy has other translational applications, such as controlling the formation of cell aggregates and spheroids (Amaral and Pasparakis 2016; Javvaji et al. 2014), promotion of cell membrane fusion (Yoshihara et al. 2020), and encapsulation of cells for transplantation (Amaral and Pasparakis 2019). In another study, by focusing on repurposing RBCs-based nanomedicine researchers developed a targeted drug delivery for cancer treatment. In this approach, they utilized sophisticated methods to alter RBCs membranes via various conjugation techniques, including lipid insertion, biotin-avidin interactions, EDC/NHS coupling, and passive adsorption (Fig. 7.3). These modification methods have enhanced the efficacy of RBCs-based nanomedicine across a range of cancer therapies (Sun et al. 2019). Lipid-insertion technique was also applied to modify the RBCs membrane by tumor-targeting peptides to bestow RBCs membrane-coated NPs with effective anti-cancer targeting (Fu et al. 2018; Chai et al. 2017; Zhang et al. 2018). Despite advancements, challenges remain in standardizing RBC engineering techniques and ensuring consistent therapeutic outcomes. Key issues include preserving RBC membrane integrity and biocompatibility. These challenges can be mitigated using gentle surface modification methods, as shown by low hemolysis rates and intact

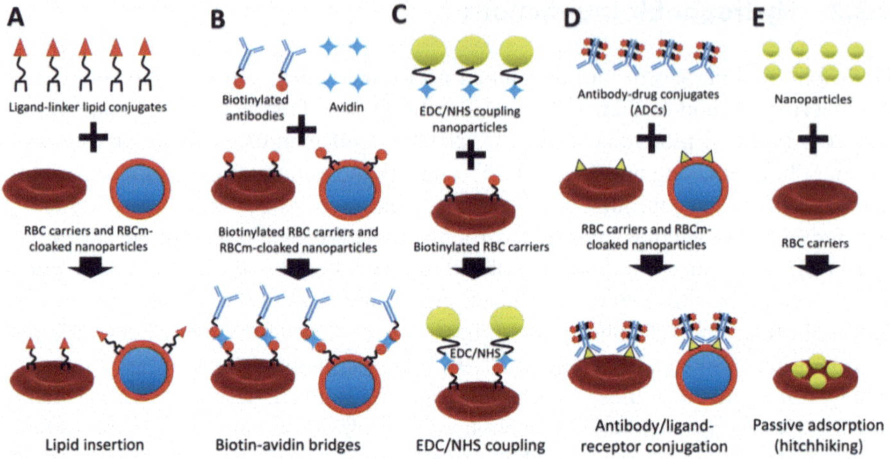

Fig. 7.3 A diagram illustrates the techniques for refunctionalizing erythrocyte-based nanomedicine, which includes: (**a**) lipid insertion, (**b**) biotin-avidin interactions, (**c**) EDC/NHS coupling, (**d**) antibody/ligand-receptor conjugation, and (**e**) passive adsorption (hitchhiking). (Reprinted with permission from Sun et al. (2019), copyright 2019, Ivyspring International Publisher)

RBC structures (Sun et al. 2015). A common issue with lipid anchoring is the potential internalization of functional moieties caused by the hydrophobic nature of the conjugates. To prevent this, researchers use hydrophilic biopolymers (e.g., PEG) between hydrophobic and functional moieties (Shi and Wang 2021; Kim and Kim 2022).

Despite cells, it is possible to engineer EVs by this strategy for multiple biomedical applications such as improving the tumor-targeting drug delivery (Kooijmans et al. 2016; Qi et al. 2016) and facilitating the uptake of cell and increasing endosomal escape of EVs-loaded drugs (Nakase and Futaki 2015).

7.3.3 Electrostatic Interaction

Due to the presence of different negative charged components, such as sialic acid residues on glycoproteins, carboxyl groups of proteins, and phosphate groups on phospholipids, the outer-surface membrane has a negatively charged nature. Therefore, cationic functionalizing nanoparticles and biomaterials such as chitosan, poly-L-lysine (PLL), and poly(ethylenimine) (PEI), can be installed to the outer-surface membrane through electrostatic interaction as a noncovalent physical bioconjugation. This simple and cost-effective strategy is utilized widely to achieve cell assembly in the tissue engineering approach via the polyelectrolyte method (Almeida-Pinto et al. 2023; Zhang et al. 2017b). This simple and inexpensive modifying method can improve the physical and chemical features of the cell membrane, increase the functionalities of cells, and also confer more mechanical support to the cells against external stimuli (Almeida-Pinto et al. 2023). The polyelectrolyte-modified cell assembly is achieved through electrostatic interaction. To achieve this, cationic biomaterials bind onto the negatively charged outer-surface membrane until the cell membrane gains a positive charge. After layer-by-layer (LbL) assembly, due to the cationic biomaterial deposition, a thin polyelectrolyte multilayer (PEM) film is formed around the cell surface. PEM can be used as a suitable environment for cell differentiation and proliferation (Almeida-Pinto et al. 2023). This method has been specifically utilized in cell therapy applications such as the immunoisolation barrier, sustained release carriers of therapeutic molecules (Abbina et al. 2017; Wilson et al. 2011), and the construction of multi-layered tissue-like structures (Almeida-Pinto et al. 2023). In comparison to other modifying methods, the major disadvantage of electrostatic interaction is the possibility of cellular damage and decreasing cell viability due to the internalization of polycations with high charge density (Davis et al. 2019; Maciel et al. 2021). To address this problem, researchers utilize spacers, typically PEG, to constrain the direct contact between the cationic biomaterials and the cell surface. Moreover, choosing suitable biocompatible biopolymers for modifying cell surface is another strategy for minimizing cytotoxicity (Lee et al. 2018; Wang and Wang 2022). Alongside, although this strategy was successfully demonstrated, there are multiple challenges in this strategy for achieving cell membrane functionalization such as a time-consuming process (e.g., LbL approach), decreasing cell bioactivity, low stability in complex biological environments, and lacking cell targeting ability (Almeida-Pinto et al. 2023; Ma et al. 2022).

7.3.4 Membrane Fusion

EVs are cell-derived nanovesicles that are very popular among researchers due to their wide biomedical applications. However, despite their valuable functionalities, natural EVs have multiple challenges including lacking targeted ability, low drug loading ability, and complex-low-yield isolation method (Rankin-Turner et al. 2021). To address these urgent issues, membrane fusion strategy is utilized to develop hybrid membrane nanovesicles (HMNVs) by fusing of homologous and heterologous membranes originating from natural or artificial sources (Sun et al. 2023). To achieve cell/membrane fusion *in vitro*, researchers usually utilize the electricity-induced method, the virus-induced method, and the PEG-based chemically induced method (Yoshihara et al. 2020). Cell membrane fusion progresses through key steps, including membrane contact, lipid mixing, and the complete integration of inner contents. Initially, membrane or vesicle contact is driven by van der Waals interactions; however, strong repulsive forces arising from hydration, electrostatic, and steric effects hinder close membrane approximation. To overcome this barrier, membrane engineering strategies can be employed to reduce lipid bilayer repulsion, either by decreasing the number of water-binding sites or by modulating membrane charge, thereby facilitating closer membrane interactions and promoting fusion (Yoshihara et al. 2020; Sauvola and Littleton 2021; Mondal Roy and Sarkar 2011; Jahn et al. 2024). The process of membrane fusion can be monitored through various methods, including imaging techniques like transmission electron microscopy, scanning electron microscopy, confocal laser scanning microscopy, and freeze-fracture microscopy, as well as through assays such as lipid mixing and content mixing assays (Mondal Roy and Sarkar 2011).

7.3.4.1 Classification of HMNVs According to Membrane Sources

HMNVs have more stability, specific targeting abilities, circulation time, and drug loading capability than EVs, therefore they are considered as alternative tools instead of EVs for biomedical applications. As shown in Fig. 7.4 based on their membrane origins, HMNVs are divided into three major groups, which include (1) HMNVs created by the fusion of homologous or heterologous membranes (2) HMNVs created by a fusion of liposome and cell membrane 3) HMNVs created by fusion of cell membrane and bacterial membrane (Sun et al. 2023).

The liposome fusion technique allows for precise regulation of the composition and density of functional agents on the cell membrane, allowing for the creation of customized cell interfaces with specific, user-defined characteristics (Shi et al. 2016). Liposome-based synthetic fusion models are divided into non-targeted vesicle fusion (e.g., metal ion- and peptide-induced fusion) and targeted vesicle fusion (e.g., DNA- and peptide-based models). The fusion recognition mechanisms are primarily designed using DNA, carbohydrates, peptides, nucleic acids, and ions. Each fusion model has distinct applications based on its structural features and biological characteristics (Mazur and Chandrawati 2021; Marsden et al. 2011). For instance, due to large amounts of diol sugars on the cell membrane, carbohydrate-binding acid model platforms, such as boronic acid/cis-diol recognition-based models, are among the best choice for developing targeted drug delivery systems (Dowlut and Hall 2006). Alongside several membrane engineering methods, the

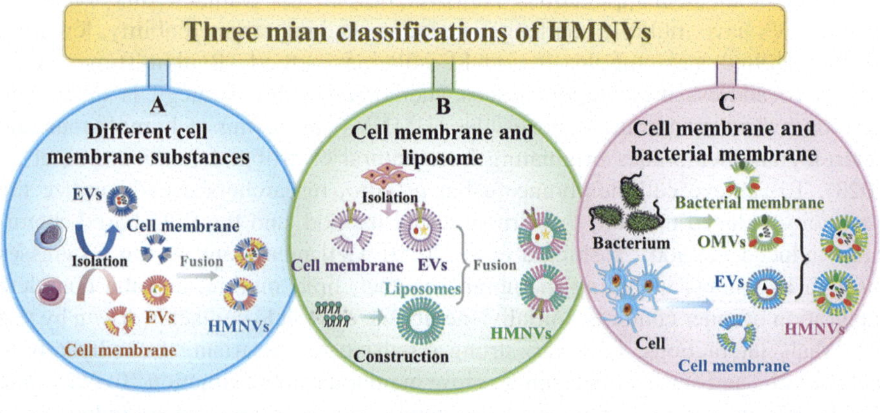

Fig. 7.4 Schematic diagram of HMNVs classification according to their membrane sources, which includes (**a**) HMNVs formed by the fusion of either homologous or heterologous cell membranes. (**b**) HMNVs generated through the fusion of synthetic liposomes and cell membranes, and (**c**) HMNVs formed by the fusion of mammalian cell membrane and bacterial membrane. (Reprinted with permission from Sun et al. (2023), copyright 2023, Wiley-VCH GmbH)

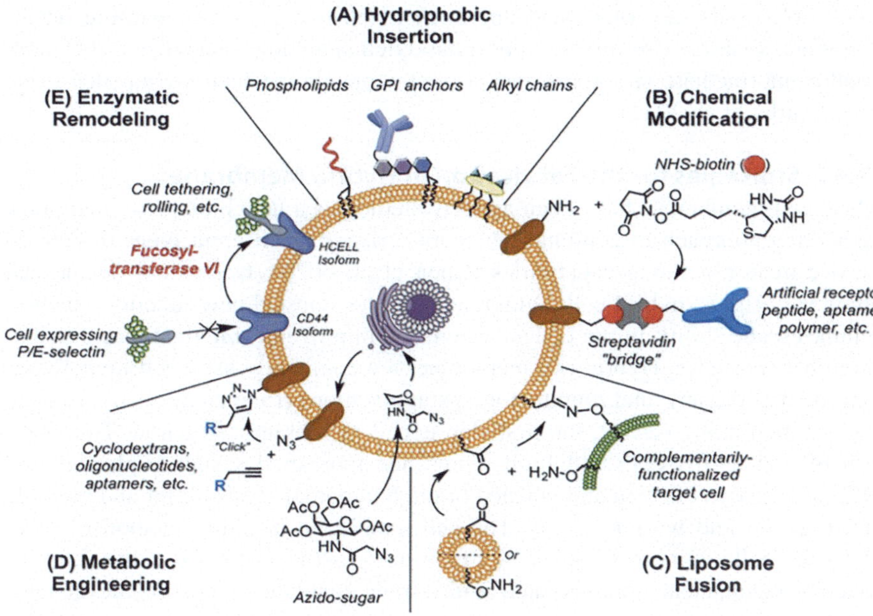

Fig. 7.5 Cell membrane engineering strategies to manage cellular interactions. Several biological membrane engineering methods have been developed to direct intercellular interactions. (**a**) Any ligand can be tethered to the cell surface by hydrophobic insertion via different strategies. (**b**) Cell membrane functional groups can be decorated with many ligands by covalent reactions. In many cases, the reactions between biotin and streptavidin are used to facilitate this method. (**c**) Liposome fusion techniques decorate the cell surface by its chemical groups with reactive modifications. (**d**) Culture cells supplemented with synthetic sugar analogs, enabling the decoration of cell surface glycoproteins with bioorthogonal groups, which can then be modified by exogenous ligands. (**e**) Enzyme remodeling strategy is utilized to modify cell membrane proteins and glycans and/or tethering ligands to the cell surface. (Reprinted with permission from Csizmar et al. (2018), copyright 2018, Elsevier B.V.)

liposome fusion technique is also utilized for directing cell-cell interactions (Fig. 7.5) (Selden et al. 2012; Cheng et al. 2012; Pulsipher 2011; Dutta et al. 2011). Liposome fusion technology has emerged as a robust method for forming targeted intercellular interactions. By fusing liposomes with specific chemical groups to cell membranes, researchers can create covalent connections between cells, facilitating the formation of multicellular structures (Luo et al. 2015). Moreover, this strategy has been used to study cellular behavior and intracellular signaling by manipulating the outer membrane. To address this, a liposome fusion technique based on hydrophobic interactions was developed to immobilize un-natural functional groups on the cell surface via membrane fusion (Lee et al. 2018; Gaspar et al. 2020; Gao and Xiao 2022; Lin et al. 2022).

In the membrane/liposome fusion technique, to achieve the optimal function lipid composition of the biological membrane/vesicle should be carefully considered because it plays a crucial role in the fate of the fusion process. For example, lysophospholipids, and oleic acid due to their intrinsic positive curvature inhibit membrane fusion. Conversely, phosphatidylethanolamine stabilizes the non-lamellar intermediate structures and promotes membrane fusion (Marsden et al. 2011; Joardar et al. 2021).

7.3.4.2 Strategies for the Fabrication of Hybrid Membranes

Hybrid cell membranes have been actively studied as a new biological membrane engineering approach to combine the characteristics of different types of cells to improve their functional capabilities (Chen et al. 2020). Depending on the cell membrane origins, hybrid cell membrane achieves boosted new functions such as immune escape ability, tumor cell adhesion, and improvement in the tissue distribution of nanoparticles. Hybrid cell membranes are commonly used in drug delivery, photothermal therapy, and immunotherapy applications (Liao et al. 2020). To form a hybrid membrane vesicle, the first step is cell membrane extraction. The membrane of nucleus-free cells with limited presence of organelles, such as platelets and RBCs, are easily collected and purified through hypotonic lysis buffer and multiple centrifugation and washing steps. The cell membrane isolation method of other eukaryotic cells is more sophisticated than nucleus-free cells. First, the cells are broken by mechanical methods, such as ultrasonic treatment and hypotonic damage. Then the cell membranes are separated by discontinuous sucrose gradient centrifugation. For bacteria, after the secretion of outer membrane vesicles from gram-negative strains, free endotoxins, and debris are removed by utilizing multiple ultrafiltration and centrifugation steps (Zhao et al. 2021). In another approach to form a hybrid cell membrane, two different cells are fused to generate fusion cells. The isolated membranes can be fused by constant stirring at 37 °C, followed by an ice bath or ultrasonic treatment. However, a drawback of this method is the need for additional purification steps to isolate fusion cells from unwanted ones (fused homologous source cells) (Dehaini et al. 2017; Liu et al. 2019b).

In general, the preparation methods of hybrid membrane biomimetic nanovesicles are classified into three major categories, including physical preparation methods, chemical induction methods, and biological transformation methods (Fig. 7.6). In physical preparation strategy coextrusion with the assistance of professional instruments such as micro-squeezer is utilized to fabricate hybrid membrane nanovesicles. It is possible to achieve uniform-sized hybrid nanovesicles by passing two different membranes through the filter membrane (Sun et al. 2023). In this context, Liao et al. summarized the biomedical applications of hybrid membrane-based nanoplatforms in which the coextrusion method is mainly utilized (Liao et al. 2020). Alongside the coextrusion strategy, some other physical methods such as freeze–thaw methods, coincubation, and ultrasonic have been utilized as hybrid membrane nanovesicle preparation strategies. By using these strategies, DC membranes have been fused into tumor cell membranes to improve cancer treatment methods through synergistic immunotherapy. By co-cultivating DCs with irradiated tumor cells, hybrid membranes are created and enable the internalization and display of tumor-associated antigens (Hao et al. 2021; Rosenthal et al. 2019). Two primary methods for generating these membranes involve fusing separately prepared dendritic and

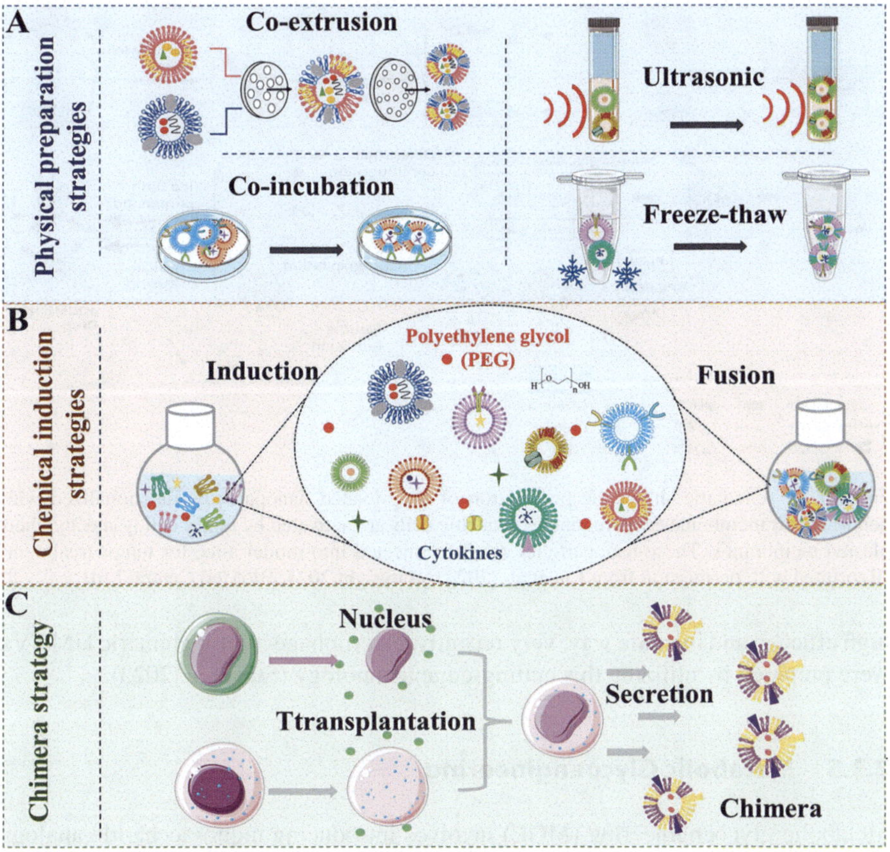

Fig. 7.6 A schematic overview of three major strategies for preparation of hybrid membrane biomimetic nanovesicles, including (**a**) Physical preparation methods, (**b**) Chemical induction methods, and (**c**) Chimera method. (Reprinted with permission from Sun et al. (2023), copyright 2023, Wiley-VCH GmbH)

tumor cell membranes via sonication or directly coalescing bone marrow-derived DCs with tumor cells in a specialized medium followed by hypotonic lysis and sonication. DC-derived membranes maintain antigen-presenting capabilities regardless of the source cell's characteristics, while NPs ensure drug concentration at tumor sites. To enhance targeting and biocompatibility, these NPs are coated with advanced DC membranes obtained through hypotonic lysis and sonication, which can be applied to PLGA-NPs or reconstituted into microcapsules, minimizing protein loss during the process (Fig. 7.7) (Kim et al. 2023; Ma et al. 2023; Sun et al. 2021). Chemical induction methods are more straightforward and have more fusion efficacy than physical preparation strategies. In this strategy, membrane fusion induction is mainly done by PEG due to the low cost, high fusion efficacy, and simple operation. In addition to these strategies, the chimera strategy is being developed, as an emerging strategy, to prepare hybrid membrane biomimetic nanovesicles with

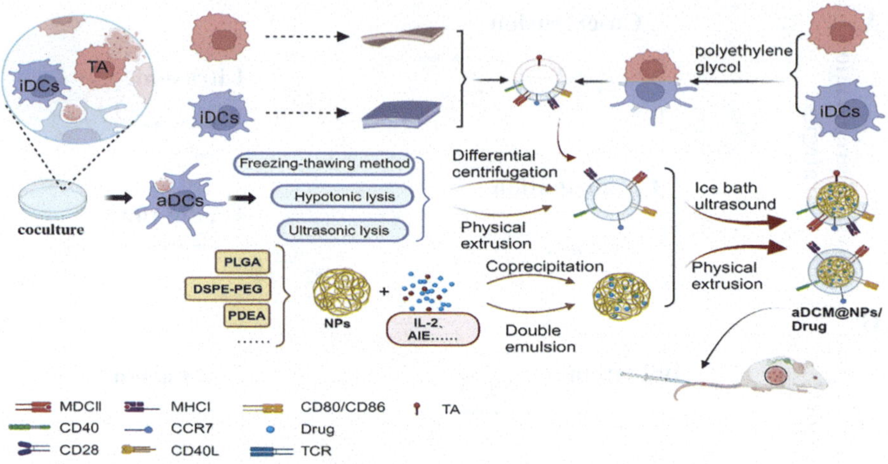

Fig. 7.7 A schematic shows the preparation of drug-loaded nanoparticles functionalized with dendritic cell membranes, where mature dendritic cells are activated by tumor cell lysate and their plasma membranes. These nanoparticles are then injected into model mice for tumor treatment. (Reprinted with permission from Liu et al. (2024), copyright 2024, BioMed Central Ltd)

high efficacy and in a safe way. Very recently, macrophage–tumor chimeric HMNVs were prepared by utilizing this cutting-edge technology (Sun et al. 2023).

7.3.5 Metabolic Glycoengineering

Metabolic glycoengineering (MGE) involves introducing monosaccharide analogs into a cell's metabolic pathways, leading to their biosynthetic incorporation into the glycocalyx on the cell membrane. This technique is valuable for modifying surface properties, enabling the manipulation of cellular function using both natural and synthetic functional groups (Nagasundaram et al. 2020; Du et al. 2009). In this method, cells are cultured with a synthetic carbohydrate analog that contains a specific chemical reporter group. The synthetic derivative is taken up by the cells, undergoes deacetylation, and is processed by the biosynthetic machinery, where it is incorporated into glycoconjugates onto membrane surface. The chemical reporter group is subsequently detected through a bioorthogonal coupling reaction (Sletten and Bertozzi 2009; Hassenrück and Wittmann 2019). Mannosamine derivatives are particularly noteworthy because they are converted into sialic acids and subsequently presented as terminal structures on cell membrane (Hassenrück and Wittmann 2019). In the past two decades, over 15 engineered mannosamine analogs suitable for MGE have been developed. These N-acetyl d-mannosamine analogs can be categorized into aliphatic analogs and bioorthogonal analogs (Wratil et al. 2016). Glycoengineering can modify cell membrane with various unnatural active groups, including ketone, azide, thiol, or alkyne-modified glycans. Among various chemical groups, azide groups are notable for their bioorthogonality, making them ideal for click chemistry in both in vitro and in vivo applications. Click chemistry encompasses a set of reactions that are characterized by high yields, rapid reaction rates, non-toxic byproducts, and the ability to occur in aqueous environments (Stöckmann et al. 2011). This has facilitated synthetic chemical conjugation on cell surfaces, within the cytosol, or throughout the body—capabilities that traditional conjugation methods could not achieve (Jewett and Bertozzi 2010). Due to these benefits, click chemistry has garnered significant interest from both organic chemists and researchers in biology and biomedicine (Boyce and Bertozzi 2011). It has been widely explored in studying biological mechanisms, targeted imaging, biomaterial development, and drug delivery. Yoon et al. innovatively combined MGE with copper-free click chemistry, using azide groups on chondrocytes to bond with near-infrared fluorescent dye-tagged DBCO-650. This technique allows efficient in vivo monitoring of labeled chondrocytes for up to four weeks with minimal impact on cartilage development, demonstrating its potential to improve cell-based therapies (Fig. 7.8) (Yoon et al. 2016).

MGE offers several advantages, including stability under physiological conditions, applicability to all cell types, safety for modified cells, and reversibility when a controlled precursor is administered (Csizmar et al. 2018). At present, cell membrane glycoengineering is an effective technique that has been extensively utilized in various fields. It enhances cells with new capabilities (Wang et al. 2020), utilizes glycoengineered membrane self-assembly as vehicles for coating nanoparticles (Rao et al. 2020), enables *in vivo* tracking of cells (Lim et al. 2019), facilitates metabolic labeling of glycans (Lim et al. 2019; Hu et al. 2018; Costa et al. 2020; Wang and Mooney 2020), improves the delivery of cells and drugs (Lim et al. 2019; Costa et al. 2020; Wang and Mooney 2020), and allows for the encapsulation of individual

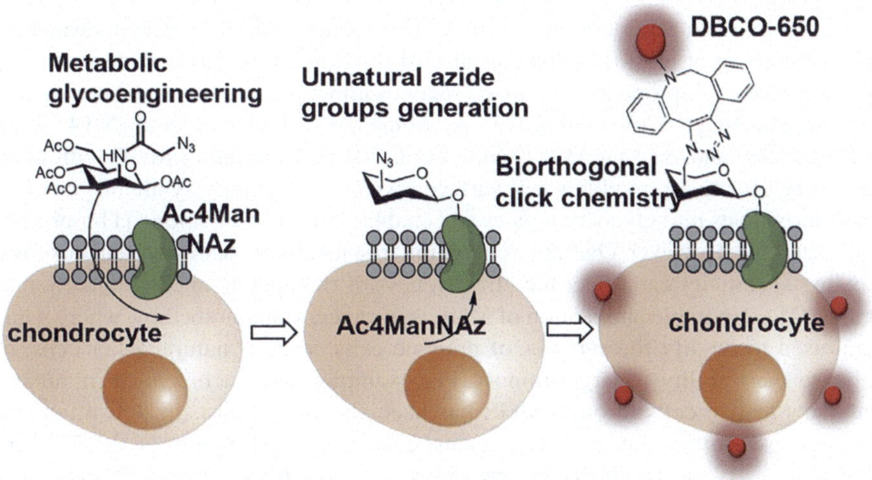

Fig. 7.8 *In vivo* labeling and tracing of transplanted chondrocytes through MGE and click chemistry. (Reprinted with permission from Yoon et al. (2016), copyright 2016, American Chemical Society)

cells (Oh et al. 2020; Kim et al. 2018). Through bioorthogonal chemistry, MGE allows for the addition of reactive groups on cell membranes, which can then be further functionalized with therapeutic agents or biomolecules (Nagahama et al. 2018; Gutmann et al. 2018a). Bioorthogonal chemistry involves fast, selective, and high-yielding reactions that take place in biological settings without disrupting endogenous functional groups (Scinto et al. 2021). It includes selective and irreversible conjugation reactions between biologically reactive groups characterized by high orthogonality and quick reaction rates. Notable bioorthogonal reactions include modified Staudinger ligation, copper(I)-catalyzed azide-alkyne cycloaddition (CuAAC), strain-promoted alkyne-azide cycloaddition (SPAAC), inverse-electron-demand Diels–Alder reaction (IEDDA), and photo-triggered bioorthogonal ligation (Scinto et al. 2021; Bird et al. 2021). Among these, SPAAC and IEDDA has shown remarkable efficiency in conjugating nanoparticles to cell surfaces, whereas CuAAC, despite its versatility, faces limitations due to the cytotoxic effects of metal catalysts (Mathew et al. 2015; Kang et al. 2014). Bioorthogonal reactions, enable versatile labeling techniques that advance the investigation of sialylation and other glycosylation processes (Scache et al. 2022). These reactions enhance drug delivery by specifically linking therapeutic agents to cell membranes, thereby reducing off-target effects (Yi et al. 2022).

Various terminal alkenes linked to sugars through amide (Niederwieser et al. 2013), carbamate (Späte et al. 2014a), and urea linkages (Dold et al. 2017) have been documented. The small size of terminal alkenes is beneficial for their recognition by the enzymes involved in the synthesis and modification of glycans. However, they demonstrate slow reactivity in the IEDDA reaction (Dold et al. 2017). On the contrary, ring-strained alkenes like norbornenes demonstrate high kinetics in the IEDDA reaction but have low incorporation efficiencies (Späte et al. 2016). Cyclopropenes, which combine rapid reaction kinetics with a small size, have proven to be ideal reporters for use in MGE (Xiong et al. 2015; Chentsova et al. 2014; Patterson et al. 2014; Späte et al. 2014b; Cole et al. 2013; Patterson et al. 2012). Three cyclopropene-derivatized mannosamine derivatives have been identified: Ac_4ManNCyc (Cole et al. 2013), Ac_4ManNCyoc (Patterson et al. 2014; Späte et al. 2014b), and Ac_4ManNCp (Xiong et al. 2015). Combining MGE with click chemistry effectively modifies cell surface glycans, integrating synthetic carbohydrate derivatives into glycoconjugates and enabling functionalization via bioorthogonal reactions. However, challenges in optimizing metabolic pathways and ensuring labeling specificity can affect the effectiveness of therapeutic applications (Scinto et al. 2021). A notable application of MGE is in cancer immunotherapy, where it has been used to modify the surfaces of immune cells, such as natural killer cells, to enhance their ability to target tumors. For example, the attachment of immunostimulants to live cells through MGE has been shown to boost the activation and effectiveness of immune responses against cancer cells (Mongis et al. 2017; Wang et al. 2021). By metabolically incorporating azide-modified monosaccharides into the cell surface, researchers can later link therapeutic antibodies or other bioactive molecules using click chemistry, thus creating a targeted strategy for cancer treatment (Altmann et al. 2021; Szponarski and Gademann 2022). Additionally, MGE has demonstrated potential in the engineering of ECM for applications in regenerative medicine. The capability to modify ECM with specific functional groups through MGE facilitates the development of biomaterials that can promote cell growth and differentiation while providing signals for tissue regeneration (Keller et al. 2020). This method not only maintains the native structure of the ECM but also improves its functionality by allowing for the controlled release of growth factors or the attachment of cells with specific phenotypes (Nellinger et al. 2019). Furthermore, cell-derived matrices (CDMs) are increasingly recognized in regenerative medicine, and a recent study showcased a targeted approach to functionalize CDMs with bioactive molecules through MGE. By integrating a tetra-acetylated azide-bearing monosaccharide into the glycan structures of ECM proteins, the modified CDMs demonstrated biocompatibility and effectively delivered a therapeutic peptide inhibitor for myostatin, underscoring their potential for future applications in tissue engineering (Fig. 7.9) (Gutmann et al. 2018b).

MGE introduces bioorthogonal groups, like azides and alkynes, into the glycocalyx of living cells, enabling click chemistry to attach therapeutic agents or imaging probes. This approach allows for specific targeting of tumor tissues in vivo and enhances the targeting abilities of nanoparticles through functionalization (Lee et al. 2014). This strategy not only boosts the effectiveness of drug delivery systems but also reduces off-target effects, which is a crucial factor in cancer treatment (Koo et al. 2012). The applications of MGE go beyond drug delivery; it is also essential for understanding and manipulating cellular interactions. By altering the glycan structures on cell surfaces, researchers can affect various aspects of cell behavior, such as adhesion, migration, and immune responses. Ying et al. emphasized the potential of MGE in regenerative medicine, where engineered surface properties of cells can be customized to improve therapeutic outcomes (Ying et al. 2022). Additionally, the integration of MGE with click chemistry has been demonstrated to facilitate the attachment of immunostimulants to live cells, thereby boosting immune

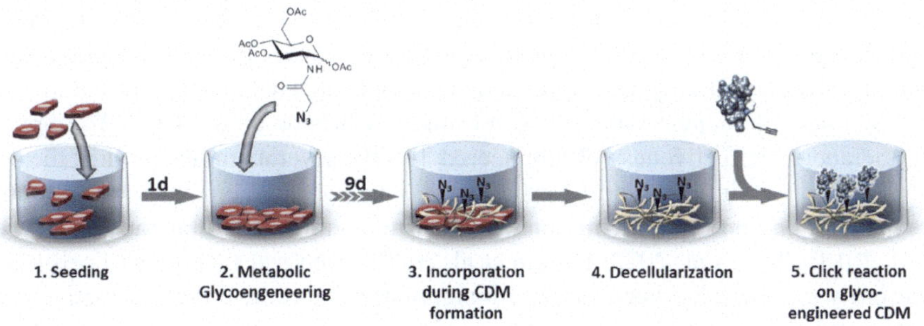

Fig. 7.9 A schematic representation of the production process for glycoengineered cell-derived matrices (CDMs) shows fibroblasts seeded and cultured in monolayers with azide-monosaccharides. After forming a network of secreted matrix proteins incorporating azide-functionalized sugars, the CDMs are obtained through decellularization. The resulting matrix components, containing azide groups, can then be modified using click chemistry techniques. (Reprinted with permission from Gutmann et al. (2018b), copyright 2018, American Chemical Society)

responses against tumors (Mongis et al. 2017). The adaptability of MGE is evident in its application across various cell types and organisms. For instance, MGE has been used to investigate skeletal precursor cells, demonstrating how modified azide/alkyne monosaccharides can be integrated into glycosylated macromolecules, providing valuable insights into skeletal development and regeneration (Altmann et al. 2021). In addition to mammalian systems, MGE has shown potential in prokaryotic systems. Saeui et al. explored its use in bacteria for producing therapeutic proteins and metabolites, indicating its broad applicability across diverse biological systems (Saeui et al. 2015). This versatility is further reinforced by the significance of click chemistry in cell engineering for precision drug delivery, illustrating its ability to modify living cells with chemical tags for various applications (Takayama et al. 2019).

7.3.6 Integrated Strategies

All cell membrane engineering strategies face intrinsic limitations and drawbacks that sometimes hinder their development and clinical applications. Integrated strategies have been developed to overcome these challenges and facilitate utilizing them. For example, metabolic oligosaccharide engineering which is a chemical reporter strategy has been developed for some biological applications such as probing sialylation, fucosylation, and mucine-typed O-linked glycosylation in living cells. In this context, transferring molecules to living target cells and lack of cell-type selectivity are the most important limitations. To overcome this limitation, some studies incorporate MGE moieties into liposomes and then tag liposomes with ligands that have the potential to target specific in vivo cells, such as cancer cells (Sun et al. 2018). In this context, Liu et al. developed an approach to enhance adoptive T cell transfer efficacy by conjugating antitumor cytokines onto T cells. They metabolically labeled T cells by adding nanoparticles containing unnatural azido sugars to the culture medium. Antitumor cytokines were then conjugated to the cell surfaces via click chemistry after washing the T cells (Liu et al. 2023). Chimeric antigen receptor T cells are valuable tools in cancer immunotherapy. Despite significant progress, this therapeutic approach faces multiple challenges, particularly due to the tumor microenvironment of solid tumors (Siriwon et al. 2018). To improve the therapeutic effectiveness of engineered T cells, several studies modify the surface of T cells mainly by binding drug-loaded nanoparticles, enzymes, and antibodies via applying multiple membrane engineering methods simultaneously (Siriwon et al. 2018; Zhao et al. 2022; Durgin et al. 2022). Both cell therapy and cell-based platforms need *ex vivo* handling and manipulation of cells, therefore, native cells with naked plasma membrane demonstrated poor viability due to high susceptibility to harmful environmental stresses (Hasturk and Kaplan 2019). To overcome this problem nanoencapsulation of cells has been developed in recent years as an emerging field in cell-membrane engineering that protects cells from *in vitro* and *in vivo* environmental stresses (Youn et al. 2020). Single-cell nanoencapsulation is achieved through various strategies, but most methods result in loosely structured encapsulation shells that fail to provide long-term cytoprotection and immunosuppression. To address this, Yang et al. developed a novel nanoencapsulation method using a polymer nanoshell, combining LbL assembly of a gelatin inner layer with cross-linked PEG outer layers via thiol-click chemistry. This approach, combining electrostatic interactions and click chemistry, enabled high-viability encapsulation of Hela cells, ensuring prolonged stability and robust resistance to external stresses (Fig. 7.10) (Yang et al. 2017).

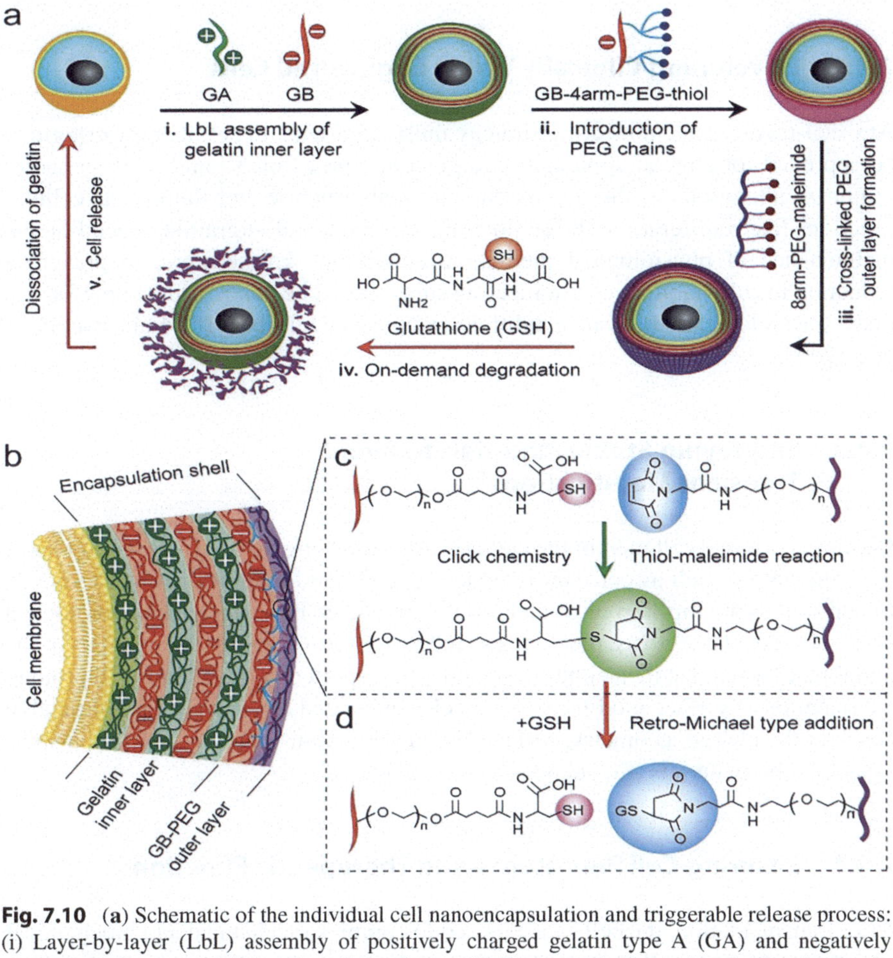

Fig. 7.10 (**a**) Schematic of the individual cell nanoencapsulation and triggerable release process: (i) Layer-by-layer (LbL) assembly of positively charged gelatin type A (GA) and negatively charged gelatin type B (GB) on the cell membrane. (ii) Electrostatic interactions lead to the attachment of PEG-modified gelatin (GB-4arm-PEG-thiol) on the cell surface. (iii) Cross-linking of the PEG coating through a thiol-maleimide reaction. (iv) GB-PEG outer layer degradation triggered by the reducing agent glutathione (GSH). (v) The release of the cell occurs as the inner gelatin layer dissociates. (**b**) Schematic representation of polymer shell on the cell membrane. (**c–d**) Thiol-maleimide based bioorthogonal chemistry reaction mechanism and degradation mechanism of the succinimide thioether group. (Reprinted with permission from Yang et al. (2017), copyright 2017, Elsevier B.V.)

7.4 Challenges of Biomaterial-Based Cell Membrane Engineering

Biomaterial modifications of cell surfaces are widely used in various therapeutic approaches and enhance therapeutic effects in cell-based therapies. However, conventional methods often face challenges related to cytotoxicity and short-term modifications, highlighting the need for more stable and efficient strategies for functionalizing cell surfaces (Rabuka et al. 2008).

7.4.1 Developing Clinically Viable Engineered Cells

Any cell-membrane engineering strategy must align with all cell therapy criteria to be approved for clinical application. Engineering protocols should be designed to minimize alterations to the membrane's normal structure and fluidity, as cellular functions like proliferation, differentiation, adhesion, and migration depend on the maintenance of physiological membrane conditions. Furthermore, to prevent a reduction in cell viability, all engineering steps should be controlled and performed under physiologically relevant conditions (Stephan and Irvine 2011; Lautenschläger et al. 2009).

7.4.2 Employing Stable Materials to Avoid Transient Modifications

Biomaterial-based cell membrane engineering is challenging due to the membrane's dynamic nature such as constant remodeling, and rapid lipid and protein turnover. In addition to the flip-flop movement of membrane components, normal biological processes like endocytosis, pinocytosis, and phagocytosis internalize portions of the membrane, which are then either degraded or transported intracellularly. To prolong the retention of surface modifications or cell-conjugated nanocarriers, it is crucial to consider the charge, geometry, surface chemistry, stability, and hybridization strategy of the biomaterials (Stephan and Irvine 2011).

7.4.3 Ensuring Cell Persistence and Therapeutic Function

After transplantation, the cells are exposed to complex mechanical and biochemical environments *in vivo*. It is important to consider the compatibility of modifications with tissue environment-related forces including shear stress and hemodynamic forces (Stephan and Irvine 2011; Espina et al. 2023). Moreover, biomaterial-modified cells are exposed to multiple compartments of the immune system when they enter the circulatory system, which may adversely affect the cell viability and its therapeutic functionality. In some cases, it is possible that phagocytes like monocytes and macrophages phagocyte surface-conjugated particulate biomaterials

(Nilsson et al. 2010; Wang et al. 2010). It was also reported that some types of negatively charged biomaterials may trigger blood vessel blockade (Ramot et al. 2010).

7.5 Conclusion and Future Prospects

Over the past two decades, cell membrane engineering strategies have been developed to enhance cell functionality by decorating cell surfaces and membrane-based platforms with therapeutic molecules, bioorthogonal chemical groups, bioactive peptides, nanomaterials, and nanoparticles. Different types of biopolymers can be tethered to the cell membrane using non-biological modifications such as covalent conjugation, electrostatic interaction, and hydrophobic interaction. Moreover, cell membrane fusion strategies and MGE methods are also among the most popular and applicable cell membrane engineering approaches. Hybrid membrane nanovesicles are prepared by the fusion of membranes or fusion of membrane to liposome by different strategies for various biomedical applications. MGE is a groundbreaking method in glycobiology, providing a powerful framework for altering glycan structures within living organisms for therapeutic applications. Cell surface engineering methods are commonly used in cancer therapy, where various cell types are modified with biomaterials. This approach enhances cancer treatment by enabling targeted drug delivery, activating immune system cells, and facilitating the movement of anti-cancer drugs across the body's natural barriers. Moreover, biomaterial-based cell surface engineering strategies could mediate and promote tissue repair and regeneration by improving anti-infection and immunomodulation reactions, alongside increasing cell proliferation/differentiation. In tissue engineering approaches, these engineering strategies play a significant role by mediating microtissue formation and directing intercellular interactions via different mechanisms. The dynamic structure of cell membranes, due to the high turnover rate of membrane building blocks, presents a significant innate challenge for this engineering method. Despite its benefits, biomaterial-based membrane engineering strategies face several challenges, including alterations to normal cellular functions, reduced cell viability, and low therapeutic functionality. We strongly believe that these challenges can be addressed by appropriate engineering design and by accurate determination of the type of lipids and/or biomaterials. Moreover, it is possible to improve the efficacy of modifications by integrating different engineering strategies.

References

Abbina S, Siren EM, Moon H, Kizhakkedathu JN (2017) Surface engineering for cell-based therapies: Techniques for manipulating mammalian cell surfaces. ACS Biomater Sci Eng 4(11):3658–3677

Agatemor C, Buettner MJ, Ariss R, Muthiah K, Saeui CT, Yarema KJ (2019) Exploiting metabolic glycoengineering to advance healthcare. Nat Rev Chem 3(10):605–620

Almeida-Pinto J, Lagarto MR, Lavrador P, Mano JF, Gaspar VM (2023) Cell surface engineering tools for programming living assemblies. Adv Sci 10(34):2304040

Altmann S, Mut J, Wolf N, Meißner-Weigl J, Rudert M, Jakob F et al (2021) Metabolic glycoengineering in hMSC-TERT as a model for skeletal precursors by using modified azide/alkyne monosaccharides. Int J Mol Sci 22(6):2820

Amaral AJ, Pasparakis G (2016) Rapid formation of cell aggregates and spheroids induced by a "smart" boronic acid copolymer. ACS Appl Mater Interfaces 8(35):22930–22941

Amaral AJ, Pasparakis G (2019) Cell membrane engineering with synthetic materials: applications in cell spheroids, cellular glues and microtissue formation. Acta Biomater 90:21–36

Bird RE, Lemmel SA, Yu X, Zhou QA (2021) Bioorthogonal chemistry and its applications. Bioconjug Chem 32(12):2457–2479

Boyce M, Bertozzi CR (2011) Bringing chemistry to life. Nat Methods 8(8):638–642

Cai F, Ren Y, Dai J, Yang J, Shi X (2023) Effects of various cell surface engineering reactions on the biological behavior of mammalian cells. Macromol Biosci 23(3):2200379

Chai Z, Hu X, Wei X, Zhan C, Lu L, Jiang K et al (2017) A facile approach to functionalizing cell membrane-coated nanoparticles with neurotoxin-derived peptide for brain-targeted drug delivery. J Control Release 264:102–111

Chen H, Teramura Y, Iwata H (2011) Co-immobilization of urokinase and thrombomodulin on islet surfaces by poly (ethylene glycol)-conjugated phospholipid. J Control Release 150(2):229–234

Chen H-Y, Deng J, Wang Y, Wu C-Q, Li X, Dai H-W (2020) Hybrid cell membrane-coated nanoparticles: a multifunctional biomimetic platform for cancer diagnosis and therapy. Acta Biomater 112:1–13

Cheng K, Shen D, Xie Y, Cingolani E, Malliaras K, Marbán E (2012) Brief report: mechanism of extravasation of infused stem cells. Stem Cells 30(12):2835–2842

Chentsova A, Kapourani E, Giannis A (2014) Synthesis of novel derivatives of 5-hydroxymethylcytosine and 5-formylcytosine as tools for epigenetics. Beilstein J Org Chem 10(1):7–11

Chung HA, Kato K, Itoh C, Ohhashi S, Nagamune T (2004) Casual cell surface remodeling using biocompatible lipid-poly (ethylene glycol)(n): development of stealth cells and monitoring of cell membrane behavior in serum-supplemented conditions. J Biomed Mater Res A 70(2):179–185

Claridge B, Lozano J, Poh QH, Greening DW (2021) Development of extracellular vesicle therapeutics: challenges, considerations, and opportunities. Front Cell Dev Biol 9:734720

Cole CM, Yang J, Šečkutė J, Devaraj NK (2013) Fluorescent live-cell imaging of metabolically incorporated unnatural cyclopropene-mannosamine derivatives. Chembiochem 14(2):205

Coronado S, Herrera J, Pino MG, Martín S, Ballesteros-Rueda L, Cea P (2024) Advancements in engineering planar model cell membranes: current techniques, applications, and future perspectives. Nanomaterials 14(18):1489

Costa AF, Campos D, Reis CA, Gomes C (2020) Targeting glycosylation: a new road for cancer drug discovery. Trends Cancer 6(9):757–766

Csizmar CM, Petersburg JR, Wagner CR (2018) Programming cell-cell interactions through non-genetic membrane engineering. Cell Chem Biol 25(8):931–940

Custódio CA, Mano JF (2016) Cell surface engineering to control cellular interactions. ChemNanoMat 2(5):376–384

Davis KA, Wu P-J, Cahall CF, Li C, Gottipati A, Berron BJ (2019) Coatings on mammalian cells: interfacing cells with their environment. J Biol Eng 13:1–28

Dehaini D, Wei X, Fang RH, Masson S, Angsantikul P, Luk BT et al (2017) Erythrocyte–platelet hybrid membrane coating for enhanced nanoparticle functionalization. Adv Mater 29(16):1606209

Dold JE, Pfotzer J, Späte AK, Wittmann V (2017) Dienophile-modified mannosamine derivatives for metabolic labeling of sialic acids: a comparative study. ChemBioChem 18(13):1242–1250

Dowlut M, Hall DG (2006) An improved class of sugar-binding boronic acids, soluble and capable of complexing glycosides in neutral water. J Am Chem Soc 128(13):4226–4227

Du J, Yarema KJ (2010) Carbohydrate engineered cells for regenerative medicine. Adv Drug Deliv Rev 62(7–8):671–682

Du J, Meledeo MA, Wang Z, Khanna HS, Paruchuri VD, Yarema KJ (2009) Metabolic glycoengineering: sialic acid and beyond. Glycobiology 19(12):1382–1401
Durgin JS, Thokala R, Johnson L, Song E, Leferovich J, Bhoj V et al (2022) Enhancing CAR T function with the engineered secretion of C. perfringens neuraminidase. Mol Ther 30(3):1201–1214
Dutta D, Pulsipher A, Luo W, Yousaf MN (2011) Synthetic chemoselective rewiring of cell surfaces: generation of three-dimensional tissue structures. J Am Chem Soc 133(22):8704–8713
Espina JA, Cordeiro MH, Milivojevic M, Pajić-Lijaković I, Barriga EH (2023) Response of cells and tissues to shear stress. J Cell Sci 136(18):jcs260985
Fang RH, Hu C-MJ, Chen KN, Luk BT, Carpenter CW, Gao W et al (2013) Lipid-insertion enables targeting functionalization of erythrocyte membrane-cloaked nanoparticles. Nanoscale 5(19):8884–8888
Fischbach MA, Bluestone JA, Lim WA (2013) Cell-based therapeutics: the next pillar of medicine. Sci Transl Med 5(179):179ps7-ps7
Fu S, Liang M, Wang Y, Cui L, Gao C, Chu X et al (2018) Dual-modified novel biomimetic nanocarriers improve targeting and therapeutic efficacy in glioma. ACS Appl Mater Interfaces 11(2):1841–1854
Gao W, Xiao Y (2022) Advances in cell membrane-encapsulated biomaterials for tissue repair and regeneration. Appl Mater Today 26:101389
Gaspar VM, Lavrador P, Borges J, Oliveira MB, Mano JF (2020) Advanced bottom-up engineering of living architectures. Adv Mater 32(6):1903975
Guan X, Xing S, Liu Y (2024) Engineered cell membrane-camouflaged nanomaterials for biomedical applications. Nanomaterials 14(5):413
Guo Y, Wang D, Song Q, Wu T, Zhuang X, Bao Y et al (2015) Erythrocyte membrane-enveloped polymeric nanoparticles as nanovaccine for induction of antitumor immunity against melanoma. ACS Nano 9(7):6918–6933
Gutmann M, Bechold J, Seibel J, Meinel L, Lühmann T (2018a) Metabolic glycoengineering of cell-derived matrices and cell surfaces: a combination of key principles and step-by-step procedures. ACS Biomater Sci Eng 5(1):215–233
Gutmann M, Braun A, Seibel J, Lühmann T (2018b) Bioorthogonal modification of cell derived matrices by metabolic glycoengineering. ACS Biomater Sci Eng 4(4):1300–1306
Han H, Chen B-T, Liu Y, Wang Y, Xing L, Wang H et al (2024) Engineered stem cell-based strategy: A new paradigm of next-generation stem cell product in regenerative medicine. J Control Release 365:981–1003
Hao W, Cui Y, Fan Y, Chen M, Yang G, Wang Y et al (2021) Hybrid membrane-coated nanosuspensions for multi-modal anti-glioma therapy via drug and antigen delivery. J Nanobiotechnol 19:1–24
Hassenrück J, Wittmann V (2019) Cyclopropene derivatives of aminosugars for metabolic glycoengineering. Beilstein J Org Chem 15(1):584–601
Hasturk O, Kaplan DL (2019) Cell armor for protection against environmental stress: Advances, challenges and applications in micro-and nanoencapsulation of mammalian cells. Acta Biomater 95:3–31
Hebb DO (2005) The organization of behavior: a neuropsychological theory: Psychology Press
Hoang DM, Pham PT, Bach TQ, Ngo AT, Nguyen QT, Phan TT et al (2022) Stem cell-based therapy for human diseases. Signal Transduct Target Ther 7(1):1–41
Holden CA, Yuan Q, Yeudall WA, Lebman DA, Yang H (2010) Surface engineering of macrophages with nanoparticles to generate a cell–nanoparticle hybrid vehicle for hypoxia-targeted drug delivery. Int J Nanomedicine 5:25–36
Hu F, Yuan Y, Wu W, Mao D, Liu B (2018) Dual-responsive metabolic precursor and light-up AIEgen for cancer cell bio-orthogonal labeling and precise ablation. Anal Chem 90(11):6718–6724
Huang D, Wang X, Wang W, Li J, Zhang X, Xia B (2024) Cell-membrane engineering strategies for clinic-guided design of nanomedicine. Biomater Sci 28:2865–2884

Ivanova A, Badertscher L, O'Driscoll G, Bergman J, Gordon E, Gunnarsson A et al (2023) Creating designer engineered extracellular vesicles for diverse ligand display, target recognition, and controlled protein loading and delivery. Adv Sci 10(34):2304389
Jahn R, Cafiso DC, Tamm LK (2024) Mechanisms of SNARE proteins in membrane fusion. Nat Rev Mol Cell Biol 25(2):101–118
Javvaji V, Dowling MB, Oh H, White IM, Raghavan SR (2014) Reversible gelation of cells using self-assembling hydrophobically-modified biopolymers: towards self-assembly of tissue. Biomater Sci 2(7):1016–1023
Jewett JC, Bertozzi CR (2010) Cu-free click cycloaddition reactions in chemical biology. Chem Soc Rev 39(4):1272–1279
Jin M, Ling Y, Jin L, Wang X, Jiang X, Li Q (2025) Cell membrane-functionalized bismuth oxyiodide nanodots as potential contrast agent for gastrointestinal tract imaging. Colloids Surf B: Biointerfaces 245:114235
Joardar A, Pattnaik GP, Chakraborty H (2021) Effect of phosphatidylethanolamine and oleic acid on membrane fusion: phosphatidylethanolamine circumvents the classical stalk model. J Phys Chem B 125(48):13192–13202
Jones RB, Mueller S, Kumari S, Vrbanac V, Genel S, Tager AM et al (2017) Antigen recognition-triggered drug delivery mediated by nanocapsule-functionalized cytotoxic T-cells. Biomaterials 117:44–53
Kang M-K, Tullman-Ercek D (2018) Engineering expression and function of membrane proteins. Methods 147:66–72
Kang S-W, Lee S, Na JH, Yoon HI, Lee D-E, Koo H et al (2014) Cell labeling and tracking method without distorted signals by phagocytosis of macrophages. Theranostics 4(4):420
Keller S, Liedek A, Shendi D, Bach M, Tovar GE, Kluger PJ et al (2020) Eclectic characterisation of chemically modified cell-derived matrices obtained by metabolic glycoengineering and re-assessment of commonly used methods. RSC Adv 10(58):35273–35286
Kim S, Kim K (2022) Lipid-mediated ex vivo cell surface engineering for augmented cellular functionalities. Biomater Adv 140:213059
Kim H, Shin K, Park OK, Choi D, Kim HD, Baik S et al (2018) General and facile coating of single cells via mild reduction. J Am Chem Soc 140(4):1199–1202
Kim HS, Ho T-C, Willner MJ, Becker MW, Kim H-W, Leong KW (2023) Dendritic cell-mimicking scaffolds for ex vivo T cell expansion. Bioact Mater 21:241–252
Koh I, Yong I, Kim B, Choi D, Hong J, Han Y-M et al (2020) Tris (2-carboxyethyl) phosphine-mediated nanometric extracellular matrix-coating method of mesenchymal stem cells. ACS Biomater Sci Eng 6(2):813–821
Koo H, Lee S, Na JH, Kim SH, Hahn SK, Choi K et al (2012) Bioorthogonal copper-free click chemistry inVivo for tumor-targeted delivery of nanoparticles. Angew Chem Int Ed 51(47):11836–11840
Kooijmans S, Fliervoet L, Van Der Meel R, Fens M, Heijnen H, van Bergen En Henegouwen PMP et al (2016) PEGylated and targeted extracellular vesicles display enhanced cell specificity and circulation time. J Control Release 224:77–85
Krishnamachari Y, Pearce ME, Salem AK (2008) Self-assembly of cell–microparticle hybrids. Adv Mater 20(5):989–993
Kufleitner M, Haiber LM, Wittmann V (2023) Metabolic glycoengineering–exploring glycosylation with bioorthogonal chemistry. Chem Soc Rev 52(2):510–535
Lanza R, Russell DW, Nagy A (2019) Engineering universal cells that evade immune detection. Nat Rev Immunol 19(12):723–733
Lautenschläger F, Paschke S, Schinkinger S, Bruel A, Beil M, Guck J (2009) The regulatory role of cell mechanics for migration of differentiating myeloid cells. Proc Natl Acad Sci 106(37):15696–15701
Lee DY, Park SJ, Nam JH, Byun Y (2006) A new strategy toward improving immunoprotection in cell therapy for diabetes mellitus: long-functioning PEGylated islets in vivo. Tissue Eng 12(3):615–623

Lee S, Koo H, Na JH, Han SJ, Min HS, Lee SJ et al (2014) Chemical tumor-targeting of nanoparticles based on metabolic glycoengineering and click chemistry. ACS Nano 8(3):2048–2063
Lee DY, Cha B-H, Jung M, Kim AS, Bull DA, Won Y-W (2018) Cell surface engineering and application in cell delivery to heart diseases. J Biol Eng 12:1–11
Li J, Xun K, Pei K, Liu X, Peng X, Du Y et al (2019) Cell-membrane-anchored DNA nanoplatform for programming cellular interactions. J Am Chem Soc 141(45):18013–18020
Liao Y, Zhang Y, Blum NT, Lin J, Huang P (2020) Biomimetic hybrid membrane-based nanoplatforms: synthesis, properties and biomedical applications. Nanoscale Horiz 5(9):1293–1302
Lim S, Yoon HY, Jang HJ, Song S, Kim W, Park J et al (2019) Dual-modal imaging-guided precise tracking of bioorthogonally labeled mesenchymal stem cells in mouse brain stroke. ACS Nano 13(10):10991–11007
Lin M, Chen Y, Zhao S, Tang R, Nie Z, Xing H (2022) Back Cover: A biomimetic approach for spatially controlled cell membrane engineering using fusogenic spherical nucleic acid (Angew. Chem. Int. Ed. 1/2022). Angew Chem Int Ed 61(1):e202116134
Liu L, He H, Liu J (2019a) Advances on non-genetic cell membrane engineering for biomedical applications. Polymers 11(12):2017
Liu WL, Zou MZ, Liu T, Zeng JY, Li X, Yu WY et al (2019b) Expandable immunotherapeutic nanoplatforms engineered from cytomembranes of hybrid cells derived from cancer and dendritic cells. Adv Mater 31(18):1900499
Liu Y, Adu-Berchie K, Brockman JM, Pezone M, Zhang DK, Zhou J et al (2023) Cytokine conjugation to enhance T cell therapy. Proc Natl Acad Sci 120(1):e2213222120
Liu H, Lu Y, Zong J, Zhang B, Li X, Qi H et al (2024) Engineering dendritic cell biomimetic membrane as a delivery system for tumor targeted therapy. J Nanobiotechnol 22(1):663
Luo W, Westcott N, Dutta D, Pulsipher A, Rogozhnikov D, Chen J et al (2015) A dual receptor and reporter for multi-modal cell surface engineering. ACS Chem Biol 10(10):2219–2226
Lutterotti A, Yousef S, Sputtek A, Stürner KH, Stellmann J-P, Breiden P et al (2013) Antigen-specific tolerance by autologous myelin peptide–coupled cells: a phase 1 trial in multiple sclerosis. Sci Transl Med 5(188):188ra75-ra75
Ma JY, Wang YX, Huang Y, Zhang Y, Cui YX, Kong DM (2022) Chemical–biological approaches for the direct regulation of cell–cell aggregation. Aggregate 3(2):e166
Ma X, Kuang L, Yin Y, Tang L, Zhang Y, Fan Q et al (2023) Tumor–antigen activated dendritic cell membrane-coated biomimetic nanoparticles with orchestrating immune responses promote therapeutic efficacy against glioma. ACS Nano 17(3):2341–2355
Maciel MM, Correia TR, Gaspar VM, Rodrigues JM, Choi IS, Mano JF (2021) Partial coated stem cells with bioinspired silica as new generation of cellular hybrid materials. Adv Funct Mater 31(29):2009619
Marsden HR, Tomatsu I, Kros A (2011) Model systems for membrane fusion. Chem Soc Rev 40(3):1572–1585
Mathew MP, Tan E, Saeui CT, Bovonratwet P, Liu L, Bhattacharya R et al (2015) Metabolic glycoengineering sensitizes drug-resistant pancreatic cancer cells to tyrosine kinase inhibitors erlotinib and gefitinib. Bioorg Med Chem Lett 25(6):1223–1227
Mazur F, Chandrawati R (2021) Membrane fusion models for bioapplications. ChemNanoMat 7(3):223–237
McMahon HT, Gallop JL (2005) Membrane curvature and mechanisms of dynamic cell membrane remodelling. Nature 438(7068):590–596
Mondal Roy S, Sarkar M (2011) Membrane fusion induced by small molecules and ions. J Lipids 2011(1):528784
Mongis A, Piller F, Piller V (2017) Coupling of immunostimulants to live cells through metabolic glycoengineering and bioorthogonal click chemistry. Bioconjug Chem 28(4):1151–1165
Murphy DE, de Jong OG, Brouwer M, Wood MJ, Lavieu G, Schiffelers RM et al (2019) Extracellular vesicle-based therapeutics: natural versus engineered targeting and trafficking. Exp Mol Med 51(3):1–12
Nacharaju P, Boctor FN, Manjula BN, Acharya SA (2005) Surface decoration of red blood cells with maleimidophenyl-polyethylene glycol facilitated by thiolation with iminothiolane: an

approach to mask A, B, and D antigens to generate universal red blood cells. Transfusion 45(3):374–383
Naderi-Meshkin H, Bahrami AR, Bidkhori HR, Mirahmadi M, Ahmadiankia N (2015) Strategies to improve homing of mesenchymal stem cells for greater efficacy in stem cell therapy. Cell Biol Int 39(1):23–34
Nagahama K, Kimura Y, Takemoto A (2018) Living functional hydrogels generated by bioorthogonal cross-linking reactions of azide-modified cells with alkyne-modified polymers. Nat Commun 9(1):2195
Nagasundaram M, Horstkorte R, Gnanapragassam VS (2020) Sialic acid metabolic engineering of breast cancer cells interferes with adhesion and migration. Molecules 25(11):2632
Nakase I, Futaki S (2015) Combined treatment with a pH-sensitive fusogenic peptide and cationic lipids achieves enhanced cytosolic delivery of exosomes. Sci Rep 5(1):10112
Nellinger S, Keller S, Southan A, Wittmann V, Volz A-C, Kluger PJ (2019) Generation of an azide-modified extracellular matrix by adipose-derived stem cells using metabolic glycoengineering. Curr Direct Biomed Eng 5(1):393–395
Niederwieser A, Späte A-K, Nguyen LD, Jüngst C, Reutter W, Wittmann V (2013) Two-color glycan labeling of live cells by a combination of Diels-Alder and click chemistry. Angew Chem Int Ed Engl 52(15):4265–4268
Nilsson B, Korsgren O, Lambris JD, Ekdahl KN (2010) Can cells and biomaterials in therapeutic medicine be shielded from innate immune recognition? Trends Immunol 31(1):32–38
Oh B, Swaminathan V, Malkovskiy A, Santhanam S, McConnell K, George PM (2020) Single-cell encapsulation via click-chemistry alters production of paracrine factors from neural progenitor cells. Adv Sci 7(8):1902573
Patterson DM, Nazarova LA, Xie B, Kamber DN, Prescher JA (2012) Functionalized cyclopropenes as bioorthogonal chemical reporters. J Am Chem Soc 134(45):18638–18643
Patterson DM, Jones KA, Prescher JA (2014) Improved cyclopropene reporters for probing protein glycosylation. Mol BioSyst 10(7):1693–1697
Peng Y, Chen B (2024) Role of cell membrane homeostasis in the pathogenicity of pathogenic filamentous fungi. Virulence 15(1):2299183
Pulsipher A (2011) Chemoselective chemistry: engineering materials and cell surfaces to control biological interactions. The University of North Carolina at Chapel Hill
Qi H, Liu C, Long L, Ren Y, Zhang S, Chang X et al (2016) Blood exosomes endowed with magnetic and targeting properties for cancer therapy. ACS Nano 10(3):3323–3333
Qian R-C, Zhou Z-R, Guo W, Wu Y, Yang Z, Lu Y (2021) Cell surface engineering using DNAzymes: metal ion mediated control of cell–cell interactions. J Am Chem Soc 143(15):5737–5744
Rabuka D, Forstner MB, Groves JT, Bertozzi CR (2008) Noncovalent cell surface engineering: incorporation of bioactive synthetic glycopolymers into cellular membranes. J Am Chem Soc 130(18):5947–5953
Ramot Y, Steiner M, Morad V, Leibovitch S, Amouyal N, Cesta MF et al (2010) Pulmonary thrombosis in the mouse following intravenous administration of quantum dot-labeled mesenchymal cells. Nanotoxicology 4(1):98–105
Rankin-Turner S, Vader P, O'Driscoll L, Giebel B, Heaney LM, Davies OG (2021) A call for the standardised reporting of factors affecting the exogenous loading of extracellular vesicles with therapeutic cargos. Adv Drug Deliv Rev 173:479–491
Rao L, Zhao SK, Wen C, Tian R, Lin L, Cai B et al (2020) Activating macrophage-mediated cancer immunotherapy by genetically edited nanoparticles. Adv Mater 32(47):2004853
Richter M, Vader P, Fuhrmann G (2021) Approaches to surface engineering of extracellular vesicles. Adv Drug Deliv Rev 173:416–426
Rosenthal R, Cadieux EL, Salgado R, Bakir MA, Moore DA, Hiley CT et al (2019) Neoantigen-directed immune escape in lung cancer evolution. Nature 567(7749):479–485
Rossi L, Fraternale A, Bianchi M, Magnani M (2019) Red blood cell membrane processing for biomedical applications. Front Physiol 10:1070

Saeui CT, Urias E, Liu L, Mathew MP, Yarema KJ (2015) Metabolic glycoengineering bacteria for therapeutic, recombinant protein, and metabolite production applications. Glycoconj J 32:425–441

Sauvola CW, Littleton JT (2021) SNARE regulatory proteins in synaptic vesicle fusion and recycling. Front Mol Neurosci 14:733138

Scache J, Rigolot V, Lion C, Mortuaire M, Lefebvre T, Biot C et al (2022) Switching azide and alkyne tags on bioorthogonal reporters in metabolic labeling of sialylated glycoconjugates: a comparative study. Sci Rep 12(1):22129

Scinto SL, Bilodeau DA, Hincapie R, Lee W, Nguyen SS, Xu M et al (2021) Bioorthogonal chemistry. Nat Rev Methods Primers 1(1):30

Selden NS, Todhunter ME, Jee NY, Liu JS, Broaders KE, Gartner ZJ (2012) Chemically programmed cell adhesion with membrane-anchored oligonucleotides. J Am Chem Soc 134(2):765–768

Shi P, Wang Y (2021) Synthetic DNA for cell-surface engineering. Angew Chem 133(21):11684–11695

Shi P, Ju E, Yan Z, Gao N, Wang J, Hou J et al (2016) Spatiotemporal control of cell–cell reversible interactions using molecular engineering. Nat Commun 7(1):13088

Siriwon N, Kim YJ, Siegler E, Chen X, Rohrs JA, Liu Y et al (2018) CAR-T cells surface-engineered with drug-encapsulated nanoparticles can ameliorate intratumoral T-cell hypofunction. Cancer Immunol Res 6(7):812–824

Sletten EM, Bertozzi CR (2009) Bioorthogonal chemistry: fishing for selectivity in a sea of functionality. Angew Chem Int Ed 48(38):6974–6998

Späte AK, Schart VF, Schöllkopf S, Niederwieser A, Wittmann V (2014a) Terminal alkenes as versatile chemical reporter groups for metabolic oligosaccharide engineering. Chemistry Eur J 20(50):16502–16508

Späte A-K, Bußkamp H, Niederwieser A, Schart VF, Marx A, Wittmann V (2014b) Rapid labeling of metabolically engineered cell-surface glycoconjugates with a carbamate-linked cyclopropene reporter. Bioconjug Chem 25(1):147–154

Späte AK, Dold JE, Batroff E, Schart VF, Wieland DE, Baudendistel OR et al (2016) Exploring the potential of norbornene-modified mannosamine derivatives for metabolic glycoengineering. ChemBioChem 17(14):1374–1383

Stephan MT, Irvine DJ (2011) Enhancing cell therapies from the outside in: cell surface engineering using synthetic nanomaterials. Nano Today 6(3):309–325

Stephan MT, Moon JJ, Um SH, Bershteyn A, Irvine DJ (2010) Therapeutic cell engineering with surface-conjugated synthetic nanoparticles. Nat Med 16(9):1035–1041

Stöckmann H, Neves AA, Stairs S, Ireland-Zecchini H, Brindle KM, Leeper FJ (2011) Development and evaluation of new cyclooctynes for cell surface glycan imaging in cancer cells. Chem Sci 2(5):932–936

Sun X, Wang C, Gao M, Hu A, Liu Z (2015) Remotely controlled red blood cell carriers for cancer targeting and near-infrared light-triggered drug release in combined photothermal–chemotherapy. Adv Funct Mater 25(16):2386–2394

Sun Y, Hong S, Xie R, Huang R, Lei R, Cheng B et al (2018) Mechanistic investigation and multiplexing of liposome-assisted metabolic glycan labeling. J Am Chem Soc 140(10):3592–3602

Sun D, Chen J, Wang Y, Ji H, Peng R, Jin L et al (2019) Advances in refunctionalization of erythrocyte-based nanomedicine for enhancing cancer-targeted drug delivery. Theranostics 9(23):6885

Sun Z, Deng G, Peng X, Xu X, Liu L, Peng J et al (2021) Intelligent photothermal dendritic cells restart the cancer immunity cycle through enhanced immunogenic cell death. Biomaterials 279:121228

Sun M, Yang J, Fan Y, Zhang Y, Sun J, Hu M et al (2023) Beyond extracellular vesicles: hybrid membrane nanovesicles as emerging advanced tools for biomedical applications. Adv Sci 10(32):2303617

Szponarski M, Gademann K (2022) Antibody recognition of cancer cells via glycan surface engineering. ChemBioChem 23(15):e202200125

Takayama Y, Kusamori K, Nishikawa M (2019) Click chemistry as a tool for cell engineering and drug delivery. Molecules 24(1):172
Tan J, Zhu C, Li L, Wang J, Xia X-H, Wang C (2024) Engineering cell membranes: from extraction strategies to emerging biosensing applications. Anal Chem 96(20):7880–7894
Teramura Y, Iwata H (2010) Cell surface modification with polymers for biomedical studies. Soft Matter 6(6):1081–1091
Vincy A, Mazumder S, Amrita BI, Hwang KC, Vankayala R (2022) Recent progress in red blood cells-derived particles as novel bioinspired drug delivery systems: challenges and strategies for clinical translation. Front Chem 10:905256
Wang H, Mooney DJ (2020) Metabolic glycan labelling for cancer-targeted therapy. Nat Chem 12(12):1102–1114
Wang W, Wang S (2022) Cell-based biocomposite engineering directed by polymers. Lab Chip 22(6):1042–1067
Wang J, Sui M, Fan W (2010) Nanoparticles for tumor targeted therapies and their pharmacokinetics. Curr Drug Metab 11(2):129–141
Wang X, Lang S, Tian Y, Zhang J, Yan X, Fang Z et al (2020) Glycoengineering of natural killer cells with CD22 ligands for enhanced anticancer immunotherapy. ACS Cent Sci 6(3):382–389
Wang X, Luo X, Tian Y, Wu T, Weng J, Li Z et al (2021) Equipping natural killer cells with cetuximab through metabolic glycoengineering and bioorthogonal reaction for targeted treatment of KRAS mutant colorectal cancer. ACS Chem Biol 16(4):724–730
Wilson JT, Chaikof EL (2008) Challenges and emerging technologies in the immunoisolation of cells and tissues. Adv Drug Deliv Rev 60(2):124–145
Wilson JT, Cui W, Kozlovskaya V, Kharlampieva E, Pan D, Qu Z et al (2011) Cell surface engineering with polyelectrolyte multilayer thin films. J Am Chem Soc 133(18):7054–7064
Wratil PR, Horstkorte R, Reutter W (2016) Metabolic glycoengineering with N-acyl side chain modified mannosamines. Angew Chem Int Ed 55(33):9482–9512
Wu H, Zhang T, Li N, Gao J (2023) Cell membrane-based biomimetic vehicles for effective central nervous system target delivery: Insights and challenges. J Control Release 360:169–184
Wu T, Jiang J, Zhang H, Liu J, Ruan H (2024a) Transcending membrane barriers: advances in membrane engineering to enhance the production capacity of microbial cell factories. Microb Cell Factories 23(1):154
Wu X, Hu JJ, Yoon J (2024b) Cell membrane as a promising therapeutic target: from materials design to biomedical applications. Angew Chem 136(18):e202400249
Xi Z, Kong H, Chen Y, Deng J, Xu W, Liang Y et al (2022) Metal-and strain-free bioorthogonal cycloaddition of o-diones with furan-2 (3H)-one as anionic cycloaddend. Angew Chem Int Ed 61(23):e202200239
Xiong D-C, Zhu J, Han M-J, Luo H-X, Wang C, Yu Y et al (2015) Rapid probing of sialylated glycoproteins in vitro and in vivo via metabolic oligosaccharide engineering of a minimal cyclopropene reporter. Org Biomol Chem 13(13):3911–3917
Yang J, Li J, Li X, Wang X, Yang Y, Kawazoe N et al (2017) Nanoencapsulation of individual mammalian cells with cytoprotective polymer shell. Biomaterials 133:253–262
Yi W, Xiao P, Liu X, Zhao Z, Sun X, Wang J et al (2022) Recent advances in developing active targeting and multi-functional drug delivery systems via bioorthogonal chemistry. Signal Transduct Target Ther 7(1):386
Ying L, Xu J, Han D, Zhang Q, Hong Z (2022) The applications of metabolic glycoengineering. Front Cell Dev Biol 10:840831
Yoon HI, Yhee JY, Na JH, Lee S, Lee H, Kang S-W et al (2016) Bioorthogonal copper free click chemistry for labeling and tracking of chondrocytes in vivo. Bioconjug Chem 27(4):927–936
Yoshihara A, Watanabe S, Goel I, Ishihara K, Ekdahl KN, Nilsson B et al (2020) Promotion of cell membrane fusion by cell-cell attachment through cell surface modification with functional peptide-PEG-lipids. Biomaterials 253:120113
Youn W, Kim JY, Park J, Kim N, Choi H, Cho H et al (2020) Single-cell nanoencapsulation: from passive to active shells. Adv Mater 32(35):1907001

Zhang X, Angsantikul P, Ying M, Zhuang J, Zhang Q, Wei X et al (2017a) Remote loading of small-molecule therapeutics into cholesterol-enriched cell-membrane-derived vesicles. Angew Chem Int Ed 56(45):14075–14079

Zhang P, Bookstaver ML, Jewell CM (2017b) Engineering cell surfaces with polyelectrolyte materials for translational applications. Polymers 9(2):40

Zhang Z, Qian H, Huang J, Sha H, Zhang H, Yu L et al (2018) Anti-EGFR-iRGD recombinant protein modified biomimetic nanoparticles loaded with gambogic acid to enhance targeting and antitumor ability in colorectal cancer treatment. Int J Nanomedicine 13:4961–4975

Zhang M, Cheng S, Jin Y, Zhang N, Wang Y (2021) Membrane engineering of cell membrane biomimetic nanoparticles for nanoscale therapeutics. Clin Transl Med 11(2):e292

Zhao Y, Li A, Jiang L, Gu Y, Liu J (2021) Hybrid membrane-coated biomimetic nanoparticles (HM@BNPs): a multifunctional nanomaterial for biomedical applications. Biomacromolecules 22(8):3149–3167

Zhao Y, Dong Y, Yang S, Tu Y, Wang C, Li J et al (2022) Bioorthogonal equipping CAR-T cells with hyaluronidase and checkpoint blocking antibody for enhanced solid tumor immunotherapy. ACS Cent Sci 8(5):603–614

Part III

Engineering Cells and Probiotics for Advanced Therapeutic Applications

Harnessing Cell Surface Engineering for Tissue Engineering and Regenerative Medicine

8

Zahid Hussain, Shah Mehmood, Ismat Ullah, Nannan Shao, and Renjun Pei

Abstract

Cells are crucial for mammalian life and perform critical functions in complex microenvironments. The heterogeneous cell surface, with variations in receptors and ligands, regulates processes such as cell adhesion, communication, immune response, and homeostasis, making it a key focus in regenerative medicine. Cell suspension therapy shows promise for intractable diseases, but heterogeneous surface markers, loss of homing signals, undesired differentiation, and poor microenvironment modulation limit its effectiveness. This chapter explores how cell surface engineering improves ligand-receptor interactions, cell homing, regulates cell fate and targeted delivery for tissue repair. Scaffold-free cellular assemblies mimic tissues, however, traditional fabrication strategies face challenges such as limited cell-cell interactions, misalignment, and difficulty in creating transplantable microtissues. This chapter explores how cell surface engineering enhances cell-cell interactions, enables precise cell patterning, and facilitates macroscale tissue fabrication. Scaffold-based tissue engineering combines cells, molecular factors, and biomaterials to create templates for tissue formation. However, traditional scaffolds face challenges such as poor cell migration, uneven distribution, and lack of bio-signals. This chapter highlights how cell surface and scaffold interface engineering modulate cell-biomaterial interactions, cellularization, and modular scaffold fabrication. Overall, this chapter emphasizes the potential of cell surface engineering in advancing tissue engineering and regenerative medicine.

Z. Hussain · S. Mehmood · I. Ullah · N. Shao · R. Pei (✉)
CAS Key Laboratory for Nano-Bio Interface, Suzhou Institute of Nano-Tech and Nano-Bionics (SINANO), Chinese Academy of Sciences (CAS), Suzhou, China
e-mail: zahid@mail.ustc.edu.cn; shahmehmood@mail.ustc.edu.cn; ullah2020@sinano.ac.cn; shaonn@ucas.ac.cn; rjpei2011@sinano.ac.cn

Z. Hussain, R. Pei (eds.), *Cell Membrane Engineering for Advancing Cell-Based and Nanotechnology-Based Therapies*, Biomaterials, Bioengineering and Sustainability 6, https://doi.org/10.1007/978-3-031-93425-4_8

Keywords

Cell engineering · Cell homing · Cell-mediated drug delivery · Scaffold-free assembly · Scaffold cellularization · Tissue reconstruction

8.1 Introduction

Cells are the cornerstones of life and perform essential functions in complex environments to support their survival and growth. Mammalian cells are protected by thin semipermeable membranes called the cell membrane. The cell membrane is a phospholipid bilayer that contains proteins, carbohydrates, glycoconjugates, and cholesterol, either embedded or spanning across it (Shaw et al. 2021). The cell membrane composition is heterogeneous, with distinct membrane proteins depending on the cell state. The cell surface is essential for cellular function, and it possesses complex structures that respond to stimuli and regulate processes such as adhesion, communication, homing, pathogen recognition, immune response, and homeostasis (Mager et al. 2011; Bausch-Fluck et al. 2019). Cellular interactions depend on the expression of membrane receptors and ligands, which can differ between healthy and diseased states, making them valuable for therapeutic applications. The cell surface is the primary point of contact for various biochemical and cellular functions, and there is increasing interest in modulating cell surfaces with biomaterials, proteins, ligands and nanoparticles to regulate cell activities for biomedical applications (Meyfour et al. 2021; Di Meo et al. 2024).

Tissue and organ reconstruction is a critical clinical challenge. Tissue engineering and regenerative medicine (TERM) combines cell biology, nanomedicine, materials science, medical sciences, and engineering principles to develop biological substitutes for repairing impaired tissues and organs. This field uses strategies such as biomaterials, scaffolds, living cells, growth factors, and nanomedicine to address clinical challenges in conditions ranging from tissue injuries to organ failure (Berthiaume et al. 2011; Salgado et al. 2013). From a technological standpoint, TERM can be categorized into cell suspension-based therapies, cell-based scaffold-free therapies, and scaffold-based therapies. These approaches have undergone significant advancements, leading to transformative changes in degenerative disease treatment and personalized therapies (Ovsianikov et al. 2018; Zelinka et al. 2022). The following sub-sections explore the principles and limitations of these approaches.

8.1.1 Cell-Suspension-Based TERM

Cell suspension-based therapies involve the infusion of autologous or allogeneic cells, offer a less invasive and quick approach with clinical translation. Cellular therapies utilize self-renewal, secretion, communication, homing, and biomodulation properties (e.g., immunomodulatory, anti-inflammatory, and proangiogenic) for regeneration (Amer et al. 2017; Wang et al. 2021). Therapeutic cells can respond to biochemical and biophysical cues for crossing biological barriers. These integrated properties are challenging to replicate with biomaterials, leading researchers to explore their potential for treating tissue defects (Buzhor et al. 2014). For example, stem cells can be obtained from various tissue sources and differentiate into various cell lineages such as chondrocytes, osteoblasts, osteocytes, adipocytes, myocytes, hepatocytes, making them promising for disease treatment (Hoang et al. 2022). However, the direct use of natural cells faces challenges like poor homing (refers to the directed migration of therapeutic cells to target tissues), low engraftment, inflammatory response, and unwanted differentiation. A significant gap remains between natural cells and effective treatment modalities, and these issues highlight the need for cell membrane engineering to improve the effectiveness of cellular therapies (Naderi et al. 2011; He and Chen 2020).

8.1.2 Scaffold-Based TERM

Scaffold-based bioengineering uses cells, molecular factors, and biomaterials to create biodegradable templates that mimic the natural extracellular matrix (ECM) and facilitate tissue formation. Scaffolds are typically populated with cells to develop functional constructs or implanted into defective sites to promote in situ cellularization. Cell-based scaffolds are preferred owing to their potential to support cell retention, survival, differentiation, migration, matrix formation, and tissue integration (Demirbag et al. 2011; Badekila et al. 2021). Recent advances in biomaterials and fabrication strategies have enhanced structural, physicochemical, mechanical, and degradation properties of scaffolds, thereby supporting neo-tissue formation and defect-bridging (Hussain et al. 2023). Hydrogels, sponges, nanofibers, cast films, aerogels, and bioprinted scaffolds have been developed to mimic natural tissue microenvironments (Eltom et al. 2019). However, scaffold-based constructs face challenges such as poor cell migration and uneven distribution owing to the absence of natural bio-signals. The interface engineering of cells and scaffolds is key to enhancing cellularization and creating living constructs with relevant cell densities (Dave and Gomes 2019).

8.1.3 Cell-Based Scaffold-Free TERM

Scaffolds have limitations such as uncontrolled degradation, unintended immune responses, and cell phenotype changes, hindering neo-tissue formation, thus highlighting the need for scaffold-free constructs (Hussain and Pei 2021). Scaffold-free strategies form stable aggregates, like cell sheets and spheroids, using the self-organization potential of cells without exogenous materials. These strategies offer high cell density, improved viability, reduced cytotoxicity, and potential elimination of sutures through adhesive matrix components (Petrenko et al. 2017). Interest in fabricating scaffold-free assemblies to support tissue integration, tissue engineering, and ex vivo tissue modelling is growing (Kim et al. 2020; Kobayashi et al. 2019). However, challenges remain in creating multilayered architectures owing to limited cell interactions and alignment issues in 3D constructs. Human tissues are heterogeneous, with various cell types in different arrangements, making it difficult to replicate them using traditional cell sheets and spheroid platforms (Laschke and Menger 2017; Ong et al. 2018). These limitations emphasize the need for cell surface engineering to enable precise cell patterning and recreating scaffold-free anatomical motifs.

Cell-cell interactions are integral to various cellular activities and processes. The cell membrane provides hotspots for engineering strategies to enhance cellular interactions, self-assembly, and biological activity. Cell surface engineering modifies the cell surface to tailor cellular functions and interactions (Amaral and Pasparakis 2019; Su et al. 2024). Various methods have been introduced for manipulating cell membrane for controlling the cell–cell and cell-scaffold interactions, including genetic engineering, covalent conjugation, metabolic glycoengineering, bioorthogonal engineering, hydrophobic insertion, physical coating, and membrane fusion. These techniques allow for decorating cell surfaces with receptors, aptamers, nanomaterials, peptides, or drugs to support therapeutic function (Abbina et al. 2017; Park et al. 2018). This chapter explores cell surface engineering strategies to (a) enhance ligand-receptor interactions for cellular homing, (b) enable cell-based targeted drug delivery, (c) modulate cell-cell interactions for scaffold-free tissue fabrication, and (d) regulate cell-biomaterial interactions for improved cellularization and modular scaffold fabrication (Fig. 8.1).

8.2 Cell Surface Engineering for Cell Homing and Drug-Assisted Therapy

Cell therapy is increasingly recognized as a promising treatment, particularly for diseases characterized by poor self-repair capacity. The systemic infusion of cells is preferred for its minimal invasiveness and repeated dosing. It shows promise for treating heart damage, brain trauma, liver dysfunction, and inflammatory disorders. However, efficient cell delivery and high engraftment to target tissues remains a significant challenge (Xia et al. 2018). Researchers have been noticed that endogenous stem cells are recruited at low rates after tissue damage. However, less than 10% of the infused cells reached the damaged site regardless of the cell type. This was likely due to heterogeneous cell surface markers, loss of homing signals, and inadequate adhesion receptors (Barbash et al. 2003; Wu et al. 2017). High stem cell doses may not improve treatment efficacy because of significant cell entrapment in the reticuloendothelial organs, resulting in poor retention of transplanted cells (Timin et al. 2018). Furthermore, many therapies require nanomedicine supplementation for modulation of diseased tissue microenvironment to support cellular responses (Dou et al. 2020; Li et al. 2023). Recent studies have suggested that engineering cell membranes with receptors and drugs could enhance cell homing and cell fate regulation for tissue repair (Table 8.1).

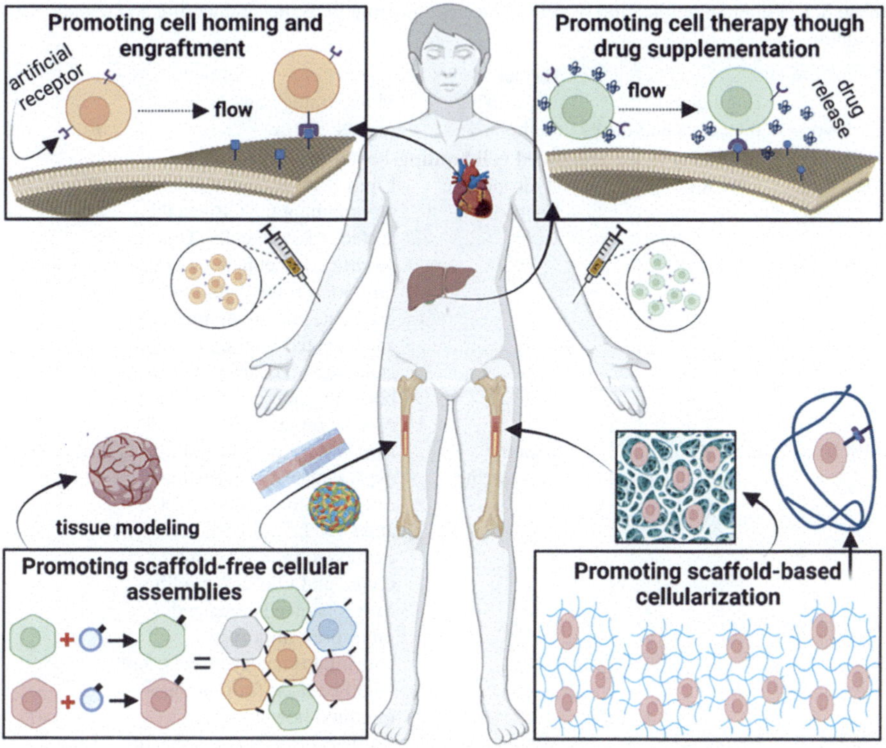

Fig. 8.1 A schematic representation showed that cell surface engineering can enhance cell homing, direct therapy via drug delivery, scaffold-free assemblies, and improved cellularization of tissue-engineered scaffolds. (Source: Created with BioRender.com)

8.2.1 Cell Homing-Based Therapies

Cardiovascular, brain, and inflammatory diseases are significant global health problems. Stem cells show great therapeutic potential towards inflammation-related diseases because of their potential to adhere to activated endothelial cells in damaged tissues, facilitating migration and repair through differentiation or paracrine signaling. Enhancing stem cell accumulation and retention at disease sites is vital for effective treatment (Ullah et al. 2019; Li et al. 2019). Studies have observed that stromal-derived factor-1 (SDF-1) expression in the ischemic myocardium is rapidly upregulated after infarction and released into the bloodstream. SDF-1 then reaches the bone marrow and recruits stem cells expressing CXC chemokine receptor 4 (CXCR4). The CXCR4/SDF-1 axis plays an essential role in mesenchymal stem cells (MSCs) homing to the infarcted myocardium, which is crucial for cardiac recovery. CXCR4 is a chemotactic receptor that recognize SDF-1 in MSCs homing to ischemic tissues. However, MSCs expansion during in vitro culture results in the downregulation of CXCR4 expression, impairing their ability to migrate to the ischemic myocardium. To address this, surface engineering strategies have been

Table 8.1 Potential application of cell surface engineering in tissue engineering and regenerative medicine

Cell types	Engineered materials and drugs	Surface engineering methods	Main strategy and critical results	Ref.
Cell surface engineering for enhanced cell homing-based therapies				
MSCs	CXCR4	Hydrophobic insertion	Lipid-PEG was used to attach recombinant CXCR4 to the surface of stem cells. This modification enhances MSCs migration towards the SDF-1 gradient, improving retention and targeted homing to the ischemic myocardium, thus boosting their potential for cardiac repair	Won et al. (2014)
MSCs	CXCR2	Genetic engineering	MSCs were engineered to express the CXCR2 using mRNA technology. CXCL2 and CXCL5 were highly expressed in the inflamed colon, and CXCR2-modified MSCs showed enhanced migration, improved anti-inflammatory effects, and could be used for treating inflamed bowel disease	Li et al. (2021a)
Cardiac stem cells	Platelets nanovesicles	Membrane fusion	Cardiac stem cells fused with platelet vesicles expressed platelet markers and enhanced adhesion to injury sites. These engineered cells showed superior binding to vascular lesions ex vivo and demonstrated improved heart retention in animal models of acute myocardial infarction	Tang et al. (2018)
MSCs	Phage-peptides	Hydrophobic insertion	Phage peptides were synthesized and conjugated with a palmitic acid tail to enhance cell membrane integration. MSCs coated with these peptides demonstrated improved localization in the heart, with retention levels correlating to the severity of myocardial infarction in a mice model	Kean et al. (2012)

(continued)

Table 8.1 (continued)

Cell types	Engineered materials and drugs	Surface engineering methods	Main strategy and critical results	Ref.
ADSCs	P-selectin binding peptide	Hydrophobic insertion	ADSCs were modified with a P-selectin binding peptide using DMPE-PEG linkers. The engineered ADSCs exhibited strong P-selectin binding, reduced inflammation, improved endothelial function, and demonstrated enhanced targeting of injured arteries, suggesting potential for vascular injury treatment	Yan et al. (2020)
MSCs	Anti-VCAM1 antibodies	Hydrophobic insertion	DNA template-directed self-assembly strategy was used to engineer polyvalent anti-VCAM1 on MSCs under physiological conditions. The engineered MSCs show enhanced adhesion to inflamed endothelial cells and improved therapeutic efficacy in mice models of inflammatory bowel disease	Ye et al. (2023)
MSCs	Vasculature binding peptides	Hydrophobic insertion	Vasculature-binding peptides were conjugated to hyperbranched polyglycerol on cell membranes to regulate cell adhesion to target tissues. It enhances peptide presentation and improves cell adhesion to inflamed blood vessels, aiding the treatment of defective vasculature	Jeong et al. (2013)
EPCs, platelets	DBCO-PEG-CD41, azide-PEG-D34	Bioorthogonal click engineering	DBCO-PEG-CD41 antibodies pretarget platelets at the myocardial infarction, and after 48 h, azide-PEG-CD34 antibodies bind to EPCs. The azide groups react with DBCO groups on platelets, redirecting EPCs to the injury, and promoting heart repair in a mice model of myocardial infarction	Li et al. (2018)

(continued)

Table 8.1 (continued)

Cell types	Engineered materials and drugs	Surface engineering methods	Main strategy and critical results	Ref.
MSCs	Heterodimerizing leucine zippers	Physical crosslinking	Cells are modified with base and complementary leucine zippers for sequential dosing, enabling layer-by-layer in situ crosslinking. This method improves the accumulation and retention of MSCs at the infarction site, thereby supporting scaffold-free cardiac repair	Jasiewicz et al. (2023)
Cell surface engineering for enhanced drug-assisted cell therapies				
MSCs	OPC-loaded liposome	Bioorthogonal click engineering	OPC-loaded liposomes were anchored to MSCs via a click reaction. This approach enabled the targeted accumulation of engineered cells at X-ray-irradiated lung sites, modulated innate immune cell polarity, promoted anti-inflammatory effects, and prevented radiation-induced pulmonary fibrosis	Zhou et al. (2023)
RBCs	Cyclodextrin, ferrocene-modified liposomes, curcumin	Hydrophobic insertion, host-guest interactions	β-Cyclodextrin-modified RBCs and ferrocene-modified liposomes were co-incubated to enable nanoparticle attachment. In inflamed lungs, high ROS trigger nanoparticle detachment, releasing therapeutic agents and significantly reduced inflammation	Li et al. (2022)
Osteoclasts	Tetracycline	Covalent conjugation	A covalent conjugation was carried out to attach tetracycline to osteoclasts. The engineered osteoclasts exhibited enhanced targeting and adhesion to ectopic calcified tissues due to their affinity for calcium minerals. They effectively removed heterotopic ossification compared to naive osteoclasts	Jin et al. (2021)

(continued)

Table 8.1 (continued)

Cell types	Engineered materials and drugs	Surface engineering methods	Main strategy and critical results	Ref.
ECFCs	TGF-β inhibitor loaded liposomal	Covalent conjugation	GDM-exposed ECFCs (GDM-ECFCs) were restored by attaching drug-loaded liposomal nanoparticles to their surface. This approach improved cell viability, normalized TAGLN expression, enhanced migration, and restored vascular progenitor function	Bui et al. (2022)
Microglia	IL-4 loaded DBCO modified liposome, platelets membrane	Membrane fusion, bioorthogonal click engineering	Platelet-hybrid microglia were created through membrane fusion. IL-4 was encapsulated in ultrasound-responsive liposomes and conjugated to microglia. This allowed microglia to adhere to injured brain tissue. Ultrasound-triggered IL-4 release promoted M2 polarization and vascular repair	Li et al. (2021b)
MSCs	Palmitic acid peptide, miR-133b	Hydrophobic insertion	Palmitic acid-peptide applied to the cell membrane directed MSCs to ischemic tissues, reducing peripheral organ accumulation and improving engraftment. Enhanced cell homing allowed MSCs to deliver miR-133b, boosting its expression in ischemic lesions	Huang et al. (2017a)
RBCs	Glc-insulin	Non-covalent interactions	Glc-insulin binds to RBC membranes via GLUT molecules. This binding is reversible under hyperglycemia, as free glucose competes with GLUT. In diabetic mice, Glc-insulin-conjugated RBCs prolonged insulin effects and maintained normal blood glucose, offering a promising strategy for diabetes management	Wang et al. (2017)

MSCs mesenchymal stem cells, *CXCR4* C-X-C chemokine receptor 4, *SDF-1* stromal cell-derived factor 1, *CXCL2* C-X-C motif chemokine ligand 2, *ADSCs* adipose-derived stem cells, *VCAM1* vascular cell adhesion molecule 1, *EPCs* endothelial progenitor cells, *DBCO* dibenzocyclooctyne, *PEG* polyethylene glycol, *RBCs* red blood cells, *ROS* reactive oxygen species, *ECFCs* endothelial colony-forming cells, *TGF-β* transforming growth factor beta, *GDM* gestational diabetes mellitus, *Glc* glucose, *GLUT* glucose transporter, *OPC* Oligomeric proanthocyanidin developed to introduce recombinant CXCR4 proteins into MSCs membranes using lipid-PEG insertion. DMPE-PEG-CXCR4 engineering on the membrane surface enhanced stem cell migration toward an SDF-1 concentration gradient, improving retention rates and homing. Systemically infused recombinant CXCR4-modified stem cells effectively migrated to the myocardial ischemia following the SDF-1 gradient, increasing their potential for cardiac repair (Fig. 8.2a) (Lee et al. 2018; Won et al. 2014).

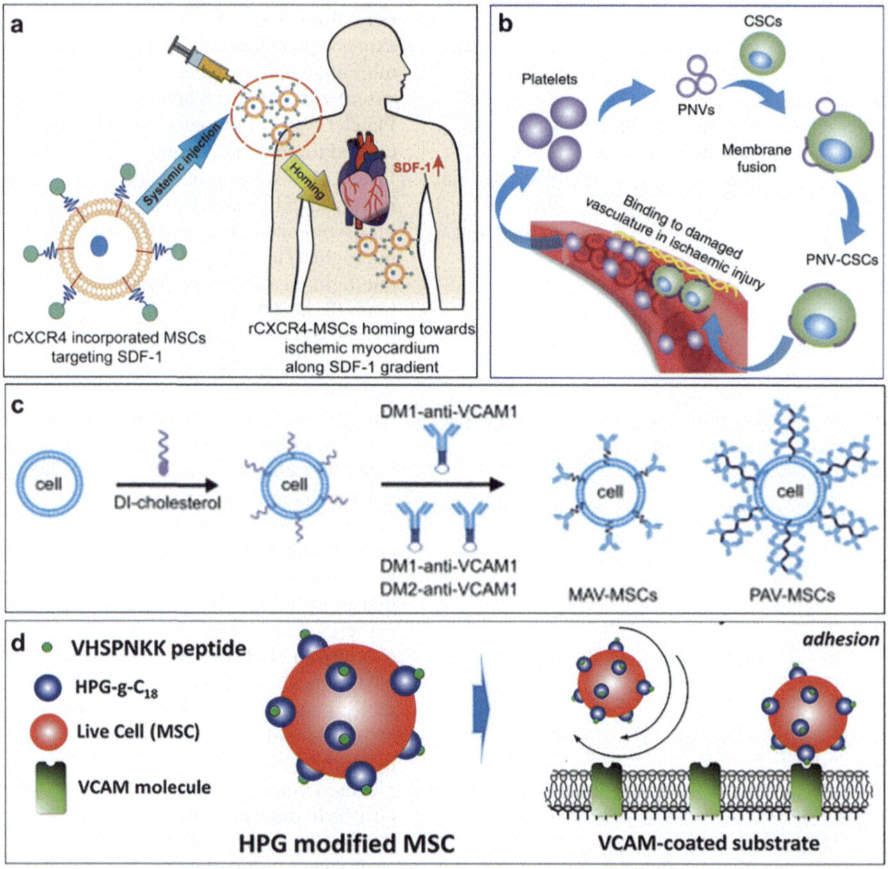

Fig. 8.2 (**a**) rCXCR4-modified MSCs, when systemically infused, migrate to the ischemic myocardium via the SDF-1 concentration gradient. (Reprinted with permission from Lee et al. (2018), copyright 2018, Springer Nature). (**b**) Platelet nanovesicles fused with cardiac stem cells harness the targeting potential of platelet surface markers, enabling engineered cells to target myocardial infarction. (Reprinted with permission from Tang et al. (2018), copyright 2018, Springer Nature). (**c**) Bottom-up assembly of polyvalent anti-VCAM1 onto MSCs surfaces to enhance cell adhesion to injured tissue vessels. (Reprinted with permission from Ye et al. (2023), copyright 2023, Springer Nature). (**d**) Peptides conjugated hyperbranched polyglycerol self-assembled onto MSCs surfaces to regulate adhesion to inflamed endothelium and VCAM-coated substrates. (Reprinted with permission from Jeong et al. (2013), copyright 2013, American Chemical Society)

MSCs have variable efficacy in treating inflammatory bowel disease, partly due to limited homing to the damaged colon. To enhance homing, MSCs have been genetically engineered to express the CXCR2 receptor. CXCL2 and CXCL5, which are significantly expressed in the inflamed colon, bind to CXCR2. MSCs with transient CXCR2 expression showed improved migration to inflamed tissue and enhanced anti-inflammatory effects. mRNA-engineered CXCR2 MSCs improve inflammatory bowel disease treatment and provide a safer and more efficient MSCs modification strategy (Li et al. 2021a). Platelet nanovesicles can fuse with cardiac stem cells via membrane fusion, enhancing their delivery to myocardial infarction sites by promoting adhesion through platelet marker expression. Engineered stem cells showed improved binding vascular lesions on denuded aortas. In animal models of myocardial infarction, these modified stem cells exhibit better accumulation in the heart tissues and reduced infarct size. This approach utilizes platelet membrane targeting and the repair capabilities of stem cells to improve delivery and retention at infarction sites, potentially enhancing angiomyogenesis in ischemic heart disease compared to unmodified stem cells (Fig. 8.2b) (Tang et al. 2018).

Peptide-based targeting was used to enhance MSCs homing to myocardial infarction (MI) sites with different MI-specific peptides (CRPPR, CRKDKC, KSTRKS, and CARSKNKDC) synthesized with palmitic acid tails for better membrane integration. After optimized the cell coating, peptide-coated MSCs were injected into myocardial infarcted mice. Treated animals showed higher MSCs retention in the heart compared to controls, with retention correlating positively with infarction severity (Kean et al. 2012). In another peptide-based study, adipose-derived stem cells (ADSCs) were modified with P-selectin binding peptide to improve targeting using DMPE-PEG linkers (DPP). DPP-modified ADSCs showed strong binding to P-selectin on activated platelets and endothelial cells, reduced inflammation, improved endothelial function, and decreased vascular smooth muscle cell migration. These DPP-ADSCs also targeted injured human arteries in ex vivo tests, supporting their potential for noninvasive vascular injury treatment (Yan et al. 2020).

Polyvalent interactions enhance stem cell targeting to injured tissue vessels compared to monovalent interactions. DNA template-directed self-assembly was used to engineer polyvalent anti-VCAM1 antibodies on MSCs. Engineered MSCs exhibit functional groups, improved adhesion to inflamed endothelial cells, and increased therapeutic efficacy in mouse models of inflammatory bowel disease (Fig. 8.2c) (Ye et al. 2023). Vasculature-binding peptides can be conjugated to hyperbranched polyglycerol on cell membranes via hydrophobic octadecyl chain insertion. Surface engineering allows precise control over the number of vasculature-binding peptides on the cell surface, thereby regulating cell adhesion to target tissues. This system can modulate the adhesion of transplanted cells to inflamed endothelial cells. The compact structure of polyglycerol minimizes intermolecular entanglements and enhances the presentation of adhesion peptides. This molecular design improves the effectiveness of cell adhesion to vascular endothelial adhesion molecules overexpressed in inflamed blood vessels, making it a promising approach for targeting defective vasculature (Fig. 8.2d) (Jeong et al. 2013).

Pretargeting and bioorthogonal engineering have been used to capture circulating stem cells and direct them to the damaged heart for functional repair. First, DBCO-PEG-CD41 antibodies were administered to the pretarget platelets at the myocardial infarction site. After 48 h, azide-modified CD34 antibodies bound to endogenous stem cells, and their azide groups reacted with dibenzocyclooctyne (DBCO) groups on platelets through a bioorthogonal chemistry mechanism. Platelets possess the inherent potential to accumulate in injured vessels and redirect engineered stem cells to the injured area. In a mice model of myocardial infarction, assessments confirmed effective heart repair facilitated by this pretargeting and bioorthogonal approach (Li et al. 2018). Layer-by-layer in situ crosslinking of cells has been investigated for cardiac tissue repair, utilizing heterodimerizing leucine zippers. MSCs are modified with either base or complementary leucine zippers, and sequential dosing of modified cells facilitates crosslinking, enhancing MSCs accumulation at the infarction site. Unlike traditional methods, this approach allows superior cell accumulation and saturation at the target site through in situ crosslinking. Sequential delivery of leucine zipper-conjugated stem cells led to a 500% increase in cell retention at the infarcted area over 10 days compared to unmodified cells (Jasiewicz et al. 2023).

8.2.2 Drug-Assisted Cell Therapies

Traditional nanoparticle-based drug delivery systems can lead to side effects due to a lack of tissue-specific targeting and personalization in treatment. Since a cell's targeting abilities mainly stem from its surface, efforts to use cell membranes for cargo delivery are gaining importance. Cell-mediated drug delivery can overcome these limitations by addressing damaged tissue microenvironments and minimizing the side effects through targeted therapies (Wang et al. 2024a). Red blood cells (RBCs), white blood cells, and stem cells are promising drug carriers owing to their biocompatibility, immune evasion, long circulation, barrier-crossing ability, and tissue targeting. Research has focused on the ability of cell carriers to modulate drug pharmacokinetics to treat degenerative diseases (Villa et al. 2016; Sun et al. 2017). Cell surface engineering for drug conjugation involves attaching active drugs or drug carriers onto the cell surface, targeting carriers' cells to specific locations, and controlling drug release. Both artificial and intrinsic ligand-receptor interactions have been used for drug loading and targeted homing to injury sites (Wang et al. 2024b).

Zhou et al. presented an innovative approach to enhancing MSCs for the treatment of radiation-induced lung injury. Oligomeric proanthocyanidin (OPC) loaded antioxidant liposomes (Lipo-OPC) were anchored to MSCs membranes through a click reaction, creating MSCs-Lipo-OPC without compromising cell function. This modification enabled MSCs-Lipo-OPC to efficiently localize in lung tissues. The modified MSCs influenced innate immune cells, such as neutrophils, macrophages, and Tregs, shifting their polarization towards anti-inflammatory responses. This strategy effectively prevented the progression of radiation-induced pulmonary fibrosis. By integrating drug-loaded nanoparticles with MSCs, this method improves the therapeutic potential of both the stem cells and the drugs, advancing their application in managing radiation-induced damage (Zhou et al. 2023). Acute pneumonia is a critical inflammatory disease that can lead to multi-organ failure. Existing anti-inflammatory treatments face challenges owing to limited systemic circulation and inadequate lung specificity. Li et al. introduced an inflammation-sensitive supramolecular drug delivery system using RBCs as carriers. By hitchhiking on RBCs, the system enhances nanomedicine circulation and targets inflamed lung tissues. β-cyclodextrin-modified RBCs and ferrocene-modified liposomes were incubated together, facilitating nanoparticle attachment through host-guest interactions. In inflamed lungs, elevated level of reactive oxygen species (ROS) trigger nanoparticle detachment, enabling targeted therapeutic release. This approach effectively delivered curcumin to the pulmonary tissue in a mouse model of acute pneumonia, thereby, significantly reduce inflammation (Li et al. 2022).

Heterotopic ossification affects over 80% of joint-revision surgery patients and frequently occurs in burn and trauma cases, posing significant medical challenges. Osteoclasts can remodel bone by dissolving organic bone matrix and mineralized bone component during bone resorption, however, naive osteoclasts struggle to remove calcified plaques owing to limited bioactivity and adhesion to ectopic sites. To address this, chemical modification of osteoclasts using tetracycline (TC) has been introduced to create engineered osteoclasts (TC-osteoclasts). TC targets calcium minerals by binding hydroxyapatite. TC-osteoclasts, with broad calcium mineral affinity, effectively targeted ectopic calcified tissues and removed heterotopic ossification, showing more significant promise than naive osteoclasts for cell therapy (Fig. 8.3a) (Jin et al. 2021). Fetal exposure to gestational diabetes increases the risks of type-2 diabetes and cardiovascular disease in children, partly due to the dysfunction of endothelial colony-forming cells (ECFCs). Current restoration methods face challenges such as drug side effects and immune responses. Bui et al. developed a strategy to rejuvenate gestational diabetes-exposed ECFCs by attaching drug-loaded liposomal particles to their surface. This approach improved cell viability, enhanced migration, and promoted vasculogenesis. These findings suggest a promising method to restore vascular progenitor function, with potential applications in regenerative medicine (Bui et al. 2022).

Stroke is a severe condition that has high mortality rate. M2 microglia show promise for reducing inflammation and promoting neuronal repair. However, their targeted delivery to stroke sites is challenging because of their sequestration in the reticuloendothelial system. Platelet-hybrid microglia-based platforms have been developed to overcome these challenges and enhance adhesion to ischemic brain tissue. IL-4 was encapsulated in ultrasound-responsive liposomes and conjugated to platelet-hybrid microglia via click reaction between azide-modified microglia and DBCO-modified liposomes. This platform enabled platelet-hybrid microglia to adhere to injured brain tissue, where ultrasound activated IL-4, inducing M2 polarization in both engineered and endogenous microglia. This approach improved neurogenesis, vascular repair, and neuroprotection, highlighting its potential for stroke recovery (Fig. 8.3b) (Li et al. 2021b). Furthermore, a peptide-based strategy was

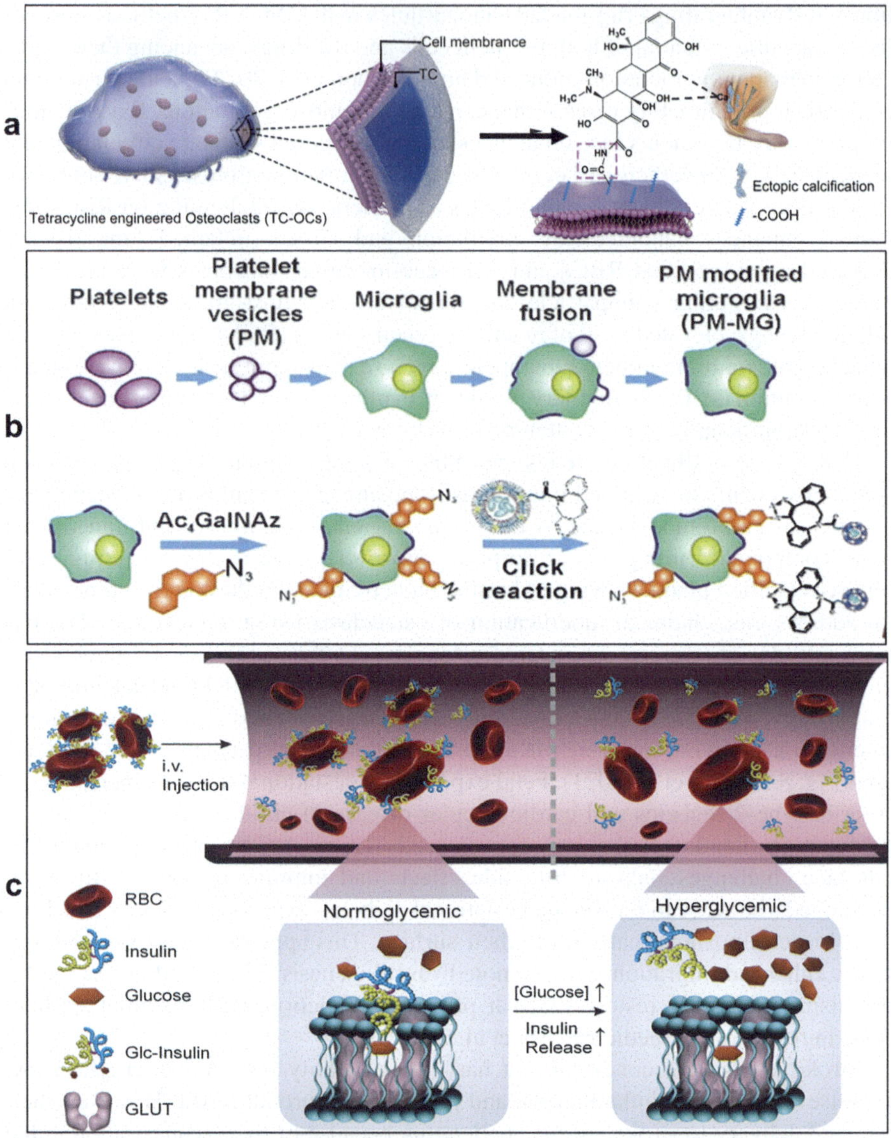

Fig. 8.3 (**a**) Schematic of tetracycline-osteoclasts as a bionic treatment for heterotopic ossification. (Reprinted with permission from Jin et al. (2021), copyright 2021, Springer Nature). (**b**) Platelet membrane vesicles were decorated onto microglia to form platelets-modified microglia. Ultrasound-responsive and IL-4 loaded liposome was conjugated to platelets-modified microglia via click conjugation for ischemia-reperfusion injury treatment. (Reprinted with permission from Li et al. (2021b), copyright 2021, Wiley-VCH GmbH). (**c**) Glucose-responsive insulin delivery based on Glc-insulin conjugated RBCs system. (Reprinted with permission from Wang et al. (2017), copyright 2017, Wiley-VCH GmbH)

developed to enhance MSCs homing to ischemic brain. By modifying the MSCs membrane with a palmitic acid–peptide, researchers have directed the cells to damaged tissues without cytotoxicity or disrupting differentiation. This approach also reduced MSCs accumulation in peripheral organs and improved engraftment in target tissues. Additionally, MSCs deliver miR-133b to ischemic lesions, thereby enhancing the therapeutic outcomes (Huang et al. 2017a).

Traditional insulin administration provides temporary control, but daily monitoring is stressful and may fail to regulate glycemia. Glucose-responsive insulin delivery using RBCs has gained attention in recent studies, suggesting that glucose-derivative-modified insulin (Glc-insulin) binds effectively to RBCs via interactions with abundant glucose transporter (GLUT) molecules on their surface. This binding is reversible under hyperglycemia, allowing insulin to be released from RBCs as free glucose competes with GLUT. In vivo experiments using a diabetic mouse model demonstrated that administration of Glc-insulin-conjugated RBCs prolonged the therapeutic effect of insulin. This approach effectively maintained blood sugar levels within the normal range. The glucose-responsive insulin delivery platform mimics the function of insulin producing β-cells and can revolutionize diabetes management (Fig. 8.3c) (Wang et al. 2017).

Significant advances in cell surface engineering using homing peptides and click molecules have created new opportunities to enhance cell homing to target areas as well as efficient drug delivery with high precision. Integrating cell therapies with drug delivery systems through cell surface engineering can expand their clinical applications (Chen et al. 2018). Drugs can be loaded into nanocarriers and subsequently conjugated to cell surfaces. Cell-based drug delivery regulates cell fate, enabling targeted delivery through cell homing or stimulus-responsive release manner. These advantages suggest cell that cell surface engineering can boost the clinical potential, making engineered cells promising for targeted therapy (Li et al. 2022; Chao et al. 2023).

8.3 Cell Surface Engineering for Scaffold-Free Cellular Assemblies

Cells in 2D cultures face increased stiffness, polarized morphology, and limited function owing to poor adhesion and lack of matrix gradients. In contrast, 3D cultures promote protein secretion, extracellular matrix production, and multi-lineage stem cell differentiation by supporting cell-cell and cell-matrix interactions (Jensen and Teng 2020). Interest in scaffold-free assemblies, like tissue sheets and spheroids, is growing due to their potential for higher cell densities and better in-vivo microenvironment mimicry, leading to improved reparative outcomes after transplantation (Cooper and Rainbow 2022). Numerous strategies, including hanging drops, nonadherent surfaces, rotary culture, micromolding, and microfluidics, have been developed to promote cell–cell aggregation, enhancing cell cohesion and reducing damage during self-assembly. Multicellular aggregates are used as ex vivo tissue models for drug efficacy screening and disease studies, and as in-vivo injectable cells for cell therapy (Decarli et al. 2021; Griffin et al. 2022). Despite progress in spheroid formation, traditional methods require long incubation times to establish cell-ECM interactions and often result in irregular, loose aggregates that do not mimic natural tissues. Creating multilayered tissues is challenging due to limited cell-cell interactions, misalignment in spheroids, and a lack of specific interactions among different cells. Given the increasing focus on lab-grown tissues, there is a need to improve the speed and consistency of cell assembly at the macro-scale (De Pieri et al. 2021; Shao et al. 2020). This section discussed how cell surface engineering enabling rapid cell aggregation and multilayered tissue formation using bottom-up tissue engineering, without relying on adhesion proteins.

8.3.1 Chemically Engineered Cells for Tissue Modeling

Chemical-based cell surface engineering enables precise cell placement and patterning to create functional scaffold-free anatomical motifs. For example, a photo-responsive thiol-ene reaction on fibroblasts allows artificial cell-cell interactions and spheroid formation. Methacryloyl groups were incorporated onto cell surface via metabolic glycoengineering with acetylated methacrylate-bearing N-mannosamine. The modified cells underwent thiol-ene reactions with thiol-terminated PEG in the presence of a photoinitiator, inducing cell-cell adhesion and cell aggregate formation via photoirradiation. This technique created millimeter-scale structures with high cell densities without a rigid scaffold and demonstrated the potential for printing heterogeneous cell populations (Fig. 8.4a) (Otaka et al. 2024). Biotin-streptavidin interactions and stress-induced rolling methods have been introduced to form multilayered tissue-like constructs. The cells were biotinylated with NHS-biotin and bound to streptavidin-coated cells to form bilayers. For instance, smooth muscle cells are biotinylated, attached to a membrane, treated with streptavidin, and overlaid with biotinylated endothelial cells to create a bilayer. This process is repeated to generate multilayered tubular structures, simulating in vivo systems like blood vessels (Gong et al. 2013).

The research group led by Muhammad N. Yousaf developed innovative strategies to modify cells with bio-orthogonal groups, facilitating rapid cellular assembly through intercellular oxime ligation. Three populations of cationic liposomes are used to deliver bioorthogonal click groups, such as ketone, oxyamine, and photo-oxyamine, to the surfaces of various cell types for modification. This approach enabled spheroid assemblies of interconnected cells via click ligation between ketone and oxyamine-engineered cells. In a scaffold-free co-culture model utilizing heart muscle cells (cardiomyocytes), blood vessel-lining cells (endothelial cells), and connective tissue cells (cardiac fibroblasts), functional cardiac tissue characterized by high cell density and generate synchronized beating (Fig. 8.4b). The photo-oxyamine lipid in the crosslinked structure responds to UV light, cleaving the photolabile oxime linkage and enabling controlled disassembly of microtissue (Fig. 8.4c). For advanced scaffold-free fabrication, surface-engineered cells are introduced into a microfluidic platform, where they flow through channels,

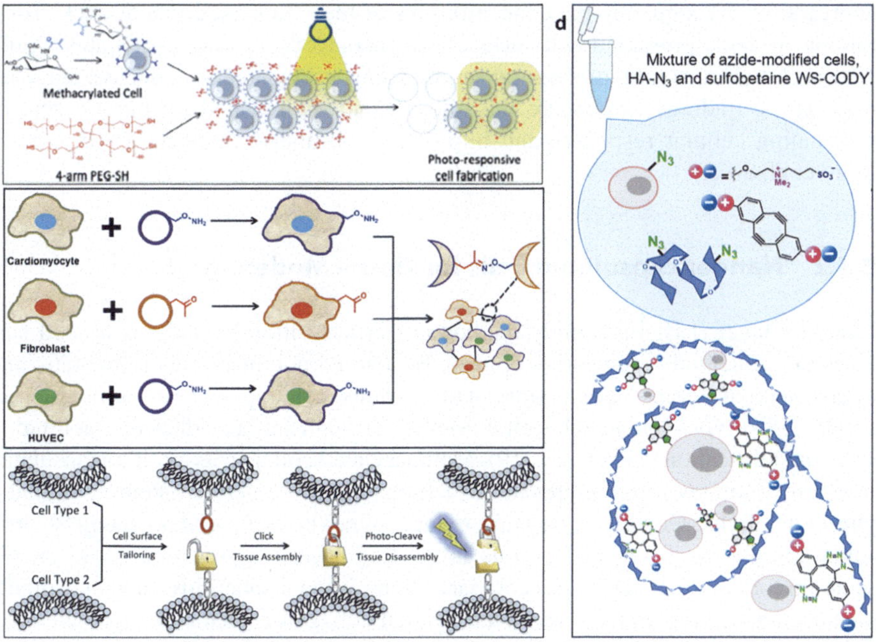

Fig. 8.4 (**a**) Photoresponsive thiol-ene reaction between thiol-terminated polyethylene glycol (PEG) and methacryloyl groups on cell membranes to form cellular aggregates. (Reprinted with permission from Otaka et al. (2024), copyright 2024, American Chemical Society). (**b**) Cells are treated with cationic liposomes to deliver bioorthogonal groups for surface modification. Ketone and oxyamine groups on the cell membrane form stable cellular assemblies through oxime linkage. (Reprinted with permission from Rogozhnikov et al. (2016), copyright 2016, Springer Nature). (**c**) Tissue assembly and disassembly at the molecular level through bio-orthogonal and photoswitchable cell membrane engineering. (Reprinted with permission from Luo et al. (2014), copyright 2014, Springer Nature). (**d**) Click cross-linking with sulfobetaine WS-CODY between azide-presenting cell surfaces and azide-modified hyaluronic acid (HA-N_3), resulting in the quick formation of cell aggregates. (Reprinted with permission from Sato et al. (2024), copyright 2024, American Chemical Society)

encounter each other, and form co-culture spheroids via oxime click chemistry. This process promotes rapid tissue assembly and enable the creation of co-culture spheroids and synthetic multilayer tissues (Dutta et al. 2011; Rogozhnikov et al. 2016; Luo et al. 2014; O'Brien et al. 2015).

Metabolic glycoengineering and bio-orthogonal engineering have been used to create scaffold-free tissue constructs. These methods utilize the biosynthetic machinery of the cells to incorporate azide group into cell membrane glycoconjugates. The introduced reporters were then crosslinked through a strain-promoted alkyne-azide cycloaddition reaction with multiarm PEG-cyclooctyne. This process rapidly assembled into aggregates, including those with different cell types, with the potential for advancing tissue modelling (Xu et al. 2023). Building on the click reaction, a robust method has been developed using zwitterionic dibenzo cyclooctadiyne (WS-CODY) and azide-conjugated hyaluronic acid as linkers to promote cell aggregation. By adjusting the concentrations of these linkers, efficient click cross-linking of azide-presented cell surfaces was achieved, forming aggregates within 10 min. A balanced ratio of these two components is essential for effective aggregation. These findings underscore the role of azide-conjugated hyaluronic acid in modulating cellular responses and improving cell-matrix interactions (Fig. 8.4d) (Sato et al. 2024).

8.3.2 Nanoencapsulated Cells for Tissue Modeling

Layer-by-layer (LBL) assembly method constructs ultrathin coatings on cell surfaces by sequentially depositing oppositely charged polymers via electrostatic and intermolecular forces. Compatible with various cell types, LBL films support single-cell survival and enable cell clustering through surface adhesion, facilitating spheroid fabrication (Liu et al. 2019). Methacrylated collagen as a cell-encapsulated material promotes rapid aggregation of human adipose-derived stem cells under ultraviolet light. Coated aggregates reduce hypoxia compared to hanging drop spheroids due to a thin and porous layer. Engineered aggregation could enable scale-up for complex 3D shapes compared to traditional spheroids. In a mice model of muscle loss, implantation of methacrylated collagen-coated cell aggregates significantly improved muscle regeneration (Chae et al. 2024).

The research group led by Mitsuru Akashi developed a bottom-up strategy using fibronectin and gelatin nano-coating to form spheroids and multilayered tissues. The LBL technique enhanced β-cell spheroid formation by creating FN-G films that promoted integrin-mediated adhesion. LBL-coated cells form dense spheroids with higher insulin secretion compared to non-coated cells. When transplanted into diabetic mice model, these assembled spheroids regulated blood glucose levels and improved glucose sensitivity, showing promise for treating type 1 diabetes mellitus (Fig. 8.5a) (Fukuda et al. 2018). Additionally, FN-G nano-coating enhances cell-cell adhesion by interacting with the α5β1 integrin receptor, enabling 3D cell assembly. This straightforward technique generated approximately 8 l of tissue after only one day of incubation. Within one-week of incubation, multilayered tissues were successfully fabricated, featuring extensive and high-density endothelial tube networks (Fig. 8.5b) (Nishiguchi et al. 2011).

LbL approach has been extended to develop a scalable system for dermal papilla spheroid formation. This approach uses gelatin as a positively charged polymer, alginate as a negatively charged polymer, and Ca^{2+} as a cross-linking agent for the nanoencapsulation of dermal papilla cells. This nanogels approach induces cell aggregation into microspheres, forming spheroids that mimic the intercellular organization of the hair follicles. Fabricated spheroid model enables high-passage dermal papilla cells, exhibit strong hair follicle inductive capacity, and can leads to successful hair follicle regeneration after implantation. This strategy is promising for generating tunable and scalable microtissues (Wang et al. 2018). Decorating cell surfaces with biomimetic materials can influence cell fate by clustering receptors and controlling aggregation. Treating cells with tris(2-carboxyethyl)phosphine

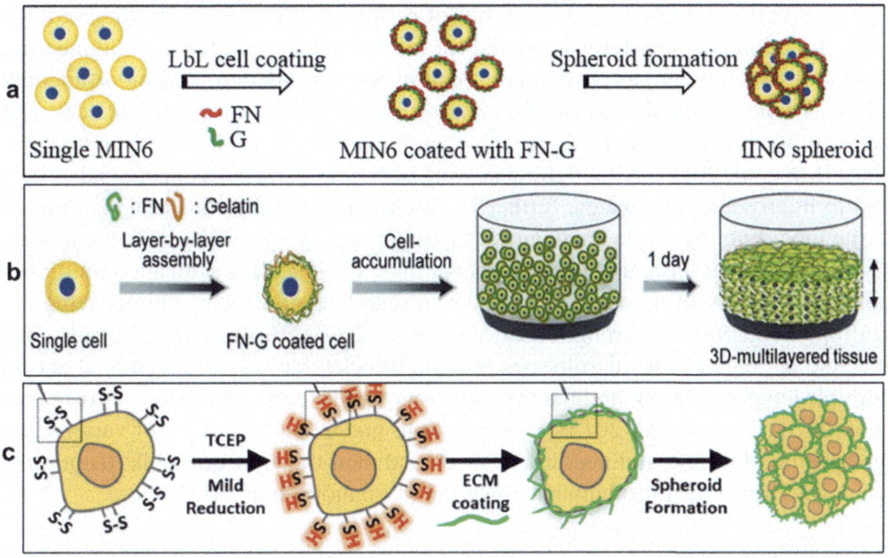

Fig. 8.5 (**a**) Fibronectin (FN)/gelatin (G) coating on cells via the layer-by-layer technique leading to rapid spheroid formation. (Reprinted with permission from Fukuda et al. (2018), copyright 2018, Elsevier B.V.). (**b**) Construction of thick tissues via self-assembly of cells coated with fibronectin/gelatin nanofilms. (Reprinted with permission from Nishiguchi et al. (2011), copyright 2011, Wiley-VCH GmbH). (**c**) Extracellular matrix (ECM) coating on cell surfaces through disulfide bond reduction and self-assembly into spheroids. (Reprinted with permission from Koh et al. (2020), copyright 2020, American Chemical Society)

(TCEP) modifies the cell surface redox status, splitting disulfide bonds to free thiol groups and enhancing anchorage to ECM. ECM-coated stem cells self-assembled into spheroids in confined microenvironments, demonstrating improved survival rates. This nanometric ECM coating shows promise for improving cell-based therapies by enhancing the retention and functionality of stem cells (Fig. 8.5c) (Koh et al. 2020).

Instant cell aggregation using lipid-conjugated heparin is achieved by adding a small concentration to a suspension of human adipose-derived stem cells. The cell coating material, composed of natural substances, enhances the bioactivity of the aggregate. Shaking of surface-engineered cells promotes cell–cell interactions and aggregation. This system allows precise control over dimensional properties of cell aggregates by adjusting the template shape, enabling the rapid formation of basic tissue structures, such as sheets and tubes. Furthermore, the co-culture of different cell types is facilitated by the sequential addition of cells with lipid-conjugated heparin. This versatile strategy can be applied to all cell types, with aggregates remaining alive and maintaining biological activity (Kim et al. 2018). Overall, cell surface engineering via nanoencapsulation effectively protects cells from harsh environments and enhances their functionalities. Conformal encapsulation directs the spatial arrangement of cells through material design, promoting secondary interactions that enhance multi-cellular organization and function.

The precise and controlled assembly of different cell types to mimic complex tissue function and structures is crucial for tissue engineering. Surface engineering techniques such as photo-crosslinking, cell nanoencapsulation, and bioorthogonal methods demonstrate that cell surface engineering could enhances artificial cell-cell interactions. It enables rapid aggregate, creating heterospheroids and layered aggregates (Ma et al. 2022). Although most work serves as proof-of-concept and is limited to microscale aggregates, further research is needed for clinical and complex tissue modeling. Multiple challenges remain in improving the controllable assembly of different cell populations that recapitulate the spatial organization and functional complexity of native tissues. Future research should focus on improving the scaffold-free construct strength (DuRaine et al. 2015), fusing minispheroids, heterotypic aggregates, or microtissues building blocks that can be assembled across multiple length scales in modular principle for complex tissue modelling (Kosheleva et al. 2020; Guven et al. 2015). It is crucial to understand the roles of various cell types within different architectural contexts and their dynamic crosstalk to optimize the development of transplantable construct (Stepanova et al. 2024).

8.4 Cell Engineering for Scaffolds Cellularization and Modular Fabrication

Tissue development relies on cell-cell and cell-matrix interactions that coordinate biochemical and biomechanical cues. Scaffolds provide structural support for therapeutic cells, preventing unintended migration, ensuring localization at target sites to enhance their effectiveness for tissue regeneration. However, most traditional scaffolding strategies face challenges such as poor cell adhesion, low cell-to-biomaterial ratios, and uneven distribution owing to the lack of bio-signals to control interactions. ECM surface modifications and incorporation of adhesion ligands can regulate cell function (Du et al. 2019). These issues limit their clinical effectiveness, highlighting the need for advanced methods to create constructs with native-like cell densities. The biomimicry gap now focuses on increasing cell-to-biomaterial ratios in engineered constructs for engineering high-cell-density tissues (Almeida-Pinto et al. 2024; Daly et al. 2021). This section examines how cell surface engineering and biomaterials enhance scaffold cellularization and modular scaffold fabrication.

8.4.1 Enhancement of Cell Adhesion

Cell-biomaterial interaction refers to the process by which adherent cells attach to a biomaterial surface. Effective cell-scaffold adhesion is driven by integrins and other cell adhesion molecules. These interactions support crucial cellular processes such as adhesion, proliferation, differentiation, and integration, which in turn drive tissue regeneration (Xing et al. 2019). Optimizing cell-scaffold interactions is crucial and many traditional scaffolds often have non-adhesive surfaces, limiting their applications in tissue engineering. Despite advances in scaffold engineering with various ligands, modulating cell-scaffold interactions remains challenging, highlighting the need for innovative strategies to enhance these interactions (Huang et al. 2017b; Cai et al. 2020).

Cationic cell-penetrating peptides (CPP) conjugated to PEG-lipid derivatives have been used for cell membrane modification. Incorporating PEG-lipid derivatives into cell membranes effectively immobilized the peptides on the cell surface. Modified cells demonstrated enhanced attachment to polystyrene scaffolds, highlighting the potential of CPP-PEG-lipid modifications in fabricating cell-based structures. The CPP-PEG-lipid material facilitated firm cell adhesion and spreading on negatively charged substrates due to the cationic CPP localization on the cell membrane. This one-step method efficiently immobilized cells onto substrates and scaffold surfaces (Teramura et al. 2017). Furthermore, bioorthogonal click strategy has been introduced for rapid in situ scaffold cellularization. A DBCO-modified PCL-PEG nanofilm was fabricated via electrospinning to provide a biocompatible environment, while azide-presenting macrophages were generated through metabolic glycoengineering. Electrospun PCL-PEG-DBCO nanofilms captured azide-labeled macrophages using copper-free click chemistry. This bioorthogonal click conjugation process promotes selective cell capture and efficient scaffold cellularization without affecting cell viability, enabling controllable in situ cellularization, with great potential for TERM applications (Fig. 8.6) (Mao et al. 2020).

Bioorthogonal reaction can be used to attach metabolically engineered stem cells to a non-cell-adhesive hydrogel surface. This approach involves metabolic glycoengineering to generate azide-presenting MSCs and the synthesis of DBCO-functionalized gellan to create DBCO-modified hydrogels. During incubation, a

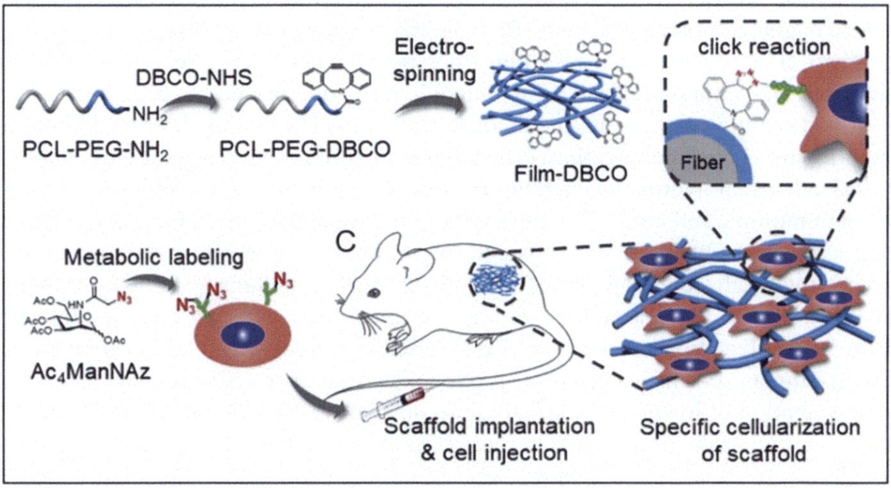

Fig. 8.6 In situ cellularization of azide-labeled macrophages on dibenzocyclooctyne (DBCO)-modified nanofibrous scaffolds. (Reprinted with permission from Mao et al. (2020), copyright 2020, Elsevier B.V.)

click reaction occurred between the azide groups on the cell surface and the cyclooctyne groups on the hydrogel. This method enhanced cell adhesion, spreading, and metabolic activity compared to un-modified or RGD peptide-decorated hydrogels (Battigelli et al. 2020). Furthermore, metabolic azide-labelled neural progenitor cells conjugated to DBCO-conjugated collagen fibers via bioorthogonal conjugation have been used to enhance cell adhesion, spreading, and differentiation. Furthermore, DBCO-conjugated lipid nanoparticles loaded with ROS scavenging drug, targeted azide-presented spinal cord tissues or azide-presented astrocytes through click conjugation, thereby improving the injury microenvironment. In a rat model, these combined strategies significantly promoted neural regeneration, emphasizing the potential of bioorthogonal cell-biomaterial interactions in tissue reconstruction (Liu et al. 2023).

8.4.2 Modular Scaffold Fabrication

Traditional cell-laden scaffolds use functional bioinks with high biomaterial-to-cell ratios, which are effective for non-cell-dens tissues such as bone. However, suboptimal cell densities limit the modelling and engineering of high-cell-density tissues such as cardiac muscle, brain, liver, and cartilage. Achieving high cell densities in scaffolds is challenging because of compromised mechanical properties and disrupted crosslinking, requiring innovative methods to organize cells into functional structures that mimic native tissues (Tiruvannamalai-Annamalai et al. 2014; Chae et al. 2021). Modular cell-dense scaffold fabrication is gaining significant attention for TERM. This strategy assembles cell-laden modules from using bottom up tissue engineering to create high cell density or tissue-like structures with soft mechanical properties, allowing precise control over the cell distribution, and material properties to engineer biological tissues (Ip et al. 2018; Kang et al. 2024).

Nam et al. developed a breathing therapeutic matrix by assembling DBCO-modified DNA microscaffolds with metabolically azide-presenting cells, that function as structural components and therapeutic agents. The hydrogel, formed through bioorthogonal click conjugation, offers a greater surface area compared to conventional cell-laden hydrogels, enabling controlled cell densities and efficient entrapment of multiple cell types. This ultra-soft, porous matrix is injectable, surgery-free, and minimizes fibrosis. It supports cell migration, angiogenesis, and tissue repair, promoting localized therapy with fewer injections. The modular scaffold overcomes challenges such as poor cellularization and integration, offering the potential for better transplantation outcomes and cell-based constructs (Fig. 8.7a) (Nam et al. 2023). Human cells have been used as dynamic crosslinkers and structural elements in the creation of living scaffolds. Hyaluronic acid was functionalized with DBCO moieties, and azide groups were introduced into cell-surface glycoproteins via metabolic glycoengineering, enabling rapid bioorthogonal click conjugation to form cell-dense hydrogels. These cell-dense hydrogels could replicate the cell-rich nature of biological tissues while maintaining hydrogel processability. Their self-assembly from liquid precursors enables micromolding into customizable geometries,

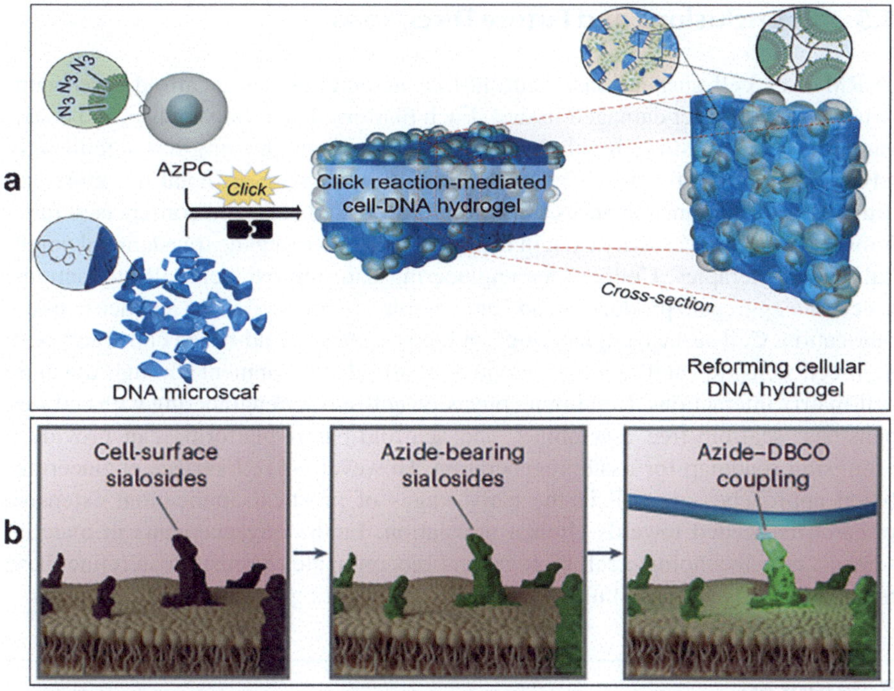

Fig. 8.7 (**a**) Click reaction-mediated binding of azide-presenting cells (AzPC) to dibenzocyclooctyne (DBCO)-modified DNA hydrogels enables precise control of cell density and efficient entrapment of multiple cell types. (Reprinted with permission from Nam et al. (2023), copyright 2023, Elsevier B.V.). (**b**) Bioorthogonal azide groups on cell-surface glycoproteins are coupled with DBCO-conjugated hyaluronic acid, allowing cells to be used as building blocks for programmable tissue-dense scaffold fabrication. (Reprinted with permission from Lavrador et al. (2025), copyright 2025, Springer Nature)

creating functional centimeter-sized constructs with relevant cell densities (Fig. 8.7b) (Lavrador et al. 2025).

Overall, surface-engineered cells, in combination with biomaterials, can improve cell-scaffold interactions and enhancing the effectiveness of cell-seeded or cell-laden scaffold. This approach enables modular scaffold fabrication with high cell densities, and minimizes biomaterial use, offering great potential for developing cell-dense macroscale constructs through assembly processes. Modular tissue engineering using human cells as material precursors and creation of macroscale engineered tissues from modular microscale units could offer a promising solution to tissue shortage (Yu et al. 2024; Ouyang et al. 2020). Further significant research is needed to precisely position mammalian cells and creations of modular microtissues with specific microarchitectural features that mimic tissue complexities. Computational simulations can aid in designing physiomimetic constructs for complex tissue reconstruction and organoid engineering (Wang et al. 2024c).

8.5 Conclusions and Future Directions

TERM uses cell suspensions, scaffold-free assemblies, and scaffold-based constructs to reconstruct damaged tissues. Each platform has its advantages and disadvantages, and this study highlights that cell surface engineering has significantly advanced these approaches. Cell surface engineering with artificial receptors and drug delivery systems enhances the homing of infused cells and confers new functions that do not necessarily exist in nature, thereby boosting the translational potential of cell therapies. Cell surface engineering can improve cell-cell interactions, accelerate aggregate production, and aid complex tissue modelling and dense tissue fabrication. Cell surface engineering can help create scaffold-based constructs with high cell densities and facilitate modular scaffold development through dynamic cell-matrix interactions. Combining precision cell surface engineering with cell suspensions, scaffold-free assemblies, and scaffold-based platforms can provide a promising roadmap for tissue regeneration. However, as cell surface engineering-based approaches are still in the early stages of practical application, extensive research is needed towards clinical translation. Future advancements in material science, nanotechnology, cell biology, and bioconjugate chemistry can refine these methods for advancing cellular therapies and fabricating tissue-mimicking grafts.

References

Abbina S, Siren EM, Moon H, Kizhakkedathu JN (2017) Surface engineering for cell-based therapies: techniques for manipulating mammalian cell surfaces. ACS Biomater Sci Eng 4(11):3658–3677

Almeida-Pinto J, Moura BS, Gaspar VM, Mano JF (2024) Advances in cell-rich inks for biofabricating living architectures. Adv Mater 36:2313776

Amaral AJ, Pasparakis G (2019) Cell membrane engineering with synthetic materials: applications in cell spheroids, cellular glues and microtissue formation. Acta Biomater 90:21–36

Amer MH, Rose FR, Shakesheff KM, Modo M, White LJ (2017) Translational considerations in injectable cell-based therapeutics for neurological applications: concepts, progress and challenges. NPJ Regen Med 2(1):23

Badekila AK, Kini S, Jaiswal AK (2021) Fabrication techniques of biomimetic scaffolds in three-dimensional cell culture: a review. J Cell Physiol 236(2):741–762

Barbash IM, Chouraqui P, Baron J, Feinberg MS, Etzion S, Tessone A, Miller L, Guetta E, Zipori D, Kedes LH, Kloner RA (2003) Systemic delivery of bone marrow-derived mesenchymal stem cells to the infarcted myocardium: feasibility, cell migration, and body distribution. Circulation 108(7):863–868

Battigelli A, Almeida B, Shukla S, Rocha AD, Shukla A (2020) Inducing mesenchymal stem cell attachment on non-cell adhesive hydrogels through click chemistry. Chem Comm 56(55):7661–7664

Bausch-Fluck D, Milani ES, Wollscheid B (2019) Surfaceome nanoscale organization and extracellular interaction networks. Curr Opin Chem Biol 48:26–33

Berthiaume F, Maguire TJ, Yarmush ML (2011) Tissue engineering and regenerative medicine: history, progress, and challenges. Annu Rev Chem Biomol Eng 2(1):403–430

Bui L, Edwards S, Hall E, Alderfer L, Round K, Owen M, Sainaghi P, Zhang S, Nallathamby PD, Haneline LS, Hanjaya-Putra D (2022) Engineering bioactive nanoparticles to rejuvenate vascular progenitor cells. Commun Biol 5(1):635

Buzhor E, Leshansky L, Blumenthal J, Barash H, Warshawsky D, Mazor Y, Shtrichman R (2014) Cell-based therapy approaches: the hope for incurable diseases. Regen Med 9(5):649–672
Cai S, Wu C, Yang W, Liang W, Yu H, Liu L (2020) Recent advance in surface modification for regulating cell adhesion and behaviors. Nanotechnol Rev 9(1):971–989
Chae S, Hong J, Hwangbo H, Kim G (2021) The utility of biomedical scaffolds laden with spheroids in various tissue engineering applications. Theranostics 11(14):6818
Chae S, Lee H, Ryu D, Kim G (2024) Macroscale pseudo-spheroids fabricated using methacrylated collagen-coated cells. Theranostics 14(3):924
Chao CJ, Zhang E, Zhao Z (2023) Engineering cells for precision drug delivery: new advances, clinical translation, and emerging strategies. Adv Drug Deliv Rev 197:114840
Chen Z, Hu Q, Gu Z (2018) Leveraging engineering of cells for drug delivery. Acc Chem Res 51(3):668–677
Cooper SM, Rainbow RS (2022) The developing field of scaffold-free tissue engineering for articular cartilage repair. Tissue Eng Part B Rev 28(5):995–1006
Daly AC, Davidson MD, Burdick JA (2021) 3D bioprinting of high cell-density heterogeneous tissue models through spheroid fusion within self-healing hydrogels. Nat Commun 12(1):753
Dave K, Gomes VG (2019) Interactions at scaffold interfaces: effect of surface chemistry, structural attributes and bioaffinity. Mater Sci Eng C 105:110078
De Pieri A, Rochev Y, Zeugolis DI (2021) Scaffold-free cell-based tissue engineering therapies: advances, shortfalls and forecast. NPJ Regen Med 6(1):18
Decarli MC, Amaral R, Dos Santos DP, Tofani LB, Katayama E, Rezende RA, da Silva JVL, Swiech K, Suazo CAT, Mota C, Moroni L (2021) Cell spheroids as a versatile research platform: formation mechanisms, high throughput production, characterization and applications. Biofabrication 13(3):032002
Demirbag B, Huri PY, Kose GT, Buyuksungur A, Hasirci V (2011) Advanced cell therapies with and without scaffolds. Biotechnol J 6(12):1437–1453
Di Meo F, Kale B, Koomen JM, Perna F (2024) Mapping the cancer surface proteome in search of target antigens for immunotherapy. Mol Ther 32(9):2892–2904
Dou Y, Li C, Li L, Guo J, Zhang J (2020) Bioresponsive drug delivery systems for the treatment of inflammatory diseases. J Control Release 327:641–666
Du Y, Guo JL, Wang J, Mikos AG, Zhang S (2019) Hierarchically designed bone scaffolds: from internal cues to external stimuli. Biomaterials 218:119334
DuRaine GD, Brown WE, Hu JC, Athanasiou KA (2015) Emergence of scaffold-free approaches for tissue engineering musculoskeletal cartilages. Ann Biomed Eng 43:543–554
Dutta D, Pulsipher A, Luo W, Yousaf MN (2011) Synthetic chemoselective rewiring of cell surfaces: generation of three-dimensional tissue structures. J Am Chem Soc 133(22):8704–8713
Eltom A, Zhong G, Muhammad A (2019) Scaffold techniques and designs in tissue engineering functions and purposes: a review. Adv Mater Sci Eng 2019(1):3429527
Fukuda Y, Akagi T, Asaoka T, Eguchi H, Sasaki K, Iwagami Y, Yamada D, Noda T, Kawamoto K, Gotoh K, Kobayashi S (2018) Layer-by-layer cell coating technique using extracellular matrix facilitates rapid fabrication and function of pancreatic β-cell spheroids. Biomaterials 160:82–91
Gong P, Zheng W, Huang Z, Zhang W, Xiao D, Jiang X (2013) A strategy for the construction of controlled, three-dimensional, multilayered, tissue-like structures. Adv Funct Mater 23(1):42–46
Griffin KH, Fok SW, Kent Leach J (2022) Strategies to capitalize on cell spheroid therapeutic potential for tissue repair and disease modeling. NPJ Regen Med 7(1):70
Guven S, Chen P, Inci F, Tasoglu S, Erkmen B, Demirci U (2015) Multiscale assembly for tissue engineering and regenerative medicine. Trends Biotechnol 33(5):269–279
He L, Chen X (2020) Cardiomyocyte induction and regeneration for myocardial infarction treatment: cell sources and administration strategies. Adv Healthc Mater 9(22):2001175
Hoang DM, Pham PT, Bach TQ, Ngo AT, Nguyen QT, Phan TT, Nguyen GH, Le PT, Hoang VT, Forsyth NR, Heke M (2022) Stem cell-based therapy for human diseases. Signal Transduct Target Ther 7(1):1–41

Huang B, Jiang XC, Zhang TY, Hu YL, Tabata Y, Chen Z, Pluchino S, Gao JQ (2017a) Peptide modified mesenchymal stem cells as targeting delivery system transfected with miR-133b for the treatment of cerebral ischemia. Int J Pharm 531(1):90–100

Huang G, Li F, Zhao X, Ma Y, Li Y, Lin M, Jin G, Lu TJ, Genin GM, Xu F (2017b) Functional and biomimetic materials for engineering of the three-dimensional cell microenvironment. Chem Rev 117(20):12764–12850

Hussain Z, Pei R (2021) Scaffold-free and scaffold-based cellular strategies and opportunities for cornea tissue engineering. Prog Biomed Eng 3(032003):3

Hussain Z, Mehmood S, Liu X, Liu Y, Wang G, Pei R (2023) Decoding bone-inspired and cell-instructive cues of scaffolds for bone tissue engineering. Eng Regen 5(1):21–44

Ip BC, Cui F, Wilks BT, Murphy J III, Tripathi A, Morgan JR (2018) Perfused organ cell-dense macrotissues assembled from prefabricated living microtissue. Adv Biosyst 2(8):1800076

Jasiewicz NE, Mei KC, Oh HM, Chansoria P, Hendy DA, Bonacquisti EE, Bachelder EM, Ainslie KM, Yin H, Qian L, Jensen BC (2023) ZipperCells exhibit enhanced accumulation and retention at the site of myocardial infarction. Adv Healthc Mater 12(4):2201094

Jensen C, Teng Y (2020) Is it time to start transitioning from 2D to 3D cell culture? Front Mol Biosci 7:33

Jeong JH, Schmidt JJ, Kohman RE, Zill AT, DeVolder RJ, Smith CE, Lai MH, Shkumatov A, Jensen TW, Schook LG, Zimmerman SC (2013) Leukocyte-mimicking stem cell delivery via in situ coating of cells with a bioactive hyperbranched polyglycerol. J Am Chem Soc 135(24):8770–8773

Jin W, Lin X, Pan H, Zhao C, Qiu P, Zhao R, Hu Z, Zhou Y, Wu H, Chen X, Ouyang H (2021) Engineered osteoclasts as living treatment materials for heterotopic ossification therapy. Nat Commun 12(1):6327

Kang D, Hong S, Kim SJ, Choi H, Kim K, Jang J (2024) Robotics-assisted modular assembly of bioactive soft materials for enhanced organ fabrication. Virtual Phys Prototyp 19(1):e2390484

Kean TJ, Duesler L, Young RG, Dadabayev A, Olenyik A, Penn M, Wagner J, Fink DJ, Caplan AI, Dennis JE (2012) Development of a peptide-targeted, myocardial ischemia-homing, mesenchymal stem cell. J Drug Target 20(1):23–32

Kim E, Kim JC, Min K, Goh M, Tae G (2018) Rapid and versatile cell aggregate formation using lipid-conjugated heparin. ACS Appl Mater Interfaces 10(29):24431–24439

Kim SJ, Kim EM, Yamamoto M, Park H, Shin H (2020) Engineering multi-cellular spheroids for tissue engineering and regenerative medicine. Adv Healthc Mater 9(23):2000608

Kobayashi J, Kikuchi A, Aoyagi T, Okano T (2019) Cell sheet tissue engineering: cell sheet preparation, harvesting/manipulation, and transplantation. J Biomed Mater Res A 107(5):955–967

Koh I, Yong I, Kim B, Choi D, Hong J, Han YM, Kim P (2020) Tris (2-carboxyethyl) phosphine-mediated nanometric extracellular matrix-coating method of mesenchymal stem cells. ACS Biomater Sci Eng 6(2):813–821

Kosheleva NV, Efremov YM, Shavkuta BS, Zurina IM, Zhang D, Zhang Y, Minaev NV, Gorkun AA, Wei S, Shpichka AI, Saburina IN (2020) Cell spheroid fusion: beyond liquid drops model. Sci Rep 10(1):12614

Laschke MW, Menger MD (2017) Life is 3D: boosting spheroid function for tissue engineering. Trends Biotechnol 35(2):133–144

Lavrador P, Moura BS, Almeida-Pinto J, Gaspar VM, Mano JF (2025) Engineered nascent living human tissues with unit programmability. Nat Mater 24(1):143–154

Lee DY, Cha BH, Jung M, Kim AS, Bull DA, Won YW (2018) Cell surface engineering and application in cell delivery to heart diseases. J Biol Eng 12(1):1–11

Li Z, Shen D, Hu S, Su T, Huang K, Liu F, Hou L, Cheng K (2018) Pretargeting and bioorthogonal click chemistry-mediated endogenous stem cell homing for heart repair. ACS Nano 12(12):12193–12200

Li Z, Hu S, Cheng K (2019) Chemical engineering of cell therapy for heart diseases. Acc Chem Res 52(6):1687–1696

Li Q, Lian Y, Deng Y, Chen J, Wu T, Lai X, Zheng B, Qiu C, Peng Y, Li W, Xiang AP (2021a) mRNA-engineered mesenchymal stromal cells expressing CXCR2 enhances cell migration and improves recovery in IBD. Mol Ther Nucleic Acids 26:222–236

Li Y, Teng X, Yang C, Wang Y, Wang L, Dai Y, Sun H, Li J (2021b) Ultrasound controlled anti-inflammatory polarization of platelet decorated microglia for targeted ischemic stroke therapy. Angew Chem Int Ed 60(10):5083–5090

Li J, Ding Y, Cheng Q, Gao C, Wei J, Wang Z, Huang Q, Wang R (2022) Supramolecular erythrocytes-hitchhiking drug delivery system for specific therapy of acute pneumonia. J Control Release 350:777–786

Li M, Yu B, Wang S, Zhou F, Cui J, Su J (2023) Microenvironment-responsive nanocarriers for targeted bone disease therapy. Nano Today 50:101838

Liu T, Wang Y, Zhong W, Li B, Mequanint K, Luo G, Xing M (2019) Biomedical applications of layer-by-layer self-assembly for cell encapsulation: current status and future perspectives. Adv Healthc Mater 8(1):1800939

Liu W, Xu B, Zhao S, Han S, Quan R, Liu W, Ji C, Chen B, Xiao Z, Yin M, Yin Y (2023) Spinal cord tissue engineering via covalent interaction between biomaterials and cells. Sci Adv 9(6):eade8829

Luo W, Pulsipher A, Dutta D, Lamb BM, Yousaf MN (2014) Remote control of tissue interactions via engineered photo-switchable cell surfaces. Sci Rep 4(1):6313

Ma JY, Wang YX, Huang Y, Zhang Y, Cui YX, Kong DM (2022) Chemical–biological approaches for the direct regulation of cell–cell aggregation. Aggregate 3(2):e166

Mager MD, LaPointe V, Stevens MM (2011) Exploring and exploiting chemistry at the cell surface. Nat Chem 3(8):582–589

Mao D, Zhang C, Liu J, Wang X, Li B, Yan H, Hu F, Kong D, Wang Z, Liu B (2020) Bio-orthogonal click reaction-enabled highly specific in situ cellularization of tissue engineering scaffolds. Biomaterials 230:119615

Meyfour A, Pahlavan S, Mirzaei M, Krijgsveld J, Baharvand H, Salekdeh GH (2021) The quest of cell surface markers for stem cell therapy. Cell Mol Life Sci 78:469–495

Naderi H, Matin MM, Bahrami AR (2011) Critical issues in tissue engineering: biomaterials, cell sources, angiogenesis, and drug delivery systems. J Biomater Appl 26(4):383–417

Nam H, Jeon H, Kim H, Yoon HY, Kim SH, Lee JB (2023) Module-assembly of injectable cellular DNA hydrogel via clickable cells and DNA scaffolds. Chem Eng J 452:139492

Nishiguchi A, Yoshida H, Matsusaki M, Akashi M (2011) Rapid construction of three-dimensional multilayered tissues with endothelial tube networks by the cell-accumulation technique. Adv Mater 23(31):3506–3510

O'Brien PJ, Luo W, Rogozhnikov D, Chen J, Yousaf MN (2015) Spheroid and tissue assembly via click chemistry in microfluidic flow. Bioconjug Chem 26(9):1939–1949

Ong CS, Zhou X, Han J, Huang CY, Nashed A, Khatri S, Mattson G, Fukunishi T, Zhang H, Hibino N (2018) In vivo therapeutic applications of cell spheroids. Biotechnol Adv 36(2):494–505

Otaka A, Hirota T, Iwasaki Y, "Direct fabrication of glycoengineered cells via photoresponsive thiol–ene reaction.," ACS Biomater Sci Eng, 2024

Ouyang L, Armstrong JP, Salmeron-Sanchez M, Stevens MM (2020) Assembling living building blocks to engineer complex tissues. Adv Funct Mater 30(26):1909009

Ovsianikov A, Khademhosseini A, Mironov V (2018) The synergy of scaffold-based and scaffold-free tissue engineering strategies. Trends Biotechnol 36(4):348–357

Park J, Andrade B, Seo Y, Kim MJ, Zimmerman SC, Kong H (2018) Engineering the surface of therapeutic "living" cells. Chem Rev 118(4):1664–1690

Petrenko Y, Syková E, Kubinová Š (2017) The therapeutic potential of three-dimensional multipotent mesenchymal stromal cell spheroids. Stem Cell Res Ther 8(1):1–9

Rogozhnikov D, O'Brien PJ, Elahipanah S, Yousaf MN (2016) Scaffold free bio-orthogonal assembly of 3-dimensional cardiac tissue via cell surface engineering. Sci Rep 6(1):39806

Salgado AJ, Oliveira JM, Martins A, Teixeira FG, Silva NA, Neves NM, Sousa N, Reis RL (2013) Tissue engineering and regenerative medicine: past, present, and future. Int Rev Neurobiol 108:1–33

Sato F, Alejandra HPL, Takemae H, Inagaki NF, Ito T, Tera M (2024) Enhancing cell aggregation and migration via double-click cross-linking with azide-modified hyaluronic acid. Bioconjug Chem 35(9):1318–1323
Shao C, Chi J, Zhang H, Fan Q, Zhao Y, Ye F (2020) Development of cell spheroids by advanced technologies. Adv Mater Technol 5(9):2000183
Shaw TR, Ghosh S, Veatch SL (2021) Critical phenomena in plasma membrane organization and function. Annu Rev Phys Chem 72(1):51–72
Stepanova D, Byrne HM, Maini PK, Alarcón T (2024) Computational modeling of angiogenesis: the importance of cell rearrangements during vascular growth. WIREs Mech Dis 16(2):e1634
Su J, Song Y, Zhu Z, Huang X, Fan J, Qiao J, Mao F (2024) Cell–cell communication: new insights and clinical implications. Signal Transduct Target Ther 9(1):196
Sun Y, Su J, Liu G, Chen J, Zhang X, Zhang R, Jiang M, Qiu M (2017) Advances of blood cell-based drug delivery systems. Eur J Pharm Sci 96:115–128
Tang J, Su T, Huang K, Dinh PU, Wang Z, Vandergriff A, Hensley MT, Cores J, Allen T, Li T, Sproul E (2018) Targeted repair of heart injury by stem cells fused with platelet nanovesicles. Nat Biomed Eng 2(1):17–26
Teramura Y, Asif S, Ekdahl KN, Gustafson E, Nilsson B (2017) Cell adhesion induced using surface modification with cell-penetrating peptide-conjugated poly (ethylene glycol)-lipid: a new cell glue for 3D cell-based structures. ACS Appl Mater Interfaces 9(1):244–254
Timin AS, Litvak MM, Gorin DA, Atochina-Vasserman EN, Atochin DN, Sukhorukov GB (2018) Cell-based drug delivery and use of nano-and microcarriers for cell functionalization. Adv Healthc Mater 7(3):1700818
Tiruvannamalai-Annamalai R, Armant DR, Matthew HW (2014) A glycosaminoglycan based, modular tissue scaffold system for rapid assembly of perfusable, high cell density, engineered tissues. PLoS One 9(1):e84287
Ullah M, Liu DD, Thakor AS (2019) Mesenchymal stromal cell homing: mechanisms and strategies for improvement. iScience 15:421–438
Villa CH, Anselmo AC, Mitragotri S, Muzykantov V (2016) Red blood cells: supercarriers for drugs, biologicals, and nanoparticles and inspiration for advanced delivery systems. Adv Drug Deliv Rev 106:88–103
Wang C, Ye Y, Sun W, Yu J, Wang J, Lawrence DS, Buse JB, Gu Z (2017) Red blood cells for glucose-responsive insulin delivery. Adv Mater 29(18):1606617
Wang J, Miao Y, Huang Y, Lin B, Liu X, Xiao S, Du L, Hu Z, Xing M (2018) Bottom-up nanoencapsulation from single cells to tunable and scalable cellular spheroids for hair follicle regeneration. Adv Healthc Mater 7(3):1700447
Wang LLW, Janes ME, Kumbhojkar N, Kapate N, Clegg JR, Prakash S, Heavey MK, Zhao Z, Anselmo AC, Mitragotri S (2021) Cell therapies in the clinic. Bioeng Transl Med 6(2):e10214
Wang LLW, Gao Y, Feng Z, Mooney DJ, Mitragotri S (2024a) Designing drug delivery systems for cell therapy. Nat Rev Bioeng 2:944–959
Wang Y, Shi J, Xin M, Kahkoska AR, Wang J, Gu Z (2024b) Cell–drug conjugates. Nat Biomed Eng 8(11):1347–1365
Wang H, Li X, You X, Zhao G (2024c) Harnessing the power of artificial intelligence for human living organoid research. Bioact Mater 42:140–164
Won YW, Patel AN, Bull DA (2014) Cell surface engineering to enhance mesenchymal stem cell migration toward an SDF-1 gradient. Biomaterials 35(21):5627–5635
Wu RX, Yin Y, He XT, Li X, Chen FM (2017) Engineering a cell home for stem cell homing and accommodation. Adv Biosyst 1(4):1700004
Xia H, Li X, Gao W, Fu X, Fang RH, Zhang L, Zhang K (2018) Tissue repair and regeneration with endogenous stem cells. Nat Rev Mater 3(7):174–193
Xing F, Li L, Zhou C, Long C, Wu L, Lei H, Kong Q, Fan Y, Xiang Z, Zhang X (2019) Regulation and directing stem cell fate by tissue engineering functional microenvironments: scaffold physical and chemical cues. Stem Cells Int 2019(1):2180925
Xu Y, Zhou A, Chen W, Ning X (2023) Scaffold-free multicellular 3D tissue constructs utilizing bio-orthogonal click strategy. Nano Lett 23(18):8770–8778

Yan H, Mi X, Midgley AC, Du X, Huang Z, Wei T, Liu R, Ma T, Zhi D, Zhu D, Wang T (2020) Targeted repair of vascular injury by adipose-derived stem cells modified with P-selectin binding peptide. Adv Sci 7(11):1903516

Ye T, Liu X, Zhong X, Yan R, Shi P (2023) Nongenetic surface engineering of mesenchymal stromal cells with polyvalent antibodies to enhance targeting efficiency. Nat Commun 14(1):5806

Yu X, Ma H, Wang Y, Hao J, Chen L, Gelinsky M, Wu C (2024) Assembled/disassembled modular scaffolds for multicellular tissue engineering. Adv Mater 36(21):2308126

Zelinka A, Roelofs AJ, Kandel RA, De Bari C (2022) Cellular therapy and tissue engineering for cartilage repair. Osteoarthr Cartil 30(12):1547–1560

Zhou H, Zhang Y, Pei P, Shen W, Yi X, Yang K (2023) Liposome-anchored mesenchymal stem cells for radiation pneumonia/fibrosis treatment. Biomaterials 300:122202

Harnessing Biomaterial-Mediated Single-Cell Nanoencapsulation for Enhanced Cellular Therapies

9

Zahid Hussain, Hewan Dawit, Shah Mehmood, Li Wang, and Renjun Pei

Abstract

Living cells, including red blood cells, insulin-producing cells, and stem cells, offer great potential for treating various diseases but face significant practical challenges. For example, the demand for blood transfusions is challenged by mismatched blood types, immune responses, and donation shortages, driving research toward the development of universal blood solutions. Similarly, transplantation of insulin-producing cells offers an alternative to frequent injections but is limited by low insulin secretion, high cell death, inflammation, and immune activation, leading to graft failure. Stem cells hold promise for various disease treatments but face significant challenges such as low survival rate, undesired differentiation, limited targeting, and lack of specific cell interactions. These challenges in blood transfusion, islet transplantation, and stem cell therapies highlight the need for single-cell or cell aggregate nanoencapsulation strategies. This chapter discusses cell encapsulation methods, summarizing strategies such as electrostatic layer-by-layer assembly, supramolecular engineering, covalent reactions, hydrogel-based coatings, and bioorthogonal glycoengineering. This chapter highlights applications such as protecting insulin-producing cells for diabetes, camouflaging red blood cells for safe transfusion, and resurfacing stem cell surfaces for tissue engineering through single-cell or cell aggregate nanoencapsulation. This chapter further outlines future research directions in single-cell and cell aggregate nanoencapsulation for biomedical applications.

Z. Hussain · H. Dawit · S. Mehmood · L. Wang · R. Pei (✉)
CAS Key Laboratory for Nano-Bio Interface, Suzhou Institute of Nano-Tech and Nano-Bionics (SINANO), Chinese Academy of Sciences (CAS), Suzhou, China
e-mail: zahid@mail.ustc.edu.cn; hewan@sinano.ac.cn; shahmehmood@mail.ustc.edu.cn; liwang2020@sinano.ac.cn; rjpei2011@sinano.ac.cn

Z. Hussain, R. Pei (eds.), *Cell Membrane Engineering for Advancing Cell-Based and Nanotechnology-Based Therapies*, Biomaterials, Bioengineering and Sustainability 6, https://doi.org/10.1007/978-3-031-93425-4_9

Keywords

Red blood cells · Insulin-producing cells · Stem cells · Immunological response · Nanoencapsulation · Biomedical applications

9.1 Introduction

Living cells, including red blood cells (RBCs), insulin-producing β-cells, and stem cells, hold significant potential in regenerative medicine. However, the clinical use of mammalian cells faces considerable practical challenges (Kwon et al. 2018; Nair et al. 2020; Zhang et al. 2024). For example, mismatched blood types, immune responses, and donation shortages hinder the demand for blood transfusions. This has spurred research on developing universal blood solutions (Liu et al. 2007). Insulin-producing cell transplantation offers a promising alternative to frequent insulin injections but is constrained by challenges such as low insulin secretion, high cell death, inflammation, and immune activation, which can ultimately lead to graft failure (Paez-Mayorga et al. 2022; Sneddon et al. 2018). Stem cells, although promising for tissue regeneration, face challenges such as low survival rates, undesired differentiation, limited targeting ability, immune rejection, and insufficient cell-to-cell interactions. (Hashemi and Kalalinia 2015; Liu et al. 2018). These challenges in universal blood transfusion, islet transplantation, and stem cell therapies underscore the need for innovative solutions.

The cell surface is a complex structure that governs numerous processes critical to biological functions. Re-engineering cell surfaces with receptors and ligands through genetic engineering can modulate interactions between cells and their environment (Stephan and Irvine 2011). However, mammalian cells are vulnerable to environmental (e.g., heat, ultraviolet (UV) radiation), biochemical (e.g., oxidative stress, inflammatory cytokines), mechanical, and immunological stresses. These stressors arise during cell manipulation, scaffold fabrication, product handling, injection, and the hostile microenvironment of diseased tissue. For example, cell-laden scaffolds are crucial for tissue engineering, but fabrication stresses like UV crosslinking, mechanical shear, toxic crosslinkers, and temperature fluctuations can compromise cell viability in scaffolds (Fig. 9.1a) (Coronel et al. 2017; Petrus-Reurer et al. 2021; Blaeser et al. 2016). These factors reduce treatment effectiveness, leading to growing interest in ensuring that transplanted cells sustain long-term functionality and acquire exogenous therapeutic properties (Farina et al. 2019).

Biomimetics studies natural entities and applies their concepts to advance engineering and medicine. Nature has evolved protective armor to safeguard species in harsh conditions, inspiring scientists to develop artificial shells based on cell wall, sporulation, seed coats, and natural biomineralized coatings (Chong et al. 2021; Wang and Guo 2014). For example, avian eggshells protect developing embryos, allowing eggs to remain viable for several weeks. Diatoms produce silica-rich shells that provide structural support and mechanical protection (Wang et al. 2018, 2024). The silk cocoon coat protects them against physical stress, toxins, and natural enemies. Some bacteria and fungi undergo sporulation, forming robust coatings that

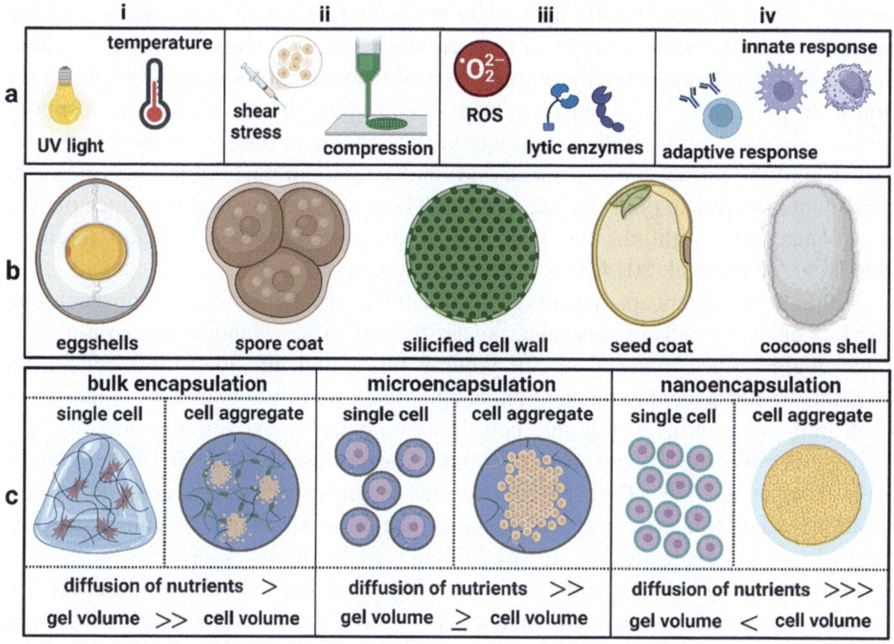

Fig. 9.1 (**a**) Vulnerability of mammalian therapeutic cells to (i) Environmental stresses, (ii) Mechanical stresses, (iii) Biochemical stresses, and (iv) Immunological stresses. (**b**) Eggshells, spore coats, silicified cell walls of diatoms, seed coats, and cocoon shells protect from various stressors and serve as inspiration for developing artificial shells to engineer cell surfaces. (**c**) Schematics of bulk encapsulation, microencapsulation, and nanoencapsulation of single-cell and cellular aggregates, highlighting features of nutrient permeation and matrix dimensions. (Source: Created with BioRender.com)

resist extreme temperatures, toxins, and desiccation. The seed coat protects the embryo by providing structural integrity and defence against pathogens, oxidative damage, and desiccation (Hong et al. 2013; Chen et al. 2012) (Fig. 9.1b). These defence mechanisms have inspired biomimetic coatings to protect living cells for potential biomedical applications.

Cell encapsulation immobilizes non-autologous cells in a semipermeable, immunoisolating membrane, enabling biological product delivery without needing immunosuppressants. The domain of single-cell and cell aggregate encapsulation has boomed, and various biomaterials have been introduced to mitigate the limitations associated with native cells (Orive et al. 2015; Youn et al. 2020). Bulk encapsulation and microencapsulation involve encapsulating cells in larger or micron-scale gel structures. These methods offer benefits like injectable delivery and immune isolation, making them promising for cell transplantation (Wu et al. 2021; Marikar et al. 2022). However, concerns such as batch variability, uncontrolled gel degradation, and poor mass transfer due to large matrix dimensions have limited their potential applications. For example, most cell encapsulation methods lack control over capsule thickness, and macro-dimensions of the coating material can increase the graft volume of encapsulated cells up to 100 times that of naked cells, exceeding the physiological diffusion distance of oxygen and nutrients, limiting clinical applications. The large volume and high proportion of blank capsules may delay the secretion of trophic factors (e.g., insulin) from encapsulated β-cells clusters in response to host glucose stimulation. Additionally, they increase transplant mass, limiting their use to the intraperitoneal space and rendering them unsuitable for portal vein transplantation due to poor vascularization. These shortcomings may reduce patient compliance and highlight the need for single-cell nanoencapsulation methods (Fig. 9.1c) (Park et al. 2014; Lee et al. 2021; Mao et al. 2017).

Single-cell nanoencapsulation is a promising surface engineering approach that involves encasing cells in nanometer-scale shell materials. Nanoencapsulation strategies could reduce material-to-cell volume ratios and aim to protect cells from mechanical stress, enable nutrient exchange, block harmful chemicals, shield them from immune responses, regulate cell division, enhance exogenous functions and degrade in response to appropriate stimuli (Hasturk and Kaplan 2019; Yang et al. 2013; Yang et al. 2023). This chapter explores key considerations in cell nanoencapsulation and biomaterial-assisted nanoencapsulation strategies to enhance transplanted cell performance. It highlights applications such as protecting insulin-producing cells for diabetes, camouflaging red blood cells for safe transfusion, and resurfacing stem cell surfaces for tissue engineering through single-cell nanoencapsulation. This chapter further outlines future research directions in single-cell nanoencapsulation for biomedical applications.

9.2 Essential Considerations in Nanoencapsulation of Cells

Single-cell encapsulation with ultrathin, robust artificial shells enhances cell survival by protecting cells from harsh conditions. Studies agree that single-cell nanoencapsulation strategies must possess following key properties (Fig. 9.2).

- **Scalability and compatibility:** To minimize immune rejection, encapsulation materials must be cell-compatible with uniform coatings. Encapsulated cells should remain stable, non-dividing, and differentiated. Non-immunogenic coatings are essential to prevent inflammation in biomedical use. Scalability is vital for transitioning from research to clinical applications.
- **Immuno-shielding:** Single-cell nanoencapsulation strategies must provide a biocompatible barrier to prevent immune cell interactions, antibody responses, and immune component infiltration. This shielding reduces the likelihood of rejection, minimizes the need for immunosuppressive drugs, and enhances the long-term effectiveness of cell-based therapies in clinical settings (Ryan et al. 2017).
- **Resilience to stressors:** Mechanical, chemical, and thermal stresses arising from injection pressure, scaffold fabrication, transplantation, and disease microenvironments can lead to cell death. Encapsulation shells must be durable enough to protect cells from these stresses, thereby enhancing the performance of transplanted grafts (Wei et al. 2024; Gupta et al. 2023).

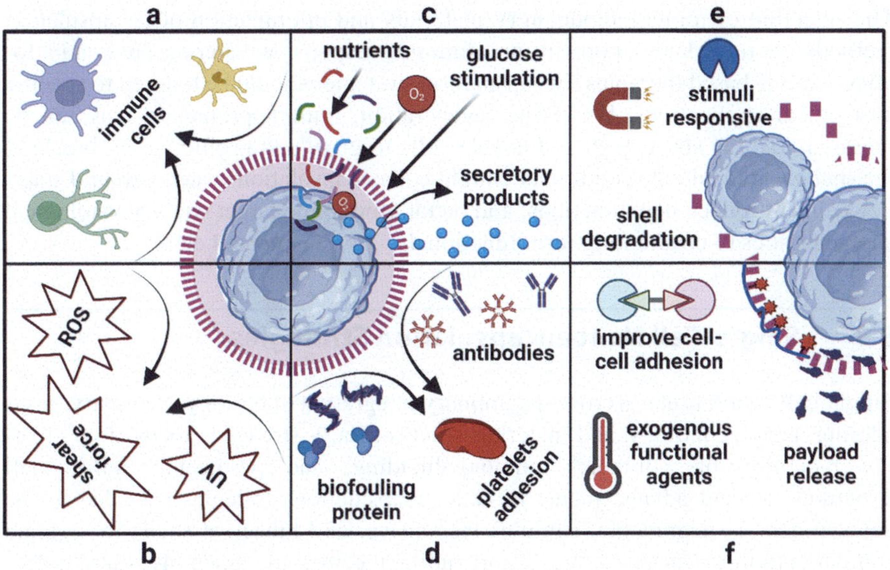

Fig. 9.2 Essential considerations for single-cell encapsulation: (**a**) creating non-recognizable surfaces to evade immune cells response. (**b**) providing chemical/mechanical protection against stressors. (**c**) allowing selective permeation of nutrients and metabolites. (**d**) blocking large macromolecules and biofouling. (**e**) ensuring stimulus-responsive shell degradation, and (**f**) incorporating exogenous capabilities for targeted functions. (Source: Created with BioRender.com)

- **Selective permeability:** The shell should be semi-permeable with optimized porosity and thickness for nutrient and oxygen exchange. It should facilitate the transport of cell-secreted metabolites essential for deriving benefits from transplanted islets. It must block harmful agents like lytic enzymes and cytotoxic molecules to maintain cell viability and enhance protection (Nafea et al. 2011; Ausländer et al. 2012).
- **Stimulus-responsive biodegradability:** The shell should degrade in a controlled manner in response to external signals or inducers to activate cell functions for drug delivery and tissue regeneration (Zhao et al. 2016; Ru et al. 2020). Minimally degradable materials are crucial for universal blood donation and islet transplantation applications. For RBCs donation, the shell must last up to 120 days to protect cells, matching their lifespan. Encapsulated islets require durable materials to maintain long-term insulin release function (Qin et al. 2023; Waeterschoot et al. 2024).
- **Exogenous functionality:** The shell should enable functionalization for cargo delivery, including anti-thrombomodulin, immunosuppressants, anti-inflammatory molecules, growth factors, and other stimulants. These components are essential for preventing coagulation, immune responses, and inflammatory issues, as well as supporting tissue regeneration (Youn et al. 2020; Pires-Santos et al. 2024).

The selection of immunomodulatory materials and optimization of encapsulation methods, shell thickness, porosity, and stimulus-responsive materials are crucial for effective cell-based therapies. Encapsulation strategies should be tailored to specific biomedical applications. For tissue regeneration, immunoprotection may not be required, and the strategy should enable cells to grow and proliferate by breaking the shells under suitable conditions. Single-cell encapsulation systems should stabilize cells, introduce new functions, and be tailored to the target site's physiological characteristics to ensure long-term functionality of transplanted cells.

9.3 Single-Cell Nanoencapsulation Strategies

Single-cell decoration is a cross-disciplinary integration of biology, chemistry, engineering, nanotechnology, and materials science. Early strategies focused on creating biocompatible, durable, immuno-shielding, and permeable cell-in-shell structures. Recent advancements in nanoencapsulation strategies have led to the development of degradable, stimulus-responsive, and bioactive shells to regulate cellular activities. Various strategies, including electrostatic assembly, supramolecular interaction, covalent conjugation, hydrogel-based coating, and bioorthogonal glycoengineering, have been developed for nano-encapsulating cells. The principles of each nanoencapsulation technique are outlined in the following sub-sections.

9.3.1 Electrostatic Layer-by-Layer Assembly

Mammalian cell membranes are negatively charged due to sialic acid residues on glycans and phosphatidylserine head groups in phospholipids. The negative charge on cell membranes attracts positively charged materials, and these electrostatic interactions are widely applied for cell surface bioengineering. Common positively charged materials employed in cell surface modification include poly(ethylenimine), poly-L-lysine (PLL), chitosan, heparin, and charged nanoparticles. Multivalent electrostatic interactions offer numerous binding sites for charged polymers, enabling adsorption under mild conditions with nanoscale precision. The electrostatic layer-by-layer (LbL) assembly technique is widely used to create a multilayered coating on cellular surfaces by assembling positively and negatively charged material layers. Shell thickness and architecture can be controlled by adjusting polyelectrolyte types and deposition steps. LbL nanofilms around cells provide mechanical support, shield against external stress, and protect from immune responses. Electrostatic functionalization has enabled nanoencapsulation of diverse mammalian cells and cell aggregates (such as pancreatic islets), thereby enhancing cellular therapies (Fig. 9.3a, b) (Liu et al. 2019; Oliveira et al. 2016).

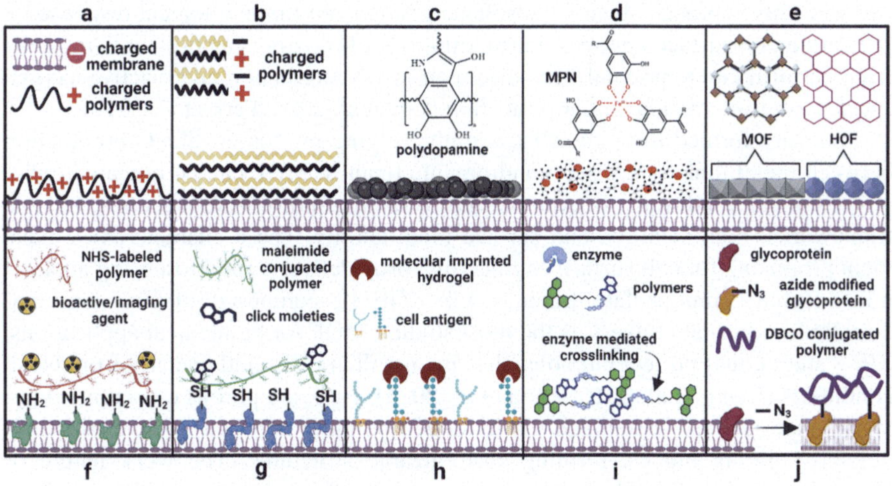

Fig. 9.3 Schematic representation of single-cell nanoencapsulation strategies: (**a**) Electrostatic functionalization, (**b**) Electrostatic layer-by-layer assembly, (**c**) Polydopamine coating, (**d**) Metal-phenolic network (MPN) complexes coating, (**e**) Metal-organic frameworks (MOF) and hydrogen organic frameworks (HOF) self-assembly, (**f**) Amine mediated covalent conjugation, (**g**) Thiol mediated covalent conjugation, (**h**) Molecular imprinting hydrogel coating, (**i**) Enzyme mediated hydrogel coating, (**j**) Bioorthogonal glycoengineering. (Source: Created with BioRender.com)

9.3.2 Supramolecular Engineering

Supramolecular engineering is a multidisciplinary field that involves designing and constructing functional systems using non-covalent interactions. It is a cornerstone of modern nanotechnology, biomaterials science, and biomedicine, as it enables the creation of dynamic and responsive systems with precise molecular-level control. Emerging systems such as polydopamine (pDA), metal-organic frameworks (MOF), hydrogen organic frameworks (HOF), and metal-phenolic networks (MPN) are notable for their ability to self-assemble on substrate surfaces, with applications in biomedicine and materials engineering (Fig. 9.3c–e) (Lin et al. 2024; Dutta et al. 2019; Lee et al. 2019; Lin and Chen 2022).

pDA coating is a rapid and cost-effective method for surface functionalization inspired by mussel adhesive proteins. pDA coatings are created through the oxidative polymerization of dopamine in mildly alkaline conditions, enabling them to coat a wide range of substrates. pDA thickness can be fine-tuned from nanometer to micrometer range by adjusting deposition cycles, dopamine concentration, and metal ions. pDA layers employ catechol-metal coordination, electrostatic interactions, $\pi - \pi$ stacking, and hydrogen bonding, facilitating bioactive functionalization. pDA-modified surfaces can modulate cellular responses and enable cell encapsulation (Alfieri et al. 2022; Yang et al. 2011). Supramolecular MPN shells onto cells surfaces could be formed upon tannic acid and ferric ion interactions. This coating could be performed in a material-independent manner onto substrate surface. This approach shields transplanted cells from immunological response by protecting cell surface epitopes. Furthermore, MPN coating can protect living cells from harmful environmental stressors, such as UV irradiation and reactive oxygen species, offering broad cellular protection (Lin et al. 2024; Park et al. 2017).

MOF are formed through self-assembly, relying on non-covalent coordination between metal ions (or clusters) and organic ligands. MOF can be prepared under physiological conditions. It exhibits micro- or mesoporous architectures, tunable frameworks, high stability, and surface modifiability. These characteristics make them promising for cell surface engineering, drug delivery, and biosensing applications. Zeolitic imidazolate framework-8 (ZIF-8), composed of Zn^{II} ions and 2-methylimidazolate linkers, is the most studied MOF for biomedical applications. ZIF-8 shell could protect biomolecules, mammalian cells, and bacteria from harsh conditions (Liang et al. 2016; Zhou et al. 2024). HOF are porous crystalline polymeric materials self-assembled via hydrogen bonding between organic or metal-organic building blocks, forming customizable structures. The weak nature of hydrogen bonds makes HOF pore structures dependent on building molecule configurations. HOF are solution-processable into membranes, providing a tunable platform for creating functional porous materials. They exhibit excellent water stability due to ligand hydrophobicity and possess sufficient thermal and chemical stability. Biocompatible HOF show great promise for encapsulating living cells and biomolecules, offering significant potential for biomedical applications (Lin and Chen 2022; Yu et al. 2022).

9.3.3 Covalent-Reactive Strategies

The cell surface has a heterogeneous biomolecular landscape with reactive groups, such as primary amines and thiols, that enable chemical modification with synthetic materials. Amine groups ($-NH_2$), found at lysine residues and the N-terminus of polypeptides, are commonly used to modify cell surfaces. N-hydroxysuccinimide (NHS) ester-labeled agents, NHS-activated carboxyl groups, and sulfo-NHS ester agents are used to bind $-NH_2$ groups on cell membranes under mild conditions. NHS ester-labeled reagents, such as NHS-DBCO, can expose bioorthogonal reactive groups on the cell membrane. Primary bioorthogonal groups introduced via covalent conjugation serve as handles for click chemistry, facilitating the attachment of exogenous polymeric materials and therapeutics to the cell membrane (Fig. 9.3f) (Abbina et al. 2017; Rossi et al. 2010).

Thiol groups (–SH) are mainly located on the cysteine residues in proteins. It can be found either as oxidized disulfide bridges or in their reduced free thiol form. Free thiols are potent nucleophiles and can be exposed to plasma membrane proteins, offering a source for chemical modification. Maleimide-containing molecules can engineer cell membranes by forming stable thioether bonds via chemoselective Michael addition, enabling precise modifications in complex microenvironments (Fig. 9.3g). However, thiol groups on the cell surface are sparse and primarily oxidized, limiting membrane engineering. Disulfide bonds can be reduced using dithiothreitol, or Traut's agent can introduce free thiol groups through thiolation. However, these modifications may produce heterogeneous products, disrupt protein active sites, and alter membrane recognition. Thus, engineering strategies should prioritize retaining modified components on the cell surface without impacting other cellular components (Abbina et al. 2017; Chen et al. 2011).

9.3.4 Hydrogel-Based Coating

Hydrogels are crosslinked polymer networks, and their degradation rate depends on factors such as polymer composition, crosslinking approach, and network density. Hydrogels can be synthesized through chemical, enzymatic, heat-induced, and light-induced polymerization. Smart hydrogels, incorporating specific chemical entities, can respond to stimuli like temperature, pH, ultrasound, ionic strength, or magnetic fields, enabling controlled drug release. Hydrogels with interactive binding sites are ideal for cell encapsulation, enhancing cell survival in semipermeable hydrogels. The biophysical characteristics of the hydrogel can be tailored based on polymer composition, molecular weight, crosslinking degree, and polymerization method (Zhang and Khademhosseini 2017; Hussain et al. 2023). Molecular imprinting and enzyme-mediated hydrogel coatings have recently been introduced for cell nanocoating (Fig. 9.3h, i).

Molecular imprinting hydrogels are designed with specific binding sites for target molecules, such as proteins or peptide epitopes, mimicking the high selectivity of antibodies and enzymes in synthetic materials. Monomers are polymerized around a template molecule in a gel matrix, and after polymerization, the molecular template is removed, leaving cavities that match the template's shape, size, and chemical properties. Molecular imprinting functions like the 'lock-and-key' model, allowing the hydrogel to rebind and recognize the target molecule. These strategies combine hydrogel versatility with molecular recognition, playing a key role in drug delivery and selective interactions (Piletsky et al. 2020; Arabi et al. 2021). A notable example of cell-recognizing hydrogel is its use in targeting blood group antigens on RBCs to develop universal blood for transfusion (Jiang et al. 2020).

Enzymatic cross-linking of hydrogels occurs through the catalytic function of exogenous enzymes, leading to covalent interactions between substrates within the bulk material. Enzymes used to initiate gelation include but are not limited to, transglutaminase, tyrosinase, phosphopantetheinyl transferase, lysyl oxidase, and peroxidases (Song et al. 2021). Enzymatically cross-linked hydrogels offer distinct advantages over chemical and physical methods, such as substrate specificity that avoids unwanted side reactions and the ability to operate under physiological conditions. Enzyme-mediated hydrogels enable cell-friendly reactions and are ideal for crosslinking natural polymers sensitive to harsh chemical conditions. Hydrogel nanofilms enable immunocamouflage of RBCs and insulin-producing β-cells, supporting universal blood transfusion and glycemic control in diabetes (Zhao et al. 2020; Kim et al. 2021).

9.3.5 Bioorthogonal Glycoengineering

Glycosylation is a post-translational process where oligosaccharide chains are added to cell-surface proteins. These modifications are essential for cellular adhesion, interactions, immune responses, and cell recognition. These significances have prompted research into metabolic glycoengineering for incorporating artificial agents into glycans for biomedical applications (Russo and Cipolla 2016; Li et al. 2023). Unnatural monosaccharide analogs (mannosamine, galactosamine, glucosamine, sialic acid) bearing a bioorthogonal group (i.e., azide, terminal alkyne) can diffuse into cells, and are metabolized into nucleotide sugars for glycosylation. These modified glycoconjugates are then transported to endosomal and plasma membranes, increasing the expression of unnatural tags on sialoglycoproteins. Metabolic changes occurring in parent cells can be mapped on cell membrane proteins, and tag quantity can be regulated by manipulating unnatural monosaccharide concentration (Critcher et al. 2021; Yoon et al. 2017). Multiple bioorthogonal groups can be attached to glycans through metabolic glycoengineering. Installed bioorthogonal groups such as azide moieties serve as handles for click reactions, such as with dibenzylcyclooctyne (DBCO), enabling the conjugation of functional polymers and therapeutic molecules to the cell surface (Fig. 9.3j) (Yi et al. 2022; Oh et al. 2020).

9.4 Biomedical Potential of Single-Cell Nanoencapsulation

9.4.1 Encapsulated RBCs for Blood Transfusion

RBCs comprise about 45% of blood cells and are crucial for transporting O_2 and CO_2 to tissues. RBCs transfusion provides life-saving treatment for patients suffering from blood loss due to traumatic injuries, surgical haemorrhage, and anaemia-related RBCs depletion. The demand for blood transfusions is increasing, but the immune rejection caused by surface glycoprotein markers exacerbates shortages. Hemolysis and cross-type agglutination from blood group mismatches can cause anemia, systemic inflammation, and organ failure (Stubbs et al. 2016; Branch 2015). For example, ABO blood groups are defined by the presence or absence of terminal oligosaccharides markers on RBCs. O-type RBCs (considered universal donors) display non-immunogenic H-antigens, while A- and B-type red blood cells present α-Gal- or GalNAc-modified epitopes within their H-antigens, respectively. RhD-negative and O-type RBCs supplies face shortages due to unclear patient blood types, severe immune responses from mismatches, and donation challenges during crises (Poole and Daniels 2007; Bjursten et al. 2016; Jajosky et al. 2023).

The challenges in ensuring safe blood transfusion have prompted researchers to develop strategies to mask RhD, A/B antigens, or rare-type antigens. This approach aims to inhibit antigen-antibody interactions, thereby eliminating immune rejection concerns caused by blood type mismatches. Strategies like gene editing to generate antigen-free RBCs and using enzymes to remove surface antigens have been explored to develop universal blood (Olsson and Clausen 2008; Rahfeld and Withers 2020). These treatments offer high specificity and conversion efficiency but face challenges in producing sufficient antigen-free RBCs cost-effectively, limiting their availability for emergency transfusions. Enzymatic removal of RBCs antigens can convert type A/B to type O but does not affect Rh antigens, as their removal disrupts the cell membrane. These shortcomings limit the practical application of these strategies in universal blood development, highlighting the need for chemical-based cell surface engineering to shield RBCs surface antigens (Liu et al. 2007; Petazzi et al. 2022; Hawksworth et al. 2018).

The LbL self-assembly of nonimmunogenic and biocompatible polyelectrolytes inhibits antibody binding to RBCs surface. The multilayered polymeric shell comprised four bilayers of alginate and chitosan-graft-phosphorylcholine, surrounded by two bilayers of alginate and PLL-grafted polyethylene glycol (PEG). The outermost negatively charged alginate coating was deposited as the final protein-repellent barrier. Coated RBCs maintained their viability and functionality, as confirmed through hemolysis assays and oxygen uptake measurements. Immune camouflage was demonstrated by reduced recognition of ABO and RhD-positive antigens by specific antibodies, marking a significant advancement toward the development of universal RBCs (Mansouri et al. 2011). Wang et al. introduced a strategy based on muscle adhesive chemistry to shield antigenic epitopes on RBCs using a polymerized network of dopamine molecules. pDA-coated RBCs (pDA-RBCs) can prevent coagulation reactions and be utilized in blood transfusion. Systemic assessment in a murine model demonstrated that pDA-RBCs exhibit excellent survival profiles, even with repeated transfusions. pDA coating caused minimal detrimental effects on RBCs structure and can produce antigenically shielded universal RBCs (Wang et al. 2014).

Park et al. introduced a "cell-in-shell" approach to mask immune-stimulating epitopes on RBCs surfaces. Each RBCs is coated with a supramolecular MPN of ferric ions (Fe^{III}) and tannic acid (TA). The galloyl groups of tannic acid form multivalent supramolecular coordination with ferric ions, and such interaction can be recognizable due to their black colorization. The universal surface-adhesion affinity of Fe^{III}/TA nanofilm on various substrates, including living cells, could produce immune-camouflaged RBCs. The thick MPN shell on the RBCs surface did not deteriorate the shape and oxygen-carrying potential of RBCs and effectively attenuated the immunoglobulin-triggered agglutination. The supramolecular-coated immune-protected RBCs could advance universal blood development (Fig. 9.4a) (Park et al. 2017). Rossi et al. created a system using nonreactive additive macromolecules to enhance the grafting rate of cell-surface-targeting hydrophilic macromolecules. Primary amine-reactive succinimidyl succinate-functionalized hyperbranched polyglycerols were used as a cell surface-reactive polymer for studying cell surface modification. Succinimidyl succinate groups interact with primary amines on lysine residues on the cell surface protein, forming stable covalent amide bonds. Cell membrane modifications were amplified tenfold by incorporating an appropriate "additive" polymer and a low concentration of a "reactive"

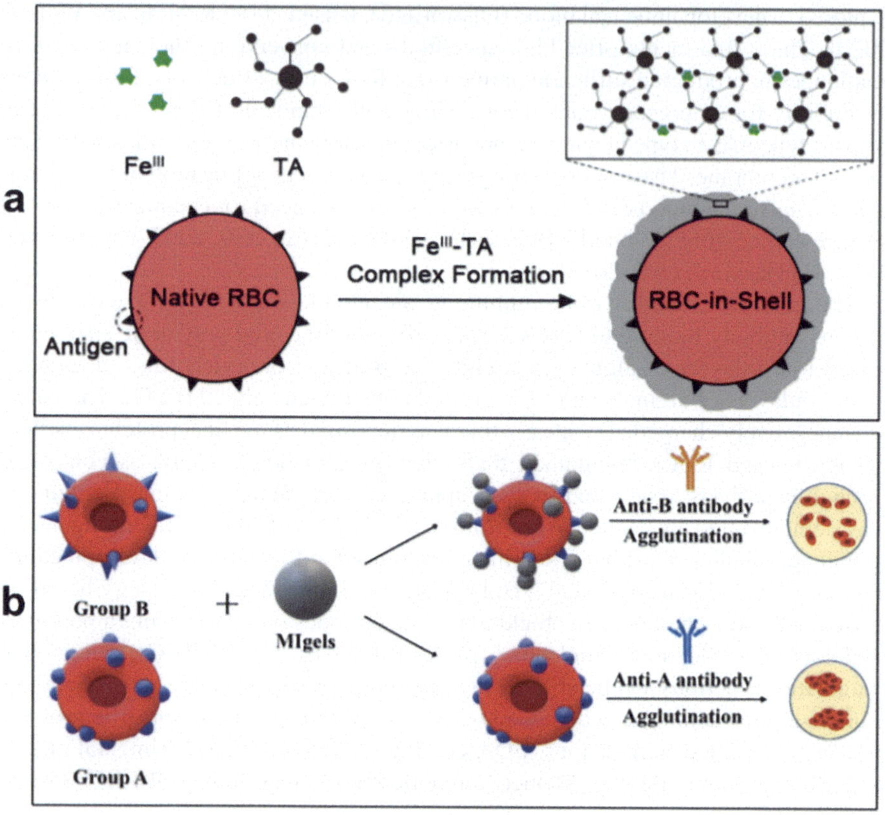

Fig. 9.4 Schematic biomaterial-assisted surface engineering for stealth red blood cells (RBCs): (**a**) Fe^{III}/TA shell formation on RBCs. (Reprinted with permission from Park et al. 2017, copyright 2017, CC-BY4.0, MDPI). (**b**) Selective blockade of antibody binding to blood group B antigen on RBCs using molecular imprinted nanogels. (Reprinted with permission from Jiang et al. 2020, copyright 2020, American Chemical Society)

macromolecule. This approach could improve RBCs surface antigen camouflage, aiding in the development of RhD-null RBCs (Rossi et al. 2010).

Traditional cell surface engineering lacks the selectivity required to create materials that specifically recognize blood group antigens. Developing biocompatible materials capable of selectively binding targeted blood group antigens to prevent recognition by corresponding antibodies is essential. Jiang et al. proposed selective cell surface engineering by developing molecularly imprinted nanogels using the epitope imprinting method, selecting the B antigen as a model. The antigen-protected RBCs effectively avoided agglutination and maintained their normal physiological functions. This selective antigen shielding by molecular imprinted nanogels improved conversion efficiency and safety, highlighting its potential in blood transfusion (Fig. 9.4b) (Jiang et al. 2020). RhD antigen, a key immunogenic determinant of RBCs, often leads to shortages of RhD-negative blood. Transitioning from RhD-positive to RhD-negative expression remains a significant challenge in transfusion medicine. Zhao et al. developed a unique cell surface engineering methodology for shielding the epitopes on RhD-positive RBCs through the surface-anchored framework of enzymatically crosslinked polysialic acid (PSA)–tyramine hydrogel. This hydrogel exhibits a flexible texture, exerts minimal physicochemical impact on cell membrane stability, and effectively conceals RhD antigens on the RBCs surface. Assessments in animal models, including mice and rabbits, confirmed the RhD-epitope shielding of the engineered RhD-positive RBCs. This approach could efficiently produce RhD-epitope stealth RBCs for universal emergency transfusions (Zhao et al. 2020).

For patients with unique blood types and allosensitization to minor RBCs antigens, antibody development complicates blood matching, causing critical donor shortages and highlighting the need for universal donor blood in transfusion medicine. The development of coating methods for nano-encapsulated RBCs requires consideration of several factors. These include the effects of in vivo shear stress, the dynamic reshaping of RBCs during tissue migration, and the risk of thrombus formation caused by platelet aggregation or activation. Additionally, preserving the oxygen-binding affinity of hemoglobin is essential due to its critical function and high concentration within the erythrocyte cytoplasm (Acker et al. 2016; Labarrere et al. 2020).

9.4.2 Encapsulated Islets for Diabetic Treatment

Type 1 diabetes (T1D) is an autoimmune disorder characterized by the depletion of insulin-producing β-cells in the pancreas by self-reactive T cells, resulting in high blood sugar levels. This destruction prevents T1D patients from producing or using insulin to regulate glucose in the bloodstream. This disease affects up to 8.4 million people and is a growing global health challenge (Gregory et al. 2022). Current treatment involves insulin administration or islet transplantation, but despite advancements, both strategies face numerous challenges. For example, insulin administration provides temporary glycemic control. It could result in several diabetes-related complications, and additionally, daily injections to maintain blood glucose within a healthy range, along with blood sugar monitoring, are stressful (Yu et al. 2017; Perkins et al. 2021). Islet transplantation offers a promising alternative to insulin injections for T1D treatment but faces challenges such as insufficient insulin secretion, high cell death, inflammation, thrombosis, immune activation, and immunosuppression side effects, which often lead to graft failure. Strategies are needed to produce large-scale β-cell aggregates that are shielded from autoimmune rejection, deleterious host responses, and environmental stressors, ensuring long-term graft acceptance without the need for immunosuppression (Bowers et al. 2019; Liu et al. 2022; Shapiro et al. 2017). In the past two decades, single-cell nanoencapsulation strategies have been developed to protect transplanted cells, enhance graft acceptance, and enable glucose-responsive insulin release without needing immunosuppression.

Successful engraftment of transplanted cells into target tissues requires mitigating the immune reactions and neutralizing the oxidative stress. Initial studies have focused on shielding grafts from autoimmune rejection and stressors. Wilson et al. demonstrated the controlled grafting of cationic PLL-grafted-PEG copolymers onto cell surfaces through electrostatic adsorption. By appropriately controlling the grafting ratio, PLL-grafted-PEG copolymers were utilized to create cytocompatible surfaces and served as "cell surface-active" carriers for ligands and reactive groups. Functionalized PEG copolymers exhibited biotin, hydrazide, and azide moieties, enabling selective binding of streptavidin, aldehyde, and cyclooctyne-labeled probes, respectively, on the islet surface. Co-adsorption of functionalized PLL-grafted-PEG copolymers enables the simultaneous presentation of multiple functional entities in a single step (Fig. 9.5) (Wilson et al. 2009, 2011). Kim et al. developed a multilayer hydrogel nanofilm to protect cells from shear stress and reduce immune responses by disrupting cell-cell interactions. The nanofilm, composed of monophenol-modified glycol chitosan (GC-T) and hyaluronic acid (HA-T), is cross-linked on the cell surface through tyrosinase-mediated reactions. The uniform positive charge on the cell membrane surface induced by GC-T enabled layer-by-layer coating with HA-T. The LbL nanofilm served as a physical barrier, withstanding external pressure, reducing acute immune responses from NK cell interactions, and protecting against cytokine attacks. Hydrogel nanofilm on mouse pancreatic β cell spheroids restored normoglycemia in a T1D mouse model, outperforming other groups. Overall, the cytoprotective, immune-protective, and glucose-monitoring properties of nanofilm underscore its potential for T1D treatment (Kim et al. 2021). Carbodiimide-based condensation of 3,5-dicarboxyphenyl glycineamide on alginate produced a branched polysaccharide with functionalizable carboxylate groups. PEG-linked azido groups were introduced to alginate, enabling nanoscale LbL coatings via Staudinger ligation with a phosphine-functionalized dendrimer. These coatings were applied to pancreatic islets, with electrostatic and covalent interactions assessed for coating efficiency, stability, and islet viability. The

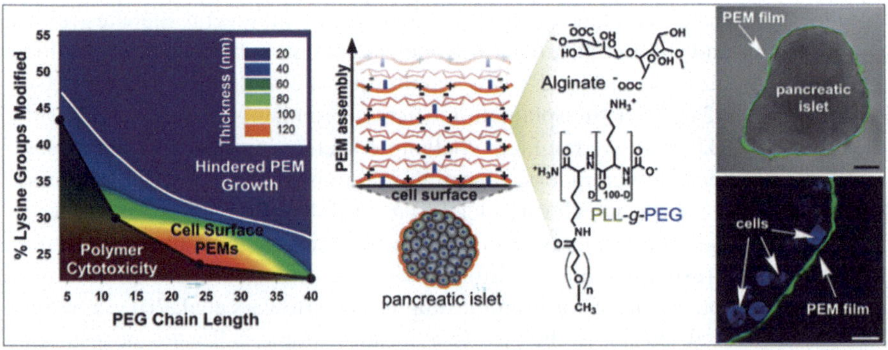

Fig. 9.5 A schematic illustrates biomaterial-assisted surface engineering using PLL-g-PEG copolymer films to create camouflage islets for immunoisolation in diabetic treatment. (Reprinted with permission from Wilson et al. 2011, copyright 2011, American Chemical Society)

hyperbranched polymeric materials provide a versatile platform for covalently stabilized thin coatings on living cells and tissues, enabling further click attachment of bioactive factors or labeling tags (Gattás-Asfura and Stabler 2013).

Thrombotic reactions in the portal vein, caused by islet contact with blood, often lead to early graft loss due to inflammatory responses and complement activation. Systemically administered immunosuppressants and anticoagulants can mitigate these complications; however, they also limit the effectiveness of islet transplantation by inhibiting beta-cell regeneration and increasing the likelihood of uncontrolled bleeding in healthy tissues. To address these limitations, significant efforts have been made to modify cell surfaces with heparin, thrombomodulin, and urokinase, localizing the anticoagulant and fibrinolytic effects. Im et al. proposed a dual-layer immune shielding strategy using multi-arm PEG-catechol (PEG8) and N-hydroxy succinimidyl-linked unfractionated heparin (UFH) to reduce inflammatory effects in transplanted islets. PEG8/UFH-coated islet surface preserved cell viability, provided anticoagulant effects, and showed significantly longer survival in an allograft model than unmodified islets. In combination with immunosuppressive drugs, shielded islets demonstrated significantly longer survival, highlighting their potential to prevent blood-mediated inflammatory reactions and immune attacks in islet transplantation (Im et al. 2013). The Mal–PEG–lipid conjugate, anchored to the islet surface through hydrophobic interactions, did not affect islet function. This surface coating was further functionalized with the fibrinolytic enzyme urokinase and anticoagulant thrombomodulin via thiol/maleimide bonding at the PEG chain end. Both enzymes retained their activity, and the results demonstrated that their local, sustained release suppressed thrombogenic reactions on transplanted islets without affecting islet volume or insulin release, thereby enhancing graft survival (Chen et al. 2011). Overall, drug-coated insulin-producing cellular aggregates inhibited thrombosis, reduced the risk of bleeding, and mitigated innate immune responses, thereby preventing coagulation and tissue inflammation.

Bhaiji et al. proposed a strategy to improve the performance of insulin-producing cells by integrating nanoencapsulation with cell cocultures. Nanoencapsulation was achieved through LbL deposition of phosphorylcholine-modified PLL and heparin, forming a nanometer-thick coating on the cell aggregate surface. Spheroids created by coculturing MIN6 cells with mesenchymal stem cells (MSCs) were tested for encapsulation. Small fractions of MSCs enhanced structural stability, insulin secretion, and resistance to cytokine-induced apoptosis. The combination of nanoencapsulation with MSCs/MIN6 coculture improved glucose responsiveness while providing the antiapoptotic and immunosuppressive benefits of pseudo islets, demonstrating the significant potential for advancing T1D treatment (Bhaiji et al. 2012). Islet transplantation monitoring relies on glycemic levels, insulin/C-peptide levels, and HbA1c levels, primarily detecting late-stage graft rejection. Magnetic resonance imaging (MRI) offers a promising non-invasive alternative. However, traditional nanoparticle labeling methods caused unpredictable toxicity, reduced viability, impaired insulin production, and lower MRI quality. The main challenge lies in developing stable chemistries for pre-labeling islets with magnetic probes. Wang et al. demonstrated that positively charged chitosan-phosphorylcholine and negatively charged heparin sequentially associate with negatively charged cell surfaces, incorporating superparamagnetic iron oxide nanoparticles during multilayer build-up. Results showed that LbL nano-coatings with superparamagnetic nanoparticles adhere to islet surfaces, remain stable post-transplantation, and enable MRI tracking of islet grafts. Combined with anti-rejection nanocoating strategies, this approach could enhance the clinical application of imaging cellular transplants (Wang et al. 2013).

Overall, biomaterial-assisted camouflage surface engineering enhances the proximity between islets and their microenvironment, which is crucial for insulin secretion. This approach limits immune cell access to islet antigens, protects transplanted cells from immune attacks, allows nutrient and gas diffusion, enhances durability under stress, and provides resistance to non-specific protein adsorption. This could eliminate the need for immunosuppression, enabling the safe transplantation of non-human islets. Chemical engineering strategies that decorate immune checkpoint ligands on islet surfaces help inhibit autoimmune responses, restore self-tolerance, and improve cellular engraftment, expanding treatment options for a wider diabetic patient population. Research is needed to address challenges like preventing fibrosis, improving vascular perfusion, overcoming insulin diffusion delays from clumping, and enhancing glucose responsiveness and insulin kinetics. Furthermore, equipping encapsulated stem cells with exogenous bioactive molecules to promote their differentiation into insulin-producing cells could represent a promising diabetes treatment.

9.4.3 Encapsulated Cells for Tissue Engineering

Cell-based therapies, characterized by their self-renewal capacity, paracrine effects, differentiation potential and bioactive functions, have demonstrated significant potential for treating cardiovascular, neurological, musculoskeletal and numerous other tissue defects (Wei et al. 2013; Zhuang et al. 2021). This approach is effective only if transplanted cells are sufficiently abundant and of high quality to survive and restore function. However, systemically injected cell suspensions often exhibit poor survival, limited retention, inadequate engraftment, lack of site specificity, suboptimal phenotypes, immune rejection, and reduced bioactivity within diseased microenvironments. Cell transplantation using scaffolds demonstrates improved survival rates; however, bulk scaffolds and other microencapsulation platforms often fail to protect transplanted cells from immune responses, thereby reducing their therapeutic efficacy in hostile diseased microenvironments (Farina et al. 2019; Hussain and Pei 2021). Consequently, the development of cell nanoencapsulation methods is crucial for protecting cells from shear stress during transplantation and providing exogenous functions afterwards, gaining significant attention for their potential to enhance cell-based therapies.

Few nanoencapsulation strategies have been proposed to modulate paracrine signaling and protect against various stressors. Transglutaminase has been used for cross-linking to create biocompatible gelatin nanoshells around cells through

LbL. This shell surrounding the mammalian cells protected them from cytotoxic enzymes, polycation attacks, and physical stress, ensuring high viability and long-term stability, which is essential for expanding their potential applications (Sun et al. 2020). Oh et al. introduce a metabolic glycoengineering and click chemistry-based single-cell nanoencapsulation method to create optimal coatings around neural progenitor cells. In brief, neural progenitor cells are treated with an unnatural metabolic precursor, such as tetraacetylated N-Azidoacetyl-mannosamine, to produce azide-modified cells. Neural progenitor cells expressing azide-modified glycan proteins are encapsulated with PEG of varying molecular weights through bioorthogonal crosslinking between the azide group on the cell membrane surface and DBCO-modified PEG. Encapsulation enhances neurotrophic factor release through the ADCY8-cAMP pathway, improving paracrine signaling and boosting the therapeutic efficacy of stem cell treatments (Oh et al. 2020).

This study presents a mild method for nanoencapsulating mammalian cells using LbL assembly of a gelatin inner layer and PEG-modified gelatin chains on the outer layer via thiol-click chemistry. Encapsulated HeLa cells demonstrated high viability, prolonged stability, resistance to external forces, and tolerance to physical stress. Furthermore, selective cleavage of the PEG layer enabled the on-demand release of active cells, an essential feature for tissue engineering applications (Yang et al. 2017). Nanoencapsulation of individual MSCs with enzyme-responsive polymer shells has been achieved using LbL assembly and click chemistry. These semipermeable nanoshells offer prolonged encapsulation, maintain high cell viability, resist physical stress, and protect against external threats and immunogenic responses. Controlled cell release was achieved through enzymatic degradation using matrix metalloproteinase-7, highlighting the potential of this approach for cell-based therapies (Yang et al. 2019).

9.4.3.1 Cardiac Tissue Engineering

Myocardial infarctions are leading global causes of death due to the heart's limited regeneration. Cell-based therapy offers promise, but challenges like poor cell survival and stability hinder its effectiveness. Innovative encapsulation strategies have been needed to improve these therapies. Zhang et al. designed a polysialic acid anchor for cell membranes (PAAM) by grafting an oleyl chain onto polysialic acid, which was then conjugated to microbial transglutaminase (mTG) using N-hydroxysuccinimide to form PAAM-mTG. The exposed mTG moiety on the cell surface catalyzed the crosslinking of surrounding gelatin molecules, leading to the fabrication of a protective nanogel coating around MSCs. This surface-anchored nanogel played a crucial role in regulating stem cell fate and enhancing anti-apoptotic capacity under hypoxic and ischemic conditions in infarcted hearts. In myocardial infarction models, engineered MSCs exhibited improved engraftment, enhanced cardiac function, reduced infarct size, and increased vascularization, highlighting the potential of this strategy for cardiac regeneration (Fig. 9.6a) (Zhang et al. 2021).

Human induced pluripotent stem cell (hiPSC)-derived cardiomyocytes are widely used in cardiac tissue regeneration, and their therapeutic potential can be

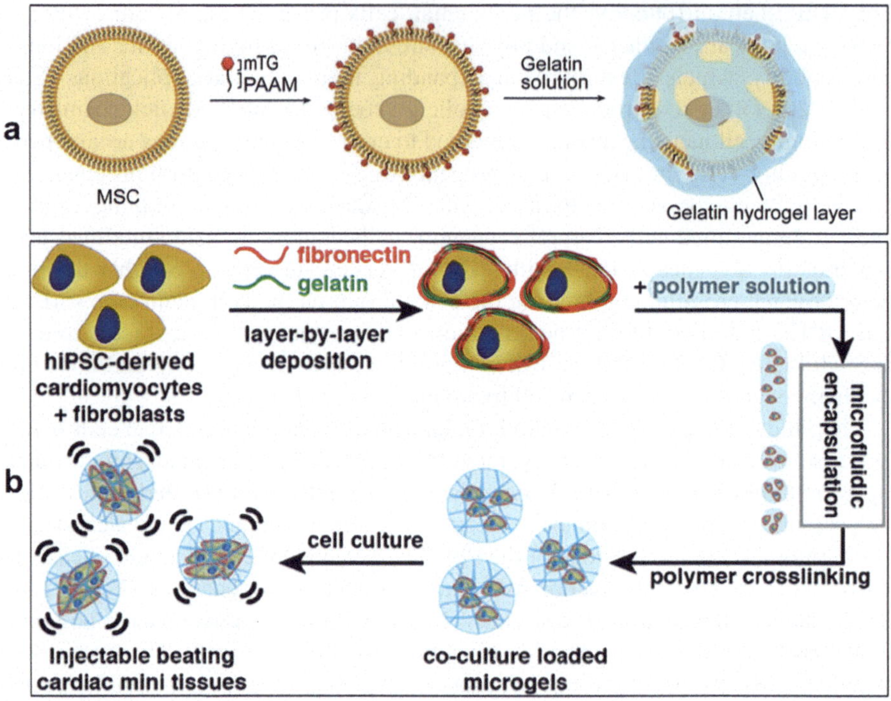

Fig. 9.6 (**a**) The PAAM-mTG molecule on the cell surface facilitated the crosslinking of neighboring gelatin molecules, resulting in the formation of a protective nanogel coating around stem cells. (Reprinted with permission from Zhang et al. 2021, copyright 2021, Wiley-VCH GmbH). (**b**) Fibronectin/gelatin nanofilm-coated dermal fibroblasts and hiPSCs are encapsulated in peptide-conjugated PEG microgels to form functional cardiac mini-tissues. (Reprinted with permission from Guerzoni et al. 2019, copyright 2019, American Chemical Society)

enhanced through advanced encapsulation strategies. To improve regenerative cell transplantation efficiency, a coculture of hiPSCs and primary dermal fibroblasts is encapsulated in injectable PEG-based gels via microfluidics. These gels are synthesized using a Michael-type addition reaction with an enzyme-sensitive peptide crosslinker and further modified with a cell-adhesive peptide. Cell viability and interactions are significantly improved by an ECM coating formed through LbL deposition, which enhances cellular microenvironments. Notably, the beating potential of encapsulated cardiomyocytes doubles compared to noncoated cells, demonstrating the effectiveness of this approach for fabricating functional cardiac mini tissues (Fig. 9.6b) (Guerzoni et al. 2019). Integrating abiotic materials with living cells has become a promising strategy for controlling cellular functions. He et al. used a simple method to partially coat BM-MSCs with a ZIF-8 nanoparticle. Partial coating acts as a protective support, maintaining cell survival in suspension and protecting against oxidative stress, and osmotic pressure. The partial exoskeleton allows cell adhesion, spreading, and signalling, enhancing cellular functions and promoting survival. During non-invasive intravenous administration, partial-coated stem cells show improved retention in infarcted myocardium, leading to enhanced heart function, angiogenesis, cellular proliferation, and reduced apoptosis (He et al. 2023).

9.4.3.2 Multi-tissue Engineering

Ru et al. embedded human embryonic stem cell-derived retinal pigment epithelial cells in polyelectrolytes gelatin and alginate shell using the LbL self-assembly strategy. The shell material significantly reduced immunogenicity without affecting intrinsic activities and optimized encapsulated cells' adhesion, survival, and function. Self-assembly at the single-cell nanoencapsulation facilitated better engraftment, reduced diffusion, and lower immunogenicity. Encapsulated cell transplantation into the subretinal space of rats significantly improved cell viability, engraftment, reduced immunogenicity, and enhanced visual function compared to control cells (Ru et al. 2020). Choi et al. coated PLL, hyaluronic acid, and arginine-glycine-aspartic acid-based nanofilm onto viable MSCs and stabilized them in a suspension state. Results indicated that nanofilm-coated stem cells exhibited increased cell survival, and coated material did not interfere with cell migration and differentiation potential. Furthermore, in vivo administration of coated stem cells prolonged their blood circulation time and enhanced stem cell recruitment to injured muscle tissue, owing to improved survival. Surface modification of stem cells with biomimetic nanofilms effectively modulated cell activity and could serve as a valuable tool to enhance the efficacy of MSCs-based treatments (Fig. 9.7a) (Choi et al. 2017). Cell encapsulation within hydrogels has emerged as a promising strategy to overcome the limitations of conventional bulk hydrogels. Huang et al. employed an LbL self-assembly approach to fabricate alginate and gelatin microgels for encapsulating individual MSCs, enabling their intradiscal injection for the treatment of disc degeneration. The microgel coating effectively suppressed pyroptosis activation under oxidative stress by maintaining mitochondrial homeostasis, thereby protecting stem cells in the harsh disc microenvironment. This preservation of essential stem cell functions, including migration, proliferation, and differentiation, contributed to enhanced therapeutic outcomes. In a rat disc degeneration model, microgel-coated MSCs demonstrated prolonged retention and superior efficacy in mitigating degeneration compared to bare MSCs (Huang et al. 2024).

Yu et al. present a novel strategy to encapsulate neural stem cells using HOFs to address issues like loss of stemness, cytomembrane damage, and prevent oxidative stress-induced apoptosis during and after transplantation. By incorporating porous carbon nanospheres into the HOF shell during mineralization, the encapsulation system establishes hierarchical hydrogen bonds within the cellular exoskeleton, enhancing resistance to oxidative stress through catalase- and superoxide dismutase-like activities. Furthermore, under near-infrared-II irradiation, temperature-responsive hydrogen bonds dissociate, enabling the controlled release of neural stem cells. In an Alzheimer's disease mouse model, stereotactic transplantation of encapsulated stem cells demonstrated improved cell viability, enhanced neurogenesis, and significant alleviation of cognitive impairment symptoms, highlighting the therapeutic potential of HOF-based nanoencapsulation (Fig. 9.7b) (Yu et al. 2022).

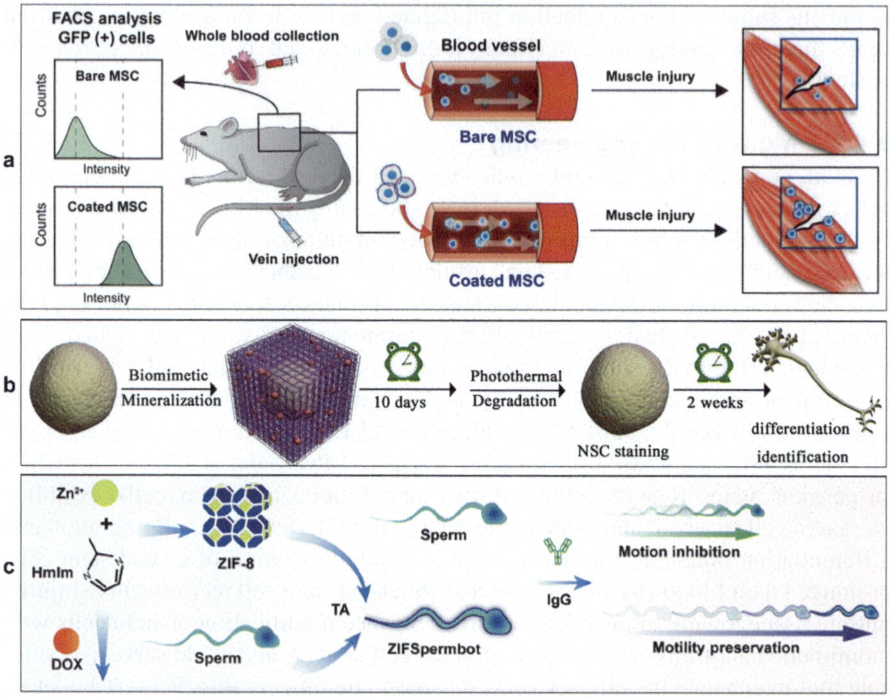

Fig. 9.7 (**a**) Nanofilm-coated stem cells injected into nude mice showed prolonged circulation and improved recruitment to injured muscle tissues. (Reprinted with permission from Choi et al. 2017, copyright 2017, American Chemical Society). (**b**) Formation and removal of porous carbon nanospheres/hydrogen organic frameworks composite shell on neural stem cells (NSC) and its role in remodeling impaired neural networks. (Reprinted with permission from Yu et al. 2022, copyright 2022, Wiley-VCH GmbH). (**c**) Biohybrid sperm microrobots enable active drug delivery and propulsion in microfluidic environments. (Reprinted with permission from Chen et al. 2021, copyright 2021, American Chemical Society)

Single-cell nanoencapsulation cytoprotective polymer shells have advanced biomedical innovations, including sperm-based microrobots. However, sperm motility is often impaired by threats like anti-sperm antibodies, which bind to surface antigens and hinder propulsion. This study introduced biohybrid sperm microrobots encapsulated with ZIF-8 nanoparticles. These ZIF-8 nanoparticle coatings, applied via tannic acid complexation, protect sperm motility. The porous coating blocks anti-sperm antibody binding, preserves propulsion, releases zinc ions to reduce oxidative damage, and supports sustained sperm movement. Biohybrid sperm microrobots enable drug delivery and propulsion in microfluidic environments, advancing adaptive cell robots for harsh biological conditions and enhanced applications (Fig. 9.7c) (Chen et al. 2021).

Overall, encapsulating stem cells within a semi-permeable shell preserves the cell phenotype, maintains cell viability, enhances resistance to shear stress, and may improve recruitment to injured tissues. Coating and functionalizing stem cell surfaces with bioactive nanoparticles and drug molecules could regulate cell differentiation and morphogenesis. However, traditionally introduced polymeric coatings are mostly nondegradable and non-responsive to external fields. On the other hand, engineering strategies that rely on covalent interactions between the coating material and cell surface raise safety concerns. Surface-independent, biodegradable, semi-permeable, and immune-protective nanometric surface engineering strategies can minimize these shortcomings and expand stem cell applications in tissue engineering (Kim et al. 2018; Jia et al. 2021).

9.5 Conclusions and Future Directions

Encapsulating living cells in functional nanoshells represents a substantial enhancement of cellular functions and a significant advancement in cell surface engineering. Significant efforts have been focused on encapsulating various types of mammalian cells using various materials and processes. Single-cell encapsulation enhances cell therapy by reducing biomaterial volume and altering the surface properties of cells compared to microencapsulation. Current nanoencapsulation strategies protect implanted cells from immune responses, shielding them from biofouling, chemical, mechanical, and oxidative stressors, while allowing nutrient and gas exchange to maintain cellular function. Surface engineering enhances cellular engraftment by incorporating targeting moieties, immune modulators, bioactive drugs, and stimulus-responsive agents, thereby imparting novel functionalities to the shell. These features are significant for universal blood transfusions, diabetes treatment, and directed cellular therapies. Despite advances, challenges persist in clinical translation, with many studies overlooking key aspects such as the use of therapeutic cells as cell models, degradation profiles, and long-term tracking of survival, biodistribution, differentiation, and other functional effects of transplanted cells or cell aggregates. The advancements in biomaterial design, cell-material interfaces, tailored chemistries, biophysical tuning, and exogenous functional customization are crucial for improving the outcomes of nanoencapsulated cells and cell aggregates in cellular therapies tailored to the specific characteristics of the patient's disease.

References

Abbina S, Siren EM, Moon H, Kizhakkedathu JN (2017) Surface engineering for cell-based therapies: techniques for manipulating mammalian cell surfaces. ACS Biomater Sci Eng 4(11):3658–3677

Acker JP, Marks DC, Sheffield WP (2016) Quality assessment of established and emerging blood components for transfusion. J Blood Transfus 2016(1):4860284

Alfieri ML, Weil T, Ng DYW, Ball V (2022) Polydopamine at biological interfaces. Adv Colloid Interf Sci 305:102689

Arabi M, Ostovan A, Li J, Wang X, Zhang Z, Choo J, Chen L (2021) Molecular imprinting: green perspectives and strategies. Adv Mater 33(30):2100543

Ausländer S, Wieland M, Fussenegger M (2012) Smart medication through combination of synthetic biology and cell microencapsulation. Metab Eng 14(3):252–260

Bhaiji T, Zhi ZL, Pickup JC (2012) Improving cellular function and immune protection via layer-by-layer nanocoating of pancreatic islet β-cell spheroids cocultured with mesenchymal stem cells. J Biomed Mater Res A 100(6):1628–1636

Bjursten H, Dardashti A, Björk J, Wierup P, Algotsson L, Ederoth P (2016) Transfusion of sex-mismatched and non–leukocyte-depleted red blood cells in cardiac surgery increases mortality. J Thorac Cardiovasc Surg 152(1):223–232

Blaeser A, Duarte Campos DF, Puster U, Richtering W, Stevens MM, Fischer H (2016) Controlling shear stress in 3D bioprinting is a key factor to balance printing resolution and stem cell integrity. Adv Healthc Mater 5(3):326–333

Bowers DT, Song W, Wang LH, Ma M (2019) Engineering the vasculature for islet transplantation. Acta Biomater 95:131–151

Branch DR (2015) Anti-a and anti-B: what are they and where do they come from? Transfusion 55(S2):S74–S79

Chen H, Teramura Y, Iwata H (2011) Co-immobilization of urokinase and thrombomodulin on islet surfaces by poly (ethylene glycol)-conjugated phospholipid. J Control Release 150(2):229–234

Chen F, Porter D, Vollrath F (2012) Structure and physical properties of silkworm cocoons. J R Soc Interface 9(74):2299–2308

Chen Q, Tang S, Li Y, Cong Z, Lu D, Yang Q, Zhang X, Wu S (2021) Multifunctional metal–organic framework exoskeletons protect biohybrid sperm microrobots for active drug delivery from the surrounding threats. ACS Appl Mater Interfaces 13(49):58382–58392

Choi D, Lee H, Kim HB, Yang M, Heo J, Won Y, Jang SS, Park JK, Son Y, Oh TI, Lee E (2017) Cytoprotective self-assembled RGD peptide nanofilms for surface modification of viable mesenchymal stem cells. Chem Mater 29(5):2055–2065

Chong LSH, Zhang J, Bhat KS, Yong D, Song J (2021) Bioinspired cell-in-shell systems in biomedical engineering and beyond: comparative overview and prospects. Biomaterials 266:120473

Coronel MM, Geusz R, Stabler CL (2017) Mitigating hypoxic stress on pancreatic islets via in situ oxygen generating biomaterial. Biomaterials 129:139–151

Critcher M, O'Leary T, Huang ML (2021) Glycoengineering: scratching the surface. Biochem J 478(4):703–719

Dutta S, Kim J, Hsieh PH, Hsu YS, Kaneti YV, Shieh FK, Yamauchi Y, Wu KCW (2019) Nanoarchitectonics of biofunctionalized metal–organic frameworks with biological macromolecules and living cells. Small Methods 3(11):1900213

Farina M, Alexander JF, Thekkedath U, Ferrari M, Grattoni A (2019) Cell encapsulation: overcoming barriers in cell transplantation in diabetes and beyond. Adv Drug Deliv Rev 139:92–115

Gattás-Asfura KM, Stabler CL (2013) Bioorthogonal layer-by-layer encapsulation of pancreatic islets via hyperbranched polymers. ACS Appl Mater Interfaces 5(20):9964–9974

Gregory GA, Robinson TI, Linklater SE, Wang F, Colagiuri S, de Beaufort C, Donaghue KC, Harding JL, Wander PL, Zhang X, Li X (2022) Global incidence, prevalence, and mortality of type 1 diabetes in 2021 with projection to 2040: a modelling study. Lancet Diabetes Endocrinol 10(10):741–760

Guerzoni LP, Tsukamoto Y, Gehlen DB, Rommel D, Haraszti T, Akashi M, De Laporte L (2019) A layer-by-layer single-cell coating technique to produce injectable beating mini heart tissues via microfluidics. Biomacromolecules 20(10):3746–3754

Gupta P, Alheib O, Shin JW (2023) Towards single cell encapsulation for precision biology and medicine. Adv Drug Deliv Rev 201:115010

Hashemi M, Kalalinia F (2015) Application of encapsulation technology in stem cell therapy. Life Sci 143:139–146

Hasturk O, Kaplan DL (2019) Cell armor for protection against environmental stress: advances, challenges and applications in micro-and nanoencapsulation of mammalian cells. Acta Biomater 95:3–31

Hawksworth J, Satchwell TJ, Meinders M, Daniels DE, Regan F, Thornton NM, Wilson MC, Dobbe JG, Streekstra GJ, Trakarnsanga K, Heesom KJ (2018) Enhancement of red blood cell transfusion compatibility using CRISPR-mediated erythroblast gene editing. EMBO Mol Med 10(6):e8454

He H, Yuan Y, Wu Y, Lu J, Yang X, Lu K, Liu A, Cao Z, Sun M, Yu M, Wang H (2023) Exoskeleton partial-coated stem cells for infarcted myocardium restoring. Adv Mater 35(52):2307169
Hong D, Park M, Yang SH, Lee J, Kim YG, Choi IS (2013) Artificial spores: cytoprotective nanoencapsulation of living cells. Trends Biotechnol 31(8):442–447
Huang G, Shen H, Xu K, Shen Y, Jin J, Chu G, Xing H, Feng Z, Wang Y (2024) Single-cell microgel encapsulation improves the therapeutic efficacy of mesenchymal stem cells in treating intervertebral disc degeneration via inhibiting pyroptosis. Research 7:0311
Hussain Z, Pei R (2021) Scaffold-free and scaffold-based cellular strategies and opportunities for cornea tissue engineering. Prog Biomed Eng 3(3):032003
Hussain Z, Ullah I, Liu X, Mehmood S, Wang L, Ma F, Ullah S, Lu Z, Wang Z, Pei R (2023) GelMA-catechol coated FeHAp nanorods functionalized nanofibrous reinforced bio-instructive and mechanically robust composite hydrogel scaffold for bone tissue engineering. Biomater Adv 155:213696
Im BH, Jeong JH, Haque MR, Lee DY, Ahn CH, Kim JE, Byun Y (2013) The effects of 8-arm-PEG-catechol/heparin shielding system and immunosuppressive drug, FK506 on the survival of intraportally allotransplanted islets. Biomaterials 34(8):2098–2106
Jajosky RP, Wu SC, Zheng L, Jajosky AN, Jajosky PG, Josephson CD, Hollenhorst MA, Sackstein R, Cummings RD, Arthur CM, Stowell SR (2023) ABO blood group antigens and differential glycan expression: perspective on the evolution of common human enzyme deficiencies. Iscience 26(1):105798
Jia HR, Zhu YX, Duan QY, Wu FG (2021) Cell surface-localized imaging and sensing. Chem Soc Rev 50(10):6240–6277
Jiang L, Liu H, Huang C, Shen X (2020) Blood group antigen shielding facilitated by selective cell surface engineering. ACS Appl Mater Interfaces 12(20):22426–22432
Kim BJ, Cho H, Park JH, Mano JF, Choi IS (2018) Strategic advances in formation of cell-in-shell structures: from syntheses to applications. Adv Mater 30(14):1706063
Kim M, Kim H, Lee YS, Lee S, Kim SE, Lee UJ, Jung S, Park CG, Hong J, Doh J, Lee DY (2021) Novel enzymatic cross-linking–based hydrogel nanofilm caging system on pancreatic β cell spheroid for long-term blood glucose regulation. Sci Adv 7(26):eabf7832
Kwon SG, Kwon YW, Lee TW, Park GT, Kim JH (2018) Recent advances in stem cell therapeutics and tissue engineering strategies. Biomater Res 22:1–8
Labarrere CA, Dabiri AE, Kassab GS (2020) Thrombogenic and inflammatory reactions to biomaterials in medical devices. Front Bioeng Biotechnol 8:123
Lee HA, Ma Y, Zhou F, Hong S, Lee H (2019) Material-independent surface chemistry beyond polydopamine coating. Acc Chem Res 52(3):704–713
Lee H, Kim N, Rheem HB, Kim BJ, Park JH, Choi IS (2021) A decade of advances in single-cell nanocoating for mammalian cells. Adv Healthc Mater 10(3):2100347
Li Y, Zhang Y, Tao Y, Huang X, Yu C, Xu H, Chen J, Xia K, Shi K, Zhang Y, Wang J (2023) Metabolic glycoengineering: a promising strategy to remodel microenvironments for regenerative therapy. Stem Cells Int 2023(1):1655750
Liang K, Richardson JJ, Cui J, Caruso F, Doonan CJ, Falcaro P (2016) Metal–organic framework coatings as cytoprotective exoskeletons for living cells. Adv Mater 28(36):7910–7914
Lin RB, Chen B (2022) Hydrogen-bonded organic frameworks: chemistry and functions. Chem 8(8):2114–2135
Lin Z, Liu H, Richardson JJ, Xu W, Chen J, Zhou J, Caruso F (2024) Metal–phenolic network composites: from fundamentals to applications. Chem Soc Rev 53:10800–10826
Liu QP, Sulzenbacher G, Yuan H, Bennett EP, Pietz G, Saunders K, Spence J, Nudelman E, Levery SB, White T, Neveu JM (2007) Bacterial glycosidases for the production of universal red blood cells. Nat Biotechnol 25(4):454–464
Liu Z, Tang M, Zhao J, Chai R, Kang J (2018) Looking into the future: toward advanced 3D biomaterials for stem-cell-based regenerative medicine. Adv Mater 30(17):1705388
Liu T, Wang Y, Zhong W, Li B, Mequanint K, Luo G, Xing M (2019) Biomedical applications of layer-by-layer self-assembly for cell encapsulation: current status and future perspectives. Adv Healthc Mater 8(1):1800939

Liu W, Flanders JA, Wang LH, Liu Q, Bowers DT, Wang K, Chiu A, Wang X, Ernst AU, Shariati K, Caserto JS (2022) A safe, fibrosis-mitigating, and scalable encapsulation device supports long-term function of insulin-producing cells. Small 18(8):2104899

Mansouri S, Merhi Y, Winnik FM, Tabrizian M (2011) Investigation of layer-by-layer assembly of polyelectrolytes on fully functional human red blood cells in suspension for attenuated immune response. Biomacromolecules 12(3):585–592

Mao AS, Shin JW, Utech S, Wang H, Uzun O, Li W, Cooper M, Hu Y, Zhang L, Weitz DA, Mooney DJ (2017) Deterministic encapsulation of single cells in thin tunable microgels for niche modelling and therapeutic delivery. Nat Mater 16(2):236–243

Marikar SN, El-Osta A, Johnston A, Such G, Al-Hasani K (2022) Microencapsulation-based cell therapies. Cell Mol Life Sci 79(7):351

Nafea EH, Marson A, Poole-Warren LA, Martens PJ (2011) Immunoisolating semi-permeable membranes for cell encapsulation: focus on hydrogels. J Control Release 154(2):110–122

Nair GG, Tzanakakis ES, Hebrok M (2020) Emerging routes to the generation of functional β-cells for diabetes mellitus cell therapy. Nat Rev Endocrinol 16(9):506–518

Oh B, Swaminathan V, Malkovskiy A, Santhanam S, McConnell K, George PM (2020) Single-cell encapsulation via click-chemistry alters production of paracrine factors from neural progenitor cells. Adv Sci 7(8):1902573

Oliveira MB, Hatami J, Mano JF (2016) Coating strategies using layer-by-layer deposition for cell encapsulation. Chem Asian J 11(12):1753–1764

Olsson ML, Clausen H (2008) Modifying the red cell surface: towards an ABO-universal blood supply. Br J Haematol 140(1):3–12

Orive G, Santos E, Poncelet D, Hernández RM, Pedraz JL, Wahlberg LU, De Vos P, Emerich D (2015) Cell encapsulation: technical and clinical advances. Trends Pharmacol Sci 36(8):537–546

Paez-Mayorga J, Lukin I, Emerich D, de Vos P, Orive G, Grattoni A (2022) Emerging strategies for beta cell transplantation to treat diabetes. Trends Pharmacol Sci 43(3):221–233

Park JH, Yang SH, Lee J, Ko EH, Hong D, Choi IS (2014) Nanocoating of single cells: from maintenance of cell viability to manipulation of cellular activities. Adv Mater 26(13):2001–2010

Park T, Kim JY, Cho H, Moon HC, Kim BJ, Park JH, Hong D, Park J, Choi IS (2017) Artificial spores: immunoprotective nanocoating of red blood cells with supramolecular ferric ion-tannic acid complex. Polymers 9(4):140

Perkins BA, Sherr JL, Mathieu C (2021) Type 1 diabetes glycemic management: insulin therapy, glucose monitoring, and automation. Science 373(6554):522–527

Petazzi P, Miquel-Serra L, Huertas S, González C, Boto N, Muñiz-Diaz E, Menéndez P, Sevilla A, Nogués N (2022) ABO gene editing for the conversion of blood type A to universal type O in Rhnull donor-derived human-induced pluripotent stem cells. Clin Transl Med 12(10):e1063

Petrus-Reurer S, Romano M, Howlett S, Jones JL, Lombardi G, Saeb-Parsy K (2021) Immunological considerations and challenges for regenerative cellular therapies. Commun Biol 4(1):798

Piletsky S, Canfarotta F, Poma A, Bossi AM, Piletsky S (2020) Molecularly imprinted polymers for cell recognition. Trends Biotechnol 38(4):368–387

Pires-Santos M, Nadine S, Mano JF (2024) Unveiling the potential of single-cell encapsulation in biomedical applications: current advances and future perspectives. Small Sci 4(5):2300332

Poole J, Daniels G (2007) Blood group antibodies and their significance in transfusion medicine. Transfus Med Rev 21(1):58–71

Qin T, Smink AM, de Vos P (2023) Enhancing longevity of immunoisolated pancreatic islet grafts by modifying both the intracapsular and extracapsular environment. Acta Biomater 167:38–53

Rahfeld P, Withers SG (2020) Toward universal donor blood: enzymatic conversion of A and B to O type. J Biol Chem 295(2):325–334

Rossi NA, Constantinescu I, Brooks DE, Scott MD, Kizhakkedathu JN (2010) Enhanced cell surface polymer grafting in concentrated and nonreactive aqueous polymer solutions. J Am Chem Soc 132(10):3423–3430

Ru L, Wu N, Wei K, Zeng Y, Li Q, Weng C, Ren C, Ren B, Huo D, Li Y, Hu X (2020) Improving cell survival and engraftment in vivo via layer-by-layer nanocoating of hESC-derived RPE cells. Stem Cell Res Ther 11:1–16

Russo L, Cipolla L (2016) Glycomics: new challenges and opportunities in regenerative medicine. Chem Eur J 22(38):13380–13388

Ryan AJ, O'Neill HS, Duffy GP, O'Brien FJ (2017) Advances in polymeric islet cell encapsulation technologies to limit the foreign body response and provide immunoisolation. Curr Opin Pharmacol 36:66–71

Shapiro AJ, Pokrywczynska M, Ricordi C (2017) Clinical pancreatic islet transplantation. Nat Rev Endocrinol 13(5):268–277

Sneddon JB, Tang Q, Stock P, Bluestone JA, Roy S, Desai T, Hebrok M (2018) Stem cell therapies for treating diabetes: progress and remaining challenges. Cell Stem Cell 22(6):810–823

Song W, Ko J, Choi YH, Hwang NS (2021) Recent advancements in enzyme-mediated crosslinkable hydrogels: in vivo-mimicking strategies. APL Bioeng 5(2):021502

Stephan MT, Irvine DJ (2011) Enhancing cell therapies from the outside in: cell surface engineering using synthetic nanomaterials. Nano Today 6(3):309–325

Stubbs JR, Zielinski MD, Jenkins D (2016) The state of the science of whole blood: lessons learned at Mayo Clinic. Transfusion 56:S173–S181

Sun J, Ren Y, Wang W, Hao H, Tang M, Zhang Z, Yang J, Zheng Y, Shi X (2020) Transglutaminase-catalyzed encapsulation of individual mammalian cells with biocompatible and cytoprotective gelatin nanoshells. ACS Biomater Sci Eng 6(4):2336–2345

Waeterschoot J, Gosselé W, Lemež Š et al (2024) Artificial cells for in vivo biomedical applications through red blood cell biomimicry. Nat Commun 15(1):2504

Wang S, Guo Z (2014) Bio-inspired encapsulation and functionalization of living cells with artificial shells. Colloids Surf B Biointerfaces 113:483–500

Wang Y, Blanco-Andujar C, Zhi ZL, So PW, Thanh NTK, Pickup JC (2013) Multilayered nanocoatings incorporating superparamagnetic nanoparticles for tracking of pancreatic islet transplants with magnetic resonance imaging. ChemComm 49(65):7255–7257

Wang B, Wang G, Zhao B, Chen J, Zhang X, Tang R (2014) Antigenically shielded universal red blood cells by polydopamine-based cell surface engineering. Chem Sci 5(9):3463–3468

Wang X, Xiao Y, Hao H, Zhang Y, Xu X, Tang R (2018) Therapeutic potential of biomineralization-based engineering. Adv Therap 1(8):1800079

Wang Q, Hu L, Wang X, Tang R (2024) Expanding from materials to biology inspired by biomineralization. Interdiscip Mater 3(2):165–188

Wei X, Yang X, Han ZP, Qu FF, Shao L, Shi YF (2013) Mesenchymal stem cells: a new trend for cell therapy. Acta Pharmacol Sin 34(6):747–754

Wei Y, Feng Y, Wang K, Wei Y, Li Q, Zuo X, Li B, Li J, Wang L, Fan C, Zhu Y (2024) Directing the encapsulation of single cells with DNA framework nucleator-based hydrogel growth. Angew Chem Int Ed 136(16):e202319907

Wilson JT, Krishnamurthy VR, Cui W, Qu Z, Chaikof EL (2009) Noncovalent cell surface engineering with cationic graft copolymers. J Am Chem Soc 131(51):18228–18229

Wilson JT, Cui W, Kozlovskaya V, Kharlampieva E, Pan D, Qu Z, Krishnamurthy VR, Mets J, Kumar V, Wen J, Song Y (2011) Cell surface engineering with polyelectrolyte multilayer thin films. J Am Chem Soc 133(18):7054–7064

Wu S, Wang L, Fang Y, Huang H, You X, Wu J (2021) Advances in encapsulation and delivery strategies for islet transplantation. Adv Healthc Mater 10(20):2100965

Yang SH, Kang SM, Lee KB, Chung TD, Lee H, Choi IS (2011) Mussel-inspired encapsulation and functionalization of individual yeast cells. J Am Chem Soc 133(9):2795–2797

Yang SH, Hong D, Lee J, Ko EH, Choi IS (2013) Artificial spores: cytocompatible encapsulation of individual living cells within thin, tough artificial shells. Small 9(2):178–186

Yang J, Li J, Li X, Wang X, Yang Y, Kawazoe N, Chen G (2017) Nanoencapsulation of individual mammalian cells with cytoprotective polymer shell. Biomaterials 133:253–262

Yang J, Yang Y, Kawazoe N, Chen G (2019) Encapsulation of individual living cells with enzyme responsive polymer nanoshell. Biomaterials 197:317–326

Yang S, Choi H, Nguyen DT, Kim N, Rhee SY, Han SY, Lee H, Choi IS (2023) Bioempowerment of therapeutic living cells by single-cell surface engineering. Adv Therap 6(7):2300037

Yi W, Xiao P, Liu X, Zhao Z, Sun X, Wang J, Zhou L, Wang G, Cao H, Wang D, Li Y (2022) Recent advances in developing active targeting and multi-functional drug delivery systems via bioorthogonal chemistry. Signal Transduct Target Ther 7(1):386

Yoon HY, Koo H, Kim K, Kwon IC (2017) Molecular imaging based on metabolic glycoengineering and bioorthogonal click chemistry. Biomaterials 132:28–36

Youn W, Kim JY, Park J, Kim N, Choi H, Cho H, Choi IS (2020) Single-cell nanoencapsulation: from passive to active shells. Adv Mater 32(35):1907001

Yu J, Qian C, Zhang Y, Cui Z, Zhu Y, Shen Q, Ligler FS, Buse JB, Gu Z (2017) Hypoxia and H2O2 dual-sensitive vesicles for enhanced glucose-responsive insulin delivery. Nano Lett 17(2):733–739

Yu D, Zhang H, Liu Z, Liu C, Du X, Ren J, Qu X (2022) Hydrogen-bonded organic framework (HOF)-based single-neural stem cell encapsulation and transplantation to remodel impaired neural networks. Angew Chem Int Ed 61(28):e202201485

Zhang YS, Khademhosseini A (2017) Advances in engineering hydrogels. Science 356(6337):eaaf3627

Zhang L, Liu G, Lv K, Xin J, Wang Y, Zhao J, Hu W, Xiao C, Zhu K, Zhu L, Nan J (2021) Surface-anchored nanogel coating endows stem cells with stress resistance and reparative potency via turning down the cytokine-receptor binding pathways. Adv Sci 8(3):2003348

Zhang X, Lin Y, Xin J, Zhang Y, Yang K, Luo Y, Wang B (2024) Red blood cells in biology and translational medicine: natural vehicle inspires new biomedical applications. Theranostics 14(1):220

Zhao S, Xu Z, Wang H, Reese BE, Gushchina LV, Jiang M, Agarwal P, Xu J, Zhang M, Shen R, Liu Z (2016) Bioengineering of injectable encapsulated aggregates of pluripotent stem cells for therapy of myocardial infarction. Nat Commun 7(1):13306

Zhao Y, Fan M, Chen Y, Liu Z, Shao C, Jin B, Wang X, Hui L, Wang S, Liao Z, Ling D (2020) Surface-anchored framework for generating RhD-epitope stealth red blood cells. Sci Adv 6(12):eaaw9679

Zhou W, Long Z, Xu C, Zhang J, Zhou X, Song X, Huo P, Guo Y, Xue W, Wang Q, Zhou C (2024) Advances in functionalized biocomposites of living cells combined with metal–organic frameworks. Langmuir 40(29):14749–14765

Zhuang WZ, Lin YH, Su LJ, Wu MS, Jeng HY, Chang HC, Huang YH, Ling TY (2021) Mesenchymal stem/stromal cell-based therapy: mechanism, systemic safety and biodistribution for precision clinical applications. J Biomed Sci 28(1):1–38

Harnessing Engineered Probiotics for Gastrointestinal Diseases Therapy

10

Ali Hasan, Saima Rani, and Ali Hussain

Abstract

The human gastrointestinal (GI) tract hosts a heterogeneous microbial consortium, the gut microbiota, which is crucial for supporting host health. Factors such as antibiotic use, poor diet, aging, and infections contribute to gut dysbiosis. Microbiota imbalance has been linked to several GI disorders, including inflammatory bowel disease, antibiotic-associated diarrhea, foodborne infections, and colorectal cancer. Probiotics, live microorganisms that mimic beneficial gut bacteria, can restore gut homeostasis and help manage gut diseases through mechanisms such as pathogen inhibition, immune modulation, digestive aid, and antimicrobial production. However, traditional probiotics lack key traits, exhibit poor colonization potential, and struggle with reduced viability during processing and exposure to environmental and GI conditions. Recent advancements in molecular biology and nanotechnology have enabled the development of engineered probiotics through genetic engineering, chemical surface modification and their combinations. Genetic engineering enhances probiotic functionality by producing therapeutic molecules, while nanoencapsulation and biomaterial immobilization improve probiotic survival, colonization, and targeted delivery in the GI tract. This chapter examines the action mechanisms of probiotics and

A. Hasan
Department of Zoology, Wildlife and Fisheries, University of Agriculture Faisalabad, Faisalabad, Pakistan
e-mail: dr.alihasan@uaf.edu.pk

S. Rani
Molecular Virology Laboratory, Department of Biosciences, COMSATS University Islamabad (CUI), Islamabad, Pakistan

A. Hussain (✉)
Institute of Zoology, University of the Punjab, Lahore, Pakistan
e-mail: ali.zool@pu.edu.pk

Z. Hussain, R. Pei (eds.), *Cell Membrane Engineering for Advancing Cell-Based and Nanotechnology-Based Therapies*, Biomaterials, Bioengineering and Sustainability 6, https://doi.org/10.1007/978-3-031-93425-4_10

factors limiting the efficacy of native probiotics, explores strategies for designing engineered probiotics, and highlights recent advancements in using engineered probiotics for managing GI-related diseases, along with future directions in this field.

Keywords

Gastrointestinal microbiota · Microbiome-associated diseases · Genetic engineering · Single-cell encapsulation · Engineered probiotics · Dysbiosis management

10.1 Introduction

The human gastrointestinal (GI) tract hosts a dynamic gut microbiome that is crucial for protecting the intestines from exogenous and opportunistic pathogens. The Human Microbiome Project has played a pivotal role in unraveling microbial diversity and provided valuable insights into the microbiome (Nash et al. 2017). Growing evidence indicates that gut microbiome participates in diverse physiological activities. The microbial consortium in the human gut outnumbers human cells, with a substantial portion colonizing the GI tract. Gut microbiota regulates metabolism, supplies nutrients, maintains homeostasis, supports intestinal function, and regulates host health (Peterson et al. 2009; Ruan et al. 2020). Gut dysbiosis, which refers to the imbalance or depletion of beneficial microbiota, can result from numerous factors such as antibiotic therapy, poor diet, aging, lifestyle changes, allergic reactions, and infections. It is linked to various gut diseases, including colorectal cancer, diarrhea, inflammatory bowel disease (IBD), opportunistic infections, celiac disease, and irritable bowel syndrome (Carding et al. 2015; Mukherjee et al. 2015). Antibiotics, used to treat gut infections, can alter microbial composition and promote the development of antibiotic-resistant microbes. The rising threat of resistant pathogens worsens dysbiosis and complicates treatment, posing significant health and economic challenges (Duan et al. 2022; Lamberte and van Schaik 2022). On the other hand, conventional drugs are often limited by high-dose toxicity and inadequate targeting of inflamed tissue, leading to severe side effects. The growing recognition of microbiomes' role in human health has sparked interest in microbiome engineering to promote gut health (Sonnenburg 2015; Foo et al. 2017).

The Human Microbiome Project highlights the therapeutic potential of microbial cells for treating gut-related diseases. Probiotics are living bacteria that mimic beneficial bacteria in the human GI tract (Peterson et al. 2009). For decades, strains from *Lactobacillus*, *Lactococcus*, *Bifidobacterium*, and *Saccharomyces* genera have been used in yogurt, milk, cheese, infant formula, and medical foods. Classified as Generally Recognized as Safe (GRAS), probiotics, when consumed in metabolically active and adequate quantities (e.g., 10^6–10^7 CFU/g), provide numerous health benefits (Gao et al. 2021; Reque and Brandelli 2021). Probiotics restore gut homeostasis through multiple mechanisms:

- Probiotics block pathogen adhesion and colonization by competing for intestinal binding sites, leading to pathogen exclusion (Lebeer et al. 2010).
- Probiotics enhance the intestinal barrier via generation of metabolic entities, support digestion, and increase short-chain fatty acids (SCFAs) to protect the mucosa (Markowiak-Kopeć and Śliżewska 2020).
- Probiotics regulate immunological responses by modulating immunoglobulins, cytokines, and other immune molecules (Yousefi et al. 2019).
- Probiotics secrete antimicrobial, anti-inflammatory, and antioxidant compounds to maintain microbial balance (Prosekov et al. 2017).
- Certain probiotics produce toxin-degrading enzymes that detoxify bacterial toxins, thereby helping to mitigate foodborne illnesses (Yao et al. 2024).

The multiple functional potential of probiotics could be crucial for protecting the intestinal barrier, regulating the ecological balance of gut microbiota, and managing gut-related diseases. However, using native probiotics presents challenges, as conventional treatments often fail to provide long-term relief. Probiotic efficacy depends on factors such as usage of specific strain selection, appropriate dosage, and protection from harsh environmental and GI conditions (Castro-López et al. 2023). Additionally, probiotics may exhibit variability in efficacy across different hosts, non-specific antimicrobial activity, and production of harmful metabolites. Probiotics typically secrete antibacterial substances effective only against specific microorganisms, limiting their broad use (Singh and Natraj 2021). The limitations of traditional probiotics have shifted the focus towards the development of designer probiotics to overcome these challenges and offer personalized and more effective treatments for GI diseases.

Engineered probiotics offer an appealing alternative to conventional treatments, which often fail to provide long-term relief. Recent advancements in bioinformatics, molecular biology and nanotechnology have enabled the creation of engineered probiotics through genetic engineering, surface modification, nanoencapsulation, and their combinations for targeted gut microbiome modulation. For example, genetically modified probiotics can secrete therapeutically important molecules, such as antimicrobial peptides, anti-inflammatory peptides, and antioxidant enzymes, to inhibit pathogen progression, reduce inflammation, and modulate the disease microenvironment (Murali and Mansell 2024; Huang et al. 2022). Chemically functionalized and encapsulated probiotics protect against environmental and stomach conditions, ensure optimal delivery in the GI tract, offer targeted therapeutic capabilities, and effectively deliver substances to treat specific conditions (Chen and Zhu 2023). These engineered strategies aim to overcome challenges of traditional antibiotics, improve microbial balance, provide health benefits with known mechanisms, maximize effectiveness, and enable probiotic-based treatments for GI diseases (Fig. 10.1). This chapter examines factors limiting the efficacy of native probiotics, explores strategies for designing engineered probiotics, and highlights recent advancements in using engineered probiotics for managing GI-related diseases. It also discusses future directions of engineered probiotics in the management of GI diseases.

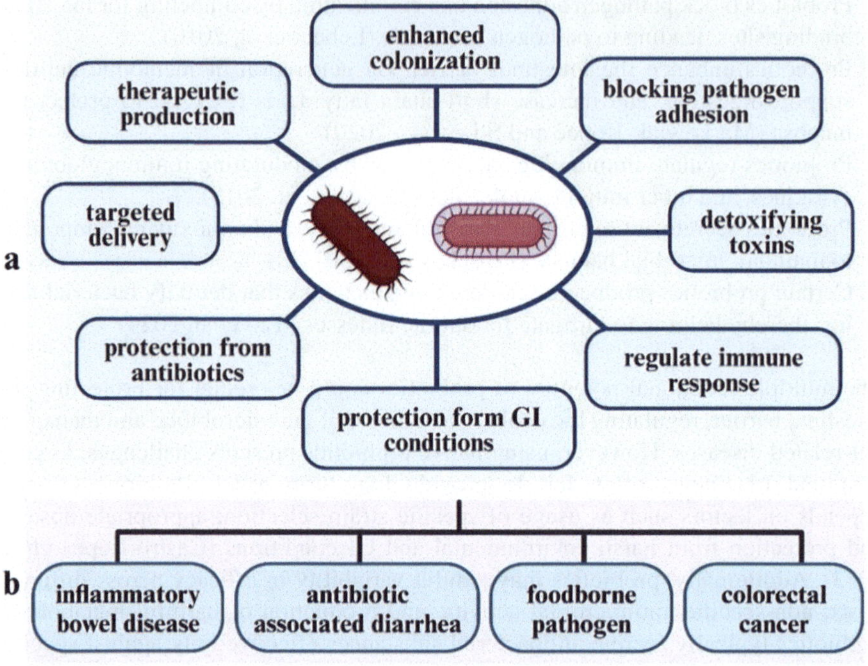

Fig. 10.1 (**a**) Mechanism of action of engineered probiotics, (**b**) Probiotics-based treatment of gastrointestinal diseases. (Source: Created with BioRender.com)

10.2 Factors Limiting Probiotic Efficacy in Therapeutic Applications

Oral probiotics hold potential for treating gastrointestinal diseases due to their various health benefits. However, challenges such as environmental factors affecting probiotic survival, the impact of gastrointestinal conditions on their viability, and the limitations of native probiotics hinder their therapeutic effectiveness (Fig. 10.2).

10.2.1 Environmental and GI Conditions Impact

The efficacy of oral probiotics is often compromised by various environmental storage, and transportation challenges, including temperature fluctuations (e.g., freezing, cooling, or heating), oxygen exposure, excessive moisture activity, light exposure, and certain processing methods (e.g., drying and encapsulation). Many probiotic strains, especially anaerobes like *Bifidobacterium* and *Lactobacillus*, are susceptible to reactive oxygen species (ROS), which damage probiotic components. Moisture is another critical factor, as elevated moisture levels can trigger microbial growth, thereby reducing the stability of probiotics. Furthermore, harsh chemicals, dehydration stress, and mechanical stress from processing methods such as

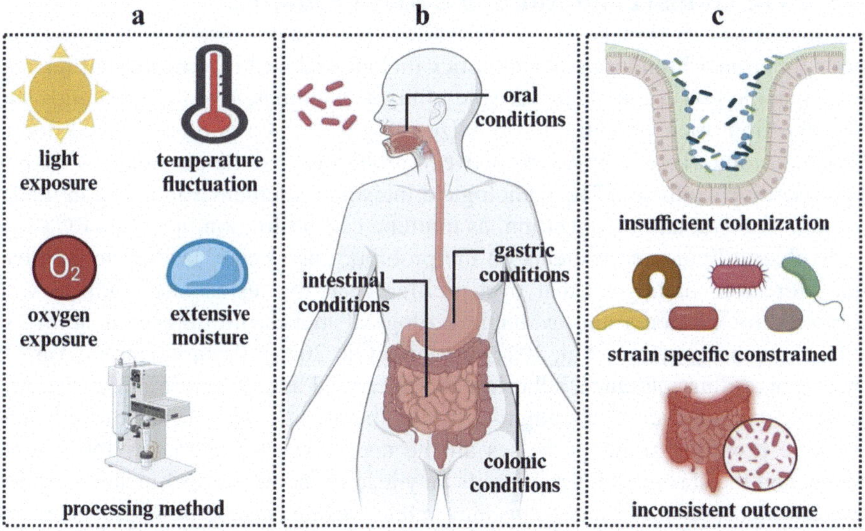

Fig. 10.2 (**a**) Environmental impacts on probiotic survival, (**b**) Gastrointestinal conditions impact on probiotic survival, and (**c**) Functional limitations of native probiotic. (Source: Created with BioRender.com)

spray-drying, extrusion, freeze-drying, and emulsification can compromise the structural integrity and viability of probiotics. Therefore, protecting probiotics from these environmental stressors during storage, transportation, and processing is essential to maintain their viability and effectiveness (Bustos et al. 2024; Misra et al. 2022).

Upon ingestion, probiotics are exposed to salivary amylase in the mouth, which begins to stress their cell membranes. In the stomach, they encounter low gastric pH and pepsin, both of which can degrade bacterial structures, necessitating acid tolerance for survival. In the small intestine, bile salts and digestive enzymes such as trypsin exert antimicrobial and protein-degrading effects, further compromising cell membrane integrity. Additionally, ROS throughout the GI tract oxidize cellular components, intensifying stress. Beyond these physiological barriers, rapid intestinal transit due to peristalsis reduces the time available for probiotics to adhere to mucosal surfaces, limiting their ability to colonize the gut and inflamed tissues. These combined stressors such as acidic conditions, antimicrobial effects of bile salts, digestive enzymes, oxidative stress, and rapid transit could impair probiotic viability, preventing effective colonization and often leading to suboptimal outcomes. Certain probiotics such as *Lactobacillus acidophilus* and *Bifidobacterium bifidum* exhibit beneficial traits such as acid tolerance, antibiotic resistance, pathogen antagonism, and metabolic activity. However, numerous wild-type strains lack these characteristics, leading to significant reductions in probiotic viability before they reach the intestines (Castro-López et al. 2023; Bustos et al. 2024).

10.2.2 Functional Limitations of Native Probiotics

Major functional limitations of probiotics include reduced colonization efficiency, strain-specific constraints, and variable clinical outcomes. Even if probiotics survive digestion, their potential to colonize the gut mucosa is often limited due to restricted biointerface interactions between probiotics and the gut mucosa, leading to competitive exclusion. The pathological intestinal microenvironment can hinder probiotic survival and colonization, as immune cell infiltration increases ROS levels, oxidizing lipids and proteins on the probiotic membrane, leading to inactivation. Moreover, pathogens that over-proliferate at the lesion site compete with probiotics for nutrients and available ecological niches, limiting their ability to establish in the gut (Han et al. 2021; Dupont et al. 2024). Probiotic strains vary in efficacy regarding tolerance, adhesion, antimicrobial activity, antibiotic resistance, and processing resilience, leading to differing therapeutic outcomes. Unfortunately, no single probiotic strain possesses all the necessary characteristics for optimal impact, and the absence of certain traits may lead to inconsistent clinical outcomes. For example, some probiotic strains may exhibit strong antimicrobial properties but poor adhesion, while others may have the opposite characteristics (Washburn et al. 2022; Kwoji et al. 2021). The effectiveness of probiotics is influenced by host genetics, gut microbiota composition, diet, and health conditions. This variability often results in inconsistent clinical outcomes, making it challenging to standardize native probiotic treatments for GI diseases across populations.

10.3 Methods for Designing Engineered Probiotics for Therapeutic Applications

Oral administration of probiotics faces challenges from environmental factors (temperature, moisture, oxidative stress), harsh GI conditions (acidic pH, digestive enzymes, bile salts, rapid transit), and functional limitations (poor colonization, rapid mucosal clearance, strain-specific constraints), all of which reduce the efficacy of probiotic-based treatments. These shortcomings have led to increased interest in engineering strategies to enhance probiotic resilience, target delivery to affected sites, modify surface properties for better colonization, and provide additional therapeutic potential for treating GI diseases (Castro-López et al. 2023; Huang et al. 2022). Genetic and chemical engineering approaches have been explored to impart specific functions to probiotics and improve their therapeutic outcomes for a wider range of GI diseases.

10.3.1 Genetically Engineered Probiotics

Genetic engineering modifies bacterial strains through exogenous gene expression, enabling wild probiotics to produce new proteins. These recombinant proteins may be intracellular, transmembrane, or secretory. Recent advancements in genome sequencing and microbial genome editing tools have led to a rapid increase in microbial genome sequences and the creation of customized probiotics. This approach aims to overcome the limitations of native probiotic treatments and enable personalized therapies (Fooladi et al. 2023; Mazhar et al. 2020). The general mechanisms of action of genetically engineered probiotics for treating GI diseases include:

- **Immunomodulatory activities**: Engineered probiotics can secrete outer membrane vesicles (OMVs) that integrate tumor antigens. Upon ingestion, these probiotics act as effective tumor vaccines in mice. Fo example, in-situ production of engineered OMVs in intestine by oral probiotic to deliver stimulatory molecules offers a promising strategy for oral vaccines against colon cancer (Yue et al. 2022).
- **Production of therapeutics**: Continuous production of therapeutic recombinant biomolecules within the intestinal tract helps combat inflammation and reprogram the gut microbiota. Compared to traditional therapies, in vivo production and delivery of functional proteins and drugs via engineered probiotics offer benefits like lower costs, non-invasiveness, and reduced need for repeat dosages (Vandenbroucke et al. 2004, 2010; Hanson et al. 2014).
- **Pathogen interaction blocking**: Engineering pathogen-specific adhesins or lectins onto probiotic surfaces strengthens adhesion to the intestinal mucosa and promote binding to host receptors. This targeted approach helps block pathogen-host interactions, protects against infection, outcompetes harmful microorganisms, and presents a novel strategy for infection prevention (Mohamadzadeh et al. 2010; Drolia et al. 2020).
- **Stress tolerance**: Modified probiotics can improve their tolerance to harsh gastrointestinal conditions and enhance stress resilience. For example, He et al. enhanced the oxidative stress tolerance of *B. longum* by introducing the *Bacillus subtilis* catalase gene. The modified strain showed increased catalase activity in the intracellular fraction, boosting their survival under oxidative stress conditions (He et al. 2012).
- **Biosensing potential:** Engineered probiotics offer a non-invasive way to monitor disease activity in IBD. Bacterial-based biosensors allow early GI diagnosis by detecting biomarkers. For instance, *Escherichia coli* Nissle (EcN) was modified to detect fecal calprotectin, a stool biomarker specific to gut inflammation, enabling accurate monitoring of IBD activity (Xia et al. 2023).

Genetic modification of probiotics represents a groundbreaking strategy for managing GI diseases. By enhancing probiotics to produce therapeutic biomolecules, block pathogen interactions, withstand environmental stress, and enable non-invasive disease monitoring, engineered probiotics offer significant advantages over traditional treatments. These advancements enhance therapeutic efficacy, and as research progresses, genetically modified probiotics may provide more efficient, cost-effective, and personalized treatments for gastrointestinal disorders, enabling adaptable healthcare solutions.

10.3.2 Chemical Surface Modification and Encapsulation

The integration of material science and microbiology has advanced oral probiotic delivery, driving significant progress in functional material-armored probiotics. Surface modification is achieved by encapsulating probiotics in protective films made from natural or synthetic polymers, boosting their resistance to environmental and GI stressors (Yang et al. 2023). Traditional microencapsulation techniques like freeze-drying, spray drying, extrusion, and emulsification protect probiotics from environmental stressors, maintain dormancy and stability at room temperature, and enhance survival against digestive enzymes, bile salts, and environmental exposure. However, the large coating material can increase the graft volume of encapsulated cells up to 100 times that of uncoated cells, exceeding the physiological diffusion distance for oxygen and nutrients. Furthermore, traditional coatings are often non-responsive to GI tract stimuli, limiting therapeutic effects and on-demand release of probiotics. This reduces intestinal mucosal adherence, impairs long-term colonization, and limits clinical applications (Centurion et al. 2021; Rajam and Subramanian 2022).

The interaction between living probiotic cells and polymeric biomaterials in protective coatings is crucial for single-cell encapsulation, providing enhanced features that overcome the limitations of traditional multicell encapsulation methods. These shells can be designed to improve resistance to environmental and GI conditions, enhance mucus adhesion, or detach upon reaching the intestine for interact with intestinal mucosa. Recently, there has been a shift toward protective, biocompatible, stimulus-responsive, and degradable nanocoating and surface functionalization technologies for individual probiotic cells. Frequently utilized coating materials are dextran, chitosan, Eudragit L100–55, alginate, cyclodextrin, silk fibroin, tannic acid, and polydopamine (Centurion et al. 2021; Xu et al. 2022). This section covers recent advancements in surface modification and encapsulation strategies, including biorthogonal conjugation, supramolecular coating, layer-by-layer (LbL) coating, and biological membranes around single probiotic cells, aimed at enhancing shelf life, delivery efficiency, and enabling targeted therapies for GI diseases.

10.3.2.1 Biorthogonal Conjugation

Unnatural monosaccharide analogs, such as mannosamine, galactosamine, and glucosamine, bearing bioorthogonal groups (e.g., azide), diffuse into cells and are metabolized into nucleotide sugars for glycosylation. The resulting modified glycoconjugates are subsequently transported to the endosomal and plasma membranes, thereby enhancing the expression of unnatural tags on sialoglycoproteins. Metabolic changes within parent cells can be mapped onto cell membrane proteins, with the quantity of tags regulated by adjusting the concentration of the unnatural monosaccharides. Through metabolic glycoengineering, various bioorthogonal groups can be incorporated into glycans. These bioorthogonal groups, such as azide moieties, serve as handles for click reactions, which facilitate the conjugation of therapeutic molecules to the cell surface. (Roy et al. 2020; Wang and Hu 2023).

The bioorthogonal-mediated bacterial delivery approach boosts probiotic engraftment by regulating bacterial adhesion between modified *Clostridium butyricum* (*C. butyricum*) and intestinal microbiota. Metabolic glycan modification process introduced azido-decorated D-alanine into the peptidoglycans of intestinal bacteria, enabling click conjugation with dibenzocyclooctyne (DBCO)-modified *C. butyricum*. Results showed that click reaction between azide group decorated D-alanine and DBCO-modified *C. butyricum* enhanced their adhesion in complex physiological conditions. DBCO-engineered *C. butyricum* exhibited better gut retention and significant disease relief in mice with dextran sodium sulfate (DSS)-induced colitis. This strategy demonstrates the potential of modified gut bacteria as binding sites for DBCO-decorated probiotics to improve bacterial colonization (Fig. 10.3) (Song et al. 2022). *E. coli Nissle* 1917 (EcN) was surface-modified with azide group by coupling N-hydroxysuccinimide ester to the bacterial cell wall's amine groups. High-molecular-weight anti-inflammatory hyaluronic acid (HA) was functionalized with DBCO and subsequently conjugated to the EcN surface via a bioorthogonal reaction. The HA-functionalized probiotic was then encapsulated in an enteric coating using calcium ion-assisted interactions. This coating protects the

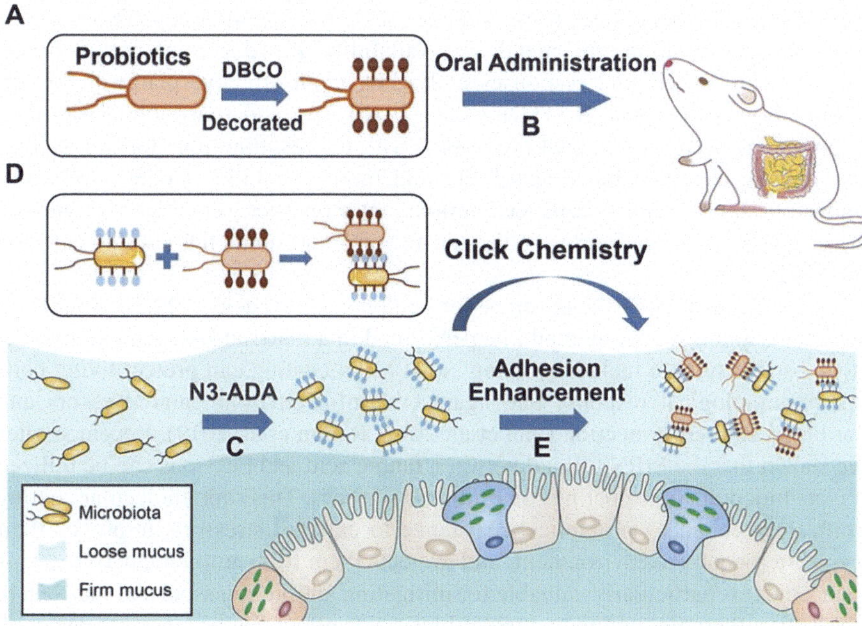

Fig. 10.3 Schematic representation of bioorthogonal-mediated enhancement of probiotic colonization. (**a**) dibenzocyclooctyne (DBCO) conjugation onto *C. butyricum*. (**b**) Oral administration of modified *C. butyricum*. (**c**) Metabolic incorporation of azido-modified D-alanine into the peptidoglycans of gut microbiota for azido decoration. (**d**) Click conjugation of the administered *C. butyricum* with gut microbiota. (**e**) Modified *C. butyricum* adhesion in the gut following the click reaction. (Reproduced with permission from Song et al. 2022, copyright 2022, American Chemical Society)

probiotics from stomach acid, dissolves in the neutral pH of the intestine, and promotes the release of immunomodulators to improve IBD therapy (Peng et al. 2024).

10.3.2.2 Supramolecular Coating

Supramolecular engineering is an interdisciplinary field focused on designing functional systems using non-covalent interactions and host-guest chemistry (Bernhard and Tibbitt 2021). Supramolecular self-assembly allows molecular components to spontaneously arrange into structured formations through specific intermolecular forces, enabling the development of biomacromolecule coatings on mammalian cells and probiotic (Cao et al. 2024; Dergham et al. 2022). Beta-cyclodextrin (β-CD) and phenolic compounds like tannic acid (TA) and dopamine exhibit various biological activities, including antioxidants, antibacterial, anti-inflammatory, and radical-scavenging effects. These compounds can be integrated onto substrate surfaces through supramolecular self-polymerization, metal-phenolic network (MPN) interactions, and host-guest interactions. Polydopamine coating is a rapid, cost-effective surface functionalization method inspired by mussel adhesive proteins. It involves the oxidative polymerization of dopamine under mildly alkaline conditions, allowing for consistent deposition of polydopamine films on various substrates, including bacterial cells. The thickness of polydopamine coatings can be adjusted from the nanometer to micrometer scale by adjusting deposition cycles, dopamine concentration, and metal ion availability. Polydopamine-modified surfaces can regulate cellular responses and serve as an effective platform for cell encapsulation (Lynge et al. 2011; Ding et al. 2016). The oxidative polymerization of dopamine has been used to coat probiotics, with the resulting polydopamine coatings showing excellent biocompatibility and biodegradability. Coated probiotics demonstrate high viability, enhanced antioxidant properties, remarkable protection against gastric acid, and strong adhesion to intestinal epithelial cells (Centurion et al. 2022).

Supramolecular MPN shells can be formed upon tannic acid and ferric ion (Fe^{3+}) interactions and this coating could be performed in a material-independent manner onto substrate surface including living cells. MPN coating can protect living cells from immunological response and against harmful environmental stressors and offer broad cellular protection (Fan et al. 2022; Rahim et al. 2019). Recent studies have shown that the MPN formed between tannic acid and Fe^{3+} ions can be utilized to create biocompatible coatings on probiotic surfaces. This supramolecular coating significantly enhances probiotics' resistance to external stressors, improves their survival in the gastric environment, and protects them from antibiotic activity. This advancement is particularly valuable for mitigating antibiotic-associated diarrhea in combination therapies and reducing antibiotic side effects in the GI tract (Pan et al. 2022). It offers a promising platform, showing that encapsulated probiotics can survive antibiotic-rich conditions, persist in the gut, release antimicrobial compounds, and help restore microbial balance (Fig. 10.4a). β-CD are macrocyclic oligosaccharides with a cone-shaped cavity, facilitating the formation of non-covalent supramolecular inclusion complexes with various drugs and polymers. Guest molecules inclusion into β-CD cone, enabling the development of novel materials for

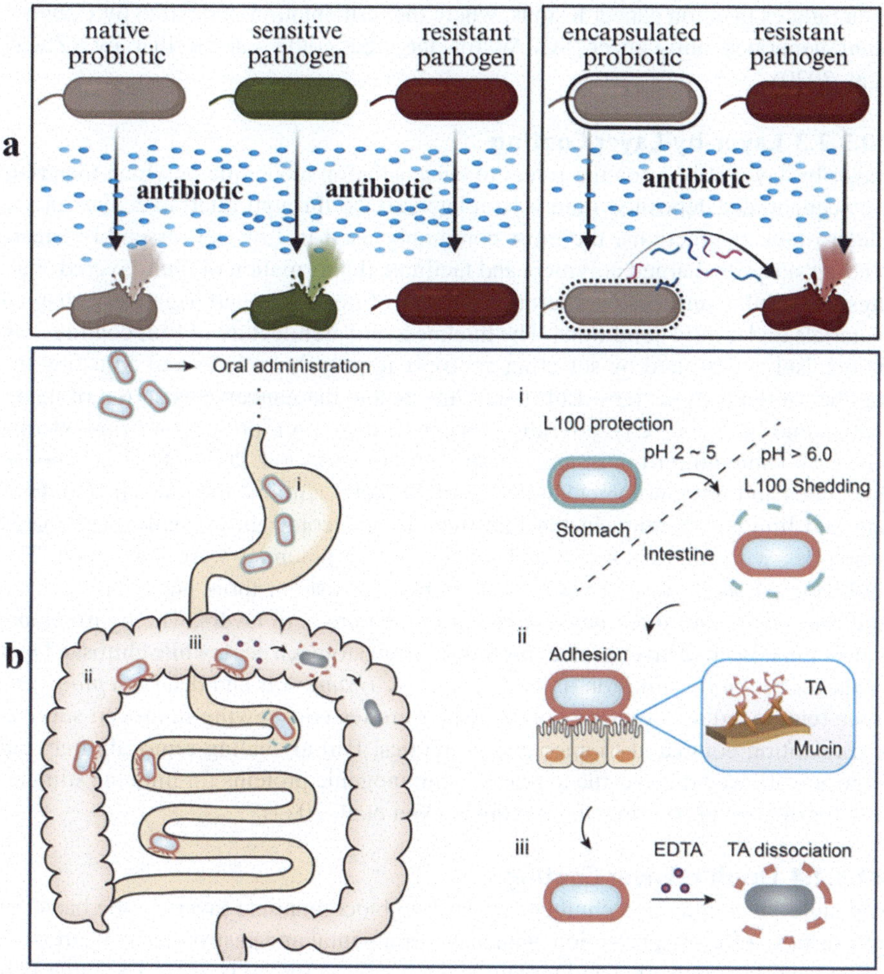

Fig. 10.4 (**a**) The schematic shows that antibiotics kill sensitive pathogens but fail against resistant strains. Native probiotics are destroyed by antibiotics, while encapsulated probiotics survive and secrete antimicrobial molecules to target resistant pathogens. (**b**) A schematic of the double-layer coating onto probiotic using tannic acid (TA) and enteric L100. The enteric layer protects the probiotic in the acidic stomach and degrades in the intestine through pH-triggered breakdown. Exposed tannic acid adheres to the mucosal epithelium, prolonging probiotic retention via tannic acid-mucin interactions. (Reproduced with permission form Liu et al. 2021, copyright 2021, Elsevier B.V.)

biomedical and pharmaceutical applications (Ullah et al. 2022). Adamantane was functionalized on the surface of *C. butyricum* via an EDC-NHS bio-conjugation reaction, followed by the attachment of β-CD-conjugated dextran to the adamantane-modified bacterial surface. This combination of probiotics bacterium (*C. butyricum*) and prebiotics (dextran) through host–guest supramolecular interaction serves as a safe microbiota-modulating material. After oral administration, it selectively accumulates in colon cancer lesions, where the fermentation of dextran by *C. butyricum* generates anti-cancer SCFAs for targeted colon cancer therapy (Zheng et al. 2020).

10.3.2.3 Layer-by-Layer Coating

Layer-by-layer (LbL) coating refers to encapsulation technique in which materials are sequentially deposited onto a template surface through intermolecular forces. Electrostatic interactions, the most commonly used in LbL self-assembly, create binding sites for charged polymers and facilitate the formation of multilayered coatings on cellular surfaces by alternating layers of positively and negatively charged materials under mild conditions. The thickness and architecture of LbL coatings can be precisely controlled by selecting appropriate polyelectrolytes and adjusting the number of deposition steps. LbL nanofilms enable the nanoencapsulation of mammalian and bacterial cells, providing mechanical support, protection from external stress, and shielding from immune responses (Borges et al. 2024). Oral delivery of probiotics and immunological agents is often inefficient due to stomach acid damage and limited retention in the intestine. To overcome these challenges, coated probiotics with pH-responsive and mucoadhesive potential were developed. The multilayered shell, created via LbL assembly, consists of three positively charged chitosan layers and three negatively trans-cinnamic acid layers. The hydrophobic trans-cinnamic acid layer protect probiotics from stomach acid, while chitosan layer enhances probiotic adhesion in the intestine. *L. rhamnosus* and dead *Streptococcus iniae* (employed as a vaccine component) were tested, showing improved survival and retention compared to uncoated probiotics. Partial coating removal after acid exposure allowed the vaccine to release immunogenic proteins for immune stimulation for disease prevention and treatment (Wei et al. 2024).

10.3.2.4 Double-Layer Coating

A double-layer coating technique applies two biocompatible layers to the bacterial cell surface, offering protection against environmental and gastrointestinal stressors while enabling controlled and targeted release at specific body sites. The outer pH-responsive layer, composed of L100 and Eudragit, protects probiotics from harsh stomach conditions and ensures their release in the intestine. The inner layer, made of mucoadhesive or antimicrobial agents, adheres to the gut mucosa, enhancing bacterial retention and promoting local therapeutic effects. The combination of the inner and outer coatings enhances the stability, functionality, and therapeutic potential of engineered bacteria (Feng et al. 2020; Liu et al. 2021). Liu et al. proposed a dual-layer coating method using tannic acid and enteric L100 to encapsulate living probiotics, overcoming the challenges associated with oral delivery. The pH-responsive L100 layer offers significant resistance against harsh environmental and GI environment, enabling selective delivery to the intestine. The tannic acid layer provides strong mucosal adhesion, prolonging probiotic retention without compromising viability, thereby enhancing the efficacy of colitis treatment. Overall, the double-layer coating significantly enhances the spatiotemporal delivery and therapeutic potential of probiotic-based GI diseases treatments (Fig. 10.4b) (Liu et al. 2021).

10.3.2.5 Biological Membrane Coating

Cell membrane coating is a bioinspired strategy that involves encapsulating nanoparticles, biomaterials, and bacterial cells with natural cell membranes to enhance their biological functionality. This method preserves the membrane's surface proteins, bioactivity, and targeting properties, enabling the coated entities to mimic natural cells and improve therapeutic efficacy (Fang et al. 2018). For example, Lin et al. describes a method to maintain intestinal microbiota homeostasis and prevent bacterial translocation through oral delivery of camouflaged probiotic bacteria into Peyer's patches. The probiotics, embedded in a yeast membrane with β-glucan through physical coextrusion. Due to specific binding, yeast membranes rich in surface glucans can be efficiently absorbed by microfold cells, which have Dectin-1 embedded in their membranes. This oral delivery method triggers mucosal immunity, enhancing secretory immunoglobulin A production and promoting gut microbiota regulation (Lin et al. 2021).

10.4 Engineered Probiotics in Gastrointestinal Diseases Management

10.4.1 Protection and Enhancement of Intestinal Barrier

The intestinal barrier maintains gut homeostasis by blocking pathogens, toxins, and antigens while allowing nutrient absorption. Strengthening this barrier helps prevent GI disorders, infections, and inflammation. Probiotics interact closely with the mucus lining of the gut, enhancing their retention in the gut, which boosts their therapeutic efficiency (Rose et al. 2021). Lin et al. describes a method to maintain intestinal microbiota homeostasis and prevent bacterial translocation through oral administration of camouflaged probiotics into Peyer's patches. The engineered probiotics, embedded in a yeast membrane with β-glucan through physical coextrusion, facilitate phagocytosis by microfold cells in the intestinal epithelium. This oral delivery method triggers mucosal immunity, enhancing secretory immunoglobulin A production and promoting gut microbiota regulation. In murine models of gut barrier impairment, encapsulated probiotics effectively prevent intestinal barrier breakdown, bacterial translocation, and systemic inflammation, offering a novel approach to defending against environmental threats (Lin et al. 2021). Centurion et al. created a catalytic method for shielding individual probiotic against stressors. The process, triggered by manganese, induces the oxidative polymerization of phenolic compounds, forming self-assembled nano-coatings onto probiotics. These coatings enhance biocompatibility, biodegradability, and gastric tolerance compared to native probiotic cells. Coated probiotics exhibit improved adhesion to intestinal epithelial cells and high ROS scavenging potential. This approach offers a novel method to enhance probiotic delivery by utilizing cell machinery to create protective and functional nano-coatings (Centurion et al. 2022).

Yang et al. engineered a modified probiotic via tannic acid and poloxamer assembling coating to enhance their intestinal colonization. The coating not only protects probiotics from harsh GI conditions but also scavenges ROS associated with inflammation, thereby improving probiotic survival. This coating promotes strong mucosal adhesion and enhances probiotic colonization in the diseased intestine. In a DSS-induced colitis mouse model, the modified probiotic exhibited significant anti-inflammatory effects, restored intestinal barrier function, and promoted a balanced gut microbiota. Additionally, in *Salmonella enterica serovar* Typhimurium (STm)-induced colitis mice, treatment with modified probiotic significantly reduced STm colonization compared to uncoated probiotics. This study highlights tannic acid and poloxamer-modified probiotics as a promising strategy for modulating the pathological microenvironment and treating microbiome-related disease (Yang et al. 2022a). Hou et al. developed a method to decorate probiotics with a therapeutic silk fibroin nanocoating to enhance their biotherapeutic potential. Probiotics were coated using an LbL procedure, taking advantage of silk fibroin's β-sheet conformation. This cytocompatible coating improved probiotic survival in simulated gastric fluid by 52-fold and enhanced colonization in the mouse intestine by 5.8 times compared to uncoated probiotics. The coating also displays natural anti-inflammatory properties, enhancing the therapeutic outcome in a mouse model of gut inflammation. This strategy boosts oral bioavailability and treatment efficacy, providing a versatile approach for probiotic-based therapies in managing GI diseases (Fig. 10.5) (Hou et al. 2021).

10.4.2 Probiotics for Inflammatory Bowel Disease

IBD refers to a long-term inflammatory condition of the GI tract, including ulcerative colitis and Crohn's disease. It is often associated with an imbalanced gut microbiome, dysregulated immune response, and excessive inflammatory microenvironment. The pathogenesis of IBD involves persistent intestinal inflammation, disruption of the mucosal barrier, immune dysregulation, and an imbalanced intestinal microbiome due to chronic inflammation. The treatment of IBD could benefit from targeted delivery of therapeutics to the intestine, however, traditional treatments are limited by low bioavailability and insufficient effectiveness (Verstockt et al. 2018; Santana et al. 2022). Engineered probiotics have been used to alleviate intestinal mucosal inflammation and restore the healthy balance of the gut microbiota.

Genetically modified EcN, overexpressing catalase and superoxide dismutase, was developed to efficiently eliminate ROS and intestinal inflammation. To enhance its gastrointestinal bioavailability, genetically modified EcN was coated with chitosan/sodium alginate films using an LbL self-assembly strategy. In a mouse model of IBD, oral delivery of the genetically engineered and chemically encapsulated EcN significantly reduced inflammation, characterized by reduced level of

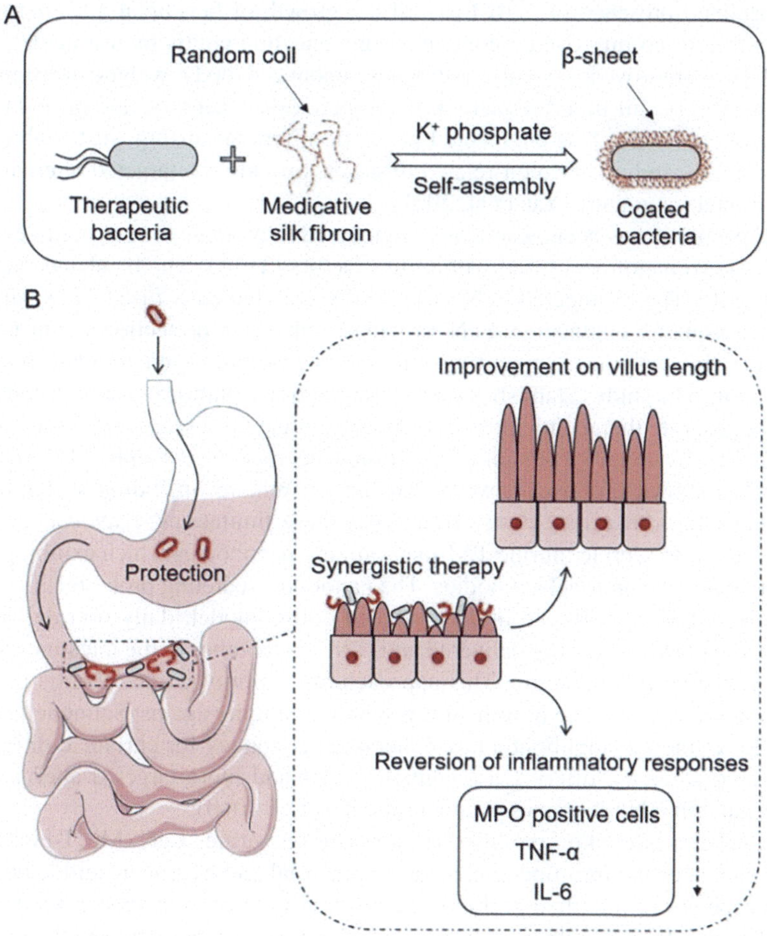

Fig. 10.5 Schematic illustration of decorating probiotic with protective and therapeutic nanocoating. (**a**) Self-assembly of medicated silk fibroin onto the probiotic surface. (**b**) Protective and therapeutic benefits of the silk fibroin coating. (Reproduced with permission from Hou et al. 2021, copyright 2021, Wiley-VCH GmbH)

pro-inflammatory cytokines and elevated level anti-inflammatory cytokines. This treatment also restored epithelial barrier integrity in the colon. Additionally, the nanocoated and genetically modified EcN regulated intestinal microbial communities, contributing to the maintenance of gut homeostasis (Zhou et al. 2022). EcN probiotic strain was engineered for sustained 3-hydroxybutyrate (3HB) production in colitis-afflicted mice. As a key ketone body, 3HB provides energy during starvation and shows therapeutic potential for inflammation and neurodegenerative diseases. Its biosynthetic pathway was integrated into the EcN genome, and engineered EcN efficiently produced 3HB under microaerobic conditions and stably colonized the mouse gut for over 14 days, significantly increasing gut 3HB and SCFA levels.

The sustained presence of 3HB fostered the growth of beneficial gut microbiota, contributing to an improved colonic environment. In a colitis mouse model, engineered EcN improved systemic outcomes, enhanced body weight maintenance, colon length, occult blood reduction, myeloperoxidase activity, and proinflammatory cytokine levels. This engineered probiotic offers promising, sustainable therapy for colitis and other colon-related diseases, providing a targeted alternative to conventional treatments (Yan et al. 2021).

EcN probiotic has been genetically engineered to produce fibrous matrices composed of curli nanofibers fused with trefoil factors (TFFs), which enhance gut barrier integrity. The engineered EcN successfully secreted curli-fused TFFs, and oral administration of engineered EcN provided enhanced protection against DSS-induced colitis in mice, supporting mucosal regeneration and immune response modulation. The study establishes a basis for creating a platform in which beneficial bacteria generate therapeutic protein matrices to treat GI diseases (Praveschotinunt et al. 2019). Systemic treatment of anti-tumor necrosis factor-alpha (TNF-α) antibodies has shown promise; however, challenges remain, including systemic side effects and high treatment costs. To address these limitations, *Lactococcus lactis* was modified to secrete murine TNF-neutralizing nanobodies which exhibit greater stability than conventional antibodies. The nanobody-secreting probiotic effectively neutralized TNF in vitro. In a DSS-induced colitis model, daily oral delivery of engineered probiotic locally delivered anti-TNF nanobodies to the intestine, significantly reducing inflammation. This approach also improved enterocolitis in IL-10 knockout mice without compromising systemic immunity against Salmonella infection. These findings highlight a novel therapeutic strategy for chronic colitis using engineered *L. lactis*, offering a potentially safer and more targeted alternative to traditional anti-TNF therapies (Vandenbroucke et al. 2010).

Yang et al. present Fe^{3+} modified montmorillonite-armed EcN (MMT-Fe@EcN), which enhances the intestinal colonization potential and hydrogen sulfide detoxification capacity of modified probiotics, contributing to the synergistic treatment of IBD. The montmorillonite layer protects EcN from environmental and GI stressors during oral delivery and improves colonization in mucus-depleted intestinal segments. MMT-Fe@EcN also manages inflammation by neutralizing hydrogen sulfide, enhancing probiotic survival rate and intestinal microbiota restoration. In DSS-induced colitis models, MMT-Fe@EcN reduces inflammation, restores the intestinal barrier, balances microbiota, and shows potential for clinical IBD treatment (Yang et al. 2024). Han et al. developed a coating around EcN using LbL technique with curcumin-loaded liposomes and a polymer blend of chitosan and gelatin. In DSS-induced colitis mice, these LbL coated EcN exhibited improved survival rates in harsh GI environments, enhanced adhesion to the intestinal lining, and reduced inflammation compared to uncoated EcN. This approach restored the intestinal barrier, balanced microbiota, and reduced bacterial motility, presenting a promising GI disease treatment (Han et al. 2024).

Peng et al. introduce a multifunctional probiotic-based approach for treating IBD with a single oral dose. EcN was engineered with azide group through the coupling of N-hydroxysuccinimide ester to amine groups on their cell wall. Functional hyaluronic acid (HA) was functionalized with DBCO and linked to EcN via bioorthogonal chemistry. The HA-conjugated EcN was then encapsulated in an enteric Eudragit L100–55 coating through calcium ion-assisted interactions. This coating protects the bacterium from stomach acidity and disintegrates in the neutral pH of the intestinal tract. Armored probiotics endure harsh conditions, release immunomodulators in the intestine, and drive pro-inflammatory macrophages toward an anti-inflammatory phenotype. In a mouse colitis model, oral EcN suppressed IBD by boosting body weight, reducing disease activity, downregulating pro-inflammatory cytokines, restoring gut barrier function, and enhancing microbiota richness and abundance (Peng et al. 2024). Inspired by probiotics' natural ability to colonize the mucus layer, a modified probiotic (EcN@TA-Ca^{2+}@Mucin) coated with tannic acid (TA) and mucin using LbL technology has been developed. This coating enhanced probiotics' resistance to harsh environmental and GI conditions while improving their adhesiveness to the intestine, promoting growth and colonization in the mucus layer. EcN@TA-Ca^{2+}@Mucin reduces inflammation through ROS scavenging, minimizes bacterial translocation in IBD, and enhances intestinal microbiota abundance. In DSS-induced colitis mice, it demonstrated excellent therapeutic and preventive efficacy. This platform presents a promising strategy for enhancing probiotic colonization and treating various GI diseases by modulating the pathological inflammatory microenvironment (Yang et al. 2022b).

Liu et al. applied norepinephrine for probiotic encapsulation, which autooxidizes to form a poly-norepinephrine film on their surface, offering protection against environmental and gastrointestinal stressors. The catecholamine group in norepinephrine provided probiotics with strong mucosal adhesion, enhancing their colonization in the intestine and thereby improving therapeutic efficacy. Building on the probiotics' ability to colonize colon tissue, hyaluronic acid/poly(propylene sulfide) nanoparticles (HPN) were attached to the engineered bacterial surface, facilitating their targeted delivery to inflamed colon tissues, balancing ROS concentration, and reducing off-target effects. This platform not only extends probiotics' retention in the intestine but also enables targeted, slow-release delivery of HPN, enhancing ROS-scavenging capabilities. It demonstrated improved prophylactic and therapeutic efficacy in a DSS-induced mouse colitis model, increasing microbiota diversity and alleviating IBDs (Liu et al. 2022). Peng et al. presents a promising strategy for treating colitis by engineered EcN and the anti-inflammatory drug 5-aminosalicylic acid within a GI-responsive alginate matrix. The alginate coating protects EcN from gastric acid, and disintegrates in the intestinal tract, releasing both probiotics and 5-aminosalicylic acid. In a DSS-induced colitis mouse model, the treatment improved microbiota richness, reduced proinflammatory cytokines, and restored intestinal barrier function, thereby demonstrating the synergistic therapeutic potential for intestinal diseases (Peng et al. 2023). Zhu et al. introduce a copolymer-modified H-silicene nanomaterial (SiH@TPGS-PEI) designed to optimize EcN probiotics for various gastrointestinal conditions. SiH nanomaterials were initially functionalized with polyethyleneimine-conjugated D-α-tocopheryl polyethylene glycol 1000 succinate (TPGS-PEI) via non-covalent hydrophobic interactions. This modification improves the dispersibility and stability of SiH nanomaterials. The SiH@TPGS-PEI coating shields EcN probiotics from stomach acidity. In the intestinal environment, the coating degrades, releasing anti-inflammatory hydrogen and exposing EcN to alleviate colitis in mice. This strategy provides new insights into developing intelligent, self-adaptive materials to enhance oral EcN probiotic delivery for managing IBD (Fig. 10.6) (Zhu et al. 2023).

Oral EcN was genetically modified to express interleukin-2 (IL-2) for the treatment of IBD. The modified probiotic regulates the intestinal microenvironment, and IL-2 promotes regulatory T cell generation and immune tolerance activation. To improve its bioavailability in the GI tract, EcN expressing IL-2 (EcN-IL2) was coated with Eudragit L100–55. This enteric coating significantly protects the probiotic against GI stressors, enhances its accumulation in the intestine, and enables controlled IL-2 secretion by the encapsulated probiotics. Coated EcN-IL2 regulated immune responses and modulated the gut microbiota. In a DSS-induced IBD mouse model, it significantly reduced inflammation and repaired the intestinal epithelial barrier (Li et al. 2024). Chen et al. developed a probiotic platform with biocatalytic and intestine-targeting coatings for managing IBD. Using a one-step bioinspired mineralization strategy, antioxidant nanocatalysts are grown on EcN probiotics, forming ferrihydrite nanostructures with catalase-like activity to scavenge ROS and reduce inflammation. Fucoidan-based shields are applied to enhance colon targeting. After administration, modified EcN show improved viability and colonization

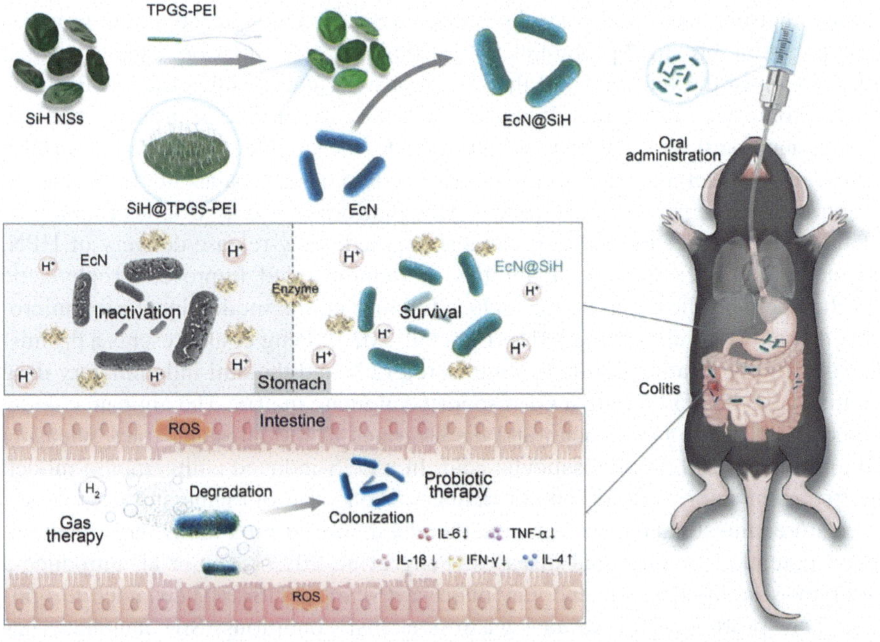

Fig. 10.6 SiH@TPGS-PEI coating shields probiotics from stomach acid, releases anti-inflammatory hydrogen in the intestine, and alleviates colitis in mice. (Reproduced with permission from Zhu et al. 2023, copyright 2023, American Chemical Society)

in the inflamed intestine, offering enhanced therapeutic effects against colitis through nanocatalysis-driven immunomodulation and microbiota reshaping (Chen et al. 2024).

10.4.3 Probiotics for Antibiotic-Associated Diarrhea

Antibiotic-associated diarrhea (AAD) is a common consequence of antibiotics, and it disrupts the gut microbiota, leading to overgrowth of opportunistic pathogens like *C. difficile*. Diarrhea severity ranges from mild to life-threatening, with symptoms such as watery diarrhea, abdominal cramping, and, in severe cases, fever and dehydration. The risk is higher with broad-spectrum antibiotics and prolonged treatment, especially in individuals with weakened immune systems (Ma et al. 2019; McFarland et al. 2016). Probiotics offer a promising strategy for preventing and treating AAD; however, antibiotics can harm native probiotics, underscoring the need for engineered solutions to enhance probiotic resilience. Pan et al. demonstrated that encapsulating probiotics within a protective supramolecular shell effectively shields them from antibiotic-induced damage. This biocompatible MPN coating, composed of tannic acid and Fe^{3+}, forms a transient barrier that protects coated bacteria from antibiotics in the GI tract. Additionally, this supramolecular coating enhances probiotic resistance to a range of antibiotics, improving their survival, colonization in the AAD-affected GI tract, and facilitating healthy microbe repopulation (Fig. 10.7) (Pan et al. 2022).

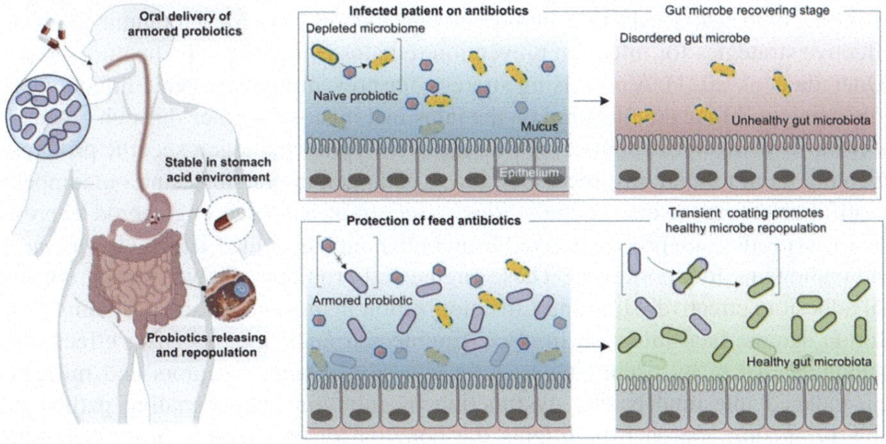

Fig. 10.7 A natural polyphenol-based single-cell coating, formed through a supramolecular metal-phenolic network between ferric ions and tannic acid, effectively protects probiotics from antibiotics in the gastrointestinal tract. (Reproduced with permission from Pan et al. 2022, copyright 2022, Springer Nature)

10.4.4 Probiotics for Opportunistic and Foodborne Pathogens

Infection occurs when pathogens or their secreted toxins are ingested and establish themselves within the host. Key bacterial pathogens include *Listeria monocytogenes* (Lm), *Helicobacter pylori*, *Vibrio cholera*, *Salmonella*, *Clostridium*, and *Shigella*. These pathogens typically colonize the gut, adhering to the mucosal layer. They release toxins and enzymes to penetrate the epithelial layer, leading to tissue injury, inflammation, and potential spread to other organs. Some opportunistic pathogens normally reside in the gut but cause disease only when the microbial balance is disturbed by stress or, more often, by antibiotics (Newell et al. 2010; Dey and Ray Chaudhuri 2023). The development of antibiotics for revolutionized treatment, providing effective and affordable solutions for infections that were once considered lethal. As microbial strains evolve resistance to multiple antibiotics, the growing prevalence of antibiotic-resistant pathogenic microorganism presents a major health challenge. The increasing prevalence of pathogenic bacteria and their infections demands innovative prevention and management strategies (Schjørring and Krogfelt 2011; Mancuso et al. 2021). In recent decades, designer probiotics have shown significant potential in combating opportunistic and food-borne pathogens that contribute to various GI diseases (Pamer 2016).

Lm is a foodborne bacterial pathogen that primarily colonizes the intestine, where it evades the innate immune defenses associated with the gut. This bacterium has the ability to penetrate the intestinal epithelial barrier, leading to serious infections in vulnerable populations such as infants and immunocompromised individuals. Furthermore, Lm can cross both the blood-brain and placental barriers, resulting in severe conditions such as meningitis, encephalitis, and, in pregnant women, the risk of abortion or stillbirth. Currently, there is no approved vaccine for preventing Lm infections, and prevention relies on proper cooking, safe food handling, and avoiding high-risk foods like unpasteurized cheeses. The development of cost-effective strategies for infection prevention remains a crucial public health concern (Quereda et al. 2021). A promising strategy for preventing pathogenic infections is enhancing probiotic colonization on the intestinal mucosal surface through bioengineering. This involves engineering probiotics to express pathogen-specific proteins, enabling them to bind host receptors, block pathogen interactions, and outcompete harmful microorganisms. *Lactobacillus casei* strain has been engineered express the Listeria adhesion protein derived from both nonpathogenic Listeria (*L. innocua*) and pathogenic foodborne Lm. These engineered probiotics showed strong intestinal adhesion, effectively limiting Lm colonization in the mucosa and providing protection against fatal infection in mice. Engineered *Lactobacillus casei* effectively outcompeted Lm by occupying Listeria adhesion protein receptors and mitigate Lm-induced intestinal barrier dysfunction by inhibiting key signaling pathways. This promising approach highlights the potential of engineering probiotics with adhesion proteins to boost protection and prevent infection (Drolia et al. 2020).

10.4.5 Probiotics for Colorectal Cancer Management

Colorectal cancer is a malignancy of the colon or rectum. It typically develops from precancerous polyps that undergo genetic and epigenetic changes, resulting in uncontrolled cell growth. Colorectal cancer is closely linked to intestinal microbiota imbalance in the colonic microenvironment (Tilg et al. 2018). Probiotic-based therapy is considered effective for treating colorectal cancer, but its safety remains a major challenge. Engineered probiotics have shown potential in regulating and treating colorectal cancer by restoring this balance. A recombinant *L. lactis* expressing the tumor metastasis-inhibiting KiSS1 peptide was successfully constructed. The secreted human KiSS1 peptide from the modified probiotic inhibited colonic epithelial cell proliferation and migration, likely through apoptosis induction and downregulation of MMP-9 expression (Zhang et al. 2016).

Yue et al. showed that commensal bacteria OMVs, when fused with a tumor-associated antigen, can be produced in the intestine and function as effective cancer vaccines in mouse models. In brief, *E. coli* was engineered to express a tumor-associated antigen fused with the cytolysin A protein on OMV surfaces, regulated by an arabinose-induced promoter. In mice, oral intake of arabinose and the modified commensal bacteria led to engineered OMV production, which crossed the intestinal lining into the lamina propria and triggered dendritic cells activation. In models of metastatic melanoma and subcutaneous colon tumors, tumor antigen-bearing bacterial vesicles effectively inhibited tumor growth. This approach of using modified commensal bacteria for in-situ production of engineered OMVs to deliver stimulatory molecules presents a promising strategy for oral colon cancer vaccines (Yue et al. 2022). Zheng et al. developed a microbiota-modulating biomaterial by combining *C. butyricum* probiotics and dextra prebiotics. Prebiotic-encapsulated probiotic spores were developed through host–guest interactions between *C. butyricum* and chemically modified prebiotic dextran. In brief, adamantane was functionalized on probiotic surface via an EDC-NHS bio-conjugation reaction, followed by the attachment of β-CD-conjugated dextran to the adamantane-modified probiotic. The study found that modified probiotic specifically accumulated in colon cancer lesions, and within these lesions, dextran was fermented by *C. butyricum*, producing anti-cancer SCFAs. Modified probiotic enhanced gut microbiota by boosting SCFA-producing bacteria and increasing microbiota richness. In subcutaneous and orthotopic tumor models, the drug-loaded modified probiotic significantly reduced tumor growth without causing significant adverse effects. This study highlights the potential of bioactive probiotics/prebiotics in regulating intestinal microbiota for treating colon cancer lesions (Zheng et al. 2020).

10.4.6 Probiotics for GI Diseases Diagnosis

Diagnosing gastrointestinal diseases is challenging due to overlapping signs such as abdominal pain, bloating, intermittent diarrhea, nausea, constipation, and rectal bleeding, often resulting in misdiagnosis. Conditions like early-stage colorectal cancer or celiac disease may be asymptomatic or have mild symptoms, complicating early detection. The clinical course of GI diseases varies among patients and is influenced by genetic and microbiome factors, adding to diagnostic complexity. Limited access to advanced invasive diagnostics, such as colonoscopy and endoscopic ultrasound, which require expertise and may cause discomfort or complications, delays timely care. As patient-reported symptoms are unreliable indicators of disease activity, biomarker-based screenings hold promise but require further validation. This highlights a clinical gap in precise diagnostic methods, emphasizing the need for advancements in non-invasive techniques, refinement of clinical protocols, and the adoption of a multidisciplinary approach to improve early detection and patient outcomes (Zhao et al. 2022; Deng et al. 2022).

The development of bacterial-based biosensors can provide a unique solution for early GI disease diagnosis by enabling timely biomarker detection. Advances in molecular biology are transforming cell-based clinical diagnostics, revolutionizing disease detection and monitoring. Probiotic EcN was genetically modified to detect calprotectin, a stool biomarker specific to gut inflammation. In short, calprotectin-sensitive genes were identified through sequencing, and their promoters were linked to a fluorescent reporter before being introduced into EcN. In murine colitis models, the reporter signal was activated during transit through the GI tract. Moreover, the engineered calprotectin-sensing probiotics accurately distinguished human patients with active IBD, those in remission, and non-IBD individuals through stool sample analysis. The reporter signal intensity correlated with clinical lab-measured calprotectin levels. This research paves the way for developing probiotics capable of detecting fecal calprotectin, enabling accurate non-invasive monitoring of intestinal disease activity (Fig. 10.8) (Xia et al. 2023). Researchers are designing intelligent gene circuits in engineered probiotics to detect disease markers, microorganisms, or metabolites. Computational processing allows for precise control of effector output, enabling probiotics to detect the presence, concentration, and frequency of signals. These designs are classified into sense-report (reporter gene expression), sense-therapy (therapeutic activation), and sense-record (signal recording). The intelligent responsive bacteria for diagnosis and therapy (i-ROBOT), based on EcN, is engineered to identify thiosulfate, an inflammatory marker, which activates a base-editing mechanism to create a heritable genomic sequence and a detectable colorimetric signal. Variations in thiosulfate also regulate the controlled release of the immunomodulator AvCystatin. When orally administered mice with colitis, i-ROBOT produced molecular recording signals in fecal and colon samples, significantly alleviating the disease. This novel approach offers a promising strategy for diagnosing and treating gastrointestinal and metabolic disorders (Zou et al. 2023).

10.5 Conclusions and Future Prospects

Significant efforts have been made to select optimal probiotic strains; however, naturally occurring probiotics often lack the full range of traits required for therapeutic efficacy. Progress in genetic engineering and chemical engineering have enabled the

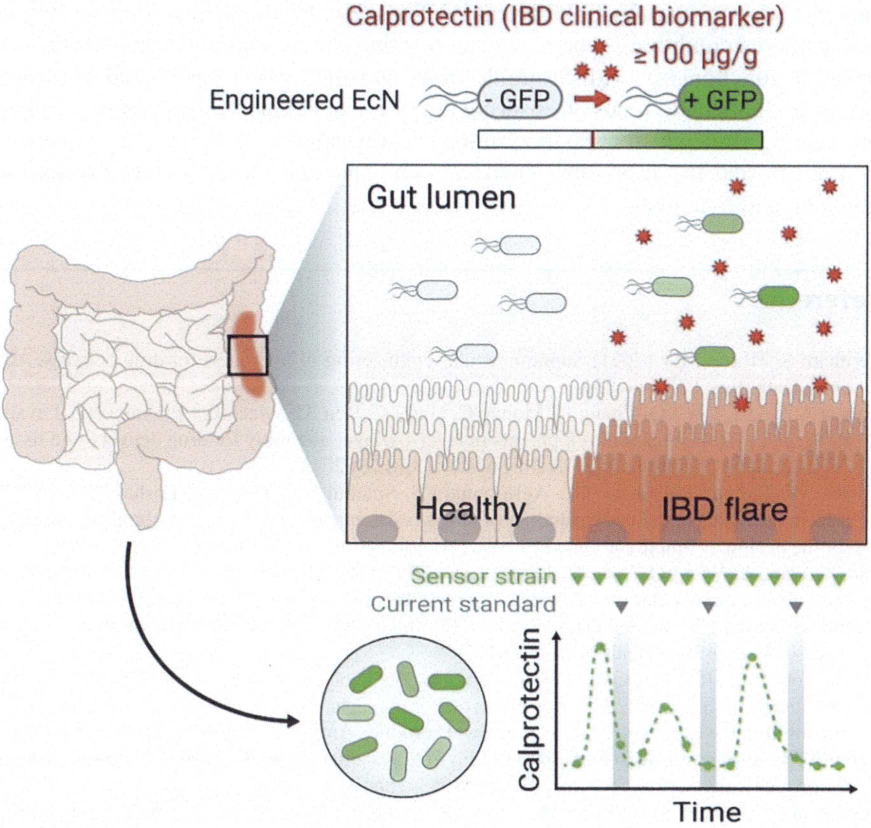

Fig. 10.8 A schematic of calprotectin-sensing *Escherichia coli* Nissle 1917 offers a reliable, non-invasive method for monitoring inflammatory bowel disease (IBD) activity. (Reproduced with permission from Xia et al. 2023, copyright 2023, Proc Natl Acad Sci)

development of designer bacterium that can produce bioactive compounds, facilitate surface conjugation of peptides and drugs, and enhance protection against environmental and GI stressors. These improvements offer significant therapeutic benefits, including enhanced intestinal adhesion, pathogen suppression, intestinal barrier reinforcement, and modulation of the pathological environment in intestinal diseases. Each engineering approach presents distinct advantages and limitations. Genetic engineering enhances probiotic function by enabling the detection of abnormal signals from the host and the production of therapeutic molecules but faces challenges such as limited gene-editing tools, potential antibiotic resistance transfer, and inadequate protection against stressors. In contrast, chemical surface modification and single-cell encapsulation strategies protect probiotics from environmental and GI stressors, improve targeted intestinal adhesion, and modulate the pathological environment through the release of antimicrobial, antioxidant, or immunomodulatory agents. However, unlike genetic engineering, these strategies may not support continuous production of therapeutic molecules. Recent studies have explored combining genetic engineering and chemical modification to enhance probiotic functionality with desirable traits and improved translational potential. Future studies should focus on ensuring the safety of engineered probiotics, refining engineering strategies for superior strains, reprogramming probiotics for personalized needs, and incorporating genetic circuits and sensors for precise GI disease detection and treatment.

References

Bernhard S, Tibbitt MW (2021) Supramolecular engineering of hydrogels for drug delivery. Adv Drug Deliv Rev 171:240–256

Borges J, Zeng J, Liu XQ, Chang H, Monge C, Garot C, Ren KF, Machillot P, Vrana NE, Lavalle P, Akagi T (2024) Recent developments in layer-by-layer assembly for drug delivery and tissue engineering applications. Adv Healthc Mater 13(8):2302713

Bustos AY, Taranto MP, Gerez CL, Agriopoulou S, Smaoui S, Varzakas T, Enshasy HA (2024) Recent advances in the understanding of stress resistance mechanisms in probiotics: relevance for the design of functional food systems. Probiotics Antimicrob Proteins 17:138–158

Cao X, Wang Y, Wang Q, Yang H, Ding Y, Mao Z (2024) Nano-armor modification of bacteria through supramolecular interactions for disease treatment. Coord Chem Rev 520:216165

Carding S, Verbeke K, Vipond DT, Corfe BM, Owen LJ (2015) Dysbiosis of the gut microbiota in disease. Microb Ecol Health Dis 26:26191

Castro-López C, Romero-Luna HE, García HS, Vallejo-Cordoba B, González-Córdova AF, Hernández-Mendoza A (2023) Key stress response mechanisms of probiotics during their journey through the digestive system: a review. Probiotics Antimicrob Proteins 15(5):1250–1270

Centurion F, Basit AW, Liu J, Gaisford S, Rahim MA, Kalantar-Zadeh K (2021) Nanoencapsulation for probiotic delivery. ACS Nano 15(12):18653–18660

Centurion F, Merhebi S, Baharfar M, Abbasi R, Zhang C, Mousavi M, Xie W, Yang J, Cao Z, Allioux FM, Harm GF (2022) Cell-mediated biointerfacial phenolic assembly for probiotic nano encapsulation. Adv Funct Mater 32(26):2200775

Chen C, Zhu Z (2023) Recent advances in the nanoshells approach for encapsulation of single probiotics. Drug Des Devel Ther 17:2763–2774

Chen T, Meng W, Li Y, Li X, Yu X, Qi J, Ding D, Li W (2024) Probiotics armed with in situ mineralized nanocatalysts and targeted biocoatings for multipronged treatment of inflammatory bowel disease. Nano Lett 24(24):7321–7331

Deng S, Gu J, Jiang Z, Cao Y, Mao F, Xue Y, Wang J, Dai K, Qin L, Liu K, Wu K (2022) Application of nanotechnology in the early diagnosis and comprehensive treatment of gastrointestinal cancer. J Nanobiotechnol 20(1):415

Dergham M, Lin S, Geng J (2022) Supramolecular self-assembly in living cells. Angew Chem 134(18):e202114267

Dey P, Ray Chaudhuri S (2023) The opportunistic nature of gut commensal microbiota. Crit Rev Microbiol 49(6):739–763

Ding YH, Floren M, Tan W (2016) Mussel-inspired polydopamine for bio-surface functionalization. Biosurf Biotribol 2(4):121–136

Drolia R, Amalaradjou MA, Ryan V, Tenguria S, Liu D, Bai X, Xu L, Singh AK, Cox AD, Bernal-Crespo V, Schaber JA (2020) Receptor-targeted engineered probiotics mitigate lethal listeria infection. Nat Commun 11:6344

Duan H, Yu L, Tian F, Zhai Q, Fan L, Chen W (2022) Antibiotic-induced gut dysbiosis and barrier disruption and the potential protective strategies. Crit Rev Food Sci Nutr 62(6):1427–1452

Dupont S, Beney L, Moundanga S, Denimal E, Hồ PH, Karbowiak T (2024) Ex-vivo investigation of probiotic bacterial adhesion to the intestinal mucus. Heliyon 10(16):e36339

Fan G, Cottet J, Rodriguez-Otero MR, Wasuwanich P, Furst AL (2022) Metal–phenolic networks as versatile coating materials for biomedical applications. ACS Appl Bio Mater 5(10):4687–4695

Fang RH, Kroll AV, Gao W, Zhang L (2018) Cell membrane coating nanotechnology. Adv Mater 30(23):1706759

Feng P, Cao Z, Wang X, Li J, Liu J (2020) On-demand bacterial reactivation by restraining within a triggerable nanocoating. Adv Mater 32(34):2002406

Foo JL, Ling H, Lee YS, Chang MW (2017) Microbiome engineering: current applications and its future. Biotechnol J 12(3):1600099

Fooladi S, Rabiee N, Iravani S (2023) Genetically engineered bacteria: a new frontier in targeted drug delivery. J Mater Chem B 11(42):10072–10087

Gao J, Li X, Zhang G, Sadiq FA, Simal-Gandara J, Xiao J, Sang Y (2021) Probiotics in the dairy industry – advances and opportunities. Compr Rev Food Sci Food Saf 20(4):3937–3982

Han S, Lu Y, Xie J, Fei Y, Zheng G, Wang Z, Liu J, Lv L, Ling Z, Berglund B, Yao M (2021) Probiotic gastrointestinal transit and colonization after oral administration: a long journey. Front Cell Infect Microbiol 11:609722

Han M, Shen N, Tan W, Wang X, Liu Y, Liang J, Li H, Gao Z (2024) Layer-by-layer coated probiotics with chitosan and liposomes exhibit enhanced therapeutic effects for DSS-induced colitis in mice. Int J Biol Macromol 269:132063

Hanson ML, Hixon JA, Li W, Felber BK, Anver MR, Stewart CA, Janelsins BM, Datta SK, Shen W, McLean MH, Durum SK (2014) Oral delivery of IL-27 recombinant bacteria attenuates immune colitis in mice. Gastroenterology 146(1):210–221

He J, Sakaguchi K, Suzuki T (2012) Acquired tolerance to oxidative stress in Bifidobacterium longum 105-a via expression of a catalase gene. Appl Environ Microbiol 78(8):2988–2990

Hou W, Li J, Cao Z, Lin S, Pan C, Pang Y, Liu J (2021) Decorating bacteria with a therapeutic nanocoating for synergistically enhanced biotherapy. Small 17(37):2101810

Huang Y, Lin X, Yu S, Chen R, Chen W (2022) Intestinal engineered probiotics as living therapeutics: chassis selection, colonization enhancement, gene circuit design, and biocontainment. ACS Synth Biol 11(10):3134–3153

Kwoji ID, Aiyegoro OA, Okpeku M, Adeleke MA (2021) Multi-strain probiotics: synergy among isolates enhances biological activities. Biology 10(4):322

Lamberte LE, van Schaik W (2022) Antibiotic resistance in the commensal human gut microbiota. Curr Opin Microbiol 68:102150

Lebeer S, Vanderleyden J, De Keersmaecker SC (2010) Host interactions of probiotic bacterial surface molecules: comparison with commensals and pathogens. Nat Rev Microbiol 8(3):171–184

Li M, Liu N, Zhu J, Wu Y, Niu L, Liu Y, Chen L, Bai B, Miao Y, Yang Y, Chen Q (2024) Engineered probiotics with sustained release of interleukin-2 for the treatment of inflammatory bowel disease after oral delivery. Biomaterials 309:122584

Lin S, Mukherjee S, Li J, Hou W, Pan C, Liu J (2021) Mucosal immunity–mediated modulation of the gut microbiome by oral delivery of probiotics into Peyer's patches. Sci Adv 7(20):eabf0677

Liu J, Li W, Wang Y, Ding Y, Lee A, Hu Q (2021) Biomaterials coating for on-demand bacteria delivery: selective release, adhesion, and detachment. Nano Today 41:101291

Liu J, Wang Y, Heelan WJ, Chen Y, Li Z, Hu Q (2022) Mucoadhesive probiotic backpacks with ROS nanoscavengers enhance the bacteriotherapy for inflammatory bowel diseases. Sci Adv 8(45):eabp8798

Lynge ME, van der Westen R, Postma A, Städler B (2011) Polydopamine – a nature-inspired polymer coating for biomedical science. Nanoscale 3(11):4916–4928

Ma H, Zhang L, Zhang Y, Liu Y, He Y, Guo L (2019) Combined administration of antibiotics increases the incidence of antibiotic-associated diarrhea in critically ill patients. Infect Drug Resist 12:1047–1054

Mancuso G, Midiri A, Gerace E, Biondo C (2021) Bacterial antibiotic resistance: the most critical pathogens. Pathogens 10(10):1310

Markowiak-Kopeć P, Śliżewska K (2020) The effect of probiotics on the production of short-chain fatty acids by human intestinal microbiome. Nutrients 12(4):1107

Mazhar SF, Afzal M, Almatroudi A, Munir S, Ashfaq UA, Rasool M, Raza H, Munir HM, Rajoka MS, Khurshid M (2020) The prospects for the therapeutic implications of genetically engineered probiotics. J Food Qual 2020(1):9676452–9676411

McFarland LV, Ozen M, Dinleyici EC, Goh S (2016) Comparison of pediatric and adult antibiotic-associated diarrhea and Clostridium difficile infections. World J Gastroenterol 22(11):3078–3104

Misra S, Pandey P, Dalbhagat CG, Mishra HN (2022) Emerging technologies and coating materials for improved probiotication in food products: a review. Food Bioprocess Technol 15(5):998–1039

Mohamadzadeh M, Durmaz E, Zadeh M, Pakanati KC, Gramarossa M, Cohran V, Klaenhammer TR (2010) Targeted expression of anthrax protective antigen by lactobacillus gasseri as an anthrax vaccine. Future Microbiol 5(8):1289–1296

Mukherjee PK, Sendid B, Hoarau G, Colombel JF, Poulain D, Ghannoum MA (2015) Mycobiota in gastrointestinal diseases. Gastroenterol Hepatol 12(2):77–87

Murali SK, Mansell TJ (2024) Next generation probiotics: engineering live biotherapeutics. Biotechnol Adv 72:108336

Nash AK, Auchtung TA, Wong MC, Smith DP, Gesell JR, Ross MC, Stewart CJ, Metcalf GA, Muzny DM, Gibbs RA, Ajami NJ (2017) The gut mycobiome of the human microbiome project healthy cohort. Microbiome 5:1–13

Newell DG, Koopmans M, Verhoef L, Duizer E, Aidara-Kane A, Sprong H, Opsteegh M, Langelaar M, Threfall J, Scheutz F, van der Giessen J (2010) Food-borne diseases – the challenges of 20 years ago still persist while new ones continue to emerge. Int J Food Microbiol 139:S3–S15

Pamer EG (2016) Resurrecting the intestinal microbiota to combat antibiotic-resistant pathogens. Science 352(6285):535–538

Pan J, Gong G, Wang Q, Shang J, He Y, Catania C, Birnbaum D, Li Y, Jia Z, Zhang Y, Joshi NS (2022) A single-cell nanocoating of probiotics for enhanced amelioration of antibiotic-associated diarrhea. Nat Commun 13(1):2117

Peng P, Feng T, Yang X, Nie C, Yu L, Ding R, Zhou Q, Jiang X, Li P (2023) Gastrointestinal microenvironment responsive nanoencapsulation of probiotics and drugs for synergistic therapy of intestinal diseases. ACS Nano 17(15):14718–14730

Peng P, Feng T, Yang X, Ding R, Wang J, Chen P, Guo Y, Li P (2024) Bioorthogonal conjugation and responsive nanocoating of probiotics for inflammatory bowel disease. J Control Release 374:538–549

Peterson J, Garges S, Giovanni M, McInnes P, Wang L, Schloss JA, Bonazzi V, McEwen JE, Wetterstrand KA, Deal C, Baker CC (2009) The NIH human microbiome project. Genome Res 19(12):2317–2323

Praveschotinunt P, Duraj-Thatte AM, Gelfat I, Bahl F, Chou DB, Joshi NS (2019) Engineered E. Coli Nissle 1917 for the delivery of matrix-tethered therapeutic domains to the gut. Nat Commun 10(1):5580

Prosekov AY, Dyshlyuk LS, Milentyeva IS, Sykhikh SA, Babich OO, Ivanova SA, Pavsky VA, Shishin MV, Matskova LV (2017) Antioxidant and antimicrobial activity of bacteriocin-producing strains of lactic acid bacteria isolated from the human gastrointestinal tract. Prog Nutr 19(1):67–80

Quereda JJ, Morón-García A, Palacios-Gorba C, Dessaux C, García-del Portillo F, Pucciarelli MG, Ortega AD (2021) Pathogenicity and virulence of listeria monocytogenes: a trip from environmental to medical microbiology. Virulence 12(1):2509–2545

Rahim MA, Kristufek SL, Pan S, Richardson JJ, Caruso F (2019) Phenolic building blocks for the assembly of functional materials. Angew Chem Int Ed 58(7):1904–1927

Rajam R, Subramanian P (2022) Encapsulation of probiotics: past, present and future. Beni-Suef Univ J Basic Appl Sci 11(1):46

Reque PM, Brandelli A (2021) Encapsulation of probiotics and nutraceuticals: applications in functional food industry. Trends Food Sci Technol 114:1–10

Rose EC, Odle J, Blikslager AT, Ziegler AL (2021) Probiotics, prebiotics and epithelial tight junctions: a promising approach to modulate intestinal barrier function. Int J Mol Sci 22(13):6729
Roy S, Cha JN, Goodwin AP (2020) Nongenetic bioconjugation strategies for modifying cell membranes and membrane proteins: a review. Bioconjug Chem 31(11):2465–2475
Ruan W, Engevik MA, Spinler JK, Versalovic J (2020) Healthy human gastrointestinal microbiome: composition and function after a decade of exploration. Dig Dis Sci 65:695–705
Santana PT, Rosas SL, Ribeiro BE, Marinho Y, de Souza HS (2022) Dysbiosis in inflammatory bowel disease: pathogenic role and potential therapeutic targets. Int J Mol Sci 23(7):3464
Schjørring S, Krogfelt KA (2011) Assessment of bacterial antibiotic resistance transfer in the gut. Int J Microbiol 2011(1):1–10
Singh TP, Natraj BH (2021) Next-generation probiotics: a promising approach towards designing personalized medicine. Crit Rev Microbiol 47(4):479–498
Song WF, Yao WQ, Chen QW, Zheng D, Han ZY, Zhang XZ (2022) In situ bioorthogonal conjugation of delivered bacteria with gut inhabitants for enhancing probiotics colonization. ACS Cent Sci 8(9):1306–1317
Sonnenburg JL (2015) Microbiome engineering. Nature 518:S10–S10
Tilg H, Adolph TE, Gerner RR, Moschen AR (2018) The intestinal microbiota in colorectal cancer. Cancer Cell 33(6):954–964
Ullah S, Hussain Z, Wang L, Ullah I, Mehmood S, Luo B, Zhang Y, Ghani MW, Pei R, Wang J (2022) Cyclodextrin-based host–guest supramolecular nanofibrous composite for biomedical applications. Chem Proc 12(1):60
Vandenbroucke K, Hans W, Van Huysse J, Neirynck S, Demetter P, Remaut E, Rottiers P, Steidler L (2004) Active delivery of trefoil factors by genetically modified Lactococcus lactis prevents and heals acute colitis in mice. Gastroenterology 127(2):502–513
Vandenbroucke K, De Haard H, Beirnaert E, Dreier T, Lauwereys M, Huyck L, Van Huysse J, Demetter P, Steidler L, Remaut E, Cuvelier C (2010) Orally administered L. lactis secreting an anti-TNF Nanobody demonstrate efficacy in chronic colitis. Mucosal Immunol 3(1):49–56
Verstockt B, Ferrante M, Vermeire S, Van Assche G (2018) New treatment options for inflammatory bowel diseases. J Gastroenterol 53:585–590
Wang Y, Hu Q (2023) Bio-orthogonal chemistry in cell engineering. Adv Nanobiomed Res 3(3):2200128
Washburn RL, Sandberg D, Stofer MA (2022) Supplementation of a single species probiotic does not affect diversity and composition of the healthy adult gastrointestinal microbiome. Hum Nutr Metab 28:200148
Wei G, Yue Feng MT, Si Z, Chan-Park MB (2024) Single-cell oral delivery platform for enhanced acid resistance and intestinal adhesion. ACS Appl Mater Interfaces 16(17):21498–21508
Xia JY, Hepler C, Tran P, Waldeck NJ, Bass J, Prindle A (2023) Engineered calprotectin-sensing probiotics for IBD surveillance in humans. Proc Natl Acad Sci 120(32):e2221121120
Xu C, Ban Q, Wang W, Hou J, Jiang Z (2022) Novel nano-encapsulated probiotic agents: encapsulate materials, delivery, and encapsulation systems. J Control Release 349:184–205
Yan X, Liu XY, Zhang D, Zhang YD, Li ZH, Liu X, Wu F, Chen GQ (2021) Construction of a sustainable 3-hydroxybutyrate-producing probiotic Escherichia coli for treatment of colitis. Cell Mol Immunol 18(10):2344–2357
Yang J, Zhang G, Yang X, Peng M, Ge S, Tan S, Wen Z, Wang Y, Wu S, Liang Y, An J (2022a) An oral "super probiotics" with versatile self-assembly adventitia for enhanced intestinal colonization by autonomous regulating the pathological microenvironment. Chem Eng J 446:137204
Yang X, Yang J, Ye Z, Zhang G, Nie W, Cheng H, Peng M, Zhang K, Liu J, Zhang Z, Shi J (2022b) Physiologically inspired mucin coated Escherichia coli Nissle 1917 enhances biotherapy by regulating the pathological microenvironment to improve intestinal colonization. ACS Nano 16(3):4041–4058
Yang X, Wang C, Wang Q, Zhang Z, Nie W, Shang L (2023) Armored probiotics for oral delivery. Smart Med 2(4):e20230019

Yang J, Ge S, Tan S, Liu H, Yang M, Liu W, Zhang K, Zhang Z, Liu J, Shi J, Wang ZH (2024) Modified montmorillonite armed probiotics with enhanced on-site mucus-depleted intestinal colonization and H2S scavenging for colitis treatment. J Control Release 374:140–153

Yao Y, Luo J, Zhang P, Wang Y, Lu B, Wu G, Zhang J, Luo X, Wang L (2024) Screening, identification and application of lactic acid bacteria for degrading mycotoxin isolated from the rumen of yaks. Microorganisms 12(11):2260

Yousefi B, Eslami M, Ghasemian A, Kokhaei P, Salek Farrokhi A, Darabi N (2019) Probiotics importance and their immunomodulatory properties. J Cell Physiol 234(6):8008–8018

Yue Y, Xu J, Li Y, Cheng K, Feng Q, Ma X, Ma N, Zhang T, Wang X, Zhao X, Nie G (2022) Antigen-bearing outer membrane vesicles as tumour vaccines produced in situ by ingested genetically engineered bacteria. Nat Biomed Eng 6(7):898–909

Zhang B, Li A, Zuo F, Yu R, Zeng Z, Ma H, Chen S (2016) Recombinant Lactococcus lactis NZ9000 secretes a bioactive kisspeptin that inhibits proliferation and migration of human colon carcinoma HT-29 cells. Microb Cell Factories 15:1–11

Zhao C, Zhang X, Chen G, Shang L (2022) Developing sensor materials for screening intestinal diseases. Mater Futures 1(2):022401

Zheng DW, Li RQ, An JX, Xie TQ, Han ZY, Xu R, Fang Y, Zhang XZ (2020) Prebiotics-encapsulated probiotic spores regulate gut microbiota and suppress colon cancer. Adv Mater 32(45):2004529

Zhou J, Li M, Chen Q, Li X, Chen L, Dong Z, Zhu W, Yang Y, Liu Z, Chen Q (2022) Programmable probiotics modulate inflammation and gut microbiota for inflammatory bowel disease treatment after effective oral delivery. Nat Commun 13(1):3432

Zhu YX, You Y, Chen Z, Xu D, Yue W, Ma X, Jiang J, Wu W, Lin H, Shi J (2023) Inorganic nanosheet-shielded probiotics: a self-adaptable oral delivery system for intestinal disease treatment. Nano Lett 23(10):4683–4692

Zou ZP, Du Y, Fang TT, Zhou Y, Ye BC (2023) Biomarker-responsive engineered probiotic diagnoses, records, and ameliorates inflammatory bowel disease in mice. Cell Host Microbe 31(2):199–212

Part IV

Engineered Cell Membrane Platforms for Immunotherapy, Vaccine Development, and Diagnostics

Cell Membrane Engineering for Advancing Cancer Immunotherapy

11

Tong Li, Bin Liu, Cheng Dong, and Yun Chang

Abstract

Immunotherapy has revolutionized cancer treatment, demonstrating remarkable clinical efficacy. A particularly promising modality is adoptive immune cell therapy, a critical component of cancer immunotherapy that provides a highly targeted, personalized strategy by harnessing the immune system to combat cancer. Its potential for long-lasting immune surveillance and its success in treating challenging cancers highlight its vital role in modern cancer therapy. Membrane engineering of adoptive immune cells further enhances the precision, efficacy, and safety of this approach. By modifying the surface characteristics of immune cells, researchers can overcome key challenges, including tumor immune evasion, limited cell persistence, and off-target effects. This membrane engineering approach demonstrates significant potential to enhance adoptive cell therapy efficacy while broadening its therapeutic scope across diverse malignancies. This chapter will explore both genetic and non-genetic methods used in immune cell membrane engineering, demonstrating how these advancements enable the development of personalized therapies. These therapies may substantially enhance outcomes for patients across a diverse array of cancer types, including those that are currently difficult to treat.

T. Li
Department of Biomedical Engineering, The Hong Kong Polytechnic University, Hong Kong, China

B. Liu · C. Dong (✉) · Y. Chang (✉)
Department of Biomedical Engineering, The Hong Kong Polytechnic University, Hong Kong, China

The Hong Kong Polytechnic University Shenzhen Research Institute, Shenzhen, China
e-mail: cheng.dong@polyu.edu.hk; yun-bme.chang@polyu.edu.hk

Z. Hussain, R. Pei (eds.), *Cell Membrane Engineering for Advancing Cell-Based and Nanotechnology-Based Therapies*, Biomaterials, Bioengineering and Sustainability 6, https://doi.org/10.1007/978-3-031-93425-4_11

Keywords

Cancer immunotherapy · Adoptive immune cells · Membrane engineering · Chimeric antigen receptor · Genetical and non-genetical engineering

11.1 Introduction

Despite significant advancements in diagnostic and therapeutic approaches, cancer remains the leading cause of disease-related mortality worldwide (Bray et al. 2018). Breakthroughs in cancer biology research have facilitated the development and approval of numerous targeted therapies, including small molecule inhibitors, monoclonal antibodies (mAbs), and genetically engineered cellular products (Wang et al. 2019). These novel agents, which either disrupt critical oncogenic pathways or enhance immune-mediated tumor clearance, have demonstrated remarkable clinical efficacy, achieving durable responses and even complete remission in certain malignancies (Wang et al. 2019).

11.1.1 Traditional Cancer Treatments and Their Drawbacks

Conventional therapeutic modalities—including surgical resection, radiation therapy, cytotoxic chemotherapy, and photodynamic treatment—have long constituted the foundation of oncologic care. However, these conventional approaches frequently induce significant toxicities and demonstrate limited efficacy against advanced-stage or metastatic malignancies (Cao et al. 2018). For example, while surgical resection remains one of the oldest and most effective approaches for localized solid tumors, its utility is limited in metastatic cancers or tumors located in critical anatomical regions. Radiotherapy, though widely used, often induces both localized and systemic adverse effects, such as damage to adjacent healthy tissues, fibrosis formation, and immunosuppression. Conventional chemotherapy employs cytotoxic agents to eliminate rapidly dividing cells throughout the body, making it a cornerstone of systemic cancer treatment. However, its clinical application is constrained by non-specific biodistribution and dose-dependent toxicities (Cao et al. 2018). Photodynamic therapy faces significant challenges due to the limited tumor specificity of most photosensitizers. Upon light activation, these agents generate reactive oxygen species (ROS) indiscriminately, potentially damaging healthy tissues and causing adverse effects (Yuan et al. 2024).

11.1.2 Cancer Immunotherapy and Its Challenges

Cancer immunotherapy has revolutionized oncology by fundamentally altering therapeutic strategies through targeted immune modulation rather than direct tumor cytotoxicity. Unlike conventional chemotherapeutic agents that indiscriminately kill rapidly dividing cells, immunotherapies specifically enhance the host's antitumor immune responses while minimizing collateral damage to healthy tissues (Riley et al. 2019). These treatments work by either activating dormant immune pathways or amplifying existing immune recognition mechanisms that tumors typically evade during malignant progression (Quesada et al. 1986; Rosenberg 2014; Ahmed and Rai 2003). This precision approach has established immunotherapy as a groundbreaking treatment modality capable of achieving durable remissions and potential cures in previously refractory malignancies.

The field of cancer immunotherapy has evolved dramatically since the introduction of recombinant interferon-α (IFNα) as the first approved biologic therapy (Quesada et al. 1986), Over the past four decades, the therapeutic arsenal has expanded to include: (1) immune checkpoint inhibitors targeting PD-1/PD-L1 pathways (Ribas and Wolchok 2018), and (2) engineered cellular therapies such as (Chimeric Antigen Receptor T Cells) CAR-T cells (June et al. 2018; Grupp et al. 2013; Porter et al. 2015; Fesnak et al. 2016). This rapid progression is evidenced by the growing list of FDA-approved immunotherapies and an extensive pipeline of preclinical candidates (Riley et al. 2019). Contemporary cancer immunotherapies comprise seven major therapeutic classes: (1) immune checkpoint inhibitors that block inhibitory pathways; (2) cytokine therapies to activate lymphocyte populations; (3) engineered cellular products including CAR-T cells; (4) co-stimulatory receptor agonists; (5) tumor antigen-specific vaccines; (6) oncolytic viral vectors; and (7) bispecific antibody constructs. Each category employs distinct immunological mechanisms to enhance antitumor responses (Riley et al. 2019). While immunotherapy represents a breakthrough in cancer treatment, its clinical implementation encounters significant limitations in therapeutic effectiveness and safety profiles. These treatments may provoke aberrant immune responses against healthy tissues, resulting in autoimmune toxicity. A prime example is observed with interleukin-2 (IL-2) therapy, which frequently induces two critical complications: (1) cytokine storm, characterized by pyrexia, refractory hypotension, and multi-organ failure; and (2) capillary leak syndrome. Such treatment-emergent adverse events may pose life-threatening risks to patients (Milling et al. 2017; Lee and Margolin 2011; June et al. 2017).

11.1.3 Adoptive Cellular Therapy

The foundational investigation of adoptive cellular therapy (ACT) for cancer began with pioneering studies in xenogeneic mouse models, establishing a critical proof of concept for immune-based tumor targeting. These studies demonstrated the potential of transferring immune cells to combat cancer, igniting widespread interest in the field (Hu et al. 2023; Ingulli 2010; Oli et al. 2022; Waldman et al. 2020). A pivotal advancement emerged in 1976 when researchers identified T-cell growth factor (later designated IL-2), revolutionizing immunotherapeutic approaches. This cytokine enabled the ex vivo expansion of T lymphocytes while preserving their effector functions, providing a critical tool for large-scale immune cell manipulation. IL-2's role in supporting T-cell proliferation and activity was a cornerstone for advancing adoptive cell therapies (Morgan et al. 1976; Ross and Cantrell 2018). In 1986, landmark *in vitro* experiments demonstrated that melanoma-derived tumor-infiltrating lymphocytes (TILs) possessed selective cytotoxicity against autologous tumor cells, establishing their therapeutic potential. These "first-generation" TILs served dual functions: establishing antigen-specific antitumor activity while simultaneously illuminating key immune regulatory mechanisms operating in neoplastic niches (Rosenberg et al. 2008). This discovery underscored the potential of using TILs as a personalized platform for personalized cancer immunotherapy against cancer. In 1988, these foundational advancements culminated in the first clinically validated demonstration of ACT. Rosenberg and colleagues used autologous TILs, expanded ex vivo with IL-2, to treat melanoma patients (Dudley et al. 2003). The therapy achieved objective regression of metastatic melanoma tumors, setting a historic milestone in oncology. This landmark achievement validated the concept of ACT and provided the groundwork for modern immunotherapy approaches, inspiring subsequent innovations in cellular membrane engineering-based therapies, such as engineered T cell receptor (TCR) and CAR-T cell technologies.

11.1.4 Cellular Membrane Engineering

Cellular membrane engineering is a novel strategy to enhancing T-cell functionality in tumor immunotherapy by modifying the cell surface to improve targeting, persistence, and overall therapeutic performance. Among these, engineered TCR and CAR-T cell technologies represent the most significant advances in T-cell membrane engineering. The conceptual foundation for CAR-T cell therapy was established in 1987 when Kuwana et al. reported the prototype chimeric antigen receptor design, marking the inception of engineered T cell therapeutics (Kuwana et al. 1987; Mitra et al. 2023). This design featured the extracellular variable region of an antibody fused to the constant intracellular region of a T-cell receptor, forming a synthetic receptor capable of recognizing tumor antigens with antibody-like specificity. Building upon this foundation, Gross et al. (1989) made significant progress by showing that engineered VHCα or VHCβ chains could successfully combine with native α/β chains in host T cells. This pairing generated functional $\alpha\beta$ heterodimers that conferred antigen-specific responsiveness, enabling programmed T cell activation against defined molecular targets (Gross et al. 1989). Researchers developed the first-generation CAR constructs by fusing an immunoglobulin-derived single-chain variable fragment (scFv) domain with the CD3ζ intracellular signaling module, resulting in functional chimeric antigen receptors (Irving and Weiss 1991; Zheng et al. 2023a). The first generation of CAR-T cells marked a theoretical milestone, but their real-world performance fell short—clinicians observed weak T-cell activation and fleeting persistence. These shortcomings revealed a critical gap: the need for stronger co-stimulation. That insight became the catalyst for Finney and colleagues' groundbreaking 1998 redesign. By integrating secondary signaling domains like CD28 into the CAR architecture, their second-generation construct finally unlocked T cells' full therapeutic potential (Honikel and Olejniczak 2022; Diehn et al. 2002; Goronzy and Weyand 2008). These modifications improved T-cell proliferation, cytokine production profiles, cellular longevity, and cytotoxic activity against malignant targets. Further improvements in 2004 added the 4-1BB co-stimulatory domain, significantly improving the *in vivo* persistence and durability of engineered T cells (Guedan et al. 2018). The emergence of third-generation CAR designs around 2010 marked a quantum leap in immunotherapy. By strategically combining multiple co-stimulatory domains—CD28, 4-1BB, OX40—researchers created a synergistic signaling architecture that outclassed previous generations. These "turbocharged" constructs didn't just improve T cell activation; they solved the persistent challenge of maintaining long-term anti-tumor activity *in vivo* (Dimitri et al. 2022). The post-2012 era ushered in a revolution with fourth-generation CARs—clinically termed TRUCKs (T cells Redirected for Universal Cytokine Killing). The "smart" T cells didn't just target tumors; they transformed the tumor microenvironment itself. By engineering them to co-express multiple classes of immunoregulators, researchers essentially created living drug factories capable of delivering: tumor-busting cytokines (IL-12, IL-15), immune cell recruiters (CCL19, CXCL10), T-cell survival boosters (IL-7, FLT3L), and microenvironment disruptors (IFN-γ, TNF-α). These CARs also incorporated safety systems, including logic-gate combinations and ON/OFF switches, to improve therapeutic control and mitigate off-target effects (Duan et al. 2021; Li et al. 2018). Complementing genetic engineering, non-genetic membrane engineering offers a flexible method to augment T-cell functionality. This approach modifies the T-cell surface without altering its genome. Functional molecules such as tumor-specific antibodies, ligands, or signaling moieties can be directly anchored into the lipid bilayer of the T-cell membrane, enabling precise targeting of tumor antigens. Techniques like click chemistry or maleimide-thiol coupling allow the attachment of functional groups or therapeutic molecules to surface proteins without disrupting T-cell viability or function. Non-genetic methods are reversible, quick to implement, and reduce the risks associated with permanent genomic modifications. These approaches are particularly valuable for testing new therapeutic designs or creating temporary enhancements for adoptive cell therapies.

The development of T cell engineering technology, from its origins in synthetic receptors to the sophisticated multi-functional structure, demonstrates the potential of cellular membrane engineering in immunotherapy. Alongside genetic modifications, non-genetic membrane engineering offers a versatile toolkit to address limitations in T cell therapies, paving the way for more effective and safer cancer treatment.

11.1.5 Antigen-Presenting Cells

Antigen-presenting cells (APCs) serve as the immune system's master conductors, expertly coordinating adaptive immunity through a sophisticated three-step process:

capturing foreign invaders, processing their proteins into recognizable fragments, and presenting these molecular fingerprints to naive T cells. Among these immune sentinels, dendritic cells stand apart as the gold standard—nature's most potent stimulators of immunity, constantly displaying antigen fragments alongside powerful co-stimulatory signals (CD80/CD86) on their major histocompatibility complex (MHC) class II complexes. While macrophages predominantly function in innate defense through phagocytic clearance, their capacity for MHC-restricted antigen presentation provides an important bridge to adaptive immunity (Kashem et al. 2017).

Cell membrane engineering is applied to APCs as well as T cells. Firstly, the expression and function of MHC molecules on the cell membrane could be regulated via cellular membrane engineering. For example, through gene editing or drug interventions, the quantity of MHC molecules on the surface of APCs can be enhanced, or their affinity for antigenic peptides can be enhanced, thereby improving the efficiency of antigen presentation. Furthermore, genetic engineering enables precise regulation of co-stimulatory and adhesion molecules, which are essential for mediating APC–T cell interactions. This modification enhances T cell activation and clonal expansion (Doan et al. 2025). Moreover, membrane engineering can modify the signaling pathways of immune cells, influencing the regulatory capacity of APCs. For example, by introducing specific signaling molecules or receptors on the cell membrane, the ability of APCs to perceive and respond to the immune microenvironment can be enhanced, allowing them to more precisely regulate immune responses. Additionally, engineering can regulate the redox state of the cell membrane, affecting the maturation and function of APCs, thus achieving fine-tuned control of immune responses (Geng et al. 2024).

11.2 Genetical Immune Cell Membrane Engineering

Genetical immune cell membrane engineering involves the modification of immune cells at the genomic level to express specific receptors, proteins, or signaling molecules on their membrane, enhancing their functionality in cancer immunotherapy. This approach enables immune cells, particularly T cells, to be precisely tailored for targeted and sustained activity against tumors. Currently, three primary types of genetical engineering techniques are commonly employed in immune cell membrane engineering: (1) viral vectors (Fig. 11.1), including retroviral, lentiviral, and adenoviral vectors; (2) non-viral vectors (Fig. 11.2), such as transposons, mRNA transduction, and DNA transduction; and (3) gene editing tools (Fig. 11.3), including CRISPR/Cas9, TALEN, and ZFN (Zhang et al. 2023).

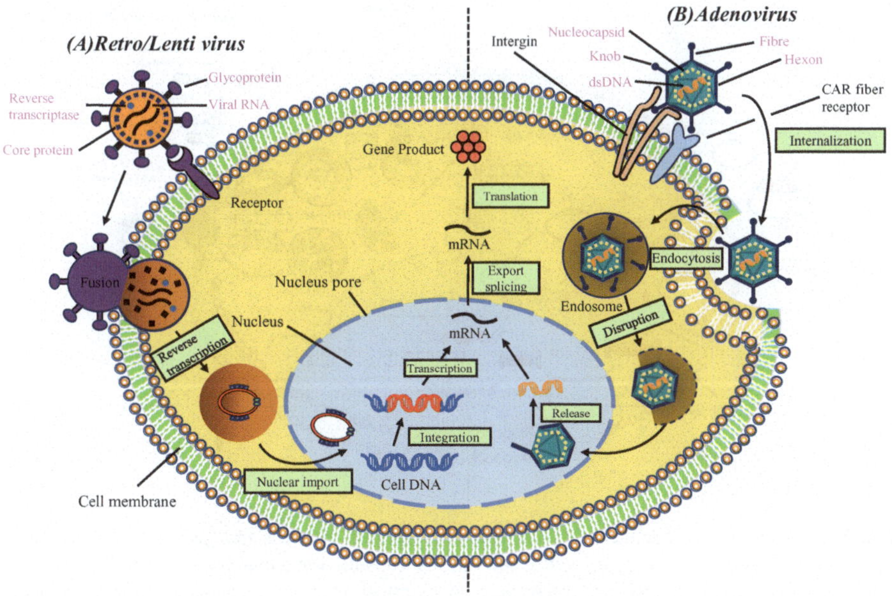

Fig. 11.1 Delivery of nucleic acids by retroviral or adenovirus particles. The mode of action for retroviruses and lentiviruses are depicted on the left, while adenoviruses are shown on the right. Retro/Lenti virus. Retroviruses or lentiviruses enter the immune cells via membrane fusion after the glycoproteins on their surface recognize specific receptors on the surface of immune cells. The viral RNA genome undergoes reverse transcription within the host cytoplasm, generating complementary DNA. This latter then traverses the nuclear pore to enter the nucleus, where it integrates into the immune cell's genome. Subsequently, the integrated viral DNA is transcribed and translated by the immune cell's own DNA polymerases and other related enzymes to express viral proteins. Adenovirus. When using adenoviruses as vectors, they bind to the coxsackievirus and adenovirus receptor on the host cell surface via their fiber proteins. With the assistance of integrins, adenoviruses enter the cell through endocytosis. Within the endosomes formed by endocytosis, adenoviruses gradually disassemble, releasing their DNA, which then traverses the nuclear pore to enter the nucleus and express the relevant proteins

11.2.1 Viral Vectors

11.2.1.1 γ-retrovirus

The retroviral family consists of seven members, with γ-retrovirus being the only one widely adopted in clinical settings. Retroviruses possess a unique capacity to permanently incorporate foreign genetic material into the host cell's DNA, which renders them highly effective for gene transfer applications. Each retrovirus encodes three essential structural genes: (1) *gag*, which encodes the viral structural proteins, including the capsid, matrix, and nucleocapsid, which form the physical framework of the virus; (2) *pol*, which encodes key enzymatic components, such as reverse transcriptase, required for converting viral RNA into DNA, and integrase, which facilitates the integration of viral DNA into the host genome; (3) *env*, which encodes the viral envelope glycoproteins that mediate the host cell recognition and entry.

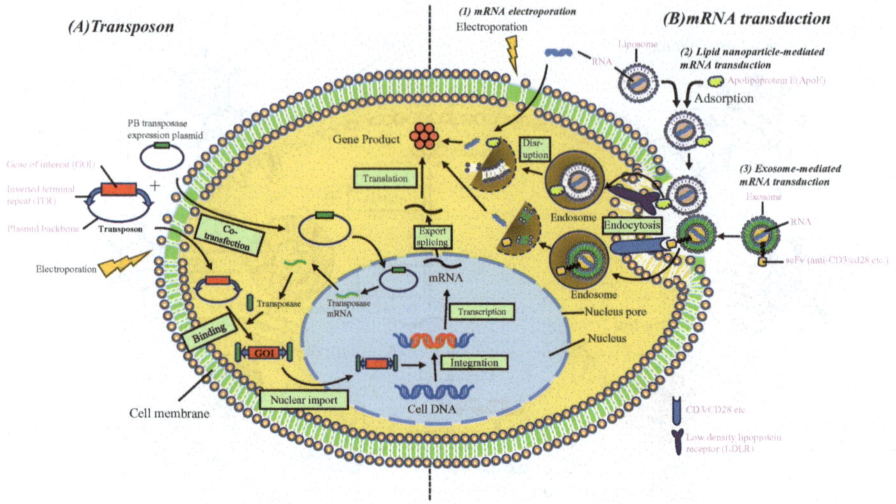

Fig. 11.2 Delivery of nucleic acids via non-viral vectors involves transposon methods, using PiggyBac transposase to integrate genes, and mRNA transduction technologies, including electroporation, lipid nanoparticles, and exosomes, to engineer immune cells with chimeric antigen receptors

The γ-retrovirus is particularly advantageous for clinical use due to its following abilities, carrying a substantial genetic payload, which enables the delivery of large therapeutic genes, such as those encoding CARs; γ-retroviruses can infect wide variety of cell types including both proliferating and select quiescent cell populations, enhancing their versatility in ex vivo gene transfer; The stable genomic integration of therapeutic transgenes facilitates persistent transgene expression, a prerequisite for achieving lasting therapeutic outcomes. Furthermore, γ-retroviruses demonstrate scalable manufacturing characteristics that meet stringent clinical production requirements. Owing to these advantageous properties, γ-retroviruses emerged as the inaugural viral vector platform to receive clinical trial authorization (Baum et al. 2006) and served as the foundational vector system for pioneering CAR-T cell therapies (Brentjens et al. 2003). Its clinical utility has been further demonstrated in trials, such as one involving 11 patients with CD19-positive tumors, where CAR-NK cells engineered using γ-retrovirus achieved a remarkable therapeutic outcome-eight patients responded, and seven achieved complete response (Liu et al. 2020). Furthermore, the γ-retrovirus has been employed in the development of approved CAR-T cell therapies, including Yescarta and Tecartus, which leverage its robust gene delivery capabilities. γ-retroviral vectors have fundamentally transformed hematologic cancer therapeutics, establishing themselves as indispensable tools in the evolution of adoptive cellular immunotherapies. The success of γ-retroviruses underscores their enduring significance in gene therapy, particularly in the production of CAR-engineered immune cells for clinical use.

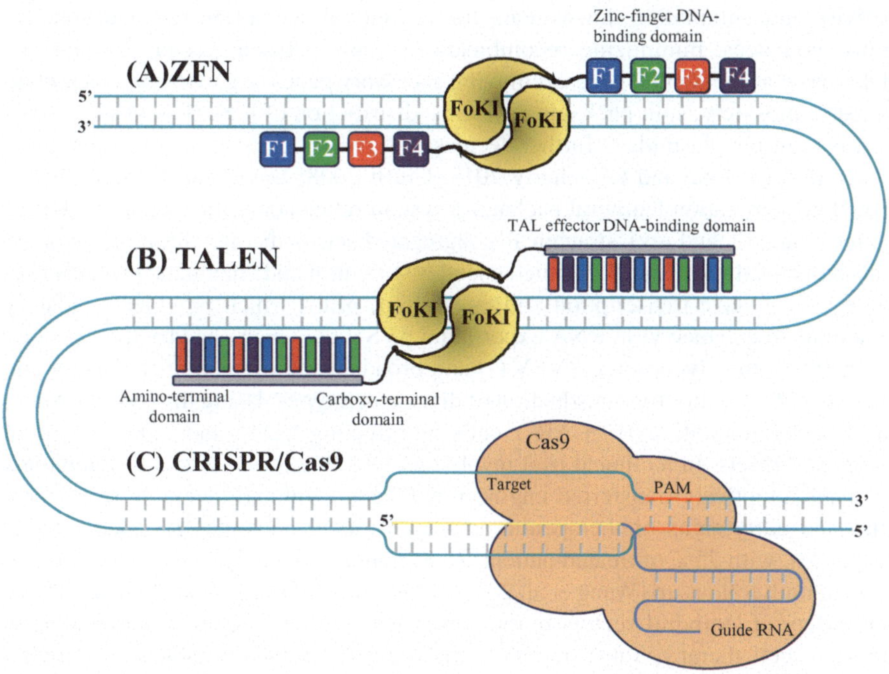

Fig. 11.3 Genome-editing platforms. Schematic representation of three major classes of programmable nucleases: zinc-finger nucleases (ZFNs), transcription activator-like effector nucleases (TALENs), and CRISPR-Cas9 systems. The illustration (not drawn to scale) comparatively depicts their distinct DNA recognition domains and cleavage mechanisms, emphasizing key structural and functional differences between each editing platform

11.2.1.2 Lentivirus

The lentiviral vector was developed through modifications to the human immunodeficiency virus-1 (HIV-1) genome, transforming it into a powerful and versatile tool for gene delivery in therapeutic applications. A defining feature of lentiviruses is the presence of a distinctive cis-acting element known as the central polypurine tract, which facilitates efficient transduction of both actively dividing and resting cells, including T cells, without the need for mitosis. This ability makes lentiviral vectors particularly valuable in engineering quiescent immune cells. The lentiviral genome includes structural (*gag*, *pol*, and *env*), regulatory (*tat, rev*), and necessary (*vif*, *vpr*, *vpu*, and *nef*) genes, all critical for viral application, cellular entry, infection, and progeny release. These genes encode essential proteins required for the natural lifecycle of the virus, including capsid formation, reverse transcription, and incorporation of viral particles into host cells (Poorebrahim et al. 2019). To improve safety and reduce the risk of replication-competent lentivirus, lentivirus vectors have undergone significant advancements, culminating in third generation systems. These vectors incorporate enhanced safety features, such as self-inactivating modifications, removal of the U3 region in the 3′ long terminal repeat, reducing promoter activity and eliminating downstream transcriptional activation; reduced homologous sequences, minimizing recombination events between vector components; deletion of non-essential genes, removing accessory genes (e.g., *vif, vpr, vpu, nef*) to enhance biosafety; split packaging systems, distribution of the essential viral genes across multiple plasmids to further decrease the risk of replication-competent virus generation (Milone and O'Doherty 2018; Ciuffi 2008; David and Doherty 2017). The third-generation lentiviral packaging system retains only three essential HIV-1 genes (gag, pol, and rev), strategically segregated across distinct plasmid vectors to enhance biosafety. pGag/Pol, which encodes structural and enzymatic proteins necessary for reverse transcription and integration; pRev, which provides regulatory functions to enhance viral RNA export; and pVSV-G, which encodes the vesicular stomatitis virus glycoprotein (VSV-G) and broadens the tropism of the vector and increases its stability for transduction of diverse cell types. Using this system, Wang et al. generated anti-CD19 CAR-T cells by infecting CD4+ and CD8+ T central memory subsets. In a clinical trial involving eight B-cell non-Hodgkin lymphoma patients, adoptively transferred engineered T cells exhibited favorable safety profiles and early signs of therapeutic activity. Clinical outcomes revealed durable responses, with 75% of treated patients (6/8) maintaining progression-free survival at 12-month follow-up (Wang et al. 2016). This study highlights the clinical efficacy and safety of lentiviral vectors in the generation of CAR-T cells, a cornerstone in adoptive cell therapy. Taken together, these attributes have established lentiviral vectors as a gold standard for genetic engineering in immunotherapy, particularly in CAR-T cell production.

11.2.1.3 Other Viral Vectors

Beyond retroviral systems, various viral vectors are utilized in gene delivery, such as adenovirus, adeno-associated virus, and herpes simplex virus (HSVs). These vectors exhibit unique properties, making them suitable for different genetic engineering applications.

Adeno-Associated Virus These vectors demonstrate tropism for both mitotically active and quiescent cell populations, while exhibiting an excellent safety profile characterized by negligible pathogenicity and low cellular toxicity. Compared to retrovirus-derived vectors, adeno-associated virus exhibits a significantly lower risk of insertional mutagenesis due to its preference for episomal maintenance rather than genomic integration. Adeno-associated virus has a constrained packaging capacity of approximately 4.7 kilobases (kb), restricting the size of the therapeutic gene or cargo they can deliver (Fernandez-Sendin et al. 2020). Zhang et al. prepared a scFv targeting phosphorylated Tau (p-Tau) and used a recombinant adeno-associated virus capable of crossing the blood-brain barrier (BBB) (rAAV/BBB) as a vector to express scFv in the mouse brain for at least 22 weeks. The results showed that the constructed scFv had high affinity for p-Tau and could bind to neuronal tangles in brain slices of Alzheimer's disease (AD) model mice (AD model mice). Moreover, rAAV/BBB could cross the blood-brain barrier, infect neuronal cells, and express scFv. This novel immunotherapy effectively delivers scFv to the brain and leads to its continuous expression *in vivo*, indicating its potential for treating AD. This new immunotherapeutic approach addresses some limitations of p-Tau immunotherapy, such as low BBB permeability, short antibody half-life, and the possibility of inflammation (Zhang et al. 2022).

Adenovirus They are highly efficient in delivering genetic material into T cells, enabling robust expression of CAR constructs. Unlike retroviruses or lentiviruses, adenoviral-based delivery systems exhibit episomal maintenance in host cells, thereby eliminating genomic integration and associated mutagenic potential. This makes them safer for clinical applications. Adenovirus-mediated gene delivery typically results in transient CAR expression, which is beneficial for proof-of-concept studies or applications requiring short-term modifications. With a packaging capacity of up to 8 kb, adenoviral vectors can deliver complex CAR constructs, including those with additional domains for co-stimulation or safety features (Syyam et al. 2022). Mathlouthi et al. explored adenovirus-vectored cancer vaccines encapsulated in extracellular vesicles as a therapeutic strategy against NRAS-mutated malignant melanoma. The research team evaluated the anti-tumor properties of this vaccine in various models, including a MUG Mel-2 melanoma xenograft model in BALB/c nude mice, a 3D co-culture model simulating the tumor microenvironment, and a humanized NSG mouse model. The results indicated that the adenovirus-based vaccine effectively activated immune responses, inhibited tumor growth, and significantly enhanced treatment efficacy and tumor-infiltrating lymphocyte infiltration through targeted delivery via extracellular vesicles. This study provided strong evidence for developing novel immunotherapies against NRAS-mutated melanoma and highlighted the potential of extracellular vesicles in cancer therapy (Mathlouthi et al. 2024).

Herpes Simplex Virus They could carry over 30 kilobases (kb) of genetic material, allowing for the delivery of complex CAR constructs, including multi-gene payloads such as co-stimulatory domains, cytokines, or safety switches. HSV efficiently transduces various cell types, including T cells, which makes it an effective vector for introducing CAR constructs into immune cells. Unlike some other viral vectors, herpes simplex virus (HSV) generally remains episomal in host cells, minimizing the potential for insertional mutagenesis and offering safer genetic modification options. HSV-mediated gene delivery supports quick and transient CAR expression, which may be beneficial for evaluating CAR-T cell functionality in preclinical studies or for short-term therapeutic applications (Chen et al. 2018). While lentiviral vectors remain the cornerstone of current CAR-T cell clinical applications, adenoviral, AAV, and HSV vectors hold significant promise for advanced research and specialized applications. Their potential is further amplified when integrated with non-viral strategies and synthetic biology approaches, enabling the development of more sophisticated and versatile CAR-T cell therapies. Bommareddy presented a study on the use of an oncolytic herpes simplex virus expressing IL-2 for the treatment of glioblastoma (GBM), a particularly lethal form of brain cancer. The team constructed G47Δ-mIL2, an IL-2-expressing oncolytic HSV variant, and evaluated its therapeutic potential in murine GBM models. The results demonstrated that G47Δ-mIL2 treatment markedly extended median survival in orthotopically implanted GBM murine models, particularly in the 005 GSC-derived model, without causing systemic IL-2-related toxicity. The therapeutic effect was linked to increased infiltration of CD8+ T cells into the TME and hinged critically on the local release of IL-2 and the presence of CD4+ T cells. The study suggested that G47Δ-mIL2 represented a promising therapeutic strategy for GBM, potentially targeting GBM stem-like cells and stimulating antitumor immunity, and warranted further exploration in clinical trials (Bommareddy et al. 2024).

11.2.2 Non-viral Vectors

11.2.2.1 Transposon

Transposons are endogenous genetic sequences that can relocate autonomously across the genome (Woodard and Wilson 2015). A transposon system consists of a transposase enzyme and target genes flanked by transposase binding sites. The transposase enzyme mediates the precise excision and genomic reintegration of target sequences by recognizing and binding to the flanking inverted terminal repeats (Izsvák et al. 2010). Transposon-based vectors have gained widespread adoption in both research and therapeutic applications owing to their substantial cargo capacity, efficient gene delivery, minimal immune response, straightforward manipulation, and cost-effective production (Magnani et al. 2020). Among transposon-based gene delivery platforms, Sleeping Beauty (SB) and PiggyBac (PB) represent the most extensively characterized systems. Both employ a dual-plasmid design, where transposase expression and gene-of-interest insertion are mediated by separate vectors. Transposition efficiency is primarily determined by transposase functionality. Through strategic engineering of terminal repeats and transposase enzymes, researchers have developed enhanced variants with improved DNA excision and integration capabilities. Notable examples include the hyperactive SB100X and SB11 mutants in the Sleeping Beauty system, as well as the optimized pB and 7pB variants in the PB system (Sato et al. 2020; Hou et al. 2015). Transposon-based gene delivery has emerged as a powerful platform for engineering T cells in cancer immunotherapy. The hyperactive SB11 transposase variant has demonstrated clinical potential in multiple therapeutic trials. In a phase I/II clinical trial, Singh et al. successfully generated CD19-targeting CAR-T cells with a CAR-positive rate exceeding 90% using the SB11 transposon, and these cells exhibited strong antitumor activity (Singh et al. 2014).

11.2.2.2 mRNA Transduction

mRNA Electroporation mRNA electroporation represents an advanced gene delivery approach where engineered mRNA molecules are first optimized for structural stability, then introduced into target cells via electrical pulse-mediated membrane permeabilization to enable transient protein expression. This innovative approach overturns conventional assumptions about mRNA's therapeutic limitations by demonstrating its viability as a stable genetic payload. Following electroporation, the mRNA directly accesses the cytoplasmic translation machinery without nuclear localization requirements. This bypasses genomic integration risks while achieving rapid, high-level protein production. Consequently, mRNA electroporation has emerged as a leading gene delivery platform, offering superior biosafety profiles compared to conventional methods. This technology provides multiple distinct advantages: (1) exceptional transfection efficacy across diverse cell types, including non-dividing primary cells and immune cell populations; (2) flexible and streamlined design parameters that facilitate rapid optimization; (3) cost-effective scalability for clinical-grade cell production; and (4) broad applicability to both research and therapeutic applications (Tavernier et al. 2011). The growing adoption of mRNA electroporation is evidenced by its expanding applications in cell therapy development. Notably, Beatty et al. achieved remarkable success by introducing mesothelin-targeting CAR mRNA into activated T lymphocytes through electroporation, generating clinical-grade CAR-T products with exceptional viability (>91%) and transduction efficiency (>98% CAR expression). The resulting CAR-T cell products demonstrated potent tumor-specific cytotoxicity in clinical evaluation, while maintaining an excellent safety profile with no observable on-target/off-tumor effects (Beatty et al. 2014). This clinical evidence establishes mRNA-electroporated CAR-T cells as both a viable and secure approach for adoptive cell therapy.

Lipid Nanoparticles-Mediated mRNA Transduction The application of lipid nanoparticles (LNPs) for mRNA delivery has gained significant attention in recent years. As a promising nanomaterial, LNPs have been widely explored in ACT due to their efficient transduction capabilities. LNPs represent a particularly promising mRNA delivery system for CAR-T cell generation. These nanoparticles enable cellular uptake through either endocytic pathways or direct membrane fusion. Compared to electroporation, LNPs achieve similar transduction rates while demonstrating significantly reduced cellular toxicity (Billingsley et al. 2020). Successful CAR-T cell generation was achieved through LNP-based mRNA delivery, with *in vitro* tumor-killing efficacy comparable to electroporation-produced CAR-T cells (Billingsley et al. 2022). However, this approach presents several limitations, including mRNA instability, suboptimal biocompatibility, inadequate degradation kinetics, and possible adverse effects in physiological systems (Si et al. 2023).

Exosome-Mediated mRNA Transduction As natural nanovesicles released by viable cells, exosomes carry diverse bioactive components that mediate molecular transport and cell-to-cell signaling via receptor-specific interactions. These nanoscale vesicles exhibit distinct biological advantages, such as: (1) robust structural integrity ensuring payload protection; (2) minimal immune recognition; (3) inherent blood-brain barrier penetrability; and (4) flexible modification potential. Notably, they demonstrate exceptional mRNA stabilization capacity against enzymatic degradation (Si et al. 2023; Yang et al. 2019). Given these advantages, exosome-based gene delivery systems have been investigated as an alternative approach for CAR-T cell engineering (Si et al. 2023). Exosomes can be engineered as dual-functional nanocarriers by simultaneously displaying anti-CD3/CD28 scFvs on their surface and encapsulating CAR-encoding mRNA. This innovative approach enables direct T cell activation and targeted mRNA delivery, and ultimately generates functional CAR-T cells with specific antitumor activity (Si et al. 2023). Comparative studies reveal that exosome-derived CAR-T cells exhibit reduced cytotoxic activity against target cells relative to their lentiviral-generated counterparts. This diminished efficacy may stem from suboptimal CAR expression levels achieved through exosome-mediated delivery.

11.2.3 Genetic Editing Tools

11.2.3.1 CRISPR/Cas9

While lentiviral and retroviral vectors remain the primary tools for gene delivery in cell therapy, their random genomic integration poses significant safety concerns. The potential insertion into critical genomic regions may compromise both cellular integrity and therapeutic performance of engineered cells (Craigie and Bushman 2014; Sakuma et al. 2012). Consequently, CRISPR/Cas9 technology has emerged as a promising alternative, enabling targeted genome modifications with substantially reduced off-target integration risks compared to conventional viral vectors.

The CRISPR/Cas9 platform represents a breakthrough in genome engineering, employing a dual-component system for site-directed DNA modification: (1) a programmable guide RNA (gRNA) that recognizes complementary genomic loci, and (2) the Cas9 nuclease that induces double-strand breaks at these predetermined targets. Directed by gRNA, Cas9 induces precise double-strand breaks (DSBs) at target sites, thereby initiating two major DNA repair pathways: non-homologous end joining (NHEJ) and homology-directed repair (HDR). In template-free conditions, NHEJ mediates direct DNA end-joining, frequently generating insertion/deletion (indel) mutations. By contrast, HDR utilizes exogenous donor DNA to enable accurate gene modifications, including targeted insertions, deletions, or point mutations (Koonin et al. 2017). The CRISPR/Cas9 platform demonstrates superior versatility beyond targeted genome editing, featuring: (1) customizable gRNA design simplicity, (2) rapid experimental deployment, (3) cost-efficient production, and (4) scalable applications across biological systems (Zhang et al. 2021). These capabilities have enabled CRISPR/Cas9's widespread adoption for diverse genomic modifications, particularly revolutionizing engineered cell therapy development. The technology has demonstrated transformative potential for CAR-T cell optimization. This system's capacity for multiplex genome editing makes it a transformative technology for developing off-the-shelf CAR-T cells. In a 2017 study, Liu and colleagues employed CRISPR/Cas9 to disrupt TRAC (responsible for the TCRα chain) and B2M (essential for MHC-I complex formation) in T cells. The resulting allogeneic CAR-T cells demonstrated potent antitumor efficacy in murine models (Liu et al. 2017). Subsequent research has expanded on these findings, with multiple investigations advancing to clinical-stage testing (Lin et al. 2021). Furthermore, CRISPR/Cas9 technology can optimize CAR-T cell performance through targeted disruption of immune checkpoint genes or site-specific insertion of CAR transgenes into designated genomic regions.

11.2.3.2 Transcription Activator-Like Effector Nucleases (TALEN)

TALEN is a synthetic restriction enzyme comprising three key structural components: (1) an N-terminal segment harboring a nuclear localization signal, (2) a central region featuring tandem TALE repeats for precise DNA binding, and (3) a C-terminal module incorporating the FokI nuclease domain. TALEN recognizes and binds to specific DNA sequences via its customizable DNA-binding domain, subsequently generating DSBs through the catalytic activity of the FokI nuclease. Like CRISPR/Cas9, this system leverages endogenous cellular repair mechanisms—either homology-directed repair (HDR) or NHEJ—to facilitate targeted gene insertions or deletions (Zhang et al. 2023). TALEN-based genome engineering has enabled the development of allogeneic CAR-T cell therapies. Jo et al. demonstrated this approach by combining TALEN-mediated knockout with AAV6-delivered gene insertion, achieving simultaneous modification of both TRAC and B2M loci to generate universal CAR-T cells. The resulting TCR- and HLA-ABC-negative T cells co-expressed both a chimeric antigen receptor and the immunomodulatory HLA-E molecule. These engineered hypoimmunogenic CAR-T cells demonstrated dual functionality: they prevented graft-versus-host disease (GvHD) while resisting elimination by host NK cells and alloreactive T lymphocytes, consequently enhancing their tumor-targeting persistence (Jo et al. 2022). These results demonstrate TALEN's significant promise for immune cell engineering applications. Importantly, TALEN-mediated genome editing achieves comparable—and in some cases superior—targeting precision relative to CRISPR/Cas9 systems.

11.2.3.3 Zinc Finger Nucleases (ZFNs)

ZFNs are chimeric proteins comprising two functional modules: (1) a customizable zinc finger array that confers sequence-specific DNA recognition, and (2) the catalytic FokI nuclease domain that mediates targeted DNA cleavage. Like TALENs, ZFNs can generate precise DSBs at designated genomic loci. These targeted breaks are subsequently repaired through either NHEJ or HDR, enabling either gene disruption or site-specific integration of exogenous sequences (Doyon et al. 2011). ZFNs have demonstrated significant utility in developing off-the-shelf CAR-T cell therapies. Pioneering work by Torikai et al. first combined the Sleeping Beauty transposon system for stable CD19-CAR integration with ZFN-mediated ablation of TCR and HLA-A loci, establishing an early proof-of-concept for allogeneic CAR-T cell engineering. This confirmed that the engineered CD19-targeting CAR-T cells maintained potent tumoricidal effects while mitigating GvHD risks and avoiding host immune rejection, thereby establishing a critical foundation for universal CAR-T cell therapeutics (Torikai et al. 2012, 2013). In a recent advancement, Brown et al. engineered glucocorticoid-resistant universal CAR-T cells by combining ZFN-mediated knockout of the glucocorticoid receptor with Ad5/

F35-vectored delivery of an anti-IL13Rα2 CAR construct into allogeneic T cells. This innovative approach yielded an off-the-shelf cellular product resistant to steroid-mediated immunosuppression (Jo et al. 2022; Brown et al. 2022). Following dexamethasone selection, the engineered cells maintained potent effector function and demonstrated dexamethasone resistance and no detectable alloreactive potential in vitro. Clinical observations demonstrated favorable tolerability of these ZFN-engineered cells, with four of six treated patients exhibiting localized tumor regression and/or necrotic changes at the administration site. These preliminary findings support both the safety profile and therapeutic potential of allogeneic ZFN-modified CAR-T cells as readily available cellular therapeutics.

11.2.4 Harnessing Genetically Modified Cells for Cancer Immunotherapy

Cancer immunotherapy has emerged as a transformative therapeutic paradigm, harnessing the host's immune defenses to selectively eliminate malignant cells. Among the most promising developments in this field are engineered cellular therapies, which have shown therapeutic effectiveness in both hematologic cancers and solid neoplasms. Through genetic engineering of immune cells to express tumor-targeting receptors, researchers have developed a highly specific anticancer strategy. This approach augments endogenous immunity while countering key mechanisms of immune escape characteristic of leukemic cells, including antigen loss and immunosuppressive signaling. The following sections present four paradigm-shifting applications of engineered cellular therapies in leukemia treatment, highlighting their innovative mechanisms and clinical translation potential. Each case study demonstrates how these advanced biotechnological interventions are redefining therapeutic outcomes in hematologic malignancies.

The development of effective therapies for acute myeloid leukemia (AML) has been hindered by significant biological challenges, particularly the disease's inherent heterogeneity. This molecular diversity frequently drives resistance to single-antigen-targeted CAR-T cell therapies and contributes to high rates of disease recurrence. Petrov et al. pioneered a bispecific cCAR-T cell platform engineered to simultaneously target CD123 and CD33—two clinically validated AML-associated surface antigens—through incorporation of distinct scFv domains in a single construct. The CAR-T cells exhibited robust and sustained cytotoxic activity toward AML cells expressing CD123 and CD33. In four distinct leukemia mouse models, *in vivo* studies confirmed enhanced survival upon cCAR administration. Furthermore, an alemtuzumab-mediated safety mechanism was integrated to enable precise regulation of cCAR T-cell function. These results indicate that simultaneous targeting of CD123 and CD33 could effectively eradicate both the bulk AML population and leukemic stem cells, thereby mitigating relapse risks associated with antigen loss or leukemic stem cells persistence. This represents a notable progression in engineered cellular therapies for AML (Petrov et al. 2018).

Preclinical glioblastoma studies indicate that CAR-T therapies targeting a single tumor antigen may result in disease relapse due to the emergence of antigen escape variants. Hegde et al. proposed that a bispecific CAR construct could reduce antigen escape while improving T-cell-mediated tumor cytotoxicity. Their design incorporated an extracellular tandem CAR(TanCAR) domain, combining an HER2-targeting scFv with an IL13Rα2-binding IL-13 mutein, paired with an intracellular CD28.ζ signaling domain. Patient-derived TanCAR T cells demonstrated high-affinity binding to both HER2 and IL13Rα2. These bispecific T cells showed activation kinetics similar to conventional single-antigen CAR T cells when stimulated by either target antigen, while maintaining potent cytotoxicity against autologous glioblastoma cells. Experimental data revealed that TanCARs mediated simultaneous HER2 and IL13Rα2 engagement through heterodimer formation, triggering synergistic T-cell activation upon dual-antigen recognition. TanCAR T cells exhibited more prolonged activity than T cells coexpressing a HER2 CAR and an IL13Rα2 CAR, T cells with a unidirectional CAR, or a pooled product, without being more exhausted. In a murine glioblastoma model, TanCAR T cells mitigated antigen escape, demonstrated enhanced antitumor efficacy, and improved animal survival rates. TanCAR T cells demonstrate enhanced glioblastoma targeting efficacy, formed bivalent immune synapses that amplify T cell activation and, mitigated tumor immune evasion through simultaneous HER2/IL13Rα2 engagement (Hegde et al. 2016).

While contemporary therapies achieve nearly 90% cure rates in pediatric B-cell acute lymphoblastic leukemia (B-ALL), patients with relapsed or chemotherapy-refractory disease continue to face dismal prognoses (Smith et al. 2014; Pui et al. 2015). Fry et al. pioneered a novel CD22-targeted CAR-T cell therapy for relapsed/refractory B-ALL, demonstrating its clinical potential in a phase I trial enrolling 21 patients, including CD19 immunotherapy-experienced cases. The phase I trial enrolled 21 participants, incorporating patients with prior CD19-targeted immunotherapeutic exposure. The trial revealed potent antileukemic efficacy of CD22-CAR-T therapy, inducing complete responses in a substantial patient cohort, including challenging CD19dim/neg B-ALL cases. These findings underscore the therapeutic value of alternative antigen targeting to circumvent antigen escape-mediated resistance, broadening B-ALL treatment paradigms and enabling development of bispecific immunotherapies (Fry et al. 2018). Besides, Zeng et al. developed an innovative dual-targeting CAR-T platform co-targeting CD19 and CD22, antigens ubiquitously expressed in B-cell malignancies. Preclinical studies confirmed the bifunctional CAR-T cells' potent cytotoxic activity against leukemia populations regardless of single or co-expressed antigens, effectively mitigating antigen-loss-mediated immune evasion. This bispecific targeting strategy significantly enhances CAR-T cell therapeutic performance while addressing a critical limitation of conventional single-antigen approaches in B-ALL treatment. The platform demonstrates the transformative potential of genetically modified cells to overcome tumor immune evasion mechanisms in hematologic malignancies (Zeng et al. 2022).

11.2.5 Fundamental Limitations and Engineering Requirements

Genetic membrane engineering has become a transformative approach in immuno-oncology, enabling precise immune cell modification for enhanced tumor recognition and elimination. However, despite significant progress, several fundamental limitations remain that challenge the full realization of its potential in clinical applications. Current challenges encompass suboptimal gene delivery efficacy, unintended off-target activity, host immune reactions, and tumor microenvironment complexity (Doudna 2020).

11.2.5.1 Gene Delivery Inefficiencies

Efficient and targeted delivery of genetic material into immune cells is a critical challenge. Viral vectors, while offering advantages of efficiency and tissue selectivity, can have limited cargo capacity and even trigger immune responses. Non-viral methods often suffer from low transfection efficiency and poor cellular uptake. For instance, Zu et al. investigation revealed critical shortcomings in non-viral gene delivery platforms, underscoring the demand for enhanced delivery systems with improved efficacy and safety profiles (Zu and Gao 2021). To improve gene delivery, researchers are engineering innovative nanoparticle-mediated delivery platforms. These systems can protect genetic material and enhance cellular uptake. Han et al. systematically reviewed cutting-edge developments in LNP-mRNA delivery platforms, highlighting innovative design optimization approaches for mRNA-based cancer therapeutics. Furthermore, the review critically analyzed existing challenges while proposing future directions to enhance the therapeutic profile of LNP-mRNA immunotherapies for oncology applications (Han et al. 2023).

11.2.5.2 Off-Target Effects

The therapeutic efficacy of genome editing hinges on two critical parameters: (1) specificity (on-target vs. off-target modification rates) and (2) fidelity (the proportion of intended on-target edits achieving the desired genetic alteration). A comprehensive review by Fu et al. detailed the off-target effects of CRISPR/Cas9 and proposed countermeasures to mitigate these adverse effects, including the use of high-fidelity Cas9 variants and optimized guide RNAs (Fu et al. 2013). Current mitigation approaches employ high-fidelity Cas9 variants and computationally designed guide RNAs. Emerging base/prime editing platforms provide alternative precision genome editing modalities that circumvent double-strand break formation, thereby minimizing off-target risks (Doudna 2020).

11.2.5.3 Immune Responses

The administration of exogenous genetic components or recombinant proteins may trigger host immune reactions, potentially resulting in the rejection of genetically modified cells or tissues. This is particularly problematic in clinical applications. Bucher et al. systematically assessed immune activation, toxicological profiles, and inflammatory responses associated with AAV-mediated retinal gene transfer. Their investigation identified key modulators of host immunity and proposed targeted approaches to optimize AAV vector immunogenicity, thereby improving the safety and efficacy of ocular gene therapy (Bucher et al. 2021; Gennari et al. 2014). To mitigate immune responses, researchers are exploring the use of immune-modulating agents and the optimization of delivery methods to reduce the exposure of foreign proteins. In ex vivo gene editing applications, the progressive degradation of residual Cas9 in modified cells reduces immunogenic risks (Doudna 2020; Wagner et al. 2019).

11.2.5.4 Complexity of the Tumor Microenvironment

The tumor microenvironment (TME) constitutes a dynamic biological niche that critically modulates response to immunotherapeutic intervention. Hypoxic conditions, acidic pH, and immunosuppressive cellular components collectively impair the functionality of engineered immune cell therapies. Pitt et al. systematically analyzed the immunosuppressive mechanisms of TME and evaluated current and emerging therapeutic strategies targeting its cellular and molecular constituents (Pitt et al. 2016). To enhance the efficacy of cancer immunotherapy, researchers are developing TME-targeting approaches. This includes the use of engineered immune cells that can withstand hypoxic conditions and advanced delivery platforms that can penetrate and modify the tumor microenvironment (Han et al. 2023; Bader et al. 2020; Sun et al. 2017).

11.3 Non-genetical Immune Cell Membrane Engineering

In response to limitations of genetic engineering, researchers have been investigating non-genetic cell surface modification approaches to precisely control cellular interactions (Abbina et al. 2017; Csizmar et al. 2018; Yao et al. 2013). These strategies are particularly suitable for applications involving immune cells due to their transient and reversible characteristics (Csizmar et al. 2018). The techniques, which fall into five main categories (Fig. 11.4), include: (1) the use of modified sugar analogs for metabolic labeling of surface glycoproteins, (2) the incorporation of hydrophobic groups into the lipid bilayer, (3) direct chemical modification of cell surface constituents, (4) enzymatic reshaping of membrane proteins, and (5) fusion of functionalized liposomes with the cell membrane. These diverse methods have been investigated to facilitate the engineering of the cell membrane in a non-genetic manner.

11.3.1 Metabolic Modification

Enzymatic modification enables precise engineering of native cell surface proteins and glycans, generating structural variants or conjugating synthetic targeting moieties. The metabolic labeling strategy for membrane protein functionalization was pioneered in 1992. It introduces biorthogonal chemical groups to the plasma membrane through cellular biosynthesis pathways. This strategy capitalizes on the

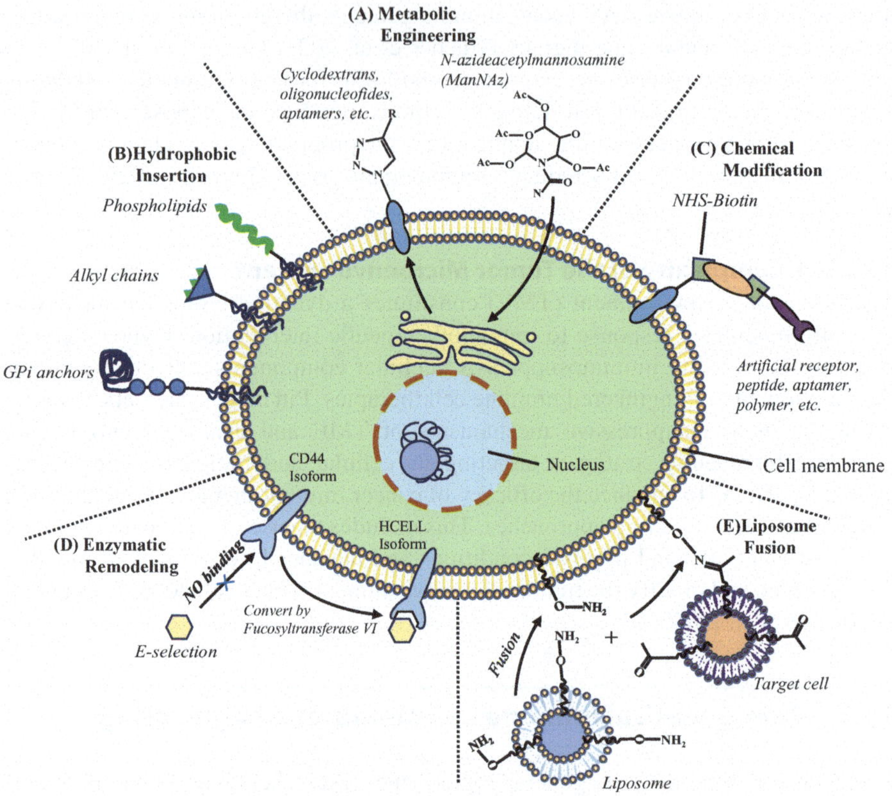

Fig. 11.4 The approaches of non-genetic membrane engineering. Several non-genetic strategies modify cell membranes to enhance intercellular interactions, including: (**a**) metabolic incorporation of biorthogonal groups, (**b**) hydrophobic anchoring of ligands, (**c**) covalent bonding via natural surface groups, (**d**) enzymatic alteration of surface biomolecules, and (**e**) liposomal fusion for targeted modifications

metabolic plasticity of native glycan biosynthesis, enabling site-specific integration of synthetic amino sugar analogs into cell surface glycoconjugates (Kayser et al. 1992).

Extending this methodology, researchers engineered Jurkat cells to metabolically incorporate N-azideacetylmannosamine, which resulted in cell surface display of azide-modified sialic acid residues through biosynthetic pathway utilization. Kim et al. subsequently achieved bioorthogonal functionalization of azide-modified CAR-T cells via copper-free click chemistry to enable stable conjugation with biodegradable fluorescent nanoparticles and preserve cellular viability (Gartner and Bertozzi 2009). The team pioneered a multimodal therapeutic platform for brain tumors, integrating fluorescent theranostic nanoparticles with CAR-T cells expressing an engineered IL-13 quadruple mutant (TQM13) targeting receptor. This variant demonstrates strong binding specificity toward glioblastoma cells expressing IL13Rα2, with minimal nonspecific cytotoxicity. The nanoparticle-T cell conjugation was achieved via azide-alkyne cycloaddition, ensuring high efficiency, selectivity, and yield. The T cell-bound nanoparticles demonstrated sustained attachment for over 8 days, confirming linkage stability and enabling sufficient duration for effective *in vivo* administration. Modified T cells carrying doxorubicin-encapsulated nanoparticles demonstrated superior tumor-killing capacity in laboratory tests relative to their unaltered counterparts. Experimental results showed that TQM13-expressing T cells effectively transported nanoparticles to target sites, leading to substantially greater nanoparticle deposition in intracranial tumors than free nanoparticle administration in both cellular and animal models (Kim et al. 2020).

While CAR-T cell therapy demonstrates remarkable efficacy against hematologic cancers, its clinical application is often limited by treatment-related toxicities. To address this challenge, Pan et al. have introduced a novel bioorthogonal CAR-T engineering approach utilizing metabolic glycan labeling to improve the therapeutic safety profile in living systems. This approach utilizes metabolically incorporated bioorthogonal handles (N3/BCN) derived from Ac4GalNAz and Ac4ManNBCN precursors, which form synthetic ligand-receptor systems capable of simultaneously engaging both CAR-T cells and malignant cells without compromising cellular integrity. The surface-anchored bioorthogonal functional groups enhance CAR-T cell-tumor interactions via click chemistry-mediated binding, which promotes targeted recognition, directional migration, and tumor-selective cytotoxicity. In vivo studies revealed that bioorthogonal CAR-T cells achieved complete lymphocyte depletion while exhibiting minimal off-target effects in hematologic malignancies, owing to their selective and efficient targeting capability. Bioorthogonal targeting markedly improves CAR-T cell tumor infiltration compared to untagged cells, facilitating enhanced tissue penetration, tumor accumulation, and ultimately superior antitumor activity in solid malignancies (Pan et al. 2021).

Wang et al. employed biomaterials to concentrate DCs *in vivo*, utilizing an innovative strategy to enable targeted immunomodulation of DCs, thereby manipulating the initiation and amplification of T-cell responses and anticancer immune reactions. DCs can be selectively modified with azide moieties through metabolic labeling, enabling precise tracking and controlled functional regulation in their native microenvironment. Azide-functionalized dendritic cells demonstrate prolonged retention in lymphoid tissues for multiple weeks and maintain the capacity to effectively internalize dibenzocyclooctyne (DBCO)-conjugated antigens and immunostimulants through bioorthogonal click conjugation. This strategy enables covalent conjugation of DBCO-functionalized cytokines (e.g., DBCO-IL-15/IL-15Rα complexes) in both experimental and physiological settings. It potentiates the activation of tumor antigen-reactive CD8+ T lymphocytes and ultimately enhances their effector functions and tumoricidal activity (Wang et al. 2020).

11.3.2 Hydrophobic Incorporation

The hydrophobic effect, which describes the tendency of hydrophobic molecules to cluster together in aqueous environments to minimize their exposure to water, serves as a key factor in the incorporation and orientation of integrated proteins. This mechanism promotes stable surface protein retention by inhibiting endocytic membrane internalization processes. In the field of cell membrane engineering, the hydrophobic effect is harnessed to insert hydrophobic molecules or lipids into the lipid bilayer, allowing these molecules to remain stably on the cell membrane surface and resist internalization. This approach can be employed to anchor specific ligands or functional molecules onto the cell membrane, facilitating engineered modifications that enhance specific cell-cell interactions (Csizmar et al. 2018).

Zhong et al. established a bioorthogonal platform enabling spatially controlled radical polymerization directly on living cell surfaces. This approach utilizes metabolic tagging strategies to confine polymerization loci and implements a cell-friendly Fenton-RAFT polymerization protocol. The membrane retention characteristics and glycan recognition patterns vary significantly depending on polymerization sites, including glycans, proteins, and lipids. Particularly, this method facilitates the direct copolymerization of glucose-sensing moieties at native cell surface glycan terminals, generating a synthetic glycocalyx with distinctive molecular recognition capabilities (Zhong et al. 2023).

Jangid et al. designed a biomaterial-mediated in vitro surface engineering technique to modify NK cells with tumor-recognizing ligand membranes, enhancing the selectivity of NK cells for cancer cells and thereby augmenting the therapeutic efficacy against tumor cells. The synthetic lipid-based biomaterial integrates three key functional elements: DSPE phospholipid moieties providing membrane anchoring, PEG chains creating an intracellular penetration barrier, and lactobionic acid (LBA) groups enabling tumor-specific targeting. The LBA-functionalized biomaterial was effectively conjugated to natural killer cell surfaces (LBA-NK) to significantly improve both tumor targeting capabilities and cytotoxic activity against asialoglycoprotein receptor (ASGPR)-overexpressing hepatocellular carcinoma models. Through a simple coating process, efficient and uniform surface editing of NK cells was achieved while preserving their inherent characteristics. The LBA-NK cell system exhibited enhanced tumor-targeting capability through specific LBA-ASGPR binding interactions, resulting in markedly improved cytotoxic effects against HepG2 hepatocellular carcinoma cell lines. Consequently, compared to currently available cell therapies, LBA-NK cells can treat hepatocellular carcinoma by promoting immunological synapse interactions, representing a novel engineering strategy (Jangid et al. 2023). Also, using Hydrophobic Incorporation to modify NK cells, Xiong et al. introduced a DNA-templated protein anchoring (DMPA) strategy for programmable cell functionalization and immunological response modulation. The researchers utilized DNA as a scaffold to anchor exogenous proteins onto living cell membranes, enabling precise control over protein density, ratio, and interactions through predictable DNA hybridization. Using the DMPA platform, NK cells were modified with sialidase to cleave immunosuppressive sialic acid residues at the immunological synapse. This modification potentiated NK cell activation and significantly boosted therapeutic efficacy in tumor immunotherapy. The study demonstrates the versatility of DMPA for membrane protein engineering, offering a new approach to enhance immune cell function and potentially revolutionize cancer immunotherapy (Xiong et al. 2023).

11.3.3 Chemical Modification

This non-genetic membrane modification approach employs direct chemical conjugation of small molecules and payloads to surface-exposed functional groups, offering a straightforward yet effective cell engineering methodology. Various ligands can be covalently linked to these native functional groups without the need for cellular or chemical pretreatment. This methodology enables universal membrane engineering through non-selective modification of diverse cellular components, encompassing proteins, lipids, and carbohydrate-based structures such as glycans (Csizmar et al. 2018).

Zhao et al. have developed a cell-based hybrid system termed "Cellnex," which facilitates the assembly of bioactive nanocomposites, including proteins, DNA, mRNA, and even viral vectors, on the surface of a vast array of immune cell adoptive cells such as red blood cells, macrophages, NK cells, and T cells through chemical modification. Cellnex offers a versatile cell engineering platform with a flexible biomolecular and cellular toolbox. The Erythrocytenex platform utilizes polyphenol-coated nanocomposites that spontaneously associate with erythrocytes, enabling selective capillary targeting. This property facilitates precise therapeutic delivery to pulmonary malignancies and persistent respiratory infections. Furthermore, the macrophage-mediated Cellnex platform, engineered with surface-conjugated PD-L1 antibodies, demonstrates deep tumor penetration in 4 T1 spheroid models via chemotactic migration. Subsequent *in vivo* studies revealed potentiated therapeutic outcomes when combined with PD-L1-targeted macrophage gene therapy. The ongoing development of novel functional molecules and therapeutic strategies positions this straightforward, scalable approach as a versatile platform for diverse cell-based therapies and biohybrid cellular engineering applications (Zhao et al. 2020).

In addition, Liao et al. reported a tumor-targeting cell drug delivery system (MAGN), which is a bioresponsive nanosponge combined with tumor-homing macrophages through surface engineering. The engineered nanosponge's exceptional porosity and hydrophilicity enable the MAGN platform to function as a versatile delivery system, achieving high payload capacity and efficient transport of multiple therapeutic modalities, including nucleic acids (DNA/RNA), biologics (monoclonal antibodies), and peptides (to disease-specific sites). These engineered nanosponges exhibit pH-responsive contraction in acidic microenvironments, triggering controlled release of encapsulated chemotherapeutic payloads. The polyphenol-based supramolecular gatekeeper enables the nanosponge to adhere to the cell surface, and the "on-off" gating capability of the iron-tannic acid (Fe^{III}-TA) gatekeeper allows for controlled drug release. Mechanistic investigations reveal that these coordination complexes function as pH-responsive gatekeepers and effectively block drug diffusion at physiological pH (7.4) via synergistic Coulombic stabilization and van der Waals interactions. Once the nanosponge reaches an acidic environment, these interactions significantly decrease, thereby allowing local drug release in TME. These results demonstrate that the MAGN system offers a versatile therapeutic platform capable of high-capacity encapsulation of diverse anticancer agents and effective treatment of advanced metastatic malignancies (Liao et al. 2023).

11.3.4 Enzymatic Remodeling

Enzymatic modification of endogenous proteins and glycans represents an unconventional yet powerful approach for non-genetic cell surface engineering. Enzymes are essential to the modification of naturally occurring proteins and glycans on the cell surface. Through these enzymatic processes, it is possible to generate diverse isoforms of proteins, each with potentially unique functional properties. Additionally, enzymes can facilitate the attachment of synthetic targeting elements to cell surface molecules. This capability is particularly valuable in biomedical applications, such as targeted drug delivery and cell-based therapies, where precise molecular targeting is essential for therapeutic efficacy (Capicciotti et al. 2017).

Shi et al. devised a novel strategy for modifying natural killer (NK) cell surfaces through supramolecular engineering. Their methodology employs aptamer-based polyvalent antibody mimics (PAMs) to achieve targeted functionalization. The researchers developed this PAM system through in situ nucleic acid assembly and hybridization on cellular membranes. Experimental data demonstrate that the polyvalent construct markedly improves NK cell adhesion to tumor targets relative to monovalent versions. Furthermore, PAM-modified NK cells exhibit significantly enhanced tumoricidal activity compared to unmodified controls. Importantly, aptamers can theoretically be identified for any cell receptor, and they can be substituted with other ligands in the development of PAM. This research successfully develops a novel platform technology to potentiate immune cell-tumor cell interactions (Shi et al. 2020).

CD22, a B-cell-specific surface marker belonging to the Siglec (sialic acid-binding immunoglobulin-like lectin) family, serves as an important therapeutic target. While CD22- and CD20-directed antibody therapies demonstrate remarkable clinical outcomes against B-cell malignancies, their non-selective action against both cancerous and healthy B cells may induce immunosuppressive side effects (Fry et al. 2018; Ramakrishna et al. 2019). To overcome this limitation, Hong's research team pioneered a chemoenzymatic glycocalyx engineering approach. Their innovative method involves incorporating high-affinity CD22-specific ligands into NK-92MI and cytokine-induced killer cells, enabling precise tumor-selective CD22 recognition. The sustained surface retention of these engineered CD22 ligands for a minimum duration of 72 hours. The engineered cells displaying CD22-targeting ligands exhibited markedly improved tumor cell adhesion and cytotoxicity *in vitro* while this kind of engineered cells-maintained specificity that spared normal B lymphocytes. To enhance the therapeutic efficacy against lymphoma in vivo, the researchers additionally engineered the CD22-targeted NK-92MI cells with sialyl Lewis X (sLeX), an E-selectin-binding glycan, to facilitate their homing to bone marrow niches. In xenograft models, the bifunctional cells effectively inhibited B-cell lymphoma (Hong et al. 2021).

Mondal et al. presented a study on the glycoengineering of CAR-T cells to enhance their binding to E-selectin, a critical factor for the homing of these cells to target tissues. The researchers utilized an enzymatic process called exofucosylation, involving the enzyme fucosyltransferase 6, to modify the surface of CAR-T cells. This biochemical conversion transforms type 2 sialyl-LacNAc substrates into sLeX, the primary molecular recognition motif for E-selectin-mediated adhesion. The enzymatic modification significantly increased the binding of CAR-T cells to E-selectin under hemodynamic shear conditions, which is essential for the cells to adhere to the vascular endothelium and home to the target tissue. The engineered CAR-T cells exhibited a tenfold increase in bone marrow infiltration capacity relative to their unmodified counterparts when administered intravenously into mice. This glycoengineering approach not only improves the tissue delivery of CAR-T cells but also has the potential to reduce the required cell dose and associated production costs, offering a promising strategy to improve the effectiveness and cost-effectiveness of CAR-T cell immunotherapy for various hematologic and non-hematologic malignancies (Mondal et al. 2019).

11.3.5 Liposomes Fusion

Lipid membrane fusion constitutes a fundamental biological phenomenon that has attracted considerable scientific investigation. Functionalized liposomes spontaneously fuse with cellular membranes during co-incubation, thereby presenting their engineered molecular components on the cell surface.

Shi et al. engineered specialized fusogenic liposomes for selective T-cell targeting, enabling modulation and quantitative assessment of T-cell function through surface redox state analysis. The liposomes were chemically modified with 2,2,6,6-tetramethylpiperidine (TEMP) moieties that effectively scavenge reactive oxygen species, protecting T lymphocytes from oxidative stress-mediated functional impairment. Additionally, the generation of paramagnetic 2,2,6,6-tetramethylpiperidine 1-oxyl radicals permits quantitative measurement of T-lymphocyte activation states via magnetic resonance imaging. Experimental studies across multiple murine models revealed that these fusogenic liposomes specifically could target T lymphocytes exhibited potent antitumor efficacy while enable early assessment of therapeutic responses to radiation treatment. This investigation utilizes a chemical targeting approach to evaluate T lymphocyte functionality within living systems for oncological theranostics, offering new insights into immune cell engineering strategies for malignant disease management (Shi et al. 2023).

Zheng et al. developed a core-shell membrane-fused liposomal system (MFL) capable of dually modulating tumor cell surface glycans through membrane fusion. This innovative platform physically introduces NK cell-activating glycans while biologically suppressing immunosuppressive glycans, effectively overcoming tumor immune evasion mediated by aberrant glycan overexpression. The incorporation of MFLs into a temperature-responsive hydrogel matrix enables both injectable administration and tumor site-specific release, as the hydrogel undergoes degradation triggered by the TME. Upon release, the MFLs undergo membrane fusion with tumor cells, achieving dual therapeutic effects: cytoplasmic delivery of sialyltransferase inhibitor payloads from the core compartment, and surface display of NK cell-activating glycans via shell integration. The MFL system achieves coordinated intracellular and extracellular modulation through its compartmentalized architecture: (1) the core component delivers sialyltransferase inhibitors to the cytosol, effectively suppressing the biosynthesis of immunosuppressive sialoglycans, while (2) the shell component deposits immunostimulatory LeX trisaccharides on the cell membrane, potently enhancing NK cell recognition and activation. This innovative approach effectively converts the tumor's immunosuppressive sialic acid shield into an immunostimulatory LeX-activated surface, significantly enhancing NK cell-mediated tumor recognition and cytotoxic elimination, ultimately promoting rapid tumor clearance (Zheng et al. 2023b).

Liu et al. designed a light-responsive platform to enhance drug delivery across the blood-brain barrier (BBB). The system consists of blue-emitting NaYF4:Yb, Tm@zeolitic imidazolate framework (UCNP@ZIF-8), a photoacid, and melatonin (MT). The UZPM system was engineered for macrophage delivery through functionalized liposome fusion, with subsequent surface modification using hydroxylamine groups. To achieve targeted specificity, researchers employed aldehyde-functionalized CTLA-4 (cytotoxic T-lymphocyte-associated protein-4) as a CAR targeting moiety. This strategic design enabled precise binding to M1-polarized microglia in the central nervous system via aldehyde-hydroxylamine condensation chemistry, forming the CAR-m-UZPM complex. Experimental validation, both in cellular models and animal studies, confirmed the therapeutic efficacy of the CAR-M-UZPM platform. This innovative delivery system successfully traverses the BBB with high specificity for activated CNS microglia. The technology demonstrates dual therapeutic mechanisms: suppression of pro-inflammatory M1 microglial polarization, and induction of prolonged immunomodulatory effects resembling therapeutic vaccination. These combined actions effectively mitigate neuroinflammation, offering preventive benefits against the initiation and progression of inflammation-mediated depressive disorders. This investigation underscores the pivotal involvement of microglial-mediated neuroinflammatory processes in depression pathogenesis (Liu et al. 2022).

11.3.6 Fundamental Limitations and Engineering Requirements

Cancer Immunotherapy has recently gained prominence as a novel anticancer strategy that utilizes the host's natural defense mechanisms. While genetic membrane engineering has been a significant focus in this field, non-genetic membrane engineering methods offer alternative strategies that avoid the complexities and potential risks associated with genetic manipulation. Here we explored the fundamental limitations of non-genetic engineering methods and discussed integrated strategies to enhance their effectiveness in cancer immunotherapy (Wang et al. 2015).

11.3.6.1 T Cell Purity and Expansion

Traditional methods of generating cancer-specific T cells often face challenges in achieving high purity and efficient expansion. The use of retrovirus-based gene transfection can be laborious and carries potential safety concerns. To address this challenge, bispecific antibodies can simultaneously bind to two different antigens, enhancing the specificity and efficacy of T cell-mediated cancer killing. This approach avoids the need for genetic engineering and can be rapidly implemented. Chen et al. developed a non-genetic engineering platform using a bispecific antibody to rapidly generate and expand cancer-specific T cells. This method achieved over 90% purity in 7 days, addressing the issue of T cell purity and expansion (Chen et al. 2023a).

11.3.6.2 Targeting Specificity

Ensuring that the immune response is specifically directed towards cancer cells without causing off-target effects is also a major challenge for non-genetic membrane engineering, which must be designed to minimize the risk of autoimmunity and collateral damage to healthy tissues. To reduce off-target effects, scFv or bispecific antibodies were anchored to specific receptors on the cell membrane to target particular cell surface receptors. To improve CAR-T efficacy against solid malignancies, this study developed a dual-targeting approach directed against both CD70 and B7-H3 tumor antigens. Yang et al. developed a novel bivalent TanCAR construct capable of simultaneously recognizing both target antigens. The research team subsequently evaluated the tumor-suppressive capacity of these engineered T cells through comprehensive laboratory and animal studies. The results revealed the concurrent expression of both CD70 and B7-H3 across diverse malignancies, particularly in renal carcinoma, mammary carcinoma, esophageal neoplasms, hepatic tumors, colorectal cancer, gliomas, and cutaneous melanoma. Compared to single-target CAR-T cells, TanCAR-T cells demonstrated superior cytotoxicity and increased cytokine production upon recognition of tumor cells co-expressing dual antigens. *In vivo* studies revealed that even minimal doses of TanCAR-T cells effectively suppressed lung cancer and melanoma xenograft growth, significantly prolonging survival in treated animals. This study demonstrated that dual-targeting TanCAR-T cells (CD70/B7-H3) not only improved tumor recognition but also increased specificity, minimizing off-target toxicity (Yang et al. 2020).

11.3.6.3 Immune Microenvironment

The TME can be immunosuppressive, limiting the effectiveness of immunotherapy. Non-genetic engineering methods must overcome this barrier to ensure that the immune response is not dampened by the TME. The integration of non-genetic modification techniques with complementary immunotherapies, including checkpoint blockade, may synergistically improve therapeutic outcomes. This synergistic approach can overcome the limitations of individual therapies and improve patient outcomes. Chen et al. proposed a novel immunotherapy approach that involves engineering CAR-T cells to secrete a bispecific single-chain variable fragment (PD-1-TREM2 scFv) to enhance the therapeutic effects on colorectal cancer. Researchers developed colorectal cancer-targeting CAR-T cells genetically modified to produce a bispecific antibody against both PD-1/PD-L1 and TREM2, and the latter being overexpressed on immunosuppressive TAMs and MDSCs. In murine models, these modified CAR-T cells demonstrated potent tumor suppression, enhanced CD8+ T cell infiltration and decreased immunosuppressive cell populations (MDSCs and M2-TAMs), ultimately reversing tumor microenvironment immunosuppression and enhancing therapeutic efficacy against solid tumors (Chen et al. 2023b).

11.4 Genetic and Non-genetic Engineering of Immune Cells for Cancer Immunotherapy

Recent advances in cancer immunotherapy have been propelled by innovative immune cell engineering strategies that synergistically integrate genetic modifications with complementary non-genetic modalities to potentiate endogenous antitumor immunity. Genetic engineering, which involves modifying the genetic material of immune cells to introduce or enhance specific functions, has been pivotal in developing CAR-T cells. These genetically modified T cells are programmed to identify and selectively attack tumor-associated antigens, enhancing both the specificity and therapeutic effectiveness of antitumor immunity (June et al. 2018). Non-genetic engineering, in contrast, focuses on modifying the physical or chemical properties of immune cells without altering their genetic makeup. This can include surface modifications or the use of nanoparticles to enhance cell function or delivery, thereby improving the cells' ability to penetrate and act within the tumor microenvironment (Csizmar et al. 2018). The integration of genetic and non-genetic engineering strategies offers a multifaceted approach to cancer treatment. By combining these methods, researchers can address the heterogeneity and resistance mechanisms of cancer more effectively. For instance, genetically engineered CAR-T cells can be further enhanced through non-genetic modifications to improve their homing to the tumor site and their resistance to the immunosuppressive TME. This dual-engineering approach not only enhances the therapeutic potential of immune cells but also provides a more robust and versatile platform for cancer immunotherapy. The following examples illustrate how these combined engineering techniques are being applied to develop innovative cancer immunotherapies, demonstrating the synergy between genetic and non-genetic modifications in enhancing the effectiveness of immune cell-based treatments.

Luo et al. introduced a novel strategy to improve CAR-T cell performance against solid tumors by creating biohybrid IL-12 nanochaperone-engineered CAR-T cell (INS-CAR T) incorporating IL-12 nanochaperones. The team employed a dual-engineering strategy, genetically reprogramming CAR-T cells for antigen-specific tumor targeting while physically enhancing them via IL-12 nanoparticle conjugation (non-genetic engineering). The IL-12 nanochaperones were engineered for antigen-responsive IL-12 release, amplifying CAR-T cell activation and expansion within the TME. The INS-CAR T biohybrids were synthesized through bioorthogonal conjugation of IL-12-loaded nanoparticles onto CAR-T cell membranes. Experimental validation revealed these constructs markedly improved tumor penetration (3-fold enhancement) and cytotoxic activity (2-fold greater tumor apoptosis) in both 2D cultures and 3D spheroid models. In vivo studies demonstrated that INS-CAR-T therapy potently inhibited tumor progression and exhibited an excellent safety profile, with no observable off-target effects or systemic toxicity in murine models. This integrated engineering approach represents a promising therapeutic platform, achieving potent tumor control while maintaining favorable safety profiles in immunotherapeutic applications (Luo et al. 2022).

Liu et al. pioneered a nanotechnology platform that augments adoptive cell therapy (ACT) effectiveness against solid tumors. Their methodology employs biorthogonal unnatural sugars to metabolically tag T cells, enabling site-specific coupling of therapeutic cytokines during ex vivo production. This innovative platform facilitates tumor-localized cytokine delivery to mitigate systemic toxicity while simultaneously enhancing T cell tumor homing. The approach further induces Type 1-polarized immune activation, stimulates broad antigen recognition, and demonstrates superior efficacy against treatment-refractory solid malignancies. This study establishes a clinically translatable approach that seamlessly integrates with existing T cell manufacturing protocols, requiring neither procedural modifications nor extended production timelines. Importantly, the technology substantially reduces the therapeutic threshold of CAR-T cells while maintaining robust antitumor efficacy. Notably, IL-12-conjugated T cells demonstrated superior solid tumor regression compared to conventional therapy, mediated through enhanced systemic persistence and improved tumor infiltration of the transferred lymphocytes. Furthermore, IL-12 conjugation induces Th1-polarized activation of endogenous immunity, broadening antitumor recognition to include non-targeted tumor-associated antigens beyond the scope of adoptive cell therapy. Within CAR-T therapy paradigms, IL-12 conjugation achieves robust tumor control at subtherapeutic cell doses while simultaneously enhancing CAR-T cell persistence in peripheral circulation post-administration. These findings demonstrate a transformative potential for advancing ACT and CAR-T platforms, and offer a streamlined yet potent strategy to amplify therapeutic efficacy across diverse clinical applications (Liu et al. 2023).

The limited clinical efficacy of CAR-T therapies against solid tumors primarily stems from two key challenges: inadequate tumor infiltration capacity and a profoundly immunosuppressive tumor niche. Zhao et al. employed biorthogonal click chemistry to functionalize CAR-T cells with both hyaluronidase (HAase) for extracellular matrix degradation and α-PD-L1 checkpoint blockade antibodies. HAase-mediated degradation of the tumor extracellular matrix significantly enhances CAR-T cell infiltration into solid tumor masses, as demonstrated through comprehensive *in vitro* assays and *in vivo* validation studies. The α-PD-L1 modification potentiates CAR-T cell antitumor efficacy through PD-1/PD-L1 axis blockade, effectively counteracting tumor-mediated immunosuppression within the microenvironment. Preclinical evaluation across dual solid tumor models revealed that these modified CAR-T cells not only achieved superior therapeutic outcomes but also demonstrated markedly increased tumor penetration and potent tumoricidal activity. Notably, HAase-modified CAR-T cells demonstrated 4.6-fold greater tumor accumulation than conventional CAR-T cells, while α-PD-L1-modified cells showed significantly increased production of effector cytokines (IFN-γ, TNF-α, and IL-2), confirming their enhanced activation status. The combination of HAase and α-PDL1 resulted in a remarkable 95.1% inhibition rate of tumor growth in one of the models. Importantly, these modifications did not cause significant systemic side effects, as indicated by minimal weight loss and no apparent organ damage in treated mice. These results establish a novel therapeutic paradigm for overcoming the limitations of CAR-T therapy in solid tumors, which demonstrate significant translational potential for clinical implementation (Zhao et al. 2022).

11.5 Conclusion and Ways Forward

In this chapter, we summarized the evolution of immune cell membrane engineering, from early studies of adoptive cellular therapy to advanced genetic and non-genetic modifications, underscores its transformative impact on tumor immunotherapy. Groundbreaking developments such as IL-2-mediated T cell expansion and the introduction of engineered receptors like CAR have revolutionized the field, enabling precise targeting, enhanced persistence, and improved functionality of immune cells. The progressive refinement of CAR-T cell therapy, from first-generation designs to multi-functional fourth-generation constructs, highlights the potential of genetic engineering to overcome therapeutic challenges and achieve durable responses in patients. Non-genetic membrane engineering complements genetic approaches by providing rapid, reversible, and flexible enhancements without the risks associated with permanent genomic alterations. Techniques such as direct membrane anchoring and click chemistry have expanded the toolkit for immune cell modification, allowing for more precise, temporary modifications that can be adapted to various clinical needs.

To advance immune cell membrane engineering for tumor immunotherapy, the following research directions should be prioritized. (1) Integrating genetic and non-genetic approaches. Combine genetic and non-genetic engineering to create immune cells with enhanced targeting specificity, controlled activation, and resistance to the immunosuppressive tumor microenvironment. (2) Addressing challenges in solid tumors. Leverage advanced membrane engineering to overcome obstacles unique to solid tumors, such as hypoxia, physical barriers, and heterogeneous antigen expression. (3) Personalized modifications. Develop patient-specific solutions, including customized CARs and tailored surface molecules, to improve therapeutic efficacy while minimizing immunogenicity. (4) Scaling up production. Optimize membrane engineering techniques for cost-effective and reliable large-scale immune cell production to expand clinical applications. (5) Synthetic circuit integration. Incorporate synthetic circuits into immune cells to enable dynamic, programmable therapeutic responses, ensuring adaptability to evolving tumor conditions. Immune cell membrane engineering remains a cornerstone of modern cancer immunotherapy, continuously enhancing the capabilities of adoptive cell therapies. Through targeted resolution of existing barriers and integration of cutting-edge approaches, the field is advancing toward realizing the complete therapeutic capacity of immune cell therapies, enabling more precise, potent, and tolerable cancer treatment paradigms.

References

Abbina S, Siren EMJ, Moon H, Kizhakkedathu JN (2017) Surface engineering for cell-based therapies: techniques for manipulating mammalian cell surfaces. ACS Biomater Sci Eng 4:3658–3677. https://doi.org/10.1021/acsbiomaterials.7b00514

Ahmed S, Rai KR (2003) Interferon in the treatment of hairy-cell leukemia. Best Pract Res Clin Haematol 16:69–81. https://doi.org/10.1016/s1521-6926(02)00084-1

Bader JE, Voss K, Rathmell JC (2020) Targeting metabolism to improve the tumor microenvironment for cancer immunotherapy. Mol Cell 78:1019–1033. https://doi.org/10.1016/j.molcel.2020.05.034

Baum C, Schambach A, Bohne J, Galla M (2006) Retrovirus vectors: toward the plentivirus? Mol Ther 13:1050–1063. https://doi.org/10.1016/j.ymthe.2006.03.007

Beatty GL, Haas AR, Maus MV, Torigian DA, Soulen MC, Plesa G et al (2014) Mesothelin-specific chimeric antigen receptor mRNA-engineered T cells induce anti-tumor activity in solid malignancies. Cancer Immunol Res 2:112–120. https://doi.org/10.1158/2326-6066.Cir-13-0170

Billingsley MM, Singh N, Ravikumar P, Zhang R, June CH, Mitchell MJ (2020) Ionizable lipid nanoparticle-mediated mRNA delivery for human CAR T cell engineering. Nano Lett 20:1578–1589. https://doi.org/10.1021/acs.nanolett.9b04246

Billingsley MM, Hamilton AG, Mai D, Patel SK, Swingle KL, Sheppard NC et al (2022) Orthogonal Design of Experiments for optimization of lipid nanoparticles for mRNA engineering of CAR T cells. Nano Lett 22:533–542. https://doi.org/10.1021/acs.nanolett.1c02503

Bommareddy PK, Wakimoto H, Martuza RL, Kaufman HL, Rabkin SD, Saha D (2024) Oncolytic herpes simplex virus expressing IL-2 controls glioblastoma growth and improves survival. J Immunother Cancer 12. https://doi.org/10.1136/jitc-2024-008880

Bray F, Ferlay J, Soerjomataram I, Siegel RL, Torre LA, Jemal A (2018) Global cancer statistics 2018: GLOBOCAN estimates of incidence and mortality worldwide for 36 cancers in 185 countries. CA Cancer J Clin 68:394–424. https://doi.org/10.3322/caac.21492

Brentjens RJ, Latouche JB, Santos E, Marti F, Gong MC, Lyddane C et al (2003) Eradication of systemic B-cell tumors by genetically targeted human T lymphocytes co-stimulated by CD80 and interleukin-15. Nat Methods 9:279–286. https://doi.org/10.1038/nm827

Brown CE, Rodriguez A, Palmer J, Ostberg JR, Naranjo A, Wagner JR et al (2022) Off-the-shelf, steroid-resistant, IL13Rα2-specific CAR T cells for treatment of glioblastoma. Neuro-Oncology 24:1318–1330. https://doi.org/10.1093/neuonc/noac024

Bucher K, Rodríguez-Bocanegra E, Dauletbekov D, Fischer MD (2021) Immune responses to retinal gene therapy using adeno-associated viral vectors – implications for treatment success and safety. Prog Retin Eye Res 83:100915. https://doi.org/10.1016/j.preteyeres.2020.100915

Cao J, Chen Z, Chi J, Sun Y, Sun Y (2018) Recent progress in synergistic chemotherapy and phototherapy by targeted drug delivery systems for cancer treatment. Artif Cells Nanomed Biotechnol 46:817–830. https://doi.org/10.1080/21691401.2018.1436553

Capicciotti CJ, Zong C, Sheikh MO, Sun T, Wells L, Boons GJ (2017) Cell-surface Glyco-engineering by exogenous enzymatic transfer using a bifunctional CMP-Neu5Ac derivative. J Am Chem Soc 139:13342–13348. https://doi.org/10.1021/jacs.7b05358

Chen YH, Keiser MS, Davidson BL (2018) Viral vectors for gene transfer. Curr Protoc Mouse Biol 8:e58. https://doi.org/10.1002/cpmo.58

Chen YJ, Chen M, Cheng TL, Tsai YS, Wang CH, Chen CY et al (2023a) A non-genetic engineering platform for rapidly generating and expanding cancer-specific armed T cells. J Biomed Sci 30:35. https://doi.org/10.1186/s12929-023-00929-z

Chen J, Zhu T, Jiang G, Zeng Q, Li Z, Huang X (2023b) Target delivery of a PD-1-TREM2 scFv by CAR-T cells enhances anti-tumor efficacy in colorectal cancer. Mol Cancer 22:131. https://doi.org/10.1186/s12943-023-01830-x

Ciuffi A (2008) Mechanisms governing lentivirus integration site selection. Curr Gene Ther 8:419–429. https://doi.org/10.2174/156652308786848021

Craigie R, Bushman FD (2014) Host factors in retroviral integration and the selection of integration target sites. Microbiol Spectr 2. https://doi.org/10.1128/microbiolspec.MDNA3-0026-2014

Csizmar CM, Petersburg JR, Wagner CR (2018) Programming cell-cell interactions through non-genetic membrane engineering. Cell Chem Biol 25:931–940. https://doi.org/10.1016/j.chembiol.2018.05.009

David RM, Doherty AT (2017) Viral vectors: the road to reducing genotoxicity. Toxicol Sci 155:315–325. https://doi.org/10.1093/toxsci/kfw220

Diehn M et al (2002) Genomic expression programs and the integration of the CD28 costimulatory signal in T cell activation. Proc Natl Acad Sci USA 99:11796–11801. https://doi.org/10.1073/pnas.092284399

Dimitri A, Herbst F, Fraietta JA (2022) Engineering the next-generation of CAR T-cells with CRISPR-Cas9 gene editing. Mol Cancer 21. https://doi.org/10.1186/s12943-022-01559-z

Doan VTH, Imai T, Kawazoe N, Chen G, Yoshitomi T (2025) Regulation of antigen presentation and interleukin 10 production in murine dendritic cells via the oxidative stimulation of cell membrane using a polycation-porphyrin-conjugate-immobilized cell culture dish. Acta Biomater. https://doi.org/10.1016/j.actbio.2025.01.004

Doudna JA (2020) The promise and challenge of therapeutic genome editing. Nature 578:229–236. https://doi.org/10.1038/s41586-020-1978-5

Doyon Y, Vo TD, Mendel MC, Greenberg SG, Wang J, Xia DF et al (2011) Enhancing zinc-finger-nuclease activity with improved obligate heterodimeric architectures. Nat Methods 8:74–79. https://doi.org/10.1038/nmeth.1539

Duan D, Wang K, Wei C, Feng D, Liu Y, He Q et al (2021) The BCMA-targeted fourth-generation CAR-T cells secreting IL-7 and CCL19 for therapy of refractory/recurrent multiple myeloma. Front Immunol 12. https://doi.org/10.3389/fimmu.2021.609421

Dudley ME, Wunderlich JR, Shelton TE, Even J, Rosenberg SA (2003) Generation of tumor-infiltrating lymphocyte cultures for use in adoptive transfer therapy for melanoma patients. J Immunother 26:332–342. https://doi.org/10.1097/00002371-200307000-00005

Fernandez-Sendin M, Tenesaca S, Vasquez M, Aranda F, Berraondo P (2020) In: Galluzzi L, Rudqvist N-P (eds) Methods in enzymology, vol 635. Academic, pp 185–203

Fesnak AD, June CH, Levine BL (2016) Engineered T cells: the promise and challenges of cancer immunotherapy. Nat Rev Cancer 16:566–581. https://doi.org/10.1038/nrc.2016.97

Fry TJ, Shah NN, Orentas RJ, Stetler-Stevenson M, Yuan CM, Ramakrishna S et al (2018) CD22-targeted CAR T cells induce remission in B-ALL that is naive or resistant to CD19-targeted CAR immunotherapy. Nat Med 24:20–28. https://doi.org/10.1038/nm.4441

Fu Y, Foden JA, Khayter C, Maeder ML, Reyon D, Joung JK et al (2013) High-frequency off-target mutagenesis induced by CRISPR-Cas nucleases in human cells. Nat Biotechnol 31:822–826. https://doi.org/10.1038/nbt.2623

Gartner ZJ, Bertozzi CR (2009) Programmed assembly of 3-dimensional microtissues with defined cellular connectivity. Proc Natl Acad Sci USA 106:4606–4610. https://doi.org/10.1073/pnas.0900717106

Geng H, Zhi S, Zhou X, Yan Y, Zhang G, Dai S et al (2024) Self-powered engineering of cell membrane receptors to on-demand regulate cellular behaviors. Nano Lett 24:7895–7902. https://doi.org/10.1021/acs.nanolett.4c01080

Gennari M, Légalité F, Zhang L, Pellegrin Y, Blart E, Fortage J et al (2014) Long-lived charge separated state in NiO-based p-type dye-sensitized solar cells with simple cyclometalated iridium complexes. J Phys Chem Lett 5:2254–2258. https://doi.org/10.1021/jz5009714

Goronzy JJ, Weyand CM (2008) T-cell co-stimulatory pathways in autoimmunity. Arthritis Res Ther 10. https://doi.org/10.1186/ar2414

Gross G, Waks T, Eshhar Z (1989) Expression of immunoglobulin-T-cell receptor chimeric molecules as functional receptors with antibody-type specificity. Proc Natl Acad Sci USA 86:10024–10028. https://doi.org/10.1073/pnas.86.24.10024

Grupp SA, Kalos M, Barrett D, Aplenc R, Porter DL, Rheingold SR et al (2013) Chimeric antigen receptor–modified t cells for acute lymphoid leukemia. N Engl J Med 368:1509–1518. https://doi.org/10.1056/NEJMoa1215134

Guedan S, Posey AD, Shaw C, Wing A, Da T, Patel PR et al (2018) Enhancing CAR T cell persistence through ICOS and 4-1BB costimulation. JCI Insight 3. https://doi.org/10.1172/jci.insight.96976

Han J, Lim J, Wang CJ, Han JH, Shin HE, Kim SN et al (2023) Lipid nanoparticle-based mRNA delivery systems for cancer immunotherapy. Nano Converg 10:36. https://doi.org/10.1186/s40580-023-00385-3

Hegde M, Mukherjee M, Grada Z, Pignata A, Landi D, Navai SA et al (2016) Tandem CAR T cells targeting HER2 and IL13Ralpha2 mitigate tumor antigen escape. J Clin Invest 126:3036–3052. https://doi.org/10.1172/JCI83416

Hong S, Yu C, Wang P, Shi Y, Cao W, Cheng B et al (2021) Glycoengineering of NK cells with glycan ligands of CD22 and selectins for B-cell lymphoma therapy. Angew Chem Int Ed Engl 60:3603–3610. https://doi.org/10.1002/anie.202005934

Honikel MM, Olejniczak SH (2022) Co-stimulatory receptor signaling in CAR-T cells. Biomol Ther 12. https://doi.org/10.3390/biom12091303

Hou X, Du Y, Deng Y, Wu J, Cao G (2015) Sleeping beauty transposon system for genetic etiological research and gene therapy of cancers. Cancer Biol Ther 16:8–16. https://doi.org/10.4161/15384047.2014.986944

Hu Y, Hu Q, Li Y, Lu L, Xiang Z, Yin Z et al (2023) Γδ T cells: origin and fate, subsets, diseases and immunotherapy. Signal Transduct Target Ther 8. https://doi.org/10.1038/s41392-023-01653-8

Ingulli E (2010) Mechanism of cellular rejection in transplantation. Pediatr Nephrol 25:61–74. https://doi.org/10.1007/s00467-008-1020-x

Irving BA, Weiss A (1991) The cytoplasmic domain of the T cell receptor zeta chain is sufficient to couple to receptor-associated signal transduction pathways. Cell 64:891–901. https://doi.org/10.1016/0092-8674(91)90314-o

Izsvák Z, Hackett PB, Cooper LJ, Ivics Z (2010) Translating sleeping beauty transposition into cellular therapies: victories and challenges. BioEssays 32:756–767. https://doi.org/10.1002/bies.201000027

Jangid AK, Kim S, Kim HJ, Kim K (2023) Biomaterial-mediated exogenous facile coating of natural killer cells for enhancing anticancer efficacy toward hepatocellular carcinoma. Bioconjug Chem 34:1789–1801. https://doi.org/10.1021/acs.bioconjchem.3c00335

Jo S, Das S, Williams A, Chretien AS, Pagliardini T, Le Roy A et al (2022) Endowing universal CAR T-cell with immune-evasive properties using TALEN-gene editing. Nat Commun 13:3453. https://doi.org/10.1038/s41467-022-30896-2

June CH, Warshauer JT, Bluestone JA (2017) Is autoimmunity the Achilles' heel of cancer immunotherapy? Nat Med 23:540–547. https://doi.org/10.1038/nm.4321

June CH, O'Connor RS, Kawalekar OU, Ghassemi S, Milone MC (2018) CAR T cell immunotherapy for human cancer. Science (New York, NY) 359:1361–1365. https://doi.org/10.1126/science.aar6711

Kashem SW, Haniffa M, Kaplan DH (2017) Antigen-presenting cells in the skin. Annu Rev Immunol 35:469–499. https://doi.org/10.1146/annurev-immunol-051116-052215

Kayser H, Geilen CC, Paul C, Zeitler R, Reutter W (1992) Incorporation of N-acyl-2-amino-2--deoxy-hexoses into glycosphingolipids of the pheochromocytoma cell line PC 12. FEBS Lett 301:137–140. https://doi.org/10.1016/0014-5793(92)81233-c

Kim GB, Aragon-Sanabria V, Randolph L, Jiang H, Reynolds JA, Webb BS et al (2020) High-affinity mutant Interleukin-13 targeted CAR T cells enhance delivery of clickable biodegradable fluorescent nanoparticles to glioblastoma. Bioactive Materials 5:624–635. https://doi.org/10.1016/j.bioactmat.2020.04.011

Koonin EV, Makarova KS, Zhang F (2017) Diversity, classification and evolution of CRISPR-Cas systems. Curr Opin Microbiol 37:67–78. https://doi.org/10.1016/j.mib.2017.05.008

Kuwana Y et al (1987) Expression of chimeric receptor composed of immunoglobulin-derived V regions and T-cell receptor-derived C regions. Biochem Biophys Res Commun 149:960–968. https://doi.org/10.1016/0006-291x(87)90502-x

Lee S, Margolin K (2011) Cytokines in cancer immunotherapy. Cancers 3:3856–3893. https://doi.org/10.3390/cancers3043856

Li J, Li W, Huang K, Zhang Y, Kupfer G, Zhao Q (2018) Chimeric antigen receptor T cell (CAR-T) immunotherapy for solid tumors: lessons learned and strategies for moving forward. J Hematol Oncol 11. https://doi.org/10.1186/s13045-018-0568-6

Liao X, Gong G, Dai M, Xiang Z, Pan J, He X et al (2023) Systemic tumor suppression via macrophage-driven automated homing of metal-phenolic-gated nanosponges for metastatic melanoma. Adv Sci 10:e2207488. https://doi.org/10.1002/advs.202207488

Lin H, Cheng J, Mu W, Zhou J, Zhu L (2021) Advances in universal CAR-T cell therapy. Front Immunol 12:744823. https://doi.org/10.3389/fimmu.2021.744823

Liu X, Zhang Y, Cheng C, Cheng AW, Zhang X, Li N et al (2017) CRISPR-Cas9-mediated multiplex gene editing in CAR-T cells. Cell Res 27:154–157. https://doi.org/10.1038/cr.2016.142

Liu E, Marin D, Banerjee P, Macapinlac HA, Thompson P, Basar R et al (2020) Use of CAR-transduced natural killer cells in CD19-positive lymphoid tumors. N Engl J Med 382:545–553. https://doi.org/10.1056/NEJMoa1910607

Liu Y, Hu P, Zheng Z, Zhong D, Xie W, Tang Z et al (2022) Photoresponsive vaccine-like CAR-M system with high-efficiency central immune regulation for inflammation-related depression. Adv Mater 34:e2108525. https://doi.org/10.1002/adma.202108525

Liu Y, Adu-Berchie K, Brockman JM, Pezone M, Zhang DKY, Zhou J et al (2023) Cytokine conjugation to enhance T cell therapy. Proc Natl Acad Sci USA 120:e2213222120. https://doi.org/10.1073/pnas.2213222120

Luo Y, Chen Z, Sun M, Li B, Pan F, Ma A et al (2022) IL-12 nanochaperone-engineered CAR T cell for robust tumor-immunotherapy. Biomaterials 281:121341. https://doi.org/10.1016/j.biomaterials.2021.121341

Magnani CF, Tettamanti S, Alberti G, Pisani I, Biondi A, Serafini M et al (2020) Transposon-based CAR T cells in acute Leukemias: where are we going? Cells 9. https://doi.org/10.3390/cells9061337

Mathlouthi S, Kuryk L, Prygiel M, Lupo MG, Zasada AA, Pesce C et al (2024) Extracellular vesicles powered cancer immunotherapy: targeted delivery of adenovirus-based cancer vaccine in humanized melanoma model. J Control Release 376:777–793. https://doi.org/10.1016/j.jconrel.2024.10.057

Milling L, Zhang Y, Irvine DJ (2017) Delivering safer immunotherapies for cancer. Adv Drug Deliv Rev 114:79–101. https://doi.org/10.1016/j.addr.2017.05.011

Milone MC, O'Doherty U (2018) Clinical use of lentiviral vectors. Leukemia 32:1529–1541. https://doi.org/10.1038/s41375-018-0106-0

Mitra A, Barua A, Huang L, Ganguly S, Feng Q, He B (2023) From bench to bedside: the history and progress of CAR T cell therapy. Front Immunol 14. https://doi.org/10.3389/fimmu.2023.1188049

Mondal N, Silva M, Castano AP, Maus MV, Sackstein R (2019) Glycoengineering of chimeric antigen receptor (CAR) T-cells to enforce E-selectin binding. J Biol Chem 294:18465–18474. https://doi.org/10.1074/jbc.RA119.011134

Morgan DA, Ruscetti FW, Gallo R (1976) Selective in vitro growth of T lymphocytes from normal human bone marrows. Sci Transl Med 10. https://doi.org/10.1126/science.181845

Oli AN, Babajide Rowaiye A, Adejumo SA, Anazodo FI, Ahmad R, Sinha S et al (2022) Classic and current opinions in human organ and tissue transplantation. Cureus. https://doi.org/10.7759/cureus.30982

Pan H, Li W, Chen Z, Luo Y, He W, Wang M et al (2021) Click CAR-T cell engineering for robustly boosting cell immunotherapy in blood and subcutaneous xenograft tumor. Bioact Mater 6:951–962. https://doi.org/10.1016/j.bioactmat.2020.09.025

Petrov JC, Wada M, Pinz KG, Yan LE, Chen KH, Shuai X et al (2018) Compound CAR T-cells as a double-pronged approach for treating acute myeloid leukemia. Leukemia 32:1317–1326. https://doi.org/10.1038/s41375-018-0075-3

Pitt JM, Marabelle A, Eggermont A, Soria JC, Kroemer G, Zitvogel L (2016) Targeting the tumor microenvironment: removing obstruction to anticancer immune responses and immunotherapy. Ann Oncol 27:1482–1492. https://doi.org/10.1093/annonc/mdw168

Poorebrahim M, Sadeghi S, Fakhr E, Abazari MF, Poortahmasebi V, Kheirollahi A et al (2019) Production of CAR-T cells by GMP-grade lentiviral vectors: latest advances and future prospects. Crit Rev Clin Lab Sci 56:393–419. https://doi.org/10.1080/10408363.2019.1633512

Porter DL, Hwang W-T, Frey NV, Lacey SF, Shaw PA, Loren AW et al (2015) Chimeric antigen receptor T cells persist and induce sustained remissions in relapsed refractory chronic lymphocytic leukemia. Sci Transl Med 7. https://doi.org/10.1126/scitranslmed.aac5415

Pui CH, Yang JJ, Hunger SP, Pieters R, Schrappe M, Biondi A et al (2015) Childhood acute lymphoblastic leukemia: Progress through collaboration. J Clin Oncol 33:2938–2948. https://doi.org/10.1200/jco.2014.59.1636

Quesada JR, Hersh EM, Manning J, Reuben J, Keating M, Schnipper E et al (1986) Treatment of hairy cell leukemia with recombinant alpha-interferon. Blood 68:493–497. https://doi.org/10.1182/blood.V68.2.493.493

Ramakrishna S, Highfill SL, Walsh Z, Nguyen SM, Lei H, Shern JF et al (2019) Modulation of target antigen density improves CAR T-cell functionality and persistence. Clin Cancer Res 25:5329–5341. https://doi.org/10.1158/1078-0432.Ccr-18-3784

Ribas A, Wolchok JD (2018) Cancer immunotherapy using checkpoint blockade. Science 359:1350–1355. https://doi.org/10.1126/science.aar4060

Riley RS, June CH, Langer R, Mitchell MJ (2019) Delivery technologies for cancer immunotherapy. Nat Rev Drug Discov 18:175–196. https://doi.org/10.1038/s41573-018-0006-z

Rosenberg SA (2014) IL-2: the first effective immunotherapy for human cancer. J Immunol 192:5451–5458. https://doi.org/10.4049/jimmunol.1490019

Rosenberg SA, Restifo NP, Yang JC, Morgan RA, Dudley ME (2008) Adoptive cell transfer: a clinical path to effective cancer immunotherapy. Nat Rev Cancer 8:299–308. https://doi.org/10.1038/nrc2355

Ross SH, Cantrell DA (2018) Signaling and function of Interleukin-2 in T lymphocytes. Annu Rev Immunol 36:411–433. https://doi.org/10.1146/annurev-immunol-042617-053352

Sakuma T, Barry MA, Ikeda Y (2012) Lentiviral vectors: basic to translational. Biochem J 443:603–618. https://doi.org/10.1042/bj20120146

Sato M, Inada E, Saitoh I, Watanabe S, Nakamura S (2020) piggyBac-based non-viral in vivo gene delivery useful for production of genetically modified animals and organs. Pharmaceutics 12. https://doi.org/10.3390/pharmaceutics12030277

Shi P, Wang X, Davis B, Coyne J, Dong C, Reynolds J et al (2020) In situ synthesis of an aptamer-based polyvalent antibody mimic on the cell surface for enhanced interactions between

immune and cancer cells. Angew Chem Int Ed 59:11892–11897. https://doi.org/10.1002/anie.202004206
Shi C, Zhang Q, Yao Y, Zeng F, Du C, Nijiati S et al (2023) Targeting the activity of T cells by membrane surface redox regulation for cancer theranostics. Nat Nanotechnol 18:86–97. https://doi.org/10.1038/s41565-022-01261-7
Si K, Dai Z, Li Z, Ye Z, Ding B, Feng S et al (2023) Engineered exosome-mediated messenger RNA and single-chain variable fragment delivery for human chimeric antigen receptor T-cell engineering. Cytotherapy 25:615–624. https://doi.org/10.1016/j.jcyt.2023.01.005
Singh H, Huls H, Kebriaei P, Cooper LJ (2014) A new approach to gene therapy using sleeping beauty to genetically modify clinical-grade T cells to target CD19. Immunol Rev 257:181–190. https://doi.org/10.1111/imr.12137
Smith MA, Altekruse SF, Adamson PC, Reaman GH, Seibel NL (2014) Declining childhood and adolescent cancer mortality. Cancer 120:2497–2506. https://doi.org/10.1002/cncr.28748
Sun Y, Su J, Liu G, Chen J, Zhang X, Zhang R et al (2017) Advances of blood cell-based drug delivery systems. Eur J Pharm Sci 96:115–128. https://doi.org/10.1016/j.ejps.2016.07.021
Syyam A, Nawaz A, Ijaz A, Sajjad U, Fazil A, Irfan S et al (2022) Adenovirus vector system: construction, history and therapeutic applications. BioTechniques 73:297–305. https://doi.org/10.2144/btn-2022-0051
Tavernier G, Andries O, Demeester J, Sanders NN, De Smedt SC, Rejman J (2011) mRNA as gene therapeutic: how to control protein expression. J Control Release 150:238–247. https://doi.org/10.1016/j.jconrel.2010.10.020
Torikai H, Reik A, Liu PQ, Zhou Y, Zhang L, Maiti S et al (2012) A foundation for universal T-cell based immunotherapy: T cells engineered to express a CD19-specific chimeric-antigen-receptor and eliminate expression of endogenous TCR. Blood 119:5697–5705. https://doi.org/10.1182/blood-2012-01-405365
Torikai H, Reik A, Soldner F, Warren EH, Yuen C, Zhou Y et al (2013) Toward eliminating HLA class I expression to generate universal cells from allogeneic donors. Blood 122:1341–1349. https://doi.org/10.1182/blood-2013-03-478255
Wagner DL, Amini L, Wendering DJ, Burkhardt LM, Akyüz L, Reinke P et al (2019) High prevalence of streptococcus pyogenes Cas9-reactive T cells within the adult human population. Nat Med 25:242–248. https://doi.org/10.1038/s41591-018-0204-6
Waldman AD, Fritz JM, Lenardo MJ (2020) A guide to cancer immunotherapy: from T cell basic science to clinical practice. Nat Rev Immunol 20:651–668. https://doi.org/10.1038/s41577-020-0306-5
Wang Q, Cheng H, Peng H, Zhou H, Li PY, Langer R (2015) Non-genetic engineering of cells for drug delivery and cell-based therapy. Adv Drug Deliv Rev 91:125–140. https://doi.org/10.1016/j.addr.2014.12.003
Wang X, Popplewell LL, Wagner JR, Naranjo A, Blanchard MS, Mott MR et al (2016) Phase 1 studies of central memory-derived CD19 CAR T-cell therapy following autologous HSCT in patients with B-cell NHL. Blood 127:2980–2990. https://doi.org/10.1182/blood-2015-12-686725
Wang F, Qin Z, Lu H, He S, Luo J, Jin C et al (2019) Clinical translation of gene medicine. J Gene Med 21:e3108. https://doi.org/10.1002/jgm.3108
Wang H, Sobral MC, Zhang DKY, Cartwright AN, Li AW, Dellacherie MO et al (2020) Metabolic labeling and targeted modulation of dendritic cells. Nat Mater 19:1244–1252. https://doi.org/10.1038/s41563-020-0680-1
Woodard LE, Wilson MH (2015) piggyBac-ing models and new therapeutic strategies. Trends Biotechnol 33:525–533. https://doi.org/10.1016/j.tibtech.2015.06.009
Xiong M, Kong G, Liu Q, Liu L, Yin Y, Liu Y et al (2023) DNA-templated anchoring of proteins for programmable cell functionalization and immunological response. Nano Lett 23:183–191. https://doi.org/10.1021/acs.nanolett.2c03928
Yang B, Chen Y, Shi J (2019) Exosome biochemistry and advanced nanotechnology for next-generation Theranostic platforms. Adv Mater 31:e1802896. https://doi.org/10.1002/adma.201802896

Yang M, Tang X, Zhang Z, Gu L, Wei H, Zhao S et al (2020) Tandem CAR-T cells targeting CD70 and B7-H3 exhibit potent preclinical activity against multiple solid tumors. Theranostics 10:7622–7634. https://doi.org/10.7150/thno.43991

Yao S, Zhu Y, Chen L (2013) Advances in targeting cell surface signalling molecules for immune modulation. Nat Rev Drug Discov 12:130–146. https://doi.org/10.1038/nrd3877

Yuan X, Zhou J-L, Yuan L, Fan J, Yoon J, Zhang X-B et al (2024) Phototherapy: progress, challenges, and opportunities. Sci China Chem. https://doi.org/10.1007/s11426-024-2411-7

Zeng W, Zhang Q, Zhu Y, Ou R, Peng L, Wang B et al (2022) Engineering novel CD19/CD22 dual-target CAR-T cells for improved anti-tumor activity. Cancer Investig 40:282–292. https://doi.org/10.1080/07357907.2021.2005798

Zhang H, Qin C, An C, Zheng X, Wen S, Chen W et al (2021) Application of the CRISPR/Cas9-based gene editing technique in basic research, diagnosis, and therapy of cancer. Mol Cancer 20:126. https://doi.org/10.1186/s12943-021-01431-6

Zhang Y, Qian L, Kuang Y, Liu J, Wang D, Xie W et al (2022) An adeno-associated virus-mediated immunotherapy for Alzheimer's disease. Mol Immunol 144:26–34. https://doi.org/10.1016/j.molimm.2022.02.006

Zhang P, Zhang G, Wan X (2023) Challenges and new technologies in adoptive cell therapy. J Hematol Oncol 16:97. https://doi.org/10.1186/s13045-023-01492-8

Zhao Z, Pan DC, Qi QM, Kim J, Kapate N, Sun T et al (2020) Engineering of living cells with polyphenol-functionalized biologically active Nanocomplexes. Adv Mater 32:e2003492. https://doi.org/10.1002/adma.202003492

Zhao Y, Dong Y, Yang S, Tu Y, Wang C, Li J et al (2022) Bioorthogonal equipping CAR-T cells with hyaluronidase and checkpoint blocking antibody for enhanced solid tumor immunotherapy. ACS Cent Sci 8:603–614. https://doi.org/10.1021/acscentsci.2c00163

Zheng Z, Li S, Liu M, Chen C, Zhang L, Zhou D (2023a) Fine-tuning through generations: advances in structure and production of CAR-T therapy. Cancers 15. https://doi.org/10.3390/cancers15133476

Zheng C, Zhong Q, Song W, Yi K, Kong H, Wang H et al (2023b) Membrane-fusion-mediated multiplex engineering of tumor cell surface Glycans for enhanced NK cell therapy. Adv Mater 35:e2206989. https://doi.org/10.1002/adma.202206989

Zhong Y, Xu L, Yang C, Xu L, Wang G, Guo Y et al (2023) Site-selected in situ polymerization for living cell surface engineering. Nat Commun 14:7285. https://doi.org/10.1038/s41467-023-43161-x

Zu H, Gao D (2021) Non-viral vectors in gene therapy: recent development, challenges, and prospects. AAPS J 23:78. https://doi.org/10.1208/s12248-021-00608-7

Cellular Membrane Engineering Platform for Capturing and Neutralizing Circulating Tumor Cells

12

Changchong Chen

Abstract

Circulating tumor cells (CTCs) are crucial biomarkers for cancer metastasis, prognosis, and treatment monitoring, but their rarity in peripheral blood and phenotypic heterogeneity pose significant challenges for isolation and analysis. Traditional capture methods, such as immunoaffinity and physical separation, often suffer from low sensitivity, loss of heterogeneous populations, and limited specificity. To overcome these limitations, biomimetic platforms using cellular membrane engineering have emerged as a promising alternative, improving both CTCs capture efficiency and functional integrity. Additionally, CTCs neutralization presents a potential strategy to prevent metastasis by disrupting survival mechanisms that facilitate immune evasion, such as interactions with platelets, endothelial adhesion, and microthrombus formation. Cellular membrane engineering not only aids in isolating CTCs but also neutralizes them by mimicking physiological interactions that interfere with these protective mechanisms. This chapter examines the biological characteristics and clinical significance of CTCs, explores current capture and neutralization strategies, and highlights the advantages of membrane-based biomimetic platforms. It discusses key advancements in membrane functionalization and their potential for both CTCs isolation and neutralization in clinical settings. Future research should focus on optimizing these platforms for clinical validation, improving reproducibility, and integrating microfluidic systems to enhance CTCs characterization and therapeutic interventions.

C. Chen (✉)
Department of Biomedical Engineering, National University of Singapore, Singapore, Singapore
e-mail: ccchen01@nus.edu.sg

Z. Hussain, R. Pei (eds.), *Cell Membrane Engineering for Advancing Cell-Based and Nanotechnology-Based Therapies*, Biomaterials, Bioengineering and Sustainability 6, https://doi.org/10.1007/978-3-031-93425-4_12

Keywords

Liquid biopsy · Circulating tumor cells (CTCs) · Cellular membrane engineering · Microfluidics · Capture efficiency · CTCs neutralization

12.1 Introduction

Circulating tumor cells (CTCs) are uncommon cancerous cells released from primary or metastatic tumors into the bloodstream. Although they are present in very low numbers, CTCs are critically involved in metastasis-the leading cause of more than 90% of cancer-related fatalities (Mina and Sledge Jr 2011; Park et al. 2022). These cells serve not only as harbingers of disease progression but also as dynamic biomarkers that reflect the molecular and phenotypic state of tumors in real time. Their transient presence in peripheral blood presents a unique opportunity to access tumor-derived information non-invasively, providing insights into tumor heterogeneity, evolution, drug resistance, and metastatic potential.

As components of liquid biopsy, CTCs serve as a promising alternative to traditional tissue biopsies, circumventing the risks, limitations, and sampling bias associated with surgical extraction (Heidrich et al. 2021; Mader and Pantel 2017; Belotti and Lim 2021). Clinically, they open avenues for personalized medicine-enabling early diagnosis, real-time monitoring, and therapeutic stratification. However, the clinical applicability of CTCs is tempered by formidable challenges. Chief among them is their extreme rarity-typically fewer than 10 CTCs per milliliter of blood, amidst billions of hematologic cells-and their pronounced biological heterogeneity. CTCs exhibit substantial phenotypic diversity not only across patients but also within the same patient over time and even within a single blood draw.

CTCs biology further complicates their detection and capture. These cells often evade immune surveillance by cloaking themselves in platelets, forming protective microthrombi that shield them from shear stress and natural killer (NK) cell-mediated cytotoxicity. They may tether to endothelial cells and secrete immunomodulatory factors that reshape the host microenvironment, promoting metastatic colonization. Notably, CTCs frequently travel as multicellular clusters, which display enhanced survival and metastatic competence compared to solitary cells, owing to intercellular communication and cooperative behavior. These complexities challenge conventional CTCs isolation methods. Traditional approaches-such as density gradient centrifugation and immunoaffinity capture targeting epithelial markers like EpCAM-often fail to recover the full spectrum of CTCs phenotypes, particularly mesenchymal or stem-like subpopulations that lack epithelial characteristics (Rushton et al. 2021; Zhang et al. 2016). Thus, there is a critical need for more versatile, robust, and physiologically relevant capture strategies. Cellular membrane engineering represents a promising frontier in addressing these limitations. This biomimetic approach harnesses the natural interactions between CTCs and host cellular components, such as immune cells or platelets, to facilitate more comprehensive and selective capture. When integrated with microfluidic technologies, membrane-engineered platforms offer dual functionality: the high-fidelity isolation of CTCs and potential therapeutic intervention through targeted neutralization (Hou et al. 2021; Xu et al. 2020; Launiere et al. 2012).

In this chapter, we first explore the biological underpinnings and clinical significance of CTCs, followed by an overview of current detection and capture technologies. We then focus on the emerging field of membrane-based biomimetic platforms, highlighting recent advances in surface functionalization, engineering strategies, and translational potential (Zhang and Wang 2024; Guan et al. 2024; Tan et al. 2024). Finally, we discuss persistent challenges and future directions in CTCs research, emphasizing the transformative potential of membrane-engineered platforms not only as diagnostic tools but also as therapeutic modalities in the fight against metastasis.

12.2 Biological and Clinical Relevance of CTCs

12.2.1 CTCs in Cancer Metastasis and Prognosis

CTCs are rare, resilient cells that play a critical role in metastasis. These cells detach from the primary tumor and enter the bloodstream through the process of intravasation. Despite encountering shear stress and immune system attacks while circulating, a small fraction of CTCs manage to survive. These surviving CTCs then exit the bloodstream through extravasation, infiltrating distant tissues where they proliferate and form secondary tumors, thereby contributing to the continued spread of cancer (Mina and Sledge Jr 2011; Klein 2008; Dujon et al. 2021).

There is a growing consensus from clinical research that CTCs in peripheral blood have significant prognostic value, linking their presence to unfavorable patient outcomes. Elevated CTCs levels have consistently been associated with more aggressive disease phenotypes, a heightened likelihood of recurrence, and poorer overall survival across cancer types, including breast, prostate, lung, and colorectal malignancies (Paterlini-Brechot and Benali 2007; Lin et al. 2021; Deng et al. 2022). In metastatic breast cancer, patients with more than 5 circulating tumor cells (CTCs) per 7.5 mL of blood show significantly worse progression-free and overall survival, underscoring the prognostic value of CTCs count in assessing disease progression (Li and Zheng 2021; Bidard et al. 2021). Similarly, in prostate cancer, increased CTCs counts correlate with advanced disease stage, elevated Gleason scores, and diminished efficacy of hormone-based therapies (Cieślikowski et al. 2021; Balázs et al. 2021).

Beyond simple enumeration, molecular profiling of CTCs provides profound insights into tumor heterogeneity and metastatic potential (Rossi et al. 2021; Khan et al. 2022; Labib and Kelley 2021). CTCs often exhibit significant phenotypic and genetic diversity, harboring mutations and expressing markers distinct from those found in the primary tumor (Xu et al. 2021; Sklias et al. 2022). This heterogeneity highlights cancer cells' adaptability and their ability to evade therapeutic interventions. Notably, epithelial-to-mesenchymal transition (EMT) is frequently observed in CTCs, characterized by the loss of epithelial markers and an increase in mesenchymal proteins such as vimentin, N-cadherin, and TWIST1 (Noubissi Nzeteu et al. 2022; Orrapin et al. 2022). This phenotypic shift enhances CTCs invasiveness and stem-like properties, facilitating metastasis. Consequently, detecting EMT-associated markers can help stratify patients for targeted therapies and refine prognostic assessments (Orrapin et al. 2022; Ahn et al. 2021).

12.2.2 Clinical Implications of CTCs Liquid Biopsy

In recent years, liquid biopsy has redefined how clinicians and researchers approach cancer diagnostics and monitoring. This minimally invasive strategy bypasses many of the limitations inherent to conventional tissue biopsies-including sampling bias, procedural complications, and the challenge of representing the spatial and temporal heterogeneity of tumors. Rather than relying on traditional methods, liquid biopsy leverages circulating biomarkers like circulating tumor DNA (ctDNA), and extracellular vesicles (EVs) to provide a real-time, comprehensive view of tumor progression (Lone et al. 2022; Zhou et al. 2022; Nikanjam et al. 2022).

One of the primary clinical applications of CTCs liquid biopsy is early cancer detection (Lin et al. 2021; Lawrence et al. 2023). Identifying CTCs in asymptomatic individuals or patients with localized disease holds significant promise for early intervention, potentially leading to improved survival outcomes. For example, in lung cancer screening, detecting CTCs in high-risk populations (e.g., long-term smokers) can complement low-dose computed tomography, improving diagnostic accuracy and reducing false positives (Li et al. 2022; Hofman 2021; Yao et al. 2023; Crook et al. 2022). Beyond early detection, CTCs analysis plays a crucial role in guiding treatment decisions and monitoring therapeutic response (Onstenk et al. 2013; Beije et al. 2019). Molecular features of CTCs enable clinicians to identify actionable genetic alterations, informing targeted therapy selection. For instance, in non-small cell lung cancer, detecting EGFR mutations in CTCs can predict potential responsiveness to tyrosine kinase inhibitors such as Osimertinib (Jiang et al. 2019; Guo et al. 2021). This non-invasive method serves as an alternative to conventional tissue biopsies for mutation analysis, enabling clinicians to personalize treatment according to the tumor's changing genetic landscape (De Luca et al. 2016; Shaw et al. 2017). Similarly, in breast cancer, identifying HER2 amplification in CTCs supports the use of HER2-targeted treatments such as trastuzumab and pertuzumab (Fehm et al. 2010; Ignatiadis et al. 2011). This is particularly valuable when primary tumor biopsies yield inconclusive results or when tumor heterogeneity complicates diagnosis. Furthermore, longitudinal CTCs analysis enables real-time tracking of tumor evolution, revealing clonal dynamics and emerging resistance mechanisms (Fabisiewicz et al. 2024; Gerratana et al. 2021). For example, in castration-resistant prostate cancer, the appearance of the androgen receptor splice variant 7 in CTCs has been associated with resistance to androgen receptor-targeted therapies like enzalutamide and abiraterone (Bastos and Antonarakis 2018; Okegawa et al. 2018). Such findings may prompt timely therapeutic adjustments, including a switch to chemotherapy or alternative targeted treatments (Miyamoto et al. 2014; Salami et al. 2019).

Beyond individual biomarkers, integrating multi-omic data from CTCs, ctDNA, and EVs offers a holistic view of tumor biology (Akhoundova and Rubin 2022; Di Sario et al. 2023). Advances in single-cell RNA sequencing and proteomic profiling provide deeper insights into the transcriptomic and proteomic landscapes of CTCs, uncovering novel therapeutic targets and resistance pathways (Xu et al. 2021; Reza et al. 2021). This comprehensive approach enhances tumor profiling resolution and supports more precise, Customized therapeutic approaches based on the specific features of a patient's tumor (Labib and Kelley 2021).

12.2.3 Challenges in CTCs Enumeration, Isolation, and Clinical Validation

Despite its promise, the widespread adoption of CTC-based liquid biopsy faces several technical and practical challenges (Ignatiadis et al. 2021). One of the foremost obstacles is the extreme rarity of CTCs in peripheral blood. In early-stage cancers, CTCs may be present at concentrations as low as one to ten cells per billion blood cells (Lawrence et al. 2023; Crook et al. 2022). This scarcity necessitates highly sensitive and specific enrichment technologies capable of isolating CTCs from a vast excess of hematopoietic cells while preserving their integrity and viability for downstream analysis (Rushton et al. 2021).

Another major challenge is the inherent heterogeneity of CTCs populations (Agnoletto et al. 2021). Conventional isolation approaches rely on epithelial markers such as EpCAM and cytokeratins. However, CTCs undergoing EMT or those with stem-like properties may downregulate these markers, rendering them undetectable by traditional methods (Wit et al. 2015; Gabriel et al. 2016). Consequently, EpCAM-negative or low-expressing CTCs are often missed, leading to an underestimation of tumor burden and an incomplete understanding of metastatic potential (Paterlini-Brechot and Benali 2007; Man et al. 2011). To address this limitation, emerging marker-independent approaches-such as size-based filtration, dielectrophoresis, and inertial microfluidics-are being developed (Philippron et al. 2021). These techniques exploit CTCs biophysical properties rather than relying solely on molecular markers, offering a more comprehensive means of capturing diverse CTCs subpopulations (Kang et al. 2023).

Standardizing CTCs enumeration and molecular profiling further complicates the clinical translation of liquid biopsy techniques (Ilie et al. 2018; Bünger et al. 2015). Currently, the CellSearch system is the only FDA-approved platform for CTCs enumeration, but its reliance on EpCAM-based enrichment limits its applicability, particularly in tumors with low or heterogeneous EpCAM expression. Alternative technologies, including the CTC-iChip, NanoVelcro, and DEPArray, have demonstrated greater sensitivity and broader capture capabilities (Alves et al. 2022). However, these methods need further validation in large-scale, multi-center clinical studies to establish their reproducibility, clinical relevance, and cost-effectiveness (Cai et al. 2023; Yaghoubi Naei et al. 2023).

Additionally, integrating CTC-based assays into routine clinical practice is hindered by cost and accessibility constraints. Many advanced CTCs isolation and analysis techniques are labor-intensive and require sophisticated instrumentation and specialized expertise, which are often limited to research centers. Developing automated, high-throughput systems, coupled with artificial intelligence-driven data interpretation, will be critical for overcoming these barriers (Beniwal et al. 2023). Standardized protocols and streamlined workflows will help reduce both the cost and operational complexity associated with CTCs analysis, making it more widely accessible (King et al. 2014; Wang et al. 2021).

To overcome these challenges, future research should focus on enhancing CTCs detection sensitivity and specificity. One promising strategy is integrating multiple capture techniques into a single platform. By combining immunoaffinity-based methods with biophysical separation and biomimetic approaches, it may be possible to capture a broader spectrum of CTCs, including those that have undergone EMT or exhibit stem-like properties (Kumar et al. 2022). Additionally, single-cell sequencing technologies hold immense potential for providing unprecedented insights into CTCs biology, facilitating the identification of new biomarkers and therapeutic targets for precision medicine (Xu et al. 2021).

Advancements in ex vivo CTCs culture techniques could also transform personalized cancer therapy. Cultivating CTCs outside the body would enable functional assays to assess drug susceptibility and resistance mechanisms in real time, allowing clinicians to tailor treatments based on a patient's individual tumor response (Pantel and Alix-Panabieres 2016). Patient-derived CTCs organoids could serve as platforms for high-throughput drug screening, revolutionizing personalized therapeutic selection (Panovska and De Smet 2022). While significant advances have been made in understanding CTCs biology and clinical relevance, continued innovation in technology development, standardization, and translational research remains critical. Addressing key challenges such as low abundance, heterogeneity, and cost will be essential for realizing the maximized potential of CTC-based liquid biopsy in the context of precision oncology.

12.3 Existing CTCs Capture Technologies

CTCs serve as vital biomarkers for cancer metastasis, disease prognosis, and therapeutic monitoring. Owing to their extreme scarcity in peripheral blood-approximately 1–10 CTCs per billion blood cells-isolating and characterizing them requires highly sensitive and specific technologies. In recent years, several technologies have been developed for the purpose of CTCs capture. These methods can generally be divided into immunoaffinity-based approaches, techniques based on physical properties, and microfluidic platforms. Each method presents distinct advantages and limitations regarding sensitivity, specificity, scalability, and clinical applicability.

12.3.1 Immunocapture Platforms and Nanomaterials

Immunoaffinity-based approaches employ specific antibodies that recognize epithelial markers, such as epithelial cell adhesion molecule (EpCAM), to selectively enrich CTCs. The most widely used immunocapture system, the FDA-approved CellSearch platform, employs EpCAM-coated magnetic beads to isolate CTCs. Following capture, cytokeratin staining confirms epithelial origin, while CD45 exclusion helps eliminate leukocytes. Although EpCAM-based approaches provide high specificity, they are inherently limited by the loss of EpCAM expression in CTCs undergoing epithelial-mesenchymal transition (EMT), leading to decreased capture efficiency for mesenchymal-like CTCs (Pantel and Alix-Panabières 2010).

To address these limitations and improve capture efficiency, nanotechnology-based strategies have been explored. For instance, magnetic nanoparticles (MNPs) conjugated with EpCAM antibodies enhance performance due to their high surface area-to-volume ratio and tunable magnetic properties. A recent study by Wang et al. (2019) demonstrated that multifunctional MNPs coated with both EpCAM and N-cadherin antibodies significantly improved the isolation of CTCs from heterogeneous tumors. Additionally, nanostructured surfaces-such as silicon nanowire arrays and graphene oxide-modified chips-have been shown to enhance CTCs adhesion through biomimetic interactions, thereby improving yield and purity (Wang et al. 2009, 2013).

Despite these advancements, immunoaffinity-based platforms face challenges related to non-specific binding, often necessitating additional purification steps to reduce background noise. Furthermore, the intrinsic heterogeneity of CTCs necessitates multi-marker targeting strategies. The incorporation of alternative biomarkers such as HER2, EGFR, and PD-L1 can facilitate the capture of diverse CTCs subpopulations, thereby improving overall detection rates (Yang et al. 2021).

12.3.2 Physical and Molecular-Based Detection Techniques

Physical separation methods exploit inherent differences between CTCs and blood cells-such as size, deformability, and density-to achieve label-free isolation. Size-based filtration systems, such as the Isolation by Size of Epithelial Tumor Cells platform, utilize microporous membranes to selectively retain larger CTCs while allowing smaller hematopoietic cells to pass through. Although this method is effective in many cases, it may fail to capture small or highly deformable CTCs with mesenchymal traits, thereby reducing overall sensitivity (Hur et al. 2011; Xu et al. 2015).

Density gradient centrifugation techniques, including those using Ficoll-Paque and OncoQuick, leverage the unique buoyant densities of CTCs to separate them from leukocytes. However, due to the overlapping density ranges of CTCs and white blood cells, these methods often yield low purity. To overcome this issue, hybrid approaches integrating both size-based and density-based separation methods have been developed, demonstrating improved CTCs isolation while minimizing contamination from non-target cells (Harouaka et al. 2014; Ajanth et al. 2020).

Molecular detection techniques complement physical and immunoaffinity-based methods by enabling genetic and transcriptomic analyses of CTCs. Polymerase chain reaction (PCR)-based assays facilitate the diagnosis of CTCs-specific mRNA markers-such as CK19, an epithelial marker, and TWIST1, a mesenchymal marker-providing insights into CTCs heterogeneity. Advanced digital droplet PCR has further enhanced sensitivity, enabling the quantification of rare mutations, including those in EGFR and KRAS, which are critical for targeted therapy selection (Huang et al. 2016; Denis et al. 2016).

Next-generation sequencing offers a high-resolution strategy to profiling the genomes of CTCs, allowing for comprehensive characterization of tumor evolution and resistance mechanisms. However, molecular techniques often require high-purity CTCs populations and involve complex processing steps, which may limit their routine clinical application (Bowes et al. 2022; Halima et al. 2022).

12.3.3 Microfluidics and Lab-on-a-Chip Approaches

Microfluidic technologies leverage precise fluid dynamics to achieve high-throughput CTCs isolation and analysis. These systems often incorporate mechanisms such as hydrodynamic forces, immunoaffinity, and dielectrophoresis to enhance capture efficiency. A notable example is the CTC-iChip, an immunoaffinity-based microfluidic chip that integrates inertial focusing with EpCAM-based capture, enabling high-throughput processing with improved purity. However, like other antibody-based methods, the CTC-iChip may fail to detect CTCs subpopulations that have undergone EMT (Descamps et al. 2022; Shi et al. 2021). To overcome the limitations of antibody-based capture, label-free microfluidic platforms have been developed. These systems exploit differences in cell size, deformability, and dielectric properties for CTCs isolation. For instance, the Vortex Chip utilizes vortex-induced inertial forces to trap CTCs on size-based differences, while deterministic lateral displacement arrays fractionate cells with high precision based on their physical characteristics (Che et al. 2016). Additionally, dielectrophoresis-based approaches have gained traction, enabling the selective manipulation of CTCs based on their electrical properties.

Lab-on-a-chip platforms further enhance CTCs analysis by integrating multiple functionalities-including capture, staining, and single-cell sequencing-into a single miniaturized system. Recent advancements in lab-on-a-chip automation have enabled real-time CTC monitoring, advancing personalized cancer management. However, challenges in standardization, scalability, and clinical translation persist due to the complexity of these devices and the need for specialized infrastructure (Surappa et al. 2023; García-Hernández et al. 2023).

12.3.4 Strengths and Limitations of Current Methods

Current CTCs capture technologies exhibit varying levels of sensitivity, specificity, and clinical feasibility. Immunoaffinity-based approaches, while highly specific, are limited by antigen loss (particularly due to EMT) and non-specific binding. Physical separation methods provide label-free isolation but often suffer from low selectivity and purity. Microfluidic devices offer superior precision and scalability, but many still require further validation before widespread adoption in clinical practice. Hybrid approaches integrating multiple modalities are emerging as a promising strategy to address these challenges. For instance, combining immunoaffinity with size-based enrichment enhances both sensitivity and selectivity. Furthermore, integrating microfluidics with molecular profiling techniques enables comprehensive characterization of CTCs at the genetic and transcriptomic levels. A recent study by Reinhardt et al. (Reinhardt et al. 2019) showed that multi-marker RNA analysis of enriched, uncultured single CTCs and cultured CTCs derived from cryopreserved diagnostic leukapheresis samples can offer valuable insights into intrapatient heterogeneity, which is closely linked to targeted therapies and treatment resistance.

In the future, progress in artificial intelligence (AI) and machine learning are anticipated to further refine CTCs detection algorithms, enhancing accuracy and clinical utility. AI-driven image processing and deep learning models have already demonstrated promise in automating CTCs identification, reducing reliance on manual assessment, and minimizing human error (Sebastian and Peter 2022). In conclusion, while CTCs capture technologies have advanced significantly, further research is essential to develop standardized, high-throughput platforms with robust clinical validation. The integration of multimodal approaches-alongside advancements in AI-driven analysis-holds great potential for advancing liquid biopsy applications in cancer diagnosis and precision medicine. Continued innovation in this field will be crucial to overcoming current limitations, ultimately enabling more accurate, efficient, and clinically relevant detection of circulating tumor cells.

12.4 Cellular Membrane Engineering for CTCs Capture

Biomimetic platforms derived from cell membranes have become a promising strategy for capturing CTCs, leveraging natural biorecognition mechanisms to enhance specificity and efficiency (Tan et al. 2024; Raza et al. 2021; Ding et al. 2020a). By integrating extracted cell membranes into functionalized capture devices, these platforms overcome limitations of traditional methods, improving CTCs isolation while preserving cellular integrity (Tan et al. 2024; Jiang et al. 2024a, b).

12.4.1 Fundamentals Structure and Function of Cell Membrane

Cell membrane also called plasma membrane, plays a vital role in maintaining cell integrity and mediating cell signal transduction, and facilitating selective interactions with the surrounding environment (Albers 2012; van Deventer et al. 2021). Composed primarily of a phospholipid bilayer interspersed with proteins, cholesterol, and carbohydrate moieties, the membrane functions as a dynamic interface between the intracellular and extracellular milieu (Ghosh and Koester 2023; Perez et al. 2005). The fluidic nature of the lipid bilayer allows for lateral movement of membrane components, which is vital for processes such as signal transduction and cellular adhesion.

Key membrane components such as integrins, selectins, and cadherins are involved in cell-cell interactions, which are crucial in physiological and pathological contexts, including cancer metastasis (Smart et al. 2021; Li et al. 2025; Schuster et al. 2021). The presence of glycoproteins and glycolipids on the outer surface of the membrane contributes to cell recognition and immune evasion, properties that are harnessed in biomimetic strategies for CTCs capture (Ghasempour and Freeman 2023; Zhang et al. 2019a). Moreover, the structural asymmetry of the bilayer, characterized by distinct lipid compositions in the inner and outer leaflets, influences interactions with circulating cells and nanoparticles engineered for CTCs isolation (Rodríguez-Hernández et al. 2020; Wang et al. 2020a; Rascol et al. 2016).

12.4.2 Cell Membrane-Based Biomimetic Platforms for Biosensing

In the context of CTCs detection, biomimetic platforms that leverage natural cell membrane properties have emerged as promising tools for enhancing sensitivity and specificity (Liu et al. 2022; Shi et al. 2022). Traditional CTCs isolation techniques, including immunoaffinity-based capture and size-exclusion filtration, often suffer from issues such as non-specific binding, low capture efficiency, and cell viability loss (Rushton et al. 2021; Deng et al. 2022). To overcome these challenges, researchers have developed biomimetic materials that incorporate extracted cell membranes into functionalized capture devices (Zhang and Wang 2024; Sun et al. 2024). The functionalization of cell membranes plays a crucial role in boosting the efficiency, specificity, and stability of CTCs capture. By leveraging biomimetic approaches, engineered cell membranes can be tailored to improve biorecognition, increase selectivity, and enable high-throughput detection of CTCs (Ding et al. 2020a). Several key strategies have emerged for cell membrane functionalization, including membrane coating of nanoparticles and microfluidic devices, hybrid biomimetic membranes, exosome-based approaches, and membrane fusion techniques.

12.4.2.1 Membrane Coating of Nanoparticles and Microfluidic Devices

Functionalization of cell membranes onto nanoparticles and microfluidic devices is a key strategy for enhancing CTCs capture efficiency. By leveraging the natural properties of cell membranes, researchers have developed hybrid platforms that integrate the advantages of both biological and synthetic materials.

Functionalizing nanoparticles with biomimetic membranes has emerged as a powerful approach to enhance CTCs capture efficiency. These nanoparticles can be modified with membranes derived from immune cells, platelets, red blood cells, or tumor cells themselves, thereby mimicking natural cell–cell interactions (Jiang et al. 2024a; Sun et al. 2024). For instance, leukocyte membrane-coated nanoparticles can exploit the adhesion properties of selectins and integrins to selectively recognize and bind to CTCs in circulation. For example, platelet membrane-coated nanoparticles have been shown to target CTCs by mimicking the natural homing behavior of platelets (Li et al. 2016a). Similarly, red blood cell membrane-coated nanoparticles can improve circulation time and reduce clearance, thereby enhancing CTCs capture efficiency (Shen et al. 2022; Zhu et al. 2018).

In an elegant display of bioengineering, Jiang et al. introduced a novel platform for capturing CTCs by cloaking magnetic nanoparticles in genetically modified leukocyte membranes. Their method, streamlined into a one-step extrusion process, cleverly mimics the native leukocyte surface, thereby reducing nonspecific interactions with white blood cells (WBCs) and allowing for high-purity CTCs isolation. What sets this system apart is the stable expression of anti-EGFR single-chain variable fragments (scFv) on the engineered membranes-enabling the nanoparticles to home in on EGFR-positive CTCs with remarkable precision, as depicted in Fig. 12.1. The impact was striking: the modified particles demonstrated significant increase in binding affinity to extracellular EGFR compared to those coated with unaltered membranes. In artificial blood models, this translated to significantly higher capture efficiency, elevating cell purity from 64.8% to an impressive 93.5%-well beyond the performance of commercial immunomagnetic tools. Most notably, the platform proved its clinical mettle, isolating high-purity CTCs from samples drawn from six cancer patients. This work not only refines how we detect CTCs but also opens a promising path forward in cancer diagnostics and monitoring (Jiang et al. 2024a).

Recognizing the potential of a multi-targeting approach to address both the heterogeneity of CTCs and their phenotypic similarities to leukocytes, L. Ou and colleagues developed a gene-editing strategy to engineer leukocyte membranes with triple-specificity toward EpCAM, EGFR, and Her2 (Fig. 12.2) (Jiang et al. 2024b). Utilizing Jurkat T cells as a platform, the team introduced a plasmid construct encoding three single-chain variable fragments (scFvs) targeting each antigen. The construct incorporated a CMV promoter to drive robust expression, an interleukin-2 (IL-2) signal sequence to facilitate membrane localization, and a platelet-derived growth factor receptor (PDGFR) transmembrane domain to stably anchor the scFvs on the cell surface. An ampicillin resistance gene enabled bacterial selection, while mCherry fluorescence served as a reporter, facilitating cell sorting via flow cytometry and fluorescence imaging.

Following stable expression, the engineered Jurkat cells were subjected to hypotonic lysis, mechanical ultrasonication, and centrifugation to isolate engineered cell membranes (E-CMs). These membranes were then used to coat Poly(lactic-co-glycolic acid) (PLGA) nanoparticles, which were subsequently incubated with tumor cell lines expressing the corresponding surface antigens. Notably,

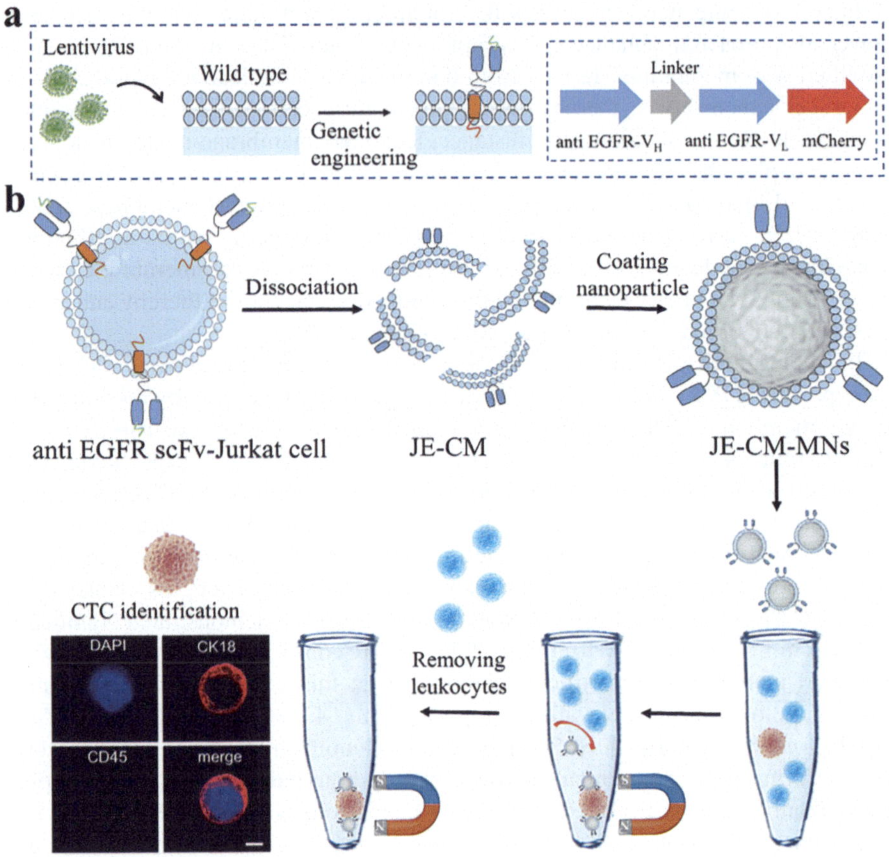

Fig. 12.1 Genetically engineered cell membrane platform for capturing circulating tumor cells (CTC). (**a**) A schematic overview depicts the generation of the engineered cellular system. (**b**) These modified cells were subsequently processed to derive Jurkat-engineered cell membrane-coated microneedles (JE-CM-MNs). Scale bar represents 5 μm. (Reproduced with permission from Ref. Jiang et al. 2024a, copyright 2024, Wiley-VCH GmbH)

E-CM-coated nanoparticles exhibited strong DiI fluorescent signals upon binding to EpCAM-positive MCF-7, EGFR-positive MDA-MB-468, and Her2-positive BT474 cells, confirming antigen-specific recognition. The fluorescence signal increased proportionally with E-CM-PLGA nanoparticle concentration, while nanoparticles coated with unmodified Jurkat membranes showed minimal binding-highlighting the specificity and efficacy of the engineered approach.

The integration of cell membrane components into microfluidic platforms is emerging as a powerful strategy for high-throughput isolation of CTCs. Within these systems, surfaces cloaked with natural membranes offer a biologically active interface that not only promotes specific cell adhesion but also reduces nonspecific binding-an essential balance in continuous-flow devices. The fidelity of CTCs capture in such systems hinges on the biochemical and biophysical properties of the

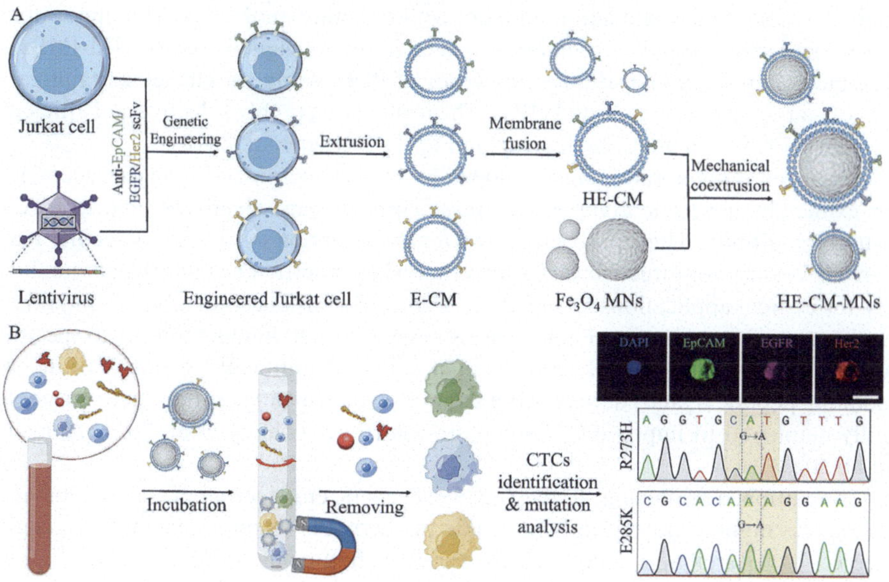

Fig. 12.2 The diagram demonstrated the promising potential of hybrid engineered cell membrane-camouflaged magnetic nanoparticles (HE-CM-MNs) for heterogeneous circulating tumor cells (CTCs) detection and downstream analysis. (Reproduced with permission from Jiang et al. 2024b, copyright 2024, Springer Nature)

capture interface. Several functionalization strategies-ranging from covalent tethering of membrane fragments to lipid insertion and electrostatic assembly-have been explored to optimize these interfaces. Notably, coating microchannels with endothelial cell membranes can mimic the native microvasculature, thereby increasing the probability of capturing CTCs under physiological conditions. Similarly, macrophage-derived membrane coatings have shown promise in selectively enhancing CTCs adhesion (Tan et al. 2024). More recently, hybrid approaches that incorporate membranes from both tumor and immune cells have yielded devices with improved selectivity and throughput for CTCs isolation (Ding et al. 2020b; Chen et al. 2019; Kang et al. 2017a).

12.4.2.2 Hybrid and Engineered Membranes for Enhanced Targetability

Hybrid membrane systems combine membranes from different cell types or engineered sources to enhance the specificity and effectiveness of CTCs capture. A notable example involves the use of leukocyte-tumor hybrid membranes, which integrate the immune-evasive properties of leukocytes with the targeting capabilities of tumor-derived membranes. This strategy enables selective recognition of CTCs while minimizing background noise from normal blood cells (Chang et al. 2020). In a noteworthy study, a biomimetic platform has been engineered by fusing platelets (PLTs) and WBC membranes, subsequently coating this hybrid membrane onto magnetic beads and adorning them with targeting antibodies. This innovative construct-termed as hybrid membrane-coated immunomagnetic beads (HM-IMBs)-capitalized on the tumor-affinity properties of PLTs while simultaneously mitigating off-target interactions with WBCs. The result was a highly selective and efficient system for CTCs capture, as depicted in Fig. 12.3 (Rao et al. 2018).

Engineered membranes are developed by integrating natural ligands, receptors, or extracellular vesicle components; these surfaces can selectively bind to CTCs with high affinity. This approach allows for the retention of critical molecular features necessary for effective CTCs targeting while improving stability and durability for clinical applications (Jiang et al. 2024a). For instance, membranes modified with aptamer-functionalized lipids have shown enhanced binding affinity to specific CTCs markers (Ding et al. 2020b; Jia et al. 2023). Additionally, membranes incorporating peptide ligands derived from tumor-associated antigens have been successfully employed to improve CTCs capture selectivity (Zhong et al. 2021; Ahmadi et al. 2023).

In addition, a promising strategy involves genetic engineering of cell membranes to express high-affinity ligands for CTCs markers. For example, membranes engineered to overexpress single-chain variable fragments (scFv) of anti-EGFR

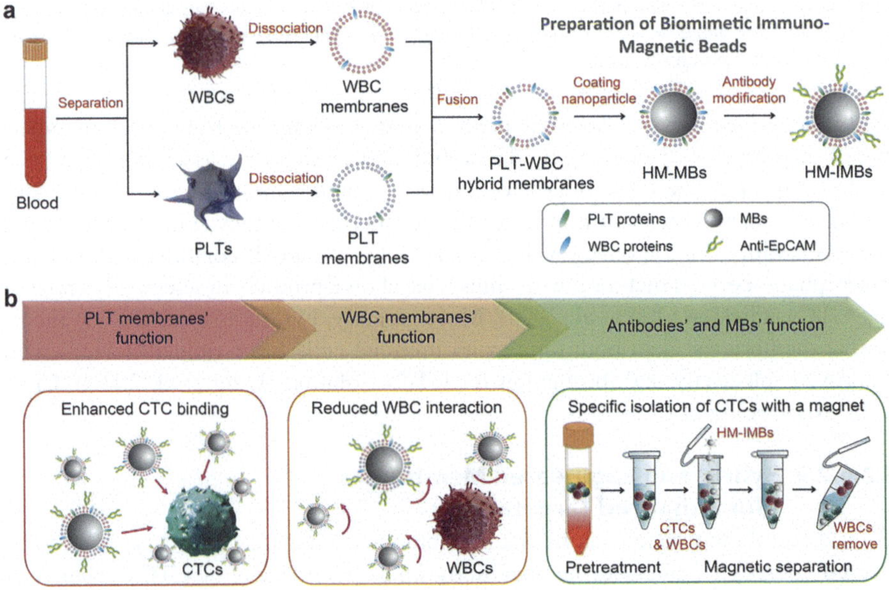

Fig. 12.3 An overview of the hybrid membrane-coated immunomagnetic beads (HM-IMBs) fabrication process for efficient circulating tumor cells (CTC) capture. (**a**) White blood cells (WBCs) and platelets (PLTs) are first individually harvested from blood samples, with their respective membranes and membrane-associated proteins carefully extracted. (**b**) These HM-IMBs exhibit a dual advantage: the PLTs-derived components enhance specific binding to CTCs, while the WBC-derived features help minimize off-target interactions with other blood cells. (Reproduced with permission from Rao et al. 2018, copyright 2018, Wiley-VCH GmbH)

antibodies have demonstrated significantly improved binding to EGFR-positive CTCs. These engineered platforms provide a high degree of customization, allowing for adaptation to different tumor types and heterogeneity in CTCs populations.

12.4.2.3 Role of Engineering Exosomes and Membrane Fusion Techniques

Engineering exosomes and other membrane vesicles play a crucial role in intercellular communication and tumor progression. These nanoscale vesicles, derived from cell membranes, carry a diverse array of surface markers and bioactive molecules that can be exploited for selective targeting (Kang et al. 2017b). By engineering exosomes to display tumor-specific ligands, researchers have created bioinspired nanoplatforms that effectively bind to and isolate CTCs from blood samples. Engineered exosome-mimicking platforms can serve as highly specific CTCs capture tools by leveraging the natural affinity between tumor-derived vesicles and CTCs.

Drawing inspiration from the innate cellular choreography of membrane fusion, researchers have begun leveraging these natural processes to advance the precision capture of CTCs. This bioinspired strategy exploits the fusion potential between synthetic liposomes and native cellular membranes to engineer hybrid vesicles-structures that merge the programmable robustness of synthetic carriers with the selective biological affinity intrinsic to cell-derived membranes (Sun et al. 2024; Ji et al. 2020; Manjunath et al. 2020). These chimeric vesicles are readily amenable to functionalization with magnetic or fluorescent nanoparticles, enabling seamless integration with downstream analytic modalities such as magnetic enrichment and optical detection.

A notable advancement in this domain was presented by Jia et al. (2025), who utilized the endogenous vesicle secretion machinery of cells to fabricate outward-facing magnetic vesicles (MVMNPs) for the rare event detection of CTCs (Fig. 12.4). Their approach involved co-incubating magnetic nanoparticles (MNPs) with donor cells, prompting the biogenesis of magnetically endowed membrane vesicles. These were further engineered using a sophisticated nucleic acid scaffold: the SYL3C aptamer, specific to EpCAM, was spatially organized via DNA tetrahedrons (TDNs) to form the TDNE complex. Upon hybridization with a complementary strand (c-EpCAM), the resulting TDNE-c was grafted onto the MVMNPs, enhancing targeting specificity via both molecular recognition and topological control.

Importantly, these vesicles inherit homotypic targeting preferences from their MCF-7 origins, allowing for increased binding efficiency with EpCAM-expressing CTCs. Furthermore, the release of c-EpCAM triggers a catalytic hairpin assembly (CHA), initiating dual-mode fluorescent and colorimetric readouts for sensitive signal amplification. The resulting biosensor platform-TDNE-c/MVMNPs-CHA-not only preserves the biofunctionality of cell membrane-coated nanoparticles (CMNPs) but also overcomes key limitations in targeting and signal transduction, marking a significant leap forward in modular, membrane-based diagnostic systems with programmable recognition interfaces.

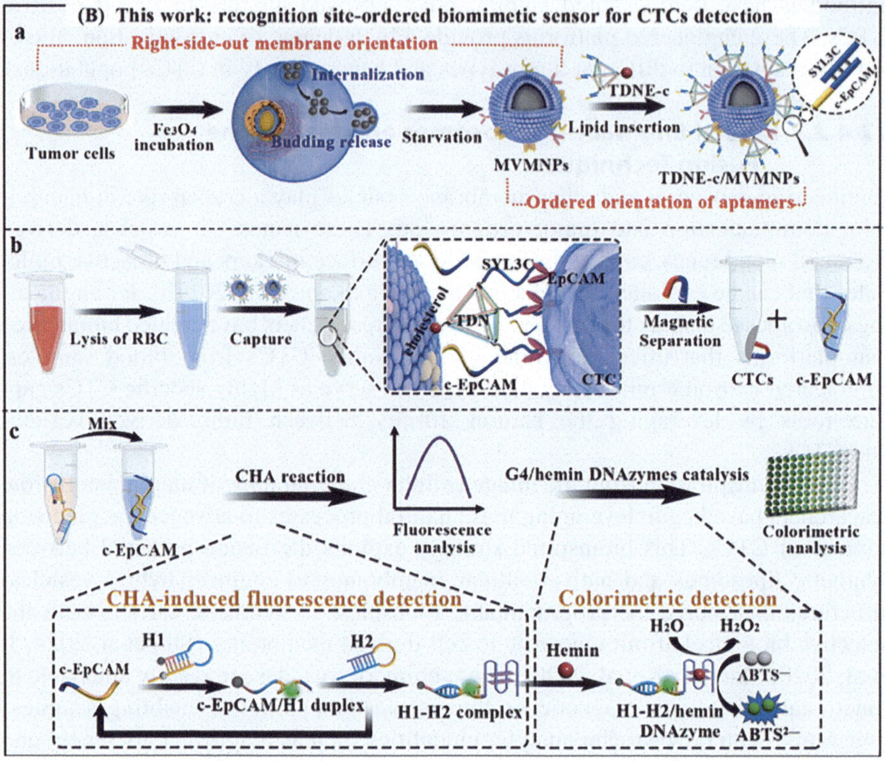

Fig. 12.4 Schematic representation depicting the fabrication of the TDNE-c/MVMNPs-CHA biomimetic sensor and its implementation for the ultrasensitive detection of circulating tumor cells (CTCs). (**a**) Overview of the synthetic pathway for TDNE-c/MVMNPs (**b**) Deployment in the sequential processes of CTCs capture, and (**c**) Signal-amplified detection (c). (Reproduced with permission from Jia et al. 2025, copyright 2025, Elsevier B.V.)

12.4.3 Advantages of Extracted Cell Membranes in CTCs Capture

One of the primary advantages of using extracted cell membranes in CTCs capture is the enhanced biorecognition capability (Vargas et al. 2022). Unlike synthetic ligands or antibodies, natural cell membranes present a complex and dynamic array of receptors, adhesion molecules, and glycoproteins that facilitate selective interactions with CTCs (Zhang and Wang 2024). This approach minimizes false positives and enhances the specificity of detection (Carneiro et al. 2024).

Another significant benefit is the improved selectivity offered by membrane-based platforms. By incorporating immune cell-derived membranes, researchers have demonstrated reduced leukocyte contamination and improved CTCs purity (Zhang and Wang 2024; Lopes et al. 2022). In addition, tumor-derived membranes can be engineered to express specific markers such as EpCAM or HER2, which are commonly overexpressed in metastatic CTCs, thus enabling targeted capture without the need for external labeling (Lin et al. 2021).

Furthermore, the integration of biomimetic membranes with microfluidic and nanotechnology-based devices has advanced the development of high-throughput systems for CTCs separation (Wu et al. 2020). These platforms can monitor in real-time, rapidly process blood samples, and efficiently retrieve viable CTCs for genomic and transcriptomic analysis (Jia et al. 2023; Zhang et al. 2021). The combination of membrane fluidity, receptor heterogeneity, and biocompatibility makes these engineered systems a powerful tool in the field of liquid biopsy (Zhang and Wang 2024; Jiang et al. 2024b).

Cellular membrane engineering offers a transformative approach to CTCs capture by leveraging naturally occurring recognition mechanisms and biophysical interactions. These biomimetic strategies provide improved selectivity, efficiency, and adaptability compared to traditional methods, enabling more effective liquid biopsy applications. Future research should focus on optimizing these platforms for clinical translation, ensuring robust performance across diverse patient populations. The integration of membrane-based technologies with multi-omics analysis and artificial intelligence-driven data interpretation holds significant promise for revolutionizing precision oncology and improving patient outcomes.

12.5 Cellular Membrane Engineering for CTCs Neutralization

Beyond their diagnostic utility in liquid biopsy, CTCs have emerged as viable therapeutic targets for curbing metastasis. By neutralizing CTCs in the bloodstream before they seed distant organs, it may be possible to interrupt the metastatic cascade and significantly improve clinical outcomes. Recent advances in cellular membrane engineering have enabled the development of biomimetic strategies that exploit natural cell-cell interactions to selectively capture and eliminate CTCs. These strategies include PLTs- and neutrophil-mimicking systems, exosome-like platforms, and immune checkpoint modulation to enhance therapeutic efficacy against metastasis-initiating cells.

12.5.1 Platelet-Mimicking Strategies for CTCs Neutralization

PLTs play a crucial role in CTCs survival by forming protective cloaks that shield CTCs from immune attack and facilitate their extravasation. To counteract this mechanism, researchers have engineered PLTs-derived biomimetic vehicles that selectively bind to CTCs and deliver cytotoxic agents (Li et al. 2016b; Ortiz-Otero et al. 2021). PLTs have been modified to express tumor necrosis factor-related apoptosis-inducing ligand (TRAIL), a cytokine that selectively induces apoptosis in cancer cells (Fig. 12.5). These TRAIL-expressing PLTs interact with CTCs, initiating tumor cell death while preserving normal PLTs function. Additionally, PLTs-mimicking synthetic nanoparticles functionalized with PLTs membranes have been

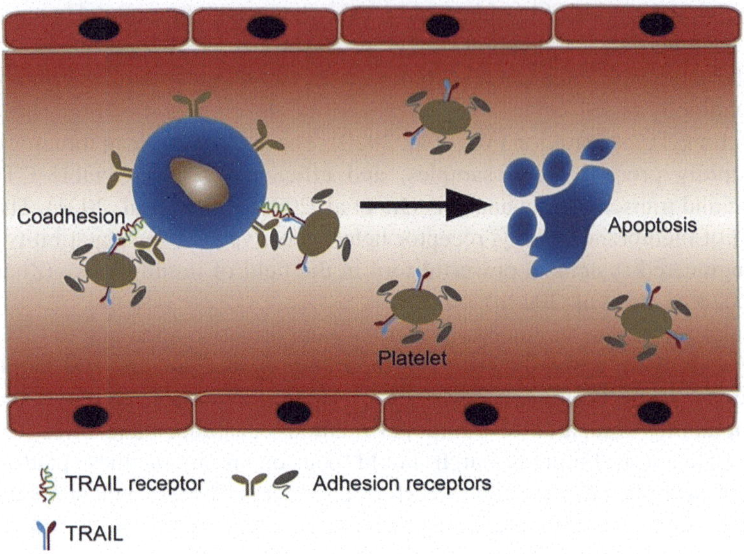

Fig. 12.5 Genetic modification of platelets to neutralize circulating tumor cells. (Reproduced with permission from Li et al. 2016a, copyright 2016, Elsevier B.V.)

developed to carry apoptosis-inducing agents, such as TRAIL, and integrate into CTC-associated microthrombi, significantly reducing metastasis in preclinical models (Li et al. 2016a; Ortiz-Otero et al. 2021). Furthermore, Zhang et al. introduced an activated PLTs membrane-based biomimetic platform (APEVs@DOX) designed for CTCs capturing while simultaneously delivering the chemotherapeutic drug doxorubicin (Zhang et al. 2025). This approach enhances drug efficacy by ensuring targeted delivery to CTCs and reducing metastatic spread. Through in vitro and in vivo evaluations, this system demonstrated improved CTCs capture efficiency and a reduction in triple-negative breast cancer metastases. In another study, Zhang et al. developed a P-selectin-targeting peptide-modified micelle system (PSN-PEG-SS-PTX4 micelle) to utilize activated PLTs as a bridge for CTCs interaction. This redox-responsive paclitaxel-loaded micelle selectively adheres to activated PLTs, facilitating CTCs capture and targeted drug delivery. The system significantly suppressed lung and liver metastases in triple-negative breast cancer models, demonstrating a promising approach for metastatic cancer treatment (Zhang et al. 2019b).

12.5.2 Neutrophil-Mimicking Strategies for CTCs Capture and Neutralization

Neutrophils naturally interact with CTCs and home to metastatic niches, making them an ideal biomimetic platform for targeted CTCs capture and neutralization. Inspired by this, Kang et al. developed neutrophil-mimicking nanoparticles

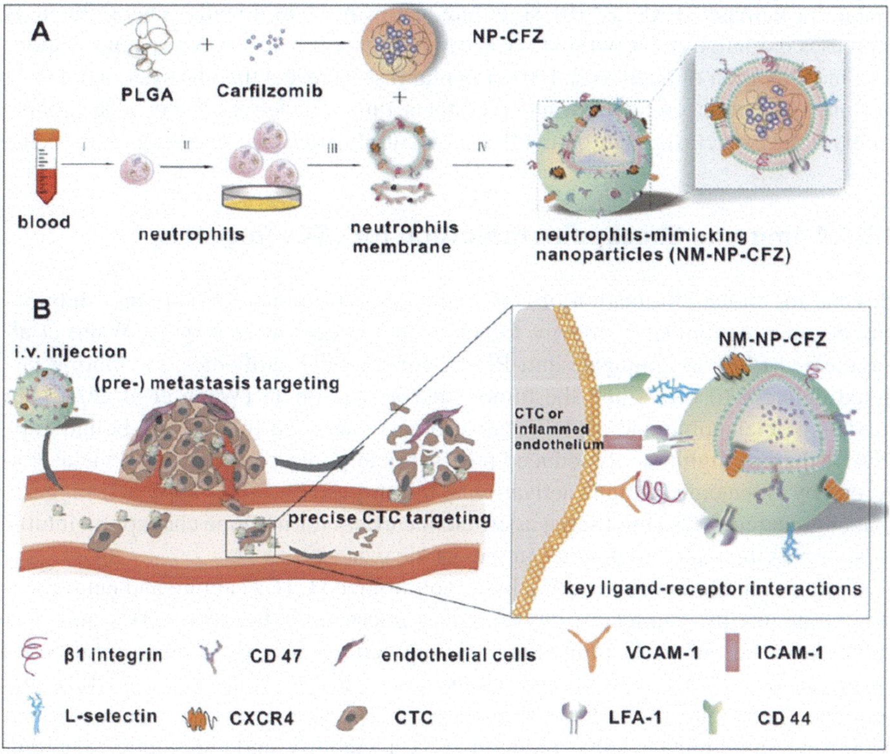

Fig. 12.6 A nanoscale neutrophil-mimicking drug delivery system (NM-NP) was created by coating PLGA nanoparticles with neutrophil membranes. This system is designed to neutralize circulating tumor cells (CTCs) and prevent the development of a metastatic niche. (Reproduced with permission from Kang et al. 2017a, copyright 2017, American Chemical Society)

(NM-NPs) by coating PLGA nanoparticles with neutrophil membranes (Fig. 12.6). This coating preserved the biobinding activity of neutrophils, enhancing cellular association, CTCs capture, and metastatic niche targeting. When loaded with carfilzomib (NM-NP-CFZ), these nanoparticles selectively depleted CTCs, effectively preventing early metastasis and inhibiting tumor progression in preclinical models (Kang et al. 2017a).

12.5.3 Exosome-Like and Nanoparticle-Based Approaches

Wang et al. to overcome the challenge of CTCs neutralization, who developed a nano-platform that mimics exosomes and delivers sequential bioactivity (Wang et al. 2020b). These nanoparticles, decorated with exosome membranes, leverage the natural affinity between cancer cell membranes and exosome-like vesicles for specific CTCs targeting. Upon CTCs uptake, the system releases cucurbitacin B to block metastasis by downregulating the focal adhesion kinase/matrix metalloproteinase (FAK/MMP) signaling pathway. Additionally, cucurbitacin B increases oxidative stress within CTCs, triggering a reactive oxygen species responsive bioactivation of paclitaxel. Experimental results reveal that the system not only prolongs systemic circulation and facilitates tumor penetration, but also significantly reduces metastasis through effective neutralization of circulating tumor cells.

12.5.4 Immune Checkpoint Inhibition for CTCs Targeting

Beyond the direct elimination of CTCs, integrating immune checkpoint inhibitors has become a promising strategy for managing cancer after surgery. Wang et al. engineered PLTs to conjugate anti-PD-L1 monoclonal antibodies, facilitating targeted delivery to CTCs and the tumor microenvironment (Wang et al. 2017). In models of melanoma and triple-negative breast cancer in mice, PLTs-bound anti-PD-L1 therapy significantly reduced post-surgical tumor recurrence and metastatic spread by enhancing immune activation at the surgical bed. This approach suggests that engineered PLTs could serve as efficient carriers for immune checkpoint inhibitors, improving patient outcomes in metastatic cancer.

Cellular membrane engineering has revolutionized CTCs capture and neutralization strategies by mimicking physiological interactions between CTCs and host cells. Advances in PLTs-mimicking (Ortiz-Otero et al. 2021), neutrophil-based (Kang et al. 2017a), exosome-like (Kang et al. 2017b; Wang et al. 2020b), and immune checkpoint inhibition (Wang et al. 2017) approaches have demonstrated promising preclinical results, reducing CTCs viability and metastatic potential. Going forward, the research emphasis of this project should center on refining these platforms, validating them clinically, ensuring their scalability, and integrating them with emerging liquid biopsy technologies to enhance real-time cancer monitoring and personalized therapeutic approaches.

12.5.5 Importance of CTCs Neutralization in Cancer Therapy

Suspended between the departure from a primary tumor and the arrival at distant organs, they serve as both messengers and harbingers of metastasis-a biological odyssey often unnoticed until its consequences become irreversible. Traditional cancer therapies, for all their power, largely aim at what is already visible: the tumor rooted in tissue. But CTCs offer something more ephemeral, and perhaps more profound-a moment, a chance, to intervene before the seeds of recurrence take hold. Intercepting circulating tumor cells (CTCs) can alter cancer progression, as lingering CTCs may persist even after tumor removal, posing a risk of metastasis. Neutralizing them in this liminal state could mean halting metastasis before it begins, reducing not only the burden of disease but the heartbreak that so often follows its return (Hu et al. 2024). The bloodstream is a challenging environment for therapeutic intervention, as CTCs often hide in plain sight, shielded by platelets or co-opted immune cells. However, by mimicking biological systems, science adapts by designing nanoparticles or engineered cells that can recognize these shields and deliver targeted treatments with remarkable precision. When combined with chemotherapy or immunotherapy, such strategies can decisively eradicate the last traces of malignancy. Meanwhile, liquid biopsies offer real-time insights into the disease's evolution, enabling timely adjustments to our approach. The promise is immense, though challenges such as safety, scalability, and regulatory review remain. Still, in the convergence of materials science, immunology, and oncology, we may be witnessing the birth of a new frontier—one where cancer's escape route is blocked before it begins.

12.6 Conclusions and Perspectives

The development of cell membrane-engineered platforms for CTCs capture marks a significant advancement in liquid biopsy and cancer diagnostics, enhancing biorecognition, selectivity, and detection efficiency. However, challenges related to standardization, scalability, and clinical validation must be overcome for successful clinical implementation. Integrating these platforms with multi-omics and personalized medicine offers immense potential for improving cancer prognosis, monitoring progression, and guiding targeted therapies. As the field evolves, collaboration across materials science, bioengineering, and oncology will be essential to unlocking the full clinical potential of CTCs capture technologies. Advances in biomaterials and biofunctionalization strategies will optimize capture efficiency, while innovations in microfluidics and biosensing will enhance detection sensitivity and enable real-time analysis. Moreover, the synergy between biomedical engineering and computational biology will be key to enhancing platform specificity and robustness, facilitating their integration into precision oncology workflows. The convergence of cell membrane-based biosensing with multi-omics technologies represents a paradigm shift in CTCs research, offering unprecedented opportunities for early cancer detection, real-time monitoring, and personalized treatment strategies. However, challenges related to regulatory approval, reproducibility, and large-scale manufacturing must be addressed to enable clinical adoption. The continued progress in biosensing technologies, automation, and AI-driven data analytics is expected to accelerate the clinical translation of these platforms, ultimately advancing cancer diagnostics, prognosis, and personalized treatment in the near future.

References

Agnoletto C, Caruso C, Garofalo C (2021) Heterogeneous circulating tumor cells in sarcoma: implication for clinical practice. Cancers 13(9):2189

Ahmadi M, Ahmadyousefi Y, Salimi Z, Mirzaei R, Najafi R, Amirheidari B, Rahbarizadeh F, Kheshti J, Safari A, Soleimani M (2023) Innovative diagnostic peptide-based technologies for cancer diagnosis: focus on EGFR-targeting peptides. ChemMedChem 18(3):e202200506

Ahn JC, Teng PC, Chen PJ, Posadas E, Tseng HR, Lu SC, Yang JD (2021) Detection of circulating tumor cells and their implications as a biomarker for diagnosis, prognostication, and therapeutic monitoring in hepatocellular carcinoma. Hepatology 73(1):422–436

Ajanth P, Sudeepthi A, Sen AK (2020) Microfluidics technology for label-free isolation of circulating tumor cells. J Inst Eng (India) Ser C 101(6):1051–1071

Akhoundova D, Rubin MA (2022) Clinical application of advanced multi-omics tumor profiling: shaping precision oncology of the future. Cancer Cell 40(9):920–938

Albers RWW (2012) Cell membrane structures and functions. In: Basic neurochemistry. Academic, pp 26–39

Alves JM, Estévez-Gómez N, Valecha M, Prado-López S, Tomás L, Alvariño P, Piñeiro R, Muinelo-Romay L, Mondelo-Macía P, Salgado M, Iglesias-Gómez A (2022) Comparative analysis of capture methods for genomic profiling of circulating tumor cells in colorectal cancer. Genomics 114(6):110500

Balázs K, Antal L, Sáfrány G, Lumniczky K (2021) Blood-derived biomarkers of diagnosis, prognosis and therapy response in prostate cancer patients. J Pers Med 11(4):296

Bastos DA, Antonarakis ES (2018) CTC-derived AR-V7 detection as a prognostic and predictive biomarker in advanced prostate cancer. Expert Rev Mol Diagn 18(2):155–163

Beije N, Martens JW, Sleijfer S (2019) Incorporating liquid biopsies into treatment decision-making: obstacles and possibilities. Drug Discov Today 24(9):1715–1719

Belotti Y, Lim CT (2021) Microfluidics for liquid biopsies: recent advances, current challenges, and future directions. Anal Chem 93(11):4727–4738

Beniwal SS, Lamo P, Kaushik A, Lorenzo-Villegas DL, Liu Y, MohanaSundaram A (2023) Current status and emerging trends in colorectal cancer screening and diagnostics. Biosensors 13(10):926

Bidard FC, Jacot W, Kiavue N, Dureau S, Kadi A, Brain E, Bachelot T, Bourgeois H, Gonçalves A, Ladoire S, Naman H (2021) Efficacy of circulating tumor cell count–driven vs clinician-driven first-line therapy choice in hormone receptor–positive, ERBB2-negative metastatic breast cancer: the STIC CTC randomized clinical trial. JAMA Oncol 7(1):34–41

Bowes AL, Tarabichi M, Pillay N, Van Loo P (2022) Leveraging single-cell sequencing to unravel intratumour heterogeneity and tumour evolution in human cancers. J Pathol 257(4):466–478

Bünger S, Zimmermann M, Habermann JK (2015) Diversity of assessing circulating tumor cells (CTCs) emphasizes need for standardization: a CTC guide to design and report trials. Cancer Metastasis Rev 34:527–545

Cai S, Deng Y, Wang Z, Zhu J, Huang C, Du L, Wang C, Yu X, Liu W, Yang C, Wang Z (2023) Development and clinical validation of a microfluidic-based platform for CTC enrichment and downstream molecular analysis. Front Oncol 13:1238332

Carneiro A, Piairo P, Matos B, Santos DA, Palmeira C, Santos LL, Lima L, Diéguez L (2024) Minimizing false positives for CTC identification. Anal Chim Acta 1288:342165

Chang ZM, Zhang R, Yang C, Shao D, Tang Y, Dong WF, Wang Z (2020) Cancer–leukocyte hybrid membrane-cloaked magnetic beads for the ultrasensitive isolation, purification, and non-destructive release of circulating tumor cells. Nanoscale 12(37):19121–19128

Che J, Yu V, Dhar M, Renier C, Matsumoto M, Heirich K, Garon EB, Goldman J, Rao J, Sledge GW, Pegram MD (2016) Classification of large circulating tumor cells isolated with ultra-high throughput microfluidic vortex technology. Oncotarget 7(11):12748–12760

Chen M, Liu A, Chen B, Zhu DM, Xie W, Deng FF, Ji LW, Chen LB, Huang HM, Fu YR, Liu W (2019) Erythrocyte-derived vesicles for circulating tumor cell capture and specific tumor imaging. Nanoscale 11(25):12388–12396

Cieślikowski WA, Antczak A, Nowicki M, Zabel M, Budna-Tukan J (2021) Clinical relevance of circulating tumor cells in prostate cancer management. Biomedicines 9(9):1179

Crook T, Leonard R, Mokbel K, Thompson A, Michell M, Page R, Vaid A, Mehrotra R, Ranade A, Limaye S, Patil D (2022) Accurate screening for early-stage breast cancer by detection and profiling of circulating tumor cells. Cancers 14(14):3341

De Luca F, Rotunno G, Salvianti F, Galardi F, Pestrin M, Gabellini S, Simi L, Mancini I, Vannucchi AM, Pazzagli M, Di Leo A (2016) Mutational analysis of single circulating tumor cells by next generation sequencing in metastatic breast cancer. Oncotarget 7(18):26107–26119

Deng Z, Wu S, Wang Y, Shi D (2022) Circulating tumor cell isolation for cancer diagnosis and prognosis. EBioMedicine 83

Denis JA, Patroni A, Guillerm E, Pépin D, Benali-Furet N, Wechsler J, Manceau G, Bernard M, Coulet F, Larsen AK, Karoui M (2016) Droplet digital PCR of circulating tumor cells from colorectal cancer patients can predict KRAS mutations before surgery. Mol Oncol 10(8):1221–1231

Descamps L, Le Roy D, Deman AL (2022) Microfluidic-based technologies for CTC isolation: a review of 10 years of intense efforts towards liquid biopsy. Int J Mol Sci 23(4):1981

Di Sario G, Rossella V, Famulari ES, Maurizio A, Lazarevic D, Giannese F, Felici C (2023) Enhancing clinical potential of liquid biopsy through a multi-omic approach: a systematic review. Front Genet 14:1152470

Ding P, Wang Z, Wu Z, Zhou Y, Sun N, Pei R (2020a) Natural biointerface based on cancer cell membranes for specific capture and release of circulating tumor cells. ACS Appl Mater Interfaces 12(18):20263–20270

Ding C, Zhang C, Cheng S, Xian Y (2020b) Multivalent aptamer functionalized Ag2S nanodots/hybrid cell membrane-coated magnetic nanobioprobe for the ultrasensitive isolation and detection of circulating tumor cells. Adv Funct Mater 30(16):1909781

Dujon AM, Capp JP, Brown JS, Pujol P, Gatenby RA, Ujvari B, Alix-Panabières C, Thomas F (2021) Is there one key step in the metastatic cascade? Cancers 13(15):3693

Fabisiewicz A, Szostakowska-Rodzos M, Grzybowska EA (2024) Improving the prognostic and predictive value of circulating tumor cell enumeration: is longitudinal monitoring the answer? Int J Mol Sci 25(19):10612

Fehm T, Müller V, Aktas B, Janni W, Schneeweiss A, Stickeler E, Lattrich C, Löhberg CR, Solomayer E, Rack B, Riethdorf S (2010) HER2 status of circulating tumor cells in patients with metastatic breast cancer: a prospective, multicenter trial. Breast Cancer Res Treat 124:403–412

Gabriel MT, Calleja LR, Chalopin A, Ory B, Heymann D (2016) Circulating tumor cells: a review of non–EpCAM-based approaches for cell enrichment and isolation. Clin Chem 62(4):571–581

García-Hernández LA, Martínez-Martínez E, Pazos-Solís D, Aguado-Preciado J, Dutt A, Chávez-Ramírez AU, Korgel B, Sharma A, Oza G (2023) Optical detection of cancer cells using lab-on-a-chip. Biosensors 13(4):439

Gerratana L, Davis AA, Zhang Q, Basile D, Rossi G, Strickland K, Franzoni A, Allegri L, Mu Z, Zhang Y, Flaum LE (2021) Longitudinal dynamics of circulating tumor cells and circulating tumor DNA for treatment monitoring in metastatic breast cancer. JCO Precis Oncol 5:943–952

Ghasempour S, Freeman SA (2023) The glycocalyx and immune evasion in cancer. FEBS J 290(1):55–65

Ghosh S, Koester DV (2023) Study of cell-cell adhesion formation and dynamics in cancer cells using a hybrid cell-lipid bilayer system. Biophys J 122(3):325a

Guan X, Xing S, Liu Y (2024) Engineered cell membrane-camouflaged nanomaterials for biomedical applications. Nano 14(5):413

Guo K, Shao C, Han L, Liu H, Ma Z, Yang Y, Feng Y, Pan M, Santarpia M, Carmo-Fonseca M, Silveira C (2021) Detection of epidermal growth factor receptor (EGFR) mutations from preoperative circulating tumor DNA (ctDNA) as a prognostic predictor for stage I–III non-small cell lung cancer (NSCLC) patients with baseline tissue EGFR mutations. Transl Lung Cancer Res 10(7):3213–3225

Halima A, Vuong W, Chan TA (2022) Next-generation sequencing: unraveling genetic mechanisms that shape cancer immunotherapy efficacy. J Clin Invest 132(12)

Harouaka R, Kang Z, Zheng SY, Cao L (2014) Circulating tumor cells: advances in isolation and analysis, and challenges for clinical applications. Pharmacol Ther 141(2):209–221

Heidrich I, Ačkar L, Mossahebi Mohammadi P, Pantel K (2021) Liquid biopsies: potential and challenges. Int J Cancer 148(3):528–545

Hofman P (2021) Liquid biopsy for lung cancer screening: usefulness of circulating tumor cells and other circulating blood biomarkers. Cancer Cytopathol 129(5):341–346
Hou J, Liu X, Zhou S (2021) Programmable materials for efficient CTCs isolation: from micro/nanotechnology to biomimicry. Viewpoints 2(6):20200023
Hu C, Long L, Lou J, Leng M, Yang Q, Xu X, Zhou X (2024) CTC-neutrophil interaction: a key driver and therapeutic target of cancer metastasis. Biomed Pharmacother 180:117474
Huang A, Zhang X, Zhou SL, Cao Y, Huang XW, Fan J, Yang XR, Zhou J (2016) Detecting circulating tumor DNA in hepatocellular carcinoma patients using droplet digital PCR is feasible and reflects intratumoral heterogeneity. J Cancer 7(13):1907–1914
Hur SC, Mach AJ, Di Carlo D (2011) High-throughput size-based rare cell enrichment using microscale vortices. Biomicrofluidics 5(2):22206
Ignatiadis M, Rothe F, Chaboteaux C, Durbecq V, Rouas G, Criscitiello C, Metallo J, Kheddoumi N, Singhal SK, Michiels S, Veys I (2011) HER2-positive circulating tumor cells in breast cancer. PLoS One 6(1):e15624
Ignatiadis M, Sledge GW, Jeffrey SS (2021) Liquid biopsy enters the clinic – implementation issues and future challenges. Nat Rev Clin Oncol 18(5):297–312
Ilie M, Hofman V, Leroy S, Cohen C, Heeke S, Cattet F, Bence C, Lalvée S, Mouroux J, Marquette CH, Hofman P (2018) Use of circulating tumor cells in prospective clinical trials for NSCLC patients–standardization of the pre-analytical conditions. Clin Chem Lab Med 56(6):980–989
Ji X, Ma Y, Liu W, Liu L, Yang H, Wu J, Zong X, Dai J, Xue W (2020) In situ cell membrane fusion for engineered tumor cells by worm-like nanocell mimics. ACS Nano 14(6):7462–7474
Jia L, Zhen X, Chen L, Feng Q, Yuan W, Bu Y, Wang S, Xie X (2023) Bioinspired nano-plate-coral platform enabled efficient detection of circulating tumor cells via the synergistic capture of multivalent aptamer and tumor cell membrane. J Colloid Interface Sci 631:55–65
Jia L, Wang R, Sun Y, Chen Y, Zhang T, Li M, Xie X, Wang S (2025) Engineered surface design of recognition site-ordered biomimetic sensor for efficient detection of circulating tumor cells. Biosens Bioelectron 269:116946
Jiang T, Zhao J, Zhao C, Li X, Shen J, Zhou J, Ren S, Su C, Zhou C, O'Brien M (2019) Dynamic monitoring and predictive value of circulating tumor cells in EGFR-mutated advanced non–small-cell lung cancer patients treated with first-line EGFR tyrosine kinase inhibitors. Clin Lung Cancer 20(2):124–133
Jiang X, Zhang X, Guo C, Ma B, Liu Z, Du Y, Wang B, Li N, Huang X, Ou L (2024a) Genetically engineered cell membrane-coated magnetic nanoparticles for high-performance isolation of circulating tumor cells. Adv Funct Mater 34(7):2304426
Jiang X, Zhang X, Guo C, Liu Z, Guo X, Tian Z, Wang Z, Yang J, Huang X, Ou L (2024b) Greatly isolated heterogeneous circulating tumor cells using hybrid engineered cell membrane-camouflaged magnetic nanoparticles. J Nanobiotechnol 22(1):231
Kang T, Zhu Q, Wei D, Feng J, Yao J, Jiang T, Song Q, Wei X, Chen H, Gao X, Chen J (2017a) Nanoparticles coated with neutrophil membranes can effectively treat cancer metastasis. ACS Nano 11(2):1397–1411
Kang YT, Kim YJ, Bu J, Cho YH, Han SW, Moon BI (2017b) High-purity capture and release of circulating exosomes using an exosome-specific dual-patterned immunofiltration (ExoDIF) device. Nanoscale 9(36):13495–13505
Kang DS, Moriarty A, Oh JM, Begum HM, Shen K, Yu M (2023) Biophysical properties and isolation of circulating tumor cells. In: Engineering and physical approaches to cancer. Springer International Publishing, Cham, pp 255–283
Keller L, Pantel K (2019) Unravelling tumour heterogeneity by single-cell profiling of circulating tumour cells. Nat Rev Cancer 19(10):553–567
Khan T, Becker TM, Po JW, Chua W, Ma Y (2022) Single-circulating tumor cell whole genome amplification to unravel cancer heterogeneity and actionable biomarkers. Int J Mol Sci 23(15):8386
King JD, Casavant BP, Lang JM (2014) Rapid translation of circulating tumor cell biomarkers into clinical practice: technology development, clinical needs and regulatory requirements. Lab Chip 14(1):24–31

Klein CA (2008) The metastasis cascade. Science 321(5897):1785–1787

Kumar J, Basak S, Kalkal A, Packirisamy G (2022) Recent advances in nanotechnology and microfluidic-based approaches for isolation and detection of circulating tumor cells (CTCs). Nano-Struct Nano-Objects 31:100886

Labib M, Kelley SO (2021) Circulating tumor cell profiling for precision oncology. Mol Oncol 15(6):1622–1646

Launiere C, Gaskill M, Czaplewski G, Myung JH, Hong S, Eddington DT (2012) Channel surface patterning of alternating biomimetic protein combinations for enhanced microfluidic tumor cell isolation. Anal Chem 84(9):4022–4028

Lawrence R, Watters M, Davies CR, Pantel K, Lu YJ (2023) Circulating tumour cells for early detection of clinically relevant cancer. Nat Rev Clin Oncol 20(7):487–500

Li B, Zheng Y (2021) Circulating tumor cell count: a reliable biomarker for treatment selection in metastatic breast cancer. Thoracic Cancer 13(2):265–266

Li J, Sharkey CC, Wun B, Liesveld JL, King MR (2016a) Genetic engineering of platelets to neutralize circulating tumor cells. J Control Release 228:38–47

Li J, Ai Y, Wang L, Bu P, Sharkey CC, Wu Q, Wun B, Roy S, Shen X, King MR (2016b) Targeted drug delivery to circulating tumor cells via platelet membrane-functionalized particles. Biomaterials 76:52–65

Li C, Wang H, Jiang Y, Fu W, Liu X, Zhong R, Cheng B, Zhu F, Xiang Y, He J, Liang W (2022) Advances in lung cancer screening and early detection. Cancer Biol Med 19(5):591–608

Li Y, Liu F, Cai Q, Deng L, Ouyang Q, Zhang XH, Zheng J (2025) Invasion and metastasis in cancer: molecular insights and therapeutic targets. Signal Transduct Target Ther 10(1):57

Lin D, Shen L, Luo M, Zhang K, Li J, Yang Q, Zhu F, Zhou D, Zheng S, Chen Y, Zhou J (2021) Circulating tumor cells: biology and clinical significance. Signal Transduct Target Ther 6(1):404

Liu Y, Li R, Zhang L, Guo S (2022) Nanomaterial-based immunocapture platforms for the recognition, isolation, and detection of circulating tumor cells. Front Bioeng Biotechnol 10:850241

Lone SN, Nisar S, Masoodi T, Singh M, Rizwan A, Hashem S, El-Rifai W, Bedognetti D, Batra SK, Haris M, Bhat AA (2022) Liquid biopsy: a step closer to transform diagnosis, prognosis and future of cancer treatments. Mol Cancer 21(1):79

Lopes J, Lopes D, Pereira-Silva M, Peixoto D, Veiga F, Hamblin MR, Conde J, Corbo C, Zare EN, Ashrafizadeh M, Tay FR (2022) Macrophage cell membrane-cloaked nanoplatforms for biomedical applications. Small methods 6(8):2200289

Mader S, Pantel K (2017) Liquid biopsy: current status and future perspectives. Oncol Res Treat 40(7–8):404–408

Man Y, Wang Q, Kemmner W (2011) Currently used markers for CTC isolation-advantages, limitations and impact on cancer prognosis. J Clin Exp Pathol 1(1):1–7

Manjunath Y, Porciani D, Mitchem JB, Suvilesh KN, Avella DM, Kimchi ET, Staveley-O'Carroll KF, Burke DH, Li G, Kaifi JT (2020) Tumor-cell–macrophage fusion cells as liquid biomarkers and tumor enhancers in cancer. Int J Mol Sci 21(5):1872

Mina LA, Sledge GW Jr (2011) Rethinking the metastatic cascade as a therapeutic target. Nat Rev Clin Oncol 8(6):325–332

Miyamoto DT, Sequist LV, Lee RJ (2014) Circulating tumour cells—monitoring treatment response in prostate cancer. Nat Rev Clin Oncol 11(7):401–412

Nikanjam M, Kato S, Kurzrock R (2022) Liquid biopsy: current technology and clinical applications. J Hematol Oncol 15(1):131

Noubissi Nzeteu GA, Geismann C, Arlt A, Hoogwater FJ, Nijkamp MW, Meyer NH, Bockhorn M (2022) Role of epithelial-to-mesenchymal transition for the generation of circulating tumors cells and cancer cell dissemination. Cancers. 14(22):5483

Okegawa T, Ninomiya N, Masuda K, Nakamura Y, Tambo M, Nutahara K (2018) AR-V7 in circulating tumor cells cluster as a predictive biomarker of abiraterone acetate and enzalutamide treatment in castration-resistant prostate cancer patients. Prostate 78(8):576–582

Onstenk W, Gratama JW, Foekens JA, Sleijfer S (2013) Towards a personalized breast cancer treatment approach guided by circulating tumor cell (CTC) characteristics. Cancer Treat Rev 39(7):691–700

Orrapin S, Udomruk S, Lapisatepun W, Moonmuang S, Phanphaisarn A, Phinyo P, Pruksakorn D, Chaiyawat P (2022) Clinical implication of circulating tumor cells expressing epithelial mesenchymal transition (EMT) and cancer stem cell (CSC) markers and their perspective in HCC: a systematic review. Cancers 14(14):3373

Ortiz-Otero N, Marshall JR, Glenn A, Matloubieh J, Joseph J, Sahasrabudhe DM, Messing EM, King MR (2021) TRAIL-coated leukocytes to kill circulating tumor cells in the flowing blood from prostate cancer patients. BMC Cancer 21:1–2

Panovska D, De Smet F (2022) Functional precision oncology: the next frontier to improve glioblastoma outcome? Int J Mol Sci 23(15):8637

Pantel K, Alix-Panabières C (2010) Circulating tumour cells in cancer patients: challenges and perspectives. Trends Mol Med 16(9):398–406

Pantel K, Alix-Panabieres C (2016) Functional studies on viable circulating tumor cells. Clin Chem 62(2):328–334

Park M, Kim D, Ko S, Kim A, Mo K, Yoon H (2022) Breast cancer metastasis: mechanisms and therapeutic implications. Int J Mol Sci 23(12):6806

Paterlini-Brechot P, Benali NL (2007) Circulating tumor cells (CTC) detection: clinical impact and future directions. Cancer Lett 253(2):180–204

Perez TD, Nelson WJ, Boxer SG, Kam L (2005) E-cadherin tethered to micropatterned supported lipid bilayers as a model for cell adhesion. Langmuir 21(25):11963–11968

Philippron A, Depypere L, Oeyen S, De Laere B, Vandeputte C, Nafteux P, De Preter K, Pattyn P (2021) Evaluation of a marker independent isolation method for circulating tumor cells in esophageal adenocarcinoma. PLoS One 16(5):e0251052

Rao L, Meng QF, Huang Q, Wang Z, Yu GT, Li A, Ma W, Zhang N, Guo SS, Zhao XZ, Liu K (2018) Platelet–leukocyte hybrid membrane-coated immunomagnetic beads for highly efficient and highly specific isolation of circulating tumor cells. Adv Funct Mater 28(34):1803531

Rascol E, Devoisselle JM, Chopineau J (2016) The relevance of membrane models to understand nanoparticles–cell membrane interactions. Nanoscale 8(9):4780–4798

Raza F, Zafar H, Zhang S, Kamal Z, Su J, Yuan WE, Mingfeng Q (2021) Recent advances in cell membrane-derived biomimetic nanotechnology for cancer immunotherapy. Adv Healthc Mater 10(6):2002081

Reinhardt F, Franken A, Meier-Stiegen F, Driemel C, Stoecklein NH, Fischer JC, Niederacher D, Ruckhaeberle E, Fehm T, Neubauer H (2019) Diagnostic leukapheresis enables reliable transcriptomic profiling of single circulating tumor cells to characterize inter-cellular heterogeneity in terms of endocrine resistance. Cancers 11(7):903

Reza KK, Dey S, Wuethrich A, Wang J, Behren A, Antaw F, Wang Y, Sina AA, Trau M (2021) In situ single cell proteomics reveals circulating tumor cell heterogeneity during treatment. ACS Nano 15(7):11231–11243

Rodríguez-Hernández AG, Vazquez-Duhalt R, Huerta-Saquero A (2020) Nanoparticle-plasma membrane interactions: thermodynamics, toxicity and cellular response. Curr Med Chem 27(20):3330–3345

Rossi T, Gallerani G, Martinelli G, Maltoni R, Fabbri F (2021) Circulating tumor cells as a tool to untangle the breast cancer heterogeneity issue. Biomedicines 9(9):1242

Rushton AJ, Nteliopoulos G, Shaw JA, Coombes RC (2021) A review of circulating tumour cell enrichment technologies. Cancers 13(5):970

Salami SS, Singhal U, Spratt DE, Palapattu GS, Hollenbeck BK, Schonhoft JD, Graf R, Louw J, Jendrisak A, Dugan L, Wang Y (2019) Circulating tumor cells as a predictor of treatment response in clinically localized prostate cancer. JCO Precis Oncol 3:1–9

Schuster E, Taftaf R, Reduzzi C, Albert MK, Romero-Calvo I, Liu H (2021) Better together: circulating tumor cell clustering in metastatic cancer. Trends Cancer 7(11):1020–1032

Sebastian AM, Peter D (2022) Artificial intelligence in cancer research: trends, challenges and future directions. Life 12(12):1991

Shaw JA, Guttery DS, Hills A, Fernandez-Garcia D, Page K, Rosales BM, Goddard KS, Hastings RK, Luo J, Ogle O, Woodley L (2017) Mutation analysis of cell-free DNA and single circulating tumor cells in metastatic breast cancer patients with high circulating tumor cell counts. Clin Cancer Res 23(1):88–96

Shen H, Su R, Peng J, Zhu L, Deng K, Niu Q, Song Y, Yang L, Wu L, Zhu Z, Yang C (2022) Antibody-engineered red blood cell interface for high-performance capture and release of circulating tumor cells. Bioact Mater 11:32–40

Shi F, Jia F, Wei Z, Ma Y, Fang Z, Zhang W, Hu Z (2021) A microfluidic chip for efficient circulating tumor cells enrichment, screening, and single-cell RNA sequencing. Proteomics 21(3–4):2000060

Shi J, Zhao C, Shen M, Chen Z, Liu J, Zhang S, Zhang Z (2022) Combination of microfluidic chips and biosensing for the enrichment of circulating tumor cells. Biosens Bioelectron 202:114025

Sklias T, Vardas V, Pantazaka E, Christopoulou A, Georgoulias V, Kotsakis A, Vasilopoulos Y, Kallergi G (2022) PARP-1 expression and BRCA1 mutations in breast cancer patients' CTCs. Cancers 14(7):1731

Smart JA, Oleksak JE, Hartsough EJ (2021) Cell adhesion molecules in plasticity and metastasis. Mol Cancer Res 19(1):25–37

Sun J, Han S, Yang R, Guo L, Li J, Li C, Xu L, Liu H, Dong B (2024) Combining hybrid cell membrane modified magnetic nanoparticles and inverted microfluidic chip for in situ CTCs capture and inactivation. Biosens Bioelectron 263:116575

Surappa S, Multani P, Parlatan U, Sinawang PD, Kaifi J, Akin D, Demirci U (2023) Integrated "lab-on-a-chip" microfluidic systems for isolation, enrichment, and analysis of cancer biomarkers. Lab Chip 23(13):2942–2958

Tan J, Zhu C, Li L, Wang J, Xia XH, Wang C (2024) Engineering cell membranes: from extraction strategies to emerging biosensing applications. Anal Chem 96(20):7880–7894

van Deventer S, Arp AB, van Spriel AB (2021) Dynamic plasma membrane organization: a complex symphony. Trends Cell Biol 31(2):119–129

Vargas E, Zhang F, Ben Hassine A, Ruiz-Valdepeñas Montiel V, Mundaca-Uribe R, Nandhakumar P, He P, Guo Z, Zhou Z, Fang RH, Gao W (2022) Using cell membranes as recognition layers to construct ultrasensitive and selective bioelectronic affinity sensors. J Am Chem Soc 144(38):17700–17708

Wang S, Wang H, Jiao J, Chen KJ, Owens GE, Kamei KI, Sun J, Sherman DJ, Behrenbruch CP, Wu H, Tseng HR (2009) Three-dimensional nanostructured substrates toward efficient capture of circulating tumor cells. Angew Chem Int Ed 48(47):8970–8973

Wang L, Asghar W, Demirci U, Wan Y (2013) Nanostructured substrates for isolation of circulating tumor cells. Nano Today 8(4):374–387

Wang C, Sun W, Ye Y, Hu Q, Bomba HN, Gu Z (2017) In situ activation of platelets with checkpoint inhibitors for post-surgical cancer immunotherapy. Nat Biom Eng 1(2):0011

Wang Z, Sun N, Liu H, Chen C, Ding P, Yue X, Zou H, Xing C, Pei R (2019) High-efficiency isolation and rapid identification of heterogeneous circulating tumor cells (CTCs) using dual-antibody-modified fluorescent-magnetic nanoparticles. ACS Appl Mater Interfaces 11(43):39586–39593

Wang L, Hartel N, Ren K, Graham NA, Malmstadt N (2020a) Effect of protein corona on nanoparticle–plasma membrane and nanoparticle–biomimetic membrane interactions. Environ Sci Nano 7(3):963–974

Wang K, Ye H, Zhang X, Wang X, Yang B, Luo C, Zhao Z, Zhao J, Lu Q, Zhang H, Kan Q (2020b) An exosome-like programmable-bioactivating paclitaxel prodrug nanoplatform for enhanced breast cancer metastasis inhibition. Biomaterials 257:120224

Wang J, Li Y, Wang R, Han C, Xu S, You T, Li Y, Xia J, Xu X, Wang D, Tang H (2021) A fully automated and integrated microfluidic system for efficient CTC detection and its application in hepatocellular carcinoma screening and prognosis. ACS Appl Mater Interfaces 13(25):30174–30186

Wit SD, Dalum GV, Lenferink AT, Tibbe AG, Hiltermann TJ, Groen HJ, van Rijn CJ, Terstappen LW (2015) The detection of EpCAM+ and EpCAM–circulating tumor cells. Sci Rep 5(1):12270

Wu L, Ding H, Qu X, Shi X, Yang J, Huang M, Zhang J, Zhang H, Song J, Zhu L, Song Y (2020) Fluidic multivalent membrane nanointerface enables synergetic enrichment of circulating tumor cells with high efficiency and viability. J Am Chem Soc 142(10):4800–4806

Xu L, Mao X, Imrali A, Syed F, Mutsvangwa K, Berney D, Cathcart P, Hines J, Shamash J, Lu YJ (2015) Optimization and evaluation of a novel size based circulating tumor cell isolation system. PLoS One 10(9):e0138032

Xu X, Jiang Z, Wang J, Ren Y, Wu A (2020) Microfluidic applications on circulating tumor cell isolation and biomimicking of cancer metastasis. Electrophoresis 41(10–11):933–951

Xu J, Liao K, Yang X, Wu C, Wu W, Han S (2021) Using single-cell sequencing technology to detect circulating tumor cells in solid tumors. Mol Cancer 20:1–7

Yaghoubi Naei V, Bordhan P, Mirakhorli F, Khorrami M, Shrestha J, Nazari H, Kulasinghe A, Ebrahimi WM (2023) Advances in novel strategies for isolation, characterization, and analysis of CTCs and ctDNA. Ther Adv Med Oncol 15:17588359231192401

Yang YP, Giret TM, Cote RJ (2021) Circulating tumor cells from enumeration to analysis: current challenges and future opportunities. Cancers 13(11):2723

Yao H, Wen L, Li Z, Xia C (2023) Analysis of diagnostic value of CTC and CTDNA in early lung cancer. Cell Mol Biol 69(6):57–62

Zhang Y, Wang J (2024) Research progress of cell membrane biomimetic nanoparticles for circulating tumour cells. Front Oncol 14:1389775

Zhang J, Chen K, Fan ZH (2016) Circulating tumor cell isolation and analysis. Adv Clin Chem 75:1–31

Zhang T, de Waard AA, Wuhrer M, Spaapen RM (2019a) The role of glycosphingolipids in immune cell functions. Front Immunol 10:90

Zhang Y, Zhu X, Chen X, Chen Q, Zhou W, Guo Q, Lu Y, Li C, Zhang Y, Liang D, Sun T (2019b) Activated platelets-targeting micelles with controlled drug release for effective treatment of primary and metastatic triple negative breast cancer. Adv Funct Mater 29(13):1806620

Zhang Z, Wuethrich A, Wang J, Korbie D, Lin LL, Trau M (2021) Dynamic monitoring of EMT in CTCs as an indicator of cancer metastasis. Anal Chem 93(50):16787–16795

Zhang H, Jiao J, Long Y, Zhou L, Lv Y, Wei W, Sun Y, Han H, Chen C, Zhu Y, Zhang W (2025) Targeting capture and eradicate circulating tumor cells by activated platelet derived vehicle for inhibiting triple-negative breast cancer metastasis. Mater Today Bio 31:101597

Zhong H, Yuan C, He J, Yu Y, Jin Y, Huang Y, Zhao R (2021) Engineering peptide-functionalized biomimetic nanointerfaces for synergetic capture of circulating tumor cells in an EpCAM-independent manner. Anal Chem 93(28):9778–9787

Zhou H, Zhu L, Song J, Wang G, Li P, Li W, Luo P, Sun X, Wu J, Liu Y, Zhu S (2022) Liquid biopsy at the frontier of detection, prognosis and progression monitoring in colorectal cancer. Mol Cancer 21(1):86

Zhu DM, Wu L, Suo M, Gao S, Xie W, Zan MH, Liu A, Chen B, Wu WT, Ji LW, Chen LB (2018) Engineered red blood cells for capturing circulating tumor cells with high performance. Nanoscale 10(13):6014–6023

Engineered Mammalian Cell and Bacterial Membrane-Based Nanovaccines and Nanoplatforms for Cancer Therapy and Infectious Disease

13

Jiaxin Ma, Yijia Xie, Lai Jiang, Xiaoliu Liang, and Gang Liu

Abstract

Cell membrane vesicles (MVs), particularly mammalian-derived vesicles (MMVs) and bacterial-derived vesicles (BMVs), have emerged as effective platforms for the treatment of infectious diseases and cancer. Through engineered modifications, the functionality of these MVs can be further enhanced, endowing them with properties such as improved targeting, stability, and biocompatibility. As research into their characteristics continues to advance, the clinical application potential of these vesicles is becoming increasingly evident. In this chapter, we focus on the application of engineered membrane vesicles (EMVs) technologies in cancer and infectious disease therapies. First, we summarize and compare various MVs functionalization methods, including physical, chemical, and

J. Ma · Y. Xie
State Key Laboratory of Vaccines for Infectious Diseases, Center for Molecular Imaging and Translational Medicine, Xiang An Biomedicine Laboratory, National Innovation Platform for Industry-Education Integration in Vaccine Research, School of Public Health, Xiamen University, Xiamen, China

L. Jiang
School of Pharmaceutical Sciences, Zhejiang Chinese Medical University, Hangzhou, China

X. Liang (✉)
College of Pharmacy, Guangxi Medical University, Nanning, China
e-mail: liangxiaoliu@sr.gxmu.edu.cn

G. Liu (✉)
State Key Laboratory of Vaccines for Infectious Diseases, Center for Molecular Imaging and Translational Medicine, Xiang An Biomedicine Laboratory, National Innovation Platform for Industry-Education Integration in Vaccine Research, School of Public Health, Xiamen University, Xiamen, China

College of Pharmacy, Guangxi Medical University, Nanning, China
e-mail: gangliu.cmitm@xmu.edu.cn

Z. Hussain, R. Pei (eds.), *Cell Membrane Engineering for Advancing Cell-Based and Nanotechnology-Based Therapies*, Biomaterials, Bioengineering and Sustainability 6, https://doi.org/10.1007/978-3-031-93425-4_13

biological approaches. We also provide an overview of MVs production and engineering techniques, discussing strategies for improving MVs yield and reducing toxicity. Next, we examine the recent progress in the use of EMVs as nanovaccines and therapeutic platforms for treating infectious diseases and cancer. Finally, to further advance the clinical application of EMVs, we address the challenges and prospects associated with their clinical translation, offering insights and recommendations to shape future research and applications in this field.

Keywords

Mammalian membrane vesicles · Bacterial membrane vesicles · Engineered membrane vesicles · Nanovaccines · Cell membrane vesicle nanoparticles · Infectious diseases

13.1 Introduction

Both eukaryotic and prokaryotic cells can actively or passively produce spherical cell membrane vesicles (MVs), such as mammalian-derived vesicles (MMVs) and bacterial-derived vesicles (BMVs). These vesicles inherit the biological characteristics of their source cells, with adjustable sizes and excellent drug-loading capacity, making them a promising platform for managing infectious diseases and cancer. MVs derived from eukaryotic or prokaryotic cells can be obtained through cell disruption methods, which preserve the membrane protein bioactivity, allowing the vesicles to retain similar biological functions as the parent cells. Additionally, live cells can actively secrete MVs (Toyofuku et al. 2019). For example, Gram-negative bacteria produce outer membrane vesicles (OMVs) containing lipopolysaccharides (LPS), outer membrane proteins, and toxins (Turnbull et al. 2016). These OMVs can effectively trigger immune responses, making them advantageous for vaccine development (Sartorio et al. 2021). Similarly, exosomes secreted by mammalian cells, due to their excellent biocompatibility, have emerged as ideal platforms for therapeutic delivery or gene therapy (Wan et al. 2022; Liang et al. 2021). Additionally, MVs-coated nanoparticles, as a nanovaccine system with enhanced targeting ability and biosafety, can be used for anti-cancer or anti-infection vaccines. The nanoparticles encapsulated within MVs not only improve the stability of the MVs-based nanovaccine system but also can carry adjuvants to enhance the therapeutic effect. Despite the progress made with naturally derived MVs for treating infectious diseases and cancer, challenges remain for BMVs and MMVs, including low yields, high toxicity, and limited targeting (Tan et al. 2018; Behrens et al. 2021; Aytar Çelik et al. 2022).

To enhance the biomedical applications of MVs, artificial modifications can be made to parent cells or MVs using physical, chemical, or biotechnological methods. These engineered membrane vesicles (EMVs) demonstrate improved targeting, immunogenicity, and stability, thereby increasing their effectiveness as vaccines or nanoplatforms (Pang et al. 2019; Liu et al. 2018, 2023). For example, engineered bacterial membrane vesicles (EBMVs), with toxic components removed or specific antigens added, can maintain their strong immunostimulatory properties while minimizing risks (Van De Waterbeemd et al. 2010; Van De Waterbeemd et al. 2012). In cancer therapy, engineered mammalian membrane vesicles (EMMVs) show significant promise. Through genetic engineering, full-length monoclonal antibodies (mAbs) targeting tumors can be expressed on the MVs surface (Liu et al. 2019a). These antibody-engineered MVs serve as tumor-targeting carriers, efficiently delivering chemotherapeutic agents or diagnostic tools to tumor sites, reducing toxicity to normal tissues, and enabling precise tumor-targeted therapy and diagnostics (Liu et al. 2019a). Additionally, viral antigen epitopes or proteins can be displayed on the vesicle surface to create safe antiviral vaccines (Zhang et al. 2015). EMVs also function as versatile nanotherapeutic platforms for neutralizing agents or immunomodulatory treatments against viral infections and cancer (Liu et al. 2018; Hou et al. 2022).

As research on cell membranes deepens, scientists have reviewed cell membrane-based nanomedicine from different perspectives in previous reports. For example, Zhang et al. reviewed fabrication of various biomimetic membrane coated nanoparticles and their preclinical assessment in evaluating anti-inflammatory effects (Zhang et al. 2021a). Wang et al. focused on the advancements of membrane coated nanoparticles in drug delivery, particularly their circulation stability and targeting efficiency, as well as their therapeutic potential in cancer and inflammation treatments (Park et al. 2021). Additionally, Ma et al. highlighted the latest progress in engineered cell membrane-based nanomedicines for anti-inflammatory applications (Ma et al. 2024). In contrast to existing reviews, this chapter focuses on the recent advancements of EMMVs and EBMVs as vaccines or nanoplatforms in the treatment of infectious diseases and cancer (Fig. 13.1). This chapter first introduces the modification strategies and purposes of EMVs, then summarizes the recent applications and insights of EMMVs and EBMVs in the treatment of infectious diseases and cancers. Finally, to gain deeper insights into the progress and future potential of EMVs, we also discuss the opportunities and challenges of EMVs in the treatment of infectious diseases and cancer.

13.2 Mammalian Cell Membrane

13.2.1 Extraction of MMVs

The extraction of MMVs involves two critical steps: cell membrane lysis and MMVs purification. These processes must be conducted under the mildest possible conditions to preserve the integrity of the membrane composition and maintain the functionality of membrane proteins. Notably, the entire extraction procedure should be performed at low temperatures, with protease inhibitors added if necessary, to prevent denaturation of functional membrane proteins (Liu et al. 2019b). For anucleate cells, such as red blood cells (RBCs), the resulting MVs are often used to

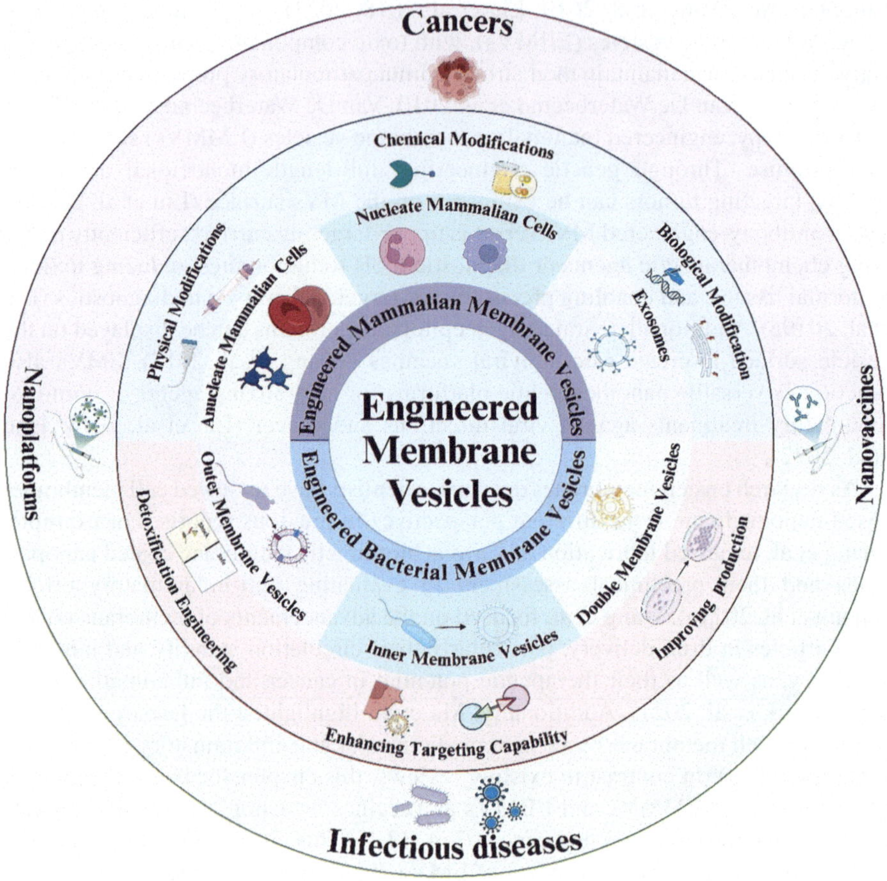

Fig. 13.1 Schematic of engineered mammalian cell and bacterial membrane-based nanovaccines and nanoplatforms for cancer therapy and infectious disease

neutralize bacterial toxins or prolong drug circulation. RBC MVs are typically extracted through isotonic treatment or repeated freeze-thaw cycles, followed by centrifugation to remove intracellular proteins. The final step is extrusion to form nanoscale MVs (Dehaini et al. 2017). Platelet MVs, which are often used to target bacteria or wound tissues, are extracted using methods similar to those for RBCs MVs (Hu et al. 2015). MVs from nucleate cells, like macrophages and neutrophils, are valued for their high affinity for bacteria and are often used to enhance nanoparticle-based pathogen targeting. After purification and *in vitro* expansion, these cells are lysed via sonication. Differential centrifugation is then employed to eliminate nuclei and cytoplasm, with nanoscale MVs obtained through sonication or extrusion (Thamphiwatana et al. 2017).

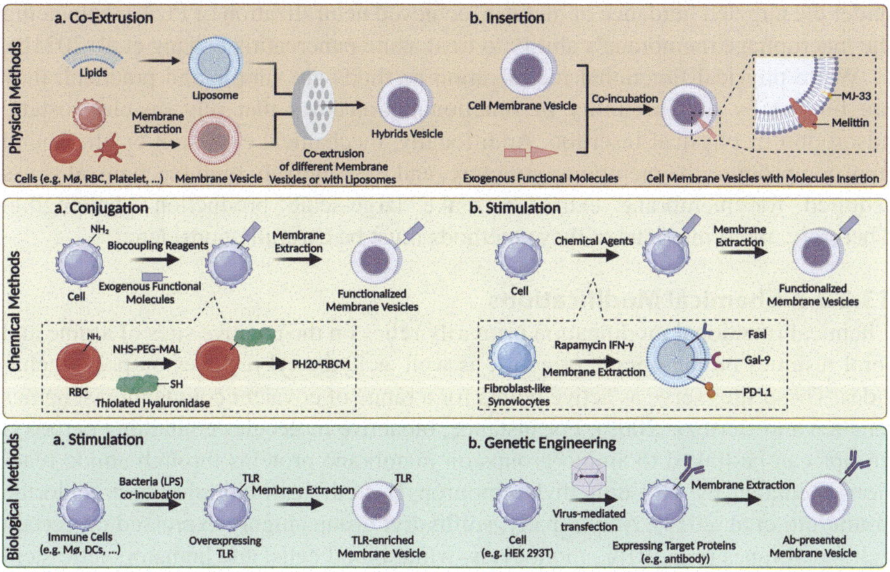

Fig. 13.2 Schematic of typical methods of engineering mammalian cell membranes. (Reproduced with permission from Ma et al. 2024, copyright 2024, Wiley-VCH GmbH)

13.2.2 Functional Modification of MMVs

MMVs have demonstrated broad potential for antibacterial, anti-inflammatory, antiviral, and antitumor applications (Ma et al. 2022; Zhang et al. 2018a; Wei et al. 2018; Chen et al. 2020). Functional modifications can further enhance these capabilities. Below, we outline three common approaches to MMV functionalization: physical, chemical, and biological modifications (Fig. 13.2).

13.2.2.1 Physical Modifications

Natural cell membranes exhibit self-assembly properties due to their phospholipid bilayer and fluidity. Hybrid MVs from different cell sources can be prepared using ultrasonication or membrane extrusion. For example, hybrid MVs combining RBC and platelet membranes have been developed, with RBC MVs neutralizing toxins and extending circulation time, while platelet MVs target vascular inflammation or bacterial infections (Dehaini et al. 2017).

Exogenous functional molecules can also be introduced into MVs via ultrasound or co-culture, bestowing new biological functions. For example, to manage acute pancreatitis resulting from premature activation of phospholipase A2 (PLA2) in pancreatic acinar cells (Lee and Papachristou 2019; Lankisch et al. 2015), researchers introduced the liposome-like PLA2 inhibitor MJ-33 into macrophage MVs via ultrasound. Following co-incubation, melittin (a bee venom peptide) was spontaneously integrated into the macrophage MVs due to its positively charged and needle-like structure. This combination significantly reduced their biological toxicity and, under the targeted guidance of melittin, achieved neutralization of PLA2, enhancing the macrophage membrane's ability to treat acute pancreatitis (Zhang et al. 2021b).

While physical functional modification methods are simple and practical, they are limited by the instability of functional molecules that rely on electrostatic adsorption or physical insertion. Additionally, mechanical extrusion or ultrasound may compromise the membrane structure, and the relatively high mechanical stress required for membrane extrusion make large-scale production challenging. Therefore, the limitations of these methods must be carefully considered.

13.2.2.2 Chemical Modifications

Chemical functional modification primarily relies on the reactive sites of amine and thiol residues on membrane proteins, as well as hydroxyl residues on polysaccharides. These sites serve as active points for a range of covalent conjugation schemes (Sletten and Bertozzi 2009). For instance, bioactive molecules containing carboxyl groups can be linked to amino groups on membrane proteins through amide reactions mediated by 1-(3-dimethylaminopropyl)-3-ethylcarbodiimide hydrochloride (Lutterotti et al. 2013). Additionally, sulfhydryl groups highly expressed on certain cell membranes (e.g., RBC, monocytes, white blood cells, and hematopoietic stem cells) can be coupled to functional molecules via maleimide (Mal) reactions (Jones et al. 2017). For instance, researchers have used Mal reactions to attach recombinant hyaluronidase to the surface of RBC MVs and encapsulate them in nanoparticles. These RBC MVs functionalized particles not only exhibit hyaluronidase activity but also have prolonged circulation time and can target hyaluronan-rich inflammatory atherosclerotic lesions (Zhou et al. 2016).

Chemical induction, as another chemical modification strategy, can upregulate the surface presentation of biologically active molecules on the plasma membrane. For example, in rheumatoid arthritis (RA), the characteristics of fibroblast-like synoviocytes (FLS) plays a crucial role. Researchers stimulated FLS cells with IFN-γ and rapamycin, resulting in high expression of programmed cell death ligand 1 (PD-L1), galectin-9, and Fas ligand on the FLS cell surface. These stimulated and modified FLS plasma membrane coated nanoparticles demonstrated better inflammation-targeting and clearance capabilities in RA mouse models (Liu et al. 2023). Despite chemical engineering approaches can improve and optimize cell membrane functions, the introduction of chemical substances during this process may cause toxic side effects and potentially inactivate other functional proteins within the membrane. Therefore, selecting less toxic, more direct, and simpler chemical reaction steps for chemical modification is a direction that needs to be explored in the future.

13.2.2.3 Biological Functional Modification Methods

Cell membranes can be biologically functionalized through bio-derived stimuli, such as live microorganism or their major components, which can induce the overexpression of certain natural receptors on the membrane surface. For example, pathogen stimulation can affect the expression and aggregation of pathogen associated molecular patterns (PAMPs) on macrophage membranes (Cao et al. 2016).

Studies have shown that macrophage membranes pre-exposed with *Staphylococcus aureus* (*S. aureus*) exhibit stronger bacterial targeting, and these biologically functionalized macrophage MVs are more effective in treating bacterial-induced osteomyelitis and local infection models compared to unstimulated MVs (Wang et al. 2018a).

Genetic engineering can also be used for biological modification. For example, Toll-like receptor 4 (TLR4) is a receptor on macrophage membranes that binds bacterial endotoxins. Through plasmid transfection, TLR4 can be overexpressed on the macrophage membrane surface, enhancing the macrophage membrane's potential to neutralize bacterial endotoxins and inhibit the secretion of inflammatory cytokines (Ou et al. 2020). Genetic engineering can also be employed to express exogenous functional proteins or antibodies on the cell membrane surface. For instance, tumor-associated mAbs can be constructed on the cell membrane surface through artificially designed fusion proteins, enabling targeted delivery of tumor drugs and antibody immunotherapy. However, this method requires careful consideration of the fusion protein design, including the orientation, conformation, and stability of the target protein, as well as the expression level and distribution of the fusion protein. Recently, a novel gene engineering functionalization method has emerged, utilizing SpyCatcher and SpyTag, a pair of synthetic proteins that spontaneously form covalent bonds, to modify target ligands on the cell membrane. The steps involve first expressing SpyCatcher on the cell membrane surface through genetic engineering, then encapsulating nanoparticles, and finally mixing the SpyTag-modified functional molecules with the nanoparticles to achieve functionalization through covalent bonding, thereby simplifying the design process of gene-engineered cell membranes (Krishnan et al. 2023).

While bio-derived stimuli enhance bacterial targeting, further research are required to assess the biosafety of live bacteria and their components. In contrast, genetic modification provides greater safety and flexibility by selectively expressing functional proteins while preserving membrane structure. However, increasing functional protein expression remains a challenge.

13.2.3 Applications of EMMVs in Cancer

13.2.3.1 Drug Delivery

Efficient drug delivery to tumor regions is crucial for the success of chemotherapy. Traditional drug delivery systems depend on the enhanced permeability and retention (EPR) effect in tumor tissues, which is a form of passive targeting with low efficiency and susceptibility to clearance by the reticuloendothelial system (Shi et al. 2020a; De Lázaro and Mooney 2021; Duan et al. 2020). In contrast, coating nanomedicines with EMMVs significantly improves tumor targeting, biocompatibility, and circulation time, enhancing therapeutic efficacy.

To improve drug circulation time, researchers have employed genetic engineering techniques to display PAS peptide sequences on cell membranes. This method has been found to significantly extend the half-life of PLGA nanoparticles coated with cell membranes *in vivo* (Krishnamurthy et al. 2019). Although CAR-T cell therapy has shown significant success in treating hematological diseases by directing T cells to target tumor cells, it has yet to achieve breakthroughs in solid tumor treatment. The primary challenge lies in the immune-suppressing microenvironment of solid tumors, characterized by nutrient deprivation, acidity, and hypoxia, which hinders effective T cell infiltration into tumor cells. To address this issue, a study constructed CAR-T cell membrane-wrapped nanoparticles loaded with IR780. These nanoparticles, capable of recognizing the hepatocellular carcinoma-specific antigen GPC3, exhibited strong tumor-targeting ability, enabling imaging and photothermal anti-tumor effects for liver cancer (Ma et al. 2020). Additionally, to achieve precise recognition of tumor cells, another study utilized genetic engineering to display human epidermal growth factor (hEGF) targeting proteins on the cell membrane surface. The results demonstrated that this novel functionalized cell membrane derived drug carrier exhibited superior bioactivity and targeting ability compared to epidermal growth factor chemically conjugated to liposomes. By loading therapeutic drugs into this carrier, small molecule drugs can be selectively enriched at tumor sites, thereby enhancing anti-tumor effects (Fig. 13.3a, b) (Zhang et al. 2018b).

Selective targeting and killing of cancer cells without harming healthy tissues is a key objective in tumor therapy. Researchers have used genetic engineering to display full-length tumor-specific monoclonal antibodies (mAbs) on cell membranes. These antibody vesicles (VAs) exhibited excellent drug-loading capacity and precise tumor targeting both *in vitro* and *in vivo*. Moreover, these VAs modulated the tumor immune microenvironment, recruiting natural killer cells and enhancing antibody-mediated cellular cytotoxicity. When combined with chemotherapeutic drugs, VAs enhanced the synergistic effects of chemotherapy and immunotherapy in tumor-bearing mice (Fig. 13.3c, d) (Liu et al. 2019a).

These novel functionalized MVs offer significant potential for targeted drug delivery to various tumor cell populations, presenting a promising alternative to natural MVs and opening new directions in targeted drug delivery.

13.2.3.2 Tumor Vaccines

Tumor vaccines aim to train the immune system for a robust attack on tumor antigens, thereby preventing or eliminating tumors without adverse effects on normal cells (Sahin and Türeci 2018; Ye et al. 2021). However, their clinical use is limited because tumor antigens often originate from normal cells, allowing them to evade immune detection through mutation or modification (Quaglino et al. 2020; Xie et al. 2023; Huang et al. 2021).

Traditional tumor vaccines rely on activating dendritic cells (DCs), which then activate T cells. However, the process of DC activation is complex, involving multiple immune signaling pathways, which can result in delayed immune responses or inefficient antigen presentation (Santos and Butterfield 2018; González et al. 2014). A promising alternative is directly presenting tumor antigens to T cells. Tumor cells possess MHC-antigen presenting peptides, but lack the co-stimulatory molecules (e.g., CD80, CD86) needed for T cell activation. Researchers addressed this by

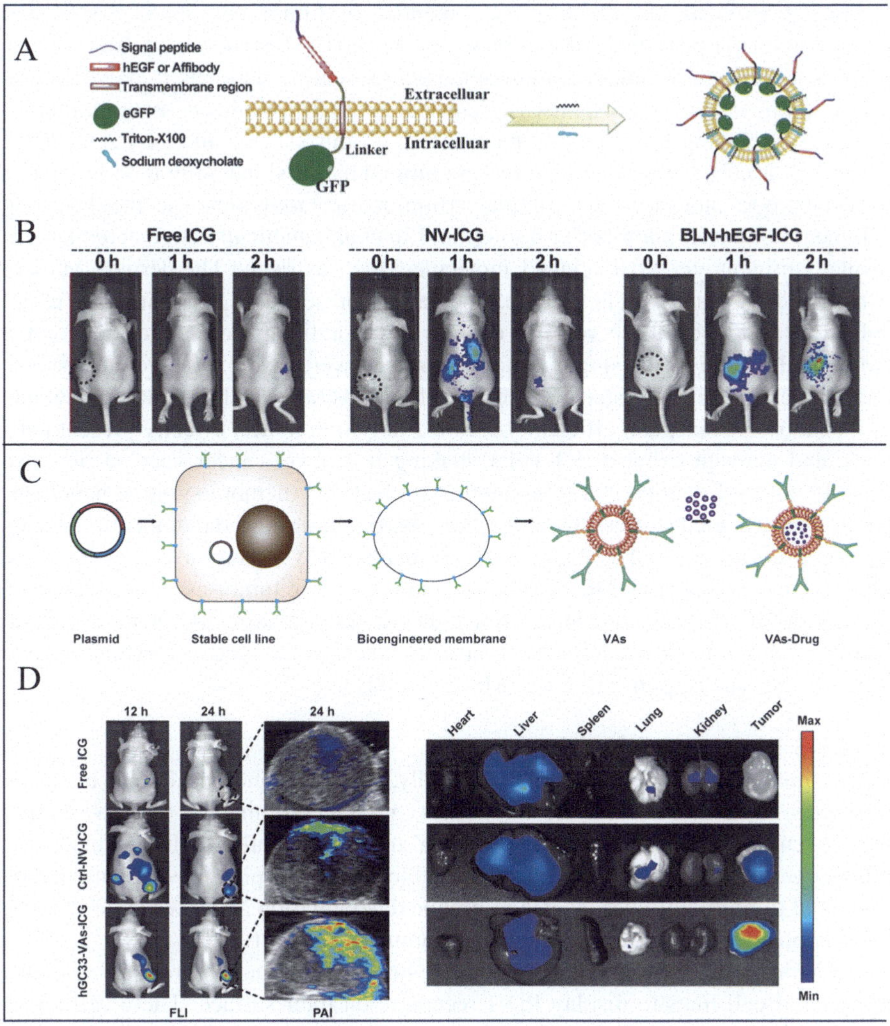

Fig. 13.3 EMMVs application in tumor drug delivery. (**a**) Schematic design of hEGF-functionalized MMVs. (**b**) hEGF-functionalized MMVs targeting drug delivery in tumor model mice. (Reproduced with permission from Zhang et al. 2018b, copyright 2018, Wiley-VCH GmbH). (**c**) Schematic construction and drug loading of antibody-functionalized MMVs. (**d**) Antibody-functionalized MMVs targeting drug delivery in tumor model mice. (Reproduced with permission from Liu et al. 2019a, copyright 2019, Wiley-VCH GmbH)

introducing the CD80 gene into mouse melanoma cells, creating B16-F10 cells that stably express CD80. MVs from these cells were used to coat nanoparticles, forming a vaccine (AP-NPs) that mimics antigen-presenting cells. AP-NPs bypass DCs, directly activating T cells, promoting T cell proliferation, and enhancing IL-2 and IFN-γ secretion. In mouse models, AP-NPs suppressed tumor progression and prolonged survival, highlighting the potential of tumor cell membrane-coated nanoparticles for personalized vaccines (Fig. 13.4a, b) (Jiang et al. 2020).

DCs are critical in tumor immunotherapy, presenting antigens to naive T cells and initiating anti-tumor responses (Wculek et al. 2020). While DC-based immunotherapies, including DC vaccines, have shown promise (Perez and De Palma 2019; Yuan et al. 2022), clinical applications are limited by DCs' low migration efficiency to lymph nodes and short survival times. Biomimetic nanovesicle vaccines based on cell membrane technology offer a solution. Liu et al. genetically engineered DCs to display antigens via MHC class I molecules, then extracted DC-derived MVs to construct a DC nanovesicle vaccine. The natural presence of co-stimulatory molecules (CD80/86) on these vesicles enables direct antigen presentation to T cells. Additionally, anti-PD-1 single chain antibodies were overexpressed on the DC membrane, effectively restoring T cell function. Overall, this DC nanovaccine, due to its nanoscale size, can efficiently localize in lymph nodes, directly present antigens, and activate exhausted T cells, making it a novel cancer vaccine platform (Fig. 13.4c, d) (Liu et al. 2022). In another study, a DC membrane-based nanocomposite was designed to cross the blood-brain barrier and showed therapeutic efficacy in a mouse glioma model. Wang et al. co-incubated lysed tumor cells with naive DCs *in vitro*, stimulating DC maturation and presenting tumor-associated antigens on their MHC class I molecules. They then extracted mature DC membranes and loaded them with nanomedicines to construct functional composites, which elicited a strong immune response in mice (Ma et al. 2023).

13.2.3.3 Immunomodulation

Tumor immunotherapy has advanced rapidly, with immune checkpoint blockade strategies, such as PD-1/PD-L1 antibodies, gaining prominence. However, challenges remain, including tumor recurrence and inflammation-induced metastasis after surgery. Although immune checkpoint blockade therapies show potential, they may bind to normal tissues, causing immune dysregulation and low response rates, making postoperative chemotherapy and immunotherapy crucial.

To address the low response rates of tumor immunotherapy, researchers have engineered cell lines to display PD-1 receptors on their surface. These cell membrane nanovesicles, produced by continuous extrusion, exhibited prolonged circulation and bound to PD-L1 on tumor cells, blocking the PD-1/PD-L1 pathway and significantly enhancing $CD8^+$ T cell infiltration into tumor tissues, thereby improving anti-tumor efficacy (Zhang et al. 2018c). Similarly, researchers genetically engineered megakaryocyte (MK) precursor cells to express PD-1 receptors. Upon stimulation, these MK cells matured and shed platelets with PD-1 receptors on their surface. These modified platelets accumulated at tumor surgical sites, recognized PD-L1 ligands on tumor cells, and blocked PD-1/PD-L1 interactions, activating $CD8^+$ T cells and reducing the risk of tumor recurrence. Additionally, researchers loaded the platelets with low-dose cyclophosphamide to specifically target and eliminate regulatory T cells (Tregs) at the tumor site, further enhancing melanoma treatment efficacy (Zhang et al. 2018d). This approach delivers drugs to target sites, minimizing off-target effects and systemic toxicity.

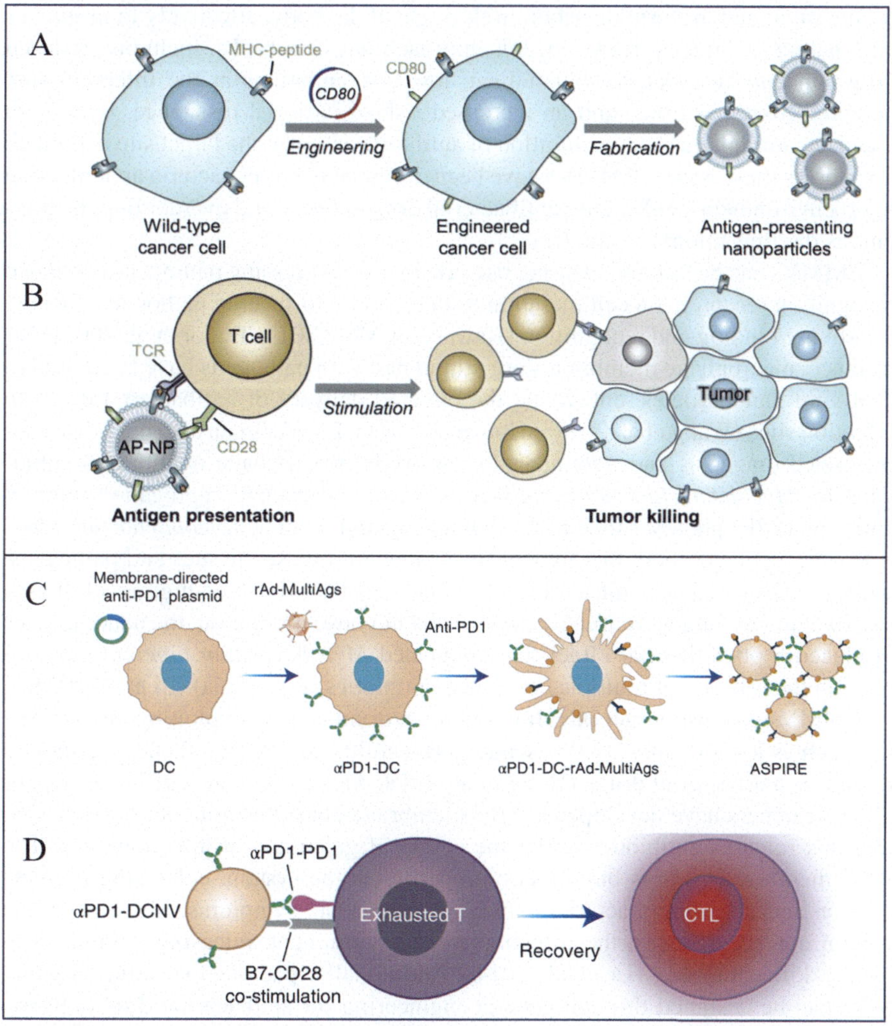

Fig. 13.4 EMMVs application as tumor vaccines. (**a**) Synthesis of CD80-expressing cell membrane-coated polymer nanoparticles (named AP-NPs). (**b**) Mechanism of direct activation of T cells by AP-NPs to generate tumor immunity. (Reproduced with permission from Jiang et al. 2020, copyright 2020, Wiley-VCH GmbH). (**c**) Schematic of genetically modified cell membrane vesicle that integrates antigen self-presentation and immunosuppression reversal (ASPIRE) for targeting tumors via immunotherapy. (**d**) Diagram of the mechanism of ASPIRE for co-delivery of B7-CD28 and anti-PD1 to promote the reversal of immunosuppression. (Reproduced with permission from Liu et al. 2022, copyright 2022, Springer Nature)

13.2.4 Applications of EMMVs in Infectious Diseases

13.2.4.1 Drug Delivery

Treating bacterial infections typically involves systemic antibiotic administration to ensure rapid and uniform distribution throughout the body, effectively inhibiting or eliminating pathogens. However, this approach has drawbacks, including reduced drug concentration due to metabolism, low concentrations at the infection site, potential toxic reactions, and an increased risk of bacterial resistance. Even with localized administration, the duration of antibiotic action at the target site is limited. To address these issues, EMMVs have been designed to target bacteria and infection microenvironments, enhancing antibacterial drug delivery and overcoming the limitations of conventional antibiotics.

EMMVs can be tailored to target bacteria by overexpressing pathogen-associated recognition receptors on cell membranes in response to specific pathogens, thereby improving binding and recognition (Kawai and Akira 2011; Taylor et al. 2005). For instance, macrophage membranes pre-stimulated with pathogens have been used to coat gold nanoparticles, enhancing the targeted delivery of antibacterial drugs to infection sites (Wang et al. 2018a; Hou et al. 2021). EMMVs can also be utilized for integrated diagnosis and treatment. In one study, macrophage membranes stimulated by *Mycobacterium marinum* were used to encapsulate aggregation-induced emission (AIE) photosensitizers for diagnosing and treating *Mycobacterium tuberculosis* infections. These membranes dual-targeted diseased tissues and pathogens, while encapsulated near-infrared II AIE photosensitizers enabled high-resolution in situ imaging of lung granulomas. Under 1064 nm laser irradiation, the nanoparticles produced a photothermal effect that eradicated *Mycobacterium tuberculosis*, surpassing the efficacy of multiple combined antibiotics (Fig. 13.5a) (Li et al. 2024a).

EMMVs can also target infection microenvironments by exploiting characteristics such as low pH, increased vascular permeability, and bacterial enzyme activity to deliver antibacterial drugs (Dong et al. 2019; Azzopardi et al. 2013). For example, researchers have developed a RBC membrane-coated gelatin nanoparticle system that releases antibiotics under the action of gelatinase at infection sites. The RBC membrane confers biomimetic properties to the system, enhancing immune evasion and absorbing bacterial exotoxins to alleviate symptoms. Meanwhile, the gelatin core degrades in the infection microenvironment, releasing antibiotics to locally kill pathogens (Li et al. 2014). Additionally, pathogen-specific targeting MVs can be designed through genetic engineering for antibacterial drug delivery. Ding et al. developed an antibody functionalized MVs (named AMV) targeting the penicillin-binding protein variant PBP2a on the surface of methicillin-resistant *S. aureus* (MRSA). AMV can effectively bind to MRSA *in vitro* and target MRSA infections in the eyes or lungs *in vivo*. When loaded with ultrasound-responsive nano-antibiotics, AMV leveraged their biomimetic and targeting properties to transport nano-antibiotics across the alveolar barrier, effectively enhancing the treatment of acute pneumonia caused by MRSA (Fig. 13.5b) (Ding et al. 2024).

Targeted antibacterial drug delivery using EMMVs enhances therapeutic efficacy by delivering nanoparticles directly to infection sites, reducing the required dosage and delaying the development of bacterial resistance while lowering antibiotic toxicity.

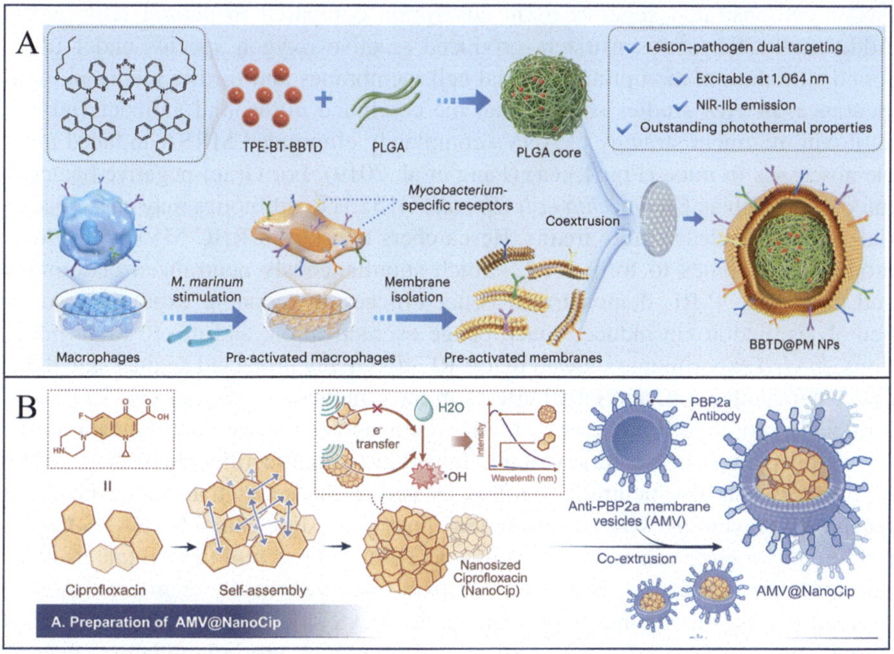

Fig. 13.5 EMMVs application in anti-infective drug delivery. (**a**) Pathogenic bacteria pre-stimulated macrophage MVs for targeted delivery of antimicrobial photothermal agents. (Reproduced with permission from Li et al. 2024a, copyright 2024, Springer Nature). (**b**) Schematic diagram of the preparation and encapsulation of nano-antibiotics by MVs designed by genetic engineering methods for PBP2a antibody presentation. (Reproduced with permission from Ding et al. 2024, copyright 2024, Wiley-VCH GmbH)

13.2.4.2 Neutralizers

EMMVs can also function as neutralizers, effectively capturing and neutralizing bacterial toxins, viruses, and inflammatory factors. Compared to traditional antibodies or chemical neutralizers, EMMVs offer multi-target neutralization while avoiding immune rejection, making them promising therapeutic agents for infectious diseases (Thamphiwatana et al. 2017).

EMMVs effectively neutralize bacterial toxins, including exotoxins and endotoxins. Gram-positive microorganisms secrete exotoxins, which are highly toxic even in small amounts, while Gram-negative bacteria release endotoxins like LPS during bacterial division or antibiotic treatment, which can lead to sepsis and severe inflammatory responses (Thamphiwatana et al. 2017; Henderson et al. 1997; Ryu et al. 2017; Beutler and Rietschel 2003). While antibiotics can kill bacteria, they cannot clear toxins and may even promote the release of more toxins. Therefore, strategies combining antibodies to neutralize bacterial toxins hold great promise in antibacterial therapy. Researchers have used genetic engineering to display mAbs against MRSA-secreted α-toxin on cell membranes, extracting these functionalized antibody MVs and encapsulating sonosensitizers within them (named ANVs).

ANVs demonstrated superior toxin adsorption compared to natural RBC MVs. Additionally, the sonosensitizers produced reactive oxygen species under ultrasound stimulation, disrupting bacterial cell membranes and further aiding in toxin clearance. *In vivo* studies showed that the combined ultrasound antibacterial and antitoxin treatment strategy of ANVs completely eliminated MRSA-induced muscle abscesses in mice (Fig. 13.6a) (Pang et al. 2019). For Gram-negative bacterial infections, such as *Escherichia coli* (*E. coli*) O157:H7, antibiotics may cause bacterial death and release more toxins. Researchers hybridized RBC MVs with PMB-modified liposomes to form P-RL, which simultaneously neutralized endotoxins and exotoxins. P-RL demonstrated better efficacy in reducing exotoxin-induced hemolysis, endotoxin-induced macrophage overactivation, and mixed toxin infections. *In vivo* experiments showed that P-RL effectively protected against skin damage, enteric infection, and lethal effects in mice induced by *E. coli* O157:H7. The artificially modified lipid layer of P-RL significantly improved detoxification efficiency, providing a new solution for treating multi-toxin infections (Jiang et al. 2021).

EMMVs can also neutralize viruses by acting as decoys that mimic host cell receptors, preventing viral invasion into host cells. For example, SARS-CoV-2 binds to the angiotensin-converting enzyme 2 (ACE2) receptor to facilitate its entry into host cell (Yang et al. 2021). Researchers fused genetically engineered ACE2-overexpressing membranes with human monocyte membranes to create nanodecoys that neutralized SARS-CoV-2 and inhibited pro-inflammatory factors, reducing immune dysfunction and lung damage in a mouse model of acute pneumonia (Rao et al. 2020a). Hepatitis B virus (HBV) drug therapies face significant obstacles, including viral heterogeneity and resistance, while prolonged medication use may lead to complications and reduced patient compliance (Iannacone and Guidotti 2022). Researchers have constructed MVs expressing HBV-specific receptor hNTCP (named hNTCP-MVs) using genetic engineering. These hNTCP-MVs retained the biological function of hNTCP and, when used as neutralizers for HBV, effectively resisted HBV infection in a humanized liver chimeric mouse model (Fig. 13.6b) (Liu et al. 2018).

EMMVs can neutralize inflammatory factors. Current treatments for inflammation consist of non-steroidal anti-inflammatory medications, corticosteroids-based treatments, disease-modifying antirheumatic medications, and mAbs that target and neutralize cytokines (Bacchi et al. 2012; Scheen 2022). However, their high cost, resistance, and inability to simultaneously neutralize multiple cytokines limit their application (Chen et al. 2023). In contrast, EMMVs, by retaining cytokine receptor characteristics, can effectively neutralize multiple inflammatory factors as neutralizers, reducing inflammatory responses. For example, neutrophil MVs coated nanoparticles can neutralize various inflammatory factors and repair joint damage caused by rheumatoid arthritis (RA) (Song et al. 2023). To improve the stability of neutrophil MVs, researchers integrated them with synthetic liposomes, thereby developing biomimetic liposomes designed for the treatment of myocardial infarction (MI). These liposomes inherited the inflammatory factor and chemokine receptors of source cells, neutralizing multiple chemokines (Chen et al. 2022). Studies have shown that genetically modified TRAIL cell membranes can target

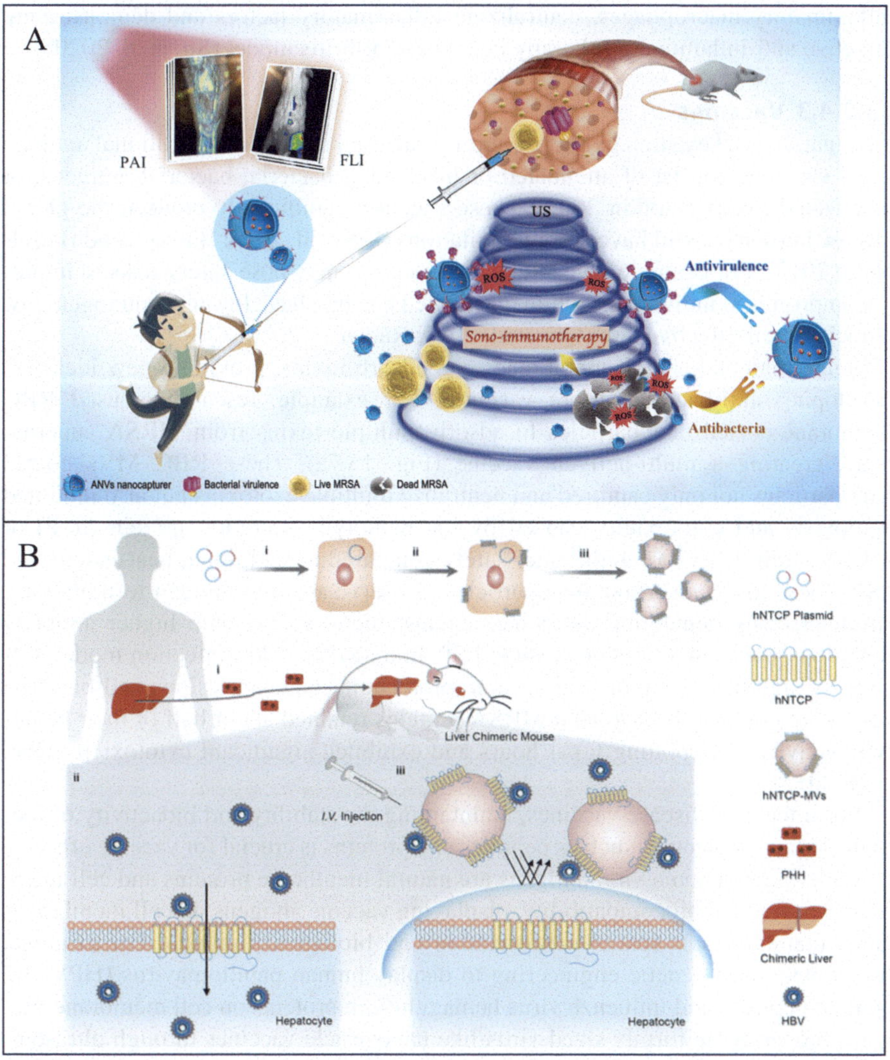

Fig. 13.6 EMMVs application as neutralizing agents for the treatment of infectious diseases. (**a**) Schematic illustration of the function of antibody-engineered MVs used to neutralize the α-toxin of MRSA in conjunction with the antimicrobial agent of the sonosensitizers. (Reproduced with permission from Pang et al. 2019, copyright 2019, Wiley-VCH GmbH). (**b**) Schematic representation of the construction of MVs expressing the hNTCP ligand on the surface and their use as a neutralizing agent for the treatment of HBV. (Reproduced with permission from Liu et al. 2018, copyright 2018, Wiley-VCH GmbH)

inflammatory macrophages, neutralizing inflammatory factors and demonstrating effective anti-inflammatory therapy in a mouse arthritis model (Shi et al. 2020b).

13.2.4.3 Vaccines

Vaccination is a key strategy for preventing bacterial infections. Traditional antibacterial vaccines consist of attenuated or inactivated bacteria, bacterial antigens, or inactivated toxins (Plotkin 2003). These vaccines significantly reduced the global disease burden but still have certain limitations (Lin et al. 2015; Barbosa and Barral-Netto 2013). For example, live attenuated vaccines may pose safety risks to immunocompromised individuals, while inactivated bacteria have lower immunogenicity, reducing their effectiveness (Qin et al. 2020; Rao et al. 2020b).

Some mammalian cell membranes can adsorb toxins, providing new ideas for developing anti-bacterial toxin vaccines. For example, researchers used RBC membrane-coated nanoparticles to adsorb multiple toxins from MRSA supernatants, creating a multi-antigen vaccine (Fig. 13.7a). These RBC MVs coated nanoparticles not only captured and neutralize multiple exotoxins but also inhibited hemolysis and cytotoxicity caused by the hemolytic secretion protein (hSP) of MRSA strain USA300, with better safety compared to traditional heat-inactivated hSP. These toxin-adsorbing nanoparticles can serve as multi-antigen formulations, simultaneously generating antibodies against these toxins, with higher antibody titers than traditional heat-inactivated hSP. In an MRSA skin infection model, this strategy significantly improved vaccine protection, reducing lesions and bacterial counts. In contrast, heat-treated MRSA proteins retained about half of their hemolytic activity after boiling for 4 hours and exhibited significant cytotoxicity (Wei et al. 2017).

For infectious disease vaccines, maintaining the stability and bioactivity of biological macromolecules such as peptides and proteins is crucial for vaccine efficacy. Considering that some viral antigens are natural membrane proteins and cell membranes have good biocompatibility, displaying vaccine antigens on cell membranes helps maintain their spatial conformation and biological function. For example, researchers used genetic engineering to display human papillomavirus (HPV) L2 epitope peptides and influenza virus hemagglutinin proteins on cell membrane vesicles, preparing uniformly sized virus-like nanovesicle vaccines through ultrasonication and surfactant treatment. This vaccine demonstrated good immunogenicity in mice, proving that this method is suitable for delivering both short-chain peptides and complex proteins, with broad application potential (Fig. 13.7b) (Zhang et al. 2015).

13.3 Bacterial Membranes

13.3.1 Types of BMVs

BMVs are derived from the bacterial cell membrane and can be categorized into three types: OMVs, inner membrane vesicles (IMVs), and double-membrane

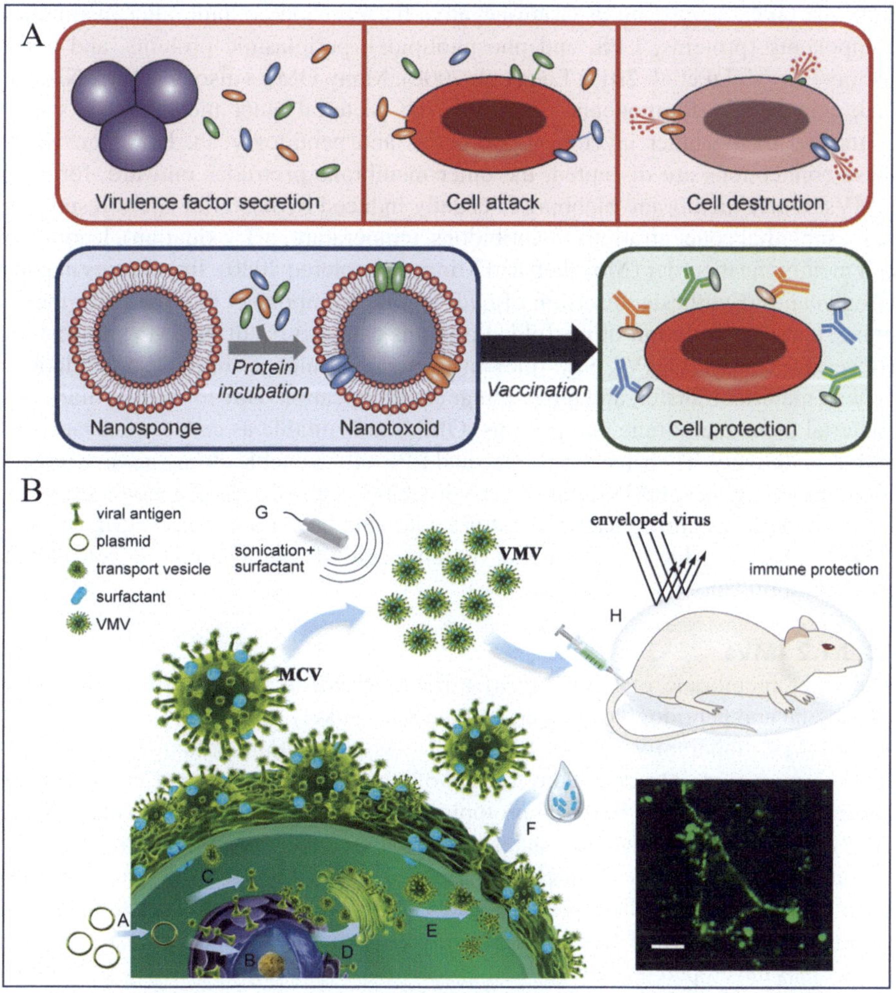

Fig. 13.7 EMMVs as vaccines against infectious diseases. (**a**) Schematic of RBC MVs adsorbed with multiple toxins coated on polymer nanoparticles for anti-bacterial toxin vaccines. (Reproduced with permission from Wei et al. 2017, copyright 2017, Wiley-VCH GmbH). (**b**) Genetically engineered virus-mimicking MVs for antiviral vaccines. (Reproduced with permission from Zhang et al. 2015, copyright 2015, Proc Natl Acad Sci U S A)

vesicles (DMVs) (Gao et al. 2022). Their formation and structure depend on the type of bacteria and preparation methods.

13.3.1.1 OMVs

OMVs are the most common and extensively studied type of bacterial vesicles, characterized by a lipid bilayer structure with diameters of 20–400 nm (Gao et al. 2020; Gerritzen et al. 2017; Hu et al. 2020). OMVs are spontaneously originate from the outer membrane of Gram-negative bacteria and contain outer membrane components (proteins, LPS, and phospholipids), periplasmic proteins, and toxic factors (Toyofuku et al. 2019; Lee et al. 2008). Many OMVs also contain DNA but lack cytoplasmic components. Generally, the bacterial outer membrane is stably connected to the inner layer by lipoproteins and peptidoglycan. However, when these connections are disrupted, the outer membrane protrudes outward, forming OMVs. Outer membrane blebbing is usually induced by bacterial stress responses (e.g., sublethal concentrations of antibiotics, temperature, pH, oxidation), leading to cell membrane disorder (Mozaheb and Mingeot-Leclercq 2020). Imbalances in peptidoglycan biosynthesis, insertion of hydrophobic compounds into the outer membrane, and accumulation of misfolded proteins can also result in OMVs formation (Beveridge 1999). OMVs play roles in bacterial virulence, immune modulation, genetic material transfer, and intercellular communication. Due to their similarity to bacterial outer membrane components, OMVs are suitable as carriers for vaccines and drug delivery. Their nanoscale size and EPR effect enable strong passive targeting to tumor tissues. OMVs can directly load hydrophilic or hydrophobic drugs or incorporate drugs during bacterial culture, allowing bacteria to secrete drug-loaded OMVs. However, due to their higher toxicity, detoxification should be considered for their application.

13.3.1.2 IMVs

The inner membrane of Gram-negative microorganisms is shielded by the outer membrane and peptidoglycan layer, making it less likely to naturally form vesicles. Since OMVs inherently contain LPS, they may cause immunotoxicity in clinical settings. Kim et al. introduced the concept of generating IMVs from Gram-negative microorganisms, termed bacterial protoplast-derived nanovesicles (Kim et al. 2015, 2017). Protoplasts are bacterial states without toxic outer membranes and peptidoglycan layers (cell walls), which are removed by lysozyme. After removing the outer membrane, protoplasts produce IMVs. IMVs were efficiently generated using continuous extrusion and developed into universal adjuvant-free vaccines. Compared to vaccines developed using OMVs, IMVs demonstrated higher efficacy and safety, inducing robust antigen-specific humoral and cellular immunological responses.

13.3.1.3 DMVs

Both OMVs and IMVs are structurally composed of a single lipid membrane, which may inherently constrain their stability and limit the capacity for efficient cargo loading (Kuerban et al. 2020; Zhang et al. 2019). DMVs, however, possess a double membrane structure, providing enhanced stability and loading capacity (Wang et al. 2018b). The outer membrane of DMVs originates from the bacterial outer membrane, while the inner membrane comes from the bacterial inner membrane, with peptidoglycan present between the two layers. DMVs formation involves both vesiculation and explosive cell lysis mechanisms. In theory, damaged Gram-negative bacterial membranes can be reassembled into new vesicles using appropriate methods, such as enzymatic digestion, high-pressure homogenization, ultrasonication, or nitrogen cavitation. Since DMVs release intracellular proteins and nucleic acids during their formation, they exhibit lower biotoxicity than OMVs while still containing many antigens required for vaccine development, retaining their immunogenic and adjuvant properties (Wang et al. 2018b; Li et al. 2021a).

13.3.2 Purpose of EBMVs

BMVs, due to their good biocompatibility and immunogenicity, have broad application prospects in drug delivery and vaccine development (Parikh et al. 2016; Haque et al. 2021; Li et al. 2021b). However, their low yield, poor tissue targeting, and biological toxicity limit their application (Yuan et al. 2018; Wang et al. 2020a; Cao and Liu 2020). Advances in bioengineering and nanotechnology enable the construction of EBMVs to address these limitations. The following sections discuss various methods and purposes for engineering BMVs.

13.3.2.1 Detoxification Engineering

Since BMVs contain antigens expressed on bacterial membranes, they can easily trigger immune responses when used as drug carriers. Therefore, reducing toxicity is a key focus of BMVs engineering. The main toxins in BMVs are LPS and toxic proteins, located on the membrane or within the vesicle. These virulence factors in BMVs can be reduced by engineering modification to improve their biosafety.

LPS, a major component of the Gram-negative bacterial outer membrane, consists of lipid A, core oligosaccharides, and O-antigen (Wang and Quinn 2010; Manissorn et al. 2020). When LPS binds to the TLR4, CD14, and MD2 receptor complex, it activates the MyD88-dependent pathway, triggering NF-κB and MAPK signaling pathways, leading to inflammation. This response can cause severe diseases such as sepsis (Dauphinee and Karsan 2006). Therefore, removing LPS from BMVs is crucial for their development as drug carriers (Needham and Trent 2013). One strategy to reduce LPS in BMVs is to control LPS expression. Studies have shown that acylation modifications of lipid A can reduce the inflammatory response induced by BMVs. Since lipid A synthesis requires enzymes, downregulating these enzymes, such as knocking out the *lpxL* and *lpxM* genes, has been shown to effectively reduce toxicity (Wang and Quinn 2010; Lee et al. 2011; Anisimov et al. 2007). Additionally, modifying the phosphorylation level of lipid A can reduce LPS toxicity, with monophosphoryl lipid A (MPL) having lower immunogenicity and FDA approval as an adjuvant (Berezow et al. 2009; Wang et al. 2020b). Early OMVs vaccines used detergent extraction to remove endotoxins, effectively preventing outbreaks of *Neisseria meningitidis* serogroup B (Matthias et al. 2020). During OMVs production, non-ionic detergents and chelating agents can reduce LPS content. Other virulence factors, such as collagenase, hyaluronidase, serine protease, and hemolysin, can also disrupt host tissues and affect BMVs safety. To reduce BMVs toxicity, selecting non-toxic or low-toxicity strains or using genetic engineering to remove toxic membrane proteins is an effective strategy.

13.3.2.2 Improving BMVs Production

BMVs production is influenced by bacterial species, growth conditions, and environmental factors, often resulting in low yields of 100–1000 μg/L, which is insufficient for clinical applications (Qiao et al. 2021). To overcome this, researchers have developed strategies such as selecting specific bacterial strains, optimizing culture conditions, and employing genetic modifications to enhance BMVs production.

Different strains produce varying amounts of OMVs. For example, *E. coli* (Mug-Opstelten and Witholt 1978; Katsui et al. 1982), *Salmonella typhimurium* (Deatherage et al. 2009; Alaniz et al. 2007), *Borrelia burgdorferi* (Charon et al. 2009), and *Campylobacter jejuni* produce more OMVs (Bager et al. 2013), while *Neisseria meningitidis* and *Francisella tularensis* produce more OMVs in later culture stages (Santos et al. 2012; Mccaig et al. 2013). *Legionella pneumophila* generates OMVs throughout its growth cycle, whereas *Helicobacter pylori* releases OMVs inversely proportional to cell growth (Galka et al. 2008; Olofsson et al. 2010). OMVs production is highly sensitive to culture conditions, with factors such as temperature (Katsui et al. 1982), amino acids (Van De Waterbeemd et al. 2013), antibiotics (Kadurugamuwa and Beveridge 1995), and bacteriophage infections affecting their release (Loeb and Kilner 1978). Studies have shown that cysteine depletion induces oxidative stress, leading to increased OMV secretion by *Neisseria meningitidis* (Van De Waterbeemd et al. 2013). Additionally, adding chemical substances to the culture medium, such as hexadecane in *Acinetobacter calcoaceticus* (Borneleit et al. 1988), glucose in *Yersinia pestis* (Kolodziejek et al. 2013), or sodium carbonate in *Vibrio cholerae* (Altindis et al. 2014), can promote OMVs production. To further increase OMVs yield, researchers use detergents or ethylenediaminetetraacetic acid to extract OMVs from bacterial membranes (Gerritzen et al. 2019a). Despite yield limitations due to culture conditions, bioreactor fermentation offers an effective solution. The sustained production of *Neisseria meningitidis* OMVs for over 600 hours, yielding 4.0×10^{14} OMVs per liter daily, demonstrates the feasibility of large-scale production for medical applications (Xia et al. 2018; Gerritzen et al. 2019b).

Genetic engineering provides greater flexibility, significantly increasing BMVs yield by screening bacterial mutants. Compared to environmental stimuli, genetic regulation of bacterial membrane stress pathways is safer. For example, knocking out stress-related genes *degS* and *degP* in *E. coli* increased OMVs yield by 100-fold (Asano et al. 2021). Additionally, manipulating membrane-related genes, such as enhancing cell membrane fluidity, reducing peptidoglycan cross-linking, and increasing outer membrane components and periplasmic protein accumulation, can promote BMVs secretion. Increasing membrane fluidity can induce membrane curvature and vesicle formation, enhancing OMVs secretion. For instance, overexpressing the *PSMα* gene significantly increased yield (Wang et al. 2018c). Knocking out the *Lgt* gene reduced lipoprotein content, enhancing membrane fluidity (Xue et al. 2022). Key targets for bacterial membrane modification include the Tol-Pal system, *Lpp*, and *OmpA*, with mutations in any of these genes leading to outer membrane defects and increased OMVs secretion (Avila-Calderón et al. 2021). Deleting

OmpA or *Lpp* genes destabilizes the outer membrane and reduces cross-linking, further increasing OMVs production (Juodeikis and Carding 2022). Additionally, *Fts* genes, *Bam* genes, *lpo* genes, and peptidoglycan hydrolases affect peptidoglycan synthesis and outer membrane stability, with gene knockouts or overexpression significantly increasing OMVs yield (Li et al. 2024b; Garde et al. 2021).

13.3.2.3 Enhancing Targeting Capability

Clinical trials have demonstrated that using attenuated bacterial strains is safe, but their anti-tumor efficacy remains limited due to insufficient active targeting capabilities. Thus, enhancing the targeting ability of BMVs is a critical area of research.

Genetic engineering can be used to develop targeted BMVs. For instance, Kim et al. fused epidermal growth factor (EGF) with the inner membrane protein PrsA to produce EGF-modified IMVs. As A549 lung cancer cells express EGF receptors (EGFR), these IMVs were able to effectively target tumors. Both *in vitro* and *in vivo* studies confirmed this hypothesis. RGD peptide motifs, which are small and exhibit tissue-specific affinity, have also been employed as novel targeting molecules (Kim et al. 2015). Gao et al. fused RGD peptides with bacterial ClyA protein and used nitrogen cavitation to prepare RGD-expressing DMVs (Gao et al. 2021). Their study demonstrated that RGD-expressing DMVs significantly enhanced tumor accumulation by simultaneously targeting immune cells, tumor blood vessels, and cancer cells. Similarly, Chen et al. functionalized RGD peptides to the outer surface of OMVs to enhance OMVs targeting potential, resulting in better accumulation of RGD-modified OMVs in tumor regions of a mouse breast xenograft model (Chen et al. 2019).

13.3.3 Application of EBMVs in Cancer Therapy

BMVs are abundant in microbial-associated molecular patterns, which confer high immunogenicity, and their nanovesicle structure provides excellent drug-loading capacity. EBMVs possess additional functionalities, offering broad prospects for cancer therapy.

13.3.3.1 Drug Delivery

Chemotherapy is one of the primary methods for treating tumors; however, the non-targeted distribution of chemotherapeutic drugs reduces their efficacy and causes severe side effects (Aghajanzadeh et al. 2022). The properties of some drugs, such as the high lipophilicity of paclitaxel or the instability of camptothecin, complicate their delivery via BMVs. To overcome these challenges, researchers are modifying OMVs to create safer, more stable, and efficient drug delivery systems.

Engineered OMVs offer a new platform for delivering potent cytotoxic drugs that may be unsuitable for traditional chemotherapy. For example, PNU-159682, a DNA topoisomerase I inhibitor, is 2000 times more potent than doxorubicin but has limited clinical application due to its high toxicity (Brahmbhatt and MacDiarmid 2022). Sagnella et al. introduced a system that integrates PNU-159682 into

Salmonella typhimurium OMVs, modified with epidermal growth factor receptor (EGFR) ligands for enhanced targeting. This system demonstrated significant antitumor activity without notable side effects. Engineered OMVs also exhibit higher drug-loading capacity, greatly improving delivery efficiency. Due to hydrogen bonding between PTX and the D-alanyl-D-alanine carboxypeptidase of bacterial OMVs, the binding affinity of PTX to OMVs is superior to that of liposomes. Under identical co-culture conditions, the packaging amount of DOX in OMVs is approximately 100 times greater than that in liposomes (Ali et al. 2020). Additionally, small molecules can be genetically engineered onto OMV surfaces to enhance their therapeutic potential. For instance, Polybia-mastoparan I, a peptide with selective toxicity toward tumor cells, was engineered onto OMVs, boosting antitumor immune responses (Ren et al. 2023).

13.3.3.2 Vaccines

An ideal cancer vaccine should contain tumor-targeted antigens, effective immune-stimulating molecules, and an efficient antigen delivery platform. Conventional materials such as liposomes and polymers are used for vaccine delivery, but these systems risk inducing autoimmune toxicity. BMVs, however, interact with the innate immune system through their PAMPs, activating antigen-presenting cells like DCs and macrophages, which process and present antigens to stimulate adaptive immune responses. Additionally, BMVs function both as carriers and adjuvants, and their immunogenicity can be enhanced through genetic modification.

Through genetic engineering, BMVs can directly produce vesicles carrying antigens, bind specific tumor antigens during growth, or express exogenous proteins within or on the surface of vesicles, thereby activating the immune system to generate specific immune responses. For example, Wang et al. used genetic recombination to express the HPV antigen E7 protein in *E. coli* OMVs, promoting DC cell uptake of tumor antigens and endosomal escape, significantly inhibiting the growth of xenograft tumors in mice (Hua et al. 2021). Adding exogenous proteins to OMVs can also maintain native antigen conformation and target specific immune responses. By fusing antigens with the hemolysin A (ClyA) protein, multiple tumor antigens can be displayed simultaneously on the OMVs surface. Cheng et al. linked SpT-Adpgk to the ClyA (CC) fusion protein on the OMVs surface, effectively alleviating the immunosuppressive microenvironment mediated by regulatory T cells and enhancing antitumor immune responses (Fig. 13.8a) (Cheng et al. 2021).

Furthermore, hybrid membrane nanovaccine carriers can be prepared by fusing bacterial cell membranes with primary tumor cell membranes. In this hybrid membrane system, BMVs act as immune adjuvants to activate innate immunity, while tumor cell membranes can transmit immune signals, serving as personalized cancer vaccines. For example, Li et al. developed an effective vesicle vaccine by fusing *E. coli* OMVs with melanoma cell membranes and coated onto PLGA nanoparticles (named BTs@PLGA). BTs@PLGA notably improved DC targeting, maturation, antigen presentation, and antigen-specific T cell responses, demonstrating excellent antitumor therapeutic effects. Additionally, vaccines prepared from BMVs can be

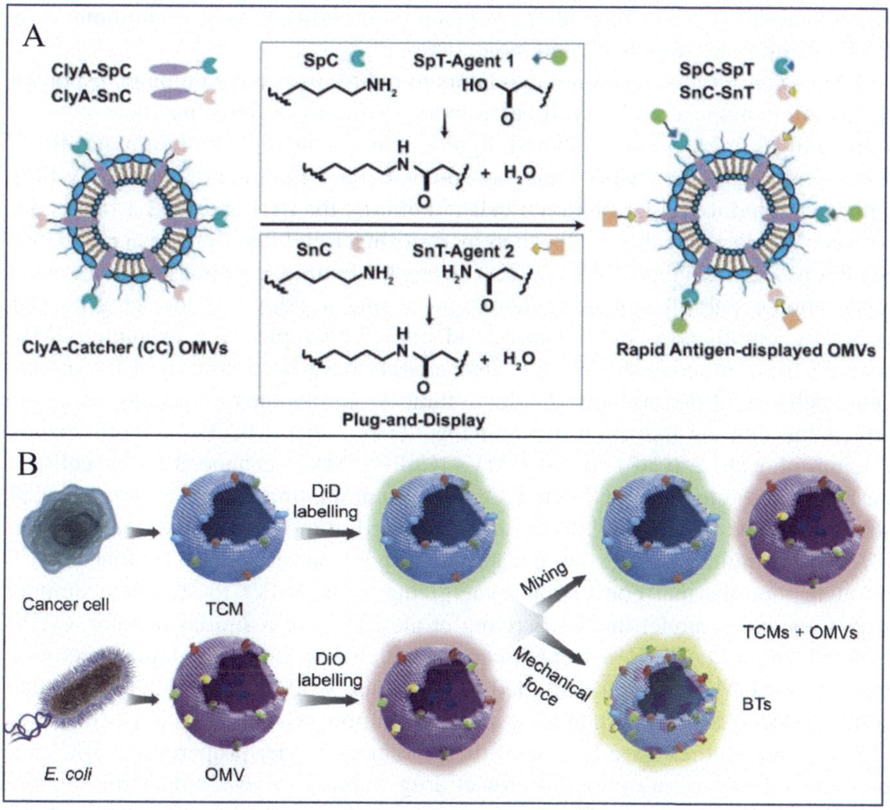

Fig. 13.8 EBMVs as multifunctional antigen-presenting platforms. (**a**) Schematic composition of OMVs expressing ClyA-Catcher (CC), SpyCatcher (SpC) and SnoopCatcher (SnC) conjugated on the surface of OMVs by fusion with cytolysin A (ClyA). (Reproduced with permission from Cheng et al. 2021, copyright 2021, Springer Nature). (**b**) Schematic of the process of hybridization of TCMs that present tumor antigens with OMVs that present PAMPs the process of OMV hybridization to form BTs is schematically illustrated. (Reproduced with permission from Li et al. 2020a, copyright 2020, Elsevier B.V.)

lyophilized and stored without a cold chain, extending shelf life and reducing vaccine application costs (Fig. 13.8b) (Li et al. 2020a).

13.3.3.3 Immunomodulation

Immunotherapy, including immune checkpoint blockade and chimeric antigen receptor T cell (CAR-T) therapy, has shown significant efficacy in certain cancers (Pardoll 2012; Fesnak et al. 2016). However, challenges such as low response rates, poor targeted delivery, and the immunosuppressive effects of the tumor microenvironment (TME) limit its broader clinical application. Factors like metabolic abnormalities, hypoxia, and low pH in the TME hinder therapeutic effectiveness (Gupta and Massagué 2006; Balkwill et al. 2012). BMVs have gained attention as immunomodulators in cancer therapy, given their ability to activate immune cells, modulate immune responses, and target tumors.

BMVs can serve as immunomodulators to induce protective immune responses. By functionalizing vesicles with heterologous proteins or antigens, their immune-stimulating properties can be altered, significantly enhancing tumor immune effects. To improve the efficiency of tumor checkpoint therapy, Li et al. prepared OMV-PD1, which can bind to PD-L1 on tumor cells, inhibiting the PD1 and PD-L1 interaction between T cells and cancer cells, thereby restoring T cell function (Li et al. 2020b). Studies have shown that OMVs possess strong host immune response activation and modulation capabilities, making them a promising strategy for enhancing the efficacy of chemotherapy, radiotherapy, and immunotherapies. For example, OMVs sourced from attenuated *Klebsiella pneumoniae* have been employed for the targeted delivery of doxorubicin (DOX) to tumors, significantly enhancing antitumor effects through the induction of immunogenic cell death (ICD). In another study, calcium phosphate (CaP)-coated OMVs (CaP@OMVs) enhanced tumor cell permeability and retention for folate receptors, accumulating at tumor sites. The CaP shell mitigated the acidic TME and polarized tumor-associated macrophages (TAMs). Simultaneously, the encapsulated photosensitizer ICG induced ICD through photothermal conversion, synergizing with OMVs to stimulate immune responses and remodel the TME (Qing et al. 2020). In a similar manner, OMVs isolated from *E. coli* DH5α were integrated with magnesium-based micromotors to form motor-OMVs, which disrupted tumor tissue while OMVs induced immune stimulation, significantly enhancing antitumor immunotherapy efficacy (Zhou et al. 2021). Some studies have leveraged the biological barrier properties of BMVs to promote substance transport and protect drug stability *in vivo*, enhancing immune stimulation. For example, engineered exosomes carrying catalase maintained stability and alleviated tumor hypoxia through intra tumoral injection. Exosomes also increased DC, NK cell, and macrophage infiltration, activated DCs, and polarized TAMs, improving the immunosuppressive TME and triggering long-term immune memory when combined with radiotherapy (Zai et al. 2021). Additionally, OMVs derived *E. coli* BL21 were engineered to encapsulate DNA damage response 1 targeting siRNA and incorporate mannose to target macrophage metabolism. Additionally, these vesicles were conjugated with chemotherapeutic drug PTX to OMVs through pH-sensitive cis-aconitic anhydride. Engineered OMVs, as immune stimulators, promoted DC maturation, increased tumor-infiltrating effector T cells, and reduced Tregs, thereby enhancing antitumor immune responses (Guo et al. 2021).

BMVs offer multiple advantages for cancer therapy, serving as natural nanocarriers with efficient drug-loading capacity and targeting ability for antigen-presenting cells in lymph nodes. Their inherent immune adjuvant properties make them ideal candidates for combination therapies, such as chemotherapy and phototherapy, to enhance tumor immunotherapy pathways. However, the presence of endotoxins in BMVs poses a biotoxicity challenge that must be addressed. Although LPS-related gene knockouts and attenuated bacterial strains have been explored to reduce toxicity, further investigation is needed to ensure the clinical safety of BMVs.

13.3.4 Applications of EBMVs in Infectious Diseases

13.3.4.1 Drug Delivery

Site-specific drug delivery systems are crucial for treating infectious diseases, as they allow precise control over the speed, timing, and location of drug release (Nazli et al. 2022). Traditional synthetic nanomaterials often struggle to interact with cells and penetrate the blood-brain barrier, leading to low delivery efficiency. In contrast, BMVs can circulate in environments of microbial disruption and inflammation, leveraging their natural affinity for biological molecules to passively or actively accumulate at infection sites, targeting pathogens and macrophages. Additionally, BMVs evade immune system recognition, resulting in high targeting efficiency.

BMVs show promise in treating intracellular bacterial infections, where bacteria survive and replicate within host cells, evading immune surveillance. Traditional antibiotics struggle to penetrate cell membranes and reach therapeutic concentrations, but BMVs can effectively enter host cells and clear intracellular bacteria. Encapsulating antibiotics in OMVs enhances nanoparticle uptake by infected cells, significantly reducing bacterial load. For example, Huang et al. demonstrated that *Acinetobacter baumannii* (*A. baumannii*) OMVs containing levofloxacin exhibited stronger antibacterial activity in mice with intestinal infections than free levofloxacin (Huang et al. 2019). Traditional nanoparticles have poor targeting ability, but BMV-encapsulated nanoparticles enhance targeting and improve antibacterial effect. Wu et al. developed a bionic nanocarrier system (Rif@MSN@OMV) comprising mesoporous silica nanoparticles (MSiNPs) loaded with rifampicin and covered with *E. coli* OMVs. This system enhanced the stability of the MSiNPs and prevented premature drug release, achieving precise targeting of *E. coli* infections (Wu et al. 2021). BMVs can also cross the blood-brain barrier, offering potential for treating central nervous system infections. For example, Lee et al. demonstrated that periodontal pathogen-derived exosomes could deliver RNA to brain immune cells, such as macrophages, leading to the induction of neuroinflammatory diseases (Ha et al. 2020). This property could be leveraged to treat conditions like meningitis. Liu et al. engineered non-inflammatory magnetic OMVs from *E. coli*, which were loaded with antibiotics and photosensitizers to synergistically treat bacterial meningitis and prevent recurrent inflammation (Shi et al. 2023).

BMVs, once isolated and purified, can deliver active molecules such as nucleic acids and proteins. They offer broad applications in drug delivery systems for various infections, including intestinal and brain infections. Additionally, BMVs provide biochemical stability, protecting biologically active molecules from degradation. However, their complex composition and unclear mechanisms of host cell endocytosis require further investigation. Future research should focus on addressing these key issues.

13.3.4.2 Vaccines

BMVs can function as vaccines or vaccine carriers for infectious diseases by stimulating MHC-I or MHC-II expression, promoting uptake by antigen-presenting cells, activating the innate immune system, and generating protective immune responses.

Their safety profile makes BMVs ideal candidates for infectious disease vaccines (Alaniz et al. 2007).

BMVs can be directly utilized as vaccines for combating homologous bacterial infections. For example, BMVs from *Streptococcus pneumoniae* (*S. pneumoniae*) and *Mycobacterium tuberculosis* contains abundant bacterial lipoproteins, which trigger humoral immunity and generate antibodies specific to these pathogens (Olaya-Abril et al. 2014; Prados-Rosales et al. 2014). Using *Salmonella* OMVs can induce specific antibody production and $CD4^+$ T cell responses, effectively protecting immunized mice from live *Salmonella* infection (Alaniz et al. 2007). Similarly, immunization with *S. aureus* BMVs can prevent *Staphylococcal* lung infections, eliciting antibody and Th1 cell-mediated T cell responses (Choi et al. 2015). Encapsulating synthetic nanoparticles with OMVs is also a strategy for designing effective antibacterial vaccines, significantly enhancing the uptake of nanoparticles by infected cells. Wei et al. demonstrated that subcutaneous administration of *E. coli* OMVs encapsulating gold nanoparticles produced more durable and potent immune responses. In a mouse model of *S. pneumoniae* infection, nasal immunization with *Salmonella* OMVs triggered strong protective responses without requiring mucosal adjuvants, significantly reducing bacterial loads (Guangzhang et al. 2023).

OMVs can also be genetically engineered to serve as carriers for heterologous antigens and proteins, thereby overcoming immune tolerance. For instance, a study utilized the ClyA-Omp22 fusion protein constructed from *E. coli* DH5α, resulting in engineered OMVs containing the *A. baumannii* Omp22 antigen. These OMVs induced a strong Omp22-specific humoral immune response, effectively protecting mice from lethal *A. baumannii* infections (Fig. 13.9a) (Huang et al. 2016). To mitigate the excessive inflammatory response induced by BMVs when used as vaccines, Jiang et al. designed biomimetic hybrid membrane vesicles (HMVs) composed of *Klebsiella pneumoniae* OMVs and alveolar macrophage MVs (Fig. 13.9b) (Jiang et al. 2024). Following intratracheal delivery, these HMVs rapidly mobilized immune cells in the lungs, promoting pathogen-targeted immunity and establishing local mucosal immunity. In an acute pneumonia mouse model, pulmonary immunization with HMVs provided significant protection, reducing bacterial colony counts in deep tissues, improving survival rates, and alleviating acute inflammatory symptoms (Jiang et al. 2024). HMVs preserved the immunogenic properties of OMVs while effectively mitigating the excessive immune response triggered by OMVs. Apart from bacterial infections, BMVs can also serve as platforms for antiviral vaccines. BMVs with reduced LPS content successfully protected mice from infections with H1N1, PR8, and H5N1 viruses, demonstrating potential in influenza virus prevention (Huang et al. 2016; Rappazzo et al. 2016; Price et al. 2016).

13.4 Challenges and Prospects

MVs derived from bacteria or mammalian cells retain the biological functions of their parent cell membranes. These vesicles can be easily engineered to enhance or introduce additional functionalities, improving their efficacy in treating cancer and

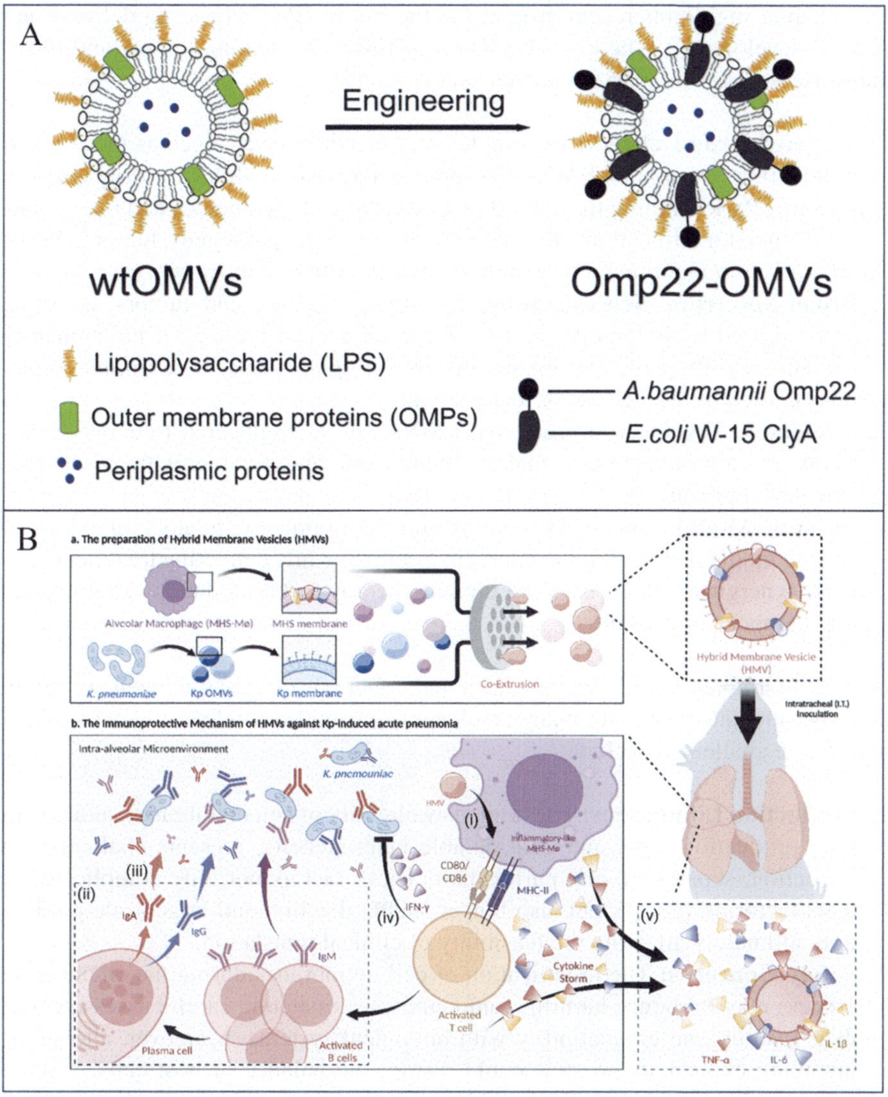

Fig. 13.9 EBMVs as vaccines against infectious diseases. (**a**) *E. coli* OMVs expressing *A. baumannii* antigens for antimicrobial vaccines. (Reproduced with permission from Huang et al. 2016, copyright 2016, Springer Nature). (**b**) Schematic of *K. pneumoniae* OMVs hybridizing with alveolar macrophage MVs to form HMVs for mucosal immunization against pneumonia. (Reproduced with permission from Jiang et al. 2024, copyright 2024, Elsevier B.V.)

infectious diseases. Engineering modifications can boost their biosafety, drug delivery targeting, vaccine immunogenicity, and immune modulation at lesion sites. Moreover, combining EMVs with organic or inorganic nanomaterials can offer complementary advantages, enhancing biocompatibility and therapeutic efficiency.

This chapter highlights recent progress in the use of EMVs for drug delivery and vaccine development. The key advantages of EMVs in treating tumors and infectious diseases are summarized as follows:

1. **Targeted Drug Delivery:** In complex microenvironments, such as those found in infections or tumors, EMVs can serve as targeted modules for drug delivery systems. By leveraging the natural properties of cell membranes and incorporating artificial modifications, they can efficiently target pathogens, tumor cells, or inflammatory environments, thereby increasing drug bioavailability.
2. **Broad-Spectrum Neutralization:** Infectious diseases and tumors are often accompanied by inflammation. EMVs can counteract a range of inflammatory mediators, pathogens, and toxins, thereby mitigating excessive immune reactions caused by the disease or its treatment.
3. **Modifiable Vaccine Immunogens:** EMVs can be engineered to enhance biosafety and immunogenicity, making them a safe and durable platform for vaccine development.
4. **Immune Modulation:** EMVs can exploit the immune regulatory functions of cells to modulate the immune microenvironment while delivering treatment, creating synergistic effects between immunotherapy and drug therapy and improving therapeutic outcomes.

Despite their potential, EMVs for cancer and infectious disease treatment are still in the experimental stage, with many challenges to overcome before clinical application. These challenges include:

1. **Production Limitations:** The limited availability of cell membranes, along with high production costs and considerable heterogeneity, presents challenges in extraction, separation, and purification. These factors not only complicate the manufacturing process but also hinder standardization and large-scale production, ultimately affecting the feasibility of clinical translation.
2. **Safety Concerns:** The modification of cell membranes during the engineering process may introduce harmful compounds, necessitating careful assessment of their toxicity and compatibility with biological systems. Moreover, alterations involving exogenous proteins could disrupt the balance of host cells, making safety evaluations more complex.
3. **Personalized Application Challenges:** Personalized treatments tailored to patient conditions could improve efficacy, with patient-derived cells being ideal. However, sourcing specific cell types from patients can be challenging, and expanding them *in vitro* to generate sufficient quantities for therapeutic use introduces further complications.
4. **Regulatory and Management Issues:** As a novel therapeutic approach, EMVs require the development of specific regulatory frameworks. Traditional regulatory systems designed for conventional pharmaceutical products may not be suitable for the clinical translation of EMVs due to their unique characteristics.

Despite these challenges, the application of EMVs is feasible. For example, CAR-T and RBC therapies (RCT) have set precedents for the clinical translation of EMVs. CAR-T therapy involves genetically engineering human T cells *in vitro* and reintroducing them into patients, with six FDA-approved therapies showing significant efficacy in treating hematologic cancers (Pagel and West 2017). Similarly, RBC-based strategies have led to the creation of unique platforms and clinical candidates (Mitra et al. 2023). For example, WTX-212, designed using third-generation RBC modification technology, efficiently delivers PD-1 antibodies to the spleen and peripheral tissues, stimulating immune effector cells to eliminate tumors and potentially overcoming immune resistance (Xiaoqian et al. 2023). WTX-212 is currently in clinical research stages. The design of CAR-T and RCT therapies shares similarities with EMVs, making them potential templates for personalized medical solutions using EMVs. Furthermore, a new method that electroporates cancer cells in liquid nitrogen for tumor microenvironment targeting and immune modulation has been developed. This simple process removes the tumorigenicity of cancer cells while maintaining their structural integrity and homing ability to tumor sites, offering a promising approach for large-scale clinical applications. Continued technological innovation and theoretical exploration are necessary to advance the clinical use of EMVs (Ci et al. 2020; Kuang et al. 2023).

13.5 Conclusion

In summary, as an advanced technology, EMVs enhance their therapeutic functions through specific engineering of cell membranes. This chapter discusses the application progress of EMVs in the treatment of cancer and infectious diseases. Due to their potential in targeted drug delivery, neutralizing therapy, vaccines, and immune regulation, EMVs have the potential to become a new strategy for treating cancer and infectious diseases. Despite existing challenges in production, quality control, personalized applications, and regulation, continuous advancements in research will help address these obstacles and facilitate the successful clinical translation of EMVs.

Acknowledgements Jiaxin Ma and Yijia Xie contributed equally to this work. This work was supported by the Major State Basic Research Development Program of China (No. 2017YFA0205201), the National Natural Science Foundation of China (Nos. 81925019, 82304366 and U1705281), the Fundamental Research Funds for the Central Universities (Nos. 20720190088 and 20720200019), the Science Foundation of Fujian Province (Nos. 2020Y4003 and 201812631010), the Program for New Century Excellent Talents in University, China (No. NCET-13-0502).

References

Aghajanzadeh M, Zamani M, Rajabi Kouchi F et al (2022) Synergic antitumor effect of photodynamic therapy and chemotherapy mediated by nano drug delivery systems. Pharmaceutics 14(2):322

Alaniz RC, Deatherage BL, Lara JC et al (2007) Membrane vesicles are immunogenic facsimiles of salmonella typhimurium that potently activate dendritic cells, prime B and T cell responses, and stimulate protective immunity in vivo. J Immunol 179(11):7692–7701

Ali MK, Liu Q, Liang K et al (2020) Bacteria-derived minicells for cancer therapy. Cancer Lett 491:11–21

Altindis E, Fu Y, Mekalanos JJ (2014) Proteomic analysis of vibrio cholerae outer membrane vesicles. Proc Natl Acad Sci USA 111(15):E1548–E1556

Anisimov AP, Shaikhutdinova RZ et al (2007) Effect of deletion of the lpxM gene on virulence and vaccine potential of Yersinia pestis in mice. J Med Microbiol 56(4):443–453

Asano Y, Onishi M, Nishi K et al (2021) Enhancement of membrane vesicle production by disrupting the degP gene in Meiothermus ruber H328. AMB Express 11(1):170

Avila-Calderón ED, Ruiz-Palma MDS, Aguilera-Arreola MG et al (2021) Outer membrane vesicles of gram-negative bacteria: an outlook on biogenesis. Front Microbiol 12:557902

Aytar Çelik P, Derkuş B, Erdoğan K et al (2022) Bacterial membrane vesicle functions, laboratory methods, and applications. Biotechnol Adv 54:107869

Azzopardi EA, Ferguson EL, Thomas DW (2013) The enhanced permeability retention effect: a new paradigm for drug targeting in infection. J Antimicrob Chemother 68(2):257–274

Bacchi S, Palumbo P, Sponta A et al (2012) Clinical pharmacology of non-steroidal anti-inflammatory drugs: a review. Antiinflamm Antiallergy Agents Med Chem 11(1):52–64

Bager RJ, Persson G, Nesta B et al (2013) Outer membrane vesicles reflect environmental cues in Gallibacterium anatis. Vet Microbiol 167(3):565–572

Balkwill FR, Capasso M, Hagemann T (2012) The tumor microenvironment at a glance. J Cell Sci 125(23):5591–5596

Barbosa T, Barral-Netto M (2013) Challenges in the research and development of new human vaccines. Braz J Med Biol Res 46(2):103–108

Behrens F, Funk-Hilsdorf TC, Kuebler WM et al (2021) Bacterial membrane vesicles in pneumonia: from mediators of virulence to innovative vaccine candidates. Int J Mol Sci 22(8)

Berezow AB, Ernst RK, Coats SR et al (2009) The structurally similar, penta-acylated lipopolysaccharides of Porphyromonas gingivalis and Bacteroides elicit strikingly different innate immune responses. Microb Pathog 47(2):68–77

Beutler B, Rietschel ET (2003) Innate immune sensing and its roots: the story of endotoxin. Nat Rev Immunol 3(2):169–176

Beveridge TJ (1999) Structures of gram-negative cell walls and their derived membrane vesicles. J Bacteriol 181(16):4725–4733

Borneleit P, Hermsdorf T, Claus R et al (1988) Effect of hexadecane-induced vesiculation on the outer membrane of acinetobacter calcoaceticus. Microbiology 134(7):1983–1992

Brahmbhatt H, MacDiarmid JA (2022) Bacterial minicells to the rescue: cyto-immunotherapy for the treatment of late stage cancers with minimal to no toxicity. Microb Biotechnol 15(1):91–94

Cao Z, Liu J (2020) Bacteria and bacterial derivatives as drug carriers for cancer therapy. J Control Release 326:396–407

Cao H, Dan Z, He X et al (2016) Liposomes coated with isolated macrophage membrane can target lung metastasis of breast cancer. ACS Nano 10(8):7738–7748

Charon NW, Goldstein SF, Marko M et al (2009) The flat-ribbon configuration of the periplasmic flagella of Borrelia burgdorferi and its relationship to motility and morphology. J Bacteriol 191(2):600–607

Chen Q, Bai H, Wu W et al (2019) Bioengineering bacterial vesicle-coated polymeric nanomedicine for enhanced cancer immunotherapy and metastasis prevention. Nano Lett 20(1):11–21

Chen HY, Deng J, Wang Y et al (2020) Hybrid cell membrane-coated nanoparticles: a multifunctional biomimetic platform for cancer diagnosis and therapy. Acta Biomater 112:1–13

Chen J, Song Y, Wang Q et al (2022) Targeted neutrophil-mimetic liposomes promote cardiac repair by adsorbing proinflammatory cytokines and regulating the immune microenvironment. J Nanobiotechnol 20(1):218

Chen Y, Zhang G, Yang Y et al (2023) The treatment of inflammatory bowel disease with monoclonal antibodies in Asia. Biomed Pharmacother 157:114081

Cheng K, Zhao R, Li Y et al (2021) Bioengineered bacteria-derived outer membrane vesicles as a versatile antigen display platform for tumor vaccination via plug-and-display technology. Nat Commun 12(1):2041

Choi SJ, Kim MH, Jeon J et al (2015) Active immunization with extracellular vesicles derived from Staphylococcus aureus effectively protects against staphylococcal lung infections, mainly via Th1 cell-mediated immunity. PLoS One 10(9):e0136021

Ci T, Li H, Chen G et al (2020) Cryo-shocked cancer cells for targeted drug delivery and vaccination. Sci Adv 6(50)

Dauphinee SM, Karsan A (2006) Lipopolysaccharide signaling in endothelial cells. Lab Investig 86(1):9–22

De Lázaro I, Mooney DJ (2021) Obstacles and opportunities in a forward vision for cancer nanomedicine. Nat Mater 20(11):1469–1479

Deatherage BL, Lara JC, Bergsbaken T et al (2009) Biogenesis of bacterial membrane vesicles. Mol Microbiol 72(6):1395–1407

Dehaini D, Wei X, Fang RH et al (2017) Erythrocyte-platelet hybrid membrane coating for enhanced nanoparticle functionalization. Adv Mater 29(16)

Ding L, Liang X, Ma J, Liu X, Zhang Y, Long Q et al (2024) Sono-triggered biomimetically nanoantibiotics mediate precise sequential therapy of MRSA-induced lung infection. Adv Mater 36(46):2403612

Dong X, Zhang CY, Jin G et al (2019) Targeting of nanotherapeutics to infection sites for antimicrobial therapy. Adv Ther (Weinh) 2(11)

Duan L, Yang L, Jin J et al (2020) Micro/nano-bubble-assisted ultrasound to enhance the EPR effect and potential theranostic applications. Theranostics 10(2):462–483

Fesnak AD, June CH, Levine BL (2016) Engineered T cells: the promise and challenges of cancer immunotherapy. Nat Rev Cancer 16(9):566–581

Galka F, Wai SN, Kusch H et al (2008) Proteomic characterization of the whole secretome of legionella pneumophila and functional analysis of outer membrane vesicles. Infect Immun 76(5):1825–1836

Gao J, Dong X, Wang Z (2020) Generation, purification and engineering of extracellular vesicles and their biomedical applications. Methods 177:114–125

Gao J, Wang S, Dong X et al (2021) RGD-expressed bacterial membrane-derived nanovesicles enhance cancer therapy *via* multiple tumorous targeting. Theranostics 11(7):3301–3316

Gao J, Su Y, Wang Z (2022) Engineering bacterial membrane nanovesicles for improved therapies in infectious diseases and cancer. Adv Drug Deliv Rev 186:114340

Garde S, Chodisetti Pavan K, Reddy M (2021) Peptidoglycan: structure, synthesis, and regulation. EcoSal Plus 9(2)

Gerritzen MJH, Martens DE, Wijffels RH et al (2017) Bioengineering bacterial outer membrane vesicles as vaccine platform. Biotechnol Adv 35(5):565–574

Gerritzen MJH, Salverda MLM, Martens DE et al (2019a) Spontaneously released Neisseria meningitidis outer membrane vesicles as vaccine platform: production and purification. Vaccine 37(47):6978–6986

Gerritzen MJH, Stangowez L, Van De Waterbeemd B et al (2019b) Continuous production of Neisseria meningitidis outer membrane vesicles. Appl Microbiol Biotechnol 103(23):9401–9410

González FE, Gleisner A, Falcón-Beas F et al (2014) Tumor cell lysates as immunogenic sources for cancer vaccine design. Hum Vaccin Immunother 10(11):3261–3269

Guangzhang C, Fangfang F, Siqian D et al (2023) Outer membrane vesicles from Escherichia coli are efficiently internalized by macrophage cells and alter their inflammatory response. Microb Pathog 175:105965

Guo Q, Li X, Zhou W et al (2021) Sequentially triggered bacterial outer membrane vesicles for macrophage metabolism modulation and tumor metastasis suppression. ACS Nano 15(8):13826–13838

Gupta GP, Massagué J (2006) Cancer metastasis: building a framework. Cell 127(4):679–695

Ha JY, Choi SY, Lee JH et al (2020) Delivery of Periodontopathogenic extracellular vesicles to brain monocytes and microglial IL-6 promotion by RNA cargo. Front Mol Biosci 7:596366

Haque S, Swami P, Khan AS (2021) Typhi derived vaccines and a proposal for outer membrane vesicles (OMVs) as potential vaccine for typhoid fever. Microb Pathog 158:105082

Henderson B, Wilson M, Wren B (1997) Are bacterial exotoxins cytokine network regulators? Trends Microbiol 5(11):454–458

Hou X, Zeng H, Chi X et al (2021) Pathogen receptor membrane-coating facet structures boost nanomaterial immune escape and antibacterial performance. Nano Lett 21(23):9966–9975

Hou M, Wei Y, Zhao Z et al (2022) Immuno-engineered nanodecoys for the multi-target anti-inflammatory treatment of autoimmune diseases. Adv Mater 34(12):e2108817

Hu CM, Fang RH, Wang KC et al (2015) Nanoparticle biointerfacing by platelet membrane cloaking. Nature 526(7571):118–121

Hu R, Liu H, Wang M et al (2020) An OMV-based nanovaccine confers safety and protection against pathogenic Escherichia coli via both humoral and predominantly Th1 immune responses in poultry. Nano 10(11):2293

Hua L, Yang Z, Li W et al (2021) A novel immunomodulator delivery platform based on bacterial biomimetic vesicles for enhanced antitumor immunity. Adv Mater 33(43):e2103923

Huang W, Wang S, Yao Y et al (2016) Employing Escherichia coli-derived outer membrane vesicles as an antigen delivery platform elicits protective immunity against Acinetobacter baumannii infection. Sci Rep 6(1):37242

Huang W, Zhang Q, Li W et al (2019) Anti-outer membrane vesicle antibodies increase antibiotic sensitivity of pan-drug-resistant Acinetobacter baumannii. Front Microbiol 10:1379

Huang X, Tang T, Zhang G et al (2021) Identification of tumor antigens and immune subtypes of cholangiocarcinoma for mRNA vaccine development. Mol Cancer 20(1):50

Iannacone M, Guidotti LG (2022) Immunobiology and pathogenesis of hepatitis B virus infection. Nat Rev Immunol 22(1):19–32

Jiang Y, Krishnan N, Zhou J et al (2020) Engineered cell-membrane-coated nanoparticles directly present tumor antigens to promote anticancer immunity. Adv Mater 32(30):e2001808

Jiang L, Zhu Y, Luan P et al (2021) Bacteria-anchoring hybrid liposome capable of absorbing multiple toxins for Antivirulence therapy of Escherichia coli infection. ACS Nano 15(3):4173–4185

Jiang L, Ding L, Long Q et al (2024) Single-dose of hybrid eukaryotic-prokaryotic nanovesicles to drive a rapid and controllable immune response against Klebsiella pneumoniae induced nosocomial pneumonia. Nano Today 54:102135

Jones RB, Mueller S, Kumari S et al (2017) Antigen recognition-triggered drug delivery mediated by nanocapsule-functionalized cytotoxic T-cells. Biomaterials 117:44–53

Juodeikis R, Carding SR (2022) Outer membrane vesicles: biogenesis, functions, and issues. Microbiol Mol Biol Rev 86(4):e00032–e00022

Kadurugamuwa JL, Beveridge TJ (1995) Virulence factors are released from Pseudomonas aeruginosa in association with membrane vesicles during normal growth and exposure to gentamicin: a novel mechanism of enzyme secretion. J Bacteriol 177(14):3998–4008

Katsui N, Tsuchido T, Hiramatsu R et al (1982) Heat-induced blebbing and vesiculation of the outer membrane of Escherichia coli. J Bacteriol 151(3):1523–1531

Kawai T, Akira S (2011) Toll-like receptors and their crosstalk with other innate receptors in infection and immunity. Immunity 34(5):637–650

Kim OY, Choi SJ, Jang SC et al (2015) Bacterial protoplast-derived Nanovesicles as vaccine delivery system against bacterial infection. Nano Lett 15(1):266–274

Kim OY, Dinh NTH, Park HT et al (2017) Bacterial protoplast-derived nanovesicles for tumor targeted delivery of chemotherapeutics. Biomaterials 113:68–79

Kolodziejek AM, Caplan AB, Bohach GA et al (2013) Physiological levels of glucose induce membrane vesicle secretion and affect the lipid and protein composition of Yersinia pestis cell surfaces. Appl Environ Microbiol 79(14):4509–4514

Krishnamurthy S, Muthukumaran P, Jayakumar MKG et al (2019) Surface protein engineering increases the circulation time of a cell membrane-based nanotherapeutic. Nanomedicine 18:169–178

Krishnan N, Jiang Y, Zhou J et al (2023) A modular approach to enhancing cell membrane-coated nanoparticle functionality using genetic engineering. Nat Nanotechnol 19:345–353

Kuang G, Zhang Q, Yu Y et al (2023) Cryo-shocked cancer cell microgels for tumor postoperative combination immunotherapy and tissue regeneration. Bioact Mater 28:326–336

Kuerban K, Gao X, Zhang H et al (2020) Doxorubicin-loaded bacterial outer-membrane vesicles exert enhanced anti-tumor efficacy in non-small-cell lung cancer. Acta Pharm Sin B 10(8):1534–1548

Lankisch PG, Apte M, Banks PA (2015) Acute pancreatitis. Lancet 386(9988):85–96

Lee PJ, Papachristou GI (2019) New insights into acute pancreatitis. Nat Rev Gastroenterol Hepatol 16(8):479–496

Lee E-Y, Choi D-S, Kim K-P et al (2008) Proteomics in gram-negative bacterial outer membrane vesicles. Mass Spectrom Rev 27(6):535–555

Lee DH, Kim S-H, Kang W et al (2011) Adjuvant effect of bacterial outer membrane vesicles with penta-acylated lipopolysaccharide on antigen-specific T cell priming. Vaccine 29(46):8293–8301

Li LL, Xu JH, Qi GB et al (2014) Core-shell supramolecular gelatin nanoparticles for adaptive and "on-demand" antibiotic delivery. ACS Nano 8(5):4975–4983

Li M, Zhou H, Jiang W et al (2020a) Nanovaccines integrating endogenous antigens and pathogenic adjuvants elicit potent antitumor immunity. Nano Today 35:101007

Li Y, Zhao R, Cheng K et al (2020b) Bacterial outer membrane vesicles presenting programmed death 1 for improved cancer immunotherapy via immune activation and checkpoint inhibition. ACS Nano 14(12):16698–16711

Li W, Hu Y, Zhang Q et al (2021a) Development of drug-resistant Klebsiella pneumoniae vaccine via novel vesicle production technology. ACS Appl Mater Interfaces 13(28):32703–32715

Li P, Wang X, Sun X et al (2021b) Recombinant pseudomonas bionanoparticles induce protection against pneumonic Pseudomonas aeruginosa infection. Infect Immun 89(11):e0039621

Li B, Wang W, Zhao L et al (2024a) Photothermal therapy of tuberculosis using targeting pre-activated macrophage membrane-coated nanoparticles. Nat Nanotechnol 19:834–845

Li J, He Y, Xu X et al (2024b) Structural insights into peptidoglycan hydrolysis by the FtsEX system in Escherichia coli during cell division. eLife Sciences Publications, Ltd. https://doi.org/10.7554/elife.94336.1

Liang Y, Duan L, Lu J et al (2021) Engineering exosomes for targeted drug delivery. Theranostics 11(7):3183–3195

Lin IY, Van TT, Smooker PM (2015) Live-attenuated bacterial vectors: tools for vaccine and therapeutic agent delivery. Vaccines (Basel) 3(4):940–972

Liu X, Yuan L, Zhang L et al (2018) Bioinspired artificial nanodecoys for hepatitis B virus. Angew Chem Int Ed Engl 57(38):12499–12503

Liu X, Liu C, Zheng Z et al (2019a) Vesicular antibodies: a bioactive multifunctional combination platform for targeted therapeutic delivery and cancer immunotherapy. Adv Mater 31(17):e1808294

Liu Y, Luo J, Chen X et al (2019b) Cell membrane coating technology: a promising strategy for biomedical applications. Nanomicro Lett 11(1):100

Liu C, Liu X, Xiang X et al (2022) A nanovaccine for antigen self-presentation and immunosuppression reversal as a personalized cancer immunotherapy strategy. Nat Nanotechnol 17(5):531–540

Liu Y, Rao P, Qian H et al (2023) Regulatory fibroblast-like Synoviocytes cell membrane coated nanoparticles: a novel targeted therapy for rheumatoid arthritis. Adv Sci (Weinh) 10(4):e2204998

Loeb MR, Kilner J (1978) Release of a special fraction of the outer membrane from both growing and phage T4-infected Escherichia coli B. Biochim et Biophys Acta (BBA) Biomembr 514(1):117–127

Lutterotti A, Yousef S, Sputtek A et al (2013) Antigen-specific tolerance by autologous myelin peptide-coupled cells: a phase 1 trial in multiple sclerosis. Sci Transl Med 5(188):188ra75

Ma W, Zhu D, Li J et al (2020) Coating biomimetic nanoparticles with chimeric antigen receptor T cell-membrane provides high specificity for hepatocellular carcinoma photothermal therapy treatment. Theranostics 10(3):1281–1295

Ma J, Jiang L, Liu G (2022) Cell membrane-coated nanoparticles for the treatment of bacterial infection. Wiley Interdiscip Rev Nanomed Nanobiotechnol 14(5):e1825

Ma X, Kuang L, Yin Y et al (2023) Tumor-antigen activated dendritic cell membrane-coated biomimetic nanoparticles with orchestrating immune responses promote therapeutic efficacy against glioma. ACS Nano 17(3):2341–2355

Ma J, Ding L, Peng X et al (2024) Recent advances of engineered cell membrane-based nanotherapeutics to combat inflammatory diseases. Small 20(28):e2308646

Manissorn J, Sitthiyotha T, Montalban JRE et al (2020) Biochemical and structural investigation of GnnA in the lipopolysaccharide biosynthesis pathway of Acidithiobacillus ferrooxidans. ACS Chem Biol 15(12):3235–3243

Matthias KA, Reveille A, Connolly KL et al (2020) Deletion of major porins from meningococcal outer membrane vesicle vaccines enhances reactivity against heterologous serogroup B Neisseria meningitidis strains. Vaccine 38(10):2396–2405

Mccaig WD, Koller A, Thanassi DG (2013) Production of outer membrane vesicles and outer membrane tubes by Francisella novicida. J Bacteriol 195(6):1120–1132

Mitra A, Barua A, Huang L et al (2023) From bench to bedside: the history and progress of CAR T cell therapy. Front Immunol 14:1188049

Mozaheb N, Mingeot-Leclercq MP (2020) Membrane vesicle production as a bacterial defense against stress. Front Microbiol 11:600221

Mug-Opstelten D, Witholt B (1978) Preferential release of new outer membrane fragments by exponentially growing Escherichia coli. Biochim et Biophys Acta (BBA) Biomembr 508(2):287–295

Nazli A, He DL, Liao D et al (2022) Strategies and progresses for enhancing targeted antibiotic delivery. Adv Drug Deliv Rev 189:114502

Needham BD, Trent MS (2013) Fortifying the barrier: the impact of lipid a remodelling on bacterial pathogenesis. Nat Rev Microbiol 11(7):467–481

Olaya-Abril A, Prados-Rosales R, McConnell MJ et al (2014) Characterization of protective extracellular membrane-derived vesicles produced by Streptococcus pneumoniae. J Proteome 106:46–60

Olofsson A, Vallström A, Petzold K et al (2010) Biochemical and functional characterization of helicobacter pylori vesicles. Mol Microbiol 77(6):1539–1555

Ou Z, Zhong H, Zhang L et al (2020) Macrophage membrane-coated nanoparticles alleviate hepatic ischemia-reperfusion injury caused by Orthotopic liver transplantation by neutralizing endotoxin. Int J Nanomedicine 15:4125–4138

Pagel JM, West HJ (2017) Chimeric antigen receptor (CAR) T-cell therapy. JAMA Oncol 3(11):1595

Pang X, Liu X, Cheng Y et al (2019) Sono-immunotherapeutic Nanocapturer to combat multidrug-resistant bacterial infections. Adv Mater 31(35):e1902530

Pardoll DM (2012) The blockade of immune checkpoints in cancer immunotherapy. Nat Rev Cancer 12(4):252–264

Parikh SR, Andrews NJ, Beebeejaun K et al (2016) Effectiveness and impact of a reduced infant schedule of 4CMenB vaccine against group B meningococcal disease in England: a national observational cohort study. Lancet 388(10061):2775–2782

Park JH, Jiang Y, Zhou J et al (2021) Genetically engineered cell membrane-coated nanoparticles for targeted delivery of dexamethasone to inflamed lungs. Sci Adv 7(25)
Perez CR, De Palma M (2019) Engineering dendritic cell vaccines to improve cancer immunotherapy. Nat Commun 10(1):5408
Plotkin SA (2003) Vaccines, vaccination, and vaccinology. J Infect Dis 187(9):1349–1359
Prados-Rosales R, Carreño Leandro J, Batista-Gonzalez A et al (2014) Mycobacterial membrane vesicles administered systemically in mice induce a protective immune response to surface compartments of mycobacterium tuberculosis. MBio 5(5):e01921–e01914. https://doi.org/10.1128/mbio.01921-14
Price NL, Goyette-Desjardins G, Nothaft H et al (2016) Glycoengineered outer membrane vesicles: a novel platform for bacterial vaccines. Sci Rep 6(1):24931
Qiao L, Rao Y, Zhu K et al (2021) Engineered remolding and application of bacterial membrane vesicles. Front Microbiol 12:729369
Qin M, Du G, Sun X (2020) Biomimetic cell-derived nanocarriers for modulating immune responses. Biomater Sci 8(2):530–543
Qing S, Lyu C, Zhu L et al (2020) Biomineralized bacterial outer membrane vesicles potentiate safe and efficient tumor microenvironment reprogramming for anticancer therapy. Adv Mater 32(47):2002085
Quaglino E, Conti L, Cavallo F (2020) Breast cancer stem cell antigens as targets for immunotherapy. Semin Immunol 47:101386
Rao L, Xia S, Xu W et al (2020a) Decoy nanoparticles protect against COVID-19 by concurrently adsorbing viruses and inflammatory cytokines. Proc Natl Acad Sci USA 117(44):27141–27147
Rao L, Tian R, Chen X (2020b) Cell-membrane-mimicking Nanodecoys against infectious diseases. ACS Nano 14(3):2569–2574
Rappazzo CG, Watkins HC, Guarino CM et al (2016) Recombinant M2e outer membrane vesicle vaccines protect against lethal influenza a challenge in BALB/c mice. Vaccine 34(10):1252–1258
Ren C, Li Y, Cong Z et al (2023) Bioengineered bacterial outer membrane vesicles encapsulated Polybia-mastoparan I fusion peptide as a promising nanoplatform for bladder cancer immune-modulatory chemotherapy. Front Immunol 14:1129771
Ryu JK, Kim SJ, Rah SH et al (2017) Reconstruction of LPS transfer Cascade reveals structural determinants within LBP, CD14, and TLR4-MD2 for efficient LPS recognition and transfer. Immunity 46(1):38–50
Sahin U, Türeci Ö (2018) Personalized vaccines for cancer immunotherapy. Science 359(6382):1355–1360
Santos PM, Butterfield LH (2018) Dendritic cell-based cancer vaccines. J Immunol 200(2):443–449
Santos S, Arauz LJD, Baruque-Ramos J et al (2012) Outer membrane vesicles (OMV) production of Neisseria meningitidis serogroup B in batch process. Vaccine 30(42):6064–6069
Sartorio MG, Pardue EJ, Feldman MF et al (2021) Bacterial outer membrane vesicles: from discovery to applications. Ann Rev Microbiol 75:609–630
Scheen AJ (2022) Anti-inflammatory drugs: from old classical ones to biotherapies and JAK inhibitors. Rev Med Liege 77(5–6):399–409
Shi Y, Van Der Meel R, Chen X et al (2020a) The EPR effect and beyond: strategies to improve tumor targeting and cancer nanomedicine treatment efficacy. Theranostics 10(17):7921–7924
Shi Y, Xie F, Rao P et al (2020b) TRAIL-expressing cell membrane nanovesicles as an anti-inflammatory platform for rheumatoid arthritis therapy. J Control Release 320:304–313
Shi R, Lv R, Dong Z et al (2023) Magnetically-targetable outer-membrane vesicles for sonodynamic eradication of antibiotic-tolerant bacteria in bacterial meningitis. Biomaterials 302:122320
Sletten EM, Bertozzi CR (2009) Bioorthogonal chemistry: fishing for selectivity in a sea of functionality. Angew Chem Int Ed Engl 48(38):6974–6998
Song W, Jia P, Ren Y et al (2023) Engineering white blood cell membrane-camouflaged nanocarriers for inflammation-related therapeutics. Bioact Mater 23:80–100
Tan K, Li R, Huang X et al (2018) Outer membrane vesicles: current status and future direction of these novel vaccine adjuvants. Front Microbiol 9:783

Taylor PR, Martinez-Pomares L, Stacey M et al (2005) Macrophage receptors and immune recognition. Annu Rev Immunol 23:901–944
Thamphiwatana S, Angsantikul P, Escajadillo T et al (2017) Macrophage-like nanoparticles concurrently absorbing endotoxins and proinflammatory cytokines for sepsis management. Proc Natl Acad Sci USA 114(43):11488–11493
Toyofuku M, Nomura N, Eberl L (2019) Types and origins of bacterial membrane vesicles. Nat Rev Microbiol 17(1):13–24
Turnbull L, Toyofuku M, Hynen AL et al (2016) Explosive cell lysis as a mechanism for the biogenesis of bacterial membrane vesicles and biofilms. Nat Commun 7:11220
Van De Waterbeemd B, Streefland M, Van Der Ley P et al (2010) Improved OMV vaccine against Neisseria meningitidis using genetically engineered strains and a detergent-free purification process. Vaccine 28(30):4810–4816
Van De Waterbeemd B, Streefland M, Van Keulen L et al (2012) Identification and optimization of critical process parameters for the production of NOMV vaccine against Neisseria meningitidis. Vaccine 30(24):3683–3690
Van De Waterbeemd B, Zomer G, Van Den Ijssel J et al (2013) Cysteine depletion causes oxidative stress and triggers outer membrane vesicle release by Neisseria meningitidis; implications for vaccine development. PLoS One 8(1):e54314
Wan T, Zhong J, Pan Q et al (2022) Exosome-mediated delivery of Cas9 ribonucleoprotein complexes for tissue-specific gene therapy of liver diseases. Sci Adv 8(37):eabp9435
Wang X, Quinn PJ (2010) Lipopolysaccharide: biosynthetic pathway and structure modification. Prog Lipid Res 49(2):97–107
Wang C, Wang Y, Zhang L et al (2018a) Pretreated macrophage-membrane-coated gold Nanocages for precise drug delivery for treatment of bacterial infections. Adv Mater 30(46):e1804023
Wang S, Gao J, Li M et al (2018b) A facile approach for development of a vaccine made of bacterial double-layered membrane vesicles (DMVs). Biomaterials 187:28–38
Wang X, Thompson CD, Weidenmaier C et al (2018c) Release of Staphylococcus aureus extracellular vesicles and their application as a vaccine platform. Nat Commun 9(1):1379
Wang X, Singh AK, Zhang X et al (2020a) Induction of protective antiplague immune responses by self-adjuvanting bionanoparticles derived from engineered Yersinia pestis. Infect Immun 88(5). https://doi.org/10.1128/iai.00081-20
Wang YQ, Bazin-Lee H, Evans JT et al (2020b) MPL adjuvant contains competitive antagonists of human TLR4. Front Immunol 11:577823
Wculek SK, Cueto FJ, Mujal AM et al (2020) Dendritic cells in cancer immunology and immunotherapy. Nat Rev Immunol 20(1):7–24
Wei X, Gao J, Wang F et al (2017) In situ capture of bacterial toxins for Antivirulence vaccination. Adv Mater 29(33)
Wei X, Zhang G, Ran D et al (2018) T-cell-mimicking nanoparticles can neutralize HIV infectivity. Adv Mater 30(45):e1802233
Wu S, Huang Y, Yan J et al (2021) Bacterial outer membrane-coated mesoporous silica nanoparticles for targeted delivery of antibiotic rifampicin against gram-negative bacterial infection in vivo. Adv Funct Mater 31(35):2103442
Xia P, He H, Kristine MS et al (2018) Therapeutic effects of recombinant human S100A6 and soluble receptor for advanced glycation end products(sRAGE) on CCl4-induced liver fibrosis in mice. Eur J Pharmacol 833:86–93
Xiaoqian N, Liu Y, Chen Y et al (2023) WTX212, an erythrocyte-anti-PD1 antibody conjugate, to demonstrate anti-tumor activities in tumor models and patients with cancer with acquired resistance to immunotherapy. J Clin Oncol 41(16_suppl):e14529-e
Xie N, Shen G, Gao W et al (2023) Neoantigens: promising targets for cancer therapy. Signal Transduct Target Ther 8(1):9
Xue K, Wang L, Liu J (2022) Bacterial outer membrane vesicles and their functionalization as vehicles for bioimaging, diagnosis and therapy. Mater Adv 3(19):7185–7197
Yang L, Xie X, Tu Z et al (2021) The signal pathways and treatment of cytokine storm in COVID-19. Signal Transduct Target Ther 6(1):255

Ye T, Li F, Ma G et al (2021) Enhancing therapeutic performance of personalized cancer vaccine via delivery vectors. Adv Drug Deliv Rev 177:113927

Yuan J, Yang J, Hu Z et al (2018) Safe staphylococcal platform for the development of multivalent nanoscale vesicles against viral infections. Nano Lett 18(2):725–733

Yuan X, Duan Y, Xiao Y et al (2022) Vitamin E enhances cancer immunotherapy by reinvigorating dendritic cells via targeting checkpoint SHP1. Cancer Discov 12(7):1742–1759

Zai W, Kang L, Dong T et al (2021) E. coli membrane vesicles as a catalase carrier for long-term tumor hypoxia relief to enhance radiotherapy. ACS Nano 15(9):15381–15394

Zhang P, Chen Y, Zeng Y et al (2015) Virus-mimetic nanovesicles as a versatile antigen-delivery system. Proc Natl Acad Sci USA 112(45):E6129–E6138

Zhang Q, Dehaini D, Zhang Y et al (2018a) Neutrophil membrane-coated nanoparticles inhibit synovial inflammation and alleviate joint damage in inflammatory arthritis. Nat Nanotechnol 13(12):1182–1190

Zhang P, Zhang L, Qin Z et al (2018b) Genetically engineered liposome-like Nanovesicles as active targeted transport platform. Adv Mater 30(7)

Zhang X, Wang C, Wang J et al (2018c) PD-1 blockade cellular vesicles for cancer immunotherapy. Adv Mater 30

Zhang X, Wang J, Chen Z et al (2018d) Engineering PD-1-presenting platelets for cancer immunotherapy. Nano Lett 18(9):5716–5725

Zhang D, Qin X, Wu T et al (2019) Extracellular vesicles based self-grown gold nanopopcorn for combinatorial chemo-photothermal therapy. Biomaterials 197:220–228

Zhang R, Wu S, Ding Q et al (2021a) Recent advances in cell membrane-camouflaged nanoparticles for inflammation therapy. Drug Deliv 28(1):1109–1119

Zhang Q, Zhou J, Zhou J et al (2021b) Lure-and-kill macrophage nanoparticles alleviate the severity of experimental acute pancreatitis. Nat Commun 12(1):4136

Zhou H, Fan Z, Lemons PK et al (2016) A facile approach to functionalize cell membrane-coated nanoparticles. Theranostics 6(7):1012–1022

Zhou J, Karshalev E, Mundaca-Uribe R et al (2021) Physical disruption of solid tumors by immunostimulatory microrobots enhances antitumor immunity. Adv Mater 33(49):2103505

Part V

Cell Membrane Engineering for Targeted Drug Delivery Across Diseases

Cell Membrane Engineering for Advancing Precision Drug Delivery in Cancer Therapy

14

Nimeet Desai, Dhwani Rana, Ragini Rai, Griva Parmar, Derajram Benival, and Lalitkumar K. Vora

Abstract

Cell membrane engineering has revolutionized the advancement of nanosystems for precision drug delivery in cancer therapy, offering unparalleled opportunities for tumor targeting. By leveraging the biological attributes of membranes derived from red blood cells, platelets, immune cells, cancer cells, exosomes, and bacterial cells, researchers have created biocompatible, immune-evasive carriers tailored for specific tumor environments. This chapter delves into the fabrication of membrane-coated nanosystems, detailing key steps such as membrane isolation, coating processes, and advanced physicochemical and biological characterization techniques. Special emphasis is placed on cutting-edge cell membrane engineering approaches, including lipid insertion, membrane hybridization, covalent conjugation, metabolic glycoengineering, and genetic modifications, which enable precise surface customization to enhance tumor interaction, payload delivery, and therapeutic efficacy. The versatility of these systems is demonstrated through their applications in chemotherapy, photothermal therapy, and immunotherapy, showcasing their ability to overcome challenges posed by the

N. Desai · D. Rana · D. Benival
National Institute of Pharmaceutical Education and Research (NIPER), Ahmedabad, Gujarat, India
e-mail: derajram@niperahm.ac.in

R. Rai
National Institute of Pharmaceutical Education and Research (NIPER), Raebareli, Uttar Pradesh, India

G. Parmar
A. R. College of Pharmacy and G. H. Patel Institute of Pharmacy, Anand, Gujarat, India

L. K. Vora (✉)
School of Pharmacy, Queen's University, Belfast, UK
e-mail: L.Vora@qub.ac.uk

Z. Hussain, R. Pei (eds.), *Cell Membrane Engineering for Advancing Cell-Based and Nanotechnology-Based Therapies*, Biomaterials, Bioengineering and Sustainability 6, https://doi.org/10.1007/978-3-031-93425-4_14

tumor microenvironment. This chapter highlights the transformative potential of engineered cell membranes to improve cancer treatment outcomes while addressing challenges in scalability and clinical translation. By integrating natural biomaterials with innovative engineering techniques, these systems offer a promising platform for advancing precision oncology and achieving enhanced therapeutic outcomes.

Keywords

Cell membrane engineering · Tumor targeting · Membrane-coated nanosystem · Cancer therapy · Precision drug delivery

Abbreviations

RBC	red blood cells
TNBC	triple-negative breast cancer
GBM	glioblastoma multiforme
NK	natural killer
AIE	aggregation-induced emission
OMVs	outer membrane vesicles
HEK293T	human embryonic kidney 293 T (cell line)
ROS	reactive oxygen species
CAR-T	chimeric antigen receptor T-cell therapy
CML	chronic myelogenous leukemia
MSC	mesenchymal stem cells
iRGD	internalizing RGD peptide

14.1 Introduction

Cancer encompasses a group of diseases marked by the uncontrolled growth of abnormal cells, which can invade adjacent tissues and often spread (metastasize) to distant organs. At the cellular and molecular levels, cancer arises due to a series of genetic and epigenetic modifications that interfere with biological regulatory mechanisms governing cell growth, differentiation, and apoptosis (Brown et al. 2023). In 2022, the global cancer burden increased to 20 million new cases and 9.7 million deaths. The most common diagnoses included breast, lung, bladder, colorectal, prostate, and stomach cancers (Bray et al. 2022). Cancer imposes significant socioeconomic and health burdens globally by straining healthcare systems and reducing workforce productivity. The high costs of treatment and long-term care lead to financial hardships for patients and their families, exacerbating inequalities and impacting economic development in many countries (Lortet-Tieulent et al. 2020).

Conventional cancer therapies often lack specificity, causing therapeutic agents to attack both malignant and healthy cells, resulting in systemic toxicity and a range of serious side effects. Moreover, the non-specific nature of these treatments can promote drug resistance in cancer cells, diminishing the efficacy of the therapy over time and necessitating higher doses that further exacerbate toxicity. These challenges highlight the need for more precise drug delivery systems that can target tumor cells specifically while minimizing harm to normal tissues (Attia et al. 2019). There is a growing shift towards personalized and targeted therapies in cancer treatment, focusing on tailoring treatments to the molecular profiles of a patient's tumor. Nanoparticles (NPs) are at the forefront of this shift due to their unique potential to be engineered for specific functions. Their surface can be modified with targeting agents, such as antibodies, aptamers, and peptides, that recognize specific markers on cancer cells, enhancing selectivity (Kandula et al. 2023). Furthermore, NPs can be designed to carry various types of cargo, including chemotherapy drugs, gene therapy agents, or imaging molecules, providing multifunctional and adaptable platforms (Li et al. 2023a).

The groundbreaking concept of "biomimetic" functionalization of NPs has garnered considerable attention in recent studies. This strategy involves coating NPs with cell-derived membranes, offering distinct advantages compared to conventional active targeting techniques. One key benefit is the biocompatibility and safety of using natural cell membranes, which allow NPs to circulate in the body for prolonged periods without eliciting an immune response. This method is particularly effective in addressing cancer heterogeneity, offering the potential for personalized treatments by selecting specific cell sources for coating (Xuan et al. 2019). Each type of cell membrane can endow the NPs with unique functionalities, leading to varied behaviours in the body. Membrane-coated NPs can evade immune detection, exhibit minimal protein corona formation and maintain their targeting abilities (Le et al. 2021). Furthermore, membranes from activated immune cells or bacterial cells can also modify immune responses within the tumor microenvironment, providing additional therapeutic functions. This membrane coating strategy simplifies the functionalization process, eliminating the need for complex chemical modifications. It allows for the straightforward selection and customization of NPs for drug delivery, promising to enhance the safety, efficiency, and adaptability of drug delivery systems (Yousefpour et al. 2023).

Besides using cell-derived membranes, extracellular vesicles like exosomes can also serve as biomimetic materials for drug delivery (Ahmed et al. 2024). Engineered extracellular vesicles are nanoscale vesicles equipped with unique membrane proteins and markers that enable targeted interactions with tumor cells (Ma et al. 2024). Recent advancements in genetic engineering and metabolic glycoengineering have further enhanced their therapeutic efficacy, allowing for precise functionalization and payload delivery (Rana et al. 2025; Desai et al. 2025). Engineered exosomes overcome challenges such as limited biodistribution and off-target effects while leveraging natural intercellular communication pathways. These attributes make them promising candidates for precision medicine and innovative cancer therapies (Makwana et al. 2025).

This chapter examines the latest advancements in cell membrane engineering for cancer treatment applications. It covers a range of cell membrane sources, including blood cells, cancer cells, exosomes, and bacterial cells, each offering unique features that improve targeted delivery and biocompatibility. The text describes various fabrication techniques for constructing membrane-coated systems, focusing on isolation, coating, and characterization processes. It further investigates advanced engineering techniques, including lipid insertion, hybridization, metabolic glycoengineering, and genetic modifications, designed to enhance the functionality of these systems and improve therapeutic efficacy. Furthermore, the chapter showcases cutting-edge applications of these engineered nanosystems in chemotherapy, phototherapy, and immunotherapy, underscoring their potential for clinical translation.

14.2 Membrane Sources and Attributes

The properties of membrane-coated nanosystem can vary based on the source of its membrane, each offering distinct advantages for drug delivery applications. Figure 14.1 illustrates different sources for obtaining functional membranes, while Table 14.1 presents a comprehensive summary of the various membrane sources along with their defining characteristics markers. The subsequent section will explore the various types of cell membrane sources and highlight their essential biomedical attributes.

14.2.1 Blood Cells

Blood cells, which originate in the bone marrow and circulate within the bloodstream, form a multifaceted and extensively utilized category of membrane sources.

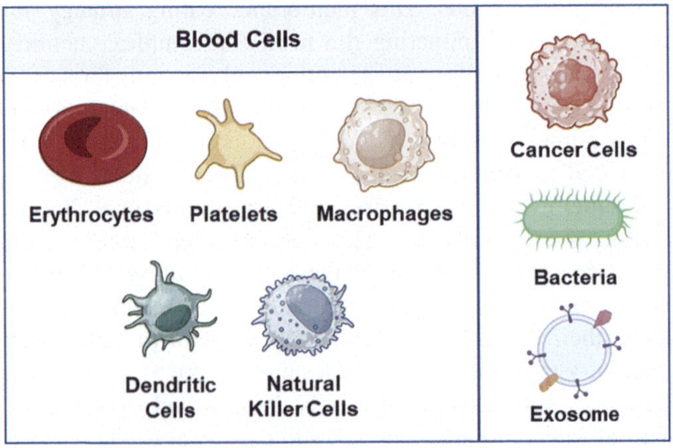

Fig. 14.1 Different cell membrane sources used in biomedical applications. (Source: Created with BioRender.com)

Table 14.1 Overview of different biological membrane sources, surface markers, and their functional attributes for therapeutic applications

Biological membrane source	Membrane-bound markers/proteins	Functional attributes	References
Red blood cells	CD47, glycophorin A	Facilitate immune evasion and circulation stability; employed in drug delivery and targeted therapeutics	Hu et al. (2012)
Platelets	CD62p, PECAM-1, CD44, GPIIb/IIIa	Exhibit robust immune evasion; utilized in targeted drug delivery, particularly for coagulation, inflammation, and tissue repair	Kunde and Wairkar (2021)
Macrophages	CCR2, VCAM-1, ICAM-1	Modulate immune responses; leveraged for targeted immune therapies and cancer treatment via immune cell trafficking	Gong et al. (2020)
Dendritic cells	MHC complex, ICAM-3, CD40, CCR7	Critical for antigen presentation and T-cell activation; used in vaccine development and immunotherapy	Cintolo et al. (2012)
Natural killer cells	NKG2D, NKp44, NKp46, DNAM-1	Recognize and eliminate tumor cells; integral in cancer immunotherapy for tumor targeting	Moretta et al. (2001)
Cancer cells	CCAM, CD44, MUC1	Display tumor antigens and promote homotypic adhesion; used in cancer diagnostics and targeted therapies	Xin et al. (2015)
Bacteria	OmpA, OmpC, flagellin, LPS	Activate immune responses through self-adjuvant properties; applied in cancer immunotherapy and drug delivery	Krishnan et al. (2022)

Abbreviations: *CD47* cluster of differentiation 47, *Glycophorin A* Glycophorin A (or Glycophorin-1), *CD62p* cluster of differentiation 62 P (P-selectin), *PECAM-1* platelet endothelial cell adhesion molecule-1, *CD44* cluster of differentiation 44, *GPIIb/IIIa* glycoprotein IIb/IIIa (Integrin αIIbβ3), *CCR2* C-C chemokine receptor type 2, *VCAM-1* vascular cell adhesion molecule-1, *ICAM-1* intercellular adhesion molecule-1, *MHC Complex* major histocompatibility complex, *ICAM-3* intercellular adhesion molecule-3, *CD40* cluster of differentiation 40, *CCR7* C-C chemokine receptor type 7, *NKG2D*, natural killer group 2 member D, *NKp44* natural killer cell receptor 44, *NKp46* natural killer cell receptor 46, *DNAM-1* DNAX accessory molecule-1, *CCAM* cancer cell adhesion molecule, *CD44* cluster of differentiation 44, *MUC1* mucin 1, *OmpA* outer membrane protein A, *OmpC* outer membrane protein C, *Flagellin* protein subunit of flagella, *LPS* lipopolysaccharide

This group encompasses several cell types, including red blood cells (RBC), platelets (PLT), and various leukocytes such as macrophages, dendritic cells (DCs), and natural killer (NK) cells.

RBC, the most prevalent cells in human blood, lack nuclei and other subcellular organelles, simplifying membrane extraction and purification. Their 120-day lifespan makes them ideal for prolonging the circulation time of encapsulated nanoparticles (Zhang 2016; Luo et al. 2023). The surface of RBC is adorned with self-identifying proteins, such as CD47, which assist in evading the immune response by interacting with the signal-regulatory protein alpha (SIRPα).

Additionally, features like glycans and acidic sialic acid residues help inhibit protein corona formation, further enhancing their functionality (Rao et al. 2015; Zhang et al. 2021a). RBCs membranes also contain membrane cofactor protein, decay-accelerating factor, and complement receptor 1, which provide resistance against complement system attacks. The first bioinspired nanosystem was developed by encapsulating biodegradable polymeric NPs within an RBC membrane. This innovation, combined with the abundance and accessibility of RBC, has significantly advanced their application in both research and medical fields (Hu et al. 2011a).

Platelets, derived from mature megakaryocytes, represent a key type of anucleate blood cell, abundant within the bloodstream with a lifespan of 7–10 days. These cells are crucial in mediating inflammatory responses and haemostasis, particularly following thrombotic events (Adhalrao et al. 2024). Research has linked platelets' ability to induce haemostasis to the progression and metastasis of cancer. Platelets can identify, engage with, and shield circulating tumor cells, facilitating their evasion from immune detection and enabling their dissemination to new sites. Key to this protective function are surface marker proteins on platelets, such as P-selectin (CD62P), platelet endothelial cell adhesion molecule-1, integrins, and CD44 receptors (Wang et al. 2020a). The unique properties of platelet membrane proteins make them promising for NP coatings, offering immune evasion and selective adherence to damaged blood vessels and tumors. Their natural attraction to inflamed tumor sites further facilitates passive accumulation, enhancing targeted delivery (Anselmo et al. 2014; Li et al. 2016).

Leukocytes defend against infections, repair tissue damage, and combat diseases by targeting and engulfing foreign particles. Their migration behavior resembles that of circulating tumor cells, and their adhesion molecules such as selectins and integrins enable efficient interaction with endothelial cells (Muller 2003). Leukocytes accumulate near endothelial walls due to the dynamics of blood flow, initiating the process of rolling adhesion. This begins with weak interactions between endothelial selectins (e.g., P-selectin) and leukocyte ligands (e.g., PSGL-1), which slows the movement of leukocytes. As they roll along the endothelium, activation signals increase the affinity of leukocyte β2 integrins, facilitating their firm binding to ICAM-1 on the endothelial cells. This interaction leads to stable adhesion, followed by transmigration, enabling leukocytes to migrate toward sites of infection or injury (Oroojalian et al. 2021). Chemoattractants, including cancer cell surface receptors and soluble chemokines in the tumor microenvironment, guide leukocyte migration toward tumors. Their ability to adhere to vascular walls and be recruited to tumor or inflammatory sites highlights their potential as sources for therapeutic membranes (Bernardini and Zabel 2019). Below are examples of key receptors and adhesion molecules involved in these interactions:

- Macrophages migrate toward inflammatory tumor sites through molecules like CCR2, VCAM-1, and ICAM-1. Targeted tumor engagement is enhanced by α4 and β1 integrins on macrophages interacting with VCAM-1 on cancer cells. Molecules such as CD45, CD11a, and glycans further assist in tumor localization and help macrophages evade phagocytosis (Cao et al. 2016).

- DCs are crucial in activating T cells by presenting antigens via diverse membrane peptide/MHC complexes. Adhesion and interaction of DCs are facilitated by molecules such as ICAM-3, CD40, CD44, and various integrins, while the CCR7 receptor aids in directing them toward lymph nodes for effective immune activation (Cheng et al. 2020).
- NK cells combat cancer by monitoring abnormal MHC-I expressions and stress proteins on cell surfaces. Despite lacking tumor-specific antigen receptors, NK cells possess alternative receptors like NKG2D, NKp44, NKp46, NKp30, and DNAM-1, which enhance their ability to recognize and attack cancer cells, improving tumor targeting and biocompatibility (Deng et al. 2018; Pitchaimani et al. 2018).

14.2.2 Cancer Cells

Cancer cells possess the capability for unlimited proliferation, a phenomenon termed immortalization, which stands as a hallmark of cancer. This trait arises from mutations and epigenetic modifications that enable cancer cells to bypass the control mechanisms of cell division and programmed cell death. For example, mutations in key genes such as p53 or RB1 disrupt normal cell cycle control, leading to unregulated proliferation and resistance to apoptosis, hallmark features of cancerous cells (Jiang et al. 2011). The capacity for limitless division not only supports the continuous proliferation of cancer cells but also makes them an excellent source for cell membrane isolation. This is particularly valuable for generating large quantities of consistent and reproducible membrane material, which is essential for experimental applications (Desai et al. 2023a).

Cancer cell membranes exhibit unique properties, such as overexpression of specific receptors and altered lipid composition, which contribute to tumor development, progression, and metastasis. For example, the overexpression of CD47 on cancer cell membranes aids in immune evasion (Lian et al. 2019). Furthermore, cancer cell adhesion molecules, including selectins, cadherins, integrins, and the immunoglobulin superfamily, play crucial roles in cell-cell adhesion, gene expression modulation, and processes like proliferation, migration, and invasion (Janiszewska et al. 2020). Moreover, tumor-associated antigens like the Thomsen–Friedenreich glycoantigen and galectin-3 play pivotal roles in facilitating the aggregation of metastatic cells (Desai et al. 2023b). The presence of these surface markers and receptors on cancer cell membranes makes them excellent candidates for the development of tumor-targeted nanosystems. For example, membrane coating from cancer cells could significantly enhance polymeric nanoparticles' uptake due to homotypic binding mechanisms, underscoring the potential for using cancer cells in localized tumor theranostics (Fang et al. 2014). Cancer cells, sourced from patient tumor biopsies allow for the extraction of membrane enriched with patient-specific cancer antigens. This personalized approach provides unique opportunities for immunotherapeutic applications, tailoring treatments to individual patients (Pan et al. 2024).

14.2.3 Exosomes

Extracellular vesicles (EVs), composed of lipids, nucleic acids, and proteins derived from the cell membrane and cytosol, play critical roles in cellular communication, maintaining homeostasis, regulating inflammation, and promoting tissue repair (Doyle and Wang 2019; Desai et al. 2021). EVs generation starts with the inward budding of the endosomal membrane, leading to the creation of intraluminal vesicles within multivesicular bodies (MVBs). These MVBs then fuse with the cell membrane, releasing exosomes into the extracellular environment (Raposo and Stahl 2019). EVs are categorized based on formation pathways, structural features, and functional capacities into three primary types: exosomes, microvesicles, and apoptotic bodies (Pitt et al. 2016). Exosomes and microvesicles hold greater biomedical potential than apoptotic bodies, as apoptotic bodies are cellular remnants resulting from cell breakdown during apoptosis (Kalluri and LeBleu 2020). Once released, exosomes and microvesicles can transfer their biological materials to nearby or distant cells, potentially altering their functions and characteristics (Batrakova and Kim 2015). EVs can be identifiable in various cell types. They usually interact with target cells via membrane-bound proteins like tetraspanins and antigen-presenting molecules (Pegtel and Gould 2019). Their membranes are designed to avoid immune detection, which enhances their suitability as drug carriers by prolonging their circulation in the bloodstream (Chavda et al. 2023). Furthermore, EVs from mesenchymal stem cells (MSCs) are characterized by surface markers like CD9, CD63, CD81, which facilitate their isolation and utilization in biomedical applications. These vesicles possess a unique blend of surface proteins that aid in their distribution within the bloodstream and enhance the longevity of encapsulated drug carriers (Ma et al. 2020). Tumor cells release EVs more abundantly than normal cells, carrying tumor-specific proteins for targeted delivery. EVs from therapeutic cells can contain immunosuppressive molecules and death receptor ligands that trigger apoptosis in tumor cells, highlighting their anticancer potential (Tian et al. 2023).

14.2.4 Bacterial Cells

The bacterial membrane differs between Gram-positive and Gram-negative microorganisms, affecting their functions. Gram-positive microorganisms have a thick peptidoglycan layer with teichoic acids but no outer membrane. On the other hand, Gram-negative bacteria have a thin peptidoglycan layer between an inner phospholipid membrane and an outer lipopolysaccharide-rich membrane (Silhavy et al. 2010). Gram-negative microorganisms produce spherical outer membrane vesicles (OMVs) by bulging and detaching the outer membrane. These vesicles facilitate gene transfer, communication, and metabolite exchange while contributing to pathogenicity. As non-replicating yet immunogenic entities, OMVs offer a safer option for therapeutic or vaccine administration (Schwechheimer and Kuehn 2015). OMVs contain structural components similar to those of the parent cell membrane, including phosphatidylglycerol and phosphatidylethanolamine. They also contain various proteins, including membrane proteins, periplasmic proteins, virulence factors, and adhesion proteins, which are essential for invading host cells (Anand and Chaudhuri 2016). OMVs display specific antigens and pathogen-associated molecular patterns (PAMPs), which activate the immune system by stimulating dendritic cells and activating T cells. Their strong immunogenicity, coupled with their potential to efficiently drain into lymph nodes and retain in tumors, makes them attractive as cancer vaccine agents (Sartorio et al. 2021). OMVs from *Salmonella* have been shown to boost the production of numerous immune cytokines, enhancing their potential in cancer therapy. Furthermore, OMVs can be modified to carry chemotherapeutic, radioactive, and photoactive molecules, enhancing their efficacy in robust antitumor treatment (Qing et al. 2019). Recent studies have indicated that gram-positive microorganism also secrete vesicle-like structures termed membrane vesicles (MVs). However, the formation of MVs in these bacteria is complex due to the thick peptidoglycan wall (Gilmore et al. 2021). Not all Gram-positive bacteria generate MVs, and their components, including lipopolysaccharides, nucleic acids, and proteins, show distinct aggregation patterns and compositional differences from the parent bacterial cell (Toyofuku et al. 2019). MVs and OMVs are crucial for biofilm formation, immune modulation, stress response, and intercellular communication. In therapeutics, they hold promise by reducing off-target effects, enhancing cellular uptake, and modulating immune response. Recently, their role in antitumor immunity has been studied, with current research focusing on minimizing batch variability, reducing immunogenicity, and optimizing cell-targeting strategies (Kaparakis-Liaskos and Ferrero 2015; Ellis and Kuehn 2010; Li et al. 2020).

14.3 Fabrication of Membrane-Coated Systems

The process for preparing membrane-coated NPs is typically categorized into three stages: first, the extraction of the biological membrane; second, the selection/preparation of the NPs core; and third, the assembly of the biological membrane around the NPs core (Zeng et al. 2023; Harris et al. 2019; Shao et al. 2023).

14.3.1 Isolation

Cell membrane isolation protocols aim to separate cell membranes from other cellular components while minimizing contamination by nuclear, mitochondrial, or cytosolic elements. Purified cell membranes are essential for creating membrane-camouflaged mimics, enhancing surface coating efficiency and functional replication. To preserve membrane proteins, ice-cold extraction buffers with protease and phosphatase inhibitors are used. Cells are pre-washed with phosphate-buffered saline to remove media residues, and isolated membranes are lyophilized and stored at −80 °C. Isolation involves two steps: first, detergent-free hypotonic lysis with physical disruption; second, membrane purification via differential or sucrose gradient centrifugation. Isolation methods differ for anucleate cells and nucleus-containing cells. In anucleate cells like RBCs and platelets, hypotonic treatment or freeze-thaw cycles followed by centrifugation effectively isolate membranes. RBC isolation includes multiple cycles to remove hemoglobin impurities, while platelets may require additional sucrose gradient purification. Bacterial membranes, isolated as OMVs, involve ultrafiltration and centrifugation to bypass challenges posed by peptidoglycan layers. For nucleus-containing cells like immune and cancer stem cells, isolation requires larger cell quantities. These cells undergo hypotonic treatment and mechanical disruption, followed by separation via differential centrifugation or sucrose gradient ultracentrifugation. Differential centrifugation removes heavier components, while sucrose gradients separate membranes based on density, ensuring minimal contamination. Maintaining low temperatures and using hypotonic buffers with divalent ions help preserve transmembrane proteins and cholesterol, ensuring membrane integrity and functionality (Chugh et al. 2021).

14.3.2 Coating

Selecting a suitable nanoparticle core is essential in cancer theranostics, as it determines the type of cargo, including drug molecules and biomolecules, that can be loaded. Its physicochemical properties, such as size, charge, and composition, significantly influence cell membrane coating efficiency and overall biocompatibility (Fang et al. 2018). After membrane isolation, NPs are coated via electrostatic interactions, forming stable core-shell structures. Physical extrusion enhances this by passing them through polycarbonate membranes with specific pore sizes. The mechanical forces disrupt the membrane, allowing NPs to integrate into the phospholipid bilayer. Composite size is determined by pore diameter, ensuring uniform distribution. This method preserves membrane protein activity but is time-intensive and relies on membrane-to-polymer ratio and nanoparticle surface charge for coating efficiency. Ultrasonication uses sonic energy to disperse cell membranes, which reassemble around nanoparticles via noncovalent interactions. It minimizes material loss, preserves protein functionality, and ensures effective dispersion. However, it may cause particle size variation, inconsistent coating, or membrane damage. Careful optimization of power, duration, and frequency is critical for balancing coating efficiency and protein preservation, making it a faster alternative to extrusion. Microfluidic electroporation generates temporary hydrophilic openings in cell membranes using high-voltage pulses, allowing nanoparticles to integrate. This method produces composites with complete coverage, uniform dispersion, and minimal protein damage while reducing material waste. Despite higher technical costs, it offers excellent repeatability and efficiency. Optimization of microfluidic chip design, voltage, and flow rate is essential for tailoring the process to specific material properties (Desai et al. 2024a). Table 14.2 provides a brief overview of different membrane coating techniques.

Table 14.2 Membrane coating techniques and their special features

Membrane coating technique	Special features	References
Physical extrusion	High precision and reproducibility Utilizes a porous membrane for co-extrusion Produces monodisperse size distribution	Cao et al. (2018), Hu et al. (2011b), Xuan et al. (2015), Chen et al. (2020a)
Sonication	Uses ultrasonic energy to fuse nanoparticles and membrane vesicles Results in higher formulation yield without material loss Influenced by factors such as power, amplitude, and suspension properties	Shao et al. (2023), Tang et al. (2017, Kang et al. (2017)
Microfluidic coating	Enables precise mixing of membrane vesicles and nanoparticle core Requires external forces (sonication or electroporation) for coating Enables high encapsulation efficiency and rapid reaction integration	Rao et al. (2017)

14.3.3 Characterization

NPs enhance drug delivery by efficiently crossing biological barriers and targeting tumors through the enhanced permeability and retention effect. Their unique properties, including catalytic abilities, mechanical strength, magnetic responsiveness, thermal properties, and optical characteristics, support diverse roles in cancer treatment (Rana et al. 2021; Desai et al. 2024b). The biological properties of NPs, such as extended circulation time, immune evasion, and targeting capabilities, can be enhanced by coating them with cell membranes. To verify the successful encapsulation of NPs within a cell membrane, a series of analytical techniques are employed. Initially, microscopic analysis and the detection of glycoproteins and sialic acids assess the integrity of the membrane coating (Liu et al. 2022). Methods like gel electrophoresis, western blotting, immunohistochemistry, fluorescent labelling, enzyme-linked immunosorbent assay, and absorbance spectroscopy can confirm the presence of biological markers on the NPs surface. For real-time assessment, dynamic light scattering measures the hydrodynamic diameter, highlighting size differences before and after coating to confirm membrane formation. Additionally, zeta potential measurements using a zeta sizer evaluate the surface charge, providing insights into particle stability and interactions with biological systems (Wang et al. 2022). Monitoring these properties over time also offers insights into the colloidal stability of the coated nanoparticles, ensuring their durability and suitability for storage and therapeutic use (Liu et al. 2021). For example, cell membrane-coated mesoporous silica NPs (CM-SiO_2 NPs) were developed, and various characterization techniques confirmed successful coating. Transmission electron microscopy revealed a distinct lipid bilayer around the NPs core, indicating membrane fusion. Dynamic light scattering showed an increase in hydrodynamic diameter from 118.6

to 139.8 nm and a slight shift in zeta potential from −41.2 to −36.0 mV. Confocal microscopy demonstrated colocalization of fluorescently labeled NPs and membranes. SDS-PAGE analysis confirmed retention of membrane proteins (Liu et al. 2021). Biomimetic magnetic NPs, Fe_3O_4-SAS@PLT, were engineered to enhance ferroptosis and mild immunogenic responses, thereby enhancing the therapeutic efficacy in non-inflamed tumors. These NPs, loaded with sulfasalazine (SAS) and camouflaged with a PLT membrane, were prepared using extrusion. The NPs showed a core/shell structure, confirming a unilamellar membrane coating. SDS-PAGE analysis detected PLT membrane proteins on the surface, and the particles exhibited excellent colloidal stability. Protein band analysis confirmed effective membrane integration (Jiang et al. 2020).

14.4 Membrane Engineering Approaches

Research on cell membrane-coated nanoparticles is advancing their application in complex biological systems, driving the need for nanoparticles with enhanced multifunctional capabilities to address the demands of advanced therapeutic and diagnostic applications. Early methods, such as conjugation techniques using amine, carboxyl, biotinylation, or thiol-based reactions, offer limited control over ligand placement and density, which can disrupt membrane integrity, cause protein aggregation, or provoke immune responses. To address these issues, more efficient, non-disruptive functionalization strategies have been developed, preserving the natural functions of the source cells. Figure 14.2 highlights these methods.

14.4.1 Functionalization via Lipid Insertion

This method uses lipid anchors conjugated with specific ligands mixed into a source membrane, exploiting the fluid nature of lipid bilayers without chemical reactions. The insertion is simple, using sonication or extrusion, and allows fine-tuned regulation of ligand density by adjusting their concentrations (Zhang et al. 2021b). Lipid anchors like DSPE-PEG are often used to embed affinity ligands like folate, mannose, cyclic peptides, and aptamers into non-cancerous source cells (Ak and Hamarat Şanlıer 2020; Guo et al. 2015; Wang et al. 2018). In one of the initial studies based on this technique, Fang et al. conjugated the nucleolin-targeting aptamer AS1411 to lipid-PEG-maleimide via maleimide–sulfhydryl coupling chemistry. The conjugate was then combined with an RBC membrane and extruded with the nanoparticle core, resulting in a twofold increase in cancer cell uptake compared to the unmodified control (Fang et al. 2013). Lipid membranes can also incorporate antibodies, achieved by linking antibodies to lipid molecules via reactive functional groups, which are then embedded into the membrane. Research has focused on cancer markers such as HER2, EGFR, and EpCAM. A key challenge is the large size of antibodies, which complicates their functional orientation on the membrane (Zhang et al. 2018a). Smaller antibody fragments offer a solution, allowing more

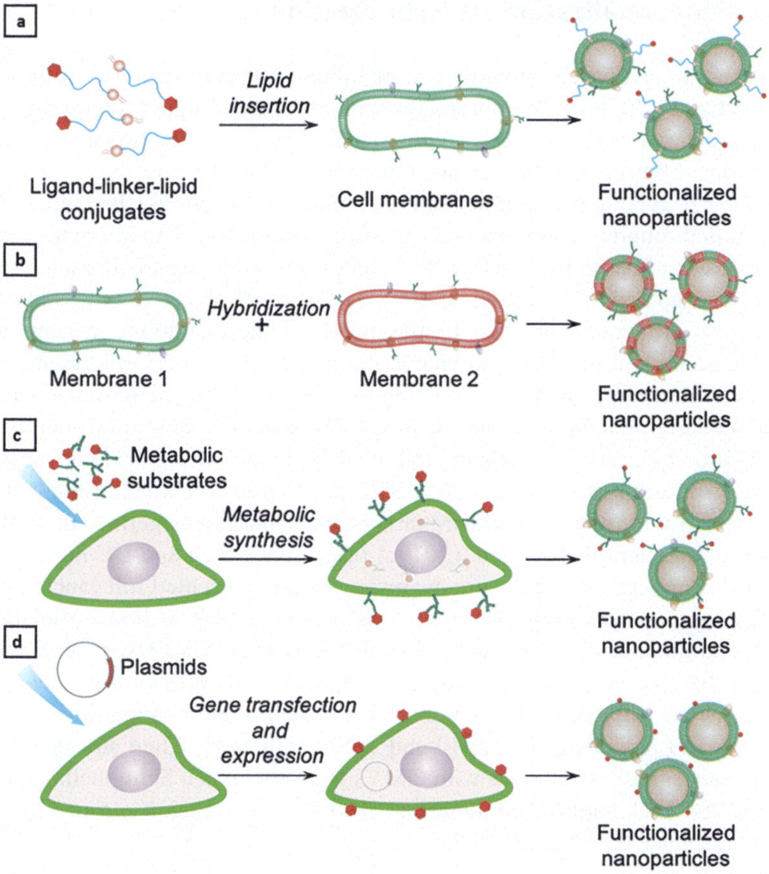

Fig. 14.2 A schematic diagram illustrating different approaches for functionalizing biological membranes, including (**a**) lipid insertion, (**b**) membrane fusion, (**c**) metabolic glycoengineering, and (**d**) genetic engineering. (Reprinted with permission from Ai et al. 2021, copyright 2021, American Chemical Society)

precise lipid conjugation and better spatial selectivity. Beyond anchoring ligands, lipids add functional properties, responding to stimuli like light, hypoxia, or pH changes (Richards et al. 2017). For example, Su et al. (Su et al. 2016) used DiR (1,1′-dioctadecyl-3,3,3′,3′-tetramethylindotricarbocyanine iodide) in RBC membranes to convert NIR light into heat, creating localized hyperthermia. Their system featured a thermosensitive lipid with a phase transition at 41.5 °C. NIR laser stimulation heats the DiR, triggering the lipid's phase change, leading to nanoparticle breakdown and rapid drug release. Similarly, Liu et al. (2019a) integrated a pH-sensitive lipid, DSPE-polyethyloxazoline, into a platelet membrane. In acidic endosomes, the polyethyloxazoline becomes protonated, causing electrostatic repulsion that destabilizes the membrane and releases the drug.

14.4.2 Functionalization via Hybridization

This technique involves combining cell membranes derived from two different cell types to create a "hybrid" membrane that possesses the unique proteins and biomimetic characteristics of the original cells. This enhances its functionality for various applications. Hybrid membranes are typically produced by isolating membranes from each cell type in proportions tailored to specific functions, then mechanically fusing them through methods such as extrusion or sonication. An alternative method involves fusing live cells together and then harvesting the membrane from the resulting hybrid cells (Chen et al. 2020b). The primary objective of this hybridization process is to overcome the limitations of single-membrane systems and to introduce additional features that enable the membrane to perform essential functions, particularly improving the targeting capabilities of the membrane-coated systems by integrating unique affinity ligands from one cell type into the membrane of another (Liu et al. 2019b). Dehaini et al. (2017) pioneered a hybrid nanosystem by merging RBC and platelet membranes. This innovation combined prolonged circulation in vivo with platelet surface markers for enhanced targeting. In context of cancer immunotherapy, Liu et al. (2019c) fused cancer cell and DC membranes to create a system that enhanced immune responses and presented full cancer antigens for targeting, incorporating DC co-stimulatory molecules to boost effectiveness. Beyond mammalian systems, bacterial membranes and OMVs also proved effective in cancer therapy. Wang et al. (Wang et al. 2020b) used OMVs from *E. coli* DH5α combined with melanoma cell membranes to encapsulate nanoparticles, targeting melanoma and increasing IL-12p40 and IFN-γ in tumors. Chen advanced this by merging cancer cell membranes with OMVs from attenuated Salmonella, creating a nano-vaccine that suppressed tumor growth by triggering a strong anti-tumor immune response (Chen et al. 2020c).

14.4.3 Functionalization via Metabolic Glycoengineering

This technique modifies cell properties by integrating functional moieties into biosynthetic pathways. Metabolic substrates introduced to cells enable non-natural conjugates to incorporate into metabolism and attach to cell surfaces (Desai et al. 2024c). Two key approaches to this concept are glycoengineering and lipid engineering. Glycoengineering produces oligosaccharides and glycoconjugates to modify cell membranes using leveraging various biochemical pathways to introduce specific glycan structures. For example, monosaccharides such as N-acetylmannosamine and N-acetylneuraminic acid are combined with functional groups to alter cell membrane characteristics. Han et al. (2019), used tetra-acetylated N-azidoacetylgalactosamine (Ac4GalNAz) to incorporate azide groups onto T cell membranes via metabolic glycoengineering. These azide-labeled T cells were then coated onto photosensitizer-loaded PLGA cores. A Bicyclo [6.1.0] nona-4-yne (BCN)-modified mannose substrate (Ac4ManN-BCN) was administered to tumors, where it expressed on tumor surfaces through metabolic glycoengineering process.

Injected azide-labeled T cells coated NPs selectively homed to the tumor via a click reaction between BCN and azide groups. This resulted in greater tumor accumulation compared to unmodified T-cell coated NPs.

Lipid engineering modifies cell membranes through natural lipid synthesis to introduce click linkers for ligand conjugation. Zhang et al. (2017) used azide-choline to add azide groups to leukocyte membranes via this pathway. These azide-labeled membranes were coated on magnetic nanoclusters and, through click chemistry, conjugated with MHC class I and anti-CD28. In a murine lymphoma model, the nanoclusters, acting as synthetic antigen-presenting cells, promoted CD8+ T cell proliferation more effectively than free anti-CD28. A single intratumoral injection significantly increased tumor-infiltrating lymphocytes and delayed tumor growth within 14 days. Zhang et al. (2018b) also used this method to create azide-labeled macrophage membranes conjugated with RGD peptides targeting integrin αvβ3, overexpressed in tumors. These engineered membranes, when coated on MNC-siRNA nanocomplexes and injected intravenously, increased tumor accumulation by 2.7-fold compared to the unmodified system. Metabolic glycoengineering can also be used to engineer microorganism membranes, as demonstrated by Price et al. (2016), who glycoengineered OMVs from non-pathogenic bacteria to express surface glycans of pathogenic bacterial strains. In a murine model, these modified OMVs induced a strong pathogen-specific IgG response. By day 42, mice treated with the engineered OMVs exhibited IgG levels comparable to those in the group receiving the pneumococcal vaccine Prevnar 13®, indicating equivalent immunogenicity. This approach, combined with an adjuvant-loaded core and tumor immunity-triggering antigen, holds potential for cancer vaccine development.

14.4.4 Functionalization via Genetic Engineering

This advanced cell functionalization method uses gene editing to modify surface proteins, upregulating native antigens or introducing new ones for added functionality. Genetically modified cells can be grown in large quantities, and their membranes extracted for nanosystem coating, potentially decreasing production costs. Efficient gene modification relies on delivering nucleic acid into the cytosol via viral vectors (e.g., lentivirus, adenovirus), chemical methods (e.g., cationic lipids), or physical techniques (e.g., electroporation), with the choice depending on feasibility, efficiency, and specificity (Krishnan et al. 2023; Bisht et al. 2024). For example, Lv et al. (2019) used genetic engineering to modify HepG2 cells to express affinity ligands, creating a membrane-coated oncolytic adenovirus. Coating the adenovirus with biological membranes reduced immunogenicity and enhanced tumor targeting through active preS1 ligand overexpression. Additionally, CRISPR technology was used to express the Asn-Gly-Arg peptide on RBC membranes, which coated adenoviruses targeting aminopeptidase N (APN) on cancer cells, improving tumor targeting in APN receptor-positive tumors like PC13 and U87. In a separate study, Krishnamurthy et al. (2019) introduced the proline-alanine-serine (PAS) peptide sequence into HEK293 cells using a plasmid vector. This approach created

PASylated nanoghosts by coating polymeric cores with the engineered membrane. The strategy resulted in a significant reduction in protein corona formation and macrophage uptake, compared to non-modified HEK293 membranes. It also improved the circulation half-life of the nanosystem, enhanced nanoparticle-target cell interactions, and minimized undesirable cellular responses.

14.4.5 Functionalization via External Stimuli

This method involves inducing the overexpression of specific membrane proteins through external stimuli. This approach enhances the therapeutic capabilities of the isolated cell membrane, facilitating targeted drug delivery. The selective upregulation of membrane proteins, which can be tailored with functionalities like targeting ligands or receptors, occurs when exposed to external stimuli. This leads to the functionalization of biological membrane by overexpressing these proteins (Rao et al. 2020). For instance, MSCs can be prompted to increase expression of the C-X-C chemokine receptor type 4 (CXCR4) under the influence of Fe_3O_4. For example, Shi et al. (2022) developed a versatile immunosuppressive nanoparticle (VIN) by combining polydopamine nanospheres loaded with A151 and MSC-derived membrane vesicles enriched with CXCR4, stimulated by Fe3O4. This approach showed significant therapeutic effects in managing overactive brain immune environments in vivo. Kang et al. (2023) developed a novel nanoplatform combining AIEgen with a hypoxia-responsive paclitaxel (PTX) prodrug, encapsulated in NPs coated with M1 macrophage membranes. Induced by external stimuli, M1 macrophages promote tumor-fighting properties. The dual approach of photodynamic therapy (PDT) and PTX chemotherapy induces immunogenic cell death, suppressing primary and metastatic tumors in 4 T1-bearing mice. This strategy leverages the natural tumor-targeting abilities and immunomodulatory effects of M1 macrophages, enhancing the efficacy of drug delivery systems.

14.5 Cutting-Edge Applications of Membrane Engineered Platforms

Optimizing therapeutic effectiveness while minimizing systemic toxicity remains a pivotal challenge in cancer treatment, making precision drug delivery a critical frontier in oncology. Cancer's cellular heterogeneity and the non-specificity of conventional therapies, which harm both healthy and malignant cells, result in severe side effects and reduced efficacy (Desai et al. 2023c). Among the innovative strategies under development, cell membrane-coated NPs have been recognized as a groundbreaking therapeutic strategy. Cloaking NPs with biological membranes enhances biocompatibility, prolongs circulation time, and enables targeted drug delivery through inherent biological interactions (Polaka et al. 2021). The membrane coating allows these NPs to evade the mononuclear phagocyte system, reducing clearance from the bloodstream and increasing accumulation at tumor sites. Moreover, functional proteins and markers present on the cell membranes facilitate the active targeting of tumor cells through homotypic binding or receptor-mediated interactions. Functionalizing the cell membranes takes this approach even further. The following section highlights recent advancements of engineered cell membrane-based systems employed in cancer treatment (summarized in Table 14.3).

14.5.1 Chemotherapy

Chemotherapy uses anti-cancer drugs to kill or inhibit the rapid growth of cancer cells as part of a standardized treatment regimen. The effectiveness of chemotherapy depends on the type of cancer and the stage at which it is diagnosed. While highly effective in treating a wide range of cancers, chemotherapy also carries a significant burden of side effects, as it can also harm healthy cells. This non-selectivity can lead to side effects ranging from hair loss and nausea to more severe health complications such as increased susceptibility to infections and anaemia. Modern approaches often combine chemotherapy with other treatments like surgery, radiation, or targeted therapies to enhance efficacy and mitigate side effects (Ahmad et al. 2019). Song et al. (2022) developed a novel RBC-biomimetic nano-drug delivery system incorporating folic acid (FA)-modified RBC membranes to enhance the delivery and efficacy of paclitaxel (PTX) for epithelial cancer treatment (Fig. 14.3a). This system leverages the tumor-targeting ability of folic acid due to its strong affinity for folate receptors, which are highly expressed on cancer cells, particularly in epithelial malignancies. The researchers synthesized a core-shell nanoparticle using a self-assembled hydrophilic and hydrophobic polymer core composed of carboxymethylcellulose sodium and stearic acid, which was coated with FA-modified RBC membranes to enhance targeting. The average nanoparticle size was approximately 226.9 ± 2.75 nm with a negative Zeta potential of −14.5 ± 0.3 mV, contributing to stability and preventing aggregation in circulation. Characterization of the membrane protein CD47 confirmed correct orientation, crucial for evading immune recognition, which reduced macrophage uptake. In vitro and in vivo studies demonstrated superior targeting efficacy, with the nanoparticles accumulating two-fold more in tumors compared to unmodified RBC nanoparticles. Furthermore, the system exhibited pH-sensitive drug release, where the acidic tumor microenvironment (pH 5.5) significantly increased PTX release. In vivo biodistribution and therapeutic efficacy studies in xenograft models (liver, lung, and melanoma) showed enhanced tumor inhibition, with the tumor weight reduction reaching 48.7% in HepG2 xenografts. Additionally, the system substantially reduced systemic toxicity, particularly renal toxicity, as evidenced by lower serum creatinine and blood urea nitrogen levels compared to free PTX treatments.

Wu et al. (2023) synthesis a photosensitizer-modified RBC membrane-coated nanoparticle system to enhance the efficacy of docetaxel (DTX)-based chemotherapy and counteract its pro-metastatic effects in triple-negative breast cancer (TNBC). The nanoparticle consists of a hybrid micelle loaded with DTX and the anti-metastatic agent calcitriol (CTL), which is encapsulated within an RBC membrane

Table 14.3 Summary of cell membrane engineering-based nanoplatforms and their applications in cancer therapy

Nanoplatform category	Cell source	Drug candidates	Cell line	In vivo model	Outcome	References
RBC-biomimetic nanodrug delivery system	RBC membranes	Paclitaxel	HepG2	Liver, lung, melanoma xenografts	Enhanced tumor inhibition, reduced systemic toxicity	Song et al. (2022)
RBC membrane-coated nanoparticle system	RBC membranes	Docetaxel, calcitriol	TNBC cell lines	TNBC xenograft mouse model	Tumor inhibition, reduced metastasis, enhanced immune response	Wu et al. (2023)
Cancer cell-mitochondria hybrid hembrane-coated nanomedicine	Cancer cell and mitochondria membranes	Gboxin	U87MG glioblastoma cells	GBM-bearing mice	Increased mitochondrial apoptosis, prolonged survival	Zou et al. (2023)
Urease-powered platelet cell motors	Platelet membranes	Gold nanoparticles	Various cancer cell lines	Murine models with radiotherapy	Enhanced radiosensitization, tumor reduction	Liu et al. (2024)
NK cell membrane-clothed virus-mimicking nanogenerator	NK cell membranes	AIPH, Fe^{III}	HepG2	HepG2 xenografts	98.9% tumor growth inhibition	Lin et al. (2022)
Genetically modified T-cell membrane-coated AIE nanoparticles	T cell membranes	Chimeric antigen receptors, AIE	Glioblastoma cells, glioblastoma stem cells	Orthotopic glioblastoma mouse models	Complete tumor ablation, recurrence prevention	Wang et al. (2023)
Hybrid cell membrane-coated nanocomposite	Brain metastatic breast cancer and glioma cells	Indocyanine green, Gambogic acid	Glioma cells (U87-MG)	Early-stage glioma-bearing mouse models	Tumor growth inhibition, effective glioma targeting	(Chi et al. 2023)
Genetically modified hybrid membrane-camouflaged nanoemulsion	OMVs and HEK293T cells	Lactate oxidase, ROS inducers	4 T1 mouse breast cancer, CT26 colorectal cancer cells	Breast cancer, and colorectal cancer models	Robust anti-tumor effects, improved immunity	Zhang et al. (2024a)
Bioorthogonal pretargeting photosensitizer system	Tumor plasma membranes	Chlorin e6	4 T1 breast cancer	4 T1 tumor models	97% tumor inhibition, enhanced imaging	Wei et al. (2021)

Bioorthogonal click-targeting nanocomposite system	Tumor cell membranes	Doxorubicin, zinc phthalocyanine	MCF-7 breast cancer	MCF-7 xenograft models	96.1% tumor inhibition, synergistic effect	Taghikhani et al. (2020)
T lymphocyte membrane-decorated nanoinducer of interferons	T lymphocyte membranes	ORY-1001	TNBC, melanoma cell lines	TNBC, melanoma, colon cancer models	Enhanced CD8+ T-cell activation, tumor growth inhibition	Zhai et al. (2021)
Genetically programmable vesicles for CAR-T therapy enhancement	HEK293T vesicles	Glutamine antagonist DON	Various tumor models	CAR-T models with solid tumors	Increased CAR-T efficacy, prolonged survival	Li et al. (2023b)
Genetically engineered membrane-based nanoengagers	Engineered membranes	Bispecific single-chain variable fragments	Pancreatic tumor cells	Murine pancreatic cancer models	Enhanced macrophage activation, prolonged survival	Zhang et al. (2024b)
IL3-R targeted exosomes (IL3L Exo)	HEK293T derived exosomes	Imatinib, BCR-ABL siRNA	CML cell lines (LAMA84, K562)	CML xenograft models	Specific targeting, tumor regression, immune response	Bellavia et al. (2017)
Genetically engineered exosomes	MSC-derived exosomes	PI3Kγ inhibitor IPI549	Pulmonary vascular endothelial cells	Postoperative lung metastasis models	Reduced metastatic lesions, prolonged survival	Han et al. (2022)
iRGD-exosomes for KRAS siRNA delivery	HEK293T derived exosomes	KRAS siRNA	A549 lung cancer cells	Lung cancer xenograft models	Effective KRAS siRNA delivery, tumor regression	Zhou et al. (2019)

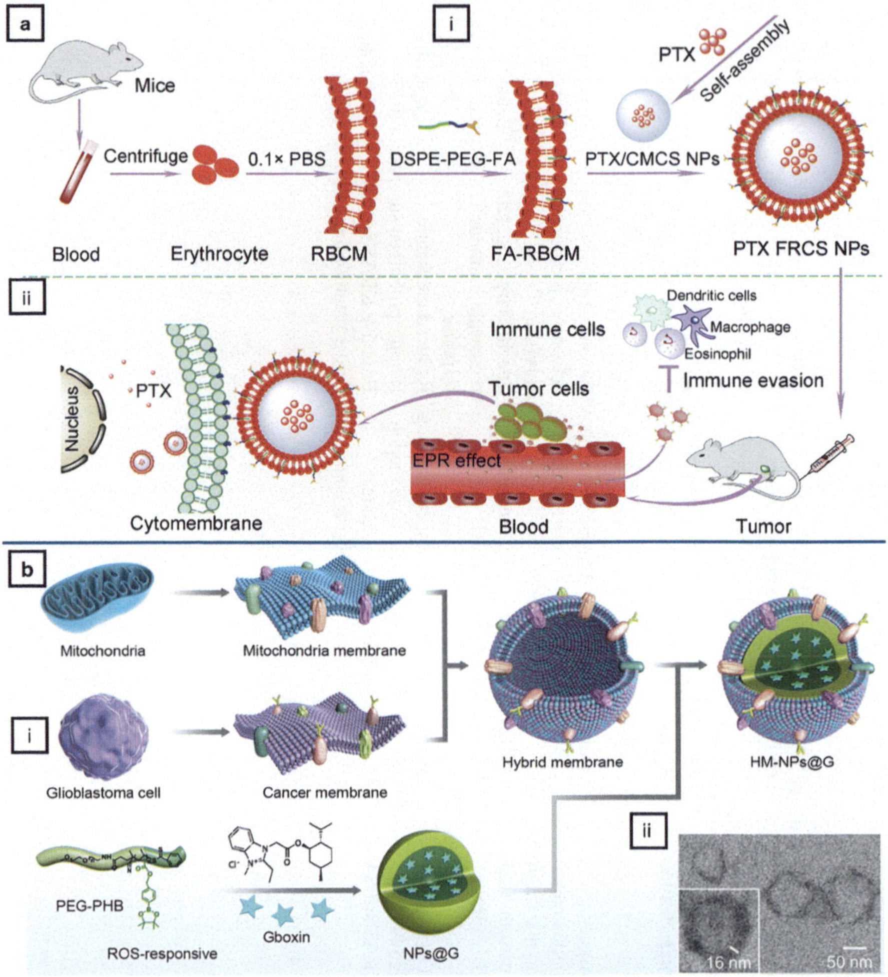

Fig. 14.3 (**a**) (i) Schematic representation of PTX FRCS NP preparation and (ii) application of these biomimetic nanoparticles for enhanced in vivo tumor chemotherapy. (Reprinted with permission from Song et al. 2022, copyright 2022, Elsevier B.V.). (**b**) Schematic synthesis of cancer cell-mitochondria hybrid membrane-coated ROS-responsive nanoparticles. (ii) Transmission electron microscopy images of HM-NPs@G. (Reprinted with permission from Zou et al. 2023, copyright 2023, Springer Nature)

shell modified with the near-infrared (NIR)-activated photosensitizer DiR. This modification enables photothermal activation, causing RBC membrane rupture under 808 nm laser irradiation, enhancing drug release and tumor penetration while inducing hyperthermia at 47.5 °C. The nanoparticle exhibited a mean size of 150.67 nm and a zeta potential of −15.27 mV, indicating successful RBC membrane coating and stability. In vivo studies showed that RBC membrane-coated nanoparticle system increased intratumoral DTX and CTL concentrations by 1.7 and 2.5

times, respectively, compared to free drugs, resulting in a tumor growth inhibition rate of 77% and an 82.9% reduction in lung metastasis after laser irradiation. Additionally, the system induced immunogenic cell death, marked by enhanced release of danger-associated molecular patterns which led to a 1.88-fold increase in intratumoral CD8+ T-cell infiltration and upregulation of mature dendritic cells (CD11c + CD80 + CD86+) to 77.3%, improving the anti-tumor immune response.

Recently, Zou et al. (2023) reported a cancer cell-mitochondria hybrid membrane-coated Gboxin-loaded nanomedicine (HM-NPs@G) for targeted glioblastoma (GBM) therapy, addressing the limitations of Gboxin, an oxidative phosphorylation inhibitor, whose therapeutic potential is hindered by rapid clearance, poor blood-brain barrier penetration, and non-specific targeting (Fig. 14.3b). The hybrid membrane combines a cancer cell membrane for homotypic tumor targeting and a mitochondrial membrane for subcellular targeting, resulting in a nanoparticle with a size of 89.1 nm, demonstrating successful membrane fusion and stability. This platform achieved 70.4% drug loading efficiency and showed ROS-responsive release, with 86.6% Gboxin release in ROS-mimicking environments, significantly enhancing the drug's availability at tumor sites. In vitro studies demonstrated two-fold higher cellular uptake in U87MG glioblastoma cells compared to unmodified nanoparticles, and Gboxin accumulation in mitochondria reached 23.4 μg/mL, significantly greater than with single membrane systems. In vivo, HM-NPs@G exhibited a prolonged circulation half-life of 4.9 hours, compared to 0.47 hours for free Gboxin, and achieved 7.73% injected dose per gram (ID/g) tumor accumulation in GBM-bearing mice, over 6.9-fold higher than free Gboxin. This system resulted in 82.9% tumor growth inhibition in orthotopic U87MG and patient-derived GBM stem cell xenografts, significantly extending median survival to 67.5 days, compared to 35 days for free Gboxin.

Liu et al. (2024) developed urease-powered platelet cell motors (PLT@Au@Urease) to enhance cancer radiosensitization by improving delivery and penetration of radiosensitizers. These motors integrate gold nanoparticles (AuNPs) into platelets (PLT@Au) and conjugate urease as the propulsion engine, enabling active movement via urea catalysis. PLT@Au@Urease achieved a sensitization enhancement ratio (SER) of 1.89, surpassing PLT@Au (SER = 1.08). In vitro, PLT@Au@Urease increased intracellular reactive oxygen species (ROS) and γ-ray-induced DNA damage, confirmed by comet assays and γ-H2AX staining, showing the longest DNA tails and highest nuclear damage among treatments. Enhanced uptake due to self-propulsion amplified radiosensitization. In vivo, PLT@Au@Urease motors demonstrated superior tumor accumulation than non-motile PLT@Au. Mice treated with PLT@Au@Urease and a moderate dose of radiation showed significant tumor volume reduction, with TUNEL and Ki67 staining confirming increased apoptosis and reduced proliferation. The motors exhibited good biocompatibility, minimal systemic toxicity, and no organ histology changes.

14.5.2 Phototherapy

Phototherapy for cancer, encompassing photothermal therapy (PTT) and PDT, utilizes the precision of light-based technologies to target and destroy cancer cells while preserving nearby healthy tissues (Overchuk et al. 2023). PDT involves a three-component system: a photosensitizer, a light source, and oxygen. It starts with the administration of a photosensitizer that accumulates preferentially in cancer cells due to their higher metabolic rates. When activated by a specific wavelength of light, usually from lasers or LEDs, the photosensitizer generates ROS. These ROS cause significant cellular damage, leading to cell death via necrosis or apoptosis, and can also impair the tumor's vasculature, causing further cell death through hypoxia (Maharjan and Bhattarai 2022). Additionally, PDT can trigger an immune response by recruiting immune cells to treated site, which may contribute to a systemic anti-tumor response (Abrahamse and Hamblin 2016).

On the other hand, PTT uses nanoparticles that absorb NIR light and convert it into heat, thus killing cancer cells through localized hyperthermia. These nanoparticles, often made from materials like gold, copper sulphide, or iron oxide, are designed to accumulate in tumors due to the increased permeability and retention effect (Chen et al. 2019). Upon NIR light exposure, they generate heat specifically at the tumor site, raising temperatures to levels that can selectively destroy cancer cells without harming surrounding tissues. This selective heating requires careful control over light exposure to achieve the desired therapeutic temperatures while avoiding damage to normal tissues (Zhi et al. 2020). Both PDT and PTT face challenges in ensuring the selective targeting of treatments to cancer cells that can be tackled by leveraging engineered cell membrane-based systems. In an interesting study, Lin et al. (2022) developed an advanced NK cell membrane-cloaked virus-mimicking nanogenerator (AsHMS-TA/Fe^{III}@NK) for synergistic thermodynamic–chemodynamic therapy, offering an innovative approach to cancer treatment (Fig. 14.4). The nanogenerator consists of a virus-like hollow mesoporous silica core (sHMS) doped with disulfide bonds and loaded with 2,2′-azobis[2-(2-imidazolin-2-yl) propane] dihydrochloride (AIPH), a thermal azo-initiator for generating toxic alkyl radicals (•C). The core is coated with a tannic acid (TA)/FeIII coordination complex, which facilitates the production of hydroxyl radicals (•OH) via Fenton reactions. The NK cell membrane cloaks the nanogenerator, providing immune evasion and enhanced tumor accumulation. Upon NIR laser irradiation, the NK membrane ruptures, inducing a shape reversal to expose virus-like surface topography, which significantly enhances cellular uptake. Simultaneously, the thermal decomposition of AIPH generates •C radicals, while intracellular glutathione-responsive disintegration of the nanogenerator releases Fe^{III} for •OH production. In vitro, the nanogenerator showed potent cytotoxicity in HepG2 cells, reducing cell viability to 28.9% under NIR irradiation. In vivo studies in a HepG2 tumor-bearing mouse model demonstrated a 98.9% tumor growth inhibition, with complete tumor elimination observed in some mice by day 21. Additionally, the NIR-triggered AsHMS-TA/Fe^{III}@NK showed prolonged circulation (half-life of 5.05 hours) and

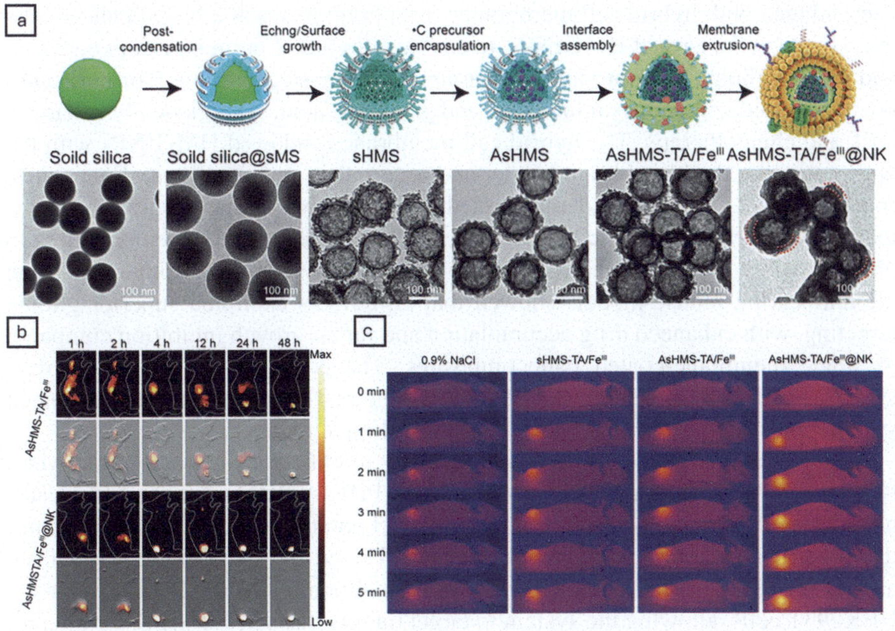

Fig. 14.4 (**a**) Synthesis method and corresponding transmission electron microscopy images of AsHMS-TA/Fe^{III}@NK. (**b**) Real-time NIR-II fluorescence imaging of HepG2 tumor-bearing nude mice following the administration of IR820-labeled AsHMS-TA/Fe^{III} and AsHMS-TA/Fe^{III}. (**c**) Thermographic scans of HepG2 tumor-bearing nude mice. (Reprinted with permission from Lin et al. 2022, copyright 2021, Wiley-VCH GmbH)

enhanced tumor accumulation, with a 1.75-fold increase in NIR fluorescence signal compared to controls.

Wang et al. (2023) developed a genetically modified T-cell membrane-coated aggregation-induced emission (AIE) nanoparticles (CM@AIE NPs) to target both glioblastoma (GBM) cells and glioblastoma stem cells (GSCs), which are pivotal in tumor recurrence. The T-cell membrane was modified to express chimeric antigen receptors against CD133 and EGFR, enabling selective targeting of GSCs and GBM cells, respectively. The CM@AIE NPs demonstrated strong stability and biocompatibility. In vitro, CM@AIE NPs efficiently crossed the blood-brain barrier with a 17% penetration efficiency, and upon laser irradiation, generated a photothermal conversion efficiency of 28.7%, raising the local temperature to 57.5 °C, which is sufficient for inducing hyperthermia. Targeting specificity was confirmed, as CM@AIE NPs showed enhanced cellular uptake in U87-MG glioblastoma cells and GSCs, significantly inhibiting GSC tumorsphere formation and self-renewal. In vivo, CM@AIE NPs accumulated in the GBM region of orthotopic mouse models, resulting in a four-fold higher fluorescence signal in the tumor site compared to controls. Upon laser activation, the treatment achieved complete tumor ablation and reduced CD133+ GSCs and SOX2 expression, effectively preventing GBM recurrence over a 60-day period. Chi et al. (2023) reported a biomimetic nanocomposite camouflaged with hybrid cell membranes from brain metastatic breast cancer cells and glioma cells, termed HMGINPs, to enable both blood-brain barrier penetration and precise glioma targeting for the treatment of early-stage glioma. The nanocomposites, loaded with indocyanine green and gambogic acid, synergistically enhanced PTT and chemotherapy. The hybrid cell membranes endowed HMGINPs with the ability to cross the blood-brain barrier and accumulate specifically at glioma sites. In vitro and in vivo studies demonstrated that HMGINPs achieved a 5.2% photothermal conversion efficiency under 808 nm laser, raising tumor temperatures to 48.8 °C, which induced effective tumor ablation. In vivo studies using early-stage glioma-bearing mouse models showed that HMGINPs exhibited superior glioma targeting, with enhanced drug accumulation and tumor growth inhibition compared to single membrane-coated nanocomposites. The combined PTT-chemotherapy synergistically minimized toxicity to healthy tissues. This hybrid membrane-coated system holds promise for targeted early-stage glioma treatment.

Zhang et al. (2024a) developed a genetically modified eukaryocyte-prokaryotic hybrid membrane-camouflaged nanoemulsion (PHL@PSHM) designed for synergistic cancer treatment combining immune checkpoint blockade, PTT, and chemodynamic therapy. The hybrid membrane was formed by integrating OMVs from attenuated *Salmonella typhimurium* with membranes from PD-1-expressing HEK293T cells, allowing the system to target tumors through PD-1/PD-L1 interactions, while leveraging the immunoadjuvant effects of bacterial OMVs. The nanoemulsion core, made of lactate oxidase, human serum albumin, and perfluorotributylamine, facilitated oxygen generation and ROS production via lactate oxidase-mediated conversion of lactic acid to hydrogen peroxide. The PHL@PSHM nanoparticles successfully upregulated PD-L1 expression in tumor cells, enhancing immune checkpoint inhibition, while also inducing RBC extravasation into the tumor microenvironment, significantly improving PTT efficacy. In vivo studies using 4 T1 breast cancer and CT26 colorectal cancer models demonstrated robust anti-tumor effects, with tumor growth inhibition rates of 92.8% and 99.3%, respectively, when combined with 808 nm laser irradiation. Additionally, the hybrid membrane system synergistically enhanced cytotoxic T cell infiltration, increasing the proportion of CD8+ T cells to 29.5%, while reducing regulatory T cells (Tregs) to 6.15%, significantly boosting anti-tumor immunity.

Wei et al. (2021) developed a click pretargeting strategy for anchoring acidity-activatable photosensitizers onto tumor cell membranes, enhancing PDT. The two-step approach uses azide-functionalized NPs (Az-NPs) to display azide groups onto tumor cell surfaces via metabolic glycoengineering process, followed by pH-responsive nanoparticles (S-NPs) containing dibenzocyclooctyne (DBCO) and chlorin e6 (Ce6). In acidic tumor environments (pH 6.5), S-NPs disassemble, exposing DBCO and Ce6, which react with azide groups to anchor photosensitizers on tumor membranes. Fluorescence intensity increased 4.9-fold under acidic conditions compared to neutral pH. This approach induced high ROS generation and tumor cell death under laser irradiation. In vivo, Az-NPs/S-NPs showed specific tumor accumulation with fluorescence retained for 72 hours, exceeding non-targeted systems. In a 4 T1 tumor model, PDT with anchored S-NPs resulted in a tumor inhibition rate of 97%, as confirmed by apoptosis and minimal damage to surrounding tissues. This strategy also enabled dual-modal fluorescence and MR imaging, improving tumor localization and therapy monitoring.

Taghikhani et al. (2020) developed a bioorthogonal click-targeting nanocomposite system (DLQ/DZ) for integrated chemo-photothermal treatment in breast cancer. The strategy integrates metabolic glycoengineering using azide-modified sugar (Ac4ManNAz) to generate high-density azide groups on tumor cell membranes, followed by tumor-targeted binding of DLQ/DZ nanocomposites via click chemistry. The nanocomposites co-encapsulate doxorubicin (DOX) and zinc phthalocyanine (ZnPc), achieving a high dual cargo-loading capacity of ~20%. The DLQ/DZ exhibited excellent stability and pH-responsive cargo release, favoring the acidic tumor microenvironment. This system demonstrated enhanced cellular uptake, resulting in synergistic cytotoxicity when combined with NIR laser irradiation. In vivo, the DLQ/DZ nanocomposites achieved a 96.1% tumor inhibition rate in MCF-7 xenograft models, significantly outperforming monotherapies or dual therapies without click-targeting. Photothermal therapy using ZnPc elevated tumor temperatures to 57 °C, enhancing drug penetration and ablation of tumor tissues.

14.5.3 Immunotherapy

Cancer immunotherapies boost the immune system's efficacy to combat cancer cells. This approach incorporates strategies like checkpoint inhibitors to enhance immune cell activity, chimeric antigen receptor T-cell (CAR-T) therapy to modify T-cells for improved cancer targeting, and cancer vaccines to activate the immune system to identify and attack cancerous cells (Rui et al. 2023). Unlike chemotherapy and radiation that target cancer cells, immunotherapy strengthens the immune system, offering a more selective approach with fewer side effects (Tan et al. 2020). This method has shown considerable effectiveness in treating melanoma and non-small cell lung cancer, with ongoing research expanding its use to more cancer types (Desai et al. 2024d).

Zhai et al. (2021) developed a T lymphocyte membrane-decorated epigenetic nanoinducer of interferons (OPEN) for enhanced cancer treatment, targeting the tumor microenvironment through simultaneous induction of type I interferons (IFNs) and immune checkpoint blockade. The core of the nanovesicles is composed of bovine serum albumin cross-linked with a reversible linker, loaded with ORY-1001, a potent inhibitor of lysine-specific histone demethylase 1, which epigenetically induces endogenous retroviruses, IFNs, and major histocompatibility complex class I (MHC-I) molecules (Fig. 14.5). The nanoparticles, coated with T lymphocyte membranes engineered to overexpress PD1, specifically target and bind to PD-L1 on tumor cells. Binding facilitates PD-L1 internalization, preventing immune evasion and enhancing IFN-mediated antitumor responses. The nanoparticles exhibited a core-shell structure, with drug release triggered by intracellular glutathione levels. In murine models of TNBC, melanoma, and colon cancer, OPEN increased intratumoral IFN-α and IFN-β levels by sixfold and threefold,

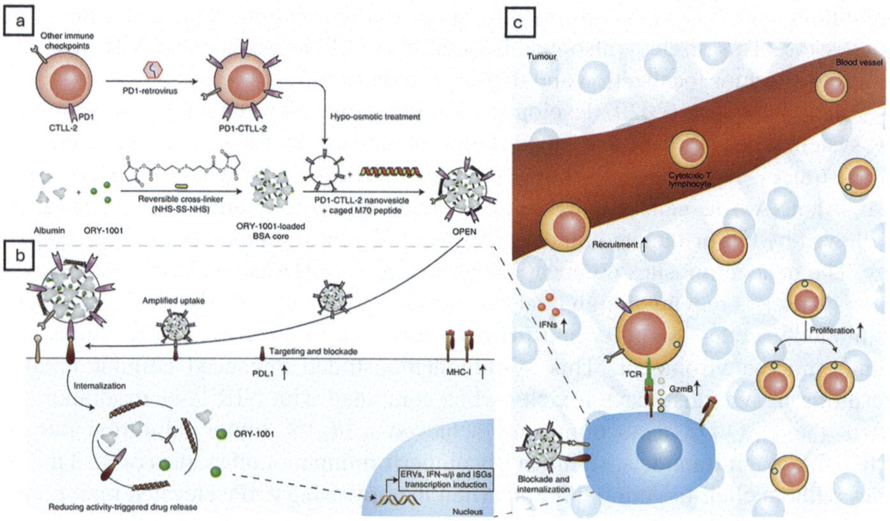

Fig. 14.5 Schematic depiction of the OPEN preparation and action mechanism. (**a**) OPEN is formulated using Macrolittin 70 (M70). (**b**) Upon administration, OPEN targets and penetrates PDL1-expressing cells, where it releases ORY-1001 aided by M70. ORY-1001 then stimulates the upregulation of interferons (IFNs), major histocompatibility complex I (MHC-I), and PDL1. The enhanced PDL1 is subsequently neutralized by additional OPEN doses. (**c**) Following intravenous administration and tumor localization, OPEN facilitates the enhancement of cytotoxic T lymphocyte (CTL) recruitment, and activity within the tumor microenvironment. (Reprinted with permission from Zhai et al. 2021, copyright 2021, Springer Nature)

respectively, and upregulated MHC-I and interferon-stimulated genes. This led to a 13-fold and 28-fold increase in active Ki-67+ and GzmB+ CD8+ T cell density in treated tumors. OPEN also inhibited tumor growth by 89% in TNBC models, outperforming traditional immune checkpoint blockade therapies by maintaining low surface PD-L1 levels through internalization.

Li et al. (2023b) developed genetically programmable vesicles expressing anti-PD-L1 single-chain variable fragments (scFv) fused with the glutamine antagonist DON (D@aPD-L1 NVs) to boost CAR-T efficacy against solid tumors, particularly addressing the limitations posed by the immunosuppressive tumor microenvironment. The engineered vesicles target PD-L1 on tumor cells, which upregulates during glutamine antagonist therapy and T cell activation, leading to CAR-T exhaustion. The D@aPD-L1 NVs achieved a homogeneous particle size of 90 nm and a zeta potential of −20.83 mV, with high DON loading efficiency. In vitro, D@aPD-L1 NVs showed a significantly higher binding affinity for PD-L1 and inhibited tumor glutamine metabolism more effectively than free DON. In vivo, these vesicles preferentially accumulated at the tumor site with an 8-hour retention rate of 27.7%, compared to 15.6% for free vesicles. Treatment with D@aPD-L1 NVs in combination with mesothelin-specific CAR-T (MSLN-mCAR-T) cells resulted in substantial tumor suppression, enhancing CAR-T infiltration and survival while reducing immune-suppressive myeloid-derived suppressor cells from 44.9% to 19.7% and

M2 macrophages, alongside increased CD8+ T-cell infiltration and effector function. The combination therapy significantly extended survival, with 83.3% of mice surviving over 100 days, demonstrating a profound improvement in CAR-T efficacy and long-term tumor resistance through enhanced immune memory response.

Recently, Zhang et al. (2024b) reported genetically modified membrane-based nanoengagers (NanoBE) designed for targeted immunotherapy against pancreatic cancer. The NanoBE system consists of PLGA nanoparticles coated with genetically engineered membranes expressing bispecific scFv targeting CD40 on macrophages and Claudin18.2 on pancreatic tumor cells. This dual-targeting mechanism promotes macrophage-mediated phagocytosis and enhances T-cell activation by facilitating the crosslinking of tumor cells and macrophages. NanoBE nanoparticles exhibited a size of ~150 nm and a high degree of bispecificity, with 85.9% of nanoparticles displaying both anti-CD40 and anti-Claudin18.2 scFv on their surface. In vitro, assays demonstrated that NanoBE significantly increased macrophage activation, upregulating CD80/CD86 expression and enhancing the phagocytosis of Claudin18.2-positive tumor cells by 6.58%, markedly higher than single-targeted controls. Additionally, NanoBE-treated macrophages showed a 200% increase in antigen cross-presentation, leading to a four-fold increase in interferon-gamma (IFN-γ) secretion from T cells, indicating improved T-cell priming. In vivo, NanoBE treatment significantly inhibited tumor growth in a murine model of Claudin18.2-expressing pancreatic cancer, extending survival to over 100 days, with the treated tumors exhibiting the highest infiltration of CD3+ and CD8+ T cells and elevated IFN-γ production.

Bellavia et al. (2017) developed Interleukin-3 receptor (IL3-R)-targeted exosomes (IL3L Exo) for precision delivery of therapeutics in chronic myelogenous leukemia (CML). Derived from HEK293T cells expressing Lamp2b fused to IL3, these exosomes selectively target IL3-R, overexpressed on CML cells. IL3L Exo, incorporated with Imatinib or BCR-ABL siRNA, showed enhanced specificity and therapeutic efficacy. Characterized at 30–60 nm, IL3L Exo efficiently integrated IL3-Lamp2b, confirmed via western blot. In vitro, Imatinib-loaded IL3L Exo reduced CML cell viability (LAMA84, K562, and K562R) at doses 37-fold lower than free Imatinib. siRNA-loaded IL3L Exo effectively knocked down BCR-ABL expression, inhibiting proliferation, including in Imatinib-resistant cells. In vivo, IL3L Exo localized to tumor sites within 1 hour, sustained for 24 hours, with minimal off-target accumulation. In xenograft models, Imatinib-loaded IL3L Exo significantly suppressed tumor growth compared to free Imatinib, while siRNA-loaded IL3L Exo caused tumor regression in Imatinib-resistant models. IL3L Exo achieved enhanced macrophage activation and antigen presentation, amplifying immune responses.

Han et al. (2022) developed genetically engineered exosomes (GExoI) to target premetastatic niches and suppress postoperative melanoma lung metastases. Derived from MSCs expressing GFE1-Lamp2b, GExoI targeted pulmonary vascular endothelial cells and delivered the PI3Kγ inhibitor IPI549. The exosomes had a size of 114.3 nm, an IPI549 encapsulation efficiency of 17.79%, and released 83.26% of the drug over 24 hours. In vivo, GExoI demonstrated enhanced pulmonary accumulation and reduced granulocytic myeloid-derived suppressor cell migration to 7.77% by inhibiting the CXCL/CXCR2/PI3Kγ axis. In a postoperative lung metastasis model, GExoI decreased metastatic lung nodules and shifted the immune microenvironment by reducing M2 macrophages and monocytic myeloid-derived suppressor cells in conjunction with increased T-cell infiltration. Histological analyses confirmed reduced metastatic lesions and minimal inflammation in GExoI-treated mice.

Zhou et al. (2019) developed iRGD-exosomes for tumor-specific delivery of KRAS siRNA to target the "undruggable" KRAS gene in lung cancer. Engineered from HEK293T cells, these exosomes expressed Lamp2b fused to iRGD peptide, which binds integrin αvβ3 on tumor cells for precise targeting. The iRGD-exosomes effectively encapsulated KRAS siRNA without compromising exosomal stability. In vitro, siRNA-loaded iRGD-exosomes significantly decreased both KRAS mRNA and protein expression in A549 lung cancer cells and inhibited proliferation, outperforming wild-type exosomes. In vivo research in xenograft mouse models demonstrated that iRGD-exosomes selectively accumulated in tumors, confirmed by fluorescence imaging, with minimal off-target distribution. Three intravenous injections of siRNA-loaded iRGD-exosomes suppressed tumor growth significantly. Histological analysis revealed reduced cellular atypia and increased apoptotic and necrotic features in treated tumors. This study establishes iRGD-exosomes as a highly effective system for precise KRAS siRNA delivery, offering a promising approach for targeting KRAS-driven cancers and other challenging oncogenes.

14.6 Conclusion

Cell membrane engineering is a promising approach for improving cancer therapeutic delivery by overcoming challenges of conventional treatments. Using biological membranes from sources like red blood cells, immune cells, exosomes, and bacterial vesicles, these engineered systems offer better compatibility, immune evasion, and targeting of tumors. Advances in methods for isolating and coating membranes, along with techniques like lipid insertion and genetic modifications, have enabled their application in therapies such as chemotherapy, phototherapy, and immunotherapy. However, challenges remain in moving these technologies to clinical use. Current methods are difficult to scale for industrial production, and variability in membrane sources affects consistency and performance. Long-term stability during storage and use also needs improvement to maintain functionality. Regulatory requirements for safety and ethical concerns, particularly for genetically engineered membranes, add further complexity. To overcome these challenges, robust testing through preclinical and clinical studies is necessary, with models that closely mimic human tumors. Standardizing protocols for membrane preparation and functionalization can improve consistency. Future research may integrate this technology with artificial intelligence to optimize design and testing. Exploring systems that combine targeting with immune modulation or multiple therapies can also enhance outcomes. With continued advancements, cell membrane-coated nanosystems could provide more effective and personalized cancer treatments.

References

Abrahamse H, Hamblin MR (2016) New photosensitizers for photodynamic therapy. Biochem J 473:347–364. https://doi.org/10.1042/BJ20150942

Adhalrao SB, Jadhav KR, Patil PL, Kadam VJ et al (2024) Engineering platelet membrane imitating nanoparticles for targeted therapeutic delivery. Curr Pharm Biotechnol 25:1230–1244. https://doi.org/10.2174/1389201024666230804140926

Ahmad A, Khan F, Mishra RK, Khan R (2019) Precision cancer Nanotherapy: evolving role of multifunctional nanoparticles for cancer active targeting. J Med Chem 62:10475–10496. https://doi.org/10.1021/acs.jmedchem.9b00511

Ahmed W, Mushtaq A, Ali S, Khan N, Liang Y, Duan L (2024) Engineering approaches for exosome cargo loading and targeted delivery: biological versus chemical perspectives. ACS Biomater Sci Eng 10:5960–5976. https://doi.org/10.1021/acsbiomaterials.4c00856

Ai X, Wang S, Duan Y, Zhang Q, Chen MS, Gao W, Zhang L (2021) Emerging approaches to functionalizing cell membrane-coated nanoparticles. Biochemistry 60:941–955. https://doi.org/10.1021/acs.biochem.0c00343

Ak G, Hamarat Şanlıer Ş (2020) Erythrocyte membrane vesicles coated biomimetic and targeted doxorubicin nanocarrier: development, characterization and in vitro studies. J Mol Struct 1205:127664. https://doi.org/10.1016/j.molstruc.2019.127664

Anand D, Chaudhuri A (2016) Bacterial outer membrane vesicles: new insights and applications. Mol Membr Biol 33:125–137. https://doi.org/10.1080/09687688.2017.1400602

Anselmo AC, Modery-Pawlowski CL, Menegatti S, Kumar S, Vogus DR, Tian LL, Chen M, Squires TM, Sen Gupta A, Mitragotri S (2014) Platelet-like nanoparticles: mimicking shape, flexibility, and surface biology of platelets to target vascular injuries. ACS Nano 8:11243–11253. https://doi.org/10.1021/nn503732m

Attia MF, Anton N, Wallyn J, Omran Z, Vandamme TF (2019) An overview of active and passive targeting strategies to improve the nanocarriers efficiency to tumour sites. J Pharm Pharmacol 71:1185–1198. https://doi.org/10.1111/jphp.13098

Batrakova EV, Kim MS (2015) Using exosomes, naturally-equipped nanocarriers, for drug delivery. J Control Release 219:396–405. https://doi.org/10.1016/j.jconrel.2015.07.030

Bellavia D, Raimondo S, Calabrese G, Forte S, Cristaldi M, Patinella A, Memeo L, Manno M, Raccosta S, Diana P, Cirrincione G, Giavaresi G, Monteleone F, Fontana S, De Leo G, Alessandro R (2017) Interleukin 3- receptor targeted exosomes inhibit *in vitro* and *in vivo* chronic Myelogenous leukemia cell growth. Theranostics 7:1333–1345. https://doi.org/10.7150/thno.17092

Bernardini G, Zabel BA (2019) Editorial: the role of Chemoattractants in the tumor microenvironment. Front Immunol 10. https://doi.org/10.3389/fimmu.2019.02671

Bisht D, Salave S, Desai N, Gogoi P, Rana D, Biswal P, Sarma G, Benival D, Kommineni N, Desai D (2024) Genome editing and its role in vaccine, diagnosis, and therapeutic advancement. Int J Biol Macromol 269:131802. https://doi.org/10.1016/j.ijbiomac.2024.131802

Bray F, Laversanne M, Sung H, Ferlay J, Siegel RL, Soerjomataram I, Jemal A (2022) GLOBOCAN estimates of incidence and mortality worldwide for 36 cancers in 185 countries. CA Cancer J Clin 74(2024):229–263. https://doi.org/10.3322/caac.21834

Brown JS, Amend SR, Austin RH, Gatenby RA, Hammarlund EU, Pienta KJ (2023) Updating the definition of cancer. Mol Cancer Res 21:1142–1147. https://doi.org/10.1158/1541-7786.MCR-23-0411

Cao H, Dan Z, He X, Zhang Z, Yu H, Yin Q, Li Y (2016) Liposomes coated with isolated macrophage membrane can target lung metastasis of breast cancer. ACS Nano 10:7738–7748. https://doi.org/10.1021/acsnano.6b03148

Cao X, Hu Y, Luo S, Wang Y, Gong T, Sun X, Fu Y, Zhang Z (2018) Neutrophil-mimicking therapeutic nanoparticles for targeted chemotherapy of pancreatic carcinoma. Acta Pharm Sin B 9:575–589. https://doi.org/10.1016/J.APSB.2018.12.009

Chavda VP, Pandya A, Kumar L, Raval N, Vora LK, Pulakkat S, Patravale V, Salwa Y, Duo BZT (2023) Exosome nanovesicles: a potential carrier for therapeutic delivery. Nano Today 49:101771. https://doi.org/10.1016/j.nantod.2023.101771

Chen J, Ning C, Zhou Z, Yu P, Zhu Y, Tan G, Mao C (2019) Nanomaterials as photothermal therapeutic agents. Prog Mater Sci 99:1–26. https://doi.org/10.1016/j.pmatsci.2018.07.005

Chen HY, Deng J, Wang Y, Wu CQ, Li X, Dai HW (2020a) Hybrid cell membrane-coated nanoparticles: a multifunctional biomimetic platform for cancer diagnosis and therapy. Acta Biomater 112:1–13. https://doi.org/10.1016/J.ACTBIO.2020.05.028

Chen H-Y, Deng J, Wang Y, Wu C-Q, Li X, Dai H-W (2020b) Hybrid cell membrane-coated nanoparticles: a multifunctional biomimetic platform for cancer diagnosis and therapy. Acta Biomater 112:1–13. https://doi.org/10.1016/j.actbio.2020.05.028

Chen Q, Huang G, Wu W, Wang J, Hu J, Mao J, Chu PK, Bai H, Tang G (2020c) A hybrid eukaryotic–prokaryotic Nanoplatform with Photothermal modality for enhanced antitumor vaccination. Adv Mater 32:e1908185. https://doi.org/10.1002/adma.201908185

Cheng S, Xu C, Jin Y, Li Y, Zhong C, Ma J, Yang J, Zhang N, Li Y, Wang C, Yang Z, Wang Y (2020) Artificial mini dendritic cells boost T cell–based immunotherapy for ovarian cancer. Adv Sci 7. https://doi.org/10.1002/advs.201903301

Chi S, Zhang L, Cheng H, Chang Y, Zhao Y, Wang X, Liu Z (2023) Biomimetic nanocomposites camouflaged with hybrid cell membranes for accurate therapy of early-stage glioma. Angew Chem Int Ed 62:e202304419. https://doi.org/10.1002/anie.202304419

Chugh V, Vijaya Krishna K, Pandit A (2021) Cell membrane-coated mimics: a methodological approach for fabrication, characterization for therapeutic applications, and challenges for clinical translation. ACS Nano 15:17080–17123

Cintolo JA, Datta J, Mathew SJ, Czerniecki BJ (2012) Dendritic cell-based vaccines: barriers and opportunities. Future Oncol 8:1273–1299. https://doi.org/10.2217/fon.12.125

Dehaini D, Wei X, Fang RH, Masson S, Angsantikul P, Luk BT, Zhang Y, Ying M, Jiang Y, Kroll AV, Gao W, Zhang L (2017) Erythrocyte–platelet hybrid membrane coating for enhanced nanoparticle functionalization. Adv Mater 29. https://doi.org/10.1002/adma.201606209

Deng G, Sun Z, Li S, Peng X, Li W, Zhou L, Ma Y, Gong P, Cai L (2018) Cell-membrane immunotherapy based on natural killer cell membrane coated nanoparticles for the effective inhibition of primary and Abscopal tumor growth. ACS Nano 12:12096–12108. https://doi.org/10.1021/acsnano.8b05292

Desai N, Gadeval A, Kathar U, Sengupta P, Kalia K, Tekade RK (2021) Emerging roles and biopharmaceutical applications of milk derived exosomes. J Drug Deliv Sci Technol 64:102577. https://doi.org/10.1016/j.jddst.2021.102577

Desai N, Katare P, Makwana V, Salave S, Vora LK, Giri J (2023a) Tumor-derived systems as novel biomedical tools – turning the enemy into an ally. Biomater Res 27:113. https://doi.org/10.1186/s40824-023-00445-z

Desai N, Rana D, Pande S, Salave S, Giri J, Benival D, Kommineni N (2023b) "Bioinspired" membrane-coated Nanosystems in cancer Theranostics: a comprehensive review. Pharmaceutics 15:1677. https://doi.org/10.3390/pharmaceutics15061677

Desai N, Hasan U et al (2023c) Biomaterial-based platforms for modulating immune components against cancer and cancer stem cells. Acta Biomater 161:1–36. https://doi.org/10.1016/j.actbio.2023.03.004

Desai N, Tambe V, Pofali P, Vora LK (2024a) Cell membrane-coated nanoparticles: a new frontier in immunomodulation. Adv Nanobiomed Res 4:2400012. https://doi.org/10.1002/ANBR.202400012

Desai N, Rana D, Salave S, Benival D, Khunt D, Prajapati BG (2024b) Achieving endo/lysosomal escape using smart nanosystems for efficient cellular delivery. Molecules 29(13):3131

Desai N, Tambe V, Pofali P, Vora LK (2024c) Cell membrane-coated nanoparticles: a new frontier in immunomodulation. Adv Nanobiomed Res 4. https://doi.org/10.1002/anbr.202400012

Desai N, Chavda V, Singh TRR, Thorat ND, Vora LK (2024d) Cancer Nanovaccines: nanomaterials and clinical perspectives. Small 20:e2401631. https://doi.org/10.1002/smll.202401631

Desai N, Pande S, Salave S, Vora LK (2025) Exosomes nanovesicles for gene delivery and vaccination. In: Exosome communication. Elsevier, pp 329–374. https://doi.org/10.1016/B978-0-443-29052-7.00013-1

Doyle LM, Wang MZ (2019) Overview of extracellular vesicles, their origin, composition, purpose, and methods for exosome isolation and analysis. Cells 8. https://doi.org/10.3390/cells8070727

Ellis TN, Kuehn MJ (2010) Virulence and immunomodulatory roles of bacterial outer membrane vesicles. Microbiol Mol Biol Rev 74:81–94. https://doi.org/10.1128/MMBR.00031-09

Fang RH, Hu C-MJ, Chen KNH, Luk BT, Carpenter CW, Gao W, Li S, Zhang D-E, Lu W, Zhang L (2013) Lipid-insertion enables targeting functionalization of erythrocyte membrane-cloaked nanoparticles. Nanoscale 5:8884–8888. https://doi.org/10.1039/c3nr03064d

Fang RH, Hu C-MJ, Luk BT, Gao W, Copp JA, Tai Y, O'Connor DE, Zhang L (2014) Cancer cell membrane-coated nanoparticles for anticancer vaccination and drug delivery. Nano Lett 14:2181–2188. https://doi.org/10.1021/nl500618u

Fang RH, Kroll AV, Gao W, Zhang L (2018) Cell membrane coating nanotechnology. Adv Mater 30:1706759. https://doi.org/10.1002/ADMA.201706759

Gilmore WJ, Johnston EL, Zavan L, Bitto NJ, Kaparakis-Liaskos M (2021) Immunomodulatory roles and novel applications of bacterial membrane vesicles. Mol Immunol 134:72–85. https://doi.org/10.1016/j.molimm.2021.02.027

Gong P, Wang Y, Zhang P, Yang Z, Deng W, Sun Z, Yang M, Li X, Ma G, Deng G, Dong S, Cai L, Jiang W (2020) Immunocyte membrane-coated nanoparticles for cancer immunotherapy. Cancers (Basel) 13:77. https://doi.org/10.3390/cancers13010077

Guo Y, Wang D, Song Q, Wu T, Zhuang X, Bao Y, Kong M, Qi Y, Tan S, Zhang Z (2015) Erythrocyte membrane-enveloped polymeric nanoparticles as Nanovaccine for induction of antitumor immunity against melanoma. ACS Nano 9:6918–6933. https://doi.org/10.1021/acsnano.5b01042

Han Y, Pan H, Li W, Chen Z, Ma A, Yin T, Liang R, Chen F, Ma Y, Jin Y, Zheng M, Li B, Cai L (2019) T cell membrane mimicking nanoparticles with bioorthogonal targeting and immune recognition for enhanced Photothermal therapy. Adv Sci 6. https://doi.org/10.1002/advs.201900251

Han X, Bi L, Wu Y, Yan J, Wu X, Zheng R, Sun Y, Zhang H, Wang Z, Wang Y, Zhang H (2022) Genetically engineered exosomes for targetedly preventing premetastatic niche formation and suppressing postoperative melanoma lung metastasis. Nano Today 46:101597. https://doi.org/10.1016/j.nantod.2022.101597

Harris JC, Scully MA, Day ES (2019) Cancer cell membrane-coated nanoparticles for cancer management. Cancers (Basel) 11. https://doi.org/10.3390/CANCERS11121836

Hu C-MJ, Zhang L, Aryal S, Cheung C, Fang RH, Zhang L (2011a) Erythrocyte membrane-camouflaged polymeric nanoparticles as a biomimetic delivery platform. Proc Natl Acad Sci 108:10980–10985. https://doi.org/10.1073/pnas.1106634108

Hu CMJ, Zhang L, Aryal S, Cheung C, Fang RH, Zhang L (2011b) Erythrocyte membrane-camouflaged polymeric nanoparticles as a biomimetic delivery platform. Proc Natl Acad Sci USA 108:10980–10985

Hu C-MJ, Fang RH, Zhang L (2012) Erythrocyte-inspired delivery systems. Adv Healthc Mater 1:537–547. https://doi.org/10.1002/adhm.201200138

Janiszewska M, Primi MC, Izard T (2020) Cell adhesion in cancer: beyond the migration of single cells. J Biol Chem 295:2495–2505. https://doi.org/10.1074/jbc.REV119.007759

Jiang Z, Jones R, Liu JC, Deng T, Robinson T, Chung PED, Wang S, Herschkowitz JI, Egan SE, Perou CM, Zacksenhaus E (2011) RB1 and p53 at the crossroad of EMT and triple-negative breast cancer. Cell Cycle 10:1563–1570. https://doi.org/10.4161/cc.10.10.15703

Jiang Q, Wang K, Zhang X, Ouyang B, Liu H, Pang Z, Yang W (2020) Platelet membrane-camouflaged magnetic nanoparticles for Ferroptosis-enhanced cancer immunotherapy. Small 16. https://doi.org/10.1002/SMLL.202001704

Kalluri R, LeBleu VS (2020) The biology, function, and biomedical applications of exosomes. Science 367. https://doi.org/10.1126/science.aau6977

Kandula S, Singh PK, Kaur GA, Tiwari A (2023) Trends in smart drug delivery systems for targeting cancer cells. Mater Sci Eng B 297:116816. https://doi.org/10.1016/j.mseb.2023.116816

Kang T, Zhu Q, Wei D, Feng J, Yao J, Jiang T, Song Q, Wei X, Chen H, Gao X, Chen J (2017) Nanoparticles coated with neutrophil membranes can effectively treat cancer metastasis. ACS Nano 11:1397–1411

Kang X, Zhang Y, Song J, Wang L, Li W, Qi J, Tang BZ (2023) A photo-triggered self-accelerated nanoplatform for multifunctional image-guided combination cancer immunotherapy. Nat Commun 14:5216. https://doi.org/10.1038/s41467-023-40996-2

Kaparakis-Liaskos M, Ferrero RL (2015) Immune modulation by bacterial outer membrane vesicles. Nat Rev Immunol 15:375–387. https://doi.org/10.1038/nri3837

Krishnamurthy S, Muthukumaran P, Jayakumar MKG, Lisse D, Masurkar ND, Xu C, Chan JM, Drum CL (2019) Surface protein engineering increases the circulation time of a cell membrane-based nanotherapeutic. Nanomedicine 18:169–178. https://doi.org/10.1016/j.nano.2019.02.024

Krishnan N, Kubiatowicz LJ, Holay M, Zhou J, Fang RH, Zhang L (2022) Bacterial membrane vesicles for vaccine applications. Adv Drug Deliv Rev 185:114294. https://doi.org/10.1016/j.addr.2022.114294

Krishnan N, Peng F-X, Mohapatra A, Fang RH, Zhang L (2023) Genetically engineered cellular nanoparticles for biomedical applications. Biomaterials 296:122065. https://doi.org/10.1016/j.biomaterials.2023.122065

Kunde SS, Wairkar S (2021) Platelet membrane camouflaged nanoparticles: biomimetic architecture for targeted therapy. Int J Pharm 598:120395. https://doi.org/10.1016/j.ijpharm.2021.120395

Le Q-V, Lee J, Lee H, Shim G, Oh Y-K (2021) Cell membrane-derived vesicles for delivery of therapeutic agents. Acta Pharm Sin B 11:2096–2113. https://doi.org/10.1016/j.apsb.2021.01.020

Li J, Ai Y, Wang L, Bu P, Sharkey CC, Wu Q, Wun B, Roy S, Shen X, King MR (2016) Targeted drug delivery to circulating tumor cells via platelet membrane-functionalized particles. Biomaterials 76:52–65. https://doi.org/10.1016/j.biomaterials.2015.10.046

Li M, Zhou H, Yang C, Wu Y, Zhou X, Liu H, Wang Y (2020) Bacterial outer membrane vesicles as a platform for biomedical applications: an update. J Control Release 323:253–268. https://doi.org/10.1016/j.jconrel.2020.04.031

Li J, Wang Q, Xia G, Adilijiang N, Li Y, Hou Z, Fan Z, Li J (2023a) Recent advances in targeted drug delivery strategy for enhancing oncotherapy. Pharmaceutics 15:2233. https://doi.org/10.3390/pharmaceutics15092233

Li X, Zhu T, Wang R, Chen J, Tang L, Huo W, Huang X, Cao Q (2023b) Genetically programmable vesicles for enhancing CAR-T therapy against solid tumors. Adv Mater 35:e2211138. https://doi.org/10.1002/adma.202211138

Lian S, Xie X, Lu Y, Lee J (2019) Checkpoint CD47 function on tumor metastasis and immune therapy. Onco Targets Ther 12:9105–9114. https://doi.org/10.2147/OTT.S220196

Lin J, Li Y, Wang P, Wu M, Zhu F, Zhang Y, Hou Z, Liu J, Liu X (2022) Natural killer cell membrane-cloaked virus-mimicking Nanogenerator with NIR-triggered shape reversal and •C/•OH storm for synergistic thermodynamic–Chemodynamic therapy. Adv Sci 9:e2103498. https://doi.org/10.1002/advs.202103498

Liu G, Zhao X, Zhang Y, Xu J, Xu J, Li Y, Min H, Shi J, Zhao Y, Wei J, Wang J, Nie G (2019a) Engineering biomimetic Platesomes for pH-responsive drug delivery and enhanced antitumor activity. Adv Mater 31:e1900795. https://doi.org/10.1002/adma.201900795

Liu W-L, Zou M-Z, Liu T, Zeng J-Y, Li X, Yu W-Y, Li C-X, Ye J-J, Song W, Feng J, Zhang X-Z (2019b) Cytomembrane nanovaccines show therapeutic effects by mimicking tumor cells and antigen presenting cells. Nat Commun 10:3199. https://doi.org/10.1038/s41467-019-11157-1

Liu W, Zou M, Liu T, Zeng J, Li X, Yu W, Li C, Ye J, Song W, Feng J, Zhang X (2019c) Expandable immunotherapeutic Nanoplatforms engineered from Cytomembranes of hybrid cells derived from cancer and dendritic cells. Adv Mater 31:e1900499. https://doi.org/10.1002/adma.201900499

Liu L, Bai X, Martikainen MV, Kårlund A, Roponen M, Xu W, Hu G, Tasciotti E, Lehto VP (2021) Cell membrane coating integrity affects the internalization mechanism of biomimetic nanoparticles. Nat Commun 12(1):1–12. https://doi.org/10.1038/s41467-021-26052-x

Liu L, Pan D, Chen S, Martikainen MV, Kårlund A, Ke J, Pulkkinen H, Ruhanen H, Roponen M, Käkelä R, Xu W, Wang J, Lehto VP (2022) Systematic design of cell membrane coating to improve tumor targeting of nanoparticles. Nat Commun 13(1):1–15. https://doi.org/10.1038/s41467-022-33889-3

Liu W, Nie H, Li H, Liu Y, Tian M, Wang S, Yang Y, Long W (2024) Engineered platelet cell motors for boosted cancer radiosensitization. J Colloid Interface Sci 658:540–552. https://doi.org/10.1016/j.jcis.2023.12.091

Lortet-Tieulent J, Georges D, Bray F, Vaccarella S (2020) Profiling global cancer incidence and mortality by socioeconomic development. Int J Cancer 147:3029–3036. https://doi.org/10.1002/ijc.33114

Luo Z, Sun L, Bian F, Wang Y, Yu Y, Gu Z, Zhao Y (2023) Erythrocyte-inspired functional materials for biomedical applications. Adv Sci 10:e2206150. https://doi.org/10.1002/advs.202206150

Lv P, Liu X, Chen X, Liu C, Zhang Y, Chu C, Wang J, Wang X, Chen X, Liu G (2019) Genetically engineered cell membrane Nanovesicles for oncolytic adenovirus delivery: a versatile platform for cancer Virotherapy. Nano Lett 19:2993–3001. https://doi.org/10.1021/acs.nanolett.9b00145

Ma Z-J, Yang J-J, Lu Y-B, Liu Z-Y, Wang X-X (2020) Mesenchymal stem cell-derived exosomes: toward cell-free therapeutic strategies in regenerative medicine. World J Stem Cells 12:814–840. https://doi.org/10.4252/wjsc.v12.i8.814

Ma Y, Dong S, Grippin AJ, Teng L, Lee AS, Kim BYS, Jiang W (2024) Engineering therapeutical extracellular vesicles for clinical translation. Trends Biotechnol 43:61–82. https://doi.org/10.1016/j.tibtech.2024.08.007

Maharjan PS, Bhattarai HK (2022) Singlet oxygen, photodynamic therapy, and mechanisms of cancer cell death. J Oncol 2022:1–20. https://doi.org/10.1155/2022/7211485

Makwana V, Desai N, Rana D, Benival D, Vora LK (2025) Challenges in exosome nanovesicle-based drug delivery and diagnostics. In: Exosome communication. Elsevier, pp 203–252. https://doi.org/10.1016/B978-0-443-29052-7.00009-X

Moretta A, Bottino C, Vitale M, Pende D, Cantoni C, Mingari MC, Biassoni R, Moretta L (2001) Activating receptors and Coreceptors involved in human natural killer cell-mediated cytolysis. Annu Rev Immunol 19:197–223. https://doi.org/10.1146/annurev.immunol.19.1.197

Muller WA (2003) Leukocyte–endothelial-cell interactions in leukocyte transmigration and the inflammatory response. Trends Immunol 24:326–333. https://doi.org/10.1016/S1471-4906(03)00117-0

Oroojalian F, Beygi M, Baradaran B, Mokhtarzadeh A, Shahbazi M (2021) Immune cell membrane-coated biomimetic nanoparticles for targeted cancer therapy. Small 17:2006484. https://doi.org/10.1002/smll.202006484

Overchuk M, Weersink RA, Wilson BC, Zheng G (2023) Photodynamic and Photothermal therapies: synergy opportunities for Nanomedicine. ACS Nano 17:7979–8003. https://doi.org/10.1021/acsnano.3c00891

Pan H, Yang S, Gao L, Zhou J, Cheng W, Chen G, Shuhang W, Li N, Veranič P, Musiol R, Cai Q, Shubhra QTH (2024) At the crossroad of nanotechnology and cancer cell membrane coating: expanding horizons with engineered nanoplatforms for advanced cancer therapy harnessing homologous tumor targeting. Coord Chem Rev 506:215712. https://doi.org/10.1016/j.ccr.2024.215712

Pegtel DM, Gould SJ (2019) Exosomes. Annu Rev Biochem 88:487–514. https://doi.org/10.1146/annurev-biochem-013118-111902

Pitchaimani A, Nguyen TDT, Aryal S (2018) Natural killer cell membrane infused biomimetic liposomes for targeted tumor therapy. Biomaterials 160:124–137. https://doi.org/10.1016/j.biomaterials.2018.01.018

Pitt JM, Kroemer G, Zitvogel L (2016) Extracellular vesicles: masters of intercellular communication and potential clinical interventions. J Clin Invest 126:1139–1143. https://doi.org/10.1172/JCI87316

Polaka S, Desai N, Kshirsagar B, Rajpoot K, Tekade RK (2021) Revamping the pharmacokinetics of poorly soluble drugs using different formulations. In: Biopharmaceutics and pharmacokinetics considerations. Elsevier, pp 387–413. https://doi.org/10.1016/B978-0-12-814425-1.00005-X

Price NL, Goyette-Desjardins G, Nothaft H, Valguarnera E, Szymanski CM, Segura M, Feldman MF (2016) Glycoengineered outer membrane vesicles: a novel platform for bacterial vaccines. Sci Rep 6:24931. https://doi.org/10.1038/srep24931

Qing G, Gong N, Chen X, Chen J, Zhang H, Wang Y, Wang R, Zhang S, Zhang Z, Zhao X, Luo Y, Liang X-J (2019) Natural and engineered bacterial outer membrane vesicles. Biophys Rep 5:184–198. https://doi.org/10.1007/s41048-019-00095-6

Rana D, Salave S, Longare S, Agarwal R, Kalia K, Benival D (2021) Nanotherapeutics in tumour microenvironment for cancer therapy. Nanosci Nanotechnol Asia 12. https://doi.org/10.2174/2210681211666210908144839

Rana D, Desai N, Salave S, Vora LK (2025) Exosomes for protein and peptide drug delivery. In: Exosome communication. Elsevier, pp 305–327. https://doi.org/10.1016/B978-0-443-29052-7.00012-X

Rao L, Bu L, Xu J, Cai B, Yu G, Yu X, He Z, Huang Q, Li A, Guo S, Zhang W, Liu W, Sun Z, Wang H, Wang T, Zhao X (2015) Red blood cell membrane as a biomimetic Nanocoating for prolonged circulation time and reduced accelerated blood clearance. Small 11:6225–6236. https://doi.org/10.1002/smll.201502388

Rao L, Cai B, Bu LL, Liao QQ, Guo SS, Zhao XZ, Dong WF, Liu W (2017) Microfluidic electroporation-facilitated synthesis of erythrocyte membrane-coated magnetic nanoparticles for enhanced imaging-guided cancer therapy. ACS Nano 11:3496–3505

Rao L, Wu L, Liu Z, Tian R, Yu G, Zhou Z, Yang K, Xiong H-G, Zhang A, Yu G-T, Sun W, Xu H, Guo J, Li A, Chen H, Sun Z-J, Fu Y-X, Chen X (2020) Hybrid cellular membrane nanovesicles amplify macrophage immune responses against cancer recurrence and metastasis. Nat Commun 11:4909. https://doi.org/10.1038/s41467-020-18626-y

Raposo G, Stahl PD (2019) Extracellular vesicles: a new communication paradigm? Nat Rev Mol Cell Biol 20:509–510. https://doi.org/10.1038/s41580-019-0158-7

Richards DA, Maruani A, Chudasama V (2017) Antibody fragments as nanoparticle targeting ligands: a step in the right direction. Chem Sci 8:63–77. https://doi.org/10.1039/C6SC02403C

Rui R, Zhou L, He S (2023) Cancer immunotherapies: advances and bottlenecks. Front Immunol 14. https://doi.org/10.3389/fimmu.2023.1212476

Sartorio MG, Pardue EJ, Feldman MF, Haurat MF (2021) Bacterial outer membrane vesicles: from discovery to applications. Ann Rev Microbiol 75:609–630. https://doi.org/10.1146/annurev-micro-052821-031444

Schwechheimer C, Kuehn MJ (2015) Outer-membrane vesicles from gram-negative bacteria: biogenesis and functions. Nat Rev Microbiol 13:605–619. https://doi.org/10.1038/nrmicro3525

Shao M, Lopes D, Lopes J, Yousefiasl S, Macário-Soares A, Peixoto D, Ferreira-Faria I, Veiga F, Conde J, Huang Y, Chen X, Paiva-Santos AC, Makvandi P (2023) Exosome membrane-coated nanosystems: exploring biomedical applications in cancer diagnosis and therapy. Matter 6:761–799. https://doi.org/10.1016/J.MATT.2023.01.012

Shi J, Yang Y, Yin N, Liu C, Zhao Y, Cheng H, Zhou T, Zhang Z, Zhang K (2022) Engineering CXCL12 biomimetic decoy-integrated versatile immunosuppressive nanoparticle for ischemic stroke therapy with Management of Overactivated Brain Immune Microenvironment. Small Methods 6. https://doi.org/10.1002/smtd.202101158

Silhavy TJ, Kahne D, Walker S (2010) The bacterial cell envelope. Cold Spring Harb Perspect Biol 2:a000414. https://doi.org/10.1101/cshperspect.a000414

Song M, Dong S, An X, Zhang W, Shen N, Li Y, Guo C, Liu C, Li X, Chen S (2022) Erythrocyte-biomimetic nanosystems to improve antitumor effects of paclitaxel on epithelial cancers. J Control Release 345:744–754. https://doi.org/10.1016/j.jconrel.2022.03.060

Su J, Sun H, Meng Q, Yin Q, Zhang P, Zhang Z, Yu H, Li Y (2016) Bioinspired nanoparticles with NIR-controlled drug release for synergetic Chemophotothermal therapy of metastatic breast cancer. Adv Funct Mater 26:7495–7506. https://doi.org/10.1002/adfm.201603381

Taghikhani A, Farzaneh F, Sharifzad F, Mardpour S, Ebrahimi M, Hassan ZM (2020) Engineered tumor-derived extracellular vesicles: potentials in cancer immunotherapy. Front Immunol 11. https://doi.org/10.3389/fimmu.2020.00221

Tan S, Li D, Zhu X (2020) Cancer immunotherapy: pros, cons and beyond. Biomed Pharmacother 124:109821. https://doi.org/10.1016/j.biopha.2020.109821

Tang J, Shen D, Caranasos TG, Wang Z, Vandergriff AC, Allen TA, Hensley MT, Dinh PU, Cores J, Li TS, Zhang J, Kan Q, Cheng K (2017) Therapeutic microparticles functionalized with biomimetic cardiac stem cell membranes and secretome. Nat Commun 8(18):1–9. https://doi.org/10.1038/ncomms13724

Tian J, Han Z, Song D, Peng Y, Xiong M, Chen Z, Duan S, Zhang L (2023) Engineered exosome for drug delivery: recent development and clinical applications. Int J Nanomedicine 18:7923–7940. https://doi.org/10.2147/IJN.S444582

Toyofuku M, Nomura N, Eberl L (2019) Types and origins of bacterial membrane vesicles. Nat Rev Microbiol 17:13–24. https://doi.org/10.1038/s41579-018-0112-2

Wang Y, Ji X, Ruan M, Liu W, Song R, Dai J, Xue W (2018) Worm-like biomimetic Nanoerythrocyte carrying siRNA for melanoma gene therapy. Small 14:e1803002. https://doi.org/10.1002/smll.201803002

Wang S, Duan Y, Zhang Q, Komarla A, Gong H, Gao W, Zhang L (2020a) Drug targeting via platelet membrane–coated nanoparticles. Small Struct 1:2000018. https://doi.org/10.1002/sstr.202000018

Wang D, Liu C, You S, Zhang K, Li M, Cao Y, Wang C, Dong H, Zhang X (2020b) Bacterial vesicle-cancer cell hybrid membrane-coated nanoparticles for tumor specific immune activation and Photothermal therapy. ACS Appl Mater Interfaces 12:41138–41147. https://doi.org/10.1021/acsami.0c13169

Wang Y, Zhao Z, Liu C, Hao M, Kong C, Zhao X, Gao Y, Zhang Y, Cui W, Zhang C, Jiang J (2022) B16 membrane-coated vesicles for combined photodynamic therapy and immunotherapy shift immune microenvironment of melanoma. Int J Nanomedicine 17:855–868. https://doi.org/10.2147/IJN.S338488

Wang W, Wu F, Mohammadniaei M, Zhang M, Li Y, Sun Y, Tang BZ (2023) Genetically edited T-cell membrane coated AIEgen nanoparticles effectively prevents glioblastoma recurrence. Biomaterials 293:121981. https://doi.org/10.1016/j.biomaterials.2022.121981

Wei R, Dong Y, Tu Y, Luo S, Pang X, Zhang W, Yao W, Tang W, Yang H, Wei X, Jiang X, Yuan Y, Yang R (2021) Bioorthogonal Pretargeting strategy for anchoring Activatable photosensitizers on plasma membranes for effective photodynamic therapy. ACS Appl Mater Interfaces 13:14004–14014. https://doi.org/10.1021/acsami.1c01259

Wu T, Lang T, Zheng C, Yan W, Li Y, Zhu R, Huang X, Xu H, Li Y, Yin Q (2023) Promote Intratumoral drug release and penetration to counteract docetaxel-induced metastasis by photosensitizer-modified red blood cell membrane-coated nanoparticle. Adv Funct Mater 33. https://doi.org/10.1002/adfm.202212109

Xin M, Dong X-W, Guo X-L (2015) Role of the interaction between galectin-3 and cell adhesion molecules in cancer metastasis. Biomed Pharmacother 69:179–185. https://doi.org/10.1016/j.biopha.2014.11.024

Xuan M, Shao J, Dai L, He Q, Li J (2015) Macrophage cell membrane camouflaged mesoporous silica Nanocapsules for in vivo cancer therapy. Adv Healthc Mater 4:1645–1652. https://doi.org/10.1002/ADHM.201500129

Xuan M, Shao J, Li J (2019) Cell membrane-covered nanoparticles as biomaterials. Natl Sci Rev 6:551–561. https://doi.org/10.1093/nsr/nwz037

Yousefpour P, Ni K, Irvine DJ (2023) Targeted modulation of immune cells and tissues using engineered biomaterials, nature reviews. Bioengineering 1:107–124. https://doi.org/10.1038/s44222-022-00016-2

Zeng S, Tang Q, Xiao M, Tong X, Yang T, Yin D, Lei L, Li S (2023) Cell membrane-coated nanomaterials for cancer therapy. Mater Today Bio 20:100633. https://doi.org/10.1016/J.MTBIO.2023.100633

Zhai Y, Wang J, Lang T, Kong Y, Rong R, Cai Y, Ran W, Xiong F, Zheng C, Wang Y, Yu Y, Zhu HH, Zhang P, Li Y (2021) T lymphocyte membrane-decorated epigenetic nanoinducer of interferons for cancer immunotherapy. Nat Nanotechnol 16:1271–1280. https://doi.org/10.1038/s41565-021-00972-7

Zhang H (2016) Erythrocytes in nanomedicine: an optimal blend of natural and synthetic materials. Biomater Sci 4:1024–1031. https://doi.org/10.1039/C6BM00072J

Zhang Q, Wei W, Wang P, Zuo L, Li F, Xu J, Xi X, Gao X, Ma G, Xie H (2017) Biomimetic Magnetosomes as versatile artificial antigen-presenting cells to potentiate T-cell-based anticancer therapy. ACS Nano 11:10724–10732. https://doi.org/10.1021/acsnano.7b04955

Zhang Z, Qian H, Huang J, Sha H, Zhang H, Yu L, Liu B, Hua D, Qian X (2018a) Anti-EGFR-iRGD recombinant protein modified biomimetic nanoparticles loaded with gambogic acid to enhance targeting and antitumor ability in colorectal cancer treatment. Int J Nanomedicine 13:4961–4975. https://doi.org/10.2147/IJN.S170148

Zhang F, Zhao L, Wang S, Yang J, Lu G, Luo N, Gao X, Ma G, Xie H, Wei W (2018b) Construction of a biomimetic Magnetosome and its application as a SiRNA carrier for high-performance anticancer therapy. Adv Funct Mater 28. https://doi.org/10.1002/adfm.201703326

Zhang J, Wei K, Shi J, Zhu Y, Guan M, Fu X, Zhang Z (2021a) Biomimetic nanoscale erythrocyte delivery system for enhancing chemotherapy via overcoming biological barriers. ACS Biomater Sci Eng 7:1496–1505. https://doi.org/10.1021/acsbiomaterials.1c00008

Zhang M, Cheng S, Jin Y, Zhang N, Wang Y (2021b) Membrane engineering of cell membrane biomimetic nanoparticles for nanoscale therapeutics. Clin Transl Med 11:e292. https://doi.org/10.1002/ctm2.292

Zhang Y, Liao Y, Bai F, Xing S, Wang M, Dang Z, Lin J, Huang P (2024a) Genetically engineered eukaryocyte-bacteria hybrid membrane-camouflaged nanoemulsion for three-pronged synergistic cancer therapy. Nano Today 54:102076. https://doi.org/10.1016/j.nantod.2023.102076

Zhang H, Li Y, Kang H, Lan J, Hou L, Chen Z, Li F, Liu Y, Zhao J, Li N, Wan Y, Zhu Y, Zhao Z, Zhang H, Zhuang J, Huang X (2024b) Genetically engineered membrane-based nanoengagers for immunotherapy of pancreatic cancer. J Nanobiotechnol 22:104. https://doi.org/10.1186/s12951-024-02369-9

Zhi D, Yang T, O'Hagan J, Zhang S, Donnelly RF (2020) Photothermal therapy. J Control Release 325:52–71. https://doi.org/10.1016/j.jconrel.2020.06.032

Zhou Y, Yuan Y, Liu M, Hu X, Quan Y, Chen X (2019) Tumor-specific delivery of KRAS siRNA with iRGD-exosomes efficiently inhibits tumor growth. ExRNA 1:28. https://doi.org/10.1186/s41544-019-0034-9

Zou Y, Sun Y, Wang Y, Zhang D, Yang H, Wang X, Zheng M, Shi B (2023) Cancer cell-mitochondria hybrid membrane coated Gboxin loaded nanomedicines for glioblastoma treatment. Nat Commun 14:4557. https://doi.org/10.1038/s41467-023-40280-3

Cell Membrane Engineering for Advancing Drug Delivery Against Infectious Diseases

15

Mahir Mohammed, Eman Hussain, Ethar Mohamed, Nourhan H. Zahran, and Yavuz Nuri Ertas

Abstract

Infectious diseases remain a significant global health challenge, exacerbated by the rise of antimicrobial resistance and the limitations of conventional drug delivery approaches. This chapter examines the potential of cell membrane engineering to enhance drug delivery strategies for combating infectious diseases. By harnessing the natural characteristics of cell membranes, researchers have developed innovative cell membrane-based nanocarriers, such as cell membrane-coated nanoparticles (CMC-NPs). These biomimetic therapeutic nanoplatforms offer enhanced biocompatibility, stability, targeted delivery, immune evasion, and treatment effectiveness, addressing critical challenges in treating infectious

M. Mohammed
Faculty of Pharmacy, University of Khartoum, Khartoum, Sudan

ERNAM–Nanotechnology Research and Application Center, Erciyes University, Kayseri, Türkiye

E. Hussain
Department of Pharmaceutics, University of Gezira, Wad Madani, Sudan

E. Mohamed
Faculty of Pharmacy, University of Khartoum, Khartoum, Sudan

N. H. Zahran
Department of Chemistry, Faculty of Science, Suez Canal University, Ismailia, Egypt

Y. N. Ertas (✉)
Department of Biomedical Engineering, Erciyes University, Kayseri, Türkiye

ERNAM–Nanotechnology Research and Application Center, Erciyes University, Kayseri, Türkiye
e-mail: yavuzertas@erciyes.edu.tr

Z. Hussain, R. Pei (eds.), *Cell Membrane Engineering for Advancing Cell-Based and Nanotechnology-Based Therapies*, Biomaterials, Bioengineering and Sustainability 6, https://doi.org/10.1007/978-3-031-93425-4_15

diseases. The chapter explores the design and fabrication of CMC-NPs, including membrane extraction and coating techniques, and classifies them into two main types: singular-sourced and hybrid CMC-NPs. It provides a detailed discussion of the drug delivery mechanisms and highlights their potential applications in treating bacterial, viral, fungal, and parasitic infections. Furthermore, this chapter discusses the challenges of scalability, heterogeneity, and immune responses while outlining future directions for optimizing drug loading, release, and clinical translation. By integrating cutting-edge nanotechnology with biological principles, cell membrane-engineered nanoplatforms hold promises to transform the treatment of infectious diseases and improve global health outcomes.

Keywords

Cell membrane types · Membrane-coated nanoparticles · Drug delivery systems · Infectious diseases therapy · Nanomedicine · Biomimetic nanocarriers

15.1 Introduction

Infectious diseases remain one of the most critical global health challenges, contributing significantly to morbidity and mortality worldwide. Recent studies highlight that these diseases account for a substantial proportion of global deaths, particularly in developing nations where healthcare infrastructure and resources are often inadequate (World Health Organization (WHO) 2023). Diseases such as tuberculosis, malaria, and lower respiratory infections continue to be leading causes of death, with their impact exacerbated by factors like antimicrobial resistance, climate change, and population displacement. Furthermore, the zoonotic origin of many emerging infectious diseases complicates prevention and control efforts, underscoring the urgent need for integrated, multidisciplinary approaches to mitigate their impact (Bloom and Cadarette 2019). Addressing the burden of these diseases and implementing effective strategies are essential for reducing their toll on global health and achieving sustainable development goals (Ellwanger et al. 2021).

The alarming rise of antimicrobial resistance is compounding this challenge, rendering many common infections progressively harder to treat (Dadgostar 2019). Over 2.8 million cases of antimicrobial-resistant infections are reported annually in the United States alone, resulting in more than 35,000 deaths. Globally, antimicrobial resistance contributed to at least 1.27 million direct deaths and nearly 5 million associated fatalities in 2019. The COVID-19 pandemic has further exacerbated this crisis by increasing antibiotic use and disrupting healthcare systems (Klein et al. 2018; Farrell et al. 2021). Without urgent action, antimicrobial resistance deaths could exceed 10 million annually by 2050, underscoring the need for innovative solutions for infectious diseases management (Naghavi et al. 2024).

Infectious diseases arise from a diverse array of pathogens, encompassing bacterial infections (e.g., methicillin-resistant *Staphylococcus aureus* (MRSA), tuberculosis), viral infections (e.g., human immunodeficiency virus (HIV), hepatitis, COVID-19), fungal infections (e.g., *Candida albicans*, cryptococcal meningitis), and parasitic infections (e.g., malaria, leishmaniasis) (McArthur 2019). Each class presents distinct challenges in diagnosis, treatment, and prevention. Bacterial infections are becoming increasingly resistant to conventional antibiotics, viral infections demand complex antiviral regimens, fungal infections remain difficult to manage due to the limited availability of antifungal agents, and parasitic infections persist as endemic threats in resource-constrained settings. Furthermore, emerging infectious diseases, including those caused by novel viruses and multidrug-resistant strains, pose a critical challenge as no fully effective treatments or vaccines currently exist. These pressing concerns highlight the urgent need for innovative therapeutic strategies capable of addressing the dynamic and evolving landscape of infectious diseases (Strasfeld and Chou 2010; Bloom and Cadarette 2019; Hossain et al. 2023; Ikuta et al. 2024).

In response to these challenges, the proliferation of pharmaceutical science and the continuous development of new therapeutic agents have significantly transformed modern medical practices. A key focus in combating infectious diseases is delivering medications effectively to targeted tissues, ensuring optimal drug release at the infection site to maximize therapeutic efficacy and minimize side effects. As a result, contemporary pharmaceutical research is increasingly directed toward designing advanced drug delivery systems capable of selectively transporting therapeutics to precise disease locations (Mitragotri et al. 2014).

Recent advancements in material science and nanotechnology have paved the way for developing innovative nanoscale formulations, including micelles, liposomes, and nanoparticles (Muzammil et al. 2018). Many of these cutting-edge delivery systems are either already in clinical use or nearing clinical application, particularly in infectious diseases, where timely and effective treatment is paramount (Farokhzad and Langer 2009). Among these, liposomes have emerged as a promising drug carrier, with over ten categories of liposomal formulations having received regulatory approval and commercialization (Omolo et al. 2019). However, despite their success, traditional delivery systems like liposomes face limitations, including stability issues and challenges in replicating the complexity of natural cell membranes (Lian and Ho 2001).

A new frontier in drug delivery has emerged, leveraging a range of biomimetic platforms, including cell membrane-coated nanoparticles (CMC-NPs), extracellular vesicles, bacterial membrane vesicles, and whole-cell-based delivery systems (Liu and Huang 2022). At the core of these systems lies the fundamental role of cell membranes, which provide structural and functional advantages. Their intricate composition and diverse molecular landscape endow them with biocompatibility, immune evasion, and precise biological interactions, making them ideal candidates for targeted drug delivery. Recent advancements in cell membrane engineering, such as genetic modifications, click chemistry, hydrophobic insertion, and membrane fusion, have further refined these platforms to optimize their therapeutic potential (Zhai et al. 2017; Huang et al. 2024). Among these, CMC-NPs have garnered significant attention for their potential to integrate nanotechnology with the biomimetic properties of natural cell membranes, offering superior biocompatibility, immune evasion, and targeting precision. By closely mimicking their source membranes, CMC-NPs effectively target drug-resistant pathogens while minimizing systemic toxicity. Their capacity to mimic natural cell-pathogen interactions positions them as a groundbreaking innovation in nanomedicine (Liu et al. 2023a, b).

This chapter will explore the foundational principles and recent advancements in cell membrane engineering for drug delivery against infectious diseases. We will examine various production techniques, unique characteristics, and applications of CMC-NPs, focusing on their role in infectious disease therapy. Additionally, we will highlight emerging opportunities and potential future developments in this field. By advancing cellular and biological membrane-based drug delivery methods, we anticipate significant improvements in treatment protocols for infectious diseases, ultimately enhancing patient outcomes and contributing to global public health.

Central to these advancements is the rise of biomimetic nanotechnology, which has revolutionized drug delivery by enabling the engineering of CMC-NPs. These innovative nanocarriers merge the distinctive attributes of NPs with the inherent characteristics of cellular membranes, including their biocompatibility, bioactivity, selective permeability, and ability to interact specifically with target cellular environments—thereby improving the stability and targeting efficiency of encapsulated therapeutics (Liu et al. 2023a, b; Zhang et al. 2024b). Integrating engineered cell membranes into NP designs has become a sophisticated strategy to enhance therapeutic efficacy, minimize systemic toxicity, and reduce immunogenic responses associated with conventional drug delivery systems (Yang et al. 2021; Imran et al. 2022). As a result, this approach has led to more efficient treatment options for various diseases, including infectious diseases (Li et al. 2023b).

15.2 Design and Fabrication of CMC-NPs

The design and fabrication of CMC-NPs are intricate, requiring a thorough understanding of cellular biology and nanotechnology applications (Li et al. 2024). Every stage of this procedure is vital so, that the resulting product delivers therapeutic payloads while retaining the beneficial features of source cell membranes. Various approaches have been used to prepare CMC-NPs. The classic preparation of CMC-NPs involves a top-down approach, which can be divided into three main steps: cell membrane extraction, inner core nanocarrier fabrication, and cell membrane coating, each key to the resulting NP functionalization (Liu et al. 2019; Liu et al. 2023a, b).

The design and preparation of CMC-NPs begin with the careful selection of a cell type for membrane isolation, which can include red blood cells, platelets, immune cells, cancer cells, or stem cells, each offering unique advantages based on their biological properties (Yaman et al. 2020). Once the cell type is chosen, the cell membranes are extracted and isolated using techniques that preserve their membrane contents and architecture (Liu et al. 2023a, b). Standard methods for membrane isolation include hypotonic lysis, freeze-thaw cycles, ultrasonic disruption, and homogenization, each followed by centrifugation or ultrafiltration to separate membranes from intracellular contents (Spanjers and Städler 2020; Chugh et al. 2021; Javed et al. 2021). Hypotonic lysis, which relies on osmotic swelling to rupture cells, is simple and widely used but is primarily effective for red blood cells and less effective for other cell types (Xia et al. 2019; Le et al. 2021; Zhang et al. 2022). Freeze-thaw cycles, involving repeated freezing and thawing, are effective for isolating platelet membranes but risk damaging membrane integrity and protein activity if not carefully controlled (Wei et al. 2016). Ultrasonic disruption uses shock waves to rupture cells but is limited to small-scale applications due to heat generation, which can alter cell structures (Liu et al. 2022; Shen et al. 2023). Membrane isolation via homogenization, another mechanical process, is versatile and applicable for lysing a wide range of cell types, such as neutrophils and carcinoma cells (Kang et al. 2017; Oroojalian et al. 2021), but is energy-intensive and unsuitable for highly viscous samples (Liu et al. 2023a, b). Combining these methods can improve membrane quality and obtain satisfactory results (Shu 2024). Subsequent purification steps, such as differential or density gradient centrifugation, are employed to isolate membrane fragments from contaminants (Jiménez-Jiménez et al. 2020). These processes ensure the production of high-quality cell membranes for coating NPs, which is critical for their biomedical applications.

CMC-NPs utilize diverse NP templates, each offering unique properties that influence permeability, functionality, and therapeutic potential. Polymeric NPs, such as those made from poly(lactic-co-glycolic acid) (PLGA) and polyethylene glycol (PEG), are highly versatile and biocompatible, enabling sustained drug release and tailored therapeutic applications (Qureshi et al. 2024). Liposomes, with their lipid bilayer structure, excel in encapsulating both water- and fat-soluble drugs, providing a dual-release system that enhances bioavailability and facilitates direct delivery to target cells (Liu et al. 2023a, b; Qureshi et al. 2024). Inorganic NPs, including gold, silver, and silica, are valued for their optical and electronic properties and ability to be functionalized for targeted delivery (Zou et al. 2020). Metal-organic frameworks (MOFs) are gaining attention for their high surface area and tunable pores, which improve drug loading and release. At the same time, cell membrane coating enhances targeting and reduces toxicity (Jiménez-Jiménez et al. 2020). Carbon-based NPs, such as carbon nanotubes and graphene oxide, offer high mechanical strength and large surface areas, making them ideal for drug delivery and imaging, especially when surface-modified for better biocompatibility (Liu et al. 2023a, b). Hybrid NPs combine different materials like polymers and inorganic components and capitalize on the strengths of each element to optimize performance. These NPs optimize stability, drug delivery efficiency, and customization for specific therapeutic needs (Liu et al. 2019; Liu et al. 2023a, b).

The production of NP cores involves techniques like solvent evaporation, coacervation, or electrospinning, followed by characterization using methods such as dynamic light scattering and scanning electron microscopy to ensure they meet application-specific requirements (Mahmood et al. 2017). Selecting the appropriate NP template is crucial for optimizing CMC-NPs, as each variant offers distinct advantages that can be leveraged to enhance drug delivery systems (Alimohammadvand et al. 2024). This versatility is key to advancing nanomedicine and improving therapeutic outcomes across various medical fields.

Various methods have been developed to merge cell membrane vesicles with core NPs, each with distinct advantages and limitations. Membrane extrusion, a widely used technique, involves passing a mixture of NPs and membrane fragments through a porous membrane to achieve uniform coating, followed by centrifugation to separate unencapsulated vesicles. This method is scalable and produces NPs of consistent size but requires precise parameter control, which can be challenging for sensitive NPs (Liu et al. 2019; Ozsoy et al. 2024). Sonication uses ultrasonic waves to rapidly combine NPs with membrane fragments, making it versatile for different NP types (Fernández-Borbolla et al. 2024). However, it may generate heat, potentially damaging membrane proteins or NPs, and does not always ensure uniform NP size, necessitating careful optimization of parameters like duration, frequency, and intensity (Shi et al. 2022; Ozsoy et al. 2024). Microfluidic electroporation, a newer technique, employs high-voltage electric pulses to create temporary openings in cell membranes, allowing NPs to enter vesicles (Rao et al. 2017; Fang et al. 2018). This method preserves membrane integrity and surface proteins while offering scalability and storage potential, although it lacks standardized specifications for industrial production (Rao et al. 2017; Zou et al. 2020). Additionally, an innovative *in situ* encapsulation approach involves exposing live cells to NP cores, such as iron-oxide NPs, gold NPs, or quantum dots, which then release vesicles containing the NPs when cultured in serum-free media (Andriola Silva et al. 2013; Guan et al. 2024b). Each technique offers unique advantages and limitations, tailored to the specific application and characteristics of CMC-NPs. Key factors such as membrane coverage, size distribution, uniformity, and protein retention must be considered to optimize production efficiency and ensure high-quality outcomes (Liu et al. 2023a, b).

15.3 Major Types of CMC-NPs and Their Potential Scope

Several cell membrane types, such as erythrocytes/red blood cells (RBCs), platelet cells, cancer cells, neutrophil, and macrophage cells, stem cells, bacterial outer membranes, and hybrid cell membranes, can be used to create CMC-NPs (Baig et al. 2024). Each cell membrane vesicle offers unique advantages based on its biological functions, facilitating targeted drug delivery and enhanced efficacy.

15.3.1 Singular-Sourced Membrane-Coated NPs

15.3.1.1 Erythrocyte Membrane-Coated Nanoparticles

NPs coated with erythrocyte membranes (EMC-NPs) represent a groundbreaking advancement in drug delivery, harnessing the unique biological properties of RBCs membranes to enhance therapeutic and diagnostic applications. These EMC-NPs encapsulate therapeutic agents within RBCs-derived membranes, significantly improving biocompatibility, circulation time, and targeted delivery. The inherent biocompatibility of erythrocyte membranes, abundant in the human body, minimizes immune recognition, allowing EMC-NPs to evade the mononuclear phagocyte system and extend their presence in the bloodstream (Zhang et al. 2021a; Guan et al. 2024a). This immune evasion is further supported by surface proteins like CD47, which act as a "don't eat me" signal to macrophages, reducing clearance and enhancing drug bioavailability (Zhang et al. 2021a). Additionally, the lipid bilayer of erythrocyte membranes functions as a diffusion barrier, enabling controlled and sustained drug release, which is particularly beneficial for therapies requiring long-lasting therapeutic effects and reduced dosing frequency (Hu et al. 2011). EMC-NPs can also be conjugated with targeting ligands to improve precision, such as binding to receptors overexpressed on cancer cells, thereby enhancing therapeutic efficacy while minimizing off-target effects (Chen et al. 2018, 2021). Furthermore, the natural composition of erythrocyte membranes stabilizes the NPs, preventing aggregation and ensuring uniform dispersion in biological fluids, which is critical for maintaining therapeutic effectiveness during intravenous administration (Guo et al. 2023). By fusing the versatility of NP technology with the biocompatibility, stability, and targeting capabilities of erythrocyte membranes, EMC-NPs offer a promising platform for advancing drug delivery systems, with potential applications in improving clinical outcomes across various medical fields.

15.3.1.2 Platelet Membrane-Coated NPs

Platelet membrane-coated NPs (PMC-NPs) represent a significant breakthrough in drug delivery systems, leveraging the natural biological functions of platelets to enhance therapeutic precision and efficacy. These NPs mimic platelets' inherent ability to target damaged blood vessels, inflamed tissues, and pathogens, making them particularly valuable in cancer therapy and infectious disease treatment (Han et al. 2022). The platelet membrane coating provides PMC-NPs with improved targeting capabilities, enhanced biocompatibility, and the ability to evade immune detection, which is critical for effective drug delivery (Wang et al. 2020b). For instance, their surface proteins enable specific interactions with subendothelial tissues and pathogens, allowing PMC-NPs to deliver therapeutic agents like siRNA to vascular damage sites with high precision, minimizing off-target effects (Wang et al. 2019; Zhuang et al. 2020). Additionally, surface markers such as CD47 help PMC-NPs avoid phagocytosis by macrophages, extending their circulation time and improving therapeutic outcomes (Wei et al. 2018a, 2019). The platelet membrane's lipid bilayer also facilitates controlled and sustained drug release, which is especially beneficial in cancer treatment, where maintaining therapeutic drug levels while reducing systemic toxicity is crucial (Zhou et al. 2021). PMC-NPs can be engineered to carry multiple therapeutic agents, such as chemotherapeutics, antibiotics, or anti-inflammatory drugs, enhancing their versatility in personalized medicine (Geng et al. 2023). Furthermore, their stability, high drug-loading capacity, and ability to incorporate additional targeting molecules make PMC-NPs a promising platform for addressing complex medical conditions, including atherosclerosis and cancer (Park et al. 2021; Feng and Zheng 2023; Fernández-Borbolla et al. 2024). By combining the natural targeting and biocompatibility of platelet membranes with the versatility of NP technology, PMC-NPs offer a transformative approach to improving drug delivery systems and clinical outcomes.

15.3.1.3 Neutrophil Membrane-Coated Nanoparticles

Neutrophil membrane-coated NPs (NMC-NPs) leverage the unique biological properties of neutrophils to create potent drug delivery systems, particularly for treating inflammatory and malignant conditions (Zhang et al. 2024a). Neutrophils, as key immune cells, possess surface markers and functional traits that enable NMC-NPs to target inflamed or infected tissues precisely (Li et al. 2023a). For instance, NMC-NPs can replicate neutrophils' ability to migrate toward chemokine signals, enhancing their localization at disease sites. This targeting is further facilitated by β2 integrins on neutrophil membranes, which bind to ICAM-1 on inflamed endothelium, improving vascular localization (Kang et al. 2017). NMC-NPs have shown promise in cancer therapy, particularly in combating metastasis, by exploiting neutrophils' natural tendency to navigate toward tumors. Their stability and therapeutic efficacy are bolstered by hydrophobic interactions and electrostatic repulsion between the neutrophil membrane and the NP core, ensuring efficient drug delivery (Kang et al. 2017; Wang et al. 2023). Additionally, NMC-NPs can cross biological barriers, such as the blood-pancreas barrier, making them valuable for treating pancreatic disorders (Xu et al. 2019). Their immunomodulatory properties, derived from neutrophil membrane components, allow them to regulate inflammation and enhance therapeutic outcomes in conditions like sepsis and chronic inflammatory diseases (Gao et al. 2017). Furthermore, the "self" properties of neutrophil membranes help NMC-NPs evade immune detection, extending their circulation time and improving drug delivery efficiency (Zhao et al. 2023). By harnessing innate targeting, immune-modulating capabilities, and prolonged circulation of neutrophils with NP technology, NMC-NPs represent a versatile and potent platform for advancing targeted and personalized therapeutic strategies.

15.3.1.4 Macrophage Membrane-Coated Nanoparticles

Macrophages, key immune cells, play a critical role in inflammation, tissue repair, and pathogen clearance, making their membrane properties ideal for biomimetic NPs. (Zhang et al. 2021c; Hasan et al. 2023). Macrophage membrane-coated NPs (MMC-NPs) represent a notable innovation in targeted drug delivery, leveraging the natural biological functions of macrophages to enhance therapeutic precision and efficacy (Wu et al. 2022). MMC-NPs benefit from the innate ability of macrophages to migrate to inflamed or tumorous tissues, enabling targeted drug delivery to disease sites (Li et al. 2022; Sha et al. 2022). Surface proteins like CD47 on macrophage membranes help MMC-NPs evade immune detection, prolonging their circulation time and reducing clearance by the reticuloendothelial system (Tang et al. 2021; Wang et al. 2024). This immune evasion is crucial for achieving sustained therapeutic effects. Additionally, MMC-NPs exhibit enhanced biocompatibility, as the natural cell membrane coating improves integration with biological systems, minimizing adverse reactions and toxicity (Wu et al. 2022). Their stability also protects encapsulated drugs from degradation, ensuring controlled release at target locations (Tang et al. 2021; Wang et al. 2024). Recent engineering advancements, such as genetic modifications and the incorporation of functional peptides, have further improved the targeting and therapeutic capabilities of MMC-NPs (Li et al. 2019; Duan et al. 2023). For instance, using M2 macrophage membranes, known for their anti-inflammatory properties, has shown promise in treating atherosclerosis and cancer (Wang et al. 2021b; Zou et al. 2024). Moreover, stimuli-responsive MMC-NPs, designed to release drugs in response to environmental cues like pH or temperature changes, offer precise control over drug delivery, reducing off-target effects (Liang et al. 2021; Feng and Zheng 2023). Thanks to the diverse spectrum of potential applications, engineering progress, and combination therapies, MMC-NPs hold immense potential for treating many diseases, including cancer, infections, and autoimmune disorders.

15.3.1.5 Cancer Cell Membrane-Coated Nanoparticles

A cutting-edge approach in drug delivery systems, cancer cell membrane-coated NPs (CCM-NPs) harnesses the distinctive biological features of cancer cell membranes to boost therapeutic compounds' targeting, effectiveness, and safety. These NPs leverage tumor-specific surface markers and receptors, enabling precise homing to cancer cells and facilitating receptor-mediated endocytosis for efficient drug release (Hisato and Saya 2022; Guo et al. 2024). By mimicking cancer cells, CCM-NPs exhibit reduced immunogenicity, allowing them to evade the immune system and prolong their circulation time. This is critical for ensuring therapeutic agents reach their targets effectively (Harris et al. 2019; Zhang et al. 2024b). Additionally, CCM-NPs can co-deliver multiple therapeutic agents, such as chemotherapeutics and siRNAs, addressing various pathways involved in cancer growth and resistance, thereby overcoming the limitations of single-drug therapies (Zhang et al. 2020). Their ability to respond to tumor microenvironment stimuli, such as pH or enzyme activity, further enhances targeted drug release, minimizing off-target effects and improving therapeutic outcomes (Zeng et al. 2023). Moreover, using patient-derived cancer cell membranes enhances biocompatibility, reducing the risk of immune rejection and improving overall treatment acceptance (Li et al. 2023b).

15.3.1.6 Stem Cell Membrane-Coated Nanoparticles

NPs coated with stem cell membranes (SCM-NPs) represent a transformative approach in drug delivery, leveraging the unique biological properties of stem cells to enhance therapeutic precision and efficacy. Stem cells' inherent ability to self-renew and migrate toward injured or diseased tissues, such as tumors, enables SCM-NPs to navigate complex biological environments effectively (Jiménez-Jiménez et al. 2020; Zhang et al. 2021c). This natural tropism, driven by chemokines and inflammatory signals, allows SCM-NPs to target tumor microenvironments with high specificity, improving cellular uptake and therapeutic outcomes (Wang and Wu 2022). The stem cell membrane coating reduces immunogenicity, enabling SCM-NPs to evade immune detection and extend their circulation time, which is critical for sustained drug delivery (Ma et al. 2019). The biocompatibility of

SCM-NPs, derived from natural stem cell membranes, minimizes the risk of adverse immune reactions, enhancing patient safety (Mei et al. 2024). These nanomaterials can encapsulate diverse therapeutic agents, including proteins, nucleic acids, and small molecules, enabling combination therapies that target multiple disease pathways simultaneously (Bose et al. 2018; Zou et al. 2023). In regenerative medicine, SCM-NPs offer a safer alternative to live stem cells by reducing tumorigenicity and immunogenicity while retaining therapeutic interactions with damaged tissues. For instance, SCM-NPs engineered to express CXCR4 have demonstrated improved targeting of ischemic tissues, showcasing their potential for precise and effective treatment (Bose et al. 2018).

15.3.2 Hybrid Cell Membrane-Coated Nanoparticles

Despite the potential of singular-sourced membrane-coated NPs, they often face limitations such as restricted functionalities, batch-to-batch variability, and limited capacity for multi-targeting, which can hinder their effectiveness in biomedical applications (Kroll et al. 2017). These limitations highlight the importance of developing hybrid cell membrane-coated NPs (HCM-NPs), which combine membranes from different cell types to enhance functionality. These hybrid platforms offer significant advantages over their counterparts by presenting a broader array of membrane ligands and receptors, improving biocompatibility, targeted delivery, and immune evasion (Yu et al. 2024). HCM-NPs expand upon the groundwork laid by CMC-NPs, which have shown significant promise in diverse biomedical applications. Scientists can develop NPs that replicate cells' inherent properties while augmenting their functional abilities by incorporating membranes from multiple cell types. For example, combining cancer cell membranes, which exhibit tumor-targeting characteristics, with erythrocyte membranes, renowned for their extended circulation periods, can produce NPs that successfully avoid immune system detection while concurrently targeting tumor locations (Dehaini et al. 2017; Zhao et al. 2021). This twofold functionality plays a vital role in enhancing the therapeutic efficacy of encapsulated drugs by reducing systemic toxicity and optimizing drug concentration at the target site.

The primary benefit of HCM-NPs is their capacity to provide NPs with a "self" signature, enabling them to bypass immune detection and extend their blood circulation time (Liao et al. 2020). For instance, coatings derived from RBCs membranes significantly reduce elimination by the mononuclear phagocyte system (Lu et al. 2023a), and when combined with other membrane types, they can introduce certain functions. Furthermore, merging various cell membranes can improve targeting accuracy. HCM-NPs, for instance, can achieve high-precision targeting by incorporating membrane proteins from cancer cells or immune cells (Han et al. 2024). Coatings derived from cancer cell membranes facilitate homotypic targeting, where the NPs preferentially attach to cancer cells of the same kind (Harris et al. 2019). Components from immune cell membranes can target inflamed or infected tissues precisely (Oroojalian et al. 2021).

The development of HCM-NPs creates opportunities for multifunctional applications by combining the characteristics of their source membranes. For instance, NPs coated with hybrid membranes from platelets (which target vascular injuries) and tumor cells (home to malignant growths) can simultaneously tackle thrombosis and cancer metastasis. Furthermore, the adaptability and inherent bioactivity of HCM-NPs are applied to various complex therapeutic approaches, such as gene therapy, immunotherapy, and targeted anti-inflammatory drug administration (Fang et al. 2018; Bahmani et al. 2021; Zeng et al. 2022). Consequently, HCM-NPs represent a groundbreaking strategy in drug delivery, merging the benefits of various biomimetic cell membranes with the flexibility of synthetic NP cores. Their exceptional abilities to avoid immune system detection, target specific cells, improve drug stability, and adapt to a range of therapeutic applications position HCM-NPs as top contenders for upcoming advancements in biomedicine.

15.4 Mechanisms of Drug Delivery Using CMC-NPs

CMC-NPs employ various mechanisms for drug delivery, which can be classified into several primary processes. These include targeted delivery methods (passive, active, and homotypic targeting), immune evasion and prolonged circulation, cellular uptake, and controlled release. Cell-specific targeting is a highly desirable characteristic that reduces unintended side effects when utilizing CMC-NPs for treating diseases. Passive targeting takes advantage of the enhanced permeability and retention effect, which enables NPs to accumulate in inflamed areas due to their porous blood vessels (Hao et al. 2021; Feng and Zheng 2023). On the other hand, active targeting involves modifying NPs with specific ligands or antibodies that bind to overexpressed receptors on target cells, such as integrins found on tumor blood vessels or ICAM-1 in inflamed tissues. This strategy significantly improves the precision of drug delivery and reduces off-target effects (Zeng and Pu 2020; Zhang et al. 2021b).

CMC-NPs also offer a unique approach to drug targeting by utilizing the natural adhesion properties of their source cells as an alternative to active targeting methods (Gao and Zhang 2015). This homotypic targeting plays a crucial role in biological processes. For instance, cancer cell-derived membranes express surface antigens that demonstrate homophilic adhesion, enabling CMC-NPs to target tumors of the exact origin (Gao and Zhang 2015). Similarly, NPs coated with platelet membranes can locate damaged blood vessels or inflammation sites due to platelets' innate attraction to injury areas (Qindeel et al. 2021). In systemic drug delivery, NPs with extended circulation times offer improved tissue targeting. Therefore, proper surface coatings are necessary to prevent the early uptake of NPs by reticuloendothelial system, prolong circulation, and effectively deliver NPs to the target tissues. CMC-NPs exploit the biomimetic features of cell membranes to evade immune detection. A critical protein on RBCs membranes, CD47, for example, interacts with macrophages' signal regulatory protein alpha to send a "don't eat me" signal (Russ et al. 2018). Furthermore, natural cell membranes delay bloodstream clearance, resulting in more extended circulation than uncoated nanocarriers. Adding PEG or specific ligands to CMC-NPs can further extend circulation by reducing protein corona formation (Suk et al. 2016; Gao et al. 2021b).

Cellular uptake is another advantageous mechanism for drug delivery using CMC-NPs. These engineered nanocarriers have the potential to deliver their therapeutic payloads by exploiting several cellular uptake and release mechanisms, including receptor-mediated endocytosis, in which surface ligands on the NPs attach to specific receptors on target cells, triggering endocytosis and internalizing the nanocarrier into the cell (Yaman et al. 2020). Sometimes, the NP coating fuses directly with the target cell membrane, evading endocytosis and delivering drugs directly into the cytoplasm. This mechanism works exceptionally well for administering medications prone to endosomal degradation (Ju et al. 2020). The controlled release mechanisms of CMC-NPs are also a significant feature. Scientists can engineer NPs to respond to specific triggers, such as changes in pH or temperature, allowing for targeted drug delivery at the intended site. For instance, NPs sensitive to pH can discharge their therapeutic contents in the acidic environment surrounding tumors or infection sites, optimizing drug availability where it is most effective (Yang et al. 2018; Chai et al. 2019). This approach enhances treatment outcomes while reducing the systemic toxicity typically associated with traditional drug delivery methods. These pathways facilitate precise drug administration and improve therapeutic results by ensuring the payload effectively reaches its cellular target.

15.5 Applications of CMC-NPs in Infectious Disease Therapy

CMC-NPs have shown great potential for functioning effectively within the body due to their exceptional ability to mimic biological systems. This makes them an ideal foundation for various therapeutic applications, especially in the fight against infectious diseases (Narain et al. 2017a). CMC-NPs offer crucial advantages in targeted and sustained antimicrobial drug delivery, primarily through enhanced circulation time and improved biodistribution (Liu et al. 2023a, b). Various CMC-NPs have recently been utilized to combat pathogen infections, leveraging the intimate interactions between pathogens and cellular membranes (Li et al. 2023b). Researchers are developing these biomimetic nanocarriers to enhance the effectiveness of diverse antimicrobial substances in combating bacteria, viruses, and other disease-causing microorganisms. The next section explores the therapeutic applications of CMC-NPs in treating bacterial, viral, fungal, and parasitic infections. Table 15.1 summarizes reported CMC-NPs, detailing membrane sources, payloads, fabrication methods, and applications.

15.5.1 CMC-NPs for Bacterial Infections

Persistent pathogenic bacteria engage with their infectious microenvironment, shaping the development of cell membrane-coating strategies for enhanced

Table 15.1 Summary of various CMC-NPs and their applications in combating infectious diseases

Cell membrane source	Incorporated payloads	Fabrication techniques	Applications	References
Bacterial infections				
Erythrocyte	Vancomycin-loaded nanogel	Membrane vesicle templated *in situ*	MRSA infections	Zhang et al. (2017)
	Tedizolid phosphate -PLGA NPs	Ultrasound method	MSSA and MRSA wound infection	Wu et al. (2021)
	Gold/polydopamine NPs	Extrusion	NIR photothermal therapy against MRSA infections	Bai et al. (2024)
Neutrophil	Sparfloxacin-loaded PCL-PEG NPs	Sonication method	MRSA pneumonia	Wang et al. (2020a)
	Antimicrobial peptide KLA-NPs	Ultrasound method	*K. Pneumoniae* lung infections	Liu et al. (2023a)
	Resolvin D1/ ceftazidime	Extrusion	*P. aeruginosa* pneumonia	Gao et al. (2021a)
	Indocyanine green/ rifampicin-loaded PLGA NPs	Co-extrusion	Photothermal-responsive therapy against *P. aeruginosa* infections	Li et al. (2023a)
	Fucoidan-loaded PLGA NPs	Sonication method	MRSA wound infection	Li et al. (2023c)
Macrophage	Silver-gold nanocage	Extrusion	Bone infection	Wang et al. (2018)
	$Ca_3(PO_4)_2/TiO_2$ magnetic nano-composite	Electroporation		Shi et al. (2021)
	Triclosan/ ciprofloxacin antimicrobial NPs	Homogenization	Intracellular *S. aureus* infections	Li et al. (2020)
	D-alanine-functionalized gold NPs	Extrusion	Cascade targeting photothermal therapy against intracellular *S. aureus* infections	Xiong et al. (2024)
	Antimicrobial peptide NPs	Co-extrusion	Bacterial sepsis	Meng et al. (2023)
	Silk fibroin NPs	Physical extrusion	Periodontitis	Deng et al. (2023)
Platelets	PLGA NPs	Sonication	Systemic *S. aureus* infections	Kim et al. (2021)
	Vancomycin-loaded Ag-MOFs	Ultrasound	MRSA infections	Huang et al. (2021)
	Fe/Zn-MOFs nanozyme	NA	*S. aureus* and *E. coli* wound infections	Shi et al. (2024)
Stem cells	FZ/MER-Ag-MOFs	Ultrasonic extrusion	Bacterial sepsis	Lu et al. (2023b)

(continued)

Table 15.1 (continued)

Cell membrane source	Incorporated payloads	Fabrication techniques	Applications	References
Macrophage/ OMVs	Gold NPs	Extrusion	Nanovaccines against *P. aeruginosa* infections	Peng et al. (2024)
Macrophage/ RBCs	Black phosphorous QDs	Co-extrusion	Photothermal therapy	Liu et al. (2024a, b)
Platelets/RBCs	Fe3O4/ cinnamaldehyde-loaded sodium alginate NPs	Ultrasound	MRSA pneumonia	Hu et al. (2023)
Viral infections				
T-cells	PLGA NPs	Sonication	HIV infections	Wei et al. (2018b)
Mosquito cell line	Gelatin NPs	Cell membrane cloaking	Zika virus	Rao et al. (2019)
RBCs Hepatocytes	CRISPR/Cas9 nanodrug	Extrusion	Hepatitis B infections	Wu et al. (2024)
Hybrid (393 T/ ACE-II cells/ human myeloid TPH1 cells	–	Sonication/ extrusion	COVID-19	Rao et al. (2020)
ACE-II-rich cells	–	Extrusion		Wang et al. (2021a, b, c)
	Quercetin-loaded PLGA NPs	Sonication		Fang et al. (2024)
Macrophage	Lopinavir-loaded PLGA NPs	Sonication		Tan et al. (2021)
Pulmonary alveolar epithelial cells	PLGA NPs	Extrusion		Chen et al. (2024)
Alveolar macrophage	PLGA NPs	Sonication		Li et al. (2021a, b)
Fungal and parasitic infections				
Streptococcus salivarius K12 membranes	Triclosan-loaded PLGA NPs	Extrusion	Oral candidiasis	Ye et al. (2023)
Vaginal epithelial cells	O_2-dissolved perfluorocarbon	Ultrasonication	*Candida albicans*	Lin et al. (2023)
Macrophage	Natamycin-loaded PLGA NPs	Extrusion	Fungal keratitis	Liu et al. (2024a, b)
Brain microvascular epithelial cells	Dihydroartemisinin-loaded MSNs	–	Cerebral malaria	Wei et al. (2022)
Parasite membrane	PLGA NPs	Sonication	*Giardia lamblia*	Zhou et al. (2022)

Abbreviations: *NPs* nanoparticles, *CMC-NPs* cell membrane-coated nanoparticles, *RBCs* red blood cells, *MRSA* methicillin-resistant *S. aureus*, *MSSA* methicillin-sensitive *S. aureus*, *PLGA* poly(lactic-co-glycolic acid), *PCL-PEG* polycaprolactone-polyethylene glycol, *MOFs* metal-organic frameworks, *Ag* silver, *Fe* iron, *QDs* quantum dots, *NIR* near-infrared, *OMVs* outer membrane vesicles, *HIV* human immunodeficiency virus, *ACE-II* angiotensin-converting enzyme II, *MSNs* mesoporous silica nanoparticles antimicrobial therapy. CMC-NPs effectively combat bacterial infections by evading immune detection, targeting pathogens, neutralizing toxins, enabling toxoid vaccine delivery, and ensuring prolonged circulation (Li et al. 2023b). This strategy positions CMC-NPs as promising candidates for targeted and sustained antibacterial therapies (Ai et al. 2018; Elhassan et al. 2022). Researchers investigated NPs enveloped in membranes extracted from various cellular sources within this domain. These membrane sources encompass erythrocytes, neutrophils, macrophages, platelets, stem cells, and hybrid cells. The potential applications of these diverse CMC-NPs in addressing bacterial infections are discussed.

Various nanosystems have been modified by applying erythrocyte membrane coatings to enhance antibiotic delivery. This approach reduces immune system clearance and extends blood circulation time, thereby improving the pharmacokinetic properties of antibiotic nanocarriers. Extensive research has been conducted to tackle the significant challenge of *S. aureus* and methicillin-resistant *S. aureus* (MRSA) infections. For example, researchers developed an RBC membrane-coated nanogel system (RBC-nanogel) to combat MRSA-associated toxins while facilitating targeted vancomycin delivery, thus enhancing treatment efficacy against this persistent pathogen. The RBC-nanogel acts as a decoy for MRSA-produced pore-forming toxins, boosting the immune response by encouraging bacterial uptake by immune cells and aiding in bacterial elimination (Zhang et al. 2017). In a similar vein, promising outcomes were demonstrated using a straightforward ultrasound technique to create RBC membrane-coated tedizolid phosphate (TR-701)-loaded PLGA NPs (TR-701-PLGA@RBC NPs) for addressing methicillin-sensitive *S. aureus* and MRSA wound infections (Wu et al. 2021). RBC membranes were recently extracted and coated onto gold/polydopamine NPs containing aptamer (Au/PDA@RBCM-aptamer) through extrusion for targeted near-infrared II photothermal therapy against *S. aureus* infections (Fig. 15.1). The encapsulation of Au/PDA@RBCM-aptamer with RBC membrane yielded the highest and most selective photothermal conversion capability and maximized localized therapeutic effect (Bai

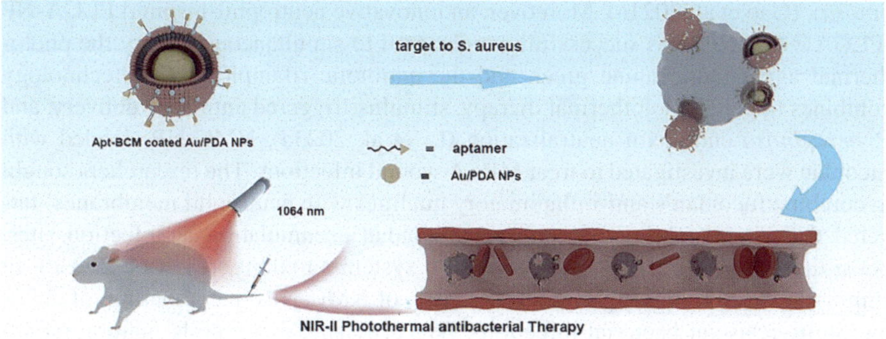

Fig. 15.1 The scheme illustrates the design of RBC membrane-coated gold/polydopamine NPs for targeted near-infrared II (NIR-II) photothermal therapy against *S. aureus* infections. (Reproduced with permission from Bai et al. 2024, copyright 2024, Elsevier B.V.)

et al. 2024). While various strategies were implemented in these investigations, all engineered NPs demonstrated enhanced biocompatibility, controlled antibiotic release, improved antibacterial effectiveness, decreased immunogenicity, and substantial toxin-neutralization capabilities. These groundbreaking studies present promising approaches to combat susceptible and resistant *S. aureus* infections by developing biomimetic erythrocyte membrane-coated NPs.

Drawing inspiration from neutrophils' natural ability to target inflamed tissues, scientists have developed NPs coated with neutrophil membranes for targeted therapy of inflammation and bacterial infections. Studies have explored the application of NMC-NPs in treating lung infections. For example, one study examined NMC-NPs as an innovative targeted drug delivery system for addressing inflammation, particularly in pneumonia and other acute lung injuries. A straightforward sonication method was employed to coat polycaprolactone-polyethylene glycol NPs loaded with sparfloxacin using neutrophil membranes (SPX/PCL-PEG@NMC-NPs). The resulting NMC-NPs exhibited preserved functional integrity of neutrophil membranes, favorable biocompatibility, controlled drug release characteristics, enhanced antibacterial activity against MRSA, and notable interaction with pulmonary inflammation sites (Wang et al. 2020a). Also, through targeted therapy, scientists critically evaluated NMC-NPs' effectiveness in treating antibiotic-resistant *Klebsiella pneumoniae* (*K. pneumoniae*) lung infections. The study involved the creation of KLA-neutrophil NPs (KLA-NMC-NPs), which integrate the antimicrobial peptide KLA with neutrophil membranes to form a targeted delivery mechanism capable of efficiently infiltrating infected cells. The engineered KLA-NMC-NPs could bypass immune system detection, achieve highly efficient targeted accumulation at infection sites, and successfully impede the progression of pneumonia caused by *K. pneumoniae* (Liu et al. 2023a). Furthermore, NMC-NPs were utilized to concurrently deliver multiple drugs to target inflammatory pathways and disease-causing bacteria. One example is using nanovesicles derived from human neutrophil membranes to simultaneously deliver Resolvin D1, a potent anti-inflammatory agent, and ceftazidime antibiotic. This approach has been introduced for enhanced antibacterial treatment of pneumonia caused by *Pseudomonas aeruginosa* (*P. aeruginosa*). (Gao et al. 2021a). Moreover, an innovative neutrophil-inspired PLGA-NP (PLGA@NM NP) was successfully engineered to simultaneously carry the photothermal agent indocyanine green and the antibiotic rifampicin. This technology combines targeted photothermal therapy, stimulus-triggered antibiotic delivery, and *P. aeruginosa* endotoxin neutralization (Li et al. 2023a). NMC-NPs loaded with fucoidan were investigated to treat MRSA wound infections. The researchers sought to combine fucoidan's anti-inflammatory qualities with neutrophil membranes' targeted delivery capabilities to enhance fucoidan accumulation at infection sites, boost therapeutic effectiveness, and reduce systemic toxicity. (Li et al. 2023c). In summary, these studies highlight the promise of NMC-NPs in combating infamous multidrug-resistant bacterial infections. By harnessing neutrophils' natural targeting abilities and integrating cutting-edge drug delivery methods, this strategy offers a prospective remedy for the escalating issue of antibiotic resistance and related inflammatory complications in infectious diseases.

The plasma membranes of macrophages are equipped with various functional proteins, including toll-like receptors and pattern recognition receptors, which play a vital role in identifying and attaching to pathogens (Li and Wu 2021). Encapsulating NPs in membranes derived from macrophages allows the coated particles to seek out inflammatory areas, identify pathogens, and adhere to infectious agents such as bacteria (Khatoon et al. 2022). The use of macrophage membrane-coated NPs (MMC-NPs) in fighting infectious diseases is diverse. These biomimetic particles excel at locating infection sites due to their natural tendency to accumulate in inflamed areas, similar to macrophages. This property can be utilized to deliver antibiotics directly to infected regions, enhancing drug effectiveness while reducing unintended effects and overall toxicity in the body (Wang et al. 2024). Leveraging these advantages, numerous studies have employed macrophage membranes to create effective nano drug delivery systems for complex diseases involving both microbial infections and immune dysregulation. To illustrate, researchers have utilized macrophage membranes to encapsulate a photothermal silver-gold nanocage (SGNC@MM) and calcium phosphate/titanium dioxide magnetic nanocomposite ($Ca_3(PO_4)_2/TiO_2$@MM-NPs), substantially improving their antibacterial effectiveness in combating bone infections. These engineered nanocarriers demonstrated remarkable biocompatibility, extended blood circulation times, enhanced bacterial adhesion, superior anti-inflammatory characteristics, and improved tissue penetration capabilities (Wang et al. 2018; Shi et al. 2021). Advancements have also been made in using MMC-NPs to treat infections caused by intracellular bacteria, particularly *S. aureus*. One example is developing a macrophage-based antimicrobial system that responds to pH changes. This system, which combines triclosan and ciprofloxacin, is taken up by macrophages harboring intracellular staphylococci and shows improved antibacterial efficacy in the acidic phagosomal environment (Li et al. 2020). Another approach involves creating a novel macrophage-based system using D-alanine-functionalized gold NPs (DAu-NPs@MM). This innovative method employs cascade-targeting photothermal therapy to target and eliminate intracellular pathogens. The DAu-NPs@MM system demonstrated strong photothermal effects and exhibited exceptional cascade-targeting capabilities, allowing for precise and localized antibacterial action (Xiong et al. 2024). Furthermore, antimicrobial peptide NPs coated with macrophage membranes (AMP-NPs@MM) have shown promise in combating bacterial sepsis. These NPs exhibited improved stability, longer circulation time within macrophages, superior antibacterial and antibiofilm properties, and decreased cytokine levels, leading to improved therapeutic outcomes (Meng et al. 2023). Recently, genetically engineered macrophage membranes functionalized with TLR4 ligands have been employed to coat silk fibroin NPs to address antibacterial treatment and immune regulation in periodontitis. These macrophage-coated nanosystems effectively targeted harmful bacteria and reduced inflammation, thereby promoting healing in periodontal tissues affected by the disease (Deng et al. 2023). These findings demonstrate the encouraging potential of MMC-NPs as a revolutionary strategy for treating numerous bacterial infections, which may pave the way for further clinical applications.

PMC-NPs have shown significant potential in combating bacterial infections. These particles perform two crucial functions: neutralize toxins produced by bacteria and improve the delivery of antibacterial substances. For example, PMC-NPs effectively attach to and neutralize virulence factors, including toxins released by *S. aureus*, a significant contributor to invasive infections. This neutralization process safeguards immune cells, such as macrophages and neutrophils, maintaining their bacteria-killing abilities while decreasing inflammation (Kim et al. 2021). Furthermore, the distinctive characteristics of platelet membranes facilitate the creation of "decoy" NPs. These decoys intercept bacterial toxins before they can harm host cells, thereby reducing systemic toxicity and enhancing survival rates in severe infections such as bacteremia. Research has demonstrated that PMC-NPs effectively neutralize MRSA toxins, decrease bacterial populations in the bloodstream, and boost host survival in experimental models (Huang et al. 2021). PMC-NPs also enhance the targeted delivery of antibiotics, addressing a significant challenge in treating infected wounds. The platelet membrane coating improves the interaction between NPs and the wound environment, advancing bacterial-targeted therapy with potent antibacterial effects against common wound-causing pathogens. A recent investigation showcased the improved delivery of metal organic framework (MOF)-based nanozyme using a platelet membrane coating approach to combat wound infections caused by *Escherichia coli* and *S. aureus*. The developed PMC-nanozymes exhibited enhanced biodegradability, favorable metabolism, non-toxic accumulation, and significant antibacterial and antibiofilm activities. They also reduced inflammatory cytokine levels, improving targeting capabilities and accelerating wound healing (Shi et al. 2024). In conclusion, the capacity of PMC-NPs to neutralize toxins and precisely deliver antibiotics positions them as a crucial component in future therapeutic approaches against drug-resistant pathogens.

Stem cell membrane-coated NPs (SCM-NPs) show great promise in combating bacterial infections, including antibiotic-resistant ones. These particles employ a two-pronged strategy: precisely delivering antimicrobial agents and influencing the host's immune response (Ramos et al. 2024). Additionally, SCMs can boost antibiotic effectiveness, creating a combined effect that enhances bacterial elimination (Song et al. 2023). The stem cell membrane coating provides several benefits to the encapsulated NPs in treating bacterial infections. Firstly, the membrane's inherent characteristics enable the NPs to evade opsonization and clearance by phagocytes, resulting in extended circulation in the bloodstream and improved accumulation at infection sites. Secondly, the membranes possess adhesion molecules that enable direct interaction with bacteria or biofilms, allowing for specific targeting of pathogens (Wang et al. 2021a). Moreover, stem cell membranes, especially those from mesenchymal stem cells, exhibit anti-inflammatory and immunomodulatory properties vital in combating infections. These membranes can reduce the excessive inflammatory reactions typically seen in bacterial infections, thus lessening tissue harm and facilitating recovery (Johnson et al. 2017). Mesenchymal SCM-NPs have demonstrated increased accumulation at bacterial infection sites in preclinical studies, enhancing drug delivery and treatment effectiveness (Wang et al. 2021a). Similarly, researchers have created bone marrow-derived mesenchymal SCM-NPs that combine anti-inflammatory and antimicrobial qualities to treat bacterial sepsis. These bioengineered NPs showed improved antibacterial action against MRSA and considerable anti-inflammatory effects through the modulation of inflammatory responses (Lu et al. 2023b).

Hybrid cell membranes (HCMs) have become a promising avenue in creating advanced materials for fighting bacterial infections. These membranes serve as multifunctional platforms that combine antibacterial, anti-inflammatory, and detoxification properties, making them essential in combating bacterial infections (Zhao et al. 2021). One example of this approach is the development of an antibacterial nanovaccine, which involves coating gold NPs (Au-NPs) with a hybrid membrane composed of bacterial outer membrane vesicles (OMVs) and macrophage membranes (Fig. 15.2). This system addresses *P. aeruginosa*-related infections by modulating immune responses and delivering effective bactericidal action (Peng et al. 2024). Furthermore, HCM-NPs were enhanced with photothermal agents to improve infection treatment. For instance, NPs that combine macrophage and RBCs membranes with black phosphorus quantum dots exhibit improved targeting of inflamed bacterial sites and enhanced bactericidal effects when exposed to near-infrared

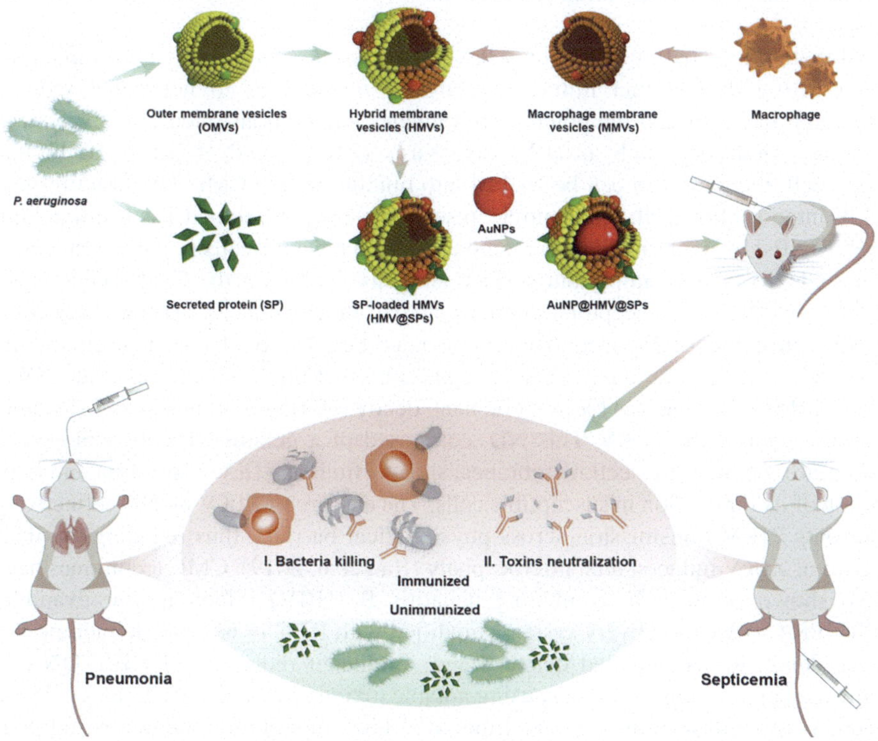

Fig. 15.2 Schematic illustration of the development and use of multi-antigenic hybrid cell membrane-based Au NPs vaccine against *P. aeruginosa* infection. (Reproduced with permission from Peng et al. 2024, copyright 2024, Wiley-VCH GmbH)

light. These strategies show potential for addressing infections caused by biofilm-forming bacteria (Liu et al. 2024a, b). Additionally, researchers have developed a ferroptosis-inducing agent based on iron oxide (Fe_3O_4) by encapsulating Fe_3O_4/cinnamaldehyde NPs within a composite membrane derived from RBCs and platelets. This approach was designed to induce ferroptosis in MRSA, disrupt bacterial communication through quorum sensing, and eradicate biofilm, ultimately providing effective treatment for severe MRSA pneumonia (Hu et al. 2023). These biomimetic NPs show potential in fighting MRSA infections by simultaneously targeting several crucial aspects of bacterial pathogenesis, including biofilm development, virulence factors, and survival mechanisms. Cell membrane coatings present a multifaceted and potent strategy for addressing bacterial infections. These systems incorporate diverse biological and chemical elements to improve targeting, counteract toxins, and deliver strong antibacterial and anti-inflammatory effects. As a result, they show great potential for clinical use in managing infections. Ongoing research in this field is expected to transform infection treatment, offering new possibilities for tackling some of the most critical global health challenges.

15.5.2 CMC-NPs for Viral Infections

Cell membrane coating technology has shown promise in fighting viral infections by creating decoys that mimic natural cell membranes to neutralize viruses. Research has explored the potential advantages and demonstrated effectiveness in treating conditions such as HIV, Zika virus (ZIKV), and Hepatitis B. In this approach, the pathogen can be tricked into binding with a CMC-NP target instead of its intended host cell. For example, researchers have developed T-cell-mimicking NPs (TNPs) by applying plasma membranes from CD4+ T cells to polymeric cores. These TNPs acquire important surface antigens, including the CD4 receptor and CCR5 or CXCR4 coreceptors, crucial for HIV attachment. By serving as decoys, TNPs can redirect HIV away from its intended targets, effectively neutralizing the virus without exerting intense selective pressure that might lead to resistance (Wei et al. 2018b). Similarly, a biomimetic nano decoy (ND) was proposed as a systemic defense against the ZIKV. This ND, composed of a gelatin NP core wrapped in mosquito medium host cell membranes, shows efficient ZIKV adsorption and inhibition of its replication in susceptible cells. The capture of ZIKV by NDs effectively prevents ZIKV transmission across physiological barriers, thus reducing the incidence of ZIKV-induced fetal microcephaly (Rao et al. 2019). CMC techniques have also shown promise in combating hepatitis B (HBV) infection. For example, CRISPR/Cas9 nanodelivery systems modified with RBC or hepatocyte membranes have demonstrated targeted delivery and significant reduction of HBV DNA in infected hepatic cells, without causing notable toxicity to the liver or kidneys. These membrane modifications enhanced the NPs' biocompatibility and decreased their elimination by the reticuloendothelial system, resulting in extended circulation time in the blood. This novel approach offers a potential therapeutic avenue for advancing HBV treatment (Wu et al. 2024).

The global outbreak of COVID-19, triggered by SARS-CoV-2, has led to unparalleled worldwide initiatives to create effective treatments and preventive measures. In this context, CMC-NPs have emerged as a promising tool for combatting viral infections. These NPs replicate the inherent characteristics of biological membranes, offering an innovative approach to enhance biointerfacing, optimize drug delivery, neutralize viruses, and regulate immune responses. This biomimetic strategy presents unique benefits for targeting SARS-CoV-2 and alleviating the severe manifestations of COVID-19 (Pereira-Silva et al. 2021). For example, CMC-NPs can function as decoys, intercepting and neutralizing SARS-CoV-2 before host cell infection. NPs coated with biological membranes from cells that express the ACE2 receptor can capture the virus, inhibiting its attachment to and entry into human cells (Rao et al. 2020; Li et al. 2021a, b; Wang et al. 2021a, b, c; Fang et al. 2024). These NPs also can absorb inflammatory cytokines, including interleukin-6 (IL-6) and granulocyte-macrophage colony-stimulating factor (GM-CSF), which are found in elevated levels during COVID-19 infections (Rao et al. 2020; Tan et al. 2021). This two-fold functionality aids in decreasing inflammation and alleviating immune disorders linked to the disease. Specific formulations, such as those utilizing pulmonary alveolar epithelial cell membranes, display antifouling characteristics that safeguard their neutralization capabilities by preventing protein corona formation in blood circulation, thus preserving their efficacy *in vivo* (Chen et al. 2024). Furthermore, NDs derived from lung spheroid cells have demonstrated the ability to remain in the lungs for extended durations when administered through inhalation, enhancing viral clearance and diminishing lung injury without observable toxicity (Li et al. 2021a, b). An additional approach of using macrophage biomimetic nanocarriers could decrease viral burdens and inflammatory responses in COVID-19 studies. The encapsulation of NPs within macrophage membrane vesicles enables targeted drug delivery to inflamed areas, improving therapeutic outcomes and minimizing systemic adverse effects (Tan et al. 2021). Certain NPs incorporate photothermal materials, allowing them to disrupt viral particles when exposed to near-infrared irradiation (NIR), offering a dual approach to antiviral and anti-inflammatory treatment (Li et al. 2021a, b).

Overall, CMC-NPs offer a multifaceted, cutting-edge solution for addressing viral infections. These nanocarriers exploit their capacity to imitate natural cell surfaces, effectively neutralizing viruses and influencing immune responses. This approach presents a promising direction for both therapeutic and preventive strategies. As scientific investigations continue, these technologies may prove instrumental in addressing present and future viral outbreaks.

15.5.3 CMC-NPs for Fungal and Parasitic Infections

Drug-resistant fungal infections present a significant obstacle in medical care. Researchers have explored biomimetic strategies, including cell membrane coating technology, to tackle this issue. One such approach involves the development of biomimetic nanoplatforms designed to target *Candida albicans*, a prevalent fungal pathogen. These platforms employ CMC-NPs to specifically attach to the pathogen, improving the delivery and effectiveness of antifungal medications while shielding healthy cells from harm. In one study, researchers created triclosan-loaded PLGA NPs encased in *Streptococcus salivarius* K12 membranes (K12/TCS@PLGA-NPs) using an extrusion method to combat *Candida albicans* and oral candidiasis (Ye et al. 2023). Another investigation led to the development of an oxygen-dissolved perfluorocarbon disguised by a photosensitizer-loaded vaginal epithelial cell membrane for treating intravaginal *Candida albicans* (Lin et al. 2023). Moreover, CMC-NPs were utilized to treat fungal keratitis (FK), a severe infectious disease of the cornea that is the primary cause of blindness. A recent investigation examined the effectiveness and safety of NPs coated with macrophage membranes for delivering natamycin (NAT) in FK therapy (Liu et al. 2024a, b). All the reported CMC-NPs demonstrated improved concentrations of antimicrobial agents in affected areas while shielding healthy tissues from toxic effects. These approaches boost the antifungal properties of the cargoes and maximize their overall therapeutic efficacy. Despite the promising outcomes of using CMC-NPs in combating fungal infections, limited research has been conducted in this area. Consequently, there is a significant need to address the substantial knowledge gap in this field to enhance the efficacy of cell membrane coating technology against fungal-induced infectious diseases.

Parasitic infections pose a major global health challenge, with conventional treatments often facing issues of drug resistance and reduced effectiveness. In this context, NPs have emerged as a promising alternative, offering innovative approaches to combat parasites. These NPs utilize various mechanisms to target parasites, such as disrupting their plasma membranes, inhibiting protein production, and generating free radicals (Bajwa et al. 2022). These strategies have shown the potential to eliminate parasitic organisms effectively. Researchers have developed specialized biomimetic CMC-NPs to combat parasitic infections. One notable example is an NP formulation designed to treat cerebral malaria, a severe condition caused by *Plasmodium falciparum*. This formulation is coated with membranes from brain microvascular endothelial cells, which enhances its ability to deliver drugs to infected red blood cells (Fig. 15.3). Additionally, it helps protect crucial organs from damage caused by the infection. Studies using experimental models showed that this approach improves survival rates (Wei et al. 2022). Furthermore, vaccines created using NPs coated with native parasite membranes have demonstrated the ability to elicit protective immunity against *Giardia lamblia*, highlighting the potential of this approach in vaccine creation (Zhou et al. 2022). Taken together, the technique of cell membrane coating shows promise for enhancing treatment efficacy and improving patient outcomes in combating parasitic infections. Nevertheless, this field of study is still in its early stages, and additional research is necessary to address the considerable knowledge gap in this area.

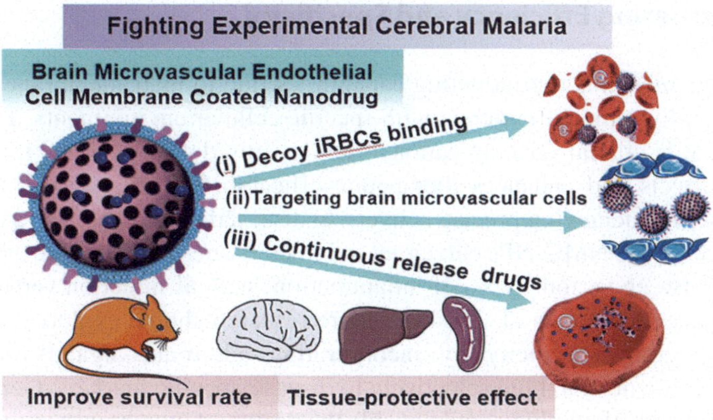

Fig. 15.3 Scheme illustrating the mechanisms and potential of the CMC-nanodrug in fighting against cerebral malaria. (Reproduced with permission from Wei et al. 2022, copyright 2022, American Chemical Society)

15.6 CMC-NPs: Advantages and Limitations

CMC-NPs have emerged as a promising platform in nanomedicine, offering unique advantages in drug delivery, immune evasion, and targeted therapy. However, their translation from laboratory research to clinical applications faces several challenges. This section comprehensively analyzes the advantages, limitations, and future directions for CMC-NPs, focusing on key thematic areas such as manufacturing, targeting, and clinical translation.

15.6.1 Manufacturing and Scalability

One of the most significant advantages of CMC-NPs lies in their biomimetic properties, which enable enhanced biocompatibility, immune evasion, and targeted drug delivery. By leveraging the natural properties of cell membranes, CMC-NPs can mimic biological systems, reducing immune system clearance and extending circulation time (Liu et al. 2019; Ai et al. 2020). However, the manufacturing process of CMC-NPs is complex and labor-intensive, involving the extraction of cell membranes and their subsequent coating onto nanoparticle cores. This complexity often leads to inconsistencies between batches, hindering large-scale production and standardization (Kroll et al. 2017; Zhao et al. 2021). Future research should prioritize scalable and reproducible manufacturing to overcome these challenges. Microfluidic technologies and automated systems could offer solutions by ensuring consistency and quality control (Rao et al. 2017; Fang et al. 2018). Additionally, advancements in membrane extraction techniques, such as improved hypotonic lysis and freeze-thaw cycles, could enhance the yield and quality of cell membranes, making the production process more efficient (Spanjers and Städler 2020; Chugh et al. 2021; Javed et al. 2021).

15.6.2 Targeting Efficiency and Specificity

CMC-NPs exhibit inherent targeting capabilities due to their membrane composition, which allows them to interact with specific cellular environments. This targeting precision is crucial for applications such as drug delivery, where it minimizes off-target effects and enhances therapeutic efficacy (Rao et al. 2017; Zou et al. 2020). However, achieving precise delivery to diseased tissues remains a challenge. For instance, while NMC-NPs can target inflamed tissues, they may struggle to differentiate between various types of inflammation, such as infection versus autoimmune responses (Zhang et al. 2024b). Future research should explore strategies to enhance targeting specificity by incorporating additional ligands or stimuli-responsive elements that bind selectively to disease-specific markers. Genetic engineering and synthetic biology could enable the design of membranes with improved targeting properties, allowing for more precise delivery of therapeutic agents (Yang et al. 2021; Imran et al. 2022). Furthermore, developing HCM-NPs, which combine membranes from different cell types, could offer multifunctional platforms with enhanced targeting capabilities (Yu et al. 2024).

15.6.3 Clinical Translation and Biocompatibility

Several factors, including potential immune responses, biological stability, and regulatory challenges, complicate the clinical translation of CMC-NPs. Although CMC-NPs are designed to evade immune detection, there remains a risk of triggering immune responses, especially when using membranes from heterologous sources (Kroll et al. 2017). Long-term biocompatibility studies are essential to ensure the safety of these systems in clinical applications (Zhang et al. 2024b). Regulatory and ethical considerations also pose significant hurdles. Using human-derived cell membranes raises concerns about donor consent, safety, and potential contamination with pathogens or oncogenic factors (Narain et al. 2017a, b). Future efforts should involve close collaboration with regulatory agencies to establish clear guidelines for the development and clinical testing of CMC-NPs. Ethical considerations, such as the sourcing of cells and the potential for misuse, must also be addressed to ensure responsible innovation (Zhai et al. 2017; Huang et al. 2024).

15.7 Conclusion and Future Outlook

CMC-NPs represent a transformative approach to drug delivery, leveraging the natural properties of cell membranes to achieve enhanced biocompatibility, immune evasion, and targeted therapy. Their ability to neutralize bacterial toxins, prolong circulation time, and precisely deliver drugs to infection sites underscores their potential in combating infectious diseases. Despite these advantages, challenges remain in scalability, immune compatibility, and clinical translation—hurdles that must be addressed to realize the promise of CMC-NPs fully. Looking ahead, the future of CMC-NPs is bright, provided research focuses on optimizing fabrication techniques (e.g., membrane isolation and coating methods) to ensure batch-to-batch consistency and therapeutic efficacy. Integrating emerging technologies, such as artificial intelligence for design optimization and advanced imaging for real-time tracking, could enhance their precision and performance. Beyond infectious diseases, expanding applications to cancer, autoimmune disorders, and regenerative medicine may unlock groundbreaking therapeutic avenues (Yu et al. 2024; Guo et al. 2024). Future studies should also explore CMC-NPs for fungal, viral, and parasitic infections, broadening their impact on global health challenges like antimicrobial resistance. Ultimately, interdisciplinary collaboration and sustained innovation will be pivotal in bridging the gap between preclinical development and clinical implementation. As CMC-NP technology evolves, it holds immense potential to revolutionize nanomedicine, offering targeted, efficient, and adaptable solutions for various diseases. By refining their design and expanding their scope, CMC-NPs could emerge as a cornerstone of next-generation drug delivery systems, reshaping the landscape of modern therapeutics.

References

Ai X, Hu M, Wang Z, Zhang W, Li J, Yang H, Lin J, Xing B (2018) Recent advances of membrane-cloaked Nanoplatforms for biomedical applications. Bioconjug Chem 29(4):838–851. https://doi.org/10.1021/acs.bioconjchem.8b00103

Ai X, Wang S, Duan Y, Zhang Q, Chen M, Gao W, Zhang L (2020) Emerging approaches to functionalizing cell membrane-coated nanoparticles. Biochemistry 60:941–955. https://doi.org/10.1021/acs.biochem.0c00343

Alimohammadvand S, Kaveh Zenjanab M, Mashinchian M, Shayegh J, Jahanban-Esfahlan R (2024) Recent advances in biomimetic cell membrane–camouflaged nanoparticles for cancer therapy. Biomed Pharmacother 177:116951. https://doi.org/10.1016/j.biopha.2024.116951

Andriola Silva AK, Di Corato R, Pellegrino T, Chat S, Pugliese G, Luciani N, Gazeau F, Wilhelm C (2013) Cell-derived vesicles as a bioplatform for the encapsulation of theranostic nanomaterials. Nanoscale 5(23):11374–11384. https://doi.org/10.1039/C3NR01541F

Bahmani B, Gong H, Luk BT, Haushalter KJ, DeTeresa E, Previti M, Zhou J, Gao W, Bui JD, Zhang L, Fang RH, Zhang J (2021) Intratumoral immunotherapy using platelet-cloaked nanoparticles enhances antitumor immunity in solid tumors. Nat Commun 12(1):1999. https://doi.org/10.1038/s41467-021-22311-z

Bai S, Yang Y, Sheng R, Qi Y, Jia Y, Wang X, Cui W, Zheng Y, Li H, Li J (2024) Blood cellular membrane-coated au/polydopamine nanoparticle-targeted NIR-II antibacterial therapy. J Colloid Interface Sci 665:855–862. https://doi.org/10.1016/j.jcis.2024.03.134

Baig MS, Ahmad A, Pathan RR, Mishra RK (2024) Precision Nanomedicine with bio-inspired Nanosystems: recent trends and challenges in mesenchymal stem cells membrane-coated bioengineered Nanocarriers in targeted Nanotherapeutics. J Xenobiot 14(3):827–872

Bajwa H, Khan M, Abbas Z, Riaz R, Rehman T, Abbas R, Aleem M, Abbas A, Almutairi M, Alshammari F, Alraey Y, Alouffi A (2022) Nanoparticles: synthesis and their role as potential drug candidates for the treatment of parasitic diseases. Life 12. https://doi.org/10.3390/life12050750

Bloom DE, Cadarette D (2019) Infectious disease threats in the twenty-first century: strengthening the global response. Front Immunol 10(MAR):549. https://doi.org/10.3389/fimmu.2019.00549

Bose RJC, Kim BJ, Arai Y, Han I, Moon JJ, Paulmurugan R, Park H, Lee S-H (2018) Bioengineered stem cell membrane functionalized nanocarriers for therapeutic targeting of severe hindlimb ischemia. Biomaterials 185:360–370. https://doi.org/10.1016/j.biomaterials.2018.08.018

Chai Z, Ran D, Lu L, Zhan C, Ruan H, Hu X, Xie C, Jiang K, Li J, Zhou J, Wang J, Zhang Y, Fang RH, Zhang L, Lu W (2019) Ligand-modified cell membrane enables the targeted delivery of drug nanocrystals to glioma. ACS Nano 13(5):5591–5601. https://doi.org/10.1021/acsnano.9b00661

Chen H, Sha H, Zhang L, Qian H, Chen F, Ding N, Ji L, Zhu A, Xu Q, Meng F, Yu L, Zhou Y, Liu B (2018) Lipid insertion enables targeted functionalization of paclitaxel-loaded erythrocyte membrane nanosystem by tumor-penetrating bispecific recombinant protein. Int J Nanomedicine 13(Null):5347–5359. https://doi.org/10.2147/IJN.S165109

Chen Z, Wang W, Li Y, Wei C, Zhong P, He D, Liu H, Wang P, Huang Z, Zhu W, Zhou Y, Qin L (2021) Folic acid-modified erythrocyte membrane loading dual drug for targeted and chemo-Photothermal synergistic cancer therapy. Mol Pharm 18(1):386–402. https://doi.org/10.1021/acs.molpharmaceut.0c01008

Chen H, Xiong X, Huang Y, Huang B, Luo X, Ke Q, Wu P, Wang S (2024) SARS-CoV-2 neutralization by cell membrane-coated antifouling nanoparticles. ACS Appl Bio Mater 7:909–917. https://doi.org/10.1021/acsabm.3c00936

Chugh V, Vijaya Krishna K, Pandit A (2021) Cell membrane-coated mimics: a methodological approach for fabrication, characterization for therapeutic applications, and challenges for clinical translation. ACS Nano 15(11):17080–17123. https://doi.org/10.1021/acsnano.1c03800

Dadgostar P (2019) Antimicrobial resistance: implications and costs. Infect Drug Resist 12:3903–3910. https://doi.org/10.2147/IDR.S234610

Dehaini D, Wei X, Fang RH, Masson S, Angsantikul P, Luk BT, Zhang Y, Ying M, Jiang Y, Kroll AV, Gao W, Zhang L (2017) Erythrocyte–platelet hybrid membrane coating for enhanced nanoparticle functionalization. Adv Mater 29(16):1606209. https://doi.org/10.1002/adma.201606209

Deng Y, Ren M, He P, Liu F, Wang X, Zhou C, Li Y, Yang S (2023) Genetically engineered cell membrane-coated nanoparticles for antibacterial and immunoregulatory dual-function treatment of ligature-induced periodontitis. Front Bioeng Biotechnol 11. https://doi.org/10.3389/fbioe.2023.1113367

Duan Y, Zhou J, Zhou Z, Zhang E, Yu Y, Krishnan N, Silva-Ayala D, Fang RH, Griffiths A, Gao W, Zhang L (2023) Extending the in vivo residence time of macrophage membrane-coated nanoparticles through genetic modification. Small 19(52):2305551. https://doi.org/10.1002/smll.202305551

Elhassan E, Devnarain N, Mohammed M, Govender T, Omolo CA (2022) Engineering hybrid nanosystems for efficient and targeted delivery against bacterial infections. J Control Release 351:598–622. https://doi.org/10.1016/j.jconrel.2022.09.052

Ellwanger JH, da Veiga ABG et al (2021) Control and prevention of infectious diseases from a one health perspective. Genet Mol Biol 44(Suppl 1):e20200256. https://doi.org/10.1590/1678-4685-GMB-2020-0256

Fang RH, Kroll AV, Gao W, Zhang L (2018) Cell membrane coating nanotechnology. Adv Mater 30(23):e1706759. https://doi.org/10.1002/adma.201706759

Fang J-Y, Huang K-Y, Wang T-H, Lin Z-C, Chen C-C, Chang S-Y, Chen E-L, Chao T-L, Yang S-C, Yang P-C, Chen C-Y (2024) Development of nanoparticles incorporated with quercetin and ACE2-membrane as a novel therapy for COVID-19. J Nanobiotechnol 22(1):169. https://doi.org/10.1186/s12951-024-02435-2

Farokhzad OC, Langer R (2009) Impact of nanotechnology on drug delivery. ACS Nano 3(1):16–20

Farrell JM, Zhao CY, Tarquinio KM, Brown SP (2021) Causes and consequences of COVID-19-associated bacterial infections. Front Microbiol 12:1911. https://doi.org/10.3389/FMICB.2021.682571

Feng M, Zheng Y (2023) Advances in the use of cell-membrane encapsulated nanoparticles to target tumor drugs. Biomater Biosensors 2(2). https://doi.org/10.58567/bab02020001

Fernández-Borbolla A, García-Hevia L, Fanarraga ML (2024) Cell membrane-coated nanoparticles for precision medicine: a comprehensive review of coating techniques for tissue-specific therapeutics. Int J Mol Sci 25(4):2071
Gao W, Zhang L (2015) Coating nanoparticles with cell membranes for targeted drug delivery. J Drug Target 23(7–8):619–626. https://doi.org/10.3109/1061186X.2015.1052074
Gao J, Wang S, Wang Z (2017) High yield, scalable and remotely drug-loaded neutrophil-derived extracellular vesicles (EVs) for anti-inflammation therapy. Biomaterials 135:62–73. https://doi.org/10.1016/j.biomaterials.2017.05.003
Gao DX, Su Y, Wang Z (2021a) Human neutrophil membrane-derived nanovesicles as a drug delivery platform for improved therapy of infectious diseases. Acta Biomater 123:354–363. https://doi.org/10.1016/j.actbio.2021.01.020
Gao Y, Zhu Y, Xu X, Wang F, Shen W, Leng X, Zhao J, Liu B, Wang Y, Liu P (2021b) Surface PEGylated cancer cell membrane-coated nanoparticles for Codelivery of curcumin and doxorubicin for the treatment of multidrug resistant esophageal carcinoma. Front Cell Dev Biol 9. https://doi.org/10.3389/fcell.2021.688070
Geng B, Chen X, Chi J, Li F, Yim WY, Wang K, Li C, Xie M, Zhu P, Fan Z, Shi J, Hu Z, Zhang Y, Dong N (2023) Platelet membrane-coated alterbrassicene a nanoparticle inhibits calcification of the aortic valve by suppressing phosphorylation P65 NF-κB. Theranostics 13(11):3781–3793. https://doi.org/10.7150/thno.85323
Guan Q, Su J, Li Y, Zhang H, Ding Z, Wang Q, Ji S, Han J, Liu M, Zhao Y (2024a) Near-infrared laser-responsive erythrocyte membrane camouflaged nanoparticles for chemo-photodynamic therapy of breast cancer. ACS Appl Nano Mater 7(11):13785–13798. https://doi.org/10.1021/acsanm.4c02669
Guan X, Xing S, Liu Y (2024b) Engineered cell membrane-camouflaged nanomaterials for biomedical applications. Nano 14(5):413
Guo K, Ren S, Zhang H, Cao Y, Zhao Y, Wang Y, Qiu W, Tian Y, Song L, Wang Z (2023) Biomimetic gold Nanorods modified with erythrocyte membranes for imaging-guided Photothermal/gene synergistic therapy. ACS Appl Mater Interfaces 15(21):25285–25299. https://doi.org/10.1021/acsami.3c00865
Guo Q, Wang S, Xu R, Tang Y, Xia X (2024) Cancer cell membrane-coated nanoparticles: a promising anti-tumor bionic platform. RSC Adv 14(15):10608–10637. https://doi.org/10.1039/d4ra01026d
Han H, Bártolo R, Li J, Shahbazi MA, Santos HA (2022) Biomimetic platelet membrane-coated nanoparticles for targeted therapy. Eur J Pharm Biopharm 172:1–15. https://doi.org/10.1016/j.ejpb.2022.01.004
Han X, Gong C, Yang Q, Zheng K, Wang Z, Zhang W (2024) Biomimetic Nano-drug delivery system: an emerging platform for promoting tumor treatment. Int J Nanomedicine 19(Null):571–608. https://doi.org/10.2147/IJN.S442877
Hao W, Cui Y, Fan Y, Chen M, Yang G, Wang Y, Yang M, Li Z, Gong W, Yang Y, Gao C (2021) Hybrid membrane-coated nanosuspensions for multi-modal anti-glioma therapy via drug and antigen delivery. J Nanobiotechnology 19(1):378. https://doi.org/10.1186/s12951-021-01110-0
Harris JC, Scully MA, Day ES (2019) Cancer cell membrane-coated nanoparticles for cancer management. Cancers (Basel) 11(12):1836
Hasan N, Imran M, Jain D, Jha SK, Nadaf A, Chaudhary A, Rafiya K, Jha LA, Almalki WH, Mohammed Y, Kesharwani P, Ahmad FJ (2023) Advanced targeted drug delivery by bioengineered white blood cell-membrane camouflaged nanoparticulate delivery nanostructures. Environ Res 238(Pt 1):117007. https://doi.org/10.1016/j.envres.2023.117007
Hisato S, Saya O (2022) Cell-derived Giant membrane vesicles enclosing the Anthracycline anticancer drug doxorubicin and their cytotoxicity. Int J Membr Sci Technol 9(1):13–19. https://doi.org/10.15379/2410-1869.2022.09.01.02
Hossain MS, Hatta T, Labony SS, Kwofie KD, Kawada H, Tsuji N, Alim MA (2023) Food- and vector-borne parasitic zoonoses: global burden and impacts. Adv Parasitol 120:87–136. https://doi.org/10.1016/bs.apar.2023.02.001

Hu C-MJ, Zhang L, Aryal S, Cheung C, Fang RH, Zhang L (2011) Erythrocyte membrane-camouflaged polymeric nanoparticles as a biomimetic delivery platform. Proc Natl Acad Sci 108(27):10980–10985. https://doi.org/10.1073/pnas.1106634108

Hu H, Hua SY, Lin X, Lu F, Zhang W, Zhou L, Cui J, Wang R, Xia J, Xu F, Chen X, Zhou M (2023) Hybrid biomimetic membrane coated particles-mediated bacterial Ferroptosis for acute MRSA pneumonia. ACS Nano 17(12):11692–11712. https://doi.org/10.1021/acsnano.3c02365

Huang R, Cai GQ, Li J, Li XS, Liu HT, Shang XL, Zhou JD, Nie XM, Gui R (2021) Platelet membrane-camouflaged silver metal-organic framework drug system against infections caused by methicillin-resistant Staphylococcus aureus. J Nanobiotechnol 19(1):229. https://doi.org/10.1186/s12951-021-00978-2

Huang D, Wang X, Wang W, Li J, Zhang X, Xia B (2024) Cell-membrane engineering strategies for clinic-guided design of nanomedicine. Biomater Sci 12(11):2865–2884. https://doi.org/10.1039/D3BM02114A

Ikuta KS, Meštrović T, Naghavi M (2024) Global incidence and mortality of severe fungal disease. Lancet Infect Dis 24(5):e268. https://doi.org/10.1016/S1473-3099(24)00102-6

Imran M, Jha LA, Hasan N, Shrestha J, Pangeni R, Parvez N, Mohammed Y, Jha SK, Paudel KR (2022) "Nanodecoys" – future of drug delivery by encapsulating nanoparticles in natural cell membranes. Int J Pharm 621:121790. https://doi.org/10.1016/j.ijpharm.2022.121790

Javed S, Alshehri S, Shoaib A, Ahsan W, Sultan MH, Alqahtani SS, Kazi M, Shakeel F (2021) Chronicles of Nanoerythrosomes: an erythrocyte-based biomimetic smart drug delivery system as a therapeutic and diagnostic tool in cancer therapy. Pharmaceutics 13(3):368

Jiménez-Jiménez C, Manzano M, Vallet-Regí M (2020) Nanoparticles coated with cell membranes for biomedical applications. Biology (Basel) 9(11):406

Johnson V, Webb T, Norman A, Coy J, Kurihara J, Regan D, Dow S (2017) Activated mesenchymal stem cells interact with antibiotics and host innate immune responses to control chronic bacterial infections. Sci Rep 7:9575. https://doi.org/10.1038/s41598-017-08311-4

Ju Y, Guo H, Edman M, Hamm-Alvarez SF (2020) Application of advances in endocytosis and membrane trafficking to drug delivery. Adv Drug Deliv Rev 157:118–141. https://doi.org/10.1016/j.addr.2020.07.026

Kang T, Zhu Q, Wei D, Feng J, Yao J, Jiang T, Song Q, Wei X, Chen H, Gao X, Chen J (2017) Nanoparticles coated with neutrophil membranes can effectively treat cancer metastasis. ACS Nano 11(2):1397–1411. https://doi.org/10.1021/acsnano.6b06477

Khatoon N, Zhang Z, Zhou C, Chu M (2022) Macrophage membrane coated nanoparticles: a biomimetic approach for enhanced and targeted delivery. Biomater Sci 10(5):1193–1208. https://doi.org/10.1039/D1BM01664D

Kim JK, Uchiyama S, Gong H, Stream A, Zhang L, Nizet V (2021) Engineered biomimetic platelet membrane-coated nanoparticles block Staphylococcus aureus cytotoxicity and protect against lethal systemic infection. Engineering (Beijing) 7(8):1149–1156. https://doi.org/10.1016/j.eng.2020.09.013

Klein EY, Van Boeckel TP, Martinez EM, Pant S, Gandra S, Levin SA, Goossens H, Laxminarayan R (2018) Global increase and geographic convergence in antibiotic consumption between 2000 and 2015. Proc Natl Acad Sci USA 115(15):E3463–E3470. https://doi.org/10.1073/PNAS.1717295115/SUPPL_FILE/PNAS.1717295115.SAPP.PDF

Kroll AV, Fang RH, Zhang L (2017) Biointerfacing and applications of cell membrane-coated nanoparticles. Bioconjug Chem 28(1):23–32. https://doi.org/10.1021/acs.bioconjchem.6b00569

Le Q-V, Lee J, Lee H, Shim G, Oh Y-K (2021) Cell membrane-derived vesicles for delivery of therapeutic agents. Acta Pharm Sin B 11(8):2096–2113. https://doi.org/10.1016/j.apsb.2021.01.020

Li D, Wu M (2021) Pattern recognition receptors in health and diseases. Signal Transduct Target Ther 6(1):291. https://doi.org/10.1038/s41392-021-00687-0

Li R, He Y, Zhu Y, Jiang L, Zhang S, Qin J, Wu Q, Dai W, Shen S, Pang Z, Wang J (2019) Route to rheumatoid arthritis by macrophage-derived microvesicle-coated nanoparticles. Nano Lett 19(1):124–134. https://doi.org/10.1021/acs.nanolett.8b03439

Li Y, Liu Y, Ren Y, Su L, Li A, An Y, Rotello V, Zhang Z, Wang Y, Liu Y, Liu S, Liu J, Laman JD, Shi L, van der Mei HC, Busscher HJ (2020) Coating of a novel antimicrobial nanoparticle with a macrophage membrane for the selective entry into infected macrophages and killing of intracellular staphylococci. Adv Funct Mater 30(48). https://doi.org/10.1002/adfm.202004942

Li B, Wang W, Song W, Zhao Z, Tan Q, Zhao Z, Tang L, Zhu T, Yin J, Bai J, Dong X, Tan S, Hu Q, Tang BZ, Huang X (2021a) Antiviral and anti-inflammatory treatment with multifunctional alveolar macrophage-like nanoparticles in a surrogate mouse model of COVID-19. Adv Sci 8

Li Z, Wang Z, Dinh P, Zhu D, Popowski K, Lutz H, Hu S, Lewis M, Cook A, Andersen H, Greenhouse J, Pessaint L, Lobo L, Cheng K (2021b) Cell-mimicking Nanodecoys neutralize SARS-CoV-2 and mitigate lung injury in a nonhuman primate model of COVID-19. Nat Nanotechnol 16:942–951. https://doi.org/10.1038/s41565-021-00923-2

Li Y, Che J, Chang L, Guo M, Bao X, Mu D, Sun X, Zhang X, Lu W, Xie J (2022) CD47- and integrin $\alpha4/\beta1$-Comodified-macrophage-membrane-coated nanoparticles enable delivery of colchicine to atherosclerotic plaque. Adv Healthc Mater 11(4):2101788. https://doi.org/10.1002/adhm.202101788

Li C, Gan Y, Li Z, Fu M, Li Y, Peng X, Yang Y, Tian G, Yang YY, Yuan P, Ding X (2023a) Neutrophil-inspired photothermo-responsive drug delivery system for targeted treatment of bacterial infection and endotoxins neutralization. Biomater Res 27(1):30. https://doi.org/10.1186/s40824-023-00372-z

Li M, Guo Q, Zhong C, Zhang Z (2023b) Multifunctional cell membranes-based nano-carriers for targeted therapies: a review of recent trends and future perspective. Drug Deliv 30(1):2288797. https://doi.org/10.1080/10717544.2023.2288797

Li J, Wang J, Zhang Z, Pan Y, Jiang Z, Hu Y, Wang L (2023c) Fucoidan-loaded, neutrophil membrane-coated nanoparticles facilitate MRSA-accompanied wound healing. Mater Des 227:111758. https://doi.org/10.1016/j.matdes.2023.111758

Li Y, Wu C, Yang R, Tang J, Li Z, Yi X, Fan Z (2024) Application and development of cell membrane functionalized biomimetic nanoparticles in the treatment of acute ischemic stroke. Int J Mol Sci 25(15):8539

Lian T, Ho RJY (2001) Trends and developments in liposome drug delivery systems. J Pharm Sci 90(6):667–680. https://doi.org/10.1002/JPS.1023

Liang T, Zhang R, Liu X, Ding Q, Wu S, Li C, Lin Y, Ye Y, Zhong Z, Zhou M (2021) Recent advances in macrophage-mediated drug delivery systems. Int J Nanomedicine 16(Null):2703–2714. https://doi.org/10.2147/IJN.S298159

Liao Y, Zhang Y, Blum NT, Lin J, Huang P (2020) Biomimetic hybrid membrane-based nanoplatforms: synthesis, properties and biomedical applications. Nanoscale Horiz 5(9):1293–1302. https://doi.org/10.1039/D0NH00267D

Lin Y, Yin Q, Tian D, Yang X, Liu S, Sun X, Chen Q, Fang B, Liang H, Li L, Zhuge D, Wang H, Weng C, Xu J, Hu C, Xie J, Zhang X, Yan L, Lu X, Wang F, Liu C, Hu Y, Chen M, Wang L, Chen Y (2023) Vaginal epithelial cell membrane-based phototherapeutic decoy confers a "three-in-one" strategy to treat against intravaginal infection of Candida albicans. ACS Nano 17(13):12160–12175. https://doi.org/10.1021/acsnano.2c12644

Liu W, Huang Y (2022) Cell membrane-engineered nanoparticles for cancer therapy. J Mater Chem B 10(37):7161–7172. https://doi.org/10.1039/D2TB00709F

Liu Y, Luo J, Chen X, Liu W, Chen T (2019) Cell membrane coating technology: a promising strategy for biomedical applications. Nanomicro Lett 11(1):100. https://doi.org/10.1007/s40820-019-0330-9

Liu Y, Liu X, Cui Y, Yuan W (2022) Ultrasound for microalgal cell disruption and product extraction: a review. Ultrason Sonochem 87:106054. https://doi.org/10.1016/j.ultsonch.2022.106054

Liu J, Chen X, Xu L, Tu F, Rui X, Zhang L, Yan Z, Liu Y, Hu R (2023a) Neutrophil membrane-coated nanoparticles exhibit increased antimicrobial activities in an anti-microbial resistant K. Pneumonia infection model. Nanomedicine 48:102640. https://doi.org/10.1016/j.nano.2022.102640

Liu H, Su YY, Jiang XC, Gao JQ (2023b) Cell membrane-coated nanoparticles: a novel multifunctional biomimetic drug delivery system. Drug Deliv Transl Res 13(3):716–737. https://doi.org/10.1007/S13346-022-01252-0

Liu H, Tang L, Yin Y, Cao Y, Fu C, Feng J, Shen Y, Wang W (2024a) Photoresponsive multirole Nanoweapon camouflaged by hybrid cell membrane vesicles for efficient antibacterial therapy of Pseudomonas aeruginosa-infected pneumonia and wound. Adv Sci 11(35):2403101. https://doi.org/10.1002/advs.202403101

Liu X, Zhang Y, Peng F, Li C, Wang Q, Wang Z, Hu L, Peng X, Zhao G, Lin J (2024b) Macrophage membrane-coated nanoparticles for the delivery of Natamycin exhibit increased antifungal and anti-inflammatory activities in fungal keratitis. ACS Appl Mater Interfaces 16(44):59777–59788. https://doi.org/10.1021/acsami.4c11366

Lu J, Gao X, Wang S, He Y, Ma X, Zhang T, Liu X (2023a) Advanced strategies to evade the mononuclear phagocyte system clearance of nanomaterials. Exploration 3(1):20220045. https://doi.org/10.1002/EXP.20220045

Lu J, Quan L, Li J, Yuan J, Nie X, Huang X, Dong H, Su Y, Huang Y, Kou Q, Liu L, Liu H, Zhou X, Gui R, Gu L (2023b) Bioengineered stem cell membrane functionalized nanoparticles combine anti-inflammatory and antimicrobial properties for sepsis treatment. J Nanobiotechnol 21(1):170. https://doi.org/10.1186/s12951-023-01913-3

Ma J, Zhang S, Liu J, Liu F, Du F, Li M, Chen AT, Bao Y, Suh HW, Avery J, Deng G, Zhou Y, Wu P, Sheth K, Wang H, Zhou J (2019) Targeted drug delivery to stroke via chemotactic recruitment of nanoparticles coated with membrane of engineered neural stem cells. Small 15(35):1902011. https://doi.org/10.1002/smll.201902011

Mahmood S, Mandal UK, Chatterjee B, Taher M (2017) Advanced characterizations of nanoparticles for drug delivery: investigating their properties through the techniques used in their evaluations. Nanotechnol Rev 6(4):355–372. https://doi.org/10.1515/ntrev-2016-0050

McArthur DB (2019) Emerging infectious diseases. Nurs Clin N Am 54(2):297–311. https://doi.org/10.1016/j.cnur.2019.02.006

Mei R, Wan Z, Yang C, Shen X, Wang R, Zhang H, Yang R, Li J, Song Y, Su H (2024) Advances and clinical challenges of mesenchymal stem cell therapy. Front Immunol 15. https://doi.org/10.3389/fimmu.2024.1421854

Meng Z, Pan L, Qian S, Yang X, Pan L, Chi R, Chen J, Pan J, Shi C (2023) Antimicrobial peptide nanoparticles coated with macrophage cell membrane for targeted antimicrobial therapy of sepsis. Mater Des 229:111883. https://doi.org/10.1016/j.matdes.2023.111883

Mitragotri S, Burke PA, Langer R (2014) Overcoming the challenges in administering biopharmaceuticals: formulation and delivery strategies. Nat Rev Drug Discov 13(9):655–672. https://doi.org/10.1038/nrd4363

Muzammil S, Hayat S, Fakhar-e-Alam M, Aslam B, Siddique MH, Nisar MA, Saqalein M, Atif M, Sarwar A, Khurshid A, Amin N, Wang Z (2018) Nanoantibiotics: future nanotechnologies to combat antibiotic resistance. Front Biosci Elite 10(2):352–374. https://doi.org/10.2741/e827

Naghavi M, Vollset SE, Ikuta KS, Swetschinski LR, Gray AP, Wool EE, Aguilar GR, Mestrovic T, Smith G, Han C, Hsu RL, Chalek J, Araki DT, Chung E, Raggi C, Hayoon AG, Weaver ND, Lindstedt PA, Smith AE, Altay U, Bhattacharjee NV, Giannakis K, Fell F, McManigal B, Ekapirat N, Mendes JA, Runghien T, Srimokla O, Abdelkader A, Abd-Elsalam S, Aboagye RG, Abolhassani H, Abualruz H, Abubakar U, Abukhadijah HJ, Aburuz S, Abu-Zaid A, Achalapong S, Addo IY, Adekanmbi V, Adeyeoluwa TE, Adnani QES, Adzigbli LA, Afzal MS, Afzal S, Agodi A, Ahlstrom AJ, Ahmad A, Ahmad S, Ahmad T, Ahmadi A, Ahmed A, Ahmed H, Ahmed I, Ahmed M, Ahmed S, Ahmed SA, Akkaif MA et al (2024) Global burden of bacterial antimicrobial resistance 1990–2021: a systematic analysis with forecasts to 2050. Lancet 404(10459):1199–1226. https://doi.org/10.1016/S0140-6736(24)01867-1

Narain A, Asawa S, Chhabria V, Patil-Sen Y (2017a) Cell membrane coated nanoparticles: next-generation therapeutics. Nanomedicine (Lond) 12(21):2677–2692. https://doi.org/10.2217/nnm-2017-0225

Narain A, Asawa S, Chhabria V, Patil-Sen Y (2017b) Cell membrane coated nanoparticles: next-generation therapeutics. Nanomedicine (Lond) 12(21):2677–2692. https://doi.org/10.2217/NNM-2017-0225

Omolo CA, Megrab NA, Kalhapure RS, Agrawal N, Jadhav M, Mocktar C, Rambharose S, Maduray K, Nkambule B, Govender T (2019) Liposomes with pH responsive 'on and off' switches for targeted and intracellular delivery of antibiotics. J Liposome Res 31(1):45–63. https://doi.org/10.1080/08982104.2019.1686517

Oroojalian F, Beygi M, Baradaran B, Mokhtarzadeh A, Shahbazi M-A (2021) Immune cell membrane-coated biomimetic nanoparticles for targeted cancer therapy. Small 17(12):2006484. https://doi.org/10.1002/smll.202006484

Ozsoy F, Mohammed M, Jan N, Lulek E, Ertas YN (2024) T cell and natural killer cell membrane-camouflaged nanoparticles for cancer and viral therapies. ACS Appl Bio Mater 7(5):2637–2659. https://doi.org/10.1021/acsabm.4c00074

Park JH, Jiang Y, Zhou J, Gong H, Mohapatra A, Heo J, Gao W, Fang RH, Zhang L (2021) Genetically engineered cell membrane–coated nanoparticles for targeted delivery of dexamethasone to inflamed lungs. Sci Adv 7(25):eabf7820. https://doi.org/10.1126/sciadv.abf7820

Peng X, Luo Y, Yang L, Yang YY, Yuan P, Chen X, Tian G-B, Ding X (2024) A multiantigenic antibacterial nanovaccine utilizing hybrid membrane vesicles for combating Pseudomonas aeruginosa infections. J Extracell Vesicles 13:e12524. https://doi.org/10.1002/jev2.12524

Pereira-Silva M, Chauhan G, Shin MD, Hoskins C, Madou MJ, Martinez-Chapa SO, Steinmetz NF, Veiga F, Paiva-Santos AC (2021) Unleashing the potential of cell membrane-based nanoparticles for COVID-19 treatment and vaccination. Expert Opin Drug Deliv 18(10):1395–1414. https://doi.org/10.1080/17425247.2021.1922387

Qindeel M, Sabir F, Sargazi S, Mohammadzadeh V, Mulla SI (2021) New insights into the application of nanoghosts as theranostic tools with an emphasis on cardiovascular diseases. J Nanopart Res 23(11):254. https://doi.org/10.1007/s11051-021-05370-7

Qureshi SA, Rafiya K, Awasthi S, Jain A, Nadaf A, Hasan N, Kesharwani P, Ahmad FJ (2024) Biomembrane camouflaged nanoparticles: a paradigm shifts in targeted drug delivery system. Colloids Surf B Biointerfaces 238:113893. https://doi.org/10.1016/j.colsurfb.2024.113893

Ramos AC, Lomanto MYW, Vuong CK, Ohneda O, Fukushige M (2024) Antibacterial effects of human mesenchymal stem cells and their derivatives: a systematic review. Front Microbiol 15. https://doi.org/10.3389/fmicb.2024.1430650

Rao L, Cai B, Bu L-L, Liao Q-Q, Guo S-S, Zhao X-Z, Dong W-F, Liu W (2017) Microfluidic electroporation-facilitated synthesis of erythrocyte membrane-coated magnetic nanoparticles for enhanced imaging-guided cancer therapy. ACS Nano 11(4):3496–3505. https://doi.org/10.1021/acsnano.7b00133

Rao L, Wang W, Meng QF, Tian M, Cai B, Wang Y, Li A, Zan M, Xiao F, Bu LL, Li G, Li A, Liu Y, Guo SS, Zhao XZ, Wang TH, Liu W, Wu J (2019) A biomimetic Nanodecoy traps Zika virus to prevent viral infection and fetal microcephaly development. Nano Lett 19(4):2215–2222. https://doi.org/10.1021/acs.nanolett.8b03913

Rao L, Xia S, Xu W, Tian R, Yu G, Gu C, Pan P, Meng QF, Cai X, Qu D, Lu L, Xie Y, Jiang S, Chen X (2020) Decoy nanoparticles protect against COVID-19 by concurrently adsorbing viruses and inflammatory cytokines. Proc Natl Acad Sci USA 117(44):27141–27147. https://doi.org/10.1073/pnas.2014352117

Russ A, Hua AB et al (2018) Blocking "don't eat me" signal of CD47-SIRPα in hematological malignancies, an in-depth review. Blood Rev 32(6):480–489. https://doi.org/10.1016/j.blre.2018.04.005

Sha X, Dai Y, Chong L, Wei M, Xing M, Zhang C, Li J (2022) Pro-efferocytic macrophage membrane biomimetic nanoparticles for the synergistic treatment of atherosclerosis via competition effect. J Nanobiotechnol 20(1):506. https://doi.org/10.1186/s12951-022-01720-2

Shen L, Pang S, Zhong M, Sun Y, Qayum A, Liu Y, Rashid A, Xu B, Liang Q, Ma H, Ren X (2023) A comprehensive review of ultrasonic assisted extraction (UAE) for bioactive components: principles, advantages, equipment, and combined technologies. Ultrason Sonochem 101:106646. https://doi.org/10.1016/j.ultsonch.2023.106646

Shi M, Shen K, Yang B, Zhang P, Lv K, Qi H, Wang Y, Li M, Yuan Q, Zhang Y (2021) An electroporation strategy to synthesize the membrane-coated nanoparticles for enhanced anti-inflammation therapy in bone infection. Theranostics 11(5):2349–2363. https://doi.org/10.7150/thno.48407

Shi Y, Qian H, Rao P, Mu D, Liu Y, Liu G, Lin Z (2022) Bioinspired membrane-based nanomodulators for immunotherapy of autoimmune and infectious diseases. Acta Pharm Sin B 12(3):1126–1147. https://doi.org/10.1016/j.apsb.2021.09.025

Shi Q, Zhao Y, Liu M, Shi F, Chen L, Xu X, Gao J, Zhao H, Lu F, Qin Y, Zhang Z, Lian M (2024) Engineering platelet membrane-coated bimetallic MOFs as biodegradable Nanozymes for efficient antibacterial therapy. Small 20(23):e2309366. https://doi.org/10.1002/smll.202309366

Shu P (2024) Application of cell membrane-coated nanoparticles in the medical field. Highl Sci Eng Technol 84:86–91. https://doi.org/10.54097/6693qm86

Song Y, Zheng X, Hu J, Subo LK, Chen J, Xu X, Lu X, Wang X (2023) Recent advances of cell membrane-coated nanoparticles for therapy of bacterial infection. Front Microbiol 14. https://doi.org/10.3389/fmicb.2023.1083007

Spanjers JM, Städler B (2020) Cell membrane coated particles. Adv Biosyst 4(11):2000174. https://doi.org/10.1002/adbi.202000174

Strasfeld L, Chou S (2010) Antiviral drug resistance: mechanisms and clinical implications. Infect Dis Clin N Am 24(2):413–437. https://doi.org/10.1016/J.IDC.2010.01.001

Suk JS, Xu Q, Kim N, Hanes J, Ensign LM (2016) PEGylation as a strategy for improving nanoparticle-based drug and gene delivery. Adv Drug Deliv Rev 99(Pt A):28–51. https://doi.org/10.1016/j.addr.2015.09.012

Tan Q, He L, Meng X, Wang W, Pan H-D, Yin W, Zhu T, Huang X, Shan H (2021) Macrophage biomimetic nanocarriers for anti-inflammation and targeted antiviral treatment in COVID-19. J Nanobiotechnol 19:173. https://doi.org/10.1186/s12951-021-00926-0

Tang W, Yang Y, Yang L, Tang M, Chen Y, Li C (2021) Macrophage membrane-mediated targeted drug delivery for treatment of spinal cord injury regardless of the macrophage polarization states. Asian J Pharm Sci 16(4):459–470. https://doi.org/10.1016/j.ajps.2021.03.005

Wang C, Wu S (2022) Research update on cell membrane camouflaged nanoparticles for cancer therapy. Front Bioeng Biotechnol 10. https://doi.org/10.3389/fbioe.2022.944518

Wang C, Wang Y, Zhang L, Miron RJ, Liang J, Shi M, Mo W, Zheng S, Zhao Y, Zhang Y (2018) Pretreated macrophage-membrane-coated gold Nanocages for precise drug delivery for treatment of bacterial infections. Adv Mater 30(46):e1804023. https://doi.org/10.1002/adma.201804023

Wang H, Wu J, Williams GR, Fan Q, Niu S, Wu J, Xie X, Zhu L-M (2019) Platelet-membrane-biomimetic nanoparticles for targeted antitumor drug delivery. J Nanobiotechnol 17(1):60. https://doi.org/10.1186/s12951-019-0494-y

Wang LY, Xia D, Xu P, Zhu T, Jiang Z, Ma Y (2020a) Neutrophil membranes coated, antibiotic agent loaded nanoparticles targeting to the lung inflammation. Colloids Surf B Biointerfaces 188:110755. https://doi.org/10.1016/j.colsurfb.2019.110755

Wang S, Duan Y, Zhang Q, Komarla A, Gong H, Gao W, Zhang L (2020b) Drug targeting via platelet membrane-coated nanoparticles. Small Struct 1(1). https://doi.org/10.1002/sstr.202000018

Wang C, Wang S, Chen Y, Zhao J, Han S, Zhao G, Kang J, Liu Y, Wang L, Wang X, Xu Y, Wang S, Huang Y, Wang J, Zhao J (2021a) Membrane nanoparticles derived from ACE2-rich cells block SARS-CoV-2 infection. ACS Nano 15(4):6340–6351. https://doi.org/10.1021/acsnano.0c06836

Wang M, Xin Y, Cao H, Li W, Hua Y, Webster TJ, Zhang C, Tang W, Liu Z (2021b) Recent advances in mesenchymal stem cell membrane-coated nanoparticles for enhanced drug delivery. Biomater Sci 9(4):1088–1103. https://doi.org/10.1039/D0BM01164A

Wang Y, Zhang K, Li T, Maruf A, Qin X, Luo L, Zhong Y, Qiu J, McGinty S, Pontrelli G, Liao X, Wu W, Wang G (2021c) Macrophage membrane functionalized biomimetic nanoparticles for targeted anti-atherosclerosis applications. Theranostics 11(1):164–180. https://doi.org/10.7150/thno.47841

Wang G, Qi Y, Xu J, Huang Y, Xiong F, Huang W, Ma X, Wang S, Wang C, Zhou Y, Wang L (2023) Neutrophil membrane-coated mesoporous silica nanoparticles loaded with hydrocortisone alleviate DSS-induced colitis in mice. ACS Appl Nano Mater 6(4):2403–2412. https://doi.org/10.1021/acsanm.2c04615

Wang C, Li C, Zhang R, Huang L (2024) Macrophage membrane-coated nanoparticles for the treatment of infectious diseases. Biomed Mater 19(4):42003. https://doi.org/10.1088/1748-605X/ad4aaa

Wei X, Gao J, Fang RH, Luk BT, Kroll AV, Dehaini D, Zhou J, Kim HW, Gao W, Lu W, Zhang L (2016) Nanoparticles camouflaged in platelet membrane coating as an antibody decoy for the treatment of immune thrombocytopenia. Biomaterials 111:116–123. https://doi.org/10.1016/j.biomaterials.2016.10.003

Wei X, Ying M, Dehaini D, Su Y, Kroll AV, Zhou J, Gao W, Fang RH, Chien S, Zhang L (2018a) Nanoparticle functionalization with platelet membrane enables multifactored biological targeting and detection of atherosclerosis. ACS Nano 12(1):109–116. https://doi.org/10.1021/acsnano.7b07720

Wei X, Zhang G, Ran D, Krishnan N, Fang RH, Gao W, Spector SA, Zhang L (2018b) T-cell-mimicking nanoparticles can neutralize HIV infectivity. Adv Mater 30

Wei W, Cheng W, Dai W, Lu F, Cheng Y, Jiang T, Ren Z, Xie Y, Xu J, Zhao Q, Yu X, Yin Y, Li J, Dong H (2022) A Nanodrug coated with membrane from brain microvascular endothelial cells protects against experimental cerebral malaria. Nano Lett 22(1):211–219. https://doi.org/10.1021/acs.nanolett.1c03514

World Health Organization (WHO) (2023) Global health estimates 2023: deaths by cause, age, sex, by country and by region, 2000–2023. https://www.who.int/data/gho. Accessed 24 Jan 2025

Wu X, Li Y, Raza F, Wang X, Zhang S, Rong R, Qiu M, Su J (2021) Red blood cell membrane-camouflaged Tedizolid phosphate-loaded PLGA nanoparticles for bacterial-infection therapy. Pharmaceutics 13(1). https://doi.org/10.3390/pharmaceutics13010099

Wu Y, Wan S, Yang S, Hu H, Zhang C, Lai J, Zhou J, Chen W, Tang X, Luo J, Zhou X, Yu L, Wang L, Wu A, Fan Q, Wu J (2022) Macrophage cell membrane-based nanoparticles: a new promising biomimetic platform for targeted delivery and treatment. J Nanobiotechnol 20(1):542. https://doi.org/10.1186/s12951-022-01746-6

Wu K, He M, Mao B, Xing Y, Wei S, Jiang D, Wang S, Alkuhali AA, Guo J, Gan Z, Li M, Li X, Chen H (2024) Enhanced delivery of CRISPR/Cas9 system based on biomimetic nanoparticles for hepatitis B virus therapy. J Control Release 374:293–311. https://doi.org/10.1016/j.jconrel.2024.08.019

Xia Q, Zhang Y, Li Z, Hou X, Feng N (2019) Red blood cell membrane-camouflaged nanoparticles: a novel drug delivery system for antitumor application. Acta Pharm Sin B 9(4):675–689. https://doi.org/10.1016/j.apsb.2019.01.011

Xiong J, Tang H, Sun L, Zhu J, Tao S, Luo J, Li J, Li J, Wu H, Yang J (2024) A macrophage cell membrane-coated cascade-targeting photothermal nanosystem for combating intracellular bacterial infections. Acta Biomater 175:293–306. https://doi.org/10.1016/j.actbio.2023.12.045

Xu Z, Cao X, Tu H, Zhang Z-R, Deng L (2019) Inflammation-targeted delivery of Celastrol via neutrophil membrane-coated nanoparticles in the Management of Acute Pancreatitis. Mol Pharm 16(3):1397–1405. https://doi.org/10.1021/acs.molpharmaceut.8b01342

Yaman S, Chintapula U, Rodriguez E, Ramachandramoorthy H, Nguyen KT (2020) Cell-mediated and cell membrane-coated nanoparticles for drug delivery and cancer therapy. Cancer Drug Resist 3(4):879–911. https://doi.org/10.20517/cdr.2020.55

Yang N, Ding Y, Zhang Y, Wang B, Zhao X, Cheng K, Huang Y, Taleb M, Zhao J, Dong W-F, Zhang L, Nie G (2018) Surface functionalization of polymeric nanoparticles with umbilical cord-derived mesenchymal stem cell membrane for tumor-targeted therapy. ACS Appl Mater Interfaces 10(27):22963–22973. https://doi.org/10.1021/acsami.8b05363

Yang Y, Wang K, Pan Y, Rao L, Luo G (2021) Engineered cell membrane-derived nanoparticles in immune modulation. Adv Sci 8(24):2102330. https://doi.org/10.1002/advs.202102330

Ye J, Liang W, Wu L, Guo R, Wu W, Yang D, Chen L (2023) Antimicrobial effect of streptococcus salivarius outer membrane-coated nanocomplexes against Candida albicans and oral candidiasis. Mater Des 233:112177. https://doi.org/10.1016/j.matdes.2023.112177

Yu Y, Peng Y, Shen WT, Zhou Z, Kai M, Gao W, Zhang L (2024) Hybrid cell membrane-coated nanoparticles for biomedical applications. Small Struct 5(5). https://doi.org/10.1002/SSTR.202300473

Zeng Z, Pu K (2020) Improving cancer immunotherapy by cell membrane-camouflaged nanoparticles. Adv Funct Mater 30(43):2004397. https://doi.org/10.1002/adfm.202004397

Zeng Y, Li S, Zhang S, Wang L, Yuan H, Hu F (2022) Cell membrane coated-nanoparticles for cancer immunotherapy. Acta Pharm Sin B 12(8):3233–3254. https://doi.org/10.1016/j.apsb.2022.02.023

Zeng S, Tang Q, Xiao M, Tong X, Yang T, Yin D, Lei L, Li S (2023) Cell membrane-coated nanomaterials for cancer therapy. Mater Today Bio 20:100633. https://doi.org/10.1016/j.mtbio.2023.100633

Zhai Y, Su J, Ran W, Zhang P, Yin Q, Zhang Z, Yu H, Li Y (2017) Preparation and application of cell membrane-camouflaged nanoparticles for cancer therapy. Theranostics 7(10):2575–2592. https://doi.org/10.7150/thno.20118

Zhang Y, Zhang J, Chen W, Angsantikul P, Spiekermann KA, Fang RH, Gao W, Zhang L (2017) Erythrocyte membrane-coated nanogel for combinatorial antivirulence and responsive antimicrobial delivery against Staphylococcus aureus infection. J Control Release 263:185–191. https://doi.org/10.1016/j.jconrel.2017.01.016

Zhang M, Du Y, Wang S, Chen B (2020) A review of biomimetic nanoparticle drug delivery systems based on cell membranes. Drug Des Devel Ther 14(Null):5495–5503. https://doi.org/10.2147/DDDT.S282368

Zhang J, Wei K, Shi J, Zhu Y, Guan M, Fu X, Zhang Z (2021a) Biomimetic nanoscale erythrocyte delivery system for enhancing chemotherapy via overcoming biological barriers. ACS Biomater Sci Eng 7(4):1496–1505. https://doi.org/10.1021/acsbiomaterials.1c00008

Zhang N, Lin J, Chew SY (2021b) Neural cell membrane-coated nanoparticles for targeted and enhanced uptake by central nervous system cells. ACS Appl Mater Interfaces 13(47):55840–55850. https://doi.org/10.1021/acsami.1c16543

Zhang R, Wu S, Ding Q, Fan Q, Dai Y, Guo S, Ye Y, Li C, Zhou M (2021c) Recent advances in cell membrane-camouflaged nanoparticles for inflammation therapy. Drug Deliv 28(1):1109–1119. https://doi.org/10.1080/10717544.2021.1934188

Zhang E, Phan P, Algarni HA, Zhao Z (2022) Red blood cell inspired strategies for drug delivery: emerging concepts and new advances. Pharm Res 39(11):2673–2698. https://doi.org/10.1007/s11095-022-03328-5

Zhang Q, Hu C, Feng J, Long H, Wang Y, Wang P, Hu C, Yue Y, Zhang C, Liu Z, Zhou X (2024a) Anti-inflammatory mechanisms of neutrophil membrane-coated nanoparticles without drug loading. J Control Release 369:12–24. https://doi.org/10.1016/j.jconrel.2024.03.030

Zhang S, Zhang X, Gao H, Zhang X, Sun L, Huang Y, Zhang J, Ding B (2024b) Cell membrane-coated biomimetic nanoparticles in cancer treatment. Pharmaceutics 16(4):531

Zhao Y, Li A, Jiang L, Gu Y, Liu J (2021) Hybrid membrane-coated biomimetic nanoparticles (HM@BNPs): a multifunctional nanomaterial for biomedical applications. Biomacromolecules 22(8):3149–3167. https://doi.org/10.1021/acs.biomac.1c00440

Zhao Y, Zhang H, Zhang Q, Tao H (2023, 1881) Research Progress of neutrophil-mediated drug delivery strategies for inflammation-related disease. Pharmaceutics 15(7)

Zhou M, Lai W, Li G, Wang F, Liu W, Liao J, Yang H, Liu Y, Zhang Q, Tang Q, Hu C, Huang J, Zhang R (2021) Platelet membrane-coated and VAR2CSA malaria protein-functionalized nanoparticles for targeted treatment of primary and metastatic cancer. ACS Appl Mater Interfaces 13(22):25635–25648. https://doi.org/10.1021/acsami.1c02581

Zhou J, Miyamoto Y, Ihara S, Kroll A, Nieskens N, Tran V, Hanson E, Fang R, Zhang L, Eckmann L (2022) Co-delivery of antigens and adjuvant in polymeric nanoparticles coated with native parasite membranes induces protective mucosal immunity against Giardia lamblia. J Infect Dis 226:319–323. https://doi.org/10.1093/infdis/jiac074

Zhuang J, Gong H, Zhou J, Zhang Q, Gao W, Fang RH, Zhang L (2020) Targeted gene silencing in vivo by platelet membrane–coated metal-organic framework nanoparticles. Sci Adv 6(13):eaaz6108. https://doi.org/10.1126/sciadv.aaz6108

Zou S, Wang B, Wang C, Wang Q, Zhang L (2020) Cell membrane-coated nanoparticles: research advances. Nanomedicine (Lond) 15(6):625–641. https://doi.org/10.2217/nnm-2019-0388

Zou D, Wu Z, Yi X, Hui Y, Yang G, Liu Y, Wang H, Brooks A, Wang H, Liu X, Xu ZP, Roberts MS, Gao H, Zhao C-X (2023) Nanoparticle elasticity regulates the formation of cell membrane-coated nanoparticles and their nano-bio interactions. Proc Natl Acad Sci 120(1):e2214757120. https://doi.org/10.1073/pnas.2214757120

Zou L, Zhang Y, Cheraga N, Abodunrin OD, Qu K-Y, Qiao L, Ma Y-Q, Hang Y, Huang N-P, Chen L-J (2024) M2 macrophage membrane-camouflaged Fe3O4-Cy7 nanoparticles with reduced immunogenicity for targeted NIR/MR imaging of atherosclerosis. Small 20(8):2304110. https://doi.org/10.1002/smll.202304110

Cell Membrane Engineering for Advanced Drug Delivery Against Neurodegenerative and Inflammatory Diseases

16

Raja Rajeswari Rajeshkumar,
Jeganpandi Senthamarai Pandi,
Murugesan Sankaranarayanan, Parasuraman Pavadai,
Panneerselvam Theivendren, and Selvaraj Kunjiappan

Abstract

One novel strategy to address challenges in drug delivery for neurodegenerative and inflammatory diseases is cell membrane engineering. Such complex pathophysiological and therapeutic challenges demand delivery systems to ensure that the drugs can overcome barriers, including the blood-brain barrier and cellular membrane selectivity. This chapter explores recent developments in cell membrane engineering and highlights how this approach can lead to improved drug delivery mediated by nanoparticles, nanovesicles, and hybrid systems that imitate natural biological membranes. This surface modification strategy can be used to target the drug to the inflamed tissue to enhance the therapeutic effects and minimize systemic side effects. This chapter summarizes cell membrane

R. R. Rajeshkumar · J. S. Pandi · S. Kunjiappan (✉)
Department of Biotechnology, Kalasalingam Academy of Research and Education, Virudhunagar, Tamil Nadu, India
e-mail: selvaraj.k@klu.ac.in

M. Sankaranarayanan
Medicinal Chemistry Research Laboratory, Department of Pharmacy, Birla Institute of Technology and Science Pilani, Pilani, Rajasthan, India
e-mail: murugesan@pilani.bits-pilani.ac.in

P. Pavadai
Department of Pharmaceutical Chemistry, Faculty of Pharmacy, M.S. Ramaiah University of Applied Sciences, Bengaluru, Karnataka, India

P. Theivendren (✉)
Department of Pharmaceutical Chemistry & Analysis, School of Pharmaceutical Sciences, Vels Institute of Science, Technology & Advanced Studies, Chennai, Tamil Nadu, India

Z. Hussain, R. Pei (eds.), *Cell Membrane Engineering for Advancing Cell-Based and Nanotechnology-Based Therapies*, Biomaterials, Bioengineering and Sustainability 6, https://doi.org/10.1007/978-3-031-93425-4_16

engineering methods, highlighting their ability to encapsulate diverse therapeutics, enhance stability, enable controlled release, and improve biocompatibility. The cell membrane engineered platform holds significant promise, there are several challenges to be addressed in terms of scalable clinical translation and long-term safety. The chapter underscores the need for further research and development to achieve clinical translation, especially concerning scalability metrics. Cell membrane engineering is revolutionizing drug delivery, offering promising, specific, efficient, and personalized approaches for neurodegenerative and inflammatory diseases, addressing unmet medical needs.

Keywords

Cell membrane engineering · Drug delivery · Blood-brain barrier · Neurodegenerative diseases · Inflammatory diseases · Biomimetic nanoparticles

16.1 Introduction

The complex pathophysiology and chronic nature of neurodegenerative and inflammatory diseases make them some of the most challenging conditions to treat with modern medicine. Alzheimer's disease, Parkinson's disease, and multiple sclerosis are neurodegenerative diseases that typically present with progressive neuronal damage, systemic inflammation, and immune system dysregulation (Amor et al. 2010; Gao and Hong 2008). Current therapeutic strategies are primarily directed toward symptomatic resolution, but disease-modifying treatments are still in their early stages. The challenge in treating these diseases is the limited efficiency of drug delivery across biological barriers, including the blood-brain barrier (BBB) or cellular membranes that prevent drugs from reaching the target site efficiently (Fig. 16.1).

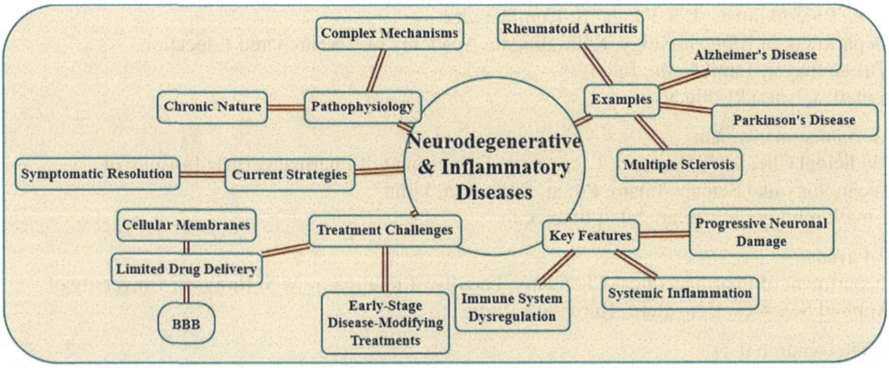

Fig. 16.1 The interconnected aspects of neurodegenerative and inflammatory diseases, including examples, key features, treatment challenges, current strategies, and the complex mechanisms influencing their pathophysiology and management

Cell membrane engineering offers an innovative strategy to overcome limitations and enhance the potential of the drug delivery system for treating neurodegenerative and inflammatory diseases. Modifying the natural properties of cell membranes, their structure, composition, and surface properties is a novel approach to engineer more effective drug carriers. Engineered cell membranes based on the intrinsic properties of natural cell membrane (biocompatibility, specific interactions with cell-type specific cellular receptors) can be utilized to improve drug delivery, enhance targeting, and circumvent physiological barriers (Leyane et al. 2022; Rekatsina et al. 2020). A key obstacle in neurodegenerative disease treatment is the BBB. The BBB closely packed endothelium of the blood vessels of the brain, hindering penetration of most therapeutic agents. Only small, lipophilic molecules or compounds with special transport mechanisms can cross this barrier efficiently (Akhtar et al. 2021; Yu et al. 2024). Many drugs are either too large, hydrophilic, and do not have the active penetration of the BBB. Therefore, strategies for drug delivery that can circumvent or transiently open the BBB are key for treating brain pathology (Fig. 16.2).

A promising approach for tackling these issues is to engineer membranes of cells to generate membrane vesicles or liposomes with similar attributes to natural cells. Focused on specific receptors or transporters on the BBB, these biomimetic carriers improve the uptake of therapeutic agents directly into the brain. The implementation of surface modification techniques, including the addition of targeting ligands or functional molecules, allows synthetic nanoparticles (NPs) to be directed toward specific brain regions or subtypes of neuronal cells, thereby enhancing the precision and efficacy of drug delivery (Ding et al. 2020; Feng et al. 2023). Similarly, in inflammatory diseases, the challenge lies in delivering drugs to inflamed sites, particularly in tissues that are difficult to reach or penetrate. Inflammatory diseases like rheumatoid arthritis or multiple sclerosis are characterized by the influx of immune

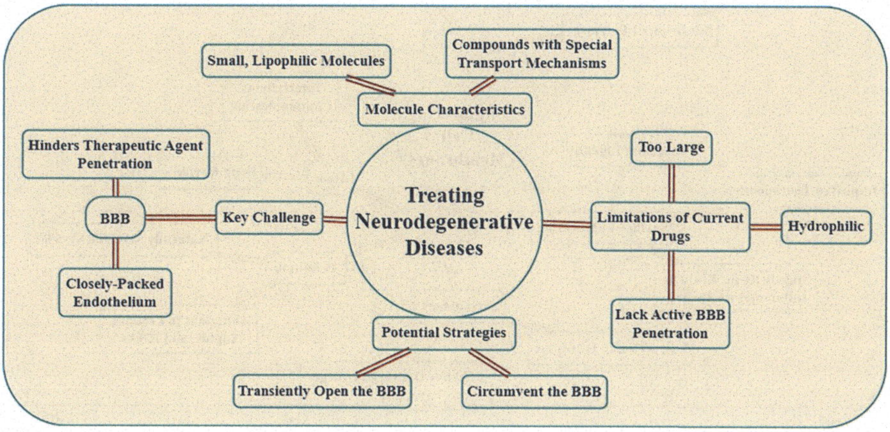

Fig. 16.2 Outlining challenges in treating neurodegenerative diseases, including BBB limitations, molecule characteristics, and current drug inefficacies, alongside potential strategies like BBB circumvention and transient opening for therapeutic delivery

cells into inflamed regions, contributing to barriers toward drug access. Membrane-engineered drug carriers can be customized to target disease site-specific markers or inflamed molecules to facilitate the accumulation of the therapeutic agent at the point of disease without off-target systemic action (Neganova et al. 2022; Patel et al. 2021).

Formation or incorporation of bio-inspired or cell-like polymers is the simple approach to engineering cell membranes; however, there are more advanced methods which can also be broadly categorized into three types employ cell-derived vesicles, modify lipid-bilayers or synthetically create polymers with optimal cell-like properties. One widely employed method is using exosomes, small vesicles that are naturally secreted by cells and are abundant in proteins, lipids, and RNAs involved in intercellular communication (Ahlawat et al. 2020; Daraban et al. 2024). Exosomes are gaining recognition as drug carriers because of their unique extracellular transport property, which enables them to interact with target cells and effectively deliver their contents. The surface of exosomes can be engineered with targeting ligands, so that they can improve therapeutic efficacy in the treatment of neurodegenerative or inflammatory diseases via control over their specificity for brain or inflammatory targets (Fig. 16.3).

A key advancement in cell membrane engineering is the development of cell fusion technologies, which involve the fusion of nanovesicles or liposomes with cell membranes to create hybrid drug carriers. These carriers integrate synthetic nanoparticles within the cell membrane, allowing for controlled drug release, enhanced bioavailability, and improved targeting and biocompatibility. This

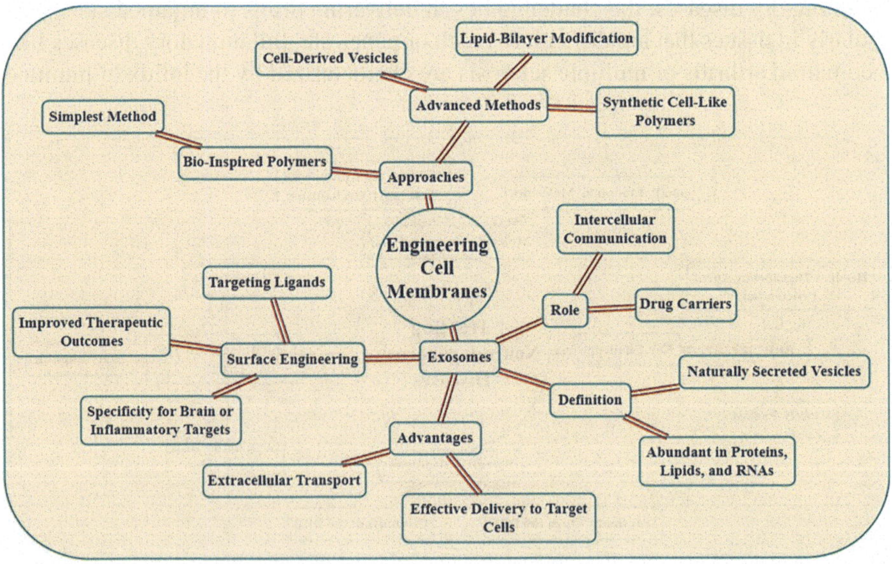

Fig. 16.3 Approaches for engineering cell membranes, highlighting exosomes' roles, advantages, and methods, including surface engineering, lipid-bilayer modifications, targeting ligands, and synthetic cell-like polymers for therapeutic applications

approach has demonstrated effectiveness in crossing the blood-brain barrier, delivering drugs to neural tissues, and neutralizing specific immune cells in inflammatory diseases (Guo et al. 2021; Liu et al. 2023). Beyond drug delivery, cell membrane engineering plays a crucial role in developing stable therapeutic agents while providing controlled drug release. Many drugs, particularly biologics such as proteins and nucleic acids, are susceptible to degradation or rapid elimination from the body. Encapsulating these therapeutics within an engineered membrane-based platform enhances their stability, regulates their release, and prolongs their therapeutic effects (Liu et al. 2021). Cell membrane engineering offers a versatile platform with broad applicability, enabling dual-targeting of neurodegenerative and inflammatory diseases or the co-delivery of multiple therapeutics within a single system (Zhang et al. 2021). Additionally, the incorporation of natural cell membranes or similar materials in the design minimizes the potential for immunogenicity, a critical issue with synthetic drug vehicle systems (Fig. 16.4).

Although cell membrane engineering has powerful prospects, there are still many challenges to overcome. We propose designing transporters that are optimized for stability and mass production. Furthermore, although there has been substantial advancement in the preclinical context, converting these technologies to be used in human clinical trials remains a challenge. A deeper understanding of the long-term biocompatibility and safety of membrane-engineered drug carriers remains essential (Baig et al. 2024). In summary, cell membrane engineering has emerged as a

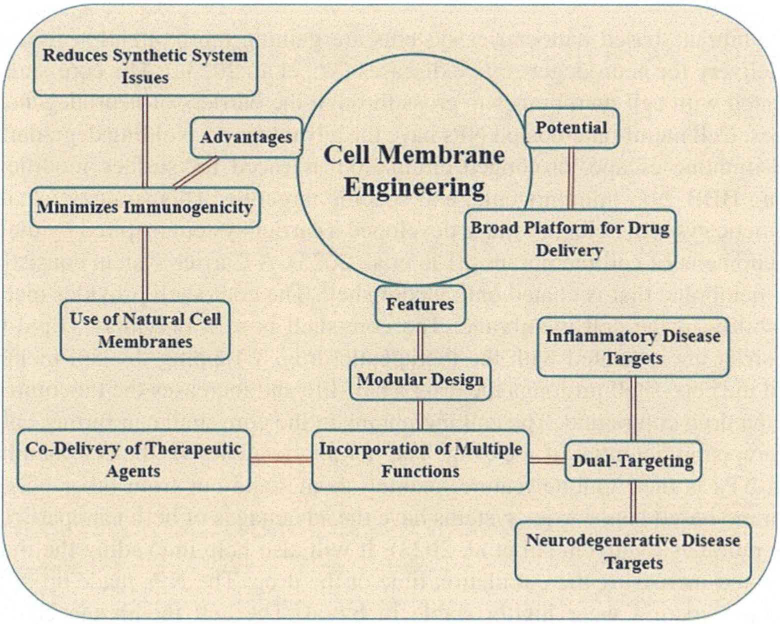

Fig. 16.4 Schematic representation of cell membrane engineering highlighting its advantages, features, and potential. It emphasizes natural membranes, modular design, multifunctionality, and dual targeting for drug delivery and disease treatment

cutting-edge technique for the solution of drug delivery in neurodegenerative/inflammatory diseases. This approach for targeting specific cells or tissues has inspired the design of therapeutic delivery systems that incorporate the unique structural and functional properties of biological membranes, thereby optimizing not just the delivery of the drug to the desired site. It also altering the pharmacokinetics and pharmacodynamics of the drug to improve its safety and efficacy. This method has the potential to revolutionize the drug delivery arena as research continues toward that goal with the hope of helping patients with these life-altering illnesses (Jan et al. 2023).

This chapter discusses current cell membrane engineering strategies to enhance drug delivery for neurodegenerative and inflammatory diseases, highlighting recent advances such as biomimetic nanoparticles (NPs), nanovesicles, hybrid cell-membrane systems, and engineered microencapsulation. These approaches facilitate BBB crossing and targeted tissue delivery (Li et al. 2020). Because of the potential for this field to redefined treatment modalities to develop more effective and personalized treatments for diseases that currently remain poorly treated, the importance of this area cannot be overstated. In this chapter, the challenges, progress, and emerging trends of drug delivery systems based on biological membranes are presented.

16.2 Cell Membrane

Cell membrane-based nanocarrier systems are gaining more attention recently in drug delivery for neurodegenerative diseases (Wu et al. 2023a). The core-shell NPs are coated with cell membranes to cross through the barriers of neurodegenerative diseases. Cell membrane-coated NPs have the advantage of avoiding degradation of cargo, immune escape, prolonged circulation, no need for surface modification, crossing BBB, non-immunogenic, and specific targeting. This strategy is called a biomimetic system, Scientists have developed a carrier system inspired by the natural phenomena of cell membranes (Liu et al. 2023). A Carrier system consisting of a cell membrane that is coated onto a core-shell. The core-shell provides mechanical stability to the cell membrane. The core-shell is of a polymeric, lipid-based nanocarrier encapsulated with the therapeutic drug. Wrapping the cell membrane around the core-shell prolongs the drug's half-life and increases the biocompatibility of the drug compound. The cell membrane in the core shell can further enhance the therapeutic potential of the drug. The homing tendency of the cell membrane-coated NPs is their unique feature, making them stand out from other NPs. Cell membrane-based nanocarrier systems have the advantages of both nanoparticle and bio membranes (Nguyen-Thi et al. 2023). It will also help in evading the immune system and increasing the circulation time of the drug. The NPs made up of a cell membrane coating were highly stable in blood. The cell membrane NPs were obtained from distinct types of cells based on the property required. The cells from which the cell membrane is derived are erythrocytes, platelets, leukocytes, macrophages, and cancer cells (Fig. 16.5). Therefore, cell membrane engineering has

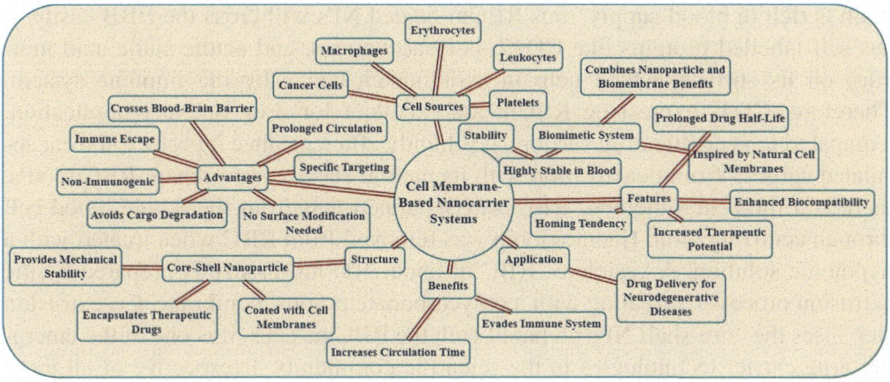

Fig. 16.5 Overview of cell membrane-based nanocarrier systems emphasizing cell sources, advantages, structure, features, and applications. Highlights include immune escape, prolonged circulation, biomimetic design, and therapeutic potential for drug delivery

become an attractive drug delivery method for neurodegenerative and inflammatory diseases. In this chapter, we summarize the recent advances in different strategies to modify and engineer cell membranes for targeted therapy. The liposome and micelle lipid-based carriers that are biocompatible, efficient for drug encapsulation, offering stability, and controlled release for the drugs (Plaza-Oliver et al. 2021; Qian et al. 2023). Biosafety of exosomal vesicles is due to their natural cell membrane origin, with intrinsic targeting capability, which enables them to cross BBB barriers. Integrated properties of different biological or synthetic systems provide localized, specific targeting through various fusion techniques, such as cell membrane coating and hybrid nanoparticles. These approaches increase drug bioavailability, reduce adverse effects, and improve therapeutic effectiveness. We summarize the latest technologies, obstacles, and perspectives in this area, with an emphasis on membrane engineering and how it could reshape drug delivery solutions for complex pathogenic conditions associated with chronic, age-related Schwann cells, such as Alzheimer's, Parkinson's, and chronic inflammatory diseases.

16.2.1 Erythrocytes

Erythrocytes are the most prevalent cells in the blood and can be easily isolated, thus helping in efficient transport. It lacks nuclei and organelles and has a high surface area/volume ratio. It has a life span of 120 day thus promising for longer circulation in the body. It lacks mitochondria and nuclei, thus it is easy to extract. One of the most frequently used nanocarriers is RBC called "carrier RBCs" or" carrier erythrocytes". RBC membrane-based NPs (RBCm NPs) can deliver the drug effectively with limited side effects. It enhances drug retention and prolongs drug circulation time. It is also biocompatible, biodegradable, and non-immunogenic. In neurodegenerative disease treatment, the most crucial part is to cross the BBB. The brain is rich in blood supply, thus RBCm-coated NPs will cross the BBB easily. It has self-labelled proteins like CD47, polysaccharides, and acidic sialic acid moieties on its surface, which help in avoiding clearance by the immune system. Therefore, CD47-expressing RBCm are excellent for drug delivery application. Compared to synthetic drug carriers, it is highly efficient since it protects the encapsulated cargoes from deactivation with its natural surface. To prepare RBCm NPs, there is a three-step method. RBC was separated first from the whole blood cell through centrifugation. Haemoglobin was removed from RBC when treated with a hypotonic solution. A vesicle of RBC of about 100 nm in size is prepared by the extrusion process by treating with a polycarbonate porous membrane. Co-extrusion step fuses the core-shell NPs prepared with the RBCm. Thus, it is one of the emerging drug carrier technologies in the scientific community. Irrespective of all these advantages RBCm NPs cannot target neurodegenerative diseases, to enhance its targeting ability it was modified by attaching a ligand (Corrons et al. 2021; Xia et al. 2019).

16.2.2 Platelet

Platelets are derived from stem cell lines called megakaryocytes, which are disc-shaped. Compared to erythrocytes, platelets have a lesser lifespan of only 7–10 days. It acts upon wound healing, vascular injury, inflammatory response, hemostasis, and thrombosis. It possesses immune evasion capabilities, facilitates subendothelial adhesion, and enables pathogen recognition. These platelet membrane-coated NPs can be used for drug delivery applications for neurodegenerative diseases with collateral vascular injury or inflammatory response. Unlike erythrocytes, platelets not only express CD47 on their surface to inhibit the immune system. It also expresses CD55 and CD59 on its surface to escape the immune system. Since it has lower immunogenicity it can circulate in blood longer than other nucleated cells (Kimmerlin et al. 2023; Liu and Welburn 2024).

16.2.3 Leukocytes

Leukocytes are white blood cells that have a diameter of 7–20 μm larger than cells derived from myeloid stem cells because of their nucleus. It includes neutrophils, eosinophils, basophils, lymphocytes, and monocytes. The surface protein CD11b present on the leukocyte membrane selectively targets activated brain capillary endothelial cells. Through the CD45 protein, it evades the mononuclear phagocytosis system, and its leukocyte receptor proteins facilitate its role as a drug delivery carrier. It helps target tumors and cells that undergo inflammation. Cytokine is released during an inflammatory response in the brain, which attracts leukocytes. Leukocyte-coated NPs exhibit chemotactic properties, enabling them to evade opsonization and delay uptake. This unique ability makes them an excellent carrier system, particularly for targeted delivery to tumors and inflammatory sites. About

5% of leukocytes are composed of neutrophiles. It responds first when inflammation occurs within 1 hour. Thus, neutrophils stand out from other leukocytes for NPs fabrication. Macrophages and monocytes function as antigen-presenting cells and are essential in regulating injury response, infection control, inflammation, and tumor development. It can evade the immune system and move within tissues for the clearance of pathogens and damaged cells. Thus, it can be used as a potential drug delivery carrier for most neurodegenerative disorders (Elhassan et al. 2022; Quarleri et al. 2021).

16.2.4 Macrophage

Macrophages are critical immune cells involved in the phagocytosis of invading pathogens, foreign particles, and apoptotic cells. They are instrumental in homeostasis and initiation of the immune response as part of the innate immune system. One interesting aspect of macrophages is that they can display characteristics of immune evasion, meaning they can change their behaviour depending on the microenvironment. This duality enables them to either enhance or inhibit immune activity, depending upon context. The intrinsic ability of dendritic cells to present antigens links innate and adaptive immunity in a manner that leads to the activation of T cells. The versatility of macrophages as defenders is due to their involvement in inflammation, tissue repair, and immune surveillance (Biscu et al. 2024; Zago et al. 2021).

16.2.5 Cancer Cells

Cancer cells can bind to homologous cancer cells through homologous adhesive molecules on the surface of the plasma membrane, such as lectins, integrins, selectors, cadherins, and proteins. Thus, it can be designed as a potential drug delivery vehicle, improving cellular uptake and specific targeting. Cancer is distinguished from other cells by its unique property of proliferation thus, it can be easily cultured in vitro. With the expression of CD47 on the cell membrane surface it evades the immune system (Zhong et al. 2023).

16.3 Cell Membrane Engineering Based Neurodegenerative-Disease Treatment

Neurodegenerative disorders are a class of progressive, irreversible disorders characterized by neuronal loss and atrophy in specific brain regions. Predominantly affecting the elderly, conditions such as Alzheimer's disease (AD), Parkinson's disease (PD), Huntington's, Pick's disease, amyotrophic lateral sclerosis, and dementia result in cognitive decline, motor impairment, reduced social engagement, and significant challenges in daily activities. Central nervous system (brain, retina, and spinal cord) neurodegenerative illnesses impact people of all ages. These can include congenital leukodystrophies that affect the white matter in children as well as those that become more common as people age, like age-related macular degeneration (AMD), PD, and AD. Degenerative mental diseases, like schizophrenia, can cause cerebral gray matter loss and symptoms of accelerated aging in their sufferers. While some neurodegenerative diseases, like Huntington's disease (HD), amyotrophic lateral sclerosis (ALS), frontotemporal dementia (FTD), corticobasal syndrome, multiple system atrophy (MSA), and progressive supranuclear palsy, are uncommon, others, like AD, AMD, and PD, are common. The causes of these illnesses could be complicated, environmental, or hereditary. Recent statistics show that 5% of adults over 65 have Parkinson's disease and up to 30% of all adults over 85 have AD (Hung et al. 2010; Przedborski et al. 2003).

Numerous neurodegenerative diseases lead to dementia, which is expected to impact 150 million people worldwide by 2050 and cost patients and their families an estimated $10 trillion. While there are presently no cures for neurodegenerative disorders, there are therapies that can help some people with their symptoms. Levodopa medication and deep brain stimulation are two effective treatments for Parkinson's disease, however they don't address the fundamental cause of the disease or its unstoppable progression (Bogetofte et al. 2020; Hariz and Blomstedt 2022) (Fig. 16.6).

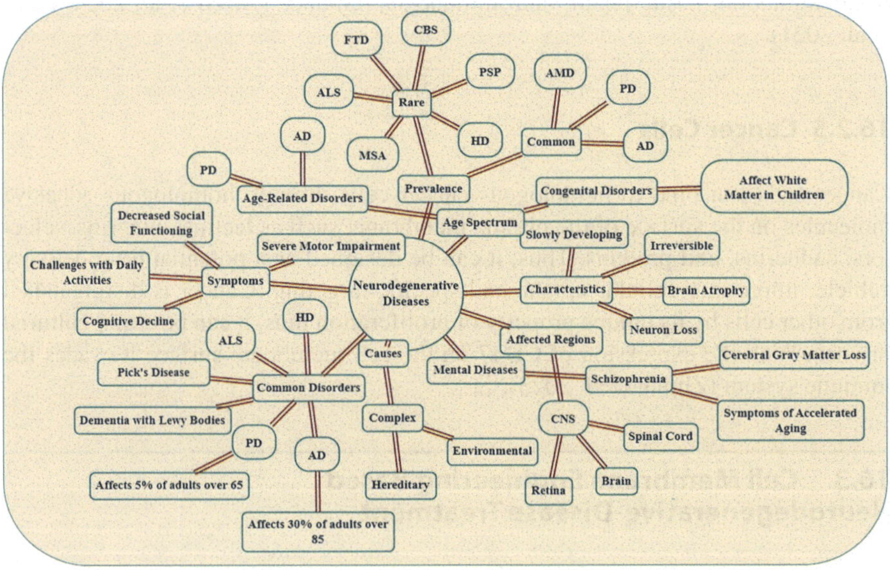

Fig. 16.6 A comprehensive depiction of neurodegenerative diseases, illustrating their prevalence, symptoms, causes, affected regions, and characteristics. Highlights include age-related and rare disorders, cognitive decline, neuron loss, and environmental and hereditary factors. Abbreviations: *FTD* frontotemporal dementia, *CBS* corticobasal syndrome, *PSP* progressive supranuclear palsy, *MSA* multiple system atrophy, *AD* Alzheimer's disease, *ALS* amyotrophic lateral sclerosis, *PD* Parkinson's disease, *HD* Huntington's disease, *CNS* central nervous system

Although the death of neurons is the hallmark of neurodegeneration, several interdependent cell types are usually involved. According to a ground-breaking study, ALS is caused by abnormalities in motor neurons and non-cell-autonomous "killing" by astrocytes, which may be made worse by microglial activation. The important roles that microglia and macroglia (oligodendrocytes and astrocytes) play in disease processes are now widely recognized. Neuropathologists have extensively demonstrated the multicellular and multisystem involvement that is characteristic of neurodegenerative disorders (Deneubourg et al. 2022; Rather et al. 2024).

Since there is currently no cure for neurodegenerative disorders, treatment focuses on reducing the disease's development, enhancing patient quality of life, and treating coexisting conditions. Over the past few decades, there has been a rise in the number of patients with identified neurodegenerative diseases. This increase has mostly been linked to longer life expectancies, which increases the likelihood that people would experience a condition in their lifetime, as well as improved early diagnostic efficiency, which increases the number of cases that healthcare systems are aware of.

16.3.1 Diagnosis

Neurodegenerative disease diagnosis is based on a multistage process that leverages the information provided by blood and cerebrospinal fluid (CSF) biomarkers as well as extensive psychological and neuropsychological evaluation. Further, biomarkers from CSF such as amyloid-β and tau proteins, as well as light chains of the neurofilaments, provide information about the underlying disease pathology and progression. Blood-based biomarkers are less invasive and more accessible alternatives. Psychological and neuropsychological evaluations assess cognitive, emotional, and behavioral functions that can help identify early symptoms and differentiate between disorders. The combined usage of diagnostic tools provides not only a more accurate representation of neurodegenerative diseases but also a more complete picture that can lead to early intervention and personalized treatment strategies for the individuals affected.

Biomarkers in Blood and Cerebrospinal Fluid Beta-amyloid and tau protein levels in cerebrospinal fluid serve as key biomarkers for differentiating neurodegenerative diseases, particularly Alzheimer's disease. The beta-amyloid 1–42 to total tau ratio enables highly accurate identification of neurodegenerative conditions and moderately distinguishes Alzheimer's from other forms of dementia (Benz and Liebner 2020).

Psychological and Neuropsychological Assessment Psychological and neuropsychological assessments, including experimental pathopsychological evaluation commonly used in post-USSR countries, help monitor disease progression and treatment effectiveness by measuring cognitive deficits linked to neurodegenerative disorders (Benz and Liebner 2020).

Neuroimaging and Connectivity The pattern of degeneration spreading across the brain is a key criterion for differentiating neurodegenerative diseases. Prognosis and clinical symptoms correlate with tau aggregation and the atrophy of specific brain regions, which can be tracked using magnetic resonance imaging (MRI) and positron emission tomography (PET) imaging (Benz and Liebner 2020). Transcranial magnetic stimulation (TMS) has shown promise in differentiating neurodegenerative diseases and has been employed to evaluate cortical connectivity in patients with frontotemporal dementia, Lewy body dementia, and Alzheimer's disease (Jan et al. 2023). The researchers then utilized an ensemble of binary decision tree classifiers to develop a predictive model, achieving exceptionally high accuracy in distinguishing different disorders. This study highlights the potential of integrating TMS with machine learning for non-invasive differentiation of neurodegenerative diseases.

Electrical Activity of the Brain Axon disease, cholinergic insufficiency, and disrupted cortical connectivity contribute to abnormal brain electrical activity in neurodegenerative disorders. While typical EEG patterns have been identified for screening, their diagnostic accuracy remains limited. However, integrating deep learning techniques into neurophysiological analysis holds promise for significantly improving diagnostic precision, and related studies are already in progress (AbuAlrob and Mesraoua 2024; Garcia Santa Cruz et al. 2023).

16.3.2 Challenges in Neurodegenerative Diseases

Treating neurodegenerative diseases is challenging due to protective barriers like the blood-brain, blood-retinal, blood-cerebrospinal fluid, and blood-spinal cord barriers, which hinder effective delivery of therapeutic agents across these interfaces.

16.3.2.1 Blood Brain Barrier

Among these barriers, the BBB is the most restrictive, consisting of astrocytes, endothelial cells, tight junctions, neurons, pericytes, and the basal membrane. Tight junctions and adherens junctions ensure the integrity of endothelial cells, with tight junctions comprising transmembrane and cytoplasmic proteins such as claudins, occludins, junction adhesion molecules, Zonula occludens, and accessory proteins. Nearly 98% of drug molecules fail to cross the BBB, making drug delivery to the brain a significant challenge. Nanotechnology-based approaches offer potential solutions to overcome this barrier. The BBB is a semipermeable membrane that selectively allows enzymes, proteins, and nutrients to pass while restricting water-soluble substances, including many therapeutic drugs and diagnostic agents. Molecules can cross the BBB through several mechanisms, including endothelial cell transport, receptor-mediated entry, cellular invagination, and transcytosis. Water-soluble molecules pass through the tight junctions, while ions and lipophilic molecules diffuse paracellularly through endothelial cells. Hydrophilic molecules, such as proteins, utilize carrier-mediated transport via specific receptors. Cellular

invagination occurs when vesicles form around molecules through caveolae, facilitating their entry. Most therapeutic molecules enter the brain via transcytosis, which occurs through two mechanisms: adsorption-mediated transcytosis (for cationic drug molecules) and receptor-mediated transcytosis (for larger drug molecules).
In neurodegenerative diseases, the BBB becomes more permeable, yet drug delivery remains challenging. Pathological conditions cause structural and functional alterations in the BBB, including loss of endothelial cell cohesiveness, disruption of integrity, and increased permeability due to inflammatory responses, protein aggregation, oxidative stress, and vascular changes. Tight junction proteins such as occludins, zonula occludens, and claudin-5 lose their integrity, weakening endothelial cohesion. The basement membrane is further disrupted by reactive oxygen species, increased matrix metalloproteinase activity, and elevated nitric oxide levels produced by endothelial nitric oxide synthase. Neuroinflammation is a hallmark of neurodegenerative diseases, leading to immune cell infiltration into the brain and further compromising BBB integrity. This inflammatory response, characterized by cytokines and other inflammatory molecules, is evident in multiple sclerosis. Additionally, protein aggregation and misfolding, as seen in Alzheimer's and Parkinson's diseases, contribute to neuroinflammation, oxidative stress, and vascular damage, further weakening the BBB (Benz and Liebner 2020; Correale and Villa 2009; Kaya and Ahishali 2021).

16.3.2.2 Blood-Retinol Barrier

The blood-retinol barrier acts as a selective permeability barrier that controls the transportation of retinoids (vitamin A derivatives) from the bloodstream to the retina. Specifically, it consists of retinal endothelial cells, retinal pigment epithelial cells, and tight junctions, which regulate the influx of retinol and its metabolites from the retinal tissue. This barrier tightly regulates the availability of retinoids that are important for visual physiology and retinal health. These changes may trigger retinal diseases such as macular degeneration and diabetic retinopathy, underscoring its role in maintaining retinal homeostasis and normal vision function. Characterisation of blood-retinol barrier is essential to devise a more focused retinal treatment (Bora et al. 2023).

16.3.2.3 Blood-Cerebrospinal Fluid Barrier

The blood-cerebrospinal fluid barrier (BCSFB) separates blood from cerebrospinal fluid (CSF) within the central nervous system (CNS), regulating substance exchange between them. Before reaching the CNS, therapeutic molecules must first pass through the BCSFB, which plays a crucial role in maintaining brain homeostasis. Located in the choroid plexus, this barrier consists of a highly vascularized stroma, a capillary network, and specialized ependymal cells within the brain's ventricles. The choroid plexus, situated near the lateral, third, and fourth ventricles, produces approximately 50% of CSF. Capillary endothelial cells permit molecular passage into the interstitial space, while tight junctions in choroid plexus epithelial cells ensure the barrier's selectivity. The molecules will enter the BCSFB through the mechanism of active transport, endocytosis, exocytosis, and enzymatic barriers. In neurodegenerative disease conditions, the tight junction of BCSFB was disrupted, dysfunction of the transporter, inflammation, and cytokine release. Tight junction disruption allows all the toxic substances into the cerebrospinal fluid like neurotoxic molecules, inflammatory cytokines, immune cells, and inflammatory cytokines. Impaired nutrient supply and waste clearance occurred due to transporter dysfunction. Inflammation and cytokine release in neurodegenerative disease further damage the integrity (Rauf et al. 2022).

16.3.3 Neurodegenerative Diseases Treatment

Neurodegenerative diseases are of five classifications based on the pathophysiological condition such as memory loss and dementia, motor impairment, vision loss, memory loss and dementia combined with motor impairment, motor impairment combined with vision loss. AD and frontotemporal dementia come under memory loss and dementia category. Whereas Parkinson's disease and amyotrophic lateral sclerosis (ALS) come under motor impairment neurodegenerative disease. Retinitis pigmentosa and AMD are categorized as vision loss neurodegenerative diseases. Within the class of memory loss and dementia combined with motor impairment neurodegenerative diseases Huntington's disease, Progressive supranuclear palsy, corticobasal syndrome, and multiple system atrophy are placed. Other diseases like multiple sclerosis, and congenital leukodystrophies (are grouped into motor impairment combined with vision loss class. Memory loss results from neuronal damage in the cerebral cortex, frontal lobe, temporal lobe, and hippocampus. Motor impairment occurs when damage affects the primary motor cortex, striatum, midbrain substantia nigra, or dopaminergic neurons. Similarly, vision loss is associated with damage to the lateral geniculate nucleus (Atapour et al. 2021).

16.3.3.1 Alzheimer's Disease

Alzheimer's disease (AD) is the leading neurodegenerative disorder, primarily marked by dementia. It predominantly affects individuals over 85 years old, with surveys indicating that 30% of AD patients fall within this age group. The disease is rare in those under 50. Dementia, marked by cognitive decline, is the primary clinical symptom, often accompanied by behavioral changes and mood swings in the early stages, progressing to severe memory loss. The hippocampus and cerebral cortex are the most affected brain regions. Pathological hallmarks include amyloid-β (Aβ) peptide aggregation, tau protein accumulation, microvascular damage, oxidative stress, and mitochondrial dysfunction, all contributing to neurodegeneration. AD is classified into familial and non-familial forms. Familial AD, often linked to early-onset cases before 65, follows an autosomal dominant inheritance pattern, primarily caused by mutations in the *APP*, *PSEN1*, and *PSEN2* genes, which promote amyloid plaque formation. Non-familial AD, the more common form, is associated with mutations in the *APOE ε4* allele, which influences Aβ deposition. Other gene mutations, such as those in *SORT1*, have been implicated, though strong evidence is lacking. Aging is the primary risk factor for non-familial AD, with women being more frequently affected than men. AD treatment focuses on pathway-targeting using US FDA-approved drugs. Cholinesterase inhibitors like donepezil, rivastigmine, and galantamine are the first-line medications, addressing cholinergic neuron loss and reduced acetylcholine levels in the brain. Enhancing acetylcholine availability helps slow cognitive decline, as cholinesterase inhibitors prevent its degradation. Donepezil, administered orally at 5–10 mg/day for mild to moderate AD, is combined with memantine for severe cases. Rivastigmine, available as a transdermal patch form, inhibits acetylcholinesterase and butyrylcholinesterase, while galantamine, given at 16–24 mg/day, also modulates nicotinic receptors. However, these treatments have limitations, including uncertain BBB penetration and minimal cognitive improvement. Excessive NMDA receptor activation by glutamate contributes to neurodegeneration, and memantine, a glutamate regulator, is used for moderate to severe AD. It preserves NMDA receptor function while preventing overactivation, improving cognitive function but not directly addressing AD pathology (Jain et al. 2021; Sweetat et al. 2024; Frewen et al. 2024; Hoogmartens et al. 2021).

16.3.3.2 Parkinson's Disease

Parkinson's disease, the second most prevalent neurodegenerative disorder, is a movement-related condition. The clinical features of PD include rest tremor, rigidity, bradykinesia, and postural instability. It rarely occurs in below age 50, a recent survey indicates out of 100,000 people 10–15 people were affected with PD. Loss of dopamine-producing neurons in the mid-brain substantia nigra is the underlying cause of PD. The pathological condition of PD includes abnormal protein aggregation, altered dopamine metabolism, deficiency in the clearance of aggregates, impaired mitochondrial function, necrosis, oxidative stress, inflammation, and accelerated apoptosis. Aggregated proteins such as α synuclein, ubiquitin, and other proteins (called Lewy bodies) are the pathological characteristics of PD. Early-onset PD is the one which occurs in people below the age of 50. This mostly happens due to genetic alteration; highly affected families have five genetic loci for PD. It was found that mutation in the gene that codes for α synuclein is responsible for autosomal dominant early-onset PD. Late-onset leucine-rich repeat kinase 2 (LRRK2 or PARK8) contains mutations associated with Parkinson's disease (PD). Parkin (PARK2), PTEN-induced putative kinase 1 (PINK1 or PARK6), and DJ-1 (DJ-1 or PARK7) have been observed in Parkinson's disease, including specific mutations. The primary therapeutic approach for Parkinson's Disease involves elevating dopamine levels in the substantia nigra region of the brain. The combination of levodopa and carbidopa is the most recognized treatment for Parkinson's disease. Levodopa, as a dopamine precursor, aids in the restoration of motor function impaired by dopamine deficiency. Levodopa, in conjunction with carbidopa, inhibited peripheral degradation of the medicine prior to its arrival in the brain. Moreover, entacapone and tolcapone inhibit the methylation of levodopa by catechol-O-methyl transferase. The dopamine agonists that mimic dopamine and stimulate the signaling pathways include apomorphine hydrochloride, pergolide dihydrochloride, pramipexole, ropinirole hydrochloride, and rotigotine. Dopamine loss occurs through oxidative deamination of dopamine. Monoamine oxidase like Salegiline and Rasagiline inhibitors inhibits oxidative deamination of dopamine. Dementia-related symptoms are managed with rivastigmine, Donepezil, and Galantamine. Other psychosis-related symptoms are treated by Clozapine, Quetiapine, and Pimavanserin. Melatonin or Clonazepam are prescribed for sleep-related symptoms (Chew et al. 2023; Lamptey et al. 2022; Naoi et al. 2022).

16.3.3.3 Amyotrophic Lateral Sclerosis

ALS is a motor disorder with loss of motor neurons in the anterior horns of the spinal cord and the cerebral cortex. Like other neurodegenerative diseases, it also has both familial and non-familial classification. Non-familial ALS accounts for 1–2 cases per 100,000 in a year, and men are more likely to be affected than women. ALS occurs in elder people and certainly leads to death within 2 to 3 years of disease onset. Neuronal death in ALS occurs due to excitotoxicity mediated by glutamine and elevated calcium ions (Ca^{2+}). Neuronal muscle atrophy occurs followed by neuronal cell death, which leads to reduced muscular strength, fasciculation, and hyperreflexia. Respiratory muscle damage can lead to pulmonary infection eventually paralysis occurs due to amyotrophy. The histology of ALS features intracytoplasmic inclusion bodies made up of neurofilaments and spheroids containing ubiquitinated copper and zinc superoxide dismutase (SOD1). It is an enzyme that catalyzes the conversion of superoxide free radicals to hydrogen peroxide. It is noted that in 20% of familial ALS cases mutation in the SOD1 gene is present, but its expression is found in 10% of non-familial ALS. Mutated SOD1 leads to insufficient clearance of oxidative stress and also misfolded SOD1 protein will aggregate and form toxic inclusion bodies. Mutation in alsin (ALS2) gene is associated with familial ALS. The risk factors of ALS are smoking, pesticides, heavy metals, electric shock, electromagnetic field, and agriculture chemicals. Riluzole and Edaravone are the two approved drugs for ALS, which have pathophysiological symptoms such as frontotemporal dementia, behavioral changes, and cognitive decline. Riluzole is one of the kinds of glutamate-receptor antagonists that was prescribed for ALS patients with a dose of 100 mg/day. For delaying the progression of the disease patients were suggested to treat with Edaravone. It is a free-radical scavenger that was given to patients 60 mg/day (Kim et al. 2021; Liu et al. 2020).

16.4 Cell Membrane Engineering Based Inflammatory Diseases Treatment

Inflammation is the defense response by the human system against dangerous triggers such as infections, cell death, or toxins. It also involves restoring normal function, repairing tissues, and eliminating the cause of injury. Almost all organs in the human body can suffer damage due to inflammatory diseases. The major types of inflammatory disorders are inflammatory infections, autoimmune diseases, and chronic inflammatory diseases. Once the body's immune system mistakenly attacks its own healthy tissues, it leads to conditions of autoimmune disorders and chronic

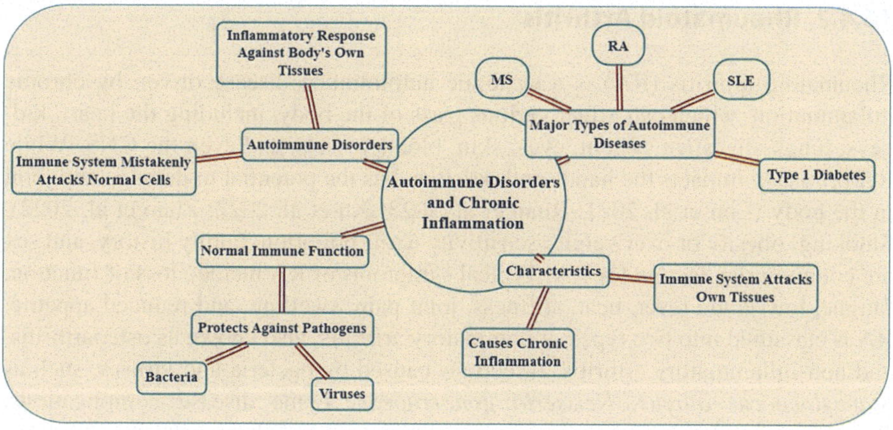

Fig. 16.7 Overview of autoimmune disorders and chronic inflammation, highlighting major types, immune dysfunction, and characteristics. Emphasizes immune system attacks on normal tissues, chronic inflammation, and protective functions against pathogens. Abbreviations: *MS* multiple sclerosis, *RA* rheumatoid arthritis, *SLE* systemic lupus erythematosus

inflammation. Autoimmune disorders cause chronic inflammation that is usually brought about by the attack of the body's immune system on its own healthy tissues. Normally, the immune system identifies and attacks foreign matters to protect the body from the harmful pathogens (viruses and bacteria). The immune system, therefore, mistakenly identifies the human body's normal cells as threats in autoimmune infections, producing an inflammatory response against them. The major types of autoimmune and inflammatory diseases include rheumatoid arthritis (RA), Systemic lupus erythematosus (SLE), multiple sclerosis (MS), and type 1 diabetes (Fig. 16.7).

16.4.1 Systemic Lupus Erythematosus

SLE is a chronic and fatal autoimmune disease that can affect any organ, including the CNS. It results from a combination of environmental, genetic, and immunological factors, leading to tissue and organ damage. Recent advancements in the treatment and management of lupus have improved care, but patients with the disease still face significant risks of illness and mortality. For example, approximately 10% of individuals with lupus nephritis progress to end-stage kidney failure or complications. In SLE, autoantibodies produced by the immune system attack the body's own tissues, leading to inflammation and organ damage. While SLE cannot be cured, therapy helps manage symptoms, reduce pain, and prevent organ failure. Diagnosing SLE can be challenging and typically involves a combination of tests, including microscopic examination, biopsy, platelet count, blood tests for antinuclear antibodies, urinalysis, and a physical examination of the patient (Adebowale et al. 2024; Ou et al. 2023; Wang et al. 2015, 2023).

16.4.2 Rheumatoid Arthritis

Rheumatoid arthritis (RA) is a systemic autoimmune disease driven by chronic inflammation, which can affect various parts of the body, including the heart, kidneys, lungs, digestive system, eyes, skin, blood vessels, and even the CNS. While RA primarily impacts the hands and wrists, it has the potential to damage any joint in the body (Lim et al. 2021; Ruan et al. 2023; Xu et al. 2022; Zhao et al. 2022). Smoking, obesity or overweight, sensitivity to air pollution, family history, and sex are common risk factors for RA. Typical symptoms of RA include loss of function, fatigue, low-grade fever, heat, stiffness, joint pain, swelling, and reduced appetite. RA is classified into two types: inflammatory arthritis, also known as osteoarthritis, and non-inflammatory arthritis. Infections caused by bacteria and viruses, such as *Staphylococcus aureus*, *Neisseria gonorrhoeae*, Lyme disease complications, Parvovirus, and Enterovirus, can also contribute to RA. Other causes include autoimmune processes and crystal deposition, such as pseudogout and basic calcium phosphate disease. RA can be diagnosed through laboratory tests, including rheumatoid factor, anti-citrullinated protein antibodies, erythrocyte sedimentation rate, and C-reactive protein. Imaging techniques like X-rays are used to detect joint damage, while ultrasound and MRI can identify early signs. Clinical evaluation by a rheumatologist, based on the patient's symptoms, is essential for diagnosis. Although no cure exists for RA, early detection and appropriate treatment can help patients control their symptoms and prevent severe joint damage (Frazzei et al. 2023; Radu and Bungau 2021).

16.4.3 Multiple Sclerosis

MS is a significant inflammatory, chronic neurological disease affecting the CNS. Once the body launches an attack on the self, the immune responses inadvertently target the myelin coating that surrounds nerve fibers. Different degrees of myelin and axon damage can occur from targeting myelinated axons in the CNS (Alvarez-Erviti et al. 2011; Wu et al. 2022; Yang et al. 2017). The primary etiology of MS is probably complex gene-environment interactions, which are still unknown. Among the symptoms of multiple sclerosis are tremors, visual impairment, mental diseases, distress, bowel/bladder malfunction, sex dysfunction, fatigue, weakness, issues with walking, and dysphagia/dysarthria. Primary progressive MS, secondary progressive MS, relapsing-remitting MS, and progressive-relapsing MS are some of the subtypes of MS. Multiple sclerosis is associated with low serum levels of vitamin D deficiency, smoking, childhood obesity, and Epstein-Barr virus infection. MS is a disorder that affects young adults and is more common in women than in males (Choi et al. 2020; Park et al. 2021; Zheng et al. 2022; Zhong et al. 2021). Multiple sclerosis (MS) can be diagnosed by MRI of the brain, and McDonald criteria have been compared favorably to MS diagnostic criteria after being evaluated for utility, sensitivity, and specificity. There are FDA-approved medicines for multiple sclerosis, even though there is currently no cure. The goal of the symptoms-based treatment is to increase quality of life and function.

16.4.4 Inflammatory Bowel Diseases

Inflammatory bowel disease (IBD) comprises a collection of chronic, spontaneous inflammatory disorders. IBD is induced by genetic and environmental factors that affect immune responses. Chronic IBD, which induce inflammation and digestive complications in the gastrointestinal tract, encompass Crohn's disease and ulcerative colitis. These diseases are linked to elevated rates of morbidity and mortality. Crohn's disease can affect the entire gastrointestinal tract, whereas ulcerative colitis primarily affects the mucosal layer of the colon (You et al. 2021; Zhao et al. 2019). Ulcerative colitis exclusively affects the colon and rectum; it cannot damage the small intestine. Bloody diarrhea, arthritis, stomach discomfort, fatigue, rectal bleeding, fever, decreased appetite, and anemia are typical symptoms of Crohn's disease and ulcerative colitis. Sometimes there could be joint pain, skin rashes, and eye irritation. The diagnostic technique used to identify IBD disorders includes blood count tests (anemia and thrombocytosis), stool testing, endoscopic, histological, and radiographic investigations, as well as the history and physical examination. Through medicine, lifestyle modifications, and even surgery, people with IBD can control their condition and lives. Laparoscopic surgery is a safe treatment option for patients with IBD (He et al. 2024; Wang et al. 2021).

16.4.5 Type 1 Diabetes

In type 1 diabetes, the immune system misidentifies and damage the insulin-producing β-cells in the pancreas. The attack is planned by CD4 cells, and glucose homeostasis is upset when CD8 cells instantly destroy the beta cells. The common symptoms of type 1 diabetes often experienced increased appetite, thirst, frequent urination, kidney failure, foot ulcers, heart disease, stroke, and vision impairment. Type 1 diabetes is frequently caused by environmental triggers, such as viral infections, and genetic predispositions that cause the immune system to target the pancreas. Cardiovascular disease, neuropathy, nephropathy, retinopathy, and an elevated risk of infection are among the many conditions that can be affected by type 1 diabetes. Insulin medication, a healthy diet, regular exercise, and regular blood glucose checks are the mainstays of controlling type 1 diabetes (Csizmar et al. 2018; Kapate et al. 2023; Li et al. 2021, 2022). A long-sought approach to treating inflammatory and neurodegenerative disorders involves leveraging mechanistic advances in complex decision strategies. NPs and hydrogel-based drug delivery systems offer an effective solution by enabling precise and controlled drug delivery to target sites. These systems can effectively penetrate the BBB, a major challenge in neurodegenerative disease treatment, while facilitating localized and sustained release of anti-inflammatory agents, neuroprotective compounds, or gene therapies. Given the critical role of inflammation in disease progression, developing targeted treatment strategies remains essential. Novel approaches, including monoclonal antibodies and small-molecule inhibitors, are being designed to modulate key inflammatory pathways such as NF-κB, TNF-α, and IL-6 signaling, potentially mitigating chronic inflammation that drives neurodegeneration. Additionally, cell-based therapies using mesenchymal stem cells and induced pluripotent stem cells are emerging as promising strategies for neuronal repair and immune modulation. The gut-brain axis is also gaining attention in neurodegenerative disease research, with probiotics, prebiotics, and microbial metabolites being explored for their ability to restore gut microbiota homeostasis, reduce systemic inflammation, and support neuronal health. Furthermore, advancements in precision medicine, particularly through genetic and epigenetic profiling, now allow for tailored therapies based on individual patient profiles. Machine learning and artificial intelligence algorithms further enhance drug discovery by identifying high-affinity drug candidates for specific molecular targets. Collectively, these innovative approaches hold great potential to transform the therapeutic landscape for inflammatory and neurodegenerative disorders.

16.5 Current Drug Delivery Approaches and Their Failures

Drug delivery across the barriers of the brain is difficult but it can be accomplished through various strategies. The current drug delivery approaches are classified majorly into three types: non-invasive, invasive, and miscellaneous techniques. The chemical, biological, and colloidal drug carrier strategies are placed under non-invasive techniques-chemical methods in which a drug compound is delivered across the BBB by attaching a chemical substance. A water-soluble drug is conjugated or encapsulated with a lipid-soluble moiety. Chimeric peptides and cationic proteins are examples of chemical-based approaches. The biological method is with the help of polymerase chain reaction, protein/peptide engineering, and recombinant DNA technology new grafts were synthesized. Colloidal drug carrier. The invasive technique consists of Intracebral implants, Intraventricular/ Intrathecal delivery, Interstitial delivery, and BBB disruption. Through Intracerebral and intrathecal delivery, only lipid-soluble nutrients and limited number of peptides can pass through the barrier by using specific transport mechanisms or passive diffusion. Intracerebral delivery is done by direct injection into the targeted part by continuous intraventricular infusion. While intrathecal delivery is the process of directly administrating the drug into the cisterna magma of the brain through the intrathecal route. Biochemical distribution of BBB is another technique in which the drug is delivered effectively inside the brain through certain energy or chemicals. Other miscellaneous techniques also show effective treatment methods for drug delivery. Intranasal delivery is one such approach in which a drug is delivered through an intranasal route into the blood passing the nasal mucosa (Fig. 16.8).

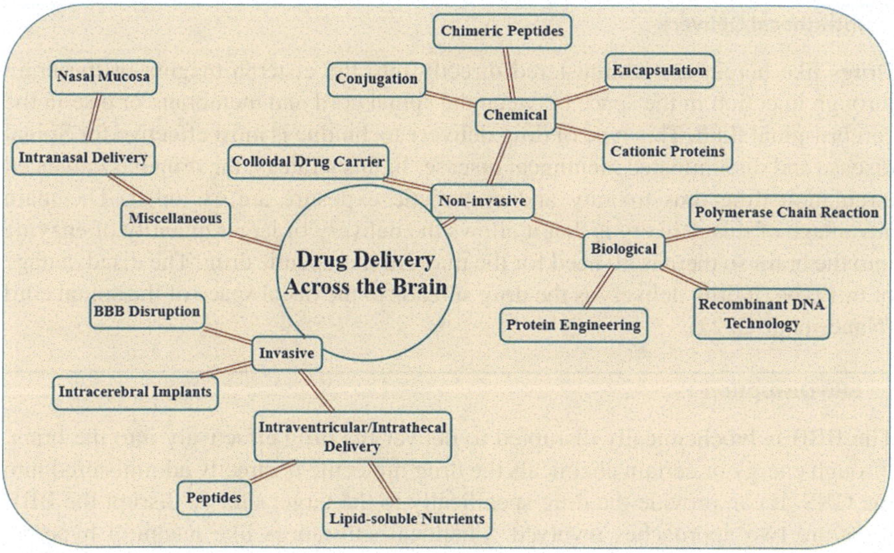

Fig. 16.8 Illustration of drug delivery methods across the brain, highlighting invasive and non-invasive techniques. Includes intranasal delivery, intracerebral implants, colloidal carriers, biological approaches like protein engineering, and chemical conjugation

16.5.1 Invasive Approach

An invasive approach is nothing but the administration of drugs directly into the brain with open-skull surgery through intracerebral and intrathecal delivery. Both ways deliver large and small molecules into the brain. Another way of invasive approach is intranasal delivery, in this method, pathogens and viruses get into the CNS. Along with this, there are a few other disadvantages like antiacne agents enter the brain and also cause physiological stress.

Intracerebral Route

To specifically deliver the drug into the brain direct injection is a more efficient technique. The drug is injected directly into the target site through stereotactic coordinates and continuous intraventricular infusion. The implanted device was made based on biodegradability and biocompatibility. It is mostly made up of a polymeric matrix in which the drug molecule is encapsulated. The skull was opened in the desired targeted part and the implanted devices were inserted. In this technique, the drug molecule was circulated in slow motion to nearer cells. So, there is a requirement for a higher dosage. The drug's physiological behavior determines the penetration capability. Through this technique, only a limited number of drugs can be delivered.

Intrathecal Delivery

Drugs like herein are administered directly into the cisterna magma of the brain through injection in the space between the spinal cord and membrane or else in the cerebrospinal fluid. This type of drug delivery technique is most effective for Spinal disease and disseminated meningeal disease. In this strategy, the drug has enhanced circulation time thus toxicity and systemetic exposure are excluded. The main advantage of this delivery is that it allows the delivery of larger quantity of enzyme into the brain so there is no need for the massive therapeutic drug. The disadvantage of this type of drug delivery is the drug spreads to the distal space of the spinal cord (Nance et al. 2022).

BBB Disruption

The BBB is biochemically disrupted to deliver the drug effectively into the brain, through energy or certain chemicals the drug molecule is directly administered into the CNS. It can provide the drug specifically to the target site. To disrupt the BBB there are two approaches involved. Chemical substances like mannitol hyperosmotic solution on application improve the delivery of chemotherapeutic agents into brain tumors. Ultrasound and electromagnetic radiation shrink the endothelial cells and cause BBB disruption. Intravenous injection of microbubbles reduces the power required to disrupt BBB (Zhan 2020).

Intranasal Delivery

It delivers the drug passing the nasal mucosa for systematic action and directly into the blood. This has a link to the brain through trigeminal nerves in the nasal mucosa. The major advantage is that it requires only a small dosage, is self-administered, and has no need for sterile conditions. It evades gastrointestinal degradation and protein binding during circulation. The disadvantage is the administration is not done through professionals and drug components are degraded by enzymes in the mucous, since a small dosage of drug reaches the brain (Islam et al. 2020).

16.5.2 Non-invasive Approach

This approach enhances drug distribution across brain blood vessels. Microbubbles combined with ultrasound irradiation reduce the expression of tight junctions in the BBB, temporarily opening it without causing damage to the brain. The method primarily involves receptor-mediated drug delivery, chemical drug delivery, prodrugs strategies, lipophilic analogs, and carrier-assisted drug delivery (Tahar et al. 2024).

Chemical Based Approach

There are two fundamental ways to deliver the drug across the BBB. Firstly, the polar function of the water-soluble drugs is conjugated with lipid-soluble moieties.

Next, the water-soluble drug is conjugated with a lipid-soluble drug carrier. Examples of these approaches are chimeric peptides and cationic peptides (Wu et al. 2023b).

Biotechnology-Based Approach

Using protein engineering, and recombinant DNA, neural grafts were created and implanted into the damaged areas of the brain. However, a major drawback of this approach is that the grafts do not persist for extended periods (Feng et al. 2022; Shahlaei et al. 2020; Zhou et al. 2023). The effectiveness of current therapies has been severely hampered by these challenges, leading to an urgent need for new solutions such as the engineering of the cell membrane. In neurodegenerative diseases, small molecules and biologics are examples of traditional treatments with limited transport across physiological barriers, including BBB (Csizmar et al. 2018; Kapate et al. 2023; Li et al. 2021, 2022). As a result, only a small fraction of the delivered drug reaches the target site, leading to unsatisfactory therapeutic effects. For inflammatory diseases such as RA and IBD, many existing drugs lack specificity and cause systemic immunosuppression along with broader side effects. Additionally, current therapies often fail to address the dynamic and complex nature of these diseases, such as the neuroinflammation associated with neurodegeneration. Multiple drug administrations are frequently needed, increasing the risk of toxicity and decreasing patient compliance. These limitations underscore the need for innovative drug delivery strategies, and cell membrane engineering has emerged as a promising alternative (You et al. 2021; Zhao et al. 2019).

16.6 Discussion

A recent review on cell membrane engineering for enhanced drug delivery in neurodegenerative and inflammatory diseases, overcoming challenges like the BBB and cellular uptake barriers. The obstacle of cellular drug delivery systems and their routes through biological barriers led to enhancements regarding the biological fitness of the drugs by mimicking the cell on drug delivery systems through efficient engineering of the cell membrane. Among newer approaches mentioned are the biomimetic nanoparticles and exosomes which utilize the innate cell-like characteristics of biological membranes to aid with translation delivery. Such engineered systems can be designed for targeted drug delivery, adding greater specificity which yields therapeutic properties of drug molecules, ranging from biologics through DNA and small molecules, without affecting other normal body cells. In the case of neurodegenerative diseases, these constructs can be tailored for crossing the BBB, to directly release therapeutic agents in the compromised neural tissues. Such membrane-engineered systems can also be modified to target immune cells or inflamed tissues in ICI and inflammatory diseases respectively to facilitate drug accumulation while minimizing systemic toxicity. In addition to the synthetic NPs mentioned, hybrids are also covered in this review, meaning those that woven natural cell

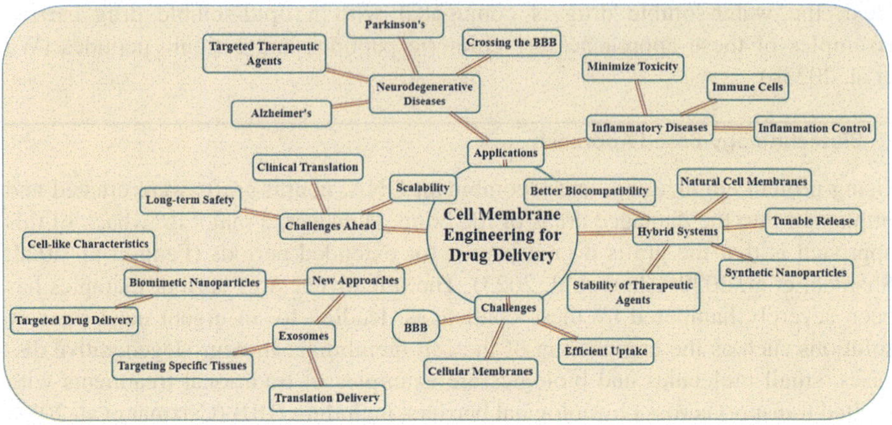

Fig. 16.9 Overview of cell membrane engineering for drug delivery, highlighting applications, challenges, and new approaches. Emphasizes targeted drug delivery, biocompatibility, scalability, neurodegenerative disease treatment, hybrid systems, and translation delivery techniques

membranes into synthetic ones, potentially allowing for better biocompatibility, finely tunable release, and better stability of therapeutic agents. Despite the enormous strides that have been made, the article underscored those issues regarding scalability, long-term safety and clinical translation remain. Despite these challenges, cell membrane engineering is emerging as a promising approach for next-generation drug delivery, with the potential to significantly enhance drug therapy for complex diseases (Fig. 16.9).

16.7 Conclusion

Next-generation therapeutic delivery systems utilize cell membrane coating technology to enhance treatment efficacy for neurodegenerative and inflammatory diseases. These disreputable disorders are defined by intricate pathophysiological pathways and substantial therapeutic hurdles like BBB and immune system limitations, which have significantly restricted the effectiveness of traditional therapies. Engineered drug delivery systems, on the other hand, exploit the inherent features of cell membranes, including biocompatibility, immune evasion, and disease specificity, to circumvent these drawbacks and deliver therapeutic agents specifically and efficaciously to target sites of action. In neurodegenerative diseases, these cell membrane-coated NPs can pass through the BBB and directly deliver the drugs to the brain, targeting the disease origins such as Alzheimer's and Parkinson's disease. This localized targeting is also advantageous in inflammatory diseases, like RA and IBD, which results in reduced systemic side effects and greater therapeutic benefit. Cell membrane engineering enables the co-delivery of multiple therapeutics, facilitating synergistic treatments and theranostic applications for therapy and disease monitoring. The application of hybrid systems and their functionalization with disease-specific ligands represents an additional area of innovation for this technology. Cell membrane engineering enables targeted drug delivery, advancing precision medicine by overcoming existing limitations, particularly in treating neurodegenerative and inflammatory diseases. Mimicking natural membrane properties enables efficient, safe delivery of therapeutics across barriers like the blood-brain barrier. Despite these advancements, challenges remain, including scalability, reproducibility, safety, and regulatory approval for clinical use. Overcoming these obstacles will require interdisciplinary approach, focusing on biomaterials, nanotechnology, and translational research. However, with ongoing research and technological progress, cell membrane engineering has the potential to revolutionize complex disease treatment, enabling more efficient, personalized, and targeted therapies.

References

AbuAlrob MA, Mesraoua B (2024) Harnessing artificial intelligence for the diagnosis and treatment of neurological emergencies: a comprehensive review of recent advances and future directions. Front Neurol 15:1485799

Adebowale K, Liao R, Suja VC et al (2024) Materials for cell surface engineering. Adv Mater 36(43):2210059

Ahlawat J, Guillama Barroso G, Masoudi Asil S et al (2020) Nanocarriers as potential drug delivery candidates for overcoming the blood–brain barrier: challenges and possibilities. Acs Omega 5(22):12583–12595

Akhtar A, Andleeb A, Waris TS et al (2021) Neurodegenerative diseases and effective drug delivery: a review of challenges and novel therapeutics. J Control Release 330:1152–1167

Alvarez-Erviti L, Seow Y, Yin H et al (2011) Delivery of siRNA to the mouse brain by systemic injection of targeted exosomes. Nat Biotechnol 29(4):341–345

Amor S, Puentes F, Baker D et al (2010) Inflammation in neurodegenerative diseases. Immunology 129(2):154–169

Atapour N, Worthy KH, Rosa MG (2021) Neurochemical changes in the primate lateral geniculate nucleus following lesions of striate cortex in infancy and adulthood: implications for residual vision and blindsight. Brain Struct Funct 226(9):2763–2775

Baig MS, Ahmad A, Pathan RR et al (2024) Precision nanomedicine with bio-inspired nanosystems: recent trends and challenges in mesenchymal stem cells membrane-coated bioengineered nanocarriers in targeted nanotherapeutics. J Xenobiot 14(3):827–872

Benz F, Liebner S (2020) Structure and function of the blood–brain barrier (BBB). In: Physiology, pharmacology and pathology of the blood-brain barrier. Springer, pp 3–31

Biscu F, Zouzaf A, Cicia D et al (2024) Innate immunity champions: the diverse functions of macrophages. Eur J Immunol 54(12):2451139

Bogetofte H, Alamyar A, Blaabjerg M et al (2020) Levodopa therapy for Parkinson's disease: history, current status and perspectives. CNS & Neurol Disord-Drug Targets-CNS & Neurolo Disord 19(8):572–583

Bora K, Kushwah N, Maurya M et al (2023) Assessment of inner blood–retinal barrier: animal models and methods. Cells 12(20):2443

Chew ZX, Lim CL, Ng KY et al (2023) The role of monoamine oxidase B inhibitors in the treatment of Parkinson's disease-an update. CNS & Neurol Disord-Drug Targets-CNS & Neurolo Disord 22(3):329–352

Choi H, Kim Y, Mirzaaghasi A et al (2020) Exosome-based delivery of super-repressor IκBα relieves sepsis-associated organ damage and mortality. Sci Adv 6(15):eaaz6980

Correale J, Villa A (2009) Cellular elements of the blood-brain barrier. Neurochem Res 34:2067–2077

Corrons JV, Casafont LB, Frasnedo EF (2021) Concise review: how do red blood cells born, live, and die? Ann Hematol 100(10):2425–2433

Csizmar CM, Petersburg JR, Wagner CR (2018) Programming cell-cell interactions through non-genetic membrane engineering. Cell Chem Biol 25(8):931–940

Daraban BS, Popa AS, Stan MS (2024) Latest perspectives on Alzheimer's disease treatment: the role of blood-brain barrier and antioxidant-based drug delivery systems. Molecules 29(17):4056

Deneubourg C, Ramm M, Smith LJ et al (2022) The spectrum of neurodevelopmental, neuromuscular and neurodegenerative disorders due to defective autophagy. Autophagy 18(3):496–517

Ding S, Khan AI, Cai X et al (2020) Overcoming blood–brain barrier transport: advances in nanoparticle-based drug delivery strategies. Mater Today 37:112–125

Elhassan TA, Rahim MSM, Swee TT et al (2022) Segmentation of white blood cells in acute myeloid leukemia microscopic images: a review, prognostic models in healthcare: AI and statistical approaches. Springer Nature, Singapore, pp 1–24

Feng Y, Li Y, Shen P-P et al (2022) Gene-modified stem cells for spinal cord injury: a promising better alternative therapy. Stem Cell Rev Rep 18(8):2662–2682

Feng J, Zheng Y, Guo M et al (2023) Oxidative stress, the blood–brain barrier and neurodegenerative diseases: the critical beneficial role of dietary antioxidants. Acta Pharm Sin B 13(10):3988–4024

Frazzei G, Musters A, de Vries N et al (2023) Prevention of rheumatoid arthritis: a systematic literature review of preventive strategies in at-risk individuals. Autoimmun Rev 22(1):103217

Frewen P, Vincent A, Olff M (2024) Childhood trauma histories in men and women assessed by the childhood attachment and relational trauma screen (CARTS) and the global psychotrauma screen (GPS): results from the global collaboration on traumatic stress (GC-TS). Child Abuse Negl 149:106610

Gao H-M, Hong J-S (2008) Why neurodegenerative diseases are progressive: uncontrolled inflammation drives disease progression. Trends Immunol 29(8):357–365

Garcia Santa Cruz B, Husch A, Hertel F (2023) Machine learning models for diagnosis and prognosis of Parkinson's disease using brain imaging: general overview, main challenges, and future directions. Front Aging Neurosci 15:1216163

Guo M, Xia C, Wu Y et al (2021) Research progress on cell membrane-coated biomimetic delivery systems. Front Bioeng Biotechnol 9:772522

Hariz M, Blomstedt P (2022) Deep brain stimulation for Parkinson's disease. J Intern Med 292(5):764–778

He J, Wu X, Li L et al (2024) Construction of curcumin-loaded macrophage and HUVECs membrane-derived vesicles for drug delivery in cardiovascular inflammatory. J Drug Deliv Sci Technol 95:105611

Hoogmartens J, Cacace R, Van Broeckhoven C (2021) Insight into the genetic etiology of Alzheimer's disease: a comprehensive review of the role of rare variants. Alzheimer's Dementia: Diagnosis Assessment Disease Monitoring 13(1):e12155

Hung C-W, Chen Y-C, Hsieh W-L et al (2010) Ageing and neurodegenerative diseases. Ageing Res Rev 9:S36–S46

Islam SU, Shehzad A, Ahmed MB et al (2020) Intranasal delivery of nanoformulations: a potential way of treatment for neurological disorders. Molecules 25(8):1929

Jain A, Norton N, Bruno KA et al (2021) Sex differences, genetic and environmental influences on dilated cardiomyopathy. J Clin Med 10(11):2289

Jan N, Madni A, Khan S et al (2023) Biomimetic cell membrane-coated poly (lactic-co-glycolic acid) nanoparticles for biomedical applications. Bioeng Transl Med 8(2):e10441

Kapate N, Dunne M, Kumbhojkar N et al (2023) A backpack-based myeloid cell therapy for multiple sclerosis. Proc Natl Acad Sci 120(17):e2221535120

Kaya M, Ahishali B (2021) Basic physiology of the blood-brain barrier in health and disease: a brief overview. Tissue Barriers 9(1):1840913

Kim M, Byun J, Chung Y et al (2021) Reactive oxygen species scavenger in acute intracerebral hemorrhage patients: a multicenter, randomized controlled trial. Stroke 52(4):1172–1181

Kimmerlin Q, Strassel C, Eckly A et al (2023) The tubulin code in platelet biogenesis. Seminars in Cell & Developmental Biology 137:63–73

Lamptey RN, Chaulagain B, Trivedi R et al (2022) A review of the common neurodegenerative disorders: current therapeutic approaches and the potential role of nanotherapeutics. Int J Mol Sci 23(3):1851

Leyane TS, Jere SW, Houreld NN (2022) Oxidative stress in ageing and chronic degenerative pathologies: molecular mechanisms involved in counteracting oxidative stress and chronic inflammation. Int J Mol Sci 23(13):7273

Li Y, Gan Y, Li C et al (2020) Cell membrane-engineered hybrid soft nanocomposites for biomedical applications. J Mater Chem B 8(26):5578–5596

Li Y, Teng X, Yang C et al (2021) Ultrasound controlled anti-inflammatory polarization of platelet decorated microglia for targeted ischemic stroke therapy. Angew Chem Int Ed 60(10):5083–5090

Li J, Ding Y, Cheng Q et al (2022) Supramolecular erythrocytes-hitchhiking drug delivery system for specific therapy of acute pneumonia. J Control Release 350:777–786

Lim GT, You DG, Han HS et al (2021) Bioorthogonally surface-edited extracellular vesicles based on metabolic glycoengineering for CD44-mediated targeting of inflammatory diseases. J Extracell Vesicles 10(5):e12077

Liu H, Welburn JP (2024) A circle of life: platelet and megakaryocyte cytoskeleton dynamics in health and disease. Open Biol 14(6):240041

Liu Y-H, Li K, Tian H-Q (2020) Renoprotective effects of a new free radical scavenger, XH-003, against cisplatin-induced nephrotoxicity. Oxidative Med Cell Longev 2020(1):9820168–9820112

Liu W, Yan Q, Xia C et al (2021) Recent advances in cell membrane coated metal–organic frameworks (MOFs) for tumor therapy. J Mater Chem B 9(22):4459–4474

Liu H, Su Y-Y, Jiang X-C et al (2023) Cell membrane-coated nanoparticles: a novel multifunctional biomimetic drug delivery system. Drug Deliv Transl Res 13(3):716–737

Nance E, Pun SH, Saigal R et al (2022) Drug delivery to the central nervous system. Nat Rev Mater 7(4):314–331

Naoi M, Maruyama W, Shamoto-Nagai M (2022) Neuroprotective function of rasagiline and selegiline, inhibitors of type B monoamine oxidase, and role of monoamine oxidases in synucleinopathies. Int J Mol Sci 23(19):11059

Neganova ME, Aleksandrova YR, Sukocheva OA et al (2022) Benefits and limitations of nanomedicine treatment of brain cancers and age-dependent neurodegenerative disorders. Seminars in cancer biology 86:805–833

Nguyen-Thi P-T, Nguyen TT, Phan HL et al (2023) Cell membrane-based nanomaterials for therapeutics of neurodegenerative diseases. Neurochem Int 105612:105612

Ou A, Wang Y, Zhang J et al (2023) Living cells and cell-derived vesicles: a trojan horse technique for brain delivery. Pharmaceutics 15(4):1257

Park JH, Jiang Y, Zhou J et al (2021) Genetically engineered cell membrane–coated nanoparticles for targeted delivery of dexamethasone to inflamed lungs. Sci Adv 7(25):eabf7820

Patel V, Chavda V, Shah J (2021) Nanotherapeutics in neuropathologies: obstacles, challenges and recent advancements in CNS targeted drug delivery systems. Curr Neuropharmacol 19(5):693–710

Plaza-Oliver M, Santander-Ortega MJ, Lozano MV (2021) Current approaches in lipid-based nanocarriers for oral drug delivery. Drug Deliv Transl Res 11:471–497

Przedborski S, Vila M, Jackson-Lewis V (2003) Series introduction: neurodegeneration: what is it and where are we? J Clin Invest 111(1):3–10

Qian J, Guo Y, Xu Y et al (2023) Combination of micelles and liposomes as a promising drug delivery system: a review. Drug Deliv Transl Res 13(11):2767–2789

Quarleri J, Cevallos C, Delpino MV (2021) Apoptosis in infectious diseases as a mechanism of immune evasion and survival. Adv Protein Chem Struct Biol 125:1–24

Radu A-F, Bungau SG (2021) Management of rheumatoid arthritis: an overview. Cells 10(11):2857
Rather MA, Khan A, Javed H et al (2024) Neuropathology of neurological disorders. In: Mechanism and genetic susceptibility of neurological disorders. Springer, pp 1–33
Rauf A, Badoni H, Abu-Izneid T et al (2022) Neuroinflammatory markers: key indicators in the pathology of neurodegenerative diseases. Molecules 27(10):3194
Rekatsina M, Paladini A, Piroli A et al (2020) Pathophysiology and therapeutic perspectives of oxidative stress and neurodegenerative diseases: a narrative review. Adv Ther 37:113–139
Ruan H, Li Y, Wang C et al (2023) Click chemistry extracellular vesicle/peptide/chemokine nanocarriers for treating central nervous system injuries. Acta Pharm Sin B 13(5):2202–2218
Shahlaei M, Asl SM, Saeidifar M (2020) Nanotechnology in gene delivery for neural regenerative medicine. In: Neural regenerative nanomedicine. Elsevier, pp 123–157
Sweetat S, Shabat MB, Theotokis P et al (2024) Ovariectomy and high fat-sugar-salt diet induced Alzheimer's disease/vascular dementia features in mice. Aging Dis 15(5):2284–2300
Tahar A, Zrour H, Dupont S et al (2024) Non-invasive approaches to hydration assessment: a literature review. Urolithiasis 52(1):132
Wang Q, Cheng H, Peng H et al (2015) Non-genetic engineering of cells for drug delivery and cell-based therapy. Adv Drug Deliv Rev 91:125–140
Wang D, Jiang S, Zhang F et al (2021) Cell membrane vesicles with enriched CXCR4 display enhances their targeted delivery as drug carriers to inflammatory sites. Adv Sci 8(23):2101562
Wang Y, Li Z, Mo F et al (2023) Chemically engineering cells for precision medicine. Chem Soc Rev 52(3):1068–1102
Wu W-C, Tian J, Xiao D et al (2022) Engineered extracellular vesicles encapsulated Bryostatin-1 as therapy for neuroinflammation. Nanoscale 14(6):2393–2410
Wu H, Zhang T, Li N et al (2023a) Cell membrane-based biomimetic vehicles for effective central nervous system target delivery: insights and challenges. J Control Release 360:169–184
Wu D, Chen Q, Chen X et al (2023b) The blood–brain barrier: structure, regulation, and drug delivery. Signal Transduct Target Ther 8(1):217
Xia Q, Zhang Y, Li Z et al (2019) Red blood cell membrane-camouflaged nanoparticles: a novel drug delivery system for antitumor application. Acta Pharm Sin B 9(4):675–689
Xu F, Wu Y, Yang Q et al (2022) Engineered extracellular vesicles with SHP2 high expression promote mitophagy for Alzheimer's disease treatment. Adv Mater 34(49):2207107
Yang J, Zhang X, Chen X et al (2017) Exosome mediated delivery of miR-124 promotes neurogenesis after ischemia. Mol Ther Nucleic Acids 7:278–287
You DG, Lim GT, Kwon S et al (2021) Metabolically engineered stem cell–derived exosomes to regulate macrophage heterogeneity in rheumatoid arthritis. Sci Adv 7(23):eabe0083
Yu S, Chen X, Yang T et al (2024) Revealing the mechanisms of blood–brain barrier in chronic neurodegenerative disease: an opportunity for therapeutic intervention. Rev Neurosci 35:895–916
Zago G, Saavedra PH, Keshari KR et al (2021) Immunometabolism of tissue-resident macrophages–an appraisal of the current knowledge and cutting-edge methods and technologies. Front Immunol 12:665782
Zhan W (2020) Effects of focused-ultrasound-and-microbubble-induced blood-brain barrier disruption on drug transport under liposome-mediated delivery in brain tumour: a pilot numerical simulation study. Pharmaceutics 12(1):69
Zhang M, Cheng S, Jin Y et al (2021) Membrane engineering of cell membrane biomimetic nanoparticles for nanoscale therapeutics. Clin Transl Med 11(2):e292
Zhao Y-Z, ZhuGe D-L, Tong M-Q et al (2019) Ulcerative colitis-specific delivery of keratinocyte growth factor by neutrophils-simulated liposomes facilitates the morphologic and functional recovery of the damaged colon through alleviating the inflammation. J Control Release 299:90–106
Zhao Y, Haney MJ, Fallon JK et al (2022) Using extracellular vesicles released by GDNF-transfected macrophages for therapy of Parkinson disease. Cells 11(12):1933
Zheng J, Lu C, Yang M et al (2022) Lung-targeted delivery of cepharanthine by an erythrocyte-anchoring strategy for the treatment of acute lung injury. Pharmaceutics 14(9):1820

Zhong Y, Qin X, Wang Y et al (2021) "Plug and play" functionalized erythrocyte nanoplatform for target atherosclerosis management. ACS Appl Mater Interfaces 13(29):33862–33873
Zhong X, Na Y, Yin S et al (2023) Cell membrane biomimetic nanoparticles with potential in treatment of Alzheimer's disease. Molecules 28(5):2336
Zhou H, He Y, Xiong W et al (2023) MSC based gene delivery methods and strategies improve the therapeutic efficacy of neurological diseases. Bioact Mater 23:409–437

The manufacturer's authorised representative in the EU is Springer Nature Customer Service Centre GmbH, Europaplatz 3, 69115 Heidelberg, Germany. If you have any concerns regarding our products, please contact ProductSafety@springernature.com

Printed and bound by CPI Group (UK) Ltd, Croydon, CR0 4YY
30/09/2026
02225564-0018